Springer Texts in Business and Economics

Springer Texts in Business and Economics (STBE) delivers high-quality instructional content for undergraduates and graduates in all areas of Business/Management Science and Economics. The series is comprised of self-contained books with a broad and comprehensive coverage that are suitable for class as well as for individual self-study. All texts are authored by established experts in their fields and offer a solid methodological background, often accompanied by problems and exercises.

More information about this series at http://www.springer.com/series/10099

Eduardo Souza de Cursi

Uncertainty Quantification and Stochastic Modelling with EXCEL

Springer

Editors
Eduardo Souza de Cursi
Department Mechanics / Civil Engineering
INSA Rouen Normandie
Saint-Etienne du Rouvray, France

ISSN 2192-4333 ISSN 2192-4341 (electronic)
Springer Texts in Business and Economics
ISBN 978-3-030-77756-2 ISBN 978-3-030-77757-9 (eBook)
https://doi.org/10.1007/978-3-030-77757-9

This Springer imprint is published by the registered company Springer Nature Switzerland AG
The registered company address is: Gewerbestrasse 11, 6330 Cham, Switzerland

Introduction

This book presents a collection of methods of Uncertainty Quantification (UQ), that is, a collection of methods for the analysis of numerical data, namely when uncertainty or variability is involved.

The general aim of UQ is to characterize the observed variability in a quantity X by using a random variable U. In the ideal situation, the connection between X and U is perfectly known and the random variable U has a known distribution. Unfortunately, such a situation may be unrealistic in practice and we must also consider situations where this knowledge is imperfect or even non-existent, for example, situations where U is simply unknown: variability is observed without precise knowledge of the cause.

UQ tries to use all the available information about (X, U) to construct an explanation of X by U into a form which will be useful for use in numerical calculations involving X. The information may be, for instance, an equation, a numerical problem involving both the variables, or samples.

The methods of UQ are general and may be applied to a wide range of situations. They generally belong to the large and well-supplied family of methods based on functional representations, that is, on expansions of the unknowns in series of functions – we find these approaches particularly in Fourier analysis, spectral methods, finite elements, Bayesian optimization, and quantum algorithms. It is a large family with numerous and very diversified applications.

Our objective is to present the practical use of UQ techniques under EXCEL®. We assume that you are a mean user of this software, that is to say you are able to perform standard tasks without difficulties, but we do not assume that you have an expert knowledge of EXCEL® and VBA. Some of the worksheets proposed by this book contain VBA code, but we do not assume that you are a VBA programmer. Obviously, if you are an expert in EXCEL® and VBA, you will find a large amount of improvements in our codes and worksheets: do not hesitate in making your own enhancements and, eventually, in sharing them.

EXCEL® is a part of Microsoft Office®. It is an application that works with spreadsheets – which are a kind of tables having a large number of lines and columns on which we may perform mathematical operations. EXCEL® offers also the possibility of object programming using Visual Basic for

Applications (VBA). EXCEL® offers wide possibilities for data analysis and scientific calculations. There are many books devoted to the use of EXCEL®, going from the basic use to the advanced programming. You may also find a lot of books concerning specific applications of EXCEL®, namely in Finance, Management, Operations Research. If you are familiar with EXCEL, you probably know the extent of the possibilities offered by this software. If, in addition, you are familiar with Visual Basic for Applications (VBA), then you can appreciate the full range of possibilities of EXCEL®. The community of the users of EXCEL® and some commercial software provide you with many useful add-ins, such as linear algebra, numerical calculus, probability, statistics, and simulation.

For instance, you may use the free add-in MATRIX, which provides some useful macros for linear algebra and works with recent versions of version of EXCEL®. There is a free powerful extension of MATRIX, known as XNUMBERS, which includes many numerical methods – XNUMBERS works with recent versions of EXCEL®, but you need to use VBA to effectively use it. Another free add-in is MATHLAYER ®, which is a MATLAB®/OCTAVE-like extension of EXCEL®, offering the possibility of programming analogously to MATLAB®/OCTAVE: programs may be written in separate files or directly in EXCEL® worksheets – in addition, MATHLAYER® allows the use of MATLAB®/Octave programs in EXCEL® Workbooks. A simple-to-use add-in is SDAT, designed for chemistry, but performing general tasks such as regression, differentiation, and statistics. If you are mainly interested in operations research, you may use the Jensen Library, an old solver proposing many classical methods for this field. These are a few examples among a lot of resources available for the users of EXCEL® – we cannot cite all the existing add-ins and we apologize for the non-cited existing contributions.

In this book, we use the add-ins proposed by EXCEL®: the SOLVER, the ANALYSIS TOOLPACK, and VBA, which are standard complements of EXCEL®. This choice is made by reasons of concentration on the purpose of this book: we don't want to lose the reader in the forest of add-ins, but – on the contrary – we want to keep the reader on the central topic all the time. Of course, you may use your preferred add-in instead of the standard EXCEL® tools.

In the next chapter, we describe some basic operations that are necessary to use all the power of EXCEL®.

Contents

1 Some Tips to Use EXCEL®

EXCEL® is a powerful software: a complete exploration of its possibilities cannot be made here. In this chapter, we present some tips that will be useful in the sequel.

1.1 How to Activate the SOLVER and the ANALYSIS TOOLPACK

Before any use, you must activate the SOLVER and ANALYSIS TOOLPACK as follows:

1. Open Excel and click on *Options* (at the very bottom, to the left)

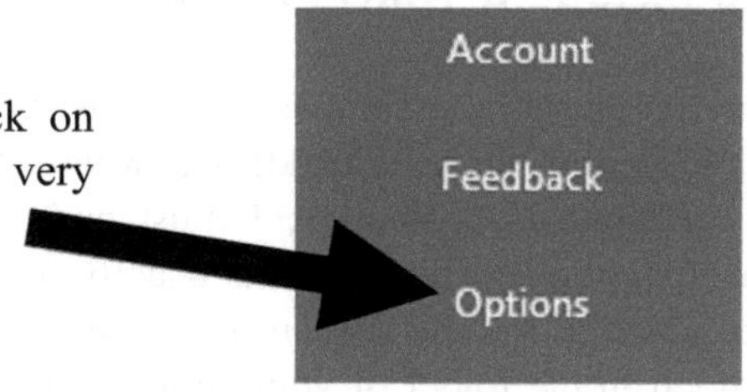

2. The window "Excel Options" is open. Click on *Add-ins*

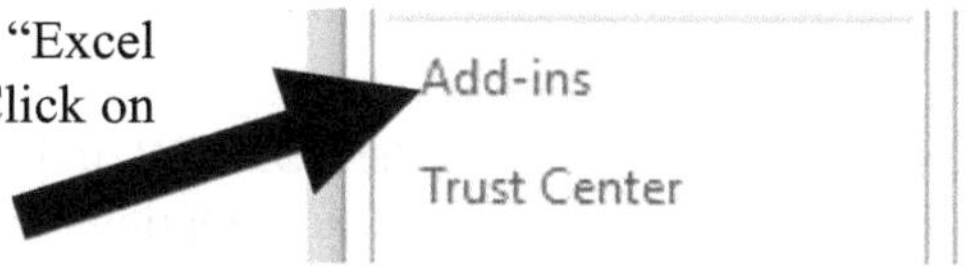

3. At the bottom, verify that "Excel Add-ins" is shown and click on *Go*

E. Souza de Cursi, *Uncertainty Quantification and Stochastic Modelling with EXCEL*,
Springer Texts in Business and Economics, https://doi.org/10.1007/978-3-030-77757-9_1

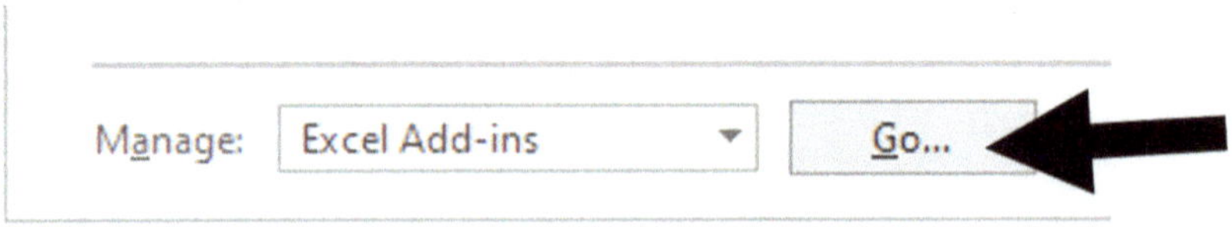

4. Check all the boxes and click *OK*

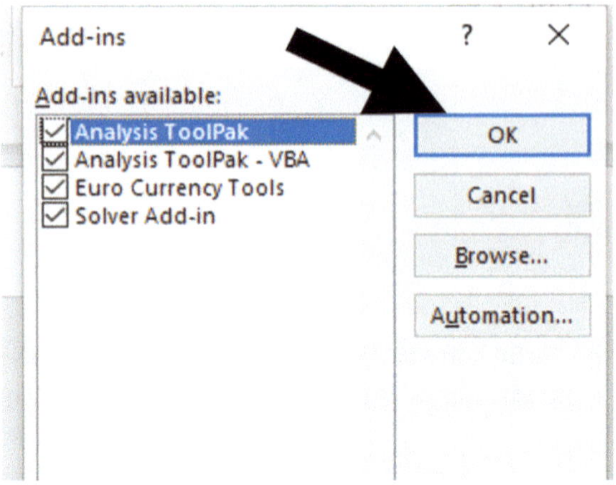

1.2 How to Include a Third-Part Add-In

If you want to use an add-in, the simplest way consists in copying it to the folder "C:\Users\your account\AppData\Roaming\Microsoft\AddIns", where "your account" is your username. For instance, if my username is "eduardo", I must copy the add-in to "C:\Users\eduardo\AppData\Roaming\Microsoft\AddIns". This folder is usually hidden, so that you must make it visible, by modifying the parameters for the display of folders and files. As an alternative way, you may download and save it in the folder of your choice (where you want to keep it permanently). Then you activate it:

1. Make again the first 3 steps to activate the SOLVER and ANALYSIS TOOLPACK: Open Excel and click on *Options*; in the window "Excel Options" which opens, click on *Add-ins*; at the bottom, verify that "Excel Add-ins" is shown and click on *Go*.

2. If SOLVER and ANALYSIS TOOLPACK are activated, all the boxes are checked (if not, check them). Click on *Browse*

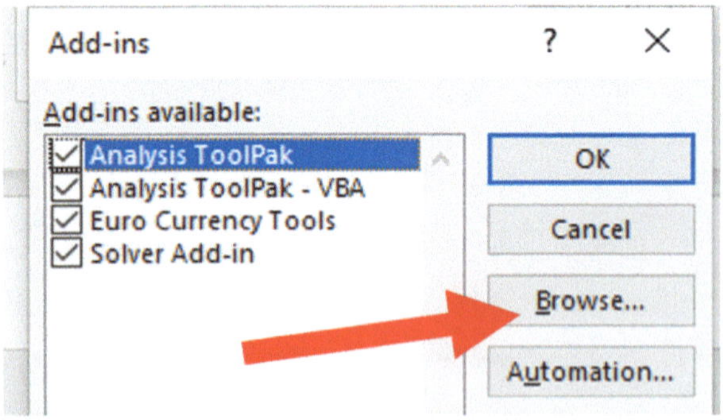

3. The file tree opens. Browse to the folder where the add-in is located, select it and click *OK*. For instance, to include the add-in RandomNumbers.xlam, you must navigate to the folder where it is located, select it and click *OK*

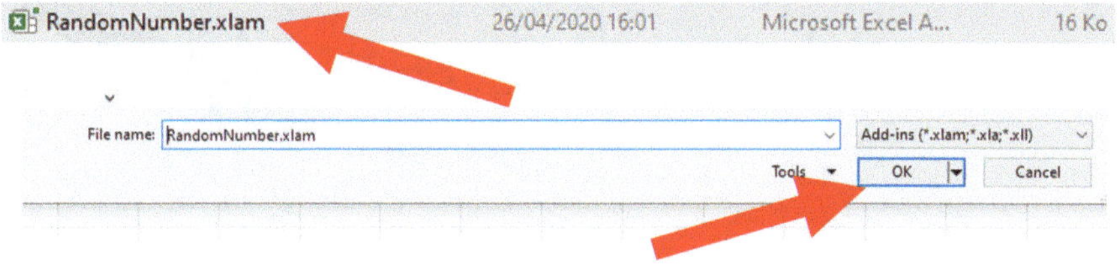

4. Now, your add-in appears in the list with a checked box. Click *OK* to validate.

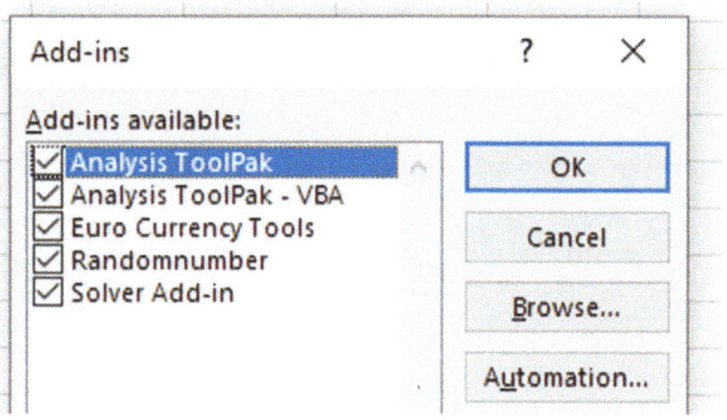

1.3 Disabling Warnings on the Add-Ins

If the add-in is not in the folder "C\Users\. . .\AddIns", it is useful to make its folder as "trusted" – otherwise, EXCEL® will give you a security warning at each time you open a workbook using the add-ins. Examples of such a message are given below:

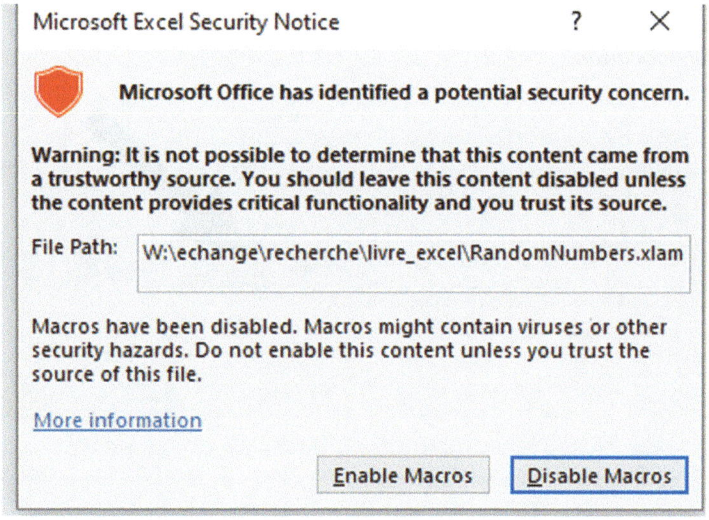

These messages ask if you authorize EXCEL to use the contents of the add-in (if yes, click *Enable*). If you do not want to see this message at each workbook's opening, you must go to the *Trust Center*, just below the *Add-ins* in the "Excel Options":	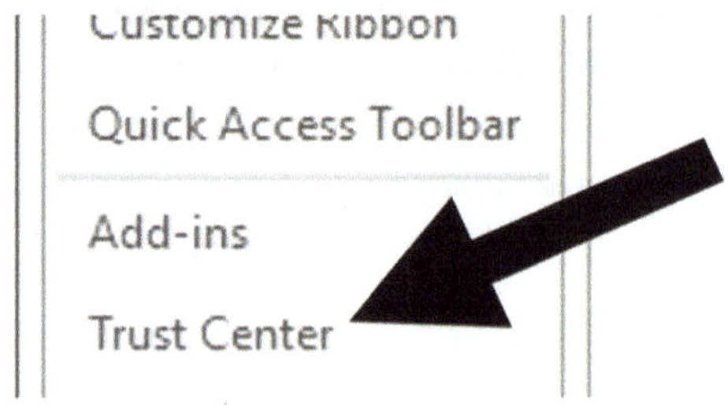

Then, click on *Trust Center Settings*

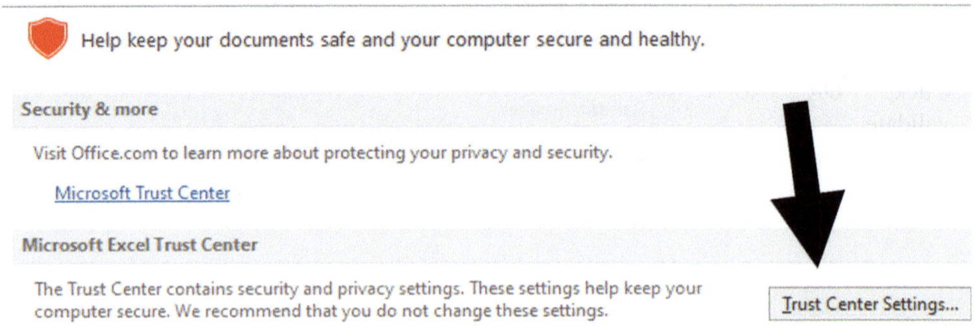

Open *Trusted Locations*

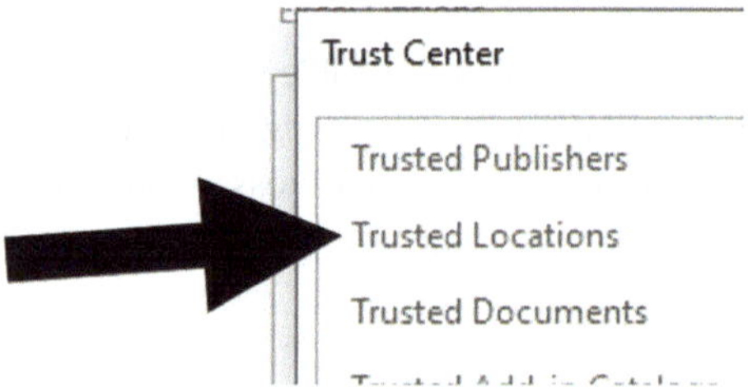

Click on *Add new location*

Then, click on *Browse*

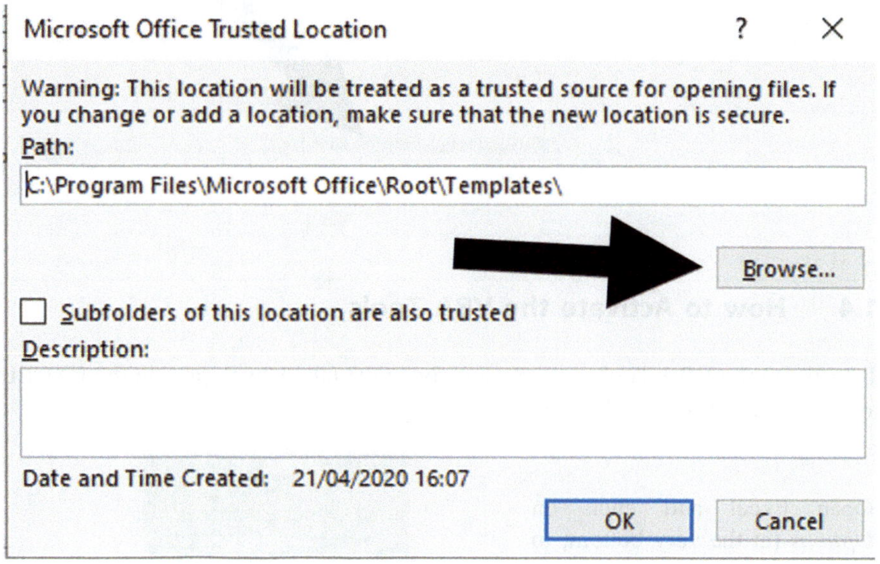

Navigate to the folder containing the add-in and click *OK*. The path is modified: you must see the path to the folder containing the add-in. Check the *Subfolders* option and click *OK*.

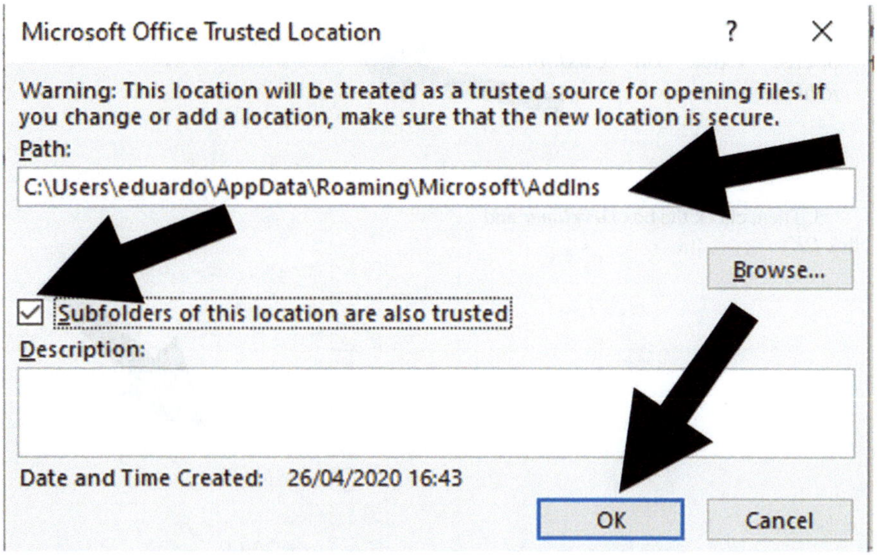

The new location appears in the list of the trusted locations. Click *OK* to validate and come back to your workbook. Now EXCEL® considers the location as safe and will allow the use of the add-in without supplementary questions.

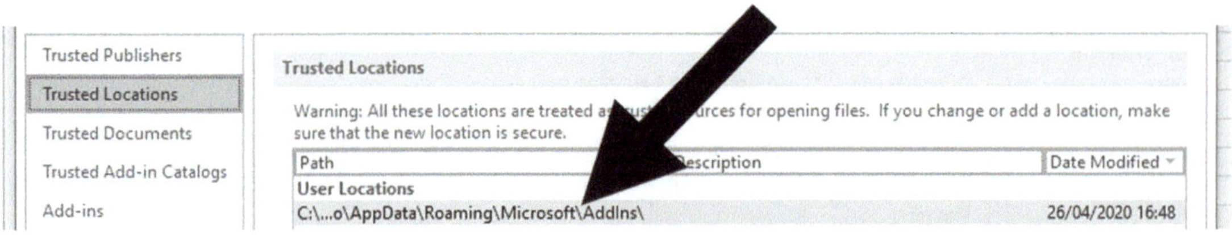

1.4 How to Activate the VBA Tools

To activate the VBA Tools, you must activate the tab “Developer” in the EXCEL® Ribbon, by going to *Options* > *Customize Ribbon*:

1. Open Excel and click on *Options* (at the very bottom, to the left)

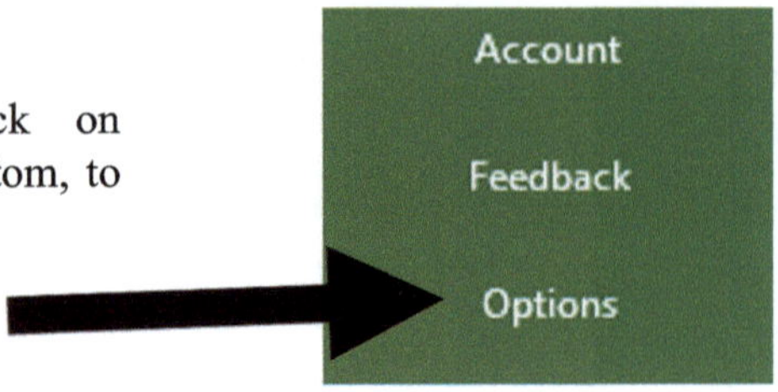

2. The window “Excel Options” opens. Click on *Customize Ribbon*

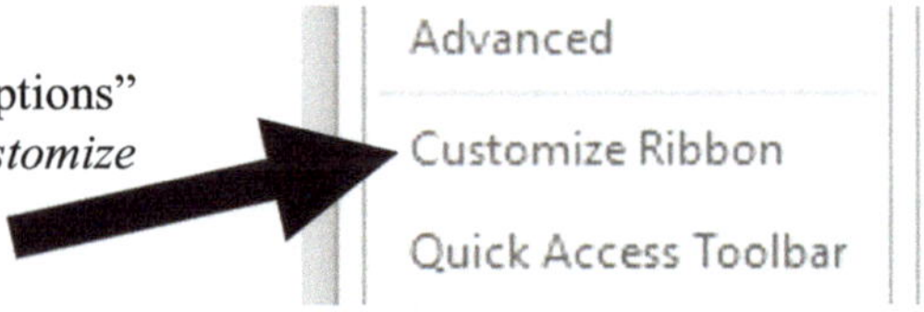

3. Then, check the box *Developer* and click *OK*

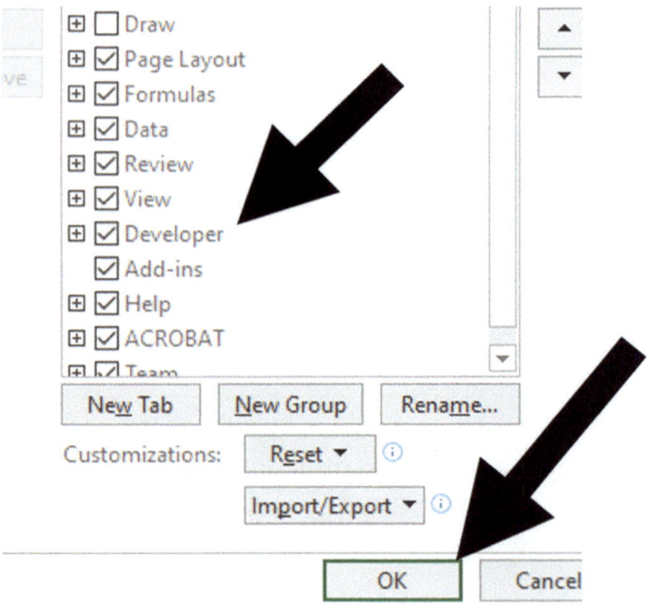

1.5 How to Insert a VBA Module

VBA modules require your workbook to be saved under the extension ".xlsm". To insert a VBA module, you must go to the tab *Developer*, click on *Visual Basic*

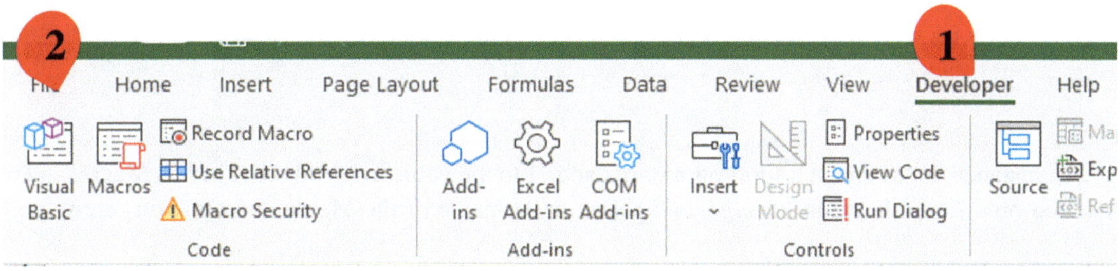

Then right click on *VBAProject(name of your workbook)* and go to *Insert* > *Module*. If you see several *VBAProject*, choose the one having the name of your workbook. For instance, the case where the workbook is *Book1* is illustrated below

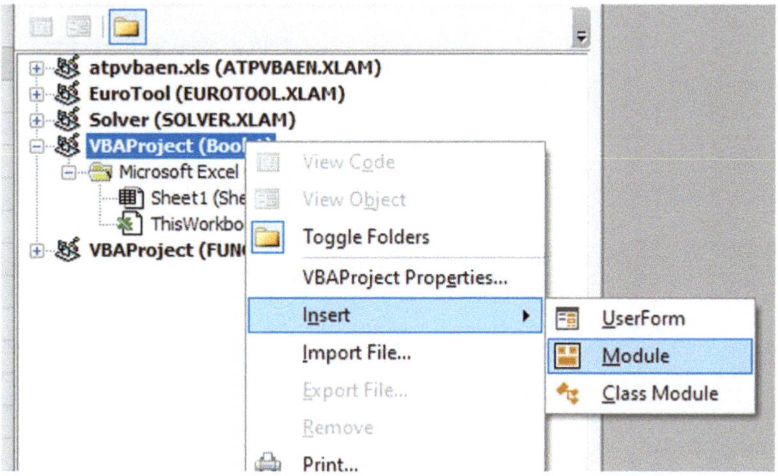

The window for the code of the module appears and you may type your code. You can change the name of your module at the "Properties"

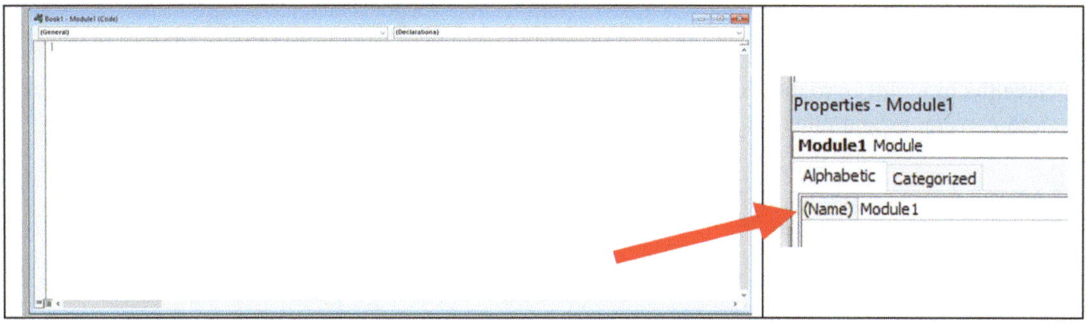

Recall that you must save your workbook with the extension ".xlsm". As an alternative, you may go to *Developer* and click on *Macros*

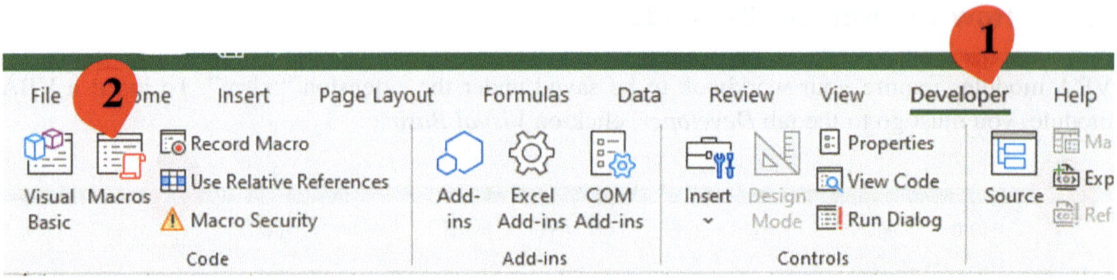

For instance, let us create a subprogram that adds 1 to the value in a Cell: clicking on Macros opens a dialog box. Enter the name *change_cell_value* in the name of the Macro and click on "`create`".

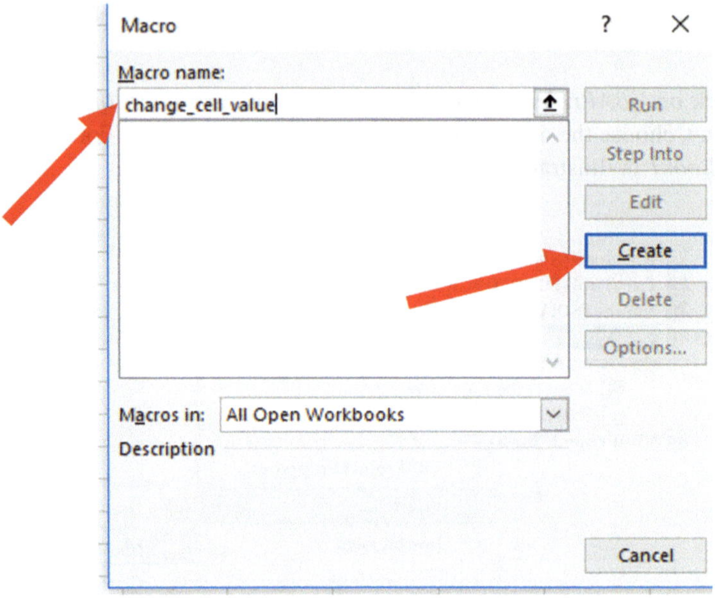

Then, enter the code below:

```
(General)
Sub change_cell_value()
ActiveCell.Value = ActiveCell.Value + 1
End Sub
```

Save the module, go to the workbook, and select cell `A1`. Enter the value "`2`". Reselect cell `A1`. Go back to *Developer* and click on *Macros*. Now, the button "Run" may be clicked:

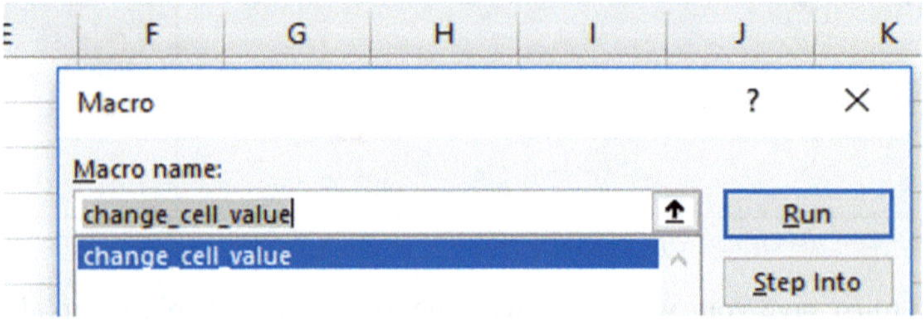

Click "Run" and the macro will be executed: the value of cell `A1` becomes `3`.

In VBA, Modules may contain many subprograms. In addition, subprograms may be Subroutines (`Sub`) or functions (`Function`). Subroutines execute instructions to manipulate and eventually modify variables, cells, sheets, workbooks. Functions calculate values of quantities or, more generally, objects: a function returns an output. Subroutines and Functions may use inputs – their arguments, which may be furnished by reference (if you want to modify them in the subprogram) or by value (if you want to use their value). Macros are Subroutines with no arguments. Subroutines with arguments do not appear in the list of Macros.

To illustrate these ideas, let us create a Function that receives as argument a range of cells and returns a vector containing the sine of each element of the given range.

Such a function may be written as follows:

```
Function funtest(r As Range) As Variant
Dim x() As Variant
x = r
Dim sinx() As Double
ReDim sinx(LBound(x, 1) To UBound(x, 1), LBound(x, 2) To UBound(x, 2))
For i = LBound(x, 1) To UBound(x, 1)
    For j = LBound(x, 2) To UBound(x, 2)
        sinx(i, j) = Sin(x(i, j))
    Next j
Next i
funtest = sinx
End Function
```

Open a blank workbook, add this function as a module. Enter the data below, select cell `A2` and enter "`=funtest(A1:D1)`".

	A	B	C	D
1	1	2	3	4
2	=funtest(A1:D1)			

The result is the following

	A	B	C	D
1	1	2	3	4
2	0,841471	0,909297	0,14112	-0,7568

Now, let us create a Subroutine that receives as argument a range of cells, calculates the sine of the range and shows the result in a region below the range:

```
Public Sub subtest(r As Range)
Dim sht As Worksheet
Set sht = ActiveSheet
xx = r.Value
m = UBound(xx, 1)
n = UBound(xx, 2)
ii = r(1, 1).Row
jj = r(1, 1).Column
For i = 1 To m
    For j = 1 To n
        sht.Cells(ii + m + i - 1, jj + j - 1) = Sin(xx(i, j))
    Next j
Next i
End Sub
```

As observed in the preceding, subroutines requesting parameters *are not* Macros and cannot be executed directly in a worksheet: it is necessary to provide a Macro for its execution. For instance, we may write a Macro that executes subtest for a range of cells having a name (see Sect. 1.10) or for a range of selected cells. For the last one, an example of code is:

```
Public Sub run_on_selection()
Dim r As Range
Set r = Selection
Call subtest(r)
End Sub
```

Open a blank workbook, add this function as a module. Enter the data below, select cells `A1:D1`

	A	B	C	D
1	1	2	3	4
2				

Go to *Developer* > *Macros* and execute the Macro `run_on_selection` (see page 8). The result is

	A	B	C	D
1	1	2	3	4
2	0,841471	0,909297	0,14112	-0,7568

1.6 How to Import a VBA Module

If you have a VBA module and you want to use it, you may load the module to your workbook. To do this, go to *Developer* > *Visual Basic* > *VBAProject*. Right Click and choose *Import File*. Your file tree opens: navigate to the folder where you can find the module, select it and click *Open*. For instance, we illustrate in the Figure below the importation of the module *RandomNumbers.bas*

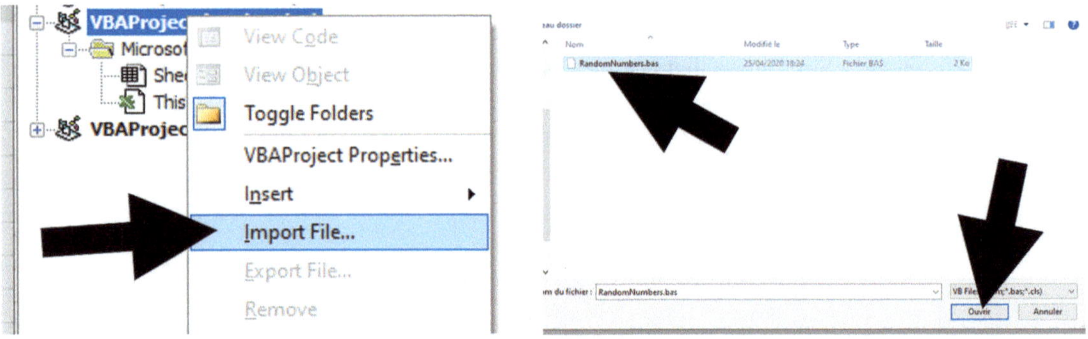

After these steps, you see a folder *Modules* in your VBA Project (it is created if not existing). Open it by a double click and you will see the module imported. For the importation of *RandomNumbers.bas* in the workbook *Book1*, you may see the result in the illustration below:

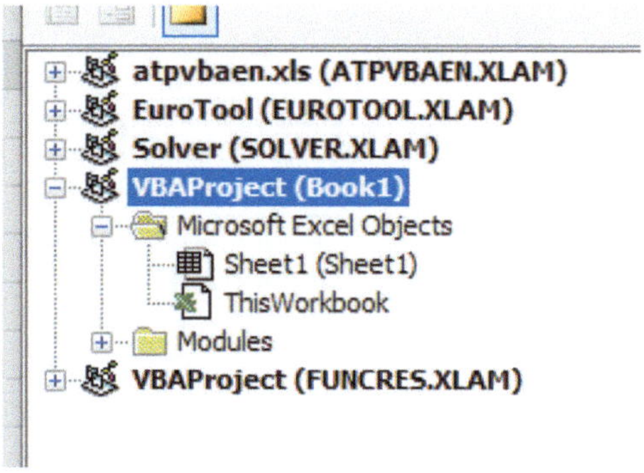

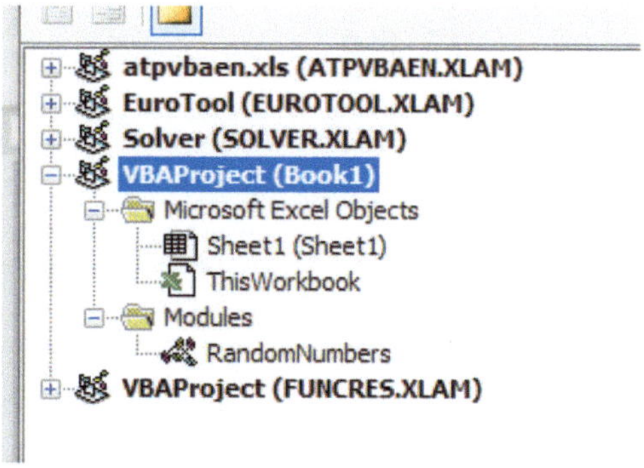

Now, you may use the module in your workbook. Recall that you must save it using the extension "`.xlsm`". Modules export to `.bas` files.

1.7 Matrix Formulas

This book refers to the recent versions of EXCEL, where Matrix Formulas are automatically executed. In the older versions of EXCEL, you must use the combination CTRL + SHIFT + ENTER to make matrix calculations: ENTER is pressed while holding CTRL and SHIFT simultaneously pressed.

CTRL: 

SHIFT: 

ENTER: 

Unnecessary for Matrix calculations in the recent versions of EXCEL

1.8 Fixing Volatile Formulas

Some functions of EXCEL are *volatile*, id est, they are recalculated each time something changes in any cell of an open workbook. For instance, random numbers generated by built-in functions are volatile: the values generated are modified each time the workbook is opened or any cell is modified. To get fixed values, we must disable such a recalculation: go to *File > Options > Formulas*

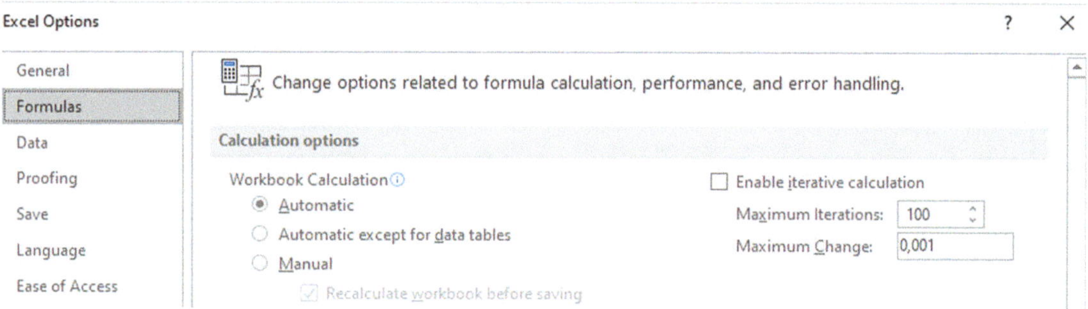

and modify the "Calculation Options" to disable the automatic recalculation. Modify the options as follows:

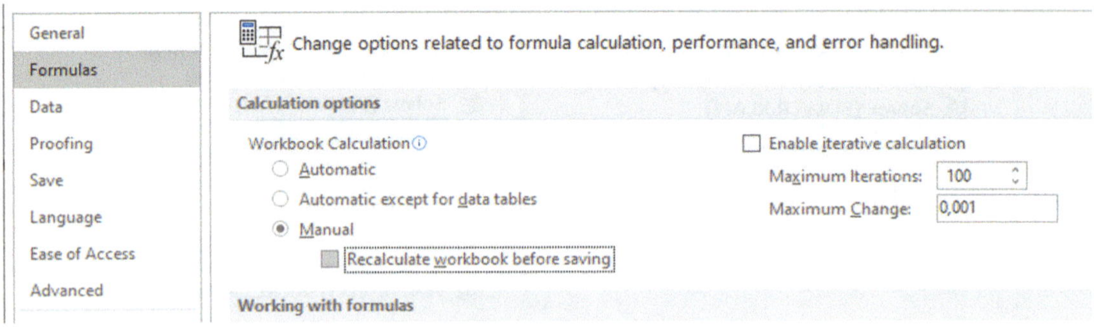

Now, the calculation is not automatic: if you desire to recalculate your workbook, you must make it manually by clicking on *Formulas > Calculate Now*

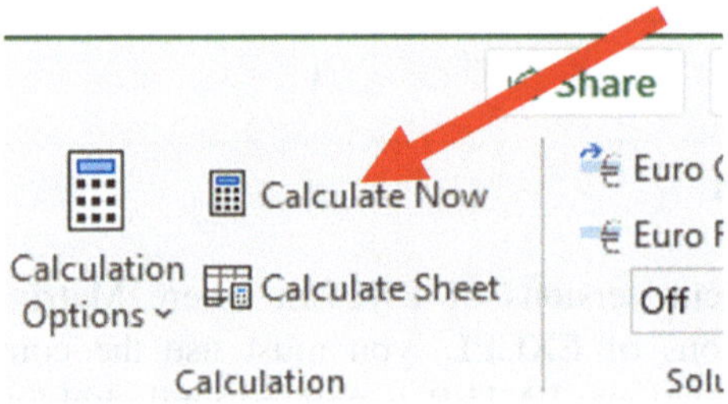

1.9 Using Addresses of Cells

Under EXCEL®, Cells may be referred

- either by the classical style "A1", where a cell is referred by letters and numbers: A2, B4, C1 and so on.
- Or by the style "R1C1", where a cell is referred by a pair (i,j) = (row number, column number). For instance, A2 = (2,1), B4 = (4,2), C1 = (1,3).
  You may set your choice in *Options > Formulas* by checking or unchecking *R1C1 Reference Style*

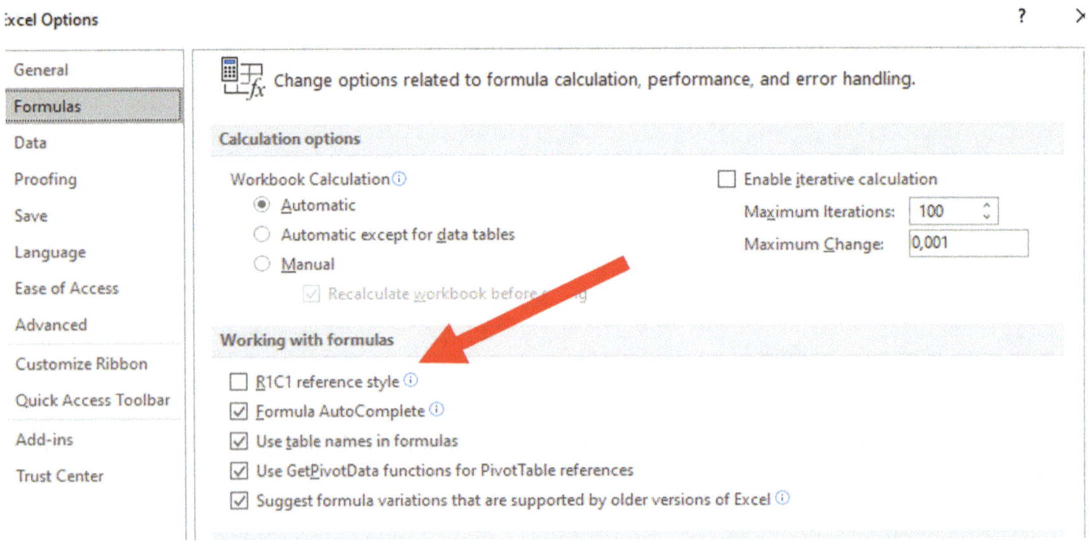

A collection of cells forms a *Range* of cells. In VBA, *Range* is used to refer to a collection of cells. For instance, `r = Range("A2")` or `r = Range(Cells(2,1).Address)` create a range referring to the cell A2. For multiple cells, use `r = Range("A2:A4")` or `r = Range(Cells(2,1).Address, Cells(4,1).Address)`. Both these instructions refer to the cells A2, A3, A4. Analogously, `r = Range("A2:B4")` or `r = Range(Cells(2,1).Address, Cells(4,2).Address)` refer to the cells A2, A3, A4, B2, B3, B4. You may also assign a name to the collection of cells to manipulate it (see next section).

By default, EXCEL® uses relative addresses in formulas typed in the worksheet. For instance, if you type in cell C1 the formula "= `A1 + B1`", copy it and paste it in the cell G3, you will get "= `E3 + F3`"

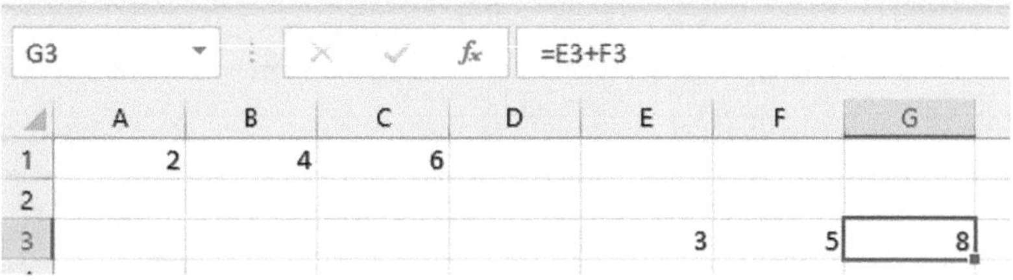

To keep absolute addresses, you must use a "$" before the index that you want to get fixed. For instance, "= `$A1 + B1`" keeps the column A fixed; "= `A$1 + B1`" keeps the row 1 fixed; "= `$A$1 + B1`" keeps the cell A1 fixed. If you copy the formula in C3 and you past it in G3, you will get different results in each case: "= `$A3 + F3`", "= `E$1 + F3`", "= `$A$1 + F3`", respectively (Figures below).

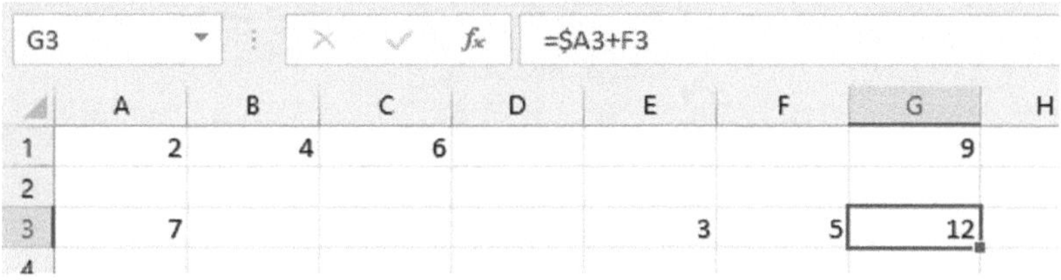

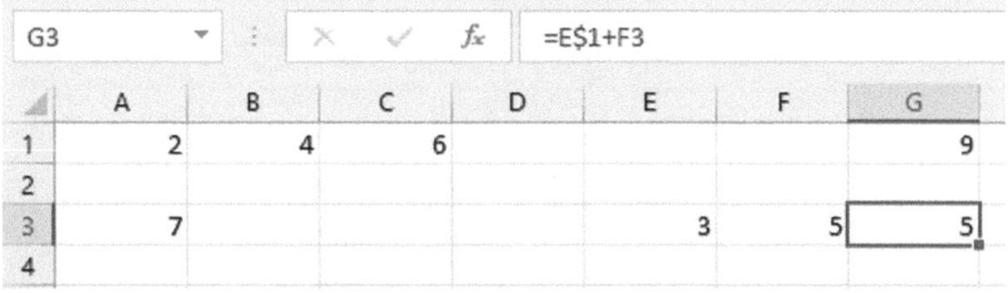

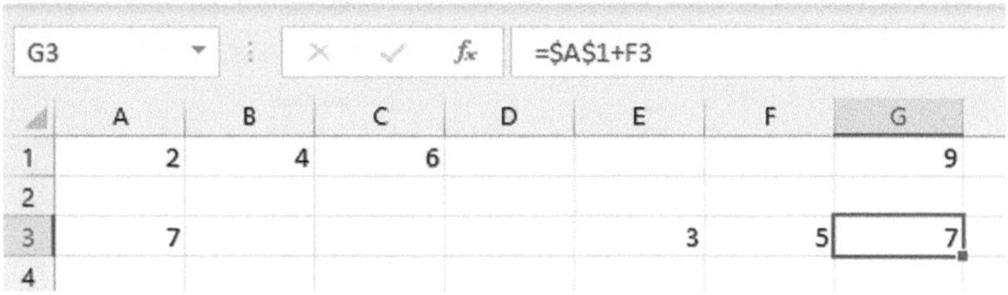

1.10 Using Names in EXCEL®

A practical way to manipulate data in EXCEL® is the use of names. To assign a name to a group of cells, you must select the group of cells. Then, **either** use the right click to select *Define Name . . .* **or** go to *Formulas > Define Name* and, then give a name to the selected cells. Let us illustrate the procedure: open a workbook and

Enter the data and select the cells A1:B2

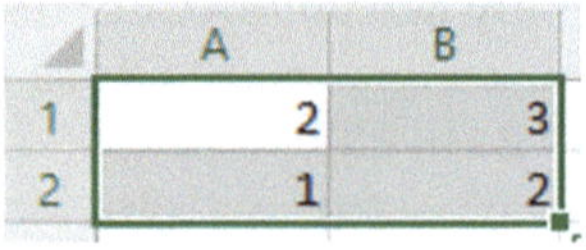

Go to *Formulas > Define Name*

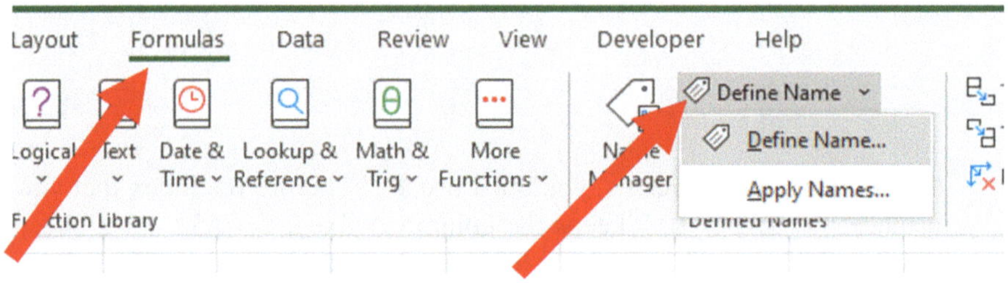

Or use the right click and select *Define Name . . .*

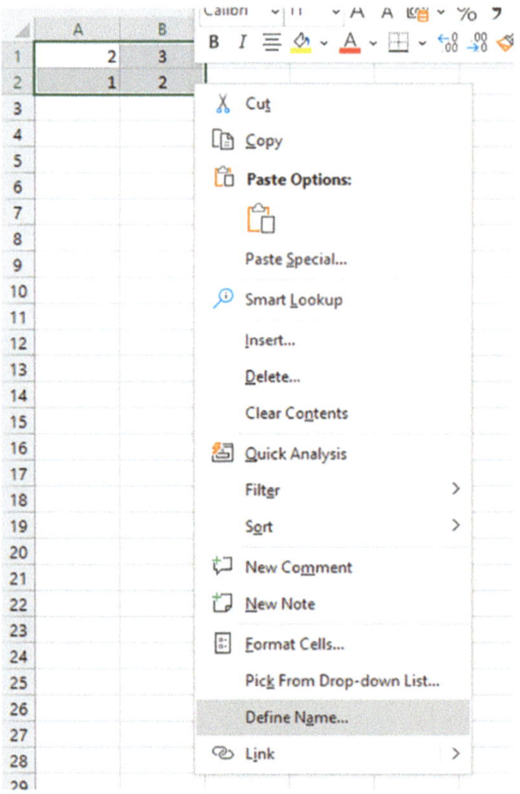

Give a name to the group of cells (here, A). You may add a comment. Then, click *OK*.

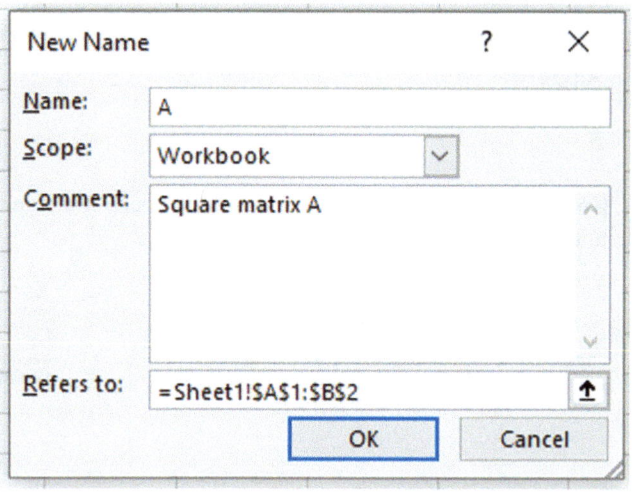

Now, you can manipulate all the cells by using the name chosen: select the cell `D1` and type "= `A/SUM(A)`". You will get the results below:

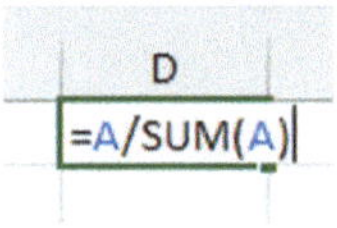

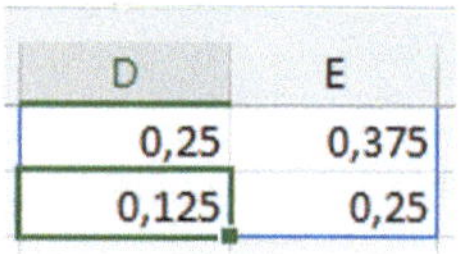

Select the cell G1 and enter "= `MINVERSE(A)`". You will get the inverse of A:

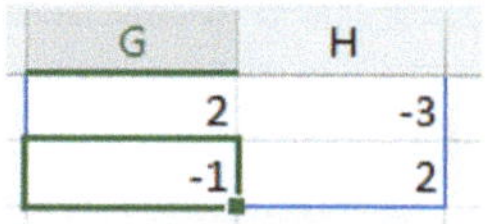

Select the cell `A4` and enter "= `INDEX(A;1;0)`". You will get the first line of `A`:

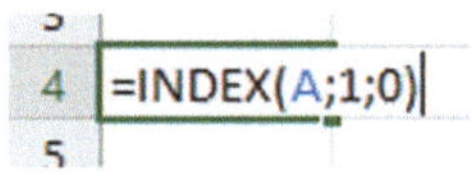

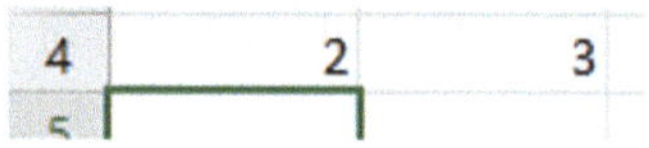

Analogously, select cell `A6` and enter "= `INDEX(A;0;2)`". You will get the second column of `A`:

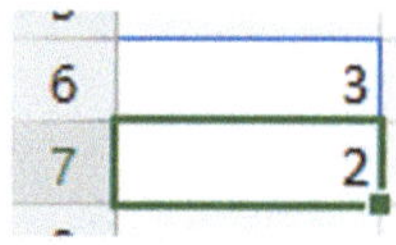

Select cell `J1` and enter "= `INDEX(A;2;1)`". You will get the value of `A(2,1)`:

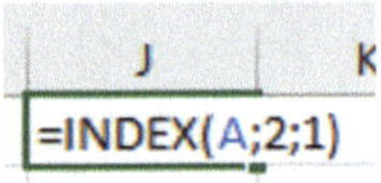

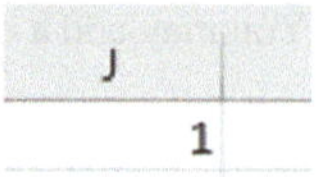

Going to *Formulas* > *Name manager*, you can manage the names: see the names defined, modify them or delete names.

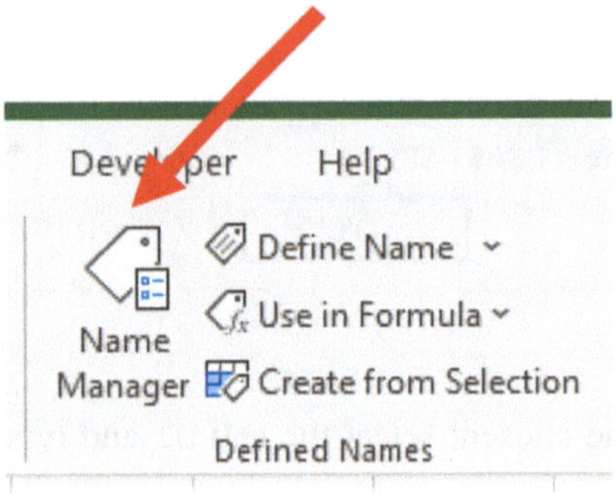

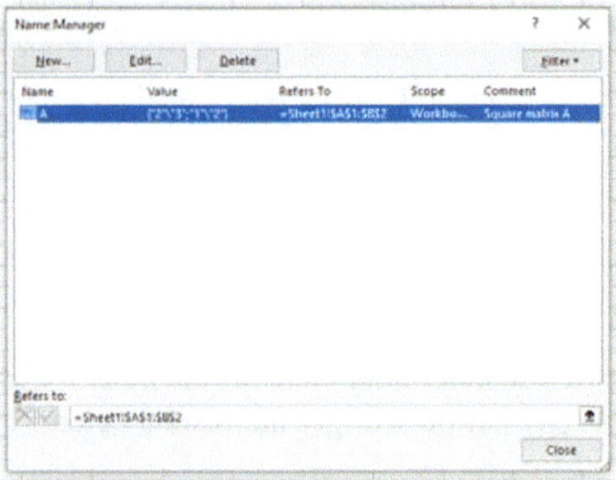

Recall that a name is a range of values, not a table or matrix. To refer to an element of, we must use `INDEX`: using `A(i,j)` instead `INDEX(A;i;j)` will produce an error. To transform `A` a into a matrix, you need to use VBA:

```
nr = "[" & ActiveWorkbook.name & "]" & ActiveSheet.name & "!" & "A"
Dim a As Variant
a = Range(nra)
```

Then, `a` is a matrix that can be manipulated by VBA.

1.11 How to Run the SOLVER

The SOLVER minimizes or maximizes a target cell by changing the values of variable cells. Constraints may be introduced. A Dialog Box allows the definition of the parameters. As an example, let us look for the solution of the following problem:

$$X = \arg\min\left\{\sum_{i=1}^{3} x_i^2 : x_1 + x_2 = 0, x_2 + x_3 \geq 3\right\}.$$

This problem has as exact solution $x_1 = -1$, $x_2 = 1, x_3 = 2$. Open a blank workbook and enter the data below: `1,2,3` in cells `A1` to `A3`; give the name `X` to these cells; "=`A1 + A2`", in `B1`; "=`A1 + A2`" in `B2`; "= `SUMSQ(X)`" in cell `C1`.

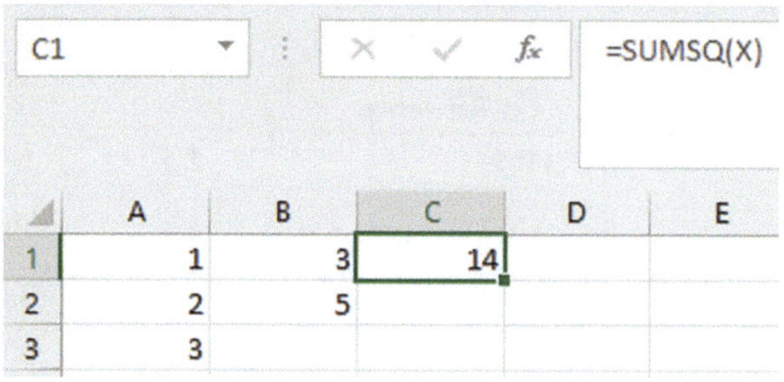

Then, go to *Data* > *Solver*: clicking on "*Solver*" opens the Dialog:

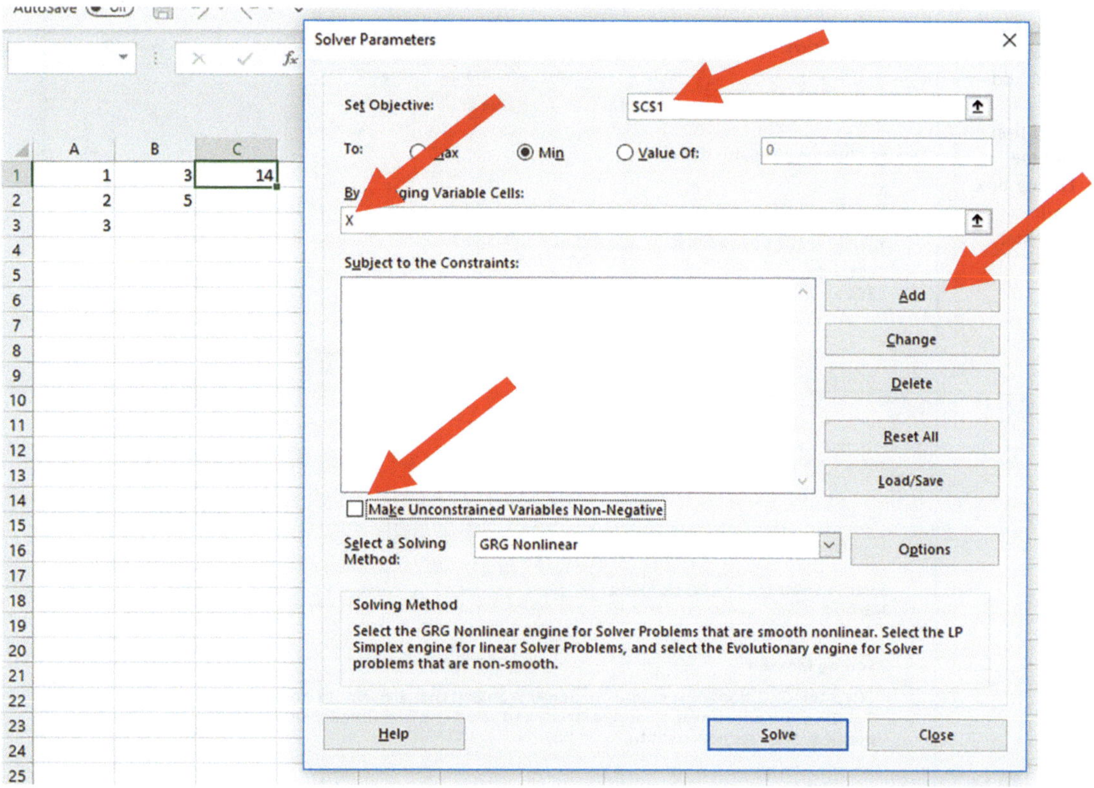

Enter the required information: Objective is the target cell `C1`; Variable cells are `X`. Uncheck "Make Unconstrained Variables Non-Negative" (except if you are looking for non-negative solutions). To introduce the constraints, click on "Add", select the cells that evaluate the constraints and enter the value of the constraint. For instance, the restriction $x_1 + x_2 = 0$ corresponds to cell `B1` equal to zero:

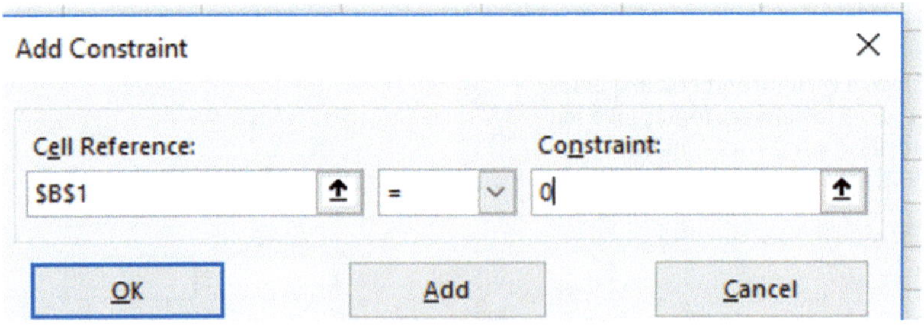

The condition $x_2 + x_3 \geq 3$ corresponds to cell B2 $\geq$ 3. Add this condition and the Dialog box will become:

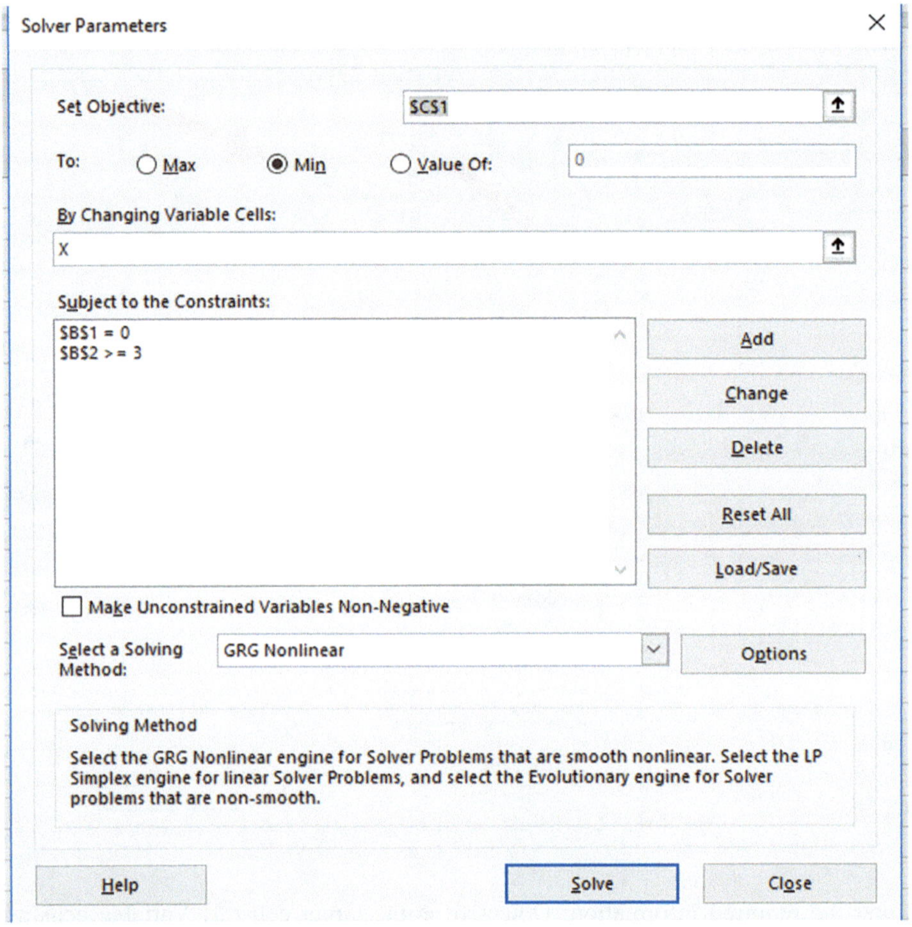

Then, click on "Solve". A new Dialog Box will inform that the Solver has found a solution. Click on OK and you will have the solution in cells `A1:A3`.

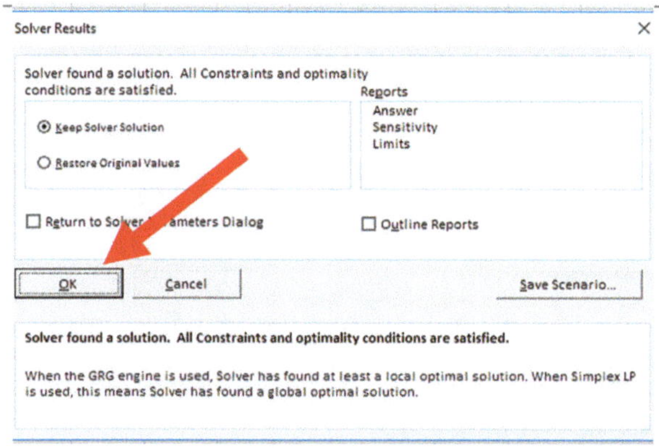

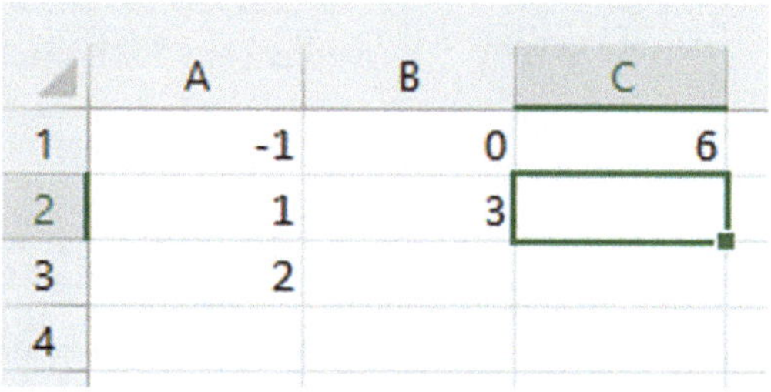

The SOLVER may be invoked in Modules. For instance, insert a module as shown in page 7. Then, enter the code below:

```
Sub solve_problem()
rT = ActiveCell.Address
ActiveCell.FormulaArray = "=SUMSQ(X)"
SolverOk SetCell:=rT, MaxMinVal:=2, ValueOf:=0, ByChange:="$A$1:$A$3",
        Engine:=1, EngineDesc:="GRG Nonlinear"
    SolverAdd CellRef:="$B$1", Relation:=2, FormulaText:="0"
    SolverAdd CellRef:="$B$2", Relation:=3, FormulaText:="3"
    SolverOptions AssNonNeg:=False
    SolverSolve True
End Sub
```

To run this code, it is mandatory to include the SOLVER in the references of the VBA project: go to *Developer> Tools>References*

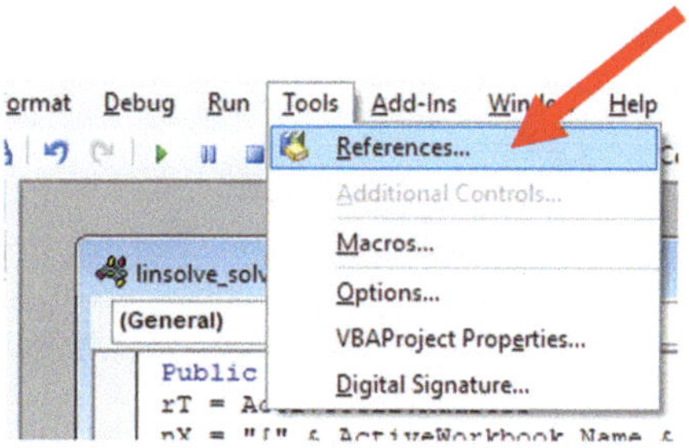

Click on reference to open the Dialog Box, check "Solver" and click OK.	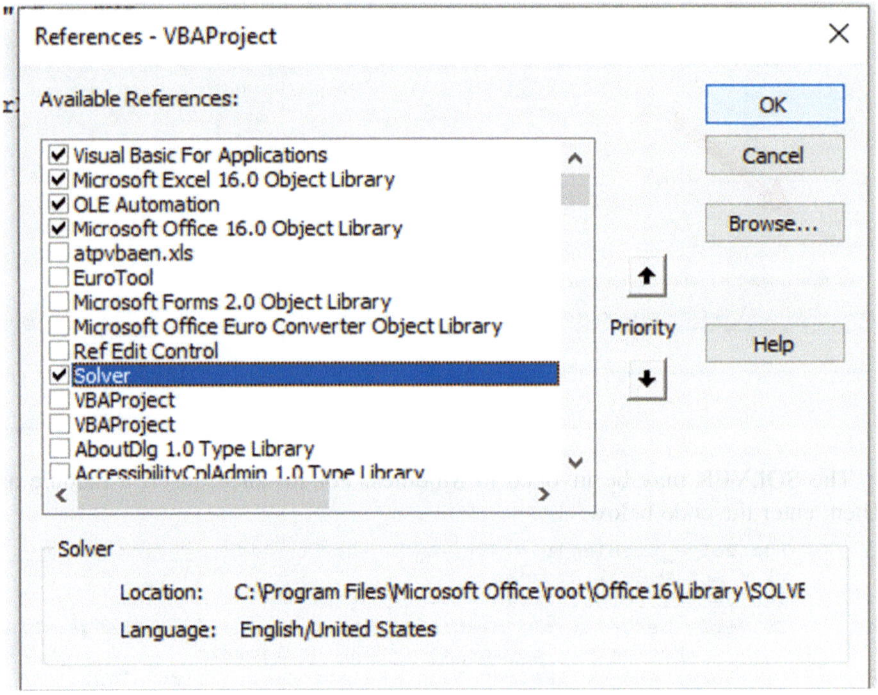

Go to the workbook, select any empty cell. Then, go to *Developer* > *Macros*. Click on *Macros* to open the Dialog Box and Run the macro	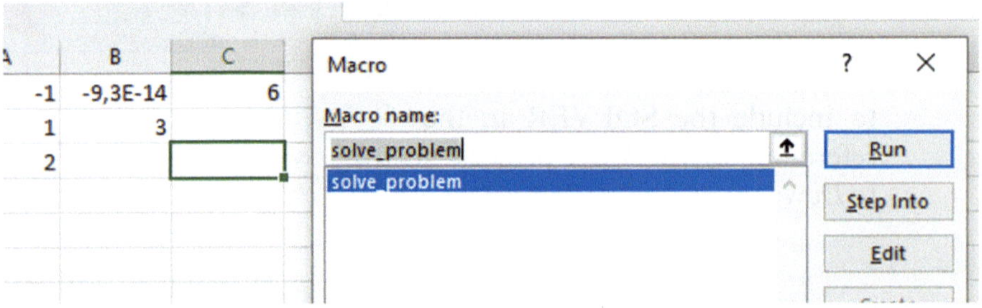

The results appear in the cells `A1:A3`	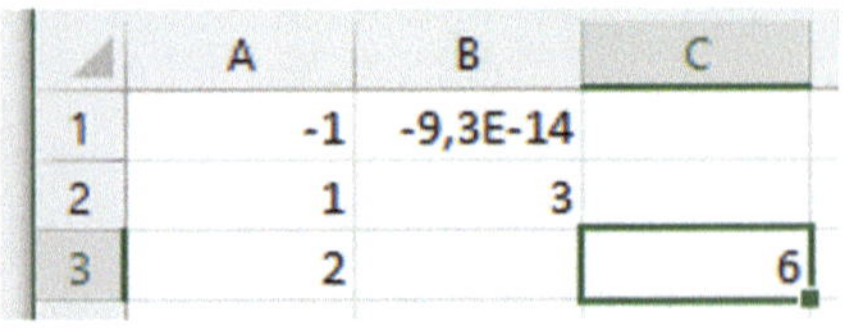

The SOLVER offers the option of using multistart methods and evolutionary algorithms. To use a multistart method, you must select it in the options, indicated by an arrow in the Figure above. Clicking on "options" opens the dialog shown at left, which allows to select "multistart" (or other options).

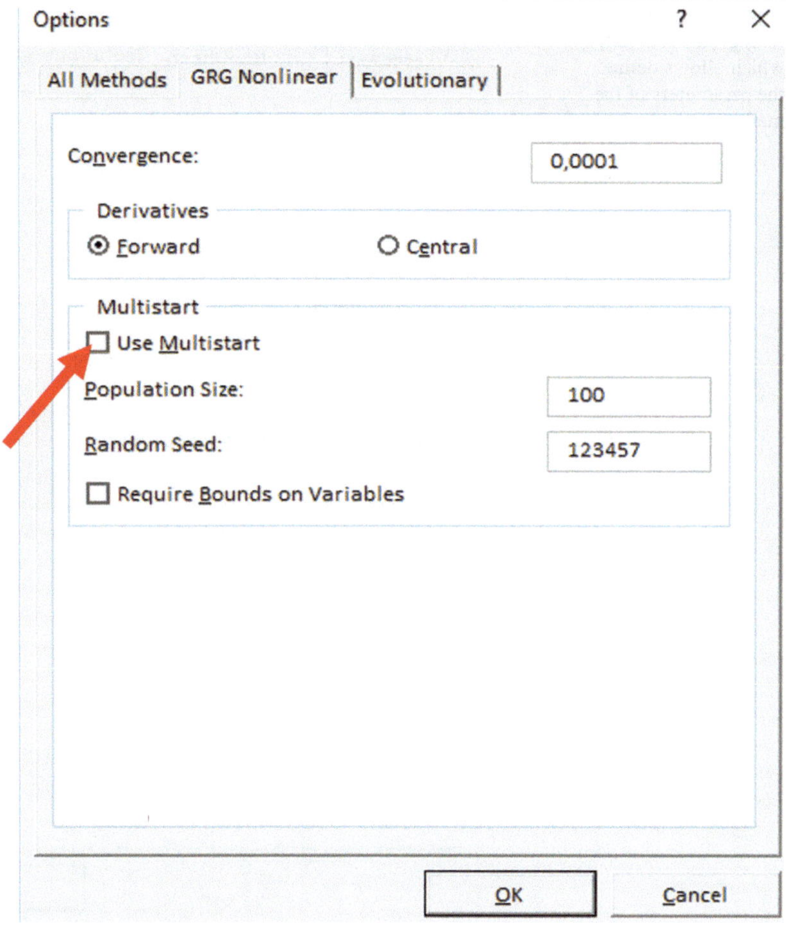

To use an evolutionary algorithm, the selection is made at "Select a Solving Method".

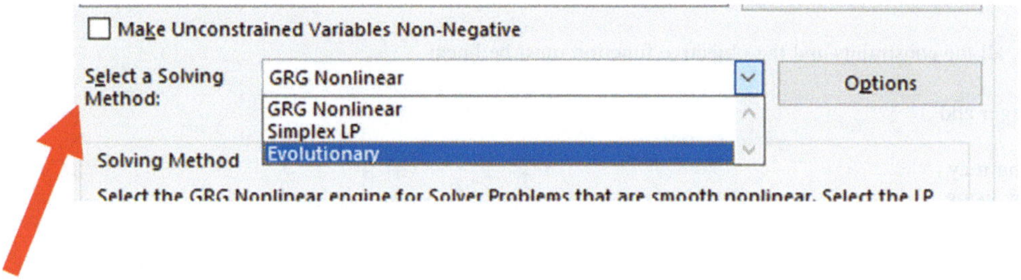

Clicking on "options" opens the dialog shown at left, which allows define the parameters of the method.

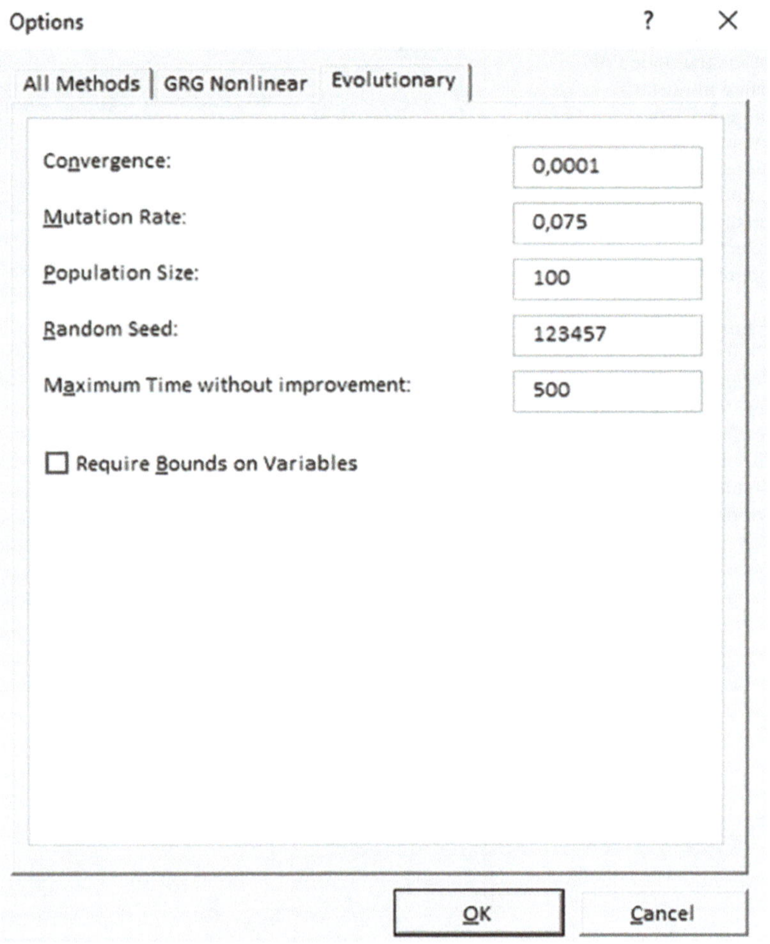

The SOLVER also proposes the Simplex Method for Linear Programming:

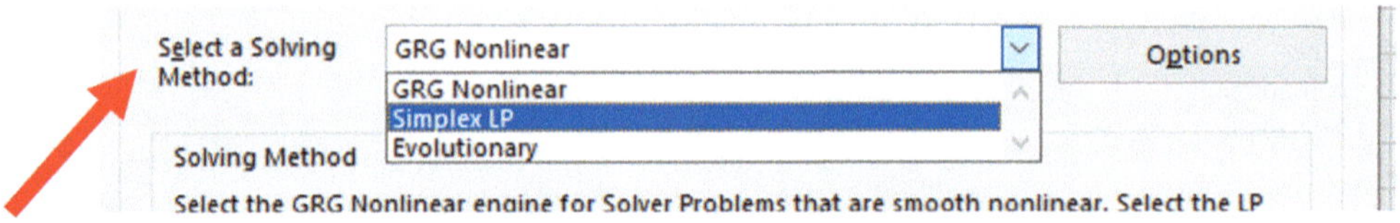

In this case, all the constraints and the objective function must be linear.

Binary, Integer and mixed programming may be solved by using adequate constraints.

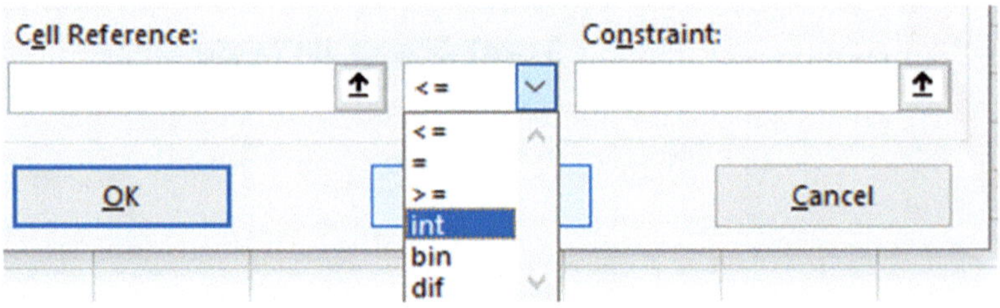

1.12 How to Include Iterative Calculations in Your Workbook

Iterative calculations are often used to solve numerical problems. An iterative method usually involves an iteration function ψ and generates a sequence $x_{n+1} = \psi(x_n)$, starting from a given x_0. Thus, an iterative method is implemented as follows:

1. Choose the initial point x_0, the maximal iteration number n_{max} and a precision *prec*: the iterations stop if $n \geq n_{max}$ or $|x_{n+1} - x_n| \leq prec$;
2. Set $n = 0$, $x_n = x_0$.
3. Evaluate $x_{n+1} = \psi(x_n)$
4. Increment n: $n \leftarrow n + 1$, evaluate $d = |x_{n+1} - x_n|$.
5. Set $x_n = x_{n+1}$.
6. If $d \leq prec$ or $n \geq n_{max}$ then stop, else go to 3.

Under EXCEL®, iterative methods may be implemented in VBA or directly on a worksheet. To use VBA, you must add a module which implement the iterations. For instance, the code at right:

```
Function iteration(x0, nmax, prec)
xn = x0
n = 0
d = prec + 1
Do Until n > nmax Or d < prec
    xnew = psi(xn)
    n = n + 1
    d = Abs(xnew - xn)
    xn = xnew
Loop
iteration = xn
End Function
```

Let us exemplify the use with the classical Newton's method for the solution of the equation $f(x) = 0$: in this case, $\psi(x) = x - f(x)/f'(x)$. For instance, consider $f(x) = x^3 - 27$ – the solution is $x = 3$. You must add the modules evaluating ψ, f and f', as shown at right.

```
Function psi(x)
psi = x - f(x) / df(x)
End Function
```

```
Function f(x)
f = x * x * x - 27
End Function
```

```
Function df(x)
df = 3 * x * x
End Function
```

Open a blank workbook, include these modules, and enter the data below in the first sheet of the workbook. Give the names `x0` to `B1`, `nmax` to `B2`, `prec` to `B3`.

	A	B
1	x0	2
2	nmax	100
3	prec	1,00E-10
4		

Then, select an empty cell and enter "=iteration(x0; nmax;prec)". You will get the result.

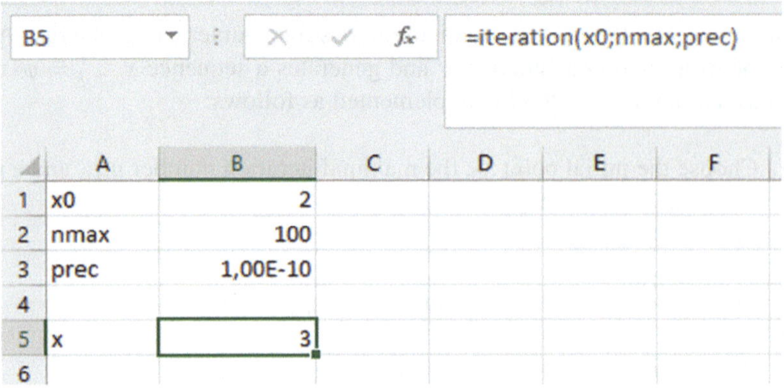

To make the calculations directly in the worksheet, you must activate the iterations: open a blank workbook, go to *File > Options > Formulas*, Check *"Enable Iterative Calculation"*, enter the values of *nmax* and *prec*, click OK.

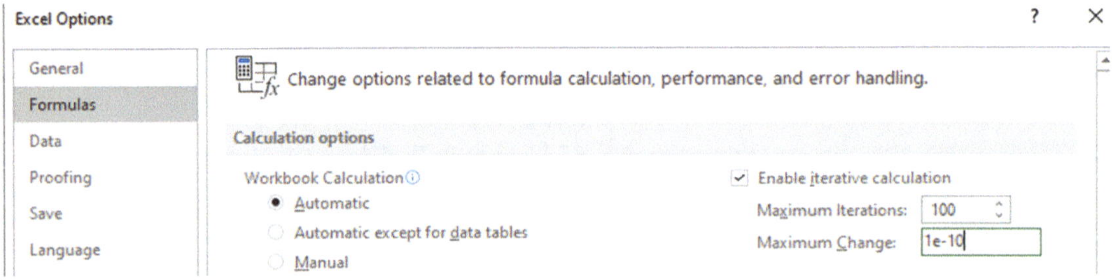

Go to the first Sheet an enter the data at right. Give the name x to B2. Cell B2 contains the formula "=B1". Cell B3 contains "=x-(x^3-27)/(3*x^2)"

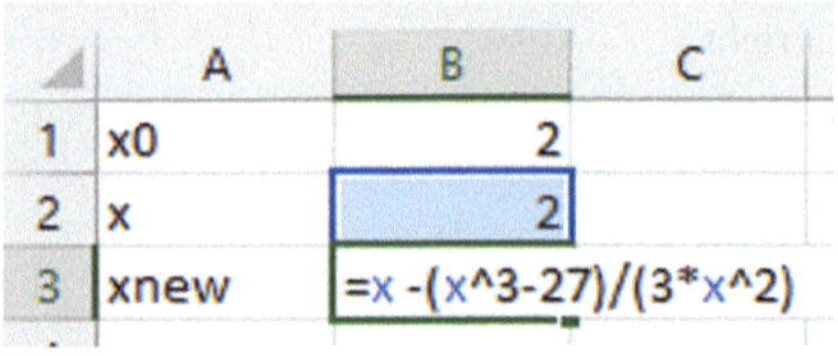

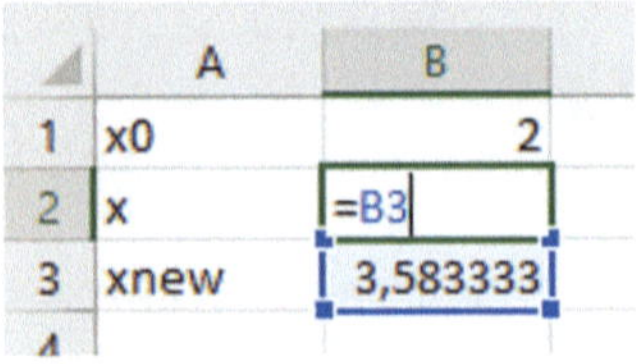

Then, Enter "=B3" in cell B2 (at left) You will get the result at right.

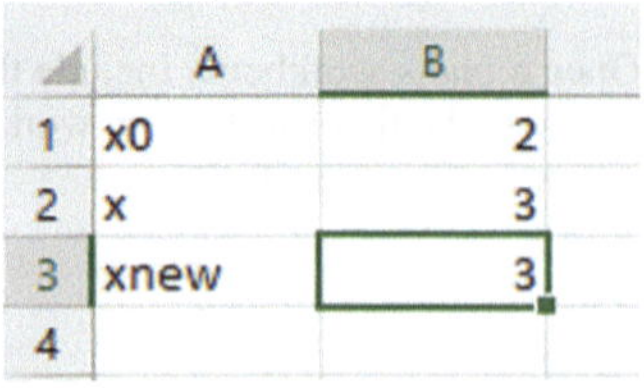

1.13 How to Include a Control in Your Workbook

EXCEL® offers the possibility of adding COM Controls, such as buttons. For instance, you may include:

A button that runs a Macro when clicked

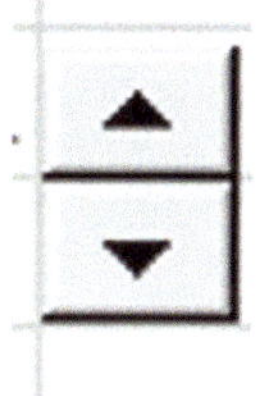

A scroll bar (at left) or a spin button (at right) that automatically increases or decreases the value of a cell, with a given step.

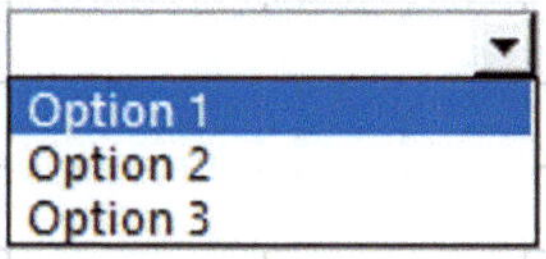

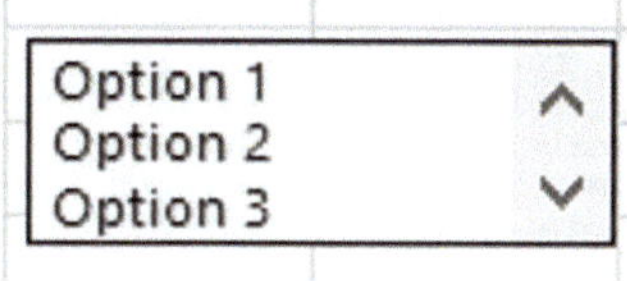

Combo (at left) and List (at right) Boxes to provide a way to select options listed in cells. In a Combo, a single option must be selected. In a List, you may authorize or not multiselection.

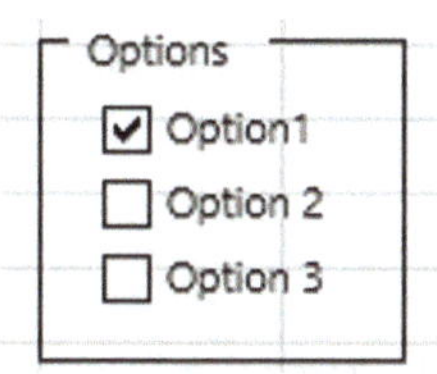

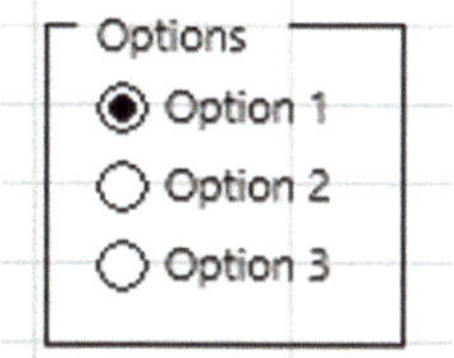

A Check Box (at left) or a panel of Radio Buttons (at right) provide an alternative way of choosing options. Check Boxes allow multiselection and give the value "TRUE" or "FALSE" to specified cells. Radio Buttons impose a single selection and give a numerical value to a cell, correponding to the chosen option.

Let us illustrate the inclusion of a button that gives the name X to a vector in line 1, gives the name Y to the vector in line 2, then calculates the scalar product of X and Y and shows the result in cell C3. The button will be placed over cells A3:B3 (controls cannot be inserted into cells). Open a blank workbook, enter the data below, go to *Developer > Insert > Form Controls* and select a Button:

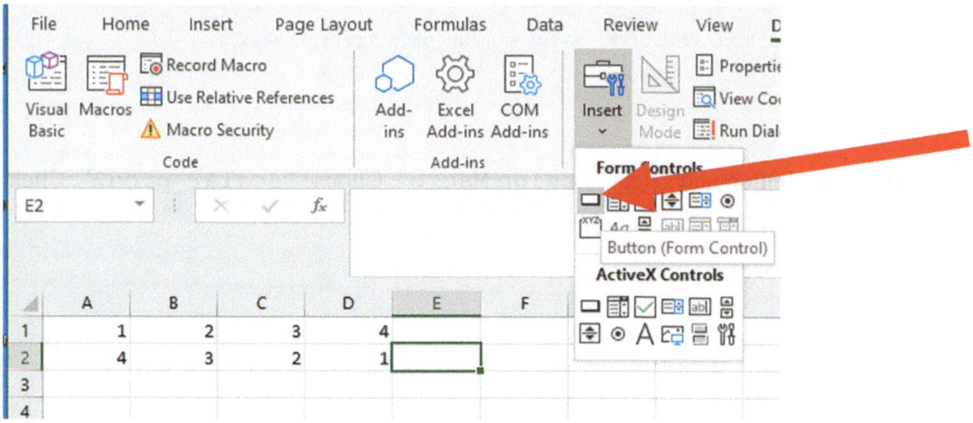

Go to cell A3, hold a left click and draw the button on A3:B3. A Dialog Box opens: change the name of the Macro and click "New".

The VBA window for Module insertion appears. Enter the code below, save it and come back to the workbook.

```
Sub scalar_prod()
l1 = Cells(1, Columns.Count).End(xlToLeft).Column
l2 = Cells(2, Columns.Count).End(xlToLeft).Column
Names.Add Name:="X", RefersToR1C1:="=" & ActiveSheet.Name & "!R1C1:R1C" & CStr(l1)
Names.Add Name:="Y", RefersToR1C1:="=" & ActiveSheet.Name & "!R2C1:R2C" & CStr(l2)
Cells(3, 3).FormulaR1C1 = "=SUMPRODUCT(X,Y)"
End Sub
```

The button shadows and the text "`Button 1`" appears. A right click on the button allows to edit the text: change it to "`scalar product:`". click on the button: the result appears in cell `C3`.

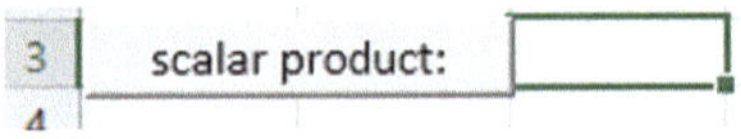

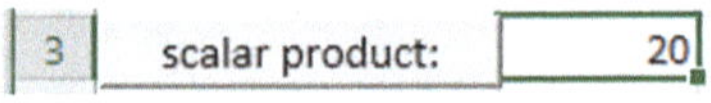

Go to *Developer > VBA*, edit the code of Module 1. After the "`End Sub`" of "`scalar_prod`", add a new sub as follows:

```
End Sub
Sub scalar_prod2()
l1 = Cells(1, Columns.Count).End(xlToLeft).Column
l2 = Cells(2, Columns.Count).End(xlToLeft).Column
Names.Add Name:="X", RefersToR1C1:="=" & ActiveSheet.Name & "!R1C1:R1C" & CStr(l1)
Names.Add Name:="Y", RefersToR1C1:="=" & ActiveSheet.Name & "!R2C1:R2C" & CStr(l2)
nX = "[" & ActiveWorkbook.Name & "]" & ActiveSheet.Name & "!" & "X"
rX = Range(nX)
nY = "[" & ActiveWorkbook.Name & "]" & ActiveSheet.Name & "!" & "Y"
rY = Range(nY)
s = WorksheetFunction.MMult(rX, WorksheetFunction.Transpose(rY))
Cells(3, 4).Value = s
End Sub
```

Right click on the button and select "`Assign Macro`". Choose "`scalar_prod2`", Click "`OK`".

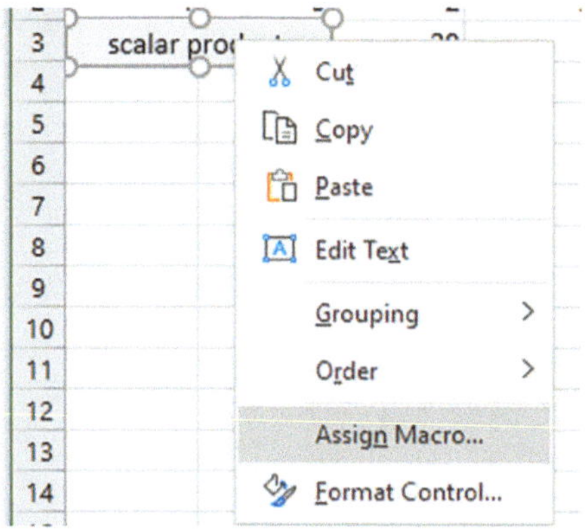

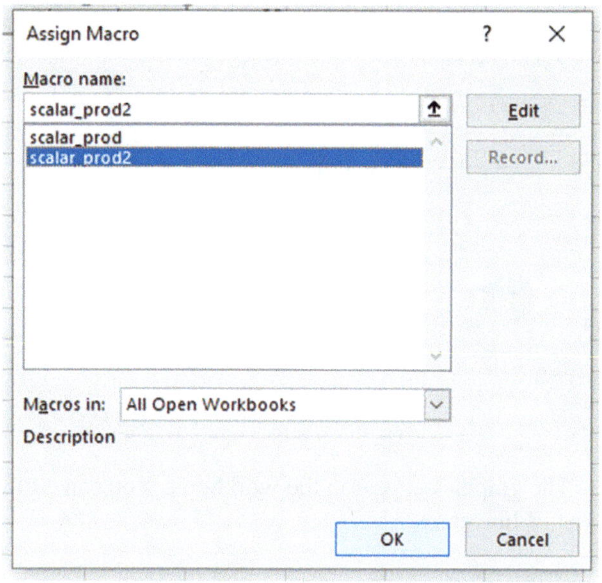

Click Again on the button. The same result appears in Cell D3.

1.14 How to Include a Chart in Your Workbook

Graphics may be inserted directly by VBA modules or by using the assistant in the workbook. In both the cases, you must define the data (x_i, y_i), $1 \leq i \leq np$ to be used. The data may also be defined directly by VBA modules or be taken from the workbook.

Let us exemplify the process by inserting a simple graphics of the function $y = x^2$:open a blank workbook and enter the data below. Giving the name "x" to `B1:L1` allows to enter "=x^2" in `B2`.

	A	B	C	D	E	F	G	H	I	J	K	L	M
1	x	-1	-0,8	-0,6	-0,4	-0,2	0	0,2	0,4	0,6	0,8	1	
2	y	1	0,64	0,36	0,16	0,04	0	0,04	0,16	0,36	0,64	1	

Select `A1:L2`, go to Insert and click on "`Recommended Charts`"

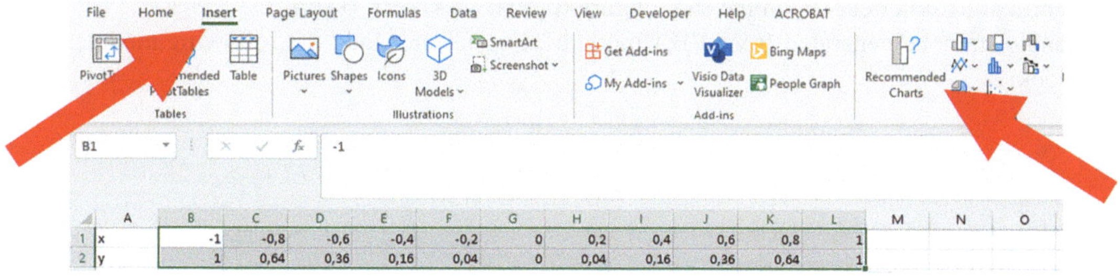

If you see a convenient type of chart, select it and click "OK". Otherwise, click on "all charts" to see all the possibilities.

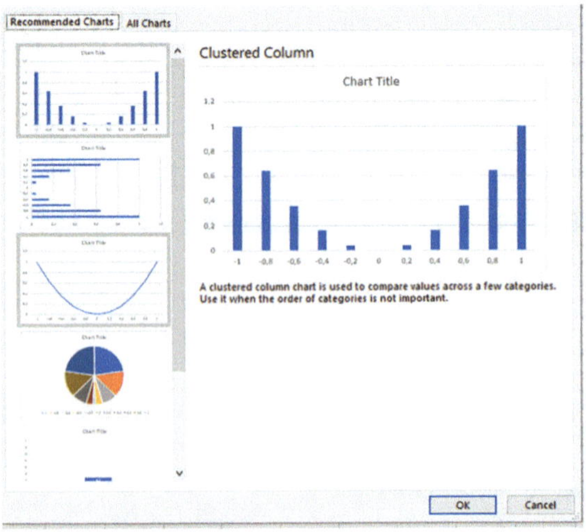

The graphics appear in the workbook. You can edit it, to add supplementary data, change color or type of line etc.

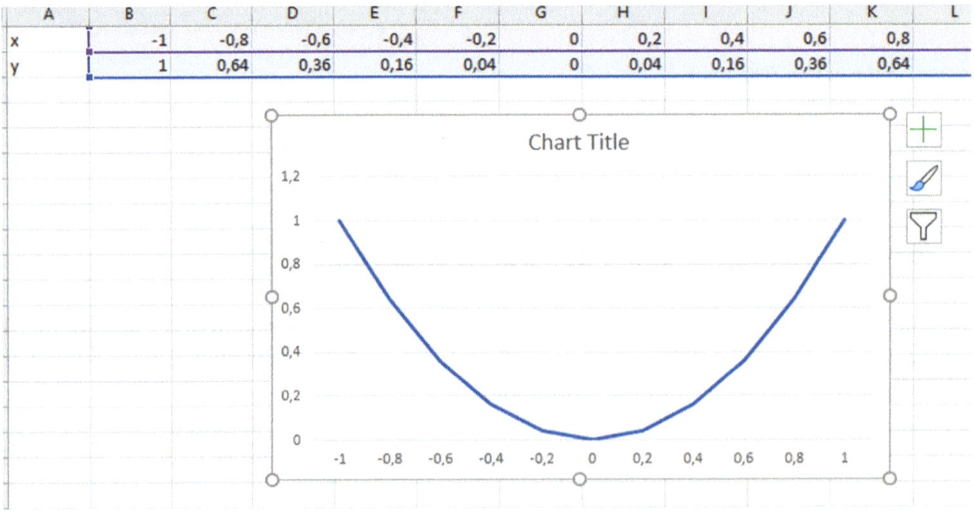

Click on the chart and you will see at the right side three options: Chart Elements, Chart Styles, Chart Filters. Chart Elements will modify Title, axes properties, legend, labels – it may also add a trendline (linear, logarithmical, polynomial etc.). Chart Styles provides modifications of the line parameters, color, markers etc. Chart Filter gives the possibility of modifying, including, or removing points or adding other curves on the same figure (option "`select data`").

Graphics may be generated from VBA modules. For instance, add the following modules to the workbook:

```
Sub graph1(gdata As Variant, ttl As String)
ActiveCell.Select
ActiveSheet.Shapes.AddChart2(227, xlLine).Select
Application.CutCopyMode = False
Application.CutCopyMode = False
For i = 1 To UBound(gdata, 1)
    ActiveChart.SeriesCollection.NewSeries
    ActiveChart.FullSeriesCollection(i).Values = gdata(i, 2)
    ActiveChart.FullSeriesCollection(i).XValues = gdata(i, 1)
    ActiveChart.FullSeriesCollection(i).Name = gdata(i, 3)
Next i
ActiveChart.HasLegend = True
ActiveChart.Legend.Position = xlLegendPositionBottom
ActiveChart.ChartTitle.Select
ActiveChart.ChartTitle.Text = ttl
End Sub
```

```
Function readline(line, j1, Optional ns)
Dim x() As Double
If IsMissing(ns) Then
    ns = Cells(line, Columns.Count).End(xlToLeft).Column
End If
ReDim x(1 To ns - j1 + 1)
For j = j1 To ns
    x(j - j1 + 1) = Cells(line, j).Value
Next j
readline = x
End Function
```

```
Sub addgraph()
Set ws = ActiveSheet
x = readline(1, 2)
y = readline(2, 2)
ngraph = 2
Dim gdata() As Variant
ReDim gdata(1 To ngraph, 3)
gdata(1, 1) = x
gdata(1, 2) = y
gdata(1, 3) = "y = x^2"
gdata(2, 1) = x
gdata(2, 2) = x
gdata(2, 3) = "y = x"
Dim ttl As String
ttl = "two functions of x"
Call graph1(gdata, ttl)
End Sub
```

Function `readline(i,j)` reads the data contained in line `i` starting at column `j`. Optionally, you may indicate the last column `ns` to be read.

Function `graph1` adds a chart containing a variable number of graphs to the active worksheet. The title of the graph is a string passed by the variable `ttl`. The series of data are passed by the argument `gdata`. Each line of `gdata` contains a series of data. `gdata(i,1)` contains the values of the abscissa x, `gdata(i,2)` contains the ordinates y, `gdata(i,3)` contains the name of the series, as a string.

The macro `addgraph` reads the data in the lines 1 and 2 and adds a chart containing two graphs (parameter `ngraph`) to the Active Worksheet, corresponding to two series of data: $y = x^2$ and $y = x$. The result is shown in the figure below.

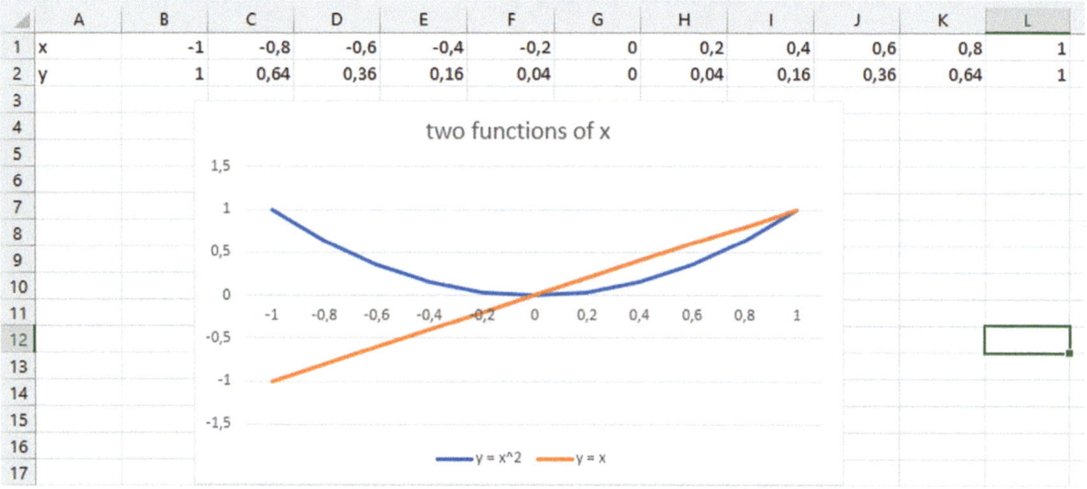

By adding supplementary instructions, you may control the type of line, its color and width ad many features of the chart. In addition, x and y may be generated by other modules, instead of reading the data in the lines 1 and 2 of the workbook.

1.15 How to Use a Variant to Store Anything in a Variable

VBA offers various types of variables, among them the type called `"Variant"`, which is useful to store other variables, such as matrices, vectors, strings etc.

The creation of a variable in VBA is made by an instruction `Dim`. For instance, `Dim k as Long` creates an integer variable named `k`; `Dim x as Double` creates an real variable named `x` and so on.

To create a vector or a matrix, VBA offers VBA offers a very convenient way with the dynamic creation in two steps: `Dim` and `ReDim`. For instance, open a blank workbook and insert the module at right. It creates a vector `x` with indexes from −2 to 4 and fills it with the values 1 to 7. The values are printed in cells `A1` to `A7`.

```
Sub test()
Dim x()
n = 4
ReDim x(-2 To n)
For i = -2 To n
      x(i) = i + 3
Next i
k = 0
For i = LBound(x, 1) To UBound(x, 1)
  k = k + 1
  Cells(1, k).Value = x(i)
Next i
End Sub
```

Running the macro test produces the result below:

	A	B	C	D	E	F	G
1	1	2	3	4	5	6	7

Add a second `Sub` to your workbook as shown at right: this `Sub` creates a matrix `x(i,j)` having indexes from 2 to 3 for `i` and 0 to 4 for `j`.

```
Sub test2()
Dim x()
n = 4
m = 3
ReDim x(2 To m, 0 To n)
For i = 2 To m
   For j = 0 To n
    x(i, j) = i + j
   Next j
Next i
k = 0
For i = LBound(x, 1) To UBound(x, 1)
  For j = LBound(x, 2) To UBound(x, 2)
     Cells(i + 1, j + 1).Value = x(i, j)
  Next j
Next i
End Sub
```

Run the macro `test2`. You will get the result below:

	A	B	C	D	E	F	G
1	1	2	3	4	5	6	7
2							
3	2	3	4	5	6		
4	3	4	5	6	7		

By the same method, we can create vectors, matrices and tensors of the type `Variant`. These variables may be used to store and manipulate any type of variables – including Variants.

Open a blank workbook and add the code at right. It creates vectors of variants `v1`, `v2`, `v3`. Notice that
`v3(-1)(2)` = `v2(2)` = `v1`,
`v3(-1)(2)(2)` = `v1(2)` = `x`,
`v3(-1)(2)(2)(-1)` = `x(-1)` = 2.
Analogously,
`v3(-2)(2)(0)` = `v1(2)` = `x(0)` = −1;
`v2(2)(0)` = `v1(0)` = `name1`;
`v3(-1)(2)(1)`=`v2(2)(1)`=`v1(1)`= `name2`
Finally notice that `v2` was not declared as Variant: VBA considers that a new variable is a Variant, except if another type is declared.

```
Sub testvariant()
name1 = "i am a cell"
name2 = "123456789"
Dim x(-1 To 1) As Double
x(-1) = 2
x(0) = -1
x(1) = 0
Dim v1(0 To 2) As Variant
v1(0) = name1
v1(1) = name2
v1(2) = x
Dim v2(1 To 2)
v2(1) = x
v2(2) = v1
Dim v3(-2 To -1) As Variant
v3(-2) = v1
v3(-1) = v2
Cells(1, 1).Value = v3(-2)(2)(0)
Cells(2, 1).Value = v2(2)(0)
Cells(3, 1).Value = v3(-1)(2)(2)(-1)
MsgBox v3(-1)(2)(1)
End Sub
```

Run the macro `testvariant`.
The result appear as at right.

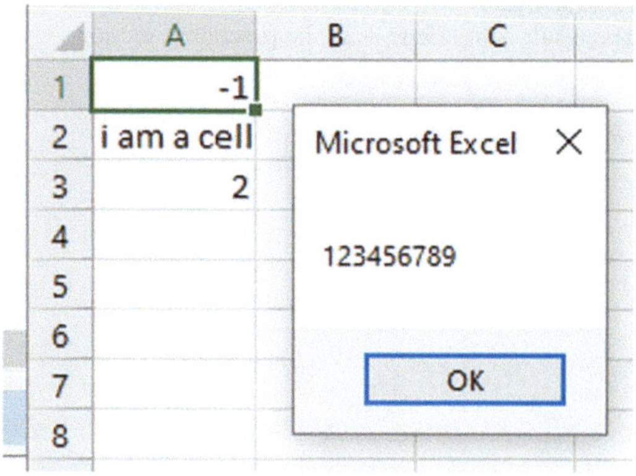

1.16 How to Use a Collection to Store Anything in a Variable

VBA offers a second manner of manipulating list of arbitrary elements through a type of variable entitled `"Collection"`, which is simply a numbered set of objects. You may add or remove an object in a collection. You may also use a string (`"blablabla"`) to refer the elements – otherwise, the elements are numbered 1,2,3, etc. To refer to the element number `i` of the collection `c`, use `c(i)` or `c.Item(i)`. The number of elements in the collection is `c.Count`. Let us remake the preceding example using collections:

Open a blank workbook and add the code at right. It creates collections `v1`, `v2`, `v3`. Notice that
`v3(2)=v2`
`v3(2)(2) = v2(3) = v1`,
`v3(2)(2)(3) = v1(3) = x`.
`v3(2)(2)(3)(−1) = x(−1) = 2`.
Analogously,
`v3(1)(3)(0) = v1(3)(0) = x(0) = −1`;
`v2(2)(2) = v1(1) = name1`;
`v3(2)(2)(2)=v2(2)(2)=v1(2)= name2`
Finally notice that `v2` was not declared as Variant: VBA considers that a new variable is a Variant, except if another type is declared.

```
Sub testcollection()
name1 = "i am a cell"
name2 = "123456789"
Dim x(-1 To 1) As Double
x(-1) = 2
x(0) = -1
x(1) = 0
Dim v1 As New Collection
v1.Add name1
v1.Add name2
v1.Add x
Dim v2 As New Collection
v2.Add x
v2.Add v1
Dim v3 As New Collection
v3.Add v1
v3.Add v2
Cells(1, 1).Value = v3(1)(3)(0)
Cells(2, 1).Value = v2(2)(1)
Cells(3, 1).Value = v3(2)(2)(3)(-1)
MsgBox v3(2)(2)(2)
End Sub
'
```

Run the macro `testcollection`.
The result is the same as in the preceding section.

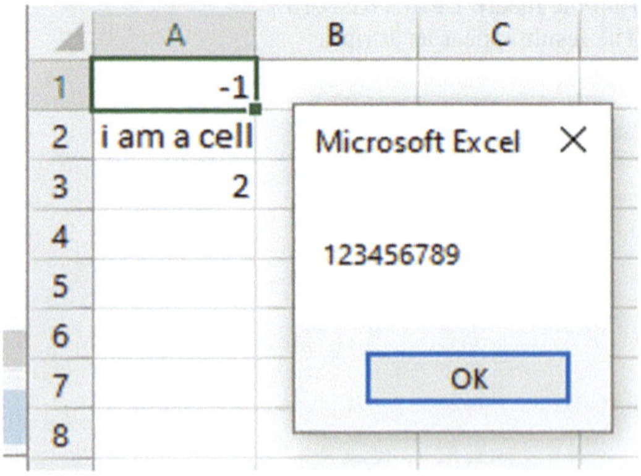

1.17 How to Include a Class in Your Workbook

If you are not familiar with object programming, it is probable that you are not using classes. It is even possible that this word may scare you a little bit. However, the use of classes can be useful and, in the situations considered in this book, is relatively easy to implement.

A class in VBA defines something that may be easily replicated at will and of which several copies can exist at the same time.

Let us exemplify it with a simple class, which defines a point in $\mathbb{R}^2$: the point is characterized by two coordinates $\boldsymbol{x} = (x_1, x_2)$. These coordinates are properties of the point.

Open a blank workbook, go to the tab *Developer*, click on *Visual Basic* and insert a Class Module, as shown at right (see also Sect. 1.5).

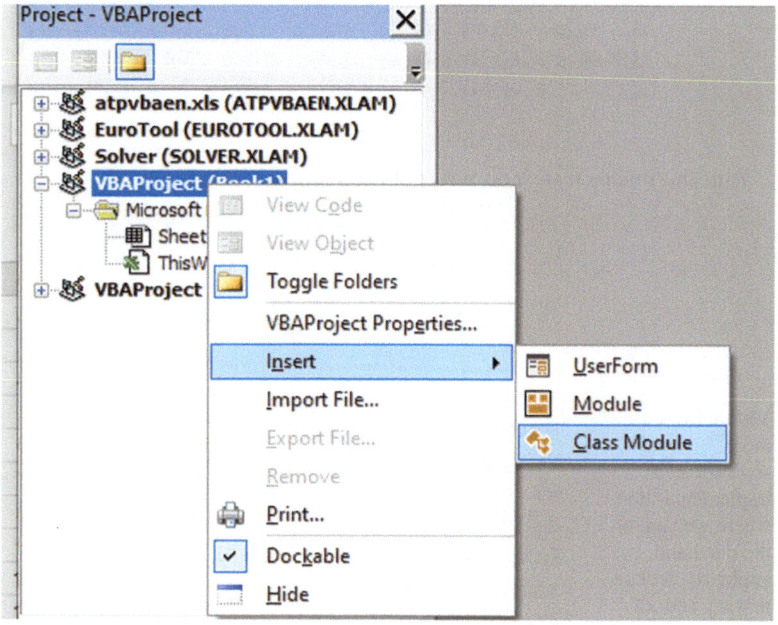

Enter the code below. `x1` is the value of x_1, `x2` the value of x_2, `dist2` evaluates the distance from the point itself to point `p2` and `dist` evaluates the distance between two arbitrary points.

```
Public x1 As Double
Public x2 As Double
Function dist2(p)
dist2 = Sqr((p.x1 - x1) ^ 2 + (p.x2 - x2) ^ 2)
End Function
Function dist(p1, p2)
dist = Sqr((p1.x1 - p2.x1) ^ 2 + (p1.x2 - p2.x2) ^ 2)
End Function
```

Give the name `point_simple` to the Class Module

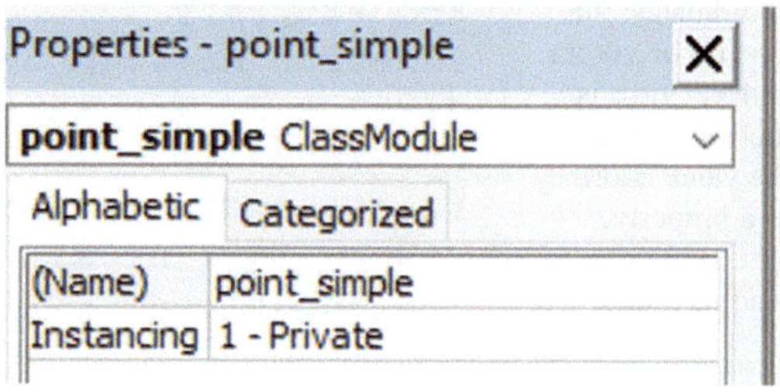

Insert a new module in your workbook and enter the code below: `d1` is the distance to `p2` (from `p1`), `d2` is the distance to `p1` (from `p2`), `d3` is again the distance between `p1` and `p2`.

```
Sub test()
Dim p1 As New point_simple
p1.x1 = 0
p1.x2 = 1
Dim p2 As New point_simple
p2.x1 = 1
p2.x2 = 0
d1 = p1.dist2(p2)
d2 = p2.dist2(p1)
d3 = p1.dist(p1, p2)
Cells(1, 1).Value = "p1"
Cells(1, 2).Value = p1.x1
Cells(1, 3).Value = p1.x2
Cells(2, 1).Value = "p2"
Cells(2, 2).Value = p2.x1
Cells(2, 3).Value = p2.x2
Cells(3, 1) = "dist"
Cells(3, 2).Value = d1
Cells(3, 3).Value = d2
Cells(3, 4).Value = d3
End Sub
```

Run the macro test. You will get the results below:

	A	B	C	D
1	p1	0	1	
2	p2	1	0	
3	dist	1,414214	1,414214	1,414214

VBA allows a more sophisticated implementa- tion, where we can, on the one hand, explicitly declare that `x1` and `x2` are properties and, on the other hand, add the vector of coordinates `coord` = `(x1, x2)` as a new property. Here, `xx` is the internal copy of the vector of coordinates. Now, we have two ways to access the coordinates: via `coord` or via `x1` and `x2`. `Let` is the way the give the value `data` to a property, `Get` is the way to read its value.

```
Private xx(1 To 2) As Double
Property Get x1() As Double
x1 = xx(1)
End Property
Property Let x1(data As Double)
xx(1) = data
End Property
Property Get x2() As Double
x2 = xx(2)
End Property
Property Let x2(data As Double)
xx(2) = data
End Property
Property Get coord() As Variant
coord = xx
End Property
Property Let coord(xp)
xx(1) = xp(1)
xx(2) = xp(2)
End Property
Function dist2(p As point)
dist2 = Sqr((p.x1 - xx(1)) ^ 2 + (p.x2 - xx(2)) ^ 2)
End Function
Function dist(p1 As point, p2 As point)
dist = Sqr((p1.x1 - p2.x1) ^ 2 + (p1.x2 - p2.x2) ^ 2)
End Function
```

Open a new blank Workbook, insert the class module above and give it the name point. Then, add a new module end enter the code at right. Run the macro `testpoint`: you will get the same results as in the preceding implementation.

	A	B	C	D
1	p1	0	1	
2	p2	1	0	
3	dist	1,414214	1,414214	1,414214

Notice that this implementation uses a vector of coordinates `coord` that may be useful.

```
Sub testpoint()
Dim p1 As New point
p1.x1 = 0
p1.x2 = 1
Dim p2 As New point
Dim xp2(1 To 2) As Double
xp2(1) = 1
xp2(2) = 0
p2.coord = xp2
d1 = p1.dist2(p2)
d2 = p2.dist2(p1)
d3 = p1.dist(p1, p2)
Cells(1, 1).Value = "p1"
Cells(1, 2).Value = p1.coord(1)
Cells(1, 3).Value = p1.coord(2)
Cells(2, 1).Value = "p2"
Cells(2, 2).Value = p2.x1
Cells(2, 3).Value = p2.x2
Cells(3, 1) = "dist"
Cells(3, 2).Value = d1
Cells(3, 3).Value = d2
Cells(3, 4).Value = d3
End Sub
```

As a second example, let us consider a class of curves in $\mathbb{R}^2$. Recall that a curve is an ordered set of points $\boldsymbol{x}_1, \ldots, \boldsymbol{x}_{np}$, where each point is characterized by two coordinates $\boldsymbol{x}_i = (x_{1,\,i}, x_{2,\,i})$. As explained in the preceding section, we may store the points in a list of type `Collection`. This list is a property of the curve.

Open a blank workbook, add a class module, name this module "`curve`" and enter the code below:

```
Private p As Collection
Public Property Let points(listpoints)
Set p = listpoints
End Property
Public Property Get points() As Collection
Set points = p
End Property
Public Sub plot()
Dim x() As Double
Dim y() As Double
ReDim x(1 To p.Count)
ReDim y(1 To p.Count)
For i = 1 To p.Count
  x(i) = p(i)(1)
  y(i) = p(i)(2)
Next i
ActiveCell.Select
ActiveSheet.Shapes.AddChart2(227, xlXYScatter).Select
Application.CutCopyMode = False
Application.CutCopyMode = False
ActiveChart.SeriesCollection.NewSeries
   ActiveChart.FullSeriesCollection(1).Values = y
   ActiveChart.FullSeriesCollection(1).XValues = x
   ActiveChart.FullSeriesCollection(1).Name = "curve"
   ActiveChart.FullSeriesCollection(1).Select
   With Selection
        .MarkerStyle = 8 'xlMarkerStyleCircle
        .MarkerSize = 2
    End With
    With Selection.Format.Line
        .Visible = msoTrue
        .Weight = 1
    End With
ActiveChart.HasLegend = True
ActiveChart.Legend.Position = xlLegendPositionBottom
ActiveChart.ChartTitle.Select
ActiveChart.ChartTitle.Text = "curve"
End Sub
```

Add a new module to your workbook and enter the code at right.

```
Sub test()
nl = 21
Dim pt(1 To 2) As Double
Dim aux As New Collection
For i = 1 To nl
  pt(1) = Cells(2, i + 1).Value
  pt(2) = Cells(3, i + 1).Value
   aux.Add (pt)
Next i
Dim cl As New curve
cl.points = aux
Call cl.plot
End Sub
```

Go to Workbook and enter the formula "= `SEQUENCE(1;21;0;PI()/10)`" in `B1`.

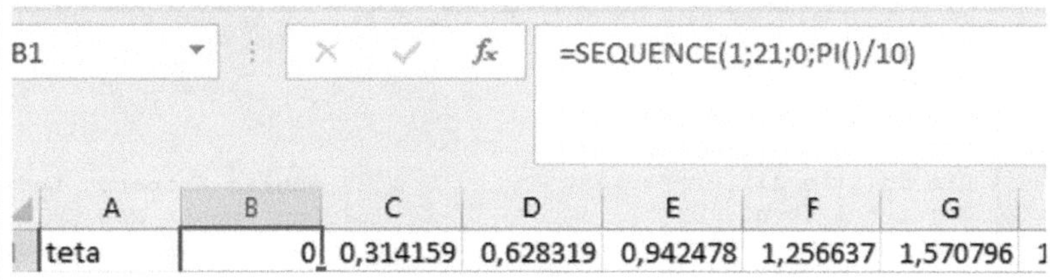

Enter the formula "`=$A2*COS(B$1`" in `B2` and "`=$A2*SIN(B$1)`" in `B3`. Copy these cells and paste them up to `V2:V3`.

	A	B	C	D	E	F	G	H	I
1	teta	0	0,314159	0,628319	0,942478	1,256637	1,570796	1,884956	2,199115	2,51
2	1	1	0,951057	0,809017	0,587785	0,309017	6,13E-17	-0,30902	-0,58779	-0,8
3		0	0,309017	0,587785	0,809017	0,951057	1	0,951057	0,809017	0,58

Run the macro test. You will get the results below:

	A	B	C	D	E	F	G	H	I
1	teta	0	0,314159	0,628319	0,942478	1,256637	1,570796	1,884956	2,19911
2	1	1	0,951057	0,809017	0,587785	0,309017	6,13E-17	-0,30902	-0,5877
3		0	0,309017	0,587785	0,809017	0,951057	1	0,951057	0,80901

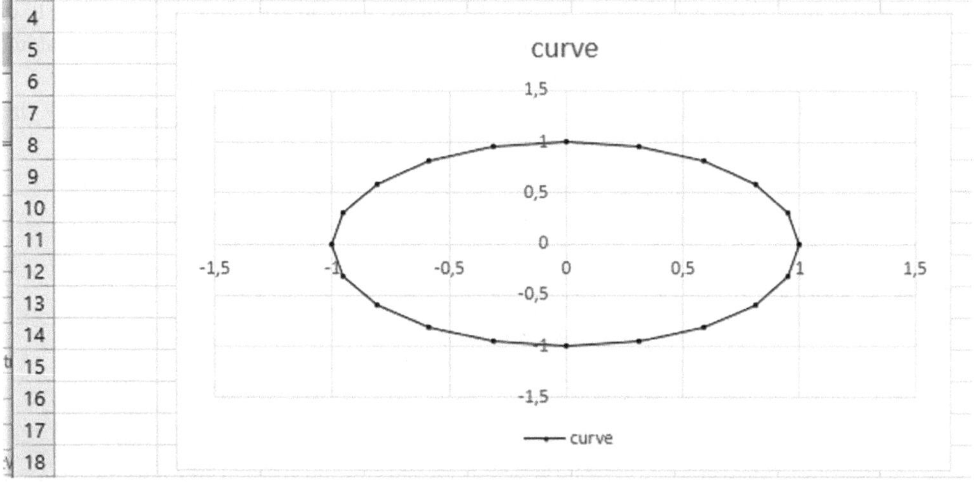

Remark

We can create a collection of objects, such as points. For instance, it is possible to write the class `curve` with a collection of points. In such a case, variable `pt` becomes a point:

```
Sub test()
nl = 21
Dim pt(1 To 2) As Double
Dim aux As New Collection
Dim pp(1 To 21) As Variant
For i = 1 To nl
  pt(1) = Cells(2, i + 1).Value
  pt(2) = Cells(3, i + 1).Value
  Set pp(i) = New Point
  pp(i).coord = pt
   aux.Add pp(i)
Next i
Dim cl As New curve
cl.points = aux
Call cl.plot
End Sub
```

```
Public Property Let points(listpoints)
Dim ppp As New Collection
For Each pp In listpoints
  ppp.Add pp
Next pp
Set p = ppp
End Property
```

In this book, we shall manipulate some classes, useful to avoid repetitive programming of some VBA modules.

For instance, our calculations use polynomials, trigonometrical functions and more generally, families of functions φ_i forming basis for expansions such as $PX = \sum_{i=0}^{k} x_i \varphi_i$. Repetitive operations are performed that use these elements: calculating their value, their derivative, a scalar product, etc. In order to prevent from having to redefine them each time we are going to make such an operation, it is possible to create a class and then call it back when necessary. It is also possible to include in the class all the usual and most commonly used operations and properties, so that we do not have to redefine them each time.

Let us illustrate this with an example: consider a family of polynomials generated by the basis $\varphi_i(x) = \left(\frac{x-a}{b-a}\right)^i$.

Open a blank workbook, go to the tab *Developer*, click on *Visual Basic* and insert a Class Module, as shown at right (see also Sect. 1.5).

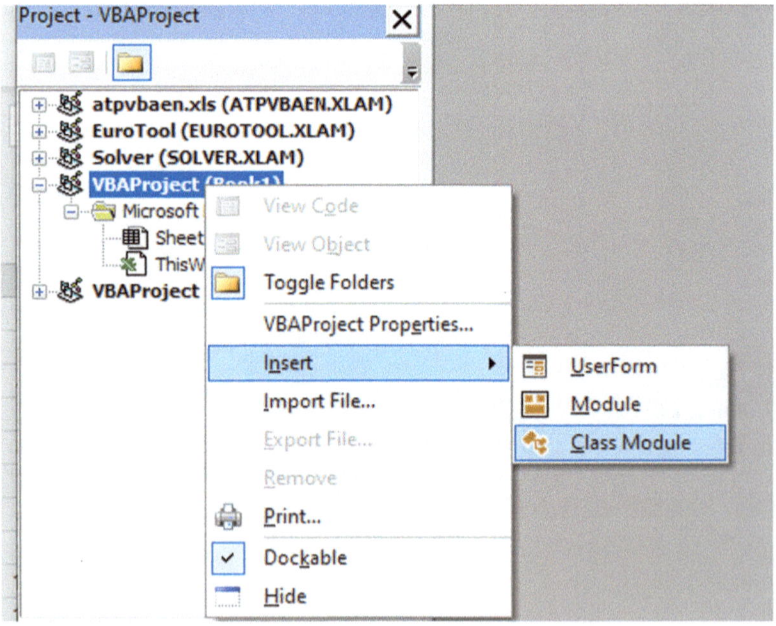

Enter the code shown at right. Function `value(i, x)` evaluates $\varphi_i(x)$ and `d1(i, x)` evaluates $\varphi_i'(x)$. Notice that the values of a and b are declared as `public` – they can be used everywhere in the module and are accessible outside.

```
Public a As Double
Public b As Double

Public Function value(i, x)
 s = (x - a) / (b - a)
 v = WorksheetFunction.Power(s, i)
 value = v
End Function

Public Function d1(i, x)
 If i < 1 Then
   v = 0
 Else
   s = (x - a) / (b - a)
   v = i * WorksheetFunction.Power(s, i - 1)
 End If
d1 = v / (b - a)
End Function
```

Change the name of the class to poly

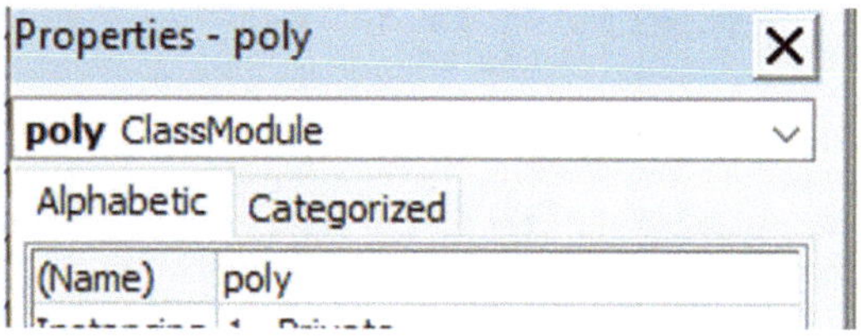

There are two methods to create a family of polynomials: use either `dim` and `new` or `dim` and `set`. Let us exemplify the methods and the use.

Add a macro to the workbook, as shown in Sect. 1.5. Enter the code at right

```
Sub create_poly()
Dim p1 As New poly
p1.a = 0
p1.b = 1
Cells(1, 1).value = p1.value(3, 0.5)
Dim p2 As poly
Set p2 = New poly
p2.a = -1
p2.b = 1
Cells(1, 2).value = p2.value(3, 0.5)
End Sub
```

Run the macro. You will get the result at right. `A1` contains the value of $p1$. *value*(3, 1/2) = 1/8 and `B1` contains the value of $p2$. *value*(3, 1/2) = (3/4) ^ 3

A	B
0,125	0,421875

You can create as many families as you need and use them in your modules. Class modules are exported to `.cls` files(for instance, `poly.cls`). They may be imported by the same way as modules (see Sect. 1.6). A more complete class `poly.cls` is given below. It takes the precaution to treat separately the values of x close to the extremities. `lb` (lower bound) corresponds to `a` and `ub` (upper bound) corresponds to `b`.

```
Public lb As Double
Public ub As Double
Public zero As Double
```

```
Public Function value(i, x)
If i = 0 Then
   v = 1
Else
   s = (x - lb) / (ub - lb)
   If Abs(s) < zero Then
      v = 0
   ElseIf Abs(s - 1) < zero Then
      v = 1
   Else
      If i = 1 Then
         v = s
      ElseIf i = 2 Then
         v = s * s
      Else
         v = WorksheetFunction.Power(s, i)
      End If
   End If
End If
value = v
End Function
```

```
Public Function d1(i, x)
If i = 0 Then
   v = 0
Else
   s = (x - lb) / (ub - lb)
   If Abs(s) < zero Then
      v = 0
   ElseIf Abs(s - 1) < zero Then
      v = i
   Else
      If i = 1 Then
         v = 1
      ElseIf i = 2 Then
         v = 2 * s
      Else
         If i = 3 Then
            v = 3 * s * s
         Else
            v = i * WorksheetFunction.Power(s, i - 1)
         End If
      End If
   End If
End If
value = v / (ub - lb)
End Function
```

This class is used in the book. It is available in the file `poly.cls`, which can be downloaded. Analogous classes `sin.cls` ($\varphi_0(x) = 1, \varphi_i(x) = \sin\left(i\left(\frac{x-a}{b-a}\right)\right)$) and `trigo.cls` ($\varphi_0(x) = 1, \varphi_{2i-1}(x) = \sin\left(i\left(\frac{x-a}{b-a}\right)\right), \varphi_{2i}(x) = \cos\left(i\left(\frac{x-a}{b-a}\right)\right)$) are available.

A second operation that will be frequent in this book is the evaluation of the expansion $\sum_{i=0}^{k} c\varphi_i(u_j), j = 1, \ldots, ns$. We create a class `expansion` to handle it:

```
Public phi As Variant
Function proj(c, u)
ns = UBound(u, 1)
n = UBound(c, 1)
k = UBound(c, 2)
Dim px() As Double
ReDim px(1 To n, 1 To ns)
For i = 1 To ns
  For r = 1 To n
    s = 0
    For j = 1 To k
      s = s + c(r, j) * phi.value(j - 1, u(i))
    Next j
    px(r, i) = s
  Next r
Next i
proj = px
End Function
```

```
Function matval(k, u)
ns = UBound(u, 1)
Dim vx() As Double
ReDim vx(1 To ns, 1 To k)
For i = 1 To ns
 For j = 1 To k
    vx(i, j) = phi.value(j - 1, u(i))
 Next j
Next i
```

One of the advantages of the use of classes is the facility in adding new features and functions: just add the new property, function or sub to the class, export the class and you can use it in all your workbooks, by importing the new file.

Some Useful Numerical Methods 2

As indicated in the Introduction, the reader may find on the net free and commercial add-ins to make Numerical Calculus with EXCEL®. We cited some of them: MATRIX, XNUMBERS, MATHLAY ER®, the Jensen Library.

For the sake of completeness of the text, we present in this chapter some numerical methods used in what follows. The reader will find here the procedure for solving some classical problems of Numerical Analysis using EXCEL. We present both calculations with the add-in SOLVER and with VBA programs. The programs presented below are far from having the sophistication of the above-mentioned add-ins, but they are sufficient for our purpose.

2.1 Linear Systems

The resolution of linear systems is one of the most important topics in Numerical Analysis. Indeed, linear systems are often found in practice, so it is essential to master tools and methods for their resolution. UQ is not an exception: the solution of linear systems is often requested, namely for the determination of approximations of random variables. To fix the ideas, let us consider a linear system

$$\boldsymbol{AX} = \boldsymbol{B}. \tag{2.1}$$

where $\boldsymbol{A}$ is a $m \times n$-matrix, $\boldsymbol{X}$ is a $n \times 1$-matrix, $\boldsymbol{B}$ is a $m \times 1$-matrix. In this book, we shall manipulate two kinds of linear systems: either $m = n$ and $\det(A) \neq 0$ or $m > n$, but $rank(A) = m$. Notice that

- If $m > n$, the system is **overdetermined**. In general, overdetermined systems do not have exact solutions. We must find adequate approximated solutions, such as, for instance, least-square solutions or, more generally, minimum norm solutions.
- If $m = n$ and $\det(A) \neq 0$, the linear system has a unique solution.

In this section, we present methods fgor the solution this last type of linear systems. Overdetermined systems are considered in Sect. 2.3.

Supplementary Information The online version of this chapter (https://doi.org/10.1007/978-3-030-77757-9_2) contains supplementary material, which is available to authorized users.

E. Souza de Cursi, *Uncertainty Quantification and Stochastic Modelling with EXCEL*, Springer Texts in Business and Economics, https://doi.org/10.1007/978-3-030-77757-9_2

2.1.1 Using the Inverse Matrix

If $m = n$ and $\det(A) \neq 0$, the simplest method to solve a linear system consists in the use of the formula

$$X = A^{-1}B\,. \tag{2.2}$$

EXCEL® has built-in functions to determine A^{-1}(`MINVERSE`) and for the product of matrices (`MMULT`). For instance, open a blank workbook and enter the data in the Figure below, at left. Give the name "`A`" to `A1:B2`. Select D1 and enter "=`MINVERSE(A)`". The results are shown at left.

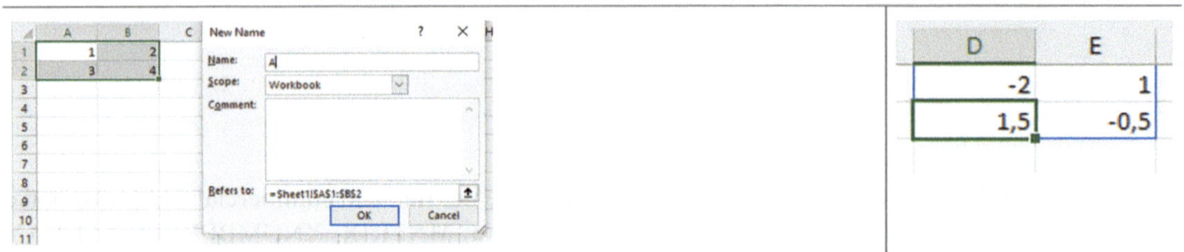

Indeed,

$$A = \begin{pmatrix} 1 & 2 \\ 3 & 4 \end{pmatrix} \Leftrightarrow A^{-1} = \begin{pmatrix} -2 & 1 \\ \frac{3}{2} & -\frac{1}{2} \end{pmatrix}$$

Now, enter the data in the Figure below, at left. Give the name "`B`" to `A4:A5`. Select C4 and enter "=`MMULT(MINVERSE(A);B)`". The results are shown at left.

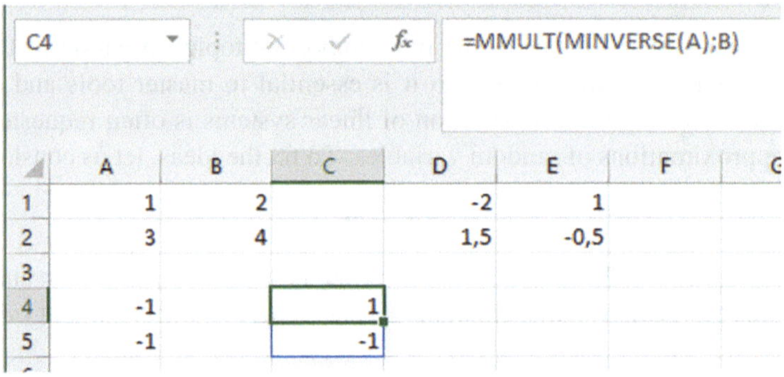

Indeed,

$$\begin{cases} x_1 + 2x_2 = -1 \\ 3x_1 + 4x_2 = -1 \end{cases} \Leftrightarrow x_1 = 1, x_2 = -1.$$

These functions may be used in VBA: go to *Developer>Visual Basic* and insert a module as indicated page 7. Then, enter the code below:

```
linsolve_inv.xlsm - Module1 (Code)
General)

Function inverse(A)
inverse = WorksheetFunction.MInverse(A)
End Function
Function linsolve_inv(A, B)
aux = inverse(A)
linsolve_inv = WorksheetFunction.MMult(aux, B)
End Function
```

Now, select A7 and enter “=`inverse(A)`”. Then, select `D1` and enter “=`linsolve_inv(A; B)`”. The results are shown below.

6				
7	-2	1		1
8	1,5	-0,5		-1
9				

2.1.2 Using the SOLVER

Linear equations may also be solved by calling the SOLVER, which performs the numerical minimization of a function. For instance, open a blank workbook and enter the data in the Figure below.

	A	B	C
1	1	2	
2	3	4	
3			
4	-1		5
5	-1		6

Give the names “A” to `A1:B2`, “B” to `A4:A5`, “X” to `C4:C5`. Then, select `D1` and enter “=`SUMSQ(MMULT(A;X)-B)`”. This expression evaluates $\|AX-B\|_2^2 = \sum_{i=1}^{2}\left(\sum_{j=1}^{2} A_{ij}X_j - B_i\right)^2$.

The value in `D1` is `1924.`

`D1` is the target cell – the solution of the linear system corresponds to a value of zero in the `D1`. We apply the procedure shown in Sect. 1.11. Go to *Data>Solver* (click on Solver) and enter the information below:

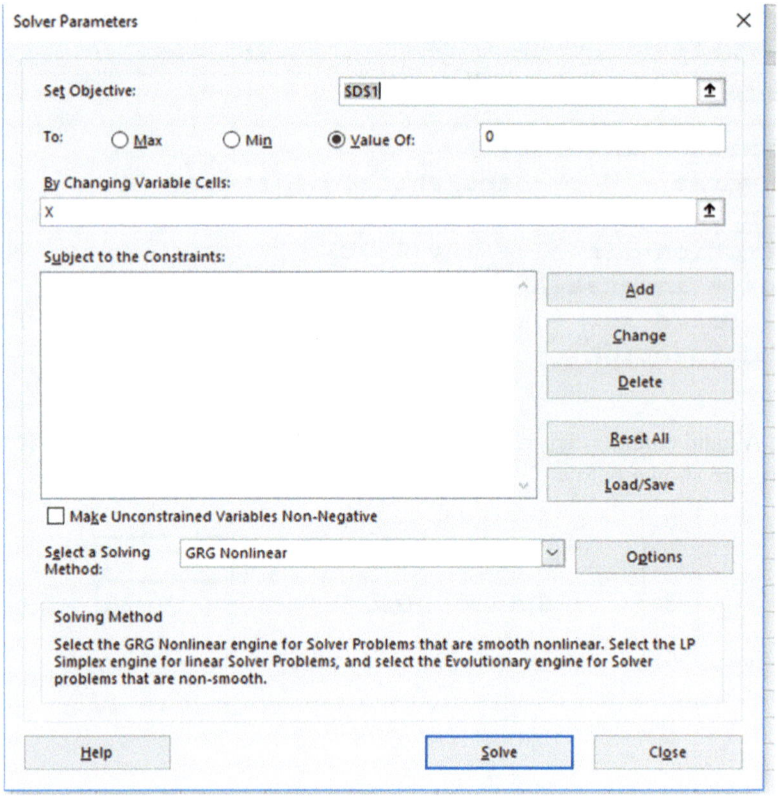

In this case, we look for the values of *X* such that the target value `D1` is equal to zero. Click on "Solve": you will new values for X, for the target cell and the Dialog Box informing that the Solver has found a solution.

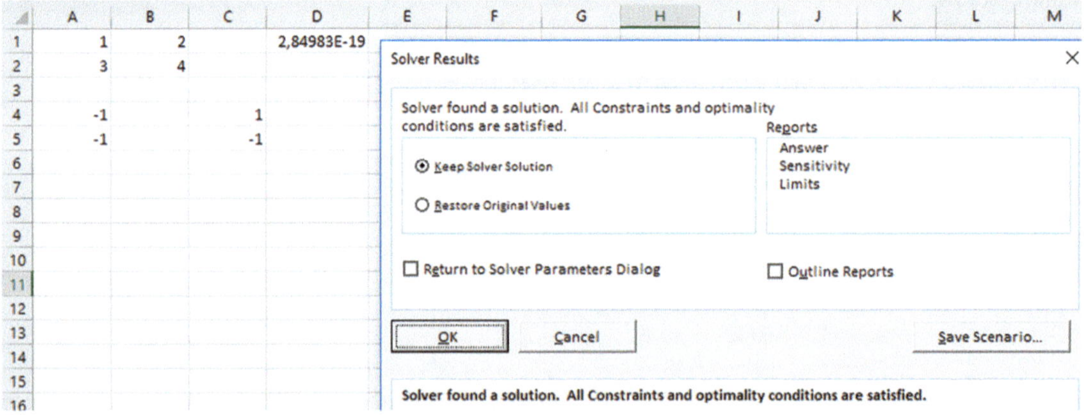

The Solver may be called inside VBA code, as shown in page 19: go to *Developer>Visual Basic* and insert a module as indicated page 7. Then, enter the code below:

```
(General)
Public Sub solvetarget()
ActiveCell.FormulaArray = "=SUMSQ(MMULT(A,X) - B)"
rT = ActiveCell.Address
nX = "[" & ActiveWorkbook.Name & "]" & ActiveSheet.Name & "!" & "X"
rX = Range(nX).Address
SolverReset
SolverOK SetCell:=rT, MaxMinVal:=3, ValueOf:=0, ByChange:=rX, _
        Engine:=1, EngineDesc:="GRG Nonlinear"
SolverOptions AssumeNonNeg:=False
SolverSolve True
End Sub
```

As indicated in the preceding, go to *Developer>Macros* and Run the macro

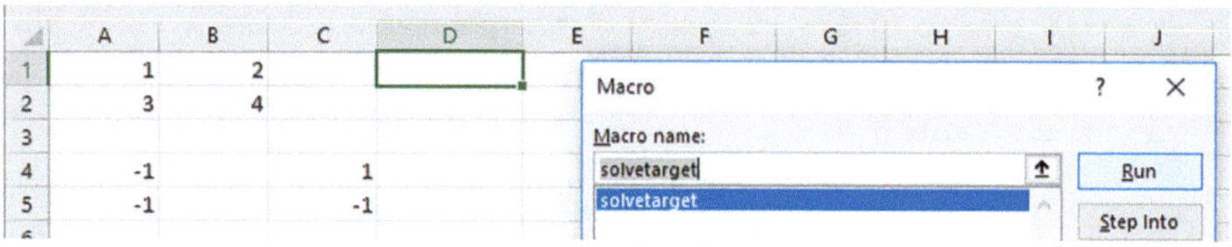

The results appear in the cells corresponding to X.

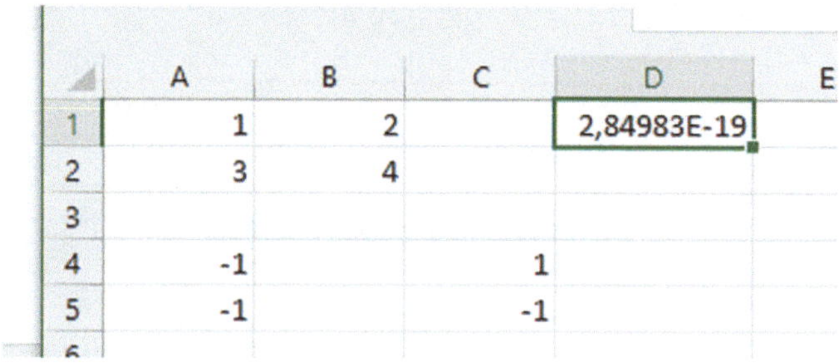

2.1.3 Using Gauss-Jordan Pivoting

Gauss-Jordan pivoting (Gauss, 1811; Jordan, 1962) is a well-known method for the solution of linear systems. We do not recall it here – the reader may refer to the literature in the field of Numerical Analysis – for instance, (Kress, 1998). In Gauss-Jordan pivoting, the linear system $\boldsymbol{AX} = \boldsymbol{B}$ is transformed into $\boldsymbol{UX} = \boldsymbol{C}$, where $\boldsymbol{U}$ is an upper triangular matrix ($U_{ij} = 0$ *if* $j < i$ – all the elements below the diagonal are null). The transformation is made by a sequence of linear operations on the lines of $\boldsymbol{A}$ and $\boldsymbol{B}$: for $i = 1, \ldots, n - 1$, lines $\boldsymbol{A}(k,:)$ with $k \geq i$ are replaced by a linear combination with line i: $\boldsymbol{A}(k,:) \longleftarrow \boldsymbol{A}(k,:) + \alpha_{ki}\boldsymbol{A}(i,:)$, with $\alpha_{ki} = -\frac{A_{ki}}{A_{ii}}$ chosen to get $A_{ki} = 0$. The same operation is made on $\boldsymbol{B}$: $B_k \longleftarrow B_k + \alpha_{ki}B_i$. The resulting upper triangular system is easily solved by backwards substitution:

$$x_n = \frac{C_n}{U_{nn}}, \quad x_k = \frac{\left(C_k - \sum_{j=k+1}^{n} U_{kj}x_j\right)}{U_{kk}} \quad (k < n) \ . \tag{2.3}$$

Gauss-Jordan elimination with total pivoting is implemented in the module *LinSys.bas* and the add-in *LinSys.xlam*, which may be used by the reader. A function `GaussSolution(a, b)` returns a vector containing the solution.

Here, `a, b` are matrices, so that it is mandatory to transform the names `A, B, X` in matrices, as indicated page 16. Such a transformation can be made by a Macro (to be included in your workbook as shown in page 8). For instance:

```
Sub Gauss()
Dim a As Variant
nA = "[" & ActiveWorkbook.Name & "]" & ActiveSheet.Name & "!" & "A"
a = Range(nA)
Dim b As Variant
nB = "[" & ActiveWorkbook.Name & "]" & ActiveSheet.Name & "!" & "B"
b = Range(nB)
nX = "[" & ActiveWorkbook.Name & "]" & ActiveSheet.Name & "!" & "X"
rxa = Range(nX).Address
x = GaussSolution(a, b)
Range(rxa).Value = x
End Sub
```

Running the macro `Gauss` above generates the solution $\boldsymbol{X}$ of the linear system $\boldsymbol{AX} = \boldsymbol{B}$ in the cells corresponding to the name `X`. You may run the macro by using buttons, as described in page 25, or directly from *Developer>Macros* (click on "`Macros`", select the macro and click "`Run`" – see Figure below).

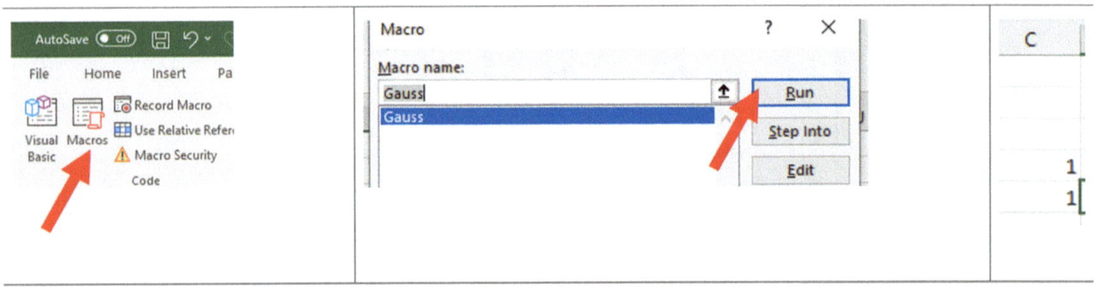

Gauss-Jordan pivoting may also be used to determine the inverse $\boldsymbol{A}^{-1}$ of the matrix $\boldsymbol{A}$: indeed, we must solve the equations $\boldsymbol{AA}^{-1} = \boldsymbol{Id}$, where $\boldsymbol{A}$ is the $n \times n$ identity matrix, so that $\boldsymbol{A}^{-1}$ may be determined by solving n linear systems $\boldsymbol{AX} = \boldsymbol{B}^j$, where $\boldsymbol{B}^j = \boldsymbol{A}(:, j)$ is the column j of $\boldsymbol{A}$.

A function `GaussInverse(a)` is implemented in the module *LinSys.bas* and the add-in *LinSys.xlam*, To use it, you must select the first cell of the range of cells destined to receive the values of $\boldsymbol{A}^{-1}$. Then, you may run a Macro:

```
Public Sub GaussInv()
Dim a As Variant
nA = "[" & ActiveWorkbook.Name & "]" & ActiveSheet.Name & "!" & "A"
a = Range(nA)
ainv = GaussInverse(a)
i = ActiveCell.Row
j = ActiveCell.Column
m = UBound(ainv, 1)
n = UBound(ainv, 2)
Range(Cells(i, j), Cells(i + m - 1, j + n - 1)).Select
Selection.Value = ainv
End Sub
```

For instance, click on Cell `D1` to select it. Then, run the macro above. The result appears in cells `D1:E2`, as the result at left.

	A	B	C	D	E
1	1	2		-2	1
2	3	4		1,5	-0,5
3					
4	-1		1		
5	-1		1		

Both the macros are implement in the workbook *test_LinSys.xlsm*.

2.1.4 Using LU Decomposition

In 1910, André Choleski (Choleski, 1910) proposed a method for the solution of linear systems involving a symmetric matrix $\boldsymbol{A}$, based on the decomposition $\boldsymbol{A} = \boldsymbol{L}\boldsymbol{L}^t$, where matrix $\boldsymbol{L}$ is lower triangular ($L_{ij} = 0 \; for \; i < j$: all the elements above the diagonal are null); The method may generalize to a $\boldsymbol{LU}$ (Lower/Upper) decomposition of matrix $\boldsymbol{A}$, by determining two triangular matrices $\boldsymbol{L}$ and $\boldsymbol{U}$, such that $\boldsymbol{A} = \boldsymbol{LU}$, with $\boldsymbol{L}$ lower triangular and $\boldsymbol{U}$ upper triangular ($U_{ij} = 0, \; if \, j < i$: all the elements below the diagonal are null). Then, the solution of the linear system is generated by solving two triangular systems: $\boldsymbol{LY} = \boldsymbol{B}$, $\boldsymbol{UX} = \boldsymbol{Y}$. The last one is solved by backwards substitution (Eq. 2.3) and the first one is solved by forwards substitution:

$$y_1 = \frac{B_1}{L_{11}}, \quad y_k = \frac{\left(B_k - \sum_{j=1}^{k-1} L_{kj} y_j \right)}{L_{kk}} \quad (k > 1). \tag{2.4}$$

In general, the decomposition $\boldsymbol{LU}$ is not unique, so that we may fix one of the coefficients: for instance, we may look for a decomposition such that $L_{ii} = 1, 1 \leq i \leq n$ – in this case, the first line of $\boldsymbol{U}$ coincides with first line of $\boldsymbol{A}$. The reader may refer to the literature to find the adequate formulae concerning this method.

LU decomposition and solution of linear systems is implemented in the module *LinSys.bas* and the add-in *LinSys.xlam*. The function is `LUSolution(a, b)`. Again, the names `A, B` must be transformed in matrices, as indicated page 16. The inverse matrix of A may be determined into an analogous way, using the function `LUInverse(a)`. Macros analogous to the preceding ones can be used by replacing `GaussSolution` by `LUSolution` and `GaussInverse` by `LUInverse`. As an example, the workbook *test_LinSys.xlsm* implements two Macros `Choleski` (to solve $\boldsymbol{AX} = \boldsymbol{B}$) and `CholeskiInv` (to find $\boldsymbol{A}^{-1}$)

Both the macros are implemented in the workbook *test_LinSys.xlsm*.

2.1.5 Using QR Decomposition

In the QR approach, matrix $\boldsymbol{A}$ is decomposed $\boldsymbol{A} = \boldsymbol{QR}$, where, on the one hand, matrix $\boldsymbol{Q}$ is orthogonal: $\boldsymbol{Q}^t\boldsymbol{Q} = \boldsymbol{Q}\boldsymbol{Q}^t = \boldsymbol{Id}$; on the other hand, $\boldsymbol{R}$ is upper triangular Two classical methods for the determination of the decomposition are Gram-Schmidt orthonormalization and Householder's reflections. Once the decomposition is determined, the linear system is solved by evaluating $\boldsymbol{y} = \boldsymbol{Q}^t\boldsymbol{B}$

and solving $\boldsymbol{Rx} = \boldsymbol{y}$ by (2.3). Householder's reflections ar and QR solution are implemented in the module *LinSys.bas* and the add-in *LinSys.xlam*. The function is `QRSolution(a, b)`. The inverse matrix of A is determined by the function `QRInverse(a)`. Macros analogous to the preceding ones are implemented in *test_LinSys.xlsm*: `Householder` (to solve $\boldsymbol{AX} = \boldsymbol{B}$) and `QRInv` (to find $\boldsymbol{A}^{-1}$)

2.1.6 Using Relaxation Iterations

The method called ***relaxation*** is used to solve problems where $\boldsymbol{A}$ is positive definite. Involves iterations: starting from an initial guess $\boldsymbol{X}^{(0)}$, we generate sequence $\boldsymbol{X}^{(1)}, \boldsymbol{X}^{(2)}, \ldots, \boldsymbol{X}^{(k)}, \ldots$ which is expected to converge to the solution of the linear system. The iterations read as

$$X_i^{(k+1)} = (1-\omega)X_i^{(k)} + \frac{\omega}{A_{ii}}\left(B_i - \sum_{j=i+1}^{n} A_{ij}X_j^{(k)} - \sum_{j=1}^{k-1} A_{ij}X_j^{(k+1)}\right), 1 \le i \le n.$$

For $\omega = 1$, relaxation iterations reduce to Gauss-Seidel iterations. For a positive definite matrix, the method converges for $0 < \omega < 2$. This method is implemented in the module *LinSys.bas* and the add-in *LinSys.xlam*. The associated function is `RelaxSolution(a, b, x0, om, tolref, nitmax)`

Here, the names `A, B, X0` must be transformed in matrices, as indicated page 16. `X0` is the initial guess $\boldsymbol{X}^{(0)}$, om is ω, `tolref` and `nitmax` are stopping conditions: the iterations stop if $k > nitmax$ or $\|\boldsymbol{X}^{(k+1)} - \boldsymbol{X}^{(k)}\| < tolref$.

A Macro analogous to the preceding ones can be used by replacing `GaussSolution` by `RelaxSolution`. This method appears as less interesting than the preceding ones for the inversion of A, so that the inversion is not implemented. As an example, the workbook *test_LinSys.xlsm* implements a Macros `RelaxSolution` to solve $\boldsymbol{AX} = \boldsymbol{B}$.

2.2 Optimization

A second class of numerical problems often found in practice is the minimization of real-valued functions of real numbers. The model situation is the one where we desire to determine

$$X = \arg\min_C f \tag{2.5}$$

id est,

$$\boldsymbol{X} \in C \subset \mathbb{R}^n : f(\boldsymbol{X}) \le f(\boldsymbol{Y}), \ \forall \boldsymbol{Y} \in C. \tag{2.6}$$

In the language of Optimization, C is the *admissible set*, $f: \mathbb{R}^n \to \mathbb{R}$ is the *objective function*, $\boldsymbol{X}$ contains the *design variables*. This problem is usually referred as *mono-objective optimization* (since $f(\boldsymbol{Y}) \in \mathbb{R}$). We shall examine multi-objective optimization in Sect. 2.6.

When $C = \mathbb{R}^n$, the optimization problem is said to be *unconstrained*. For $C \ne \mathbb{R}^n$, it is said to be *constrained*. A situation often found in the practice is the one where

$$C = \{X \in \mathbb{R}^n : \varphi_i(X) \le 0, 1 \le i \le m; \varphi_i(X) = 0, m+1 \le i \le m+p\} \tag{2.7}$$

The first m inequalities are the *inequality constraints* or *inequality restrictions*. The last p equalities are the *equality constraints* or *equality restrictions*. Constrained optimization under the formulation (2.7) involves Lagrange's multipliers $\boldsymbol{\lambda} = (\lambda_1, \ldots, \lambda_{p+q})^t$ – sometimes referred as *shadow prices* in the context of Economics. Lagrange's multipliers generate the Lagrangian $L(\boldsymbol{y};\ \boldsymbol{\lambda})$ given by

$$L(\boldsymbol{y};\boldsymbol{\lambda}) = J(\boldsymbol{y}) + \boldsymbol{\lambda}^t\boldsymbol{\Psi}(\boldsymbol{y}) = J(\boldsymbol{y}) + \sum_{i=1}^{p+q} \lambda_i \psi_i(\boldsymbol{y}). \tag{2.8}$$

They satisfy

$$\boldsymbol{\lambda}^t\boldsymbol{\Psi}(\boldsymbol{x}^*) = \boldsymbol{0} \ , \ \lambda_i \geq 0, 1 \leq i \leq p \ . \tag{2.9}$$

In addition, $(\boldsymbol{x}^*;\ \boldsymbol{\lambda})$ is a saddle point of:

$$L(\boldsymbol{x}^*;\boldsymbol{\eta}) \leq L(\boldsymbol{x}^*;\boldsymbol{\lambda}) \leq L(\boldsymbol{y};\boldsymbol{\lambda}), \forall \boldsymbol{y} \in \mathbb{R}^n, \boldsymbol{\eta} \in \mathbb{R}^{p+q}, \eta_i \geq 0, 1 \leq i \leq p. \tag{2.10}$$

Thus, we have, for $1 \leq i \leq n$, $1 \leq j \leq p$, $1 \leq k \leq q$:

$$\frac{\partial L}{\partial x_i}(\boldsymbol{x}^*;\boldsymbol{\lambda}) = 0, \lambda_j \frac{\partial L}{\partial \lambda_j}(\boldsymbol{x}^*;\boldsymbol{\lambda}) = 0, \frac{\partial L}{\partial \lambda_{k+p}}(\boldsymbol{x}^*;\boldsymbol{\lambda}) = 0. \tag{2.11}$$

These conditions are known as KKT conditions. λ_i gives a measure of the influence of an unit change of the constraint ψ_i on the variation of the optimal value $J(\boldsymbol{x}^*)$. As an example, consider $n = 2$, $J(\boldsymbol{y}) = \|\boldsymbol{y}\|^2 = y_1^2 + y_2^2$, $\psi_1(\boldsymbol{y}) = 1 - y_1 \leq 0$, $\psi_2(\boldsymbol{y}) = y_2 - 2 = 0$. The obvious solution is $\boldsymbol{x}^* = (1,\ 2)$, for which $J(\boldsymbol{x}^*) = 5$. In addition, $\lambda_1 = 2$, $\lambda_2 = -4$. These values show that the influence of the second constraint on the optimal value is the most significant: indeed, if we modify the restrictions to $1 - y_1 \leq 0$, $y_2 - 1 = 0$, the new solution is $\boldsymbol{x}^* = (1,\ 1)$, for which $J(\boldsymbol{x}^*) = 2$. The modification to $-y_1 \leq 0$, $y_2 - 2 = 0$ yields the solution $\boldsymbol{x}^* = (0,\ 2)$, for which $J(\boldsymbol{x}^*) = 4$.: the second restriction is more influent on the optimal value.

Under EXCEL®, optimization problems may be numerically solved by calling the SOLVER, as indicated in Sect. 1.11. In this section we recall this use, and we present some extensions.

2.2.1 Unconstrained Optimization Using the SOLVER in a Worksheet

To use the SOLVER in a worksheet, you must define the objective function in a cell and the variables in a range. We illustrated it in the Sect. 1.11. Here, we present the situation where the objective function is defined by a subprogram. For instance, let us consider the classical Rosenbrock's function:

$$f(x) = \sum_{i=1}^{n-1} \left(100\left(x_{i+1} - x_i^2\right)^2 + (x_i - 1)^2\right)$$

```
Function rosenbrock(x)
Dim s1 As Double
Dim s2 As Double
s1 = 0
s2 = 0
For i = 1 To UBound(x, 1) - 1
      s1 = s1 + (x(i + 1) - x(i) ^ 2) ^ 2
      s2 = s2 + (x(i) - 1) ^ 2
Next i
rosenbrock = 100 * s1 + s2
End Function
```

```
Function rose_on_range(r As Range)
x = range2vector(r)
rose_on_range = rosenbrock(x)
End Function
```

```
Function range2vector(r As Range) As Variant
vx = r.Value
m = UBound(vx, 1)
n = UBound(vx, 2)
Dim x() As Variant
If m > 1 Then
   ReDim x(1 To m)
   For i = 1 To m
      x(i) = vx(i, 1)
   Next i
Else
   ReDim x(1 To n)
   For i = 1 To n
      x(i) = vx(1, i)
   Next i
End If
range2vector = x
End Function
```

A subprogram evaluating f on a vector x of dimension $n \times 1$ is given above (`rosenbrock` at left). To execute this function on a range of cells, it is necessary to transfer the range to a vector by using the function `range2vector` given at right. The function which is called on the worksheet is `rose_on_range`

Open a blank workbook and add these modules as indicated in Sect. 1.5. Then, go to the workbook and enter the data below:

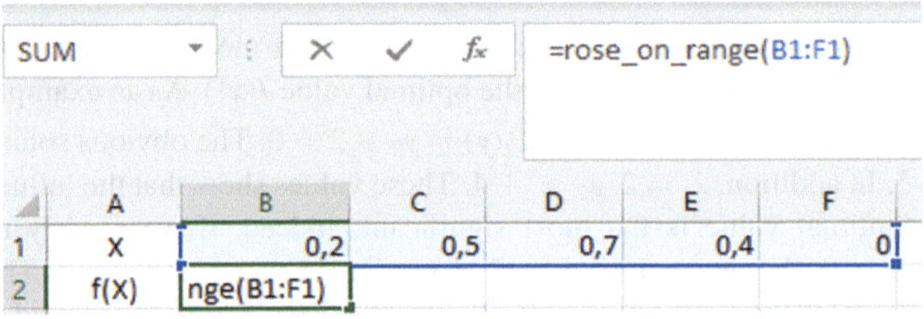

Now, we may call the solver to minimize f (see Sect. 1.11. The dialog appears at left). The initial value in cell B2 is 46,12.

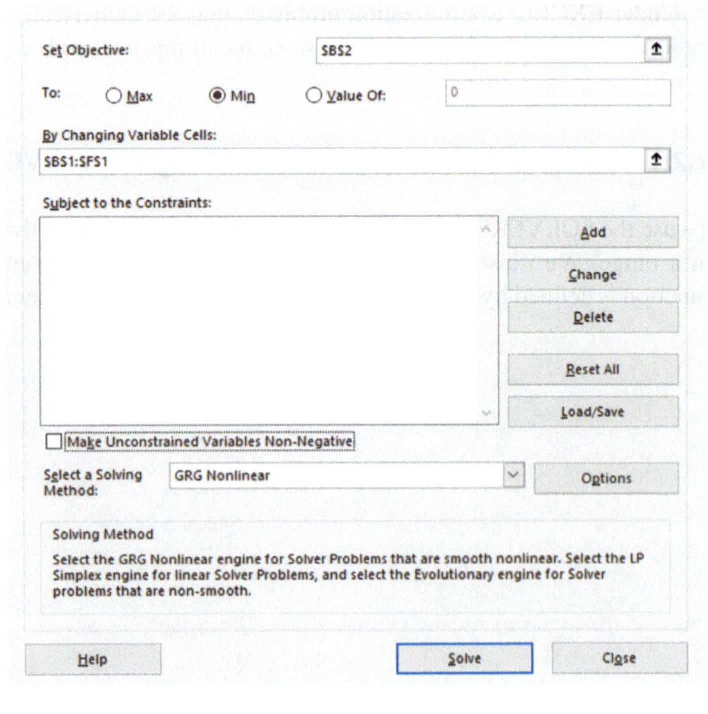

The results furnished by the SOLVER are

X	0,99992757	0,999846	0,999686	0,999369	0,998736
f(X)	5,3894E-07				

The exact point of minimum is (1, 1, 1, 1, 1), where $f = 0$.

2.2.2 Unconstrained Optimization Using the SOLVER in VBA

As previously indicated, we may invoke the solver in a VBA program. For instance, we may use a macro `fminunc` defined as follows

```
Public Sub fminunc()
Dim x As Variant
Set sht = ActiveSheet
colmax = Cells(1, Cells.Columns.Count).End(xlToLeft).Column
For i = 2 To colmax
    Cells(4, i) = Cells(1, i).Value
Next i
rx = Range(Cells(4, 2), Cells(4, colmax)).Address
fname = Range("B2").Value
rfobj = "$B$3"
strfunc = "= f_on_range(" & Chr(34) & fname & Chr(34) & ", " & rx & ")"
Cells(3, 2).Formula = strfunc
SolverReset
SolverOk SetCell:=rfobj, MaxMinVal:=2, ValueOf:=0, ByChange:=rx, _
        Engine:=1, EngineDesc:="GRG Nonlinear"
SolverOptions AssumeNonNeg:=False
SolverOptions Multistart:=True, PopulationSize:=100
SolverSolve True
End Sub
Function f_on_range(nom As String, r As Range)
x = range2vector(r)
f_on_range = Run(nom, x)
End Function
```

This macro optimizes the function defined by the string placed in cell `B2`, using the initial point given in the first line of the Sheet 1 of Workbook, starting at `A2` (the end is calculated automatically). The results appear in cell `B3` (the value of the objective function) and line 4, starting at `D2`. For instance, open a blank workbook, insert a module containing the subprograms above and the subprograms `rosenbrock` and `range2vector` of the preceding section. Enter the data below:

	A	B	C	D	E	F
1	initial point (x0)	0,2	0,5	0,7	0,4	0
2	objective function	rosenbrock				
3	value of objective					
4	solution found					

Then, go to *Developer>Macros* and run Macro `fminunc` (see page 8). The result is

	A	B	C	D	E	F
1	initial point (x0)	0,2	0,5	0,7	0,4	0
2	objective function	rosenbrock				
3	value of objective	1,1738E-13				
4	solution found	1	1	1	1	1

Here we used a single option of the SOLVER to remove the restriction on the sign of the variables. Other options may be set, such as the use of a multistart method: adding the line

`SolverOptions MultiStart:=True, PopulationSize:=100.`

makes the SOLVER use a multistart method with 100 initial points. To use an evolutionary method, we call `SolverOK` with the parameters `Engine:=3`, `EngineDesc:="Evolutionary"`. An example of the complete options of the SOLVER is

```
SolverOptions MaxTime:=0, Iterations:=0, Precision:=0.000001, Convergence:= _
    0.0001, StepThru:=False, Scaling:=False, AssumeNonNeg:=False, Derivatives:=2
SolverOptions PopulationSize:=100, RandomSeed:=0, MutationRate:=0.075, Multistart _
    :=True, RequireBounds:=False, MaxSubproblems:=0, MaxIntegerSols:=0, _
    IntTolerance:=0.1, SolveWithout:=False, MaxTimeNoImp:=30
```

2.2.3 Constrained Optimization Using the SOLVER in a Worksheet

For constrained optimization problems, both the objective function and the constraints must be defined in cells of the worksheet. For instance, let us assume that the admissible set C is given by (2.7): then we may assign a line for the restrictions $\varphi_i(X) \leq 0, 1 \leq i \leq m$ and a line for the restrictions ;$\varphi_i(X) = 0, m+1 \leq i \leq m+p$. If each of these elements is calculated by a subprogram, we may recall them in the worksheet. As an example, let us consider the minimization of the Rosenbrock's function under the restrictions

$$\varphi_1(x) = \|x\| - \sqrt{n} \leq 0, \varphi_2(x) = \sum_{i=1}^{n} x_i - n = 0$$

We introduce two subprograms `eqcons` and `ineqcons` for the evaluation of the equality and inequality constraints, respectively. Both return either a vector of values or an empty object (when the subset of constraints is empty).

For instance, open a blank workbook, add the modules `rosenbrock`, `range2vector`, `rose_on_range` defined in the preceding and include new modules `eqcons`, `ineqcons`, `eq_on_range`, `ineq_on_range`:

```
Function eqcons(x)
s = 0
For i = 1 To UBound(x, 1)
    s = s + x(i)
Next i
eqcons = s - UBound(x, 1)
End Function
Function eq_on_range(r As Range)
x = range2vector(r)
eq_on_range = eqcons(x)
End Function
```

```
Function ineqcons(x)
s = 0
For i = 1 To UBound(x, 1)
    s = s + x(i) ^ 2
Next i
inqeqcons = Sqr(s) - Sqr(UBound(x, 1))
End Function
Function ineq_on_range(r As Range)
x = range2vector(r)
ineq_on_range = ineqcons(x)
End Function
```

Then, enter the data below: select `B2` and enter "`=eq_on_range(B1:F1)`", select `B3` and enter "`=ineq_on_range(B1:F1)`", select `B4` and enter "`=rose_on_range(B1:F1)`".

	A	B	C	D	E	F
1	x	0,2	0,5	0,7	0,4	0
2	equality	-3,2				
3	inequality	0				
4	f(x)	46,12				

Then, call the solver with the constraints on the cells `B2 = 0` and `B3 <= 0;` the objective is now cell `B4`. The results obtained are

	A	B	C	D	E	F
1	x	1	1	1	1	1,000001
2	equality	-8,9E-16				
3	inequality	0				
4	f(x)	1,06E-10				

Now let us consider

$$\varphi_1(x) = \|x\| - \sqrt{n} \leq 0, \varphi_2(x) = \sum_{i=1}^{n} x_i - n \leq 0$$

In this case, `eqcons` returns an empty value and `ineqcons` returns a vector:

```
Function eq2_on_range(r As Range)
x = range2vector(r)
eq2_on_range = eqcons2(x)
End Function


Function ineq2_on_range(r As Range)
x = range2vector(r)
ineq2_on_range = ineqcons2(x)
End Function


Function eqcons2(x)
eqcons2 = Empty
End Function
```

```
Function ineqcons2(x)
Dim outs() As Double
ReDim outs(1 To 2)
s = 0
For i = 1 To UBound(x, 1)
    s = s + x(i)
Next i
outs(1) = s - UBound(x, 1)
s = 0
For i = 1 To UBound(x, 1)
    s = s + x(i) ^ 2
Next i
outs(2) = Sqr(s) - Sqr(UBound(x, 1))
ineqcons2 = outs
End Function
```

Entering the new data in the worksheet, we have:

	A	B	C	D	E	F
1	x	0,2	0,5	0,7	0,4	0
2	equality	0				
3	inequality	-1,26653	-3,2			
4	f(x)	46,12				

Calling the solver with the constraints `B2 = 0, B3 <= 0, C3 <= 0` produces the results below. The result is the same if the cell B2 is empty and the corresponding constraint is deleted in the SOLVER

	A	B	C	D	E	F
1	x	1,000129	1,000227	1,000018	0,999989	0,999637
2	equality	0				
3	inequality	1,69E-07	2,77E-07			
4	f(x)	3,1E-05				

The SOLVER also furnishes the values of the Lagrange's multipliers. To get their values, you must select the option "Sensitivity Report" in the final Dialog Box:

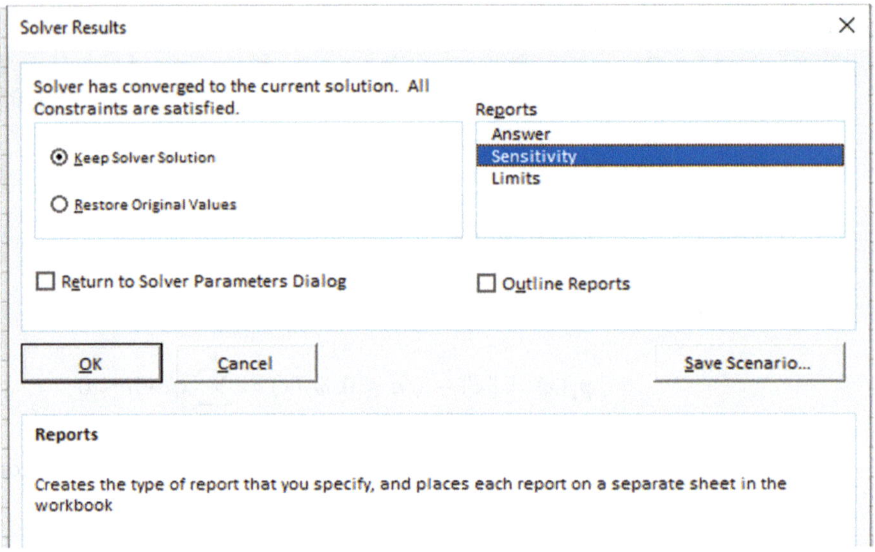

With such a selection, clicking "OK" produces a new sheet entitled "Sensitivity Report 1" containing information about the Lagrange's multipliers

Microsoft Excel 16.0 Sensitivity Report
Worksheet: [optimization_constrained_worksheet.xlsm]Sheet1
Report Created: 09/02/2021 11:34:16

Variable Cells

Cell	Name	Final Value	Reduced Gradient
$B$1	x	1,000128787	0
$C$1	x	1,000226747	0
$D$1	x	1,000018131	0
$E$1	x	0,999989314	0
$F$1	x	0,999637299	0

Constraints

Cell	Name	Final Value	Lagrange Multiplier
$B$3	inequality	1,68747E-07	3,82162177
$C$3	inequality	2,77332E-07	-1,776528257

```
Public Sub fmincon()
Dim x As Variant
Set sht = ActiveSheet
colmax = Cells(1, Cells.Columns.Count).End(xlToLeft).Column
For i = 2 To colmax
    Cells(8, i) = Cells(1, i).Value
Next i
rx = Range(Cells(8, 2), Cells(8, colmax)).Address
fname = Range("B2").Value
rfobj = "$B$3"
strfunc = "= f_on_range(" & Chr(34) & fname & Chr(34) & ", " & rx & ")"
Cells(3, 2).Formula = strfunc

eqname = Range("B4").Value
streq = "= f_on_range(" & Chr(34) & eqname & Chr(34) & ", " & rx & ")"
Cells(5, 2).Formula2 = streq

ineqname = Range("B6").Value
streq = "= f_on_range(" & Chr(34) & ineqname & Chr(34) & ", " & rx & ")"
Cells(7, 2).Formula2 = streq

SolverReset
SolverOk SetCell:=rfobj, MaxMinVal:=2, ValueOf:=0, ByChange:=rx, _
        Engine:=1, EngineDesc:="GRG Nonlinear"
SolverOptions AssumeNonNeg:=False
eqcolmax = Cells(5, Cells.Columns.Count).End(xlToLeft).Column
If eqcolmax > 1 Then
  For i = 2 To eqcolmax
      SolverAdd Cellref:=Range(Cells(5, i), Cells(5, i)).Address, Relation:=2, FormulaText:="0"
  Next i
End If
ineqcolmax = Cells(7, Cells.Columns.Count).End(xlToLeft).Column
If ineqcolmax > 1 Then
  For i = 2 To ineqcolmax
      SolverAdd Cellref:=Range(Cells(7, i), Cells(7, i)).Address, Relation:=1, FormulaText:="0"
  Next i
End If
SolverSolve True
End Sub
```

Fig. 2.1 Macro fmincon

If the option of a sensitivity report does not appear, this means that the solution found by the SOLVER does not satisfies traditional convergence conditions – it may satisfy only probabilistic conditions or, in the worst case, the calculations did not converge. In addition, sensitivity reports are available only for the Engines "`GRG NonLinear`" and "`Simplex LP`".

2.2.4 Constrained Optimization Using the SOLVER in VBA

As in the unconstrained situation, we may invoke the solver in a VBA program. For instance, we may use a macro `fmincon` defined analogously to `fminunc` (see Fig. 2.1).

To use it, open a blank workbook, add the preceding modules, and enter the data below:

	A	B	C	D	E	F
1	initial point (x0)	0,2	0,5	0,7	0,4	0
2	objective function	rosenbrock				
3	value of objective					
4	equality constraints	eqcons				
5	value of equality constraints					
6	inequality constraints	ineqcons				
7	value of inequality constraints					
8	solution found					

Running the macro `fmincon` below produces the result

	A	B	C	D	E	F
1	initial point (x0)	0,2	0,5	0,7	0,4	0
2	objective function	rosenbrock				
3	value of objective	8,3542E-17				
4	equality constraints	eqcons				
5	value of equality constraints	0				
6	inequality constraints	ineqcons				
7	value of inequality constraints	0				
8	solution found	1	1	1	1	1

Now, modify the initial data as follows:

	A	B	C	D	E	F
1	initial point (x0)	0,2	0,5	0,7	0,4	0
2	objective function	rosenbrock				
3	value of objective					
4	equality constraints	eqcons2				
5	value of equality constraints					
6	inequality constraints	ineqcons2				
7	value of inequality constraints					
8	solution found					

The result is

	A	B	C	D	E	F
1	initial point (x0)	0,2	0,5	0,7	0,4	0
2	objective function	rosenbrock				
3	value of objective	3,1328E-05				
4	equality constraints	eqcons2				
5	value of equality constraints	0				
6	inequality constraints	ineqcons2				
7	value of inequality constraints	2,1247E-07	4,06E-07			
8	solution found	1,00028021	1,0001	0,99991	0,999863	0,999846

To get the values of the Lagrange's multipliers, the simplest way consists in opening the Final Dialog Box and selecting the reports you need. To open the Final Dialog Box, you must change "`True`" in "`False`" at `SolverSolve`:

```
SolverSolve False
```

Then, the Final Dialog Box will open, and you will be able to select the reports you need, namely those with the Lagrange's multipliers. Notice that the report will be offered only if the method converges.

If you desire to make the generation of the report automatic, you can add a line after "`SolverSolve True`":

```
SolverSolve True
SolverFinish ReportArray:=Array(2)
```

The supplementary line creates a new sheet containing the sensibility report. Again, notice that *the report is created only if the criteria of the SOLVER for the convergence to a solution are fulfilled.* For instance, they are not satisfied in the examples of this subsection, although the results are awfully close to the exact solution – the SOLVER points a "probable solution", with convergence in probability.

Otherwise, the Lagrange's multipliers can be determined by solving the Eq. (2.11), which form a linear system for the values of $\boldsymbol{\lambda}$. Observe that

$$\forall i, 1 \leq i \leq p+q: \ \psi_i(\boldsymbol{x}^*) < 0 \Longrightarrow \lambda_i = 0.$$

Thus, you must start by finding the subset of indexes

$$A(\boldsymbol{x}^*) = \{i : 1 \leq i \leq p+q : \psi_i(\boldsymbol{x}^*) = 0\}$$

$A(\boldsymbol{x}^*)$ is said to be the set of the *active constraints* (those attaining the equality). In practice a tolerance *tol* is used for the inequalities: $A(\boldsymbol{x}^*)$ is formed of the indexes corresponding to the equalities ($p + 1$ *to* $p + q$) and the indexes $1 \leq i \leq p$ such that $\varphi_i(\boldsymbol{x}^*) < -tol$. Then, Eq. (2.11) show that

$$\sum_{i \in A(x^*)} \lambda_i \frac{\partial \psi_i}{\partial x_j}(\boldsymbol{x}^*) = -\frac{\partial J}{\partial x_j}(\boldsymbol{x}^*), \quad 1 \leq j \leq n.$$

Let $A(\boldsymbol{x}^*) = \{i_1, \ldots, i_\ell\}$ and denote $\boldsymbol{\lambda_A} = (\lambda_{i_1}, \lambda_{i_1}, \ldots, \lambda_{i_1})^t$. These equations form a linear system

$$\boldsymbol{M\lambda_A} = \boldsymbol{N},$$

with

$$M_{ij} = \frac{\partial \psi_j}{\partial x_i}(\boldsymbol{x}^*), N_i = -\frac{\partial J}{\partial x_i}(\boldsymbol{x}^*), 1 \leq i \leq n, j \in A(\boldsymbol{x}^*)$$

To apply this approach, you must furnish a subprogram that evaluates the gradients of the constraints and of the objective function. In addition, the linear system for $\boldsymbol{\lambda_A}$ generally must be solved in the sense of Least Squares:

$$(\boldsymbol{M}^t\boldsymbol{M})\boldsymbol{\lambda_A} = \boldsymbol{M}^t\boldsymbol{N}.$$

For instance, using this approach for the examples of this subsection, with $tol = 1E - 3$ leads to $\lambda_1 = \lambda_2 = 0$ in both the situations considered.

2.2.5 Linear Programming Using the SOLVER in a Worksheet

A general linear programming problem corresponds to

$$f(x) = c^t x = \sum_{i=1}^{n} c_i x_i$$

$$C = \{x \in \mathbb{R}^n : \ A.x \leq B, Ae.x = Be\}$$

A supplementary condition $x \geq 0$ may be considered. To solve such a problem using the solver, you must furnish the vectors c, B, Be and the matrices A, Ae. Then, you must associate cells to the values of $c^t x$, $A.\,x$, $Ae.\,x$ and pass them as parameters to the SOLVER. As an example, let us consider the problem where

$$c = (-4, -3, 5), x \geq 0$$

$$A = \begin{pmatrix} 4 & -1 & 2 \\ 2 & -1 & 1 \end{pmatrix}, B = \begin{pmatrix} 4 \\ -1 \end{pmatrix}$$

$$Ae = (1, 1, 1), \quad Be = (3)$$

Open a Blank Workbook and add 2 new sheets: the workbook will have 3 Sheets, named Sheet1, Sheet2, Sheet3 (you can rename them if you desire). Of course, all the data may be placed in a single Sheet but working with 3 Sheets will make your Worksheet more general and make easier to deal with large matrices.

The first sheet will contain c at column `A`. The result x will appear in column `C.` give the name `c` to `Sheet1!A2:A4` and the name `x` to `Sheet1!C2:C4.` Sheet 2 will contain: matrix A and vector B: give the name `A` to `A2:C3` and the name `B` to `E2:E3`. Sheet 3 will contain the equalities: matrix Ae and vector B: give the name `Ae` to `Sheet3!A2:C2` and the name `Be` to `Sheet3!E2`. The value of $f(x)$ is placed in `Sheet1!E2:` "`= MMULT(TRANSPOSE(c);x)`". Analogously, the value of $AX - B$ will appear in column `Sheet1!G`: the formula in G2 is "`=MMULT(A;x) - B`". Finally, the formula in I2 is "`=MMULT(Ae;x) - Be`": $AeX - Be$ will appear in column `Sheet1!I`: The three sheets appear as:

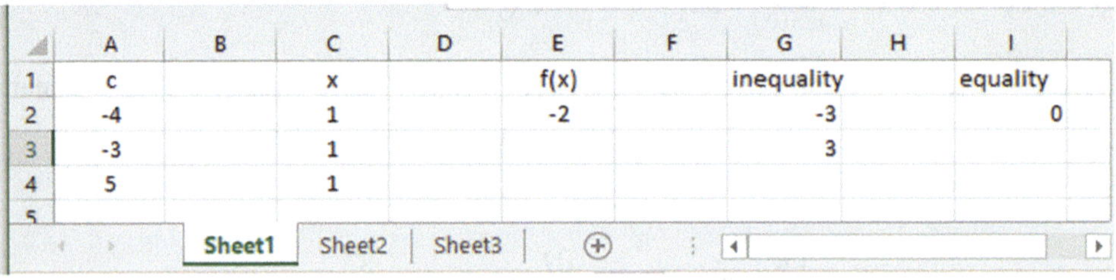

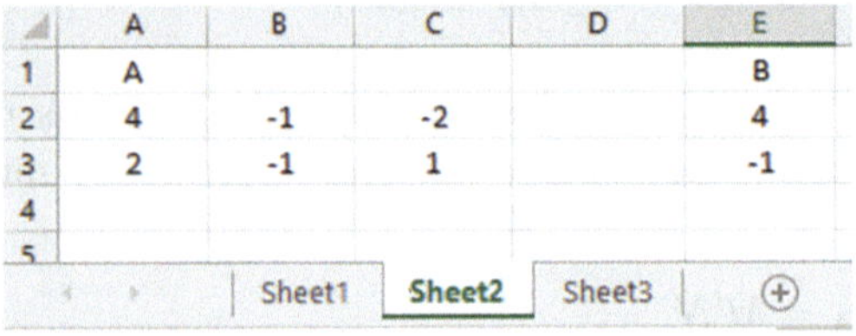

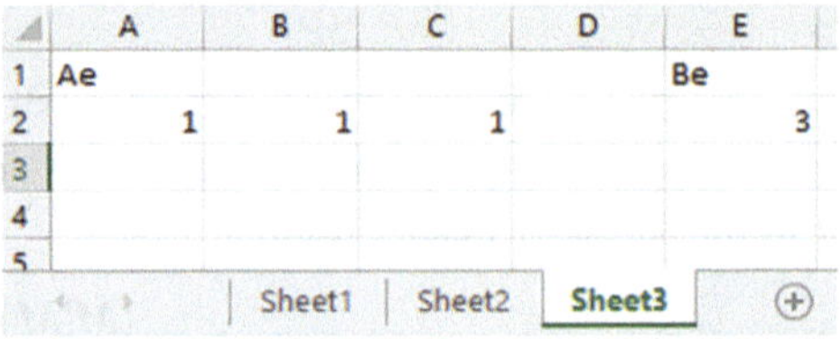

Go to Sheet1 and click on cell `E2`. Then, go to *Data>Solver* and enter the data corresponding to the objective and the constraints. Check "Make unconstrained variables non-negative". Select "Simplex LP" as Solving Method. Click on "Solve" to run the Solver.

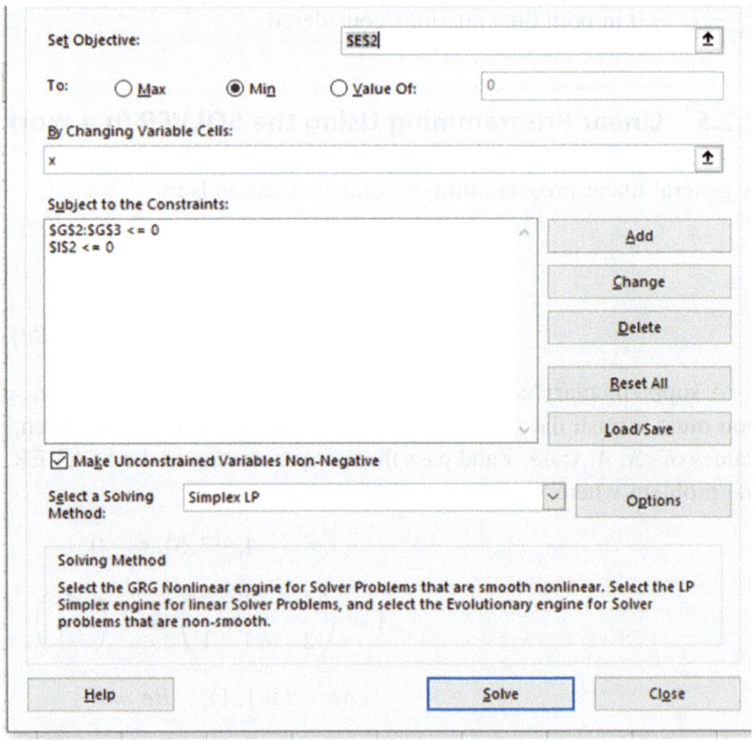

The solution appears, close to the exact one, which is (2/3,7/3):

	A	B	C	D	E	F	G	H	I
1	c		x		f(x)		inequality		equality
2	-4		0,666667		-9,66667		-3,66667		-4,4E-16
3	-3		2,333333				2,22E-16		
4	5		0						

You may get the shadow prices (Lagrange's multipliers) by checking the corresponding report ("Sensitivity Report") in the Final Dialog Box:

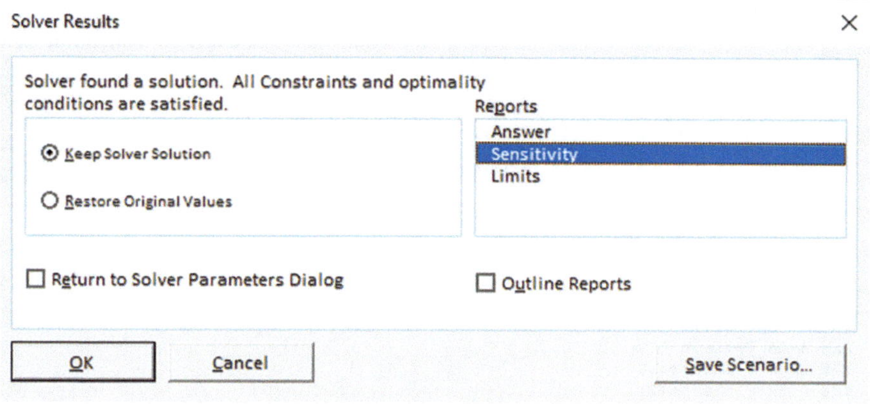

This will add a new sheet to the workbook containing the values of the shadow prices.

6	Variable Cells						
7			Final	Reduced	Objective	Allowable	Allowable
8	Cell	Name	Value	Cost	Coefficient	Increase	Decrease
9	$C$2	x	0,666666667	0	-4	1	1E+30
10	$C$3	x	2,333333333	0	-3	26	1
11	$C$4	x	0	8,666666667	5	1E+30	8,666666667
12							
13	Constraints						
14			Final	Shadow	Constraint	Allowable	Allowable
15	Cell	Name	Value	Price	R.H. Side	Increase	Decrease
16	$G$2	inequality	-3,666666667	0	0	1E+30	3,666666667
17	$G$3	inequality	0	-0,333333333	0	2,2	2
18	$I$2	equality	0	-3,333333333	0	5,5	2

2.2.6 Linear Programming Using the SOLVER in VBA

An example of macro which runs the SOLVER to get the solution is given in Fig. 2.2. Executing this Macro, we obtain:

	A	B	C	D	E	F	G	H	I
1	c		x		f(x)		inequality		equality
2	-4		0,67		-9,66667		-3,666666667		0
3	-3		2,33				2,22045E-16		
4	5		0,00						

Again, it is possible to get the Lagrange's Multipliers (shadow prices) by using "`SolverSolve False`" instead of "`SolverSolve True`".

6	Variable Cells						
7			Final	Reduced	Objective	Allowable	Allowable
8	Cell	Name	Value	Cost	Coefficient	Increase	Decrease
9	$C$2	x	0,666666667	0	-4	1	1E+30
10	$C$3	x	2,333333333	0	-3	26	1
11	$C$4	x	0	8,666666667	5	1E+30	8,666666667
12							
13	Constraints						
14			Final	Shadow	Constraint	Allowable	Allowable
15	Cell	Name	Value	Price	R.H. Side	Increase	Decrease
16	$I$2	equality	0	-3,333333333	0	5,5	2
17	$G$2	inequality	-3,666666667	0	0	1E+30	3,666666667
18	$G$3	inequality	0	-0,333333333	0	2,2	2
19							

```
Public Sub lp()
Dim x As Variant
Set sht1 = Sheets(1)
Set sht2 = Sheets(2)
Set sht3 = Sheets(3)
colmax2 = Sheets(2).Cells(2, Cells.Columns.Count).End(xlToLeft).Column
linmax2 = Sheets(2).Range("A" & Rows.Count).End(xlUp).Row
With sht2
  rA = Range(Cells(2, 1), Cells(linmax2, colmax2 - 2)).Address
  rB = Range(Cells(2, colmax2), Cells(linmax2, colmax2)).Address
End With
colmax3 = Sheets(3).Cells(2, Cells.Columns.Count).End(xlToLeft).Column
linmax3 = Sheets(3).Range("A" & Rows.Count).End(xlUp).Row
With sht3
  rAe = Range(Cells(2, 1), Cells(linmax3, colmax3 - 2)).Address
  rBe = Range(Cells(2, colmax3), Cells(linmax3, colmax3)).Address
End With
With sht1
  linmaxc = Range("A" & Rows.Count).End(xlUp).Row
  rc = Range(Cells(2, 1), Cells(linmaxc, 1)).Address
  rx = Range(Cells(2, 3), Cells(linmaxc, 3)).Address
  strf = "= SUMPRODUCT(Sheet1!" & rc & ", Sheet1!" & rx & ")"
  Range(Cells(2, 5), Cells(2, 5)).Formula = strf
  strineq = "= MMULT(Sheet2!" & rA & ", Sheet1!" & rx & ") - Sheet2!" & rB
  Range(Cells(2, 7), Cells(2, 7)).FormulaArray = strineq
  streq = "= MMULT(Sheet3!" & rAe & ", Sheet1!" & rx & ") - Sheet3!" & rBe
  Range(Cells(2, 9), Cells(2, 9)).FormulaArray = streq
  rfobj = Range(Cells(2, 5), Cells(2, 5)).Address
  SolverReset
  SolverOk SetCell:=rfobj, MaxMinVal:=2, ValueOf:=0, ByChange:=rx, _
        Engine:=2, EngineDesc:="Simplex LP"
  SolverOptions AssumeNonNeg:=True
  eqlinemax = Sheets(1).Range("I" & Rows.Count).End(xlUp).Row
  If eqlinemax > 1 Then
    For i = 2 To eqlinemax
      SolverAdd Cellref:=Range(Cells(i, 9), Cells(i, 9)).Address, Relation:=2, FormulaText:="0"
    Next i
  End If
  ineqlinemax = Sheets(1).Range("G" & Rows.Count).End(xlUp).Row
  If ineqlinemax > 1 Then
    For i = 2 To ineqlinemax
      SolverAdd Cellref:=Range(Cells(i, 7), Cells(i, 7)).Address, Relation:=1, FormulaText:="0"
    Next i
  End If
  SolverSolve True
End With
End Sub
```

Fig. 2.2 Macro lp

It is possible to generate the values of the shadow prices automatically by the same methods of Sect. 2.2.4 (see page 59): for instance, adding a line "`SolverFinish ReportArray:=Array (2)`" after "`SolverSolve True`":

```
SolverSolve True
SolverFinish ReportArray:=Array(2)
```

This command will generate a new sheet containing the Sensibility report.

2.3 Nonlinear Equations

A system of nonlinear equations reads as

$$f(x) = \mathbf{0}\,. \tag{2.12}$$

where $f: \mathbb{R}^n \longrightarrow \mathbb{R}^m$ is a function associating to vector $X \in \mathbb{R}^n$ a vector from $\mathbb{R}^m$: x is a $n \times 1$-matrix and $f(x) = (f_1(x), \ldots, f_m(x))^t$ is a $m \times 1$-matrix. Analogously to the linear systems, we shall consider $m \geq n$.

Systems of nonlinear equations may be solved by finding the minimum of $\|f(x)\|$. For instance, we may look for a least squares solution, which minimizes

$$F(x) = \left(\sum_{i=1}^{n} f_i^2(x) \right)^{\frac{1}{2}} \tag{2.13}$$

In such an approach, the solution of a nonlinear system becomes an optimization problem, which can be solved by the methods introduced in the preceding section. To solve Eq. (2.12) by minimizing F defined by Eq. (2.13) and using the SOLVER, we need just to add subprograms of a cell calculating F.

2.3.1 Nonlinear Equations Using the SOLVER in a Worksheet

Let us exemplify the approach by solving the nonlinear system of equations

$$\begin{aligned} f_1(x) &= x_1^2 + x_2^2 + x_3^2 - 3 \\ f_2(x) &= x_1 - x_2^2 \\ f_3(x) &= x_1 - x_3^2 \end{aligned} \tag{2.14}$$

Open a blank workbook and enter the data below.

	A	B	C	D	E
1	x		f		objective
2	1		11		64,66666669
3	2		-3		
4	3		-8		
5					

Formulas		
C2	=SUMSQ(A2:A4)-3	f_1
C3	=A2 - A3^2	f_2
C4	=A2 - A4^2	f_3
E2	=SUMSQ(C2:C4)/3	F

Then, run the SOLVER to minimize cell E2 by changing cells `A2:A4`. The result is close to the exact solution (1, 1, 1).

	A	B	C	D	E
1	x		f		objective
2	1		-5,2E-08		2,83672E-15
3	1		-1,2E-08		
4	1		7,54E-08		

2.3.2 Nonlinear Equations Using the SOLVER in VBA

To solve a nonlinear system of equations using VBA, you must create a VBA function that evaluates the equations to be solved and a function that evaluates their norm. For instance:

```
Function norm_2(x) As Double
Dim s As Double
s = 0
For i = 1 To UBound(x)
    s = s + x(i) ^ 2
Next i
norm_2 = Sqr(s)
End Function
```

```
Public Function nonlineq1(x)
Dim outs() As Double
ReDim outs(1 To 3)
s = 0
For i = 1 To UBound(x, 1)
    s = s + x(i) ^ 2
Next i
outs(1) = s - 3
outs(2) = x(1) - x(2) ^ 2
outs(3) = x(1) - x(3) ^ 3
nonlineq1 = outs
End Function
```

`nonlineq1` evaluates the equations and `norm_2` evaluates the least squares norm (2.13). An example of macro `fsolve` to solve the equations in column `C` by modifying the values of the variables in column `A` is given below:

```
Public Sub fsolve()
Dim x As Variant
Set sht = ActiveSheet
colx = Cells(3, Cells.Columns.Count).End(xlToLeft).Column
For i = 2 To colx
    Cells(4, i).Value = Cells(3, i).Value
Next i
rx = Range(Cells(4, 2), Cells(4, colx)).Address
eqname = Range("B1").Value
streq = "= fobjnorm(" & Chr(34) & eqname & Chr(34) & ", " & rx & ", " & CStr(Range("B2").Value) & ")"
Cells(5, 2).Formula = streq
rfobj = "$B$5"

SolverReset
SolverOk SetCell:=rfobj, MaxMinVal:=2, ValueOf:=0, ByChange:=rx, _
        Engine:=1, EngineDesc:="GRG Nonlinear"
SolverOptions AssumeNonNeg:=False
SolverSolve True
End Sub
```

```
Public Function fobjnorm(nom As String, r As Range, indnorm)
x = range2vector(r)
erreur = Application.Run(nom, x)
nrm = "norm_" & CStr(indnorm)
aux = Application.Run(nrm, erreur)
fobjnorm = aux
End Function
```

To Solve Eq. (2.14), open a blank workbook and add the modules `nonlineq1`, `norm_2`, `range2vector` and the macro `fsolve` given in the preceding. Do not forget that you must run the SOLVER, so that it must be included in the references (see Sect. 1.11, page 19). Then, enter the data below.

	A	B	C	D
1	equations	nonlineq1		
2	norm to be used	2		
3	initial point (x0)	1	2	3
4	solution found			
5	value of the norm			

The macro solves the equations defined in `nonlineq1` by changing the values in line 5, starting from column `B`. The solution is generated by minimizing the norm of the equation – the number of the norm is defined in `B2`. Here, we use the least squares norm corresponding to (2.13). Running the macro produces:

	A	B	C	D	E
1	equations	nonlineq1			
2	norm to be used	2			
3	initial point (x0)	1	2	3	
4	solution found	1,000002947	1,000001	0,999999	
5	value of the norm	8,6864E-06			

These results are close to the exact solution (1, 1, 1).

2.3.3 Nonlinear Equations Using Newton-Raphson

Newton-Raphson's method is one of the most popular ways to get numerical solutions of nonlinear systems of equations. The method starts from an initial guess $\boldsymbol{x}^{(0)}$ and generates a sequence $\boldsymbol{x}^{(1)}$, $\boldsymbol{x}^{(2)}$... such that $\boldsymbol{x}^{(i)} = \boldsymbol{x}^{(i-1)} + \Delta\boldsymbol{x}^{(i-1)}$, where $\Delta\boldsymbol{x}^{(i)}$ is the solution of the linear system

$$\boldsymbol{H}^{(i)}\,\Delta\boldsymbol{x}^{(i)} = -\boldsymbol{f}\left(\boldsymbol{x}^{(i-1)}\right), H_{jk}^{(i)} = \frac{\partial f_j}{\partial x_k}\left(\boldsymbol{x}^{(i-1)}\right)$$

Such a linear system may be solved by one of the methods exposed in Sect. 2.1. An example of Macro for Newton-Raphson's iterations is shown in Fig. 2.3 (it uses `GaussSolution` previously introduced). Its code assumes that the names of the subprograms evaluating $\boldsymbol{f}$ and $\boldsymbol{H}$ are given in `B1` and `B2`, respectively. The maximum number of iterations and the minimal tolerance between two iterations are given in `B4` and `B5`. Line 3 contains the initial guess starting at `B3`. The solution

```
Public Sub newton_raphson()
Dim x As Variant
Set sht = ActiveSheet
colx = Cells(3, Cells.Columns.Count).End(xlToLeft).Column
For i = 2 To colx
    Cells(6, i).Value = Cells(3, i).Value
Next i
eqname = Range("B1").Value
jacname = Range("B2").Value
nitmax = Range("B4").Value
tolref = Range("B5").Value
rx = Range(Cells(6, 2), Cells(6, colx)).Address
xac = range2vector(Range(rx))
nit = 0
tol = tolref + 1
Dim not_stop As Boolean
not_stop = True
While not_stop
   nit = nit + 1
   f = Application.Run(eqname, xac)
   H = Application.Run(jacname, xac)
   dx = GaussSolution(H, f)
   tol = norm_2(dx)
   For i = 1 To colx - 1
       xac(i) = xac(i) - dx(i)
   Next i
   not_stop = nit < nitmax And tol > tolref
Wend
Range(rx).Value = xac
f = Application.Run(eqname, xac)
aux = norm_2(f) / Sqr(colx - 1)
Range("B7").Value = aux
Range("B8").Value = nit
Range("B9").Value = tol
End Sub
```

Fig. 2.3 A macro for Newton-Raphson

calculated appears in line 6, starting at `B6`. `B7`, `B8`, `B9` show, respectively, the mean quadratic error $F\left(\boldsymbol{x}^{(k)}\right)/\sqrt{n}$, the final iteration number k and the final tolerance $\|\boldsymbol{x}^{(k)} - \boldsymbol{x}^{(k-1)}\|$.

For instance, let us consider Eq. (2.14). Then

$$\boldsymbol{H}(\boldsymbol{x}) = \begin{pmatrix} 2x_1 & 2x_2 & 2x_3 \\ 1 & -2x_2 & 0 \\ 1 & 0 & -2x_3 \end{pmatrix}$$

H is the jacobian matrix associated to ***f***. To use this method, you must supply subprograms evaluating ***H***. An example of subprogram that evaluates ***H*** is given at right:	see code below

```
Public Function jacobian1(x)
Dim outs() As Double
ReDim outs(1 To 3, 1 To 3)
outs(1, 1) = 2 * x(1)
outs(1, 2) = 2 * x(2)
outs(1, 3) = 2 * x(3)
outs(2, 1) = 1
outs(2, 2) = -2 * x(2)
outs(2, 3) = 0
outs(3, 1) = 1
outs(3, 2) = 0
outs(3, 3) = -2 * x(3)
jacobian1 = outs
End Function
```

The workbook appears as:

	A	B	C	D
1	equations	nonlineq1		
2	Jacobian	jacobian1		
3	initial point (x0)	3	2	1
4	maximum iterations	20		
5	desired precision	1,00E-06		
6	solution found			
7	mean quadratic error			
8	iterations attained			
9	tolerance attained			

Running the macro, we obtain

6	solution found	1,000000065	1	1
7	mean quadratic error	2,05581E-07		
8	iterations attained	16		
9	tolerance attained	4,44107E-07		

Newton-Raphson's method may converge or diverge according to the starting point $\boldsymbol{x}^{(0)}$. For instance, the initial point previously used (1, 2, 3) leads to divergence

2.3.4 Overdetermined Linear Systems

Recall that an overdetermined linear system $\boldsymbol{AX} = \boldsymbol{B}$ is a system where $\boldsymbol{A}$ is $m \times n$, with $m > n$; $\boldsymbol{X}$ is $n \times 1$; $\boldsymbol{B}$ is $m \times 1$. In general, such a system does not admit an exact solution and we look for an approximated solution that minimizes $f(X) = \|\boldsymbol{AX} - \boldsymbol{B}\|$.

Thus, the initial overdetermined linear system is solved by optimization methods: we look for the minimum of f.

If $\|\boldsymbol{v}\| = \sqrt{\boldsymbol{v}^t\boldsymbol{v}}$ (Euclidean norm), then the minimum is attained at $\boldsymbol{X}^*$ such that $\boldsymbol{A}^t\boldsymbol{AX}^* = \boldsymbol{A}^t\boldsymbol{B}$ (Least Squares Solution) – a linear system that can be solved by the methods previously introduced (Sect. 2.1).

If $\|\boldsymbol{v}\| = \max\ \{|v_i| : 1 \leq i \leq n\}$, then the minimum is attained at $\boldsymbol{X}^*$ such that $X_i^* = \frac{y_i^*}{y_{n+1}^*}$, for $1 \leq i \leq n$, where $\boldsymbol{y}^* = \left(\boldsymbol{Y}_n^*, y_{n+1}^*\right), \boldsymbol{Y}_n^* = \left(y_1^*, \ldots, y_n^*\right)$ is the solution of the following Linear Programming problem:

$$y^* = \arg\min\left\{-y_{n+1} :\ \boldsymbol{AY}_n^* - \boldsymbol{B}y_{n+1} \leq 1, -\boldsymbol{AY}_n^* + \boldsymbol{B}y_{n+1} \leq 1, y_{n+1} \geq 0\right\},$$

This LP program can be solved by the methods presented in Sects. 2.2.5 and 2.2.6.

As an example, let us consider the situation where

$$A = \begin{pmatrix} 1 & -3 & 9 & -27 \\ 1 & -2 & 4 & -8 \\ 1 & -1 & 1 & -1 \\ 1 & 0 & 0 & 0 \\ 1 & 1 & 1 & 1 \\ 1 & 2 & 4 & 8 \\ 1 & 3 & 9 & 27 \end{pmatrix}, \quad B = \begin{pmatrix} -0.14112 \\ -0.90930 \\ -0.84147 \\ 0 \\ 0.84147 \\ 0.90930 \\ 0.14112 \end{pmatrix}$$

To solve the problem, open a blank workbook and enter the data as shown at left. Give the name `A` to `A2:C8` and the name `B` to `B2:B8`.

	A	B	C	D	E	F
1	A					B
2	1	-3	9	-27		-0,14112
3	1	-2	4	-8		-0,9093
4	1	-1	1	-1		-0,84147
5	1	0	0	0		0
6	1	1	1	1		0,841471
7	1	2	4	8		0,909297
8	1	3	9	27		0,14112
9						

Then, select cell `H2` and enter =MMULT (MINVERSE (MMULT(TRANSPOSE(A);A)); MMULT(TRANSPOSE(A);B)). The Least Squares Solution appears in cells `H2: H5`.

H
LSQ
6,61E-18
0,846219
1,32E-18
-0,08942

To use Linear Programming, Enter the data in `L2:L5`, `M2` as shown at left. Give the name `Yn` to `L2:L5` and `ynp1` to `M2`. Then, go to column `J` and enter =`L2/ynp1` in `J2`. Copy the formula and paste it in `J3:J5`.

J	K	L	M
LP		Yn	ynp1
0,2		1	5
0,4		2	
0,6		3	
0,8		4	

The next step is the creation of the Linear Programming problem: go to column `N` and enter `=MMULT(A;Yn)-ynp1*B` in `N2`; `=-MMULT(A;Yn)+ynp1*B` in `O2`; `=-ynp1` in `P2`.

N	O	P
Cplus	Cminus	sign
-85,2944	85,2944	-5
-18,4535	18,45351	
2,207355	-2,20735	
1	-1	
5,792645	-5,79265	
44,45351	-44,4535	
141,2944	-141,294	

Run the Solver:

- Choose `$P$2` as objective
- Set `Yn;ynp1` as variables
- add the constraints
  `$N$2:$N$8 <= 1`
  `$O$2:$O$8 <= 1`
  `$P$2 <= 0`
- Select `Simplex LP` as method
- Uncheck "*Make Unconstrained variables non-negative*"
- Click "*Solve*"

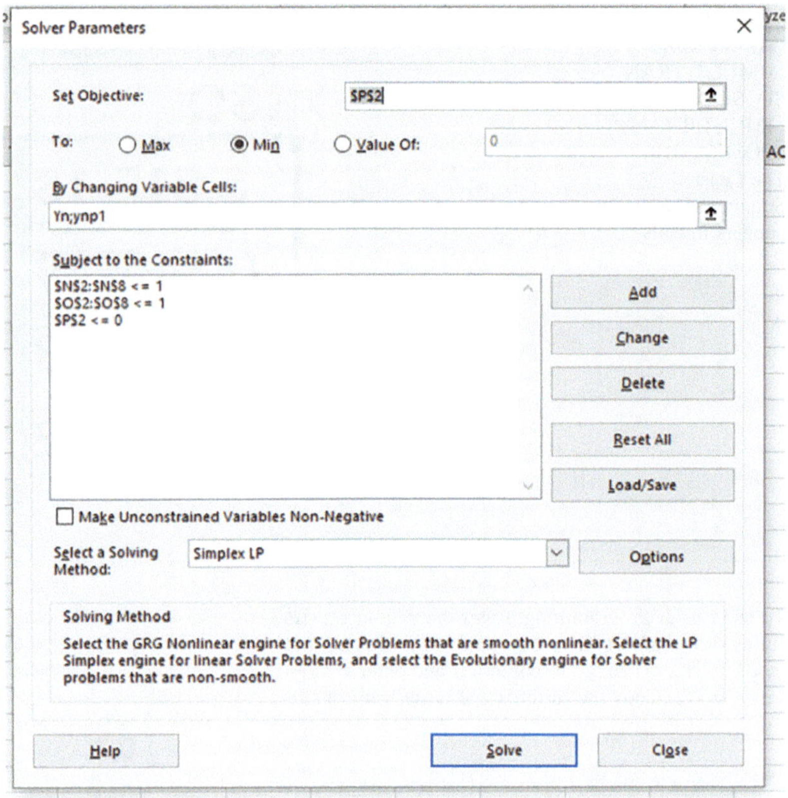

The LP solution x^* appears in `J2: J5`. `L2:L4` contains Y_n^* and `M2` contains the value of y_{n+1}^*

J	K	L	M
LP		Yn	ynp1
5E-17		0	14,059
0,864		12,1	
8E-18		0	
-0,09		-1,31	

2.4 Ordinary Differential Equations

A typical initial value problem for a first-order Ordinary Differential Equation (ODE) reads as

$$\frac{dX}{dt} = f(t, X) \quad , \quad X(0) = X_0 \quad , t \in [0, T] \tag{2.15}$$

The reader may find in the literature many methods for the numerical solution of Eq. (2.15), usually involving a discretization of the time interval (0, T) in n subintervals (t_{i-1}, t_i), $1 \le i \le n$ such that $0 = t_0 < t_1 < t_2 < \ldots < t_n = T$. For instance, we may consider a uniform grid with step $\Delta t = \frac{T}{n}$ and $t_i = i\Delta t$, $0 \le i \le n$. The solution is approximated at these discrete times: $X(t_i) \approx X_i$. The values of X_i are used to approximate the equation on each subinterval, what lead to a sequential determination of the value X_i from the preceding ones $X_0, \ldots, X_{i-1}$.

2.4.1 Runge-Kutta's Methods

A popular family of numerical methods for the solution of Eq. (2.15) is the family of Runge-Kutta (RK) methods, where

$$X_i = X_{i-1} + \Delta t \sum_{i=1}^{n} b_i Y_i; \tag{2.16}$$

$$Y_0 = X_{i-1}; \quad Y_k = f\left(t_{i-1} + c_k \Delta t, X_{i-1} + \Delta t \sum_{j=0}^{n} a_{kj} Y_j\right), 1 \le k \le n. \tag{2.17}$$

In some of these methods, the error may be estimated by

$$e_i = X(t_i) - X_i \approx \sum_{i=1}^{n} (b_i - b_i^*) Y_i$$

The coefficients are determined to ensure that the local error on each subinterval is of order n (id est, $\|e_i\| \le C\Delta t^{n+1}$). b^* are usually the coefficients of a method of order $n - 1$. The method is usually synthetized in a Table analogous to Table 2.1:

The method is explicit when $a_{ij} = 0$ for $j \ge i$. In this case, the coefficients appear as in Table 2.2.

In general, the line for b^* is given only for some methods, where the error estimate is used to determine the step (adaptive methods). An implicit method usually involves more calculations: for instance, it may be mandatory to solve a nonlinear equation to determine Y_k. Thus, implicit methods usually have a higher computational cost. Their advantage is their better numerical properties such as, for instance, a higher stability. Explicit methods are faster and simpler to use but may be less stable.

Table 2.1 Coefficients of a general Runge-Kutta's method

c_1	a_{10}	a_{11}	$\cdots$	$a_{1,n-1}$	$a_{1,n}$
c_2	a_{20}	a_{21}	$\cdots$	$a_{2,n-1}$	$a_{2,n}$
c_3	a_{30}	a_{31}	$\cdots$	$a_{3,n-1}$	$a_{3,n}$
$\vdots$	$\vdots$	$\vdots$	$\ddots$	$\vdots$	$\vdots$
c_n	a_{n0}	a_n	$\cdots$	$a_{n,n-1}$	$a_{n,n}$
	b_0	b_1	$\cdots$	b_{n-1}	b_n
	b_0^*	b_1^*	$\cdots$	b_{n-1}^*	b_n^*

The simplest method of this family is *Forward Euler's method*, which approximates

$$\frac{d\boldsymbol{X}}{dt}(t) \approx \frac{\boldsymbol{X}_i - \boldsymbol{X}_{i-1}}{\Delta t}, \ \boldsymbol{f}(t,\boldsymbol{X}) \approx \boldsymbol{f}(t_{i-1},\boldsymbol{X}_{i-1}) \ on \ (t_{i-1},t_i).$$

Thus,

$$\boldsymbol{X}_i = \boldsymbol{X}_{i-1} + \Delta t f(t_{i-1},\boldsymbol{X}_{i-1}), 1 \le i \le n.$$

Backward Euler's method is an implicit method that reads as

$$\boldsymbol{X}_i = \boldsymbol{X}_{i-1} + \Delta t f(t_{i-1} + \Delta t, \boldsymbol{X}_i), 1 \le i \le n.$$

Midpoint Euler's method is an explicit method that uses

$$\boldsymbol{X}_i = \boldsymbol{X}_{i-1} + \Delta t f\left(t_{i-1} + \frac{\Delta t}{2}, \boldsymbol{X}_i + \frac{\Delta t}{2} Y_1\right); \boldsymbol{Y}_1 = f(t_{i-1},\boldsymbol{X}_{i-1}) \ , 1 \le i \le n.$$

These Euler's methods are represented as shown in Table 2.3.

The classical RK2 (Runge-Kutta of order 2) is given in Table 2.4a. In Table 2.4b, we present the coefficients for a general 2nd order method.

RK3 and RK4 (Runge-Kutta of orders 3 and 4) as shown in Table 2.5.

Table 2.2 Coefficients of an Explicit Runge-Kutta's method

c_1	a_{10}				
c_2	a_{20}	a_{21}			
c_3	a_{30}	a_{31}			
$\vdots$	$\vdots$	$\vdots$	$\ddots$		
c_n	a_{n0}	a_{n1}	$\cdots$	$a_{n,n-1}$	
	b_0	b_1	$\cdots$	b_{n-1}	b_n
	b_0^*	b_1^*	$\cdots$	b_{n-1}^*	b_n^*

Table 2.3 Coefficients of different Euler's methods

a) Forward

0	
	1

b) Backward

1	1
	1

c) Midpoint

0		
$\frac{1}{2}$	$\frac{1}{2}$	
	0	1

Table 2.4 Tables for RK2 (Runge-Kutta order 2) and general 2nd order methods

a) RK2

0		
$\frac{1}{2}$	$\frac{1}{2}$	
	0	1

b) General 2nd_order ($a > 0$)

0		
a	a	
	$1 - \frac{1}{2a}$	$\frac{1}{2a}$

Table 2.5 Tables for RK3 and RK4 (Runge-Kutta orders 3 and 4)

a) RK3

0			
$\frac{1}{2}$	$\frac{1}{2}$		
1	-1	2	
	$\frac{1}{6}$	$\frac{4}{6}$	$\frac{1}{6}$

b) RK4

0				
$\frac{1}{2}$	$\frac{1}{2}$			
1/2	0	1/2		
1	0	0	1	
	$\frac{1}{6}$	$\frac{2}{6}$	$\frac{2}{6}$	$\frac{1}{6}$

Table 2.6 Coefficients of a Dormand-Prince's method

0							
$\frac{1}{5}$	$\frac{1}{5}$						
$\frac{3}{10}$	$\frac{3}{40}$	$\frac{9}{40}$					
$\frac{4}{5}$	$\frac{44}{45}$	$-\frac{56}{15}$	$\frac{32}{9}$				
$\frac{8}{9}$	$\frac{19372}{6561}$	$-\frac{25360}{2187}$	$\frac{64448}{6561}$	$-\frac{212}{729}$			
1	$\frac{9017}{3168}$	$-\frac{355}{33}$	$\frac{46732}{5247}$	$\frac{49}{176}$	$-\frac{5103}{18656}$		
1	$\frac{35}{384}$	0	$\frac{500}{1113}$	$\frac{125}{192}$	$-\frac{2187}{6784}$	$\frac{11}{84}$	
	$\frac{35}{384}$	0	$\frac{500}{1113}$	$\frac{125}{192}$	$-\frac{2187}{6784}$	$\frac{11}{84}$	0
	$\frac{5179}{57600}$	0	$\frac{7571}{16695}$	$\frac{393}{640}$	$-\frac{92097}{339200}$	$\frac{187}{2100}$	$\frac{1}{40}$

For instance, RK4 corresponds to

$$\boldsymbol{Y}_1 = f(t_{i-1}, \boldsymbol{X}_{i-1}) \;; \boldsymbol{Y}_2 = f\left(t_{i-1} + \frac{\Delta t}{2}, \boldsymbol{X}_{i-1} + \frac{\Delta t}{2} \boldsymbol{Y}_1\right);$$

$$\boldsymbol{Y}_3 = f\left(t_{i-1} + \frac{\Delta t}{2}, \boldsymbol{X}_{i-1} + \frac{\Delta t}{2} \boldsymbol{Y}_2\right) \;; \boldsymbol{Y}_2 = f(t_{i-1} + \Delta t, \boldsymbol{X}_{i-1} + \Delta t \boldsymbol{Y}_3);$$

$$\boldsymbol{X}_i = \boldsymbol{X}_{i-1} + \frac{\Delta t}{6} (\boldsymbol{Y}_1 + 2\boldsymbol{Y}_2 + 2\boldsymbol{Y}_3 + \boldsymbol{Y}_4).$$

A popular method from the Runge-Kutta's family is the method of *Dormand-Prince*, which is implemented in many free or commercial software. This method allows the estimation of the error by using a supplementary vector of coefficients, usually added below the last line of the coefficient's Table 2.6. The coefficients of Dormand's-Prince method are shown in Table 2.6.

A general explicit RK method may be implemented in VBA as follows:

```
Function explicitRKexec(a, b, c, tspan, x0, f)
ndim = UBound(x0, 1)
ntp = UBound(tspan, 1)
Dim solx() As Double
ReDim solx(1 To ndim, 1 To ntp)
For i = 1 To ndim
    solx(i, 1) = x0(i)
Next i
Dim x() As Double
ReDim x(1 To ndim)
For j = 2 To ntp
    For i = 1 To ndim
        x(i) = solx(i, j - 1)
    Next i
    tt = tspan(j - 1)
    h = tspan(j) - tt
    aux = explicitRKstep(a, b, c, h, tt, x, f)
    For i = 1 To ndim
        solx(i, j) = aux(i)
    Next i
Next j
Dim solRK(1 To 2) As Variant
solRK(1) = tspan
solRK(2) = solx
explicitRKexec = solRK
End Function
```

This function returns a structure formed by two elements: the first one contains the time moments where the solution was calculated (which is also the argument `tspan` of the procedure) and the second contains the solution calculated. Each column of the solution corresponds to a time moment: column i contains $\boldsymbol{X}_\mathbf{i}$. a, b, c are the coefficients of the Runge-Kutta method; `f` is a string with the name of the function that evaluates f; `x0` is a vector containing the initial point $\boldsymbol{X}_\mathbf{0}$. `explicitRKStep` performs a step of the method and reads as

```
Function explicitRKstep(a, b, c, h, t, x, f)
ndim = UBound(x, 1)
n = UBound(b, 1)
Dim y() As Double
ReDim y(1 To ndim, 0 To n)
For i = 1 To ndim
    y(i, 0) = x(i)
    For j = 2 To n
        y(i, j) = 0
    Next j
Next i
Dim yy() As Double
ReDim yy(1 To ndim)
For k = 0 To n - 1
    ck = c(k)
    ak = a(k)
    For i = 1 To ndim
        yy(i) = 0
    Next i
    tk = t + ck * h
    For j = 1 To UBound(ak, 1)
        For i = 1 To ndim
            yy(i) = yy(i) + ak(j) * y(i, j - 1)
        Next i
    Next j
    For i = 1 To ndim
        yy(i) = x(i) + h * yy(i)
    Next i
    aux = Application.Run(f, tk, yy)
    For i = 1 To ndim
        y(i, k + 1) = aux(i)
    Next i
Next k
For j = 1 To ndim
    yy(j) = 0
Next j
For i = 1 To n
    For j = 1 To ndim
        yy(j) = yy(j) + b(i) * y(j, i)
    Next j
Next i
For j = 1 To ndim
    yy(j) = x(j) + h * yy(j)
Next j
explicitRKstep = yy
End Function
```

To illustrate the use of these methods, let us consider the simple system of linear differential equations, where

$$f(t, \boldsymbol{X}) = (X_2, -X_1)^t \quad , \boldsymbol{X}_\mathbf{0} = (0, 1)^t \quad , T = 2\pi \ .$$

The solution is $\boldsymbol{X} = (sin(t), cos(t))$. We use a workbook containing the data below:

	A	B	C
1	f	ftrig	
2	T	7	
3	x0	0	1
4	nts	700	
5	number	4	

```
Function ftrig(t, x)
Dim aux() As Double
ReDim aux(1 To 2)
aux(1) = x(2)
aux(2) = -x(1)
ftrig = aux
End Function
```

Function `ftrig` evaluates f and its name is given in cell `B1`; the final time T is given in `B2`; $\boldsymbol{X}_0$ appears in `B3 : C3`; `B4` contains the number of time steps; `B5` contains the number of the method to run. Here, we use RK4. The results are shown in Fig. 2.4.

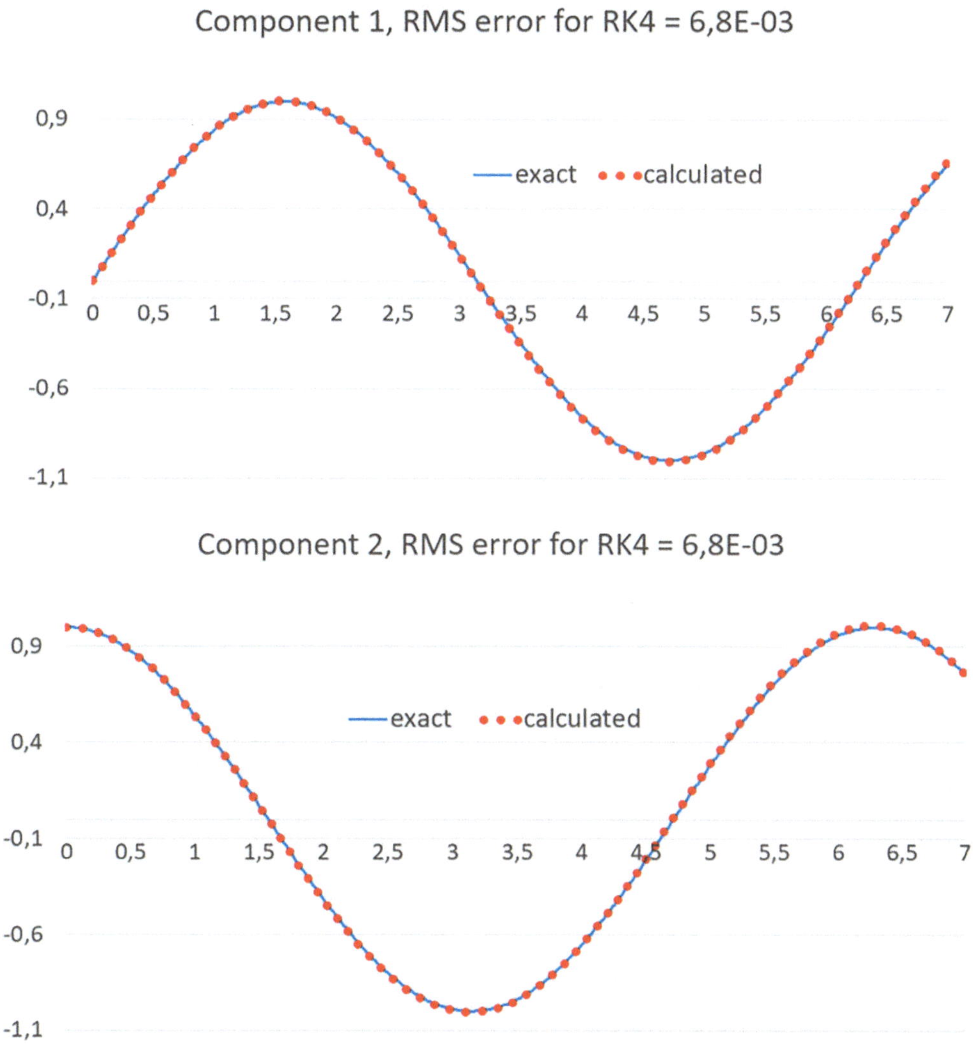

Fig. 2.4 Results furnished by RK4 for the trigonometrical example. The curves almost coincide

2.5 Numerical Integration

The simplest method to evaluate $\int_a^b f(x)dx$ is the trapezoidal rule:

$$\int_a^b f(x)dx \approx \frac{h}{2}(f(a) + f(b)) + h\sum_{i=1}^{n-1} f(a + ih), \quad h = \frac{b-a}{n}.$$

Trapezoidal rule is easily implemented in VBA. For instance,

```
Function trapezoidal(f, a, b, n) As Double
h = (b - a) / n
x = grid(a, h, n + 1)
y = spam(f, x)
s = 0.5 * (y(1) + y(n + 1))
For i = 2 To n
    s = s + y(i)
Next i
trapezoidal = h * s
End Function
```

```
Function spam(f, x) As Variant
m = LBound(x, 1)
n = UBound(x, 1)
Dim y() As Double
ReDim y(m To n)
For i = m To n
    y(i) = Application.Run(f, x(i))
Next i
spam = y
End Function
```

```
Function grid(a, h, np)
Dim x() As Double
ReDim x(1 To np)
x(1) = a
For i = 1 To np - 1
    x(i + 1) = x(i) + h
Next i
grid = x
End Function
```

Analogously to the preceding section, `f` is a string containing the name of the function that evaluates *f*. As an example, let us consider $f(x) = e^x$, $a = -2$, $b = 3$. The exact value is $\int_a^b f(x)dx = e^3 - e^{-2}$. We use a workbook containing the data below (`B5` contains *n*):

	A	B	C	D
1	f	fexp		
2	a	-2		
3	b	3		
4	method	trapezoidal		
5	number	1000		
6	calculated		exact	19,95020164
7	error			

```
Function fexp(x) As Double
fexp = Exp(x)
End Function
```

A macro runs the method and furnishes the result

```
Public Sub calculate()
f = Range("B1").Value
a = Range("B2").Value
b = Range("B3").Value
met = Range("B4").Value
nmet = Range("B5").Value
r = Application.Run(met, f, a, b, nmet)
Range("B6").Value = r
End Sub
```

6	calculated	19,9502432
7	error	-4,15629E-05

The trapezoidal rule may be improved by using Romberg's method:

$$R_{n,m} = \frac{4^m R_{n,m-1} - R_{n-1,m-1}}{4^m - 1}, R_{n,0} = trapezoidal(f, a, b, 2^n).$$

Then $\int_a^b f(x)dx \approx R_{n,n}$. Romberg's method is implemented in *integral.xlam* and *integral.bas*. For instance, we have (`B5` contains n):

	A	B	C	D
1	f	fexp		
2	a	-2		
3	b	3		
4	method	romberg		
5	number	5		
6	calculated	19,95020164	exact	19,95020164
7	error	-1,20331E-09		

Another popular method is Gauss Integration:

$$\int_a^b f(x)dx \approx \frac{b-a}{2}\sum_{i=1}^{n} w_i f\left(\frac{a+b}{2} + \xi_i \frac{(b-a)}{2}\right) .$$

Here, ξ_i, $1 \le i \le n$ are called Gauss Points and w_i, $1 \le i \le n$ are the associated weights. Tables of values of Gauss points and their weights may be found in the literature. The method is implemented in *integral.xlam* and *integral.bas*. For instance, we have (`B5` contains n):

	A	B	C	D
1	f	fexp		
2	a	-2		
3	b	3		
4	method	gauss		
5	number	6		
6	calculated	19,95020122	exact	19,95020164
7	error	4,22798E-07		

	A	B	C	D
1	f	fexp		
2	a	-2		
3	b	3		
4	method	gauss		
5	number	64		
6	calculated	19,95020164	exact	19,95020164
7	error	0		

The reader can easily find in the literature many variations of Gauss Integration, such as Gauss-Lobatto, Gauss-Tchebychev, Gauss-Laguerre, Gauss-Hermite, Gauss-Radau. Gauss-Lobatto points are available in *integral.xlam* and *integral.bas*. For instance, we have (`B5` contains n):

	A	B	C	D
1	f	fexp		
2	a	-2		
3	b	3		
4	method	gausslobatto		
5	number	6		
6	calculated	19,95020122	exact	19,95020164
7	error	4,22798E-07		

	A	B	C	D
1	f	fexp		
2	a	-2		
3	b	3		
4	method	gausslobatto		
5	number	10		
6	calculated	19,95020164	exact	19,95020164
7	error	0		

2.6 Multiobjective Optimization

In multiobjective optimization, we consider a vector of objectives $\boldsymbol{f} = (f_1, \ldots, f_m)$, with $f_i : \mathbb{R}^n \to \mathbb{R}$. In such a situation, the main concept is the ***dominance***: we say that $\boldsymbol{u} = (u_1, \ldots, u_m)$ dominates $\boldsymbol{v} = (v_1, .., v_m)$ if and only if $u_i \le v_i$, for $1 \le i \le m$ and there exists $1 \le k \le m$ such that $u_k < v_k$. $\boldsymbol{x}$ is a

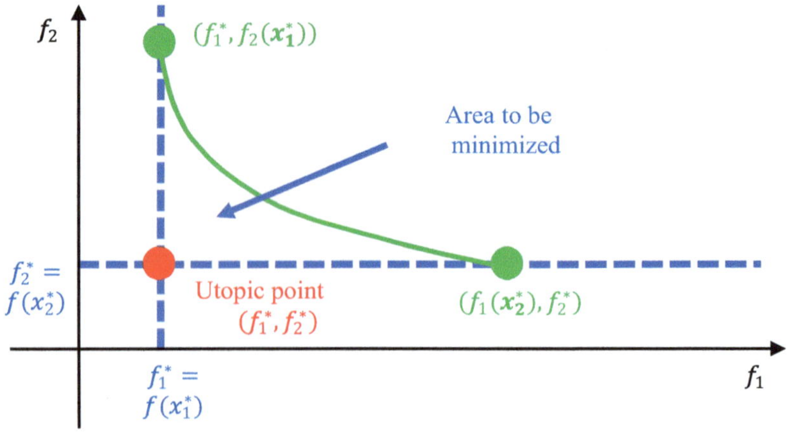

Fig. 2.5 A method for the determination of Pareto's fronts

Pareto's optimal solution if and only if there is not any $\boldsymbol{y}$ such that $\boldsymbol{f}(\boldsymbol{y})$ dominates $\boldsymbol{f}(\boldsymbol{x})$. A Pareto's optimal solution is a point for which the values of the objectives cannot be simultaneously improved: improving one of the objectives implies to deteriorate another one. It is usual to synthetize the optima using a ***Pareto's front***: the set of all the values $\boldsymbol{f}(\boldsymbol{x})$ where $\boldsymbol{x}$ is a Pareto's optimal solution. In many bidimensional situations, the Pareto's front is a curve $\boldsymbol{y} = \boldsymbol{f}(\boldsymbol{x}^*(t))$, where $t \in (a_1, b_1)$ is a parameter and $\boldsymbol{x}^*(\mathrm{t})$ is a Pareto's optimal point. Analogously, in many 3D situations, the Pareto's front is a surface $\boldsymbol{y} = \boldsymbol{f}(\boldsymbol{x}^*(t))$, $\boldsymbol{t} \in (a_1, b_1) \times (a_2, b_2)$. We may take $a_i = 0$, $b_i = 1$, with no loss of generality.

The reader may find in the literature many methods for the determination of Pareto's fronts. For instance, we may look for a curve (resp. a surface) $\boldsymbol{y} = \boldsymbol{f}(\boldsymbol{x}(\boldsymbol{t}))$ that minimizes the area (resp. the volume) limited by itself and lines (resp. planes) orthogonal to the axes and passing by the utopic point $\boldsymbol{f}^* = \left(f_1^*, \ldots, f_m^*\right)$, $f_i^* = f_i\left(\boldsymbol{x}_i^*\right) = \min f_i$ (see Fig. 2.5).

In 2D, by Stoke's theorem, it is equivalent to minimize the circulation of the vector $(-f_2(\boldsymbol{x}(t)), f_1(\boldsymbol{x}(t)))$ from $\left(f_1\left(\boldsymbol{x}_2^*\right), f_2^*\right)$ to $\left(f_1^*, f_2\left(\boldsymbol{x}_1^*\right)\right)$. In practice, we consider expansions

$$\boldsymbol{x}(t) \approx \boldsymbol{Px}(\mathrm{t}) = \sum_{i=0}^{n} \boldsymbol{x}_i \varphi_i(t), \boldsymbol{Px}(0) = \boldsymbol{x}_1^* \quad \boldsymbol{Px}(1) = \boldsymbol{x}_2^*,$$

where $\{\varphi_i : i \in \mathbb{N}\}$ is a Hilbert basis or a total family. The unknowns to be determined are the coefficients $\boldsymbol{x}_i$, $0 \le i \le n$. The solution is found by running the SOLVER with the adequate data: firstly, we determine $\boldsymbol{x}_1^*$ and $\boldsymbol{x}_2^*$. Then, we look for the coefficients. The method reads as

- Determine $\boldsymbol{x}_1^*$ by running the SOLVER
- Determine $\boldsymbol{x}_2^*$ by running the SOLVER
- Define a function $((\boldsymbol{x}_0, \ldots, \boldsymbol{x}_n), t) \to \boldsymbol{Px}(\mathrm{t})$
- Define a function $(\boldsymbol{x}_0, \ldots, \boldsymbol{x}_n) \to$ *area limited by* $\boldsymbol{f}(\boldsymbol{Px}(t))$ (for instance, use numerical integration on a grid of values of t)
- Find its point of optimum by running the SOLVER. (Do not forget the constraints, if any: they must be satisfied at each point t)

For instance, let us consider the Binh-Korn's problem:

$$f_1(\boldsymbol{x}) = 4\,x_1^2 + 4x_2^2\ ,\ \ f_2(\boldsymbol{x}) = (x_1 - 5)^2 + (x_2 - 5)^2$$

$$C = \{\boldsymbol{x} = (x_1, x_2) : \psi_i(\boldsymbol{x}) \leq 0, 1 \leq i \leq 6\},$$

$$\psi_1(\boldsymbol{x}) = (x_1 - 5)^2 + x_2^2 - 25,\ \ \psi_2(\boldsymbol{x}) = 7.7 - (x_1 - 8)^2 - (x_2 - 3)^2,$$

$$\psi_3(\boldsymbol{x}) = -x_1, \psi_4(\boldsymbol{x}) = -x_2,\ \ \psi_5(\boldsymbol{x}) = x_1 - 5,\ \ \psi_6(\boldsymbol{x}) = x_2.$$

As indicated in the procedure above, we start by determining $\boldsymbol{x}_1^*$ and $\boldsymbol{x}_2^*$. We create VBA functions `bk1, bk2, cbk` evaluating, respectively, f_1, f_2 and the 6 inequality constraints defining C. The SOLVER furnishes the solutions below:

	A	B	C	D	E	F	G	H
1	f	bk1	bk2					
2	constraints	cbk						
3	x0	0	0					
4	x1*	0	0					
5	f(x1*)	0	50					
6	constraints at x1*	0	-65,3	-5	-3	-0	-0	
7	x2*	5	3					
8	f(x2*)	136	4					
9	constraints at x2*	-16	-37,3	0	0	-5	-3	

To determine the Pareto's front, the interval (0, 1) is discretized in 100 subintervals by a uniform grid $t_i = ih, h = 0.01, 0 \leq i \leq 100$. We use $\varphi_0(t) = t, \varphi_i(t) = t^{i+1} - t, i \geq 1, n = 3$. The approximation reads as

$$\boldsymbol{x}(t) \approx \boldsymbol{P}\boldsymbol{x}(\mathrm{t}) = \boldsymbol{x}_1^* + t(\boldsymbol{x}_2^* - \boldsymbol{x}_1^*) + \boldsymbol{x}_1(t^2 - t) + \boldsymbol{x}_2(t^3 - t) + \boldsymbol{x}_3(t^4 - t).$$

A VBA function evaluating $\boldsymbol{P}\boldsymbol{x}(\mathrm{t})$ is used. T The area is evaluated by a trapezoidal rule using the values $(f_1(\boldsymbol{P}\boldsymbol{x}(t_i)), f_2(\boldsymbol{P}\boldsymbol{x}(t_i)))$ on the grid. The constraints $\boldsymbol{P}\boldsymbol{x}(t_i) \in C$ are imposed for $0 \leq i \leq 100$ (600 constraints). The solution furnished by the SOLVER is exhibited in Fig. 2.6.

As a second example, let us consider the Fleming-Fonseca's problem:

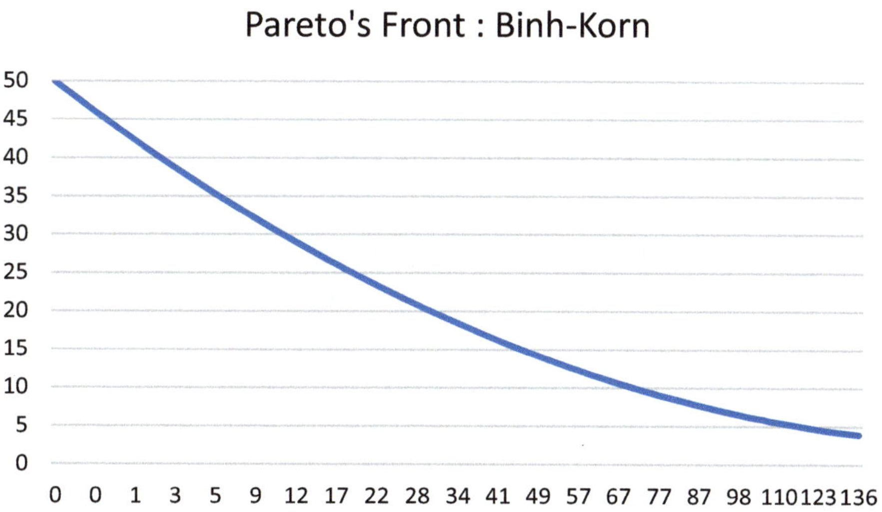

Fig. 2.6 An example of Pareto's front: Binh-Korn's problem

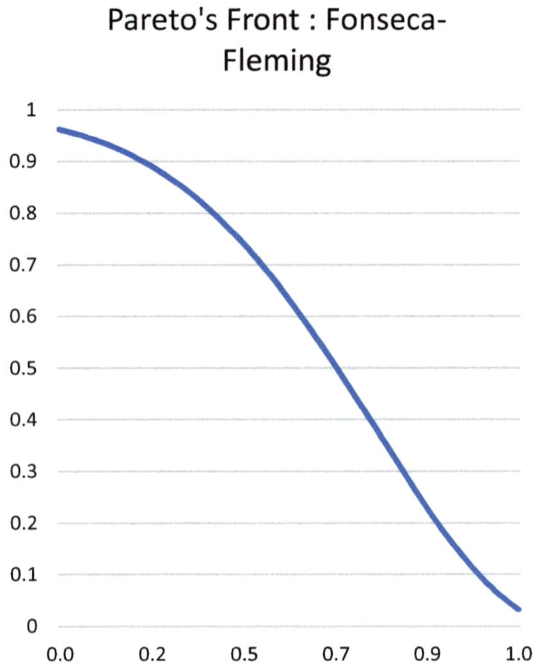

Fig. 2.7 An example of Pareto's front: Fonseca-Fleming's problem

$$f_1(\boldsymbol{x}) = 1 - \exp\left(-\sum_{i=1}^{n}\left(x_i - \frac{1}{\sqrt{n}}\right)^2\right),$$

$$f_2(\boldsymbol{x}) = 1 - \exp\left(-\sum_{i=1}^{n}\left(x_i + \frac{1}{\sqrt{n}}\right)^2\right)$$

$$C = \{\boldsymbol{x} = (x_1, x_2) : \psi_i(\boldsymbol{x}) \le 0, 1 \le i \le 2n\},$$

$$\psi_{2i-1}(\boldsymbol{x}) = 4 - x_i, \quad \psi_{2i}(\boldsymbol{x}) = x_i - 4, \quad 1 \le i \le n$$

The results furnished by the SOLVER for $n = 3$ are shown in Fig. 2.7.

2.7 Interpolation of Discrete Numerical Data

In numerical practice, we are frequently confronted with the following problem: find an expression $y(t) \approx PY(t) = \sum_{j=0}^{n} c_j \varphi_j(t)$ that represents the discrete data $(t_1, y_1), \ldots, (t_m, y_m)$. In the classical approach, the coefficients $\boldsymbol{C} = (c_0, c_1, \ldots, c_n)^t$ are determined by solving the overdetermined linear system

$$\boldsymbol{AC} = \boldsymbol{Y}, A_{ij} = \varphi_j(t_i), Y_i = y_i, 0 \le j \le n, 1 \le i \le m.$$

To get a solution to these equations, we may use the methods presented in Sect. 2.3.4.

An alternative method consists in considering that PY is the orthogonal projection of y onto the linear subspace generated by $\{\varphi_0, \ldots, \varphi_n\}$. Let us assume that the points are ranged in increasing order according the values of t: $t_1 < t_2 < \ldots < t_m$. Then, using the orthogonal projection property:

$$\int_{t_1}^{t_m} PY(t)\varphi_i(t)dt = \int_{t_1}^{t_m} y(t)\varphi_i(t)dt \ ,0 \le i \le n.$$

Thus,

$$\boldsymbol{AC} = \boldsymbol{B}, A_{ij} = \int_{t_1}^{t_m} \varphi_i(t)\varphi_j(t)dt, B_i = \int_{t_1}^{t_m} y(t)\varphi_i(t)dt, 0 \le i,j \le n$$

We have

$$B_i = \sum_{j=1}^{m-1} \int_{t_j}^{t_{j+1}} y(t)\varphi_i(t)dt.$$

And we may approximate

$$\int_{t_i}^{t_{j+1}} y(t)\varphi_i(t)dt \approx \int_{t_j}^{t_{j+1}} \left(y_j + \frac{y_{j+1} - y_j}{t_{j+1} - t_j}(t - t_j)\right)\varphi_j(t)dt.$$

The right-hand side integral may be evaluated by one of the methods introduced in Sect. 2.5.

Let us exemplify these approaches: we consider 11 points $t_i = (i - 1)\Delta t, y_i = (1 + \varepsilon u_i)e^{t_i} + \varepsilon v_i$, $\Delta t = 0.1$, u_i, v_i independent and both uniformly distributed on $(-1, 1)$, $\varepsilon \ge 0$ is the level of noise. The interpolation used the Hilbert basis $\varphi_i(t) = t^i$ and $n = 3$. PY was evaluated in 101 points $\tau_i = (i - 1)\delta t$, $\delta t = 0.01$.

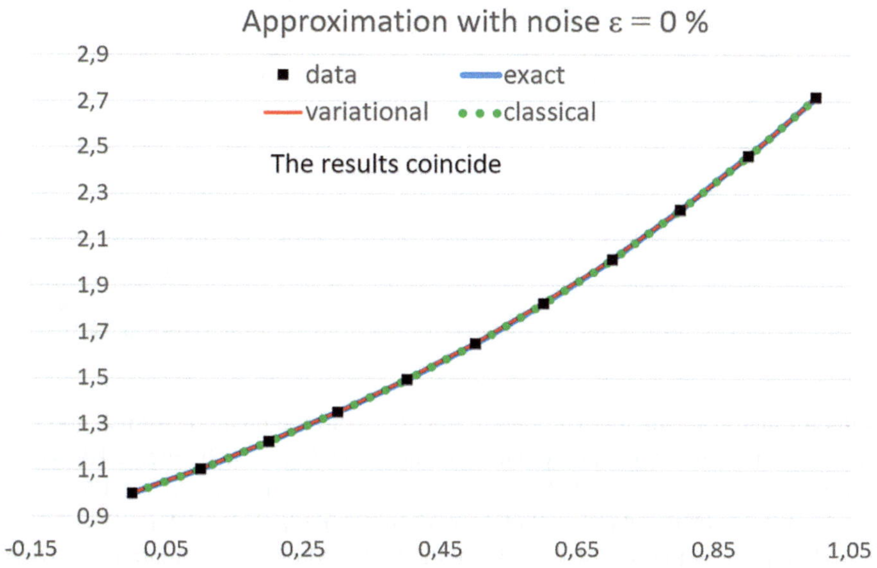

Fig. 2.8 Interpolation for unnoisy data

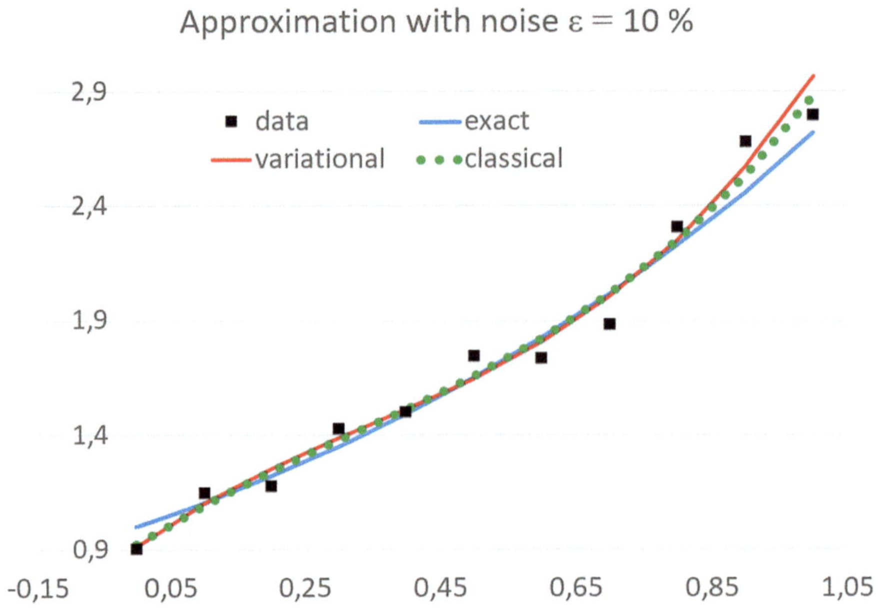

Fig. 2.9 Interpolation for noisy data (level of noise $\varepsilon = 10\ \%$)

Fig. 2.10 Interpolation for noisy data (level of noise $\varepsilon = 10\ \%$)

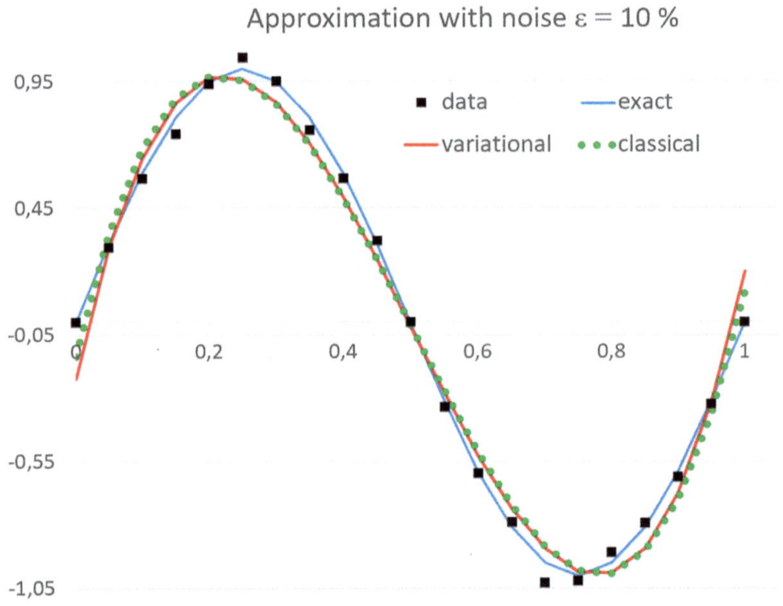

The results for unnoisy data are exhibited in Fig. 2.8: both the methods furnish good results, close to the exact solution. Results for noisy data are shown in Fig. 2.9. In Fig. 2.10, we consider the interpolation for 21 points $t_i = (i - 1)\Delta t$, $y_i = (1 + \varepsilon u_i)\sin(2\pi t_i) + \varepsilon v_i$, $\Delta t = 0.05$.

In all the situations both the classical and the variational methods furnish results of similar quality.

A second alternative is the use of particle methods, based on approximations of Dirac's delta δ, which may be interpreted as the probability density associated to a cumulative function F_T of

Heaviside type, corresponding to a random variable T which takes a single value: $P(T = 0) = 1$. Then $F_T(t) = 0$, *if* $x < 0$; $F_T(t) = 1$, *if* $t > 0$. Such a function does not have a classical derivative, but it has a variational one (see Sect. 2.8), which is δ. Dirac's delta may be approximated by probability densities $g(t, \sigma)$ of random variables having a mean equal to zero and a standard deviation σ that converges to zero, such as normal distributions $N(m, \sigma)$ when $\sigma \to 0+$. We may approximate (see, for instance, (Souza de Cursi, 2015))

$$y(t) \approx E(u(T)) = \frac{\int_{t_1}^{t_m} g(s-t,\sigma)y(s)ds}{\int_{t_1}^{t_m} g(s-t,\sigma)ds}.$$

In practice, we use Smoothed Particle Hydrodynamics (SPH) methods to approximate the integrals: we give a mass $m_i > 0$ for each t_i – referred as a particle – and we approximate (see, for instance, (Souza de Cursi, 2015))

$$y(t) \approx \sum_{i=1}^{m} \frac{m_i}{\rho_i} g(t_i - t, \sigma) y_i,$$

with

$$\rho_i = \sum_{k=1}^{m} m_k g(t_k - t_i, \sigma).$$

The value of σ is often set as a fraction or a multiple of the mean value of $|t_{i+1} - t_i|$, $1 \leq i \leq m-1$. The value of the masses is arbitrary: it is often set to 1 for all the points. In the context of SPH, the probability density g is called a kernel. The reader may find in the literature different kernels. We exemplify this method on the same parameters used in the illustration of the variational interpolation. The results are shown in Figs. 2.11, 2.12, and 2.13.

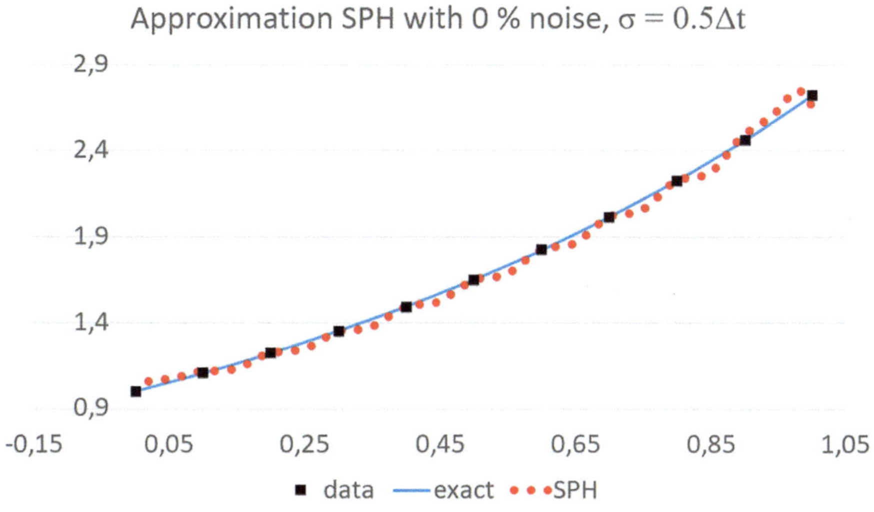

Fig. 2.11 Interpolation for unnoisy data (*SPH, Gaussian Kernel*)

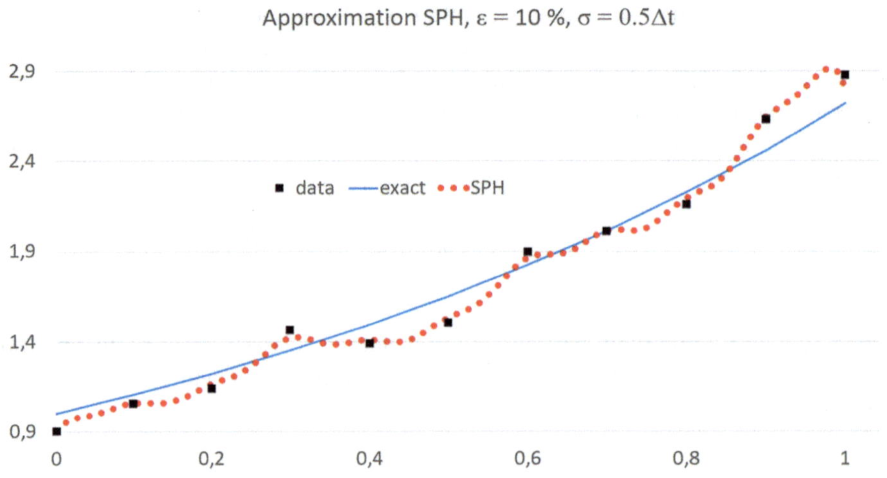

Fig. 2.12 Interpolation for noisy data (level of noise $\varepsilon = 10\ \%$, *SPH*, *Gaussian Kernel*)

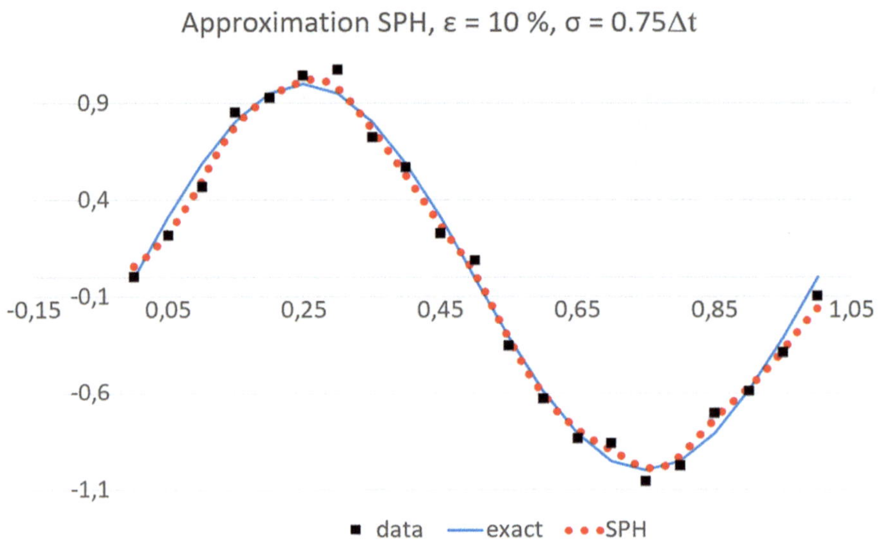

Fig. 2.13 Interpolation for noisy data (level of noise $\varepsilon = 10\ \%$, *SPH*, *Gaussian Kernel*)

2.8 Numerical Derivatives

Other frequent numerical problem is the determination of $y' = \frac{dy}{dt}$ from discrete data: analogously to the preceding situation, a finite number of couples $(t_1, y_1), \ldots, (t_m, y_m)$, such that $t_1 < t_2 < \ldots < t_m$ is given and we need to evaluate y' from these data. A simple approach consist in using finite differences:

$$y' \approx \frac{y_{i+1} - y_i}{t_{i+1} - t_i} \quad on \ \ (t_i, t_{i+1}).$$

However, such an approach is much noise sensitive. An alternative approach consists in determining the expression $y(t) \approx PY(t) = \sum_{j=0}^{n} c_j \varphi_j(t)$ by one of the approaches introduced in the previous section and then evaluating $y'(t) \approx PY'^{(t)} = \sum_{j=0}^{n} c_j \varphi'_j(t)$. We may also consider the definition of the variational derivative:

$$\int_{t_1}^{t_m} y'(t)v(t)dt = y(t_m)v(t_m) - y(t_1)v(t_1) - \int_{t_1}^{t_m} y(t)v'(t)dt, \forall v.$$

Then, taking $y'(t) \approx \sum_{j=0}^{n} c_j \varphi_j(t)$ and $=\varphi_i$, we obtain a linear system $\boldsymbol{AC} = \boldsymbol{B}$, with, for $0 \leq i, j \leq n$:

$$A_{ij} = \int_{t_1}^{t_m} \varphi_i(t)\varphi_j(t)dt, B_i = y_m\varphi_i(t_m) - y_1\varphi_i(t_1) - \int_{t_1}^{t_m} y(t)\varphi'_i(t)dt.$$

We have

$$\int_{t_1}^{t_m} y(t)\varphi'_i(t)dt = \sum_{j=1}^{m-1} \int_{t_j}^{t_{j+1}} y(t)\varphi'_i(t)dt.$$

and we may approximate

$$\int_{t_i}^{t_{j+1}} y(t)\varphi'_i(t)dt \approx \int_{t_j}^{t_{j+1}} \left(y_j + \frac{y_{j+1} - y_j}{t_{j+1} - t_j}(t - t_j)\right)\varphi'_j(t)dt.$$

id est,

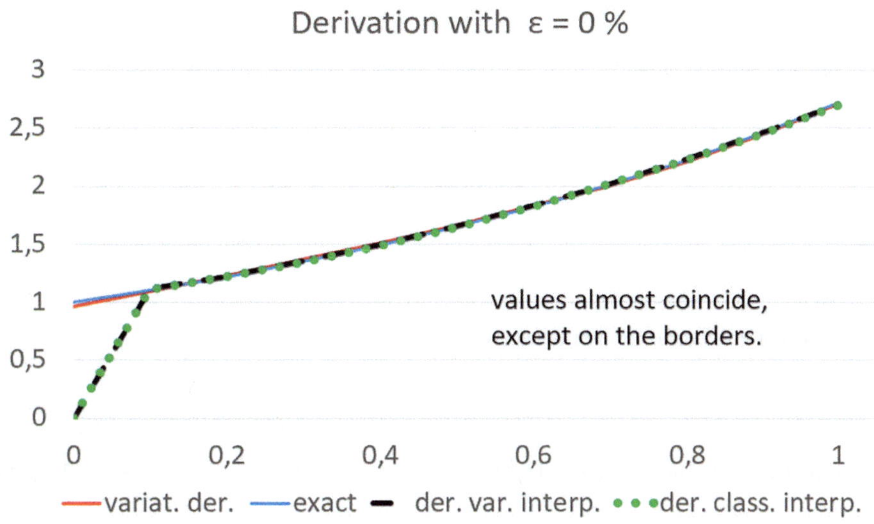

Fig. 2.14 Derivative of unnoisy data

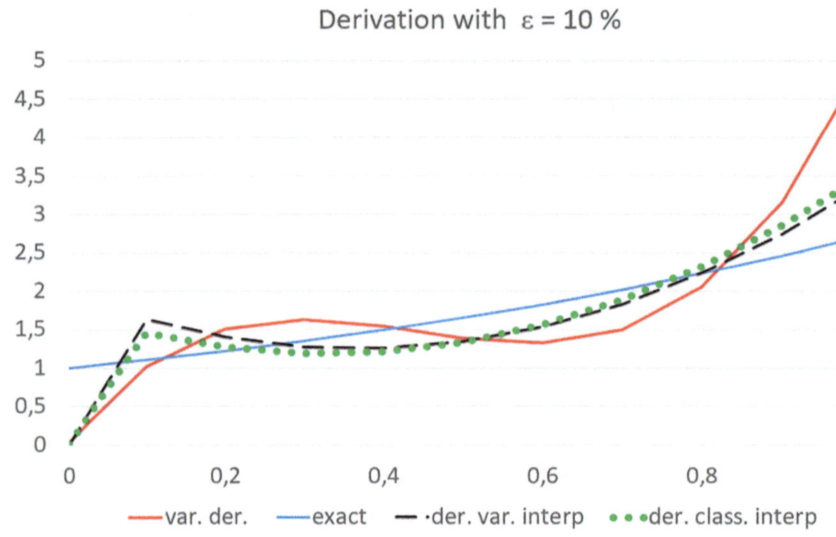

Fig. 2.15 Derivative of noisy data (level of noise $\varepsilon = 10\ \%$)

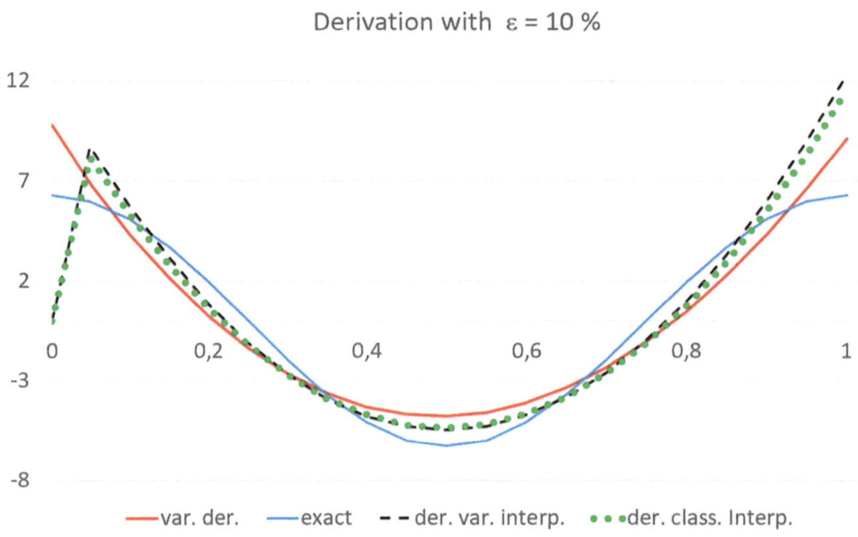

Fig. 2.16 Derivative of noisy data (level of noise $\varepsilon = 10\ \%$)

$$\int_{t_i}^{t_{j+1}} y(t)\varphi_i'(t)dt \approx y_j(\varphi(t_{j+1}) - \varphi(t_j)) + \left(y_{j+1} - y_j\right)\varphi(t_{j+1}) - \frac{y_{j+1} - y_j}{t_{j+1} - t_j} \int_{t_j}^{t_{j+1}} \varphi(t)dt.$$

Examples of derivation by these methods are shown in Figs. 2.14, 2.15, and 2.16. We use the same parameters of the preceding section.

Particle methods may be considered. Indeed, we may approximate (see (Souza de Cursi, 2015))

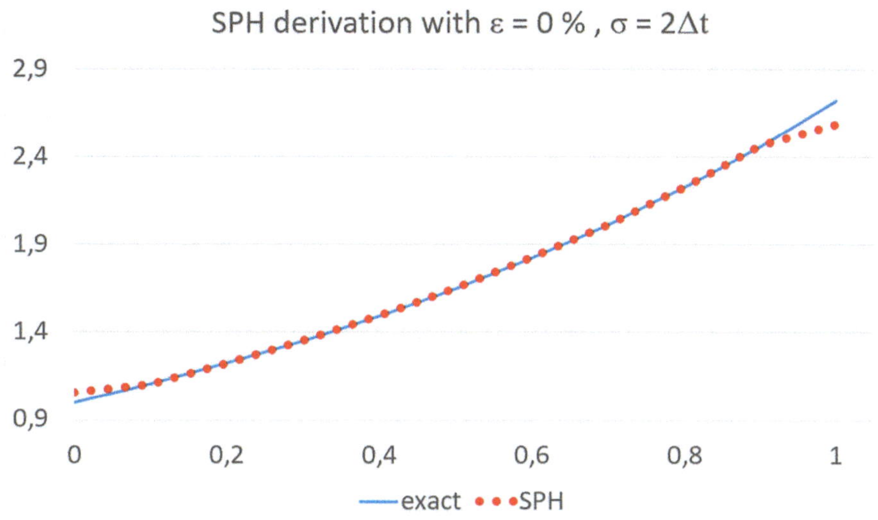

Fig. 2.17 Derivative of unnoisy data (*SPH*, *Gaussian Kernel*)

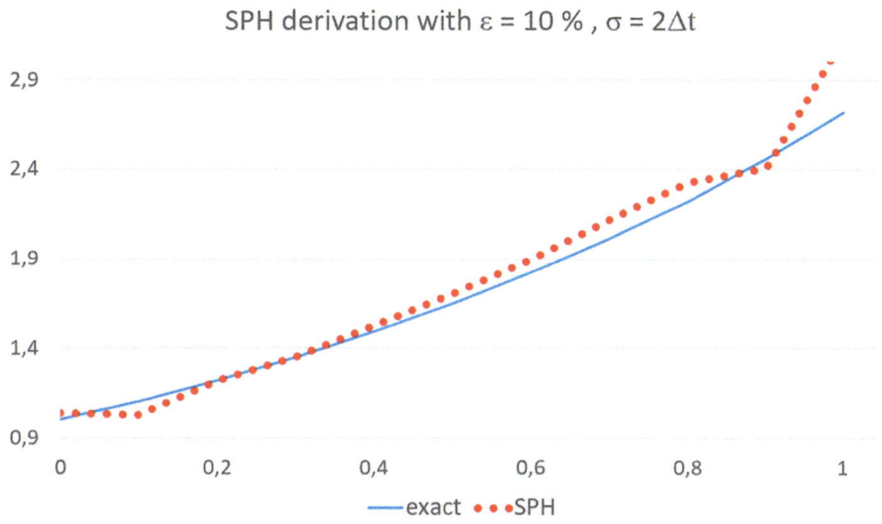

Fig. 2.18 Derivative of noisy data (level of noise $\varepsilon = 10\ \%$, *SPH*, *Gaussian Kernel*)

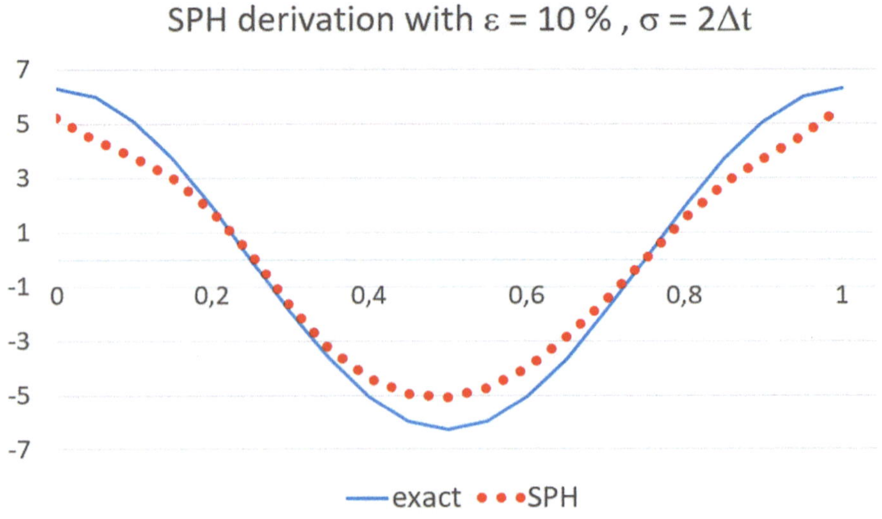

Fig. 2.19 Derivative of noisy data (level of noise $\varepsilon = 10\ \%$, *SPH*, *Gaussian Kernel*)

$$y'(\mathrm{t}) \approx \frac{\sum_{j=1}^{np} V_j \left(y_j - y(t)\right) \frac{d}{dt} g\left(t_j - t, \sigma\right)}{\sum_{j=1}^{np} \mathrm{V_j} \left(t_j - t\right) \frac{d}{dt} g\left(t_j - t, \sigma\right)}, V_j = \frac{m_j}{\rho_j} \quad .$$

Results furnished by SPH are shown in Figs. 2.17, 2.18, and 2.19.

Probabilities with EXCEL® 3

EXCEL® proposes many intrinsic functions to deal with probabilities, but, as previously observed, you may also extend their performance by using add-ins. In the sequel, we recall some basic elements of probability, statistics and the corresponding built-in functions in EXCEL®.

3.1 Probability

We recall below the main elements necessary which are used in the text. The first one is the definition of a probability (Kolmogorov, 1950):

A **probability** defined on a set Ω is an application P that, on the one hand, associates to each part $E \subset \Omega$ a value $P(E) \in (0, 1)$; and, on the other hand, it is additive for the disjoint parts of Ω, id est, if $E_1, \ldots, E_n, \ldots$ are disjoint subsets of Ω, then $P(E_1 \cup \ldots \cup E_n \cup \ldots) = P(E_1) + \ldots + P(E_n) + \ldots$ ∎

Ω is the ***universe***. The subsets of $E \subset \Omega$ are usually called ***events***. Probabilities have some well-known properties, such as

$$P(\varnothing) = 0, \quad P(\Omega) = 1$$

$$P(E \cup F) = P(E) + P(F) - P(E \cap F)$$

$$P(E - F) = P(E) - P(E \cap F)$$

$$P(E^c = \Omega - E) = 1 - P(E).$$ ∎

E. Souza de Cursi, *Uncertainty Quantification and Stochastic Modelling with EXCEL*, Springer Texts in Business and Economics, https://doi.org/10.1007/978-3-030-77757-9_3

There are two events that deserve special attention:

$E \subset \Omega$ is ***almost sure*** if and only if $P(E) = 1$; $F \subset \Omega$ is ***negligible*** if and only if $P(F) = 0$. Then for any other event $A \subset \Omega$: $P(A \cap E) = P(A)$; $P(A \cup F) = P(A)$. In addition, $P(A \cup E) = 1$; $P(A \cap F) = 0$. ∎

Notice that the complement of an event almost sure is negligible and reciprocally. Also notice that "almost sure" is not "sure" and "negligible" is not "impossible" (but may be interpreted as "almost impossible").

Example
Let us consider $\Omega = (0, 1)$ and the probability $P((a, b)) = b - a$. The event $F = \{x \in \Omega, \sin(2\pi x) = 0\}$: we have $F = \{0, 1\}$, so that $P(F) = P(\{0\}) + P(\{1\}) = (0 - 0) + (1 - 1) = 0$. This means that, if you are able to choose at random a real number on the interval (0, 1), it is almost impossible to hit on a zero of $\sin(2\pi x)$ – but it does not mean that such a zero does not exist. Analogously, the event $F = \{x \in \Omega, x \in \mathbb{Q}\}$ is negligible: hit a rational number at random is almost impossible, but the interval contains rational numbers. ∎

3.1.1 Mass Functions and Mass Densities

A convenient way to define probabilities consists in the use of mass functions or mass densities. **Mass functions** are used on finite or enumerable universes:

Let Ω be finite or enumerable. A **mass function** $\boldsymbol{\mu}$ on Ω is an application $\mu : \Omega \longrightarrow \mathbb{R}$ such that $\forall \omega \in \Omega : \mu(\omega) = P(\{\omega\})$. We have $\mu(\omega) \geq 0$ on Ω and $\sum_{\omega \in \Omega} \mu(\omega) = 1$. If useful, μ may be extended by zero for $\omega \notin \Omega$. ∎

Mass densities are used on real numbers:

Let $\Omega \subset \mathbb{R}$. A **mass density** $\boldsymbol{\mu}$ on Ω is an application $\mu : \Omega \longrightarrow \mathbb{R}$ such that $\forall (a, b) \subset \Omega :$

$P((a,b)) = \int_a^b \mu(\omega)d\omega$. We have $\mu(\omega) \geq 0$ on Ω and $\int_\Omega \mu(\omega)d\omega = 1$. If useful, μ may be extended by zero for $\omega \notin \Omega$. ∎

Example
Consider $\Omega = (\alpha, \beta)$ and the probability

$$P((a,b)) = \frac{b - a}{\beta - \alpha}, \quad \alpha < a < b < \beta$$

Then, a mass density is

$$\mu(\omega) = \frac{1}{\beta - \alpha}, \quad \alpha < \omega < \beta.$$

■

Example
Consider $\Omega = \{\omega_1, \omega_2\}$ and the probability

$$P(\omega_1) = p, P(\omega_2) = 1 - p$$

Then, a mass function is

$$\mu(\omega_1) = p, \mu(\omega_2) = 1 - p.$$

■

Exercises

1. Consider $\Omega = (0, 1)$ and the mass density $\mu(\omega) = e^{\alpha\omega + \beta}$. What are the possible values of α and β?

   TIP: Use that $P(\Omega) = 1$.

2. Consider $\Omega = \{\omega_1, \omega_2, \omega_3\}$ and the mass function $\mu(\omega_i) = \alpha i + \beta$. What are the possible values of α and β?

   TIP: Use that $P(\Omega) = 1$, and recall that $\mu(\omega_i) \geq 0, \forall\, i$.

3. Consider $\Omega = (0, 1)$ and the mass density $\mu(\omega) = \alpha\omega^{\beta}$. What are the possible values of α and β?

   TIP: Use that $P(\Omega) = 1$, and recall that $\mu(\omega) \geq 0, \forall\, \omega \in \Omega$.

4. Consider $\Omega = \{\omega_1, \omega_2, \omega_3\}$ and the mass function $\mu(\omega_i) = \alpha i^2 + \beta i$. What are the possible values of α and β?

   TIP: Use that $P(\Omega) = 1$, and recall that $\mu(\omega_i) \geq 0, \forall\, i$.

3.1.2 The Case of Finite Populations

When dealing with finite populations, id est, when $\Omega = \{\omega_1, \ldots, \omega_n\}$ is a finite set, the simplest way to define a probability on Ω is to use a mass function μ to assign a probability $\mu(\omega_i) = p_i \geq 0$ to each $\omega_i \in \Omega$: $P(\{\omega_i\}) = p_i$, with $\sum_i p_i = 1$. Then, $P(E = \{\omega_{i_1}, \ldots, \omega_{i_k}\}) = p_{i_1} + \ldots + p_{i_k}$. EXCEL® offers a built-in function PROB to evaluate, $P(\{\omega_{i_1}, \ldots, \omega_{i_1+k}\})$ from the individual probabilities p_i. The first argument of PROB is the range of values of Ω (must be numeric); the second one is the range of associated probabilities; the third one is i_1 and the last one is $i_1 + k$. For instance, open a workbook and type the data below. Then, select cell A4 and enter "`= PROB(B1:G1;B2:G2;2;4)`". You will get $P(\{\omega_2, \omega_3, \omega_4\}) = 0.6$

	A	B	C	D	E	F	G	H
1	ω_i	1	2	3	4	5	6	
2	p_i	0,1	0,2	0,25	0,15	0,1	0,2	
3								
4	=PROB(B1:G1;B2:G2;2;4)							
5								

Notice that $\sum_i p_i = 1$. Otherwise, EXCEL® returns an error #NUM!.

Exercises

1. Consider again $\Omega = (0, 1)$ and the probability $P((a, b)) = b - a$(mass density $\mu(\omega) = 1$). Find the probabilities of the following events

   (a) $A = \left\{x \in \Omega : x < \frac{1}{3}\right\}$;
   (b) $B = \left\{x \in \Omega : x^2 < \frac{1}{2}\right\}$;
   (c) $C = \left\{x \in \Omega : x^2 > \frac{1}{4}\right\}$;
   (d) $D = \{x \in \Omega : 9x^2 - 5x + 1 \leq 0\}$.

   TIP: transform all the events in intervals (a, b).

2. Consider $\Omega = \{\omega_1, \omega_2, \omega_3\}$ and the events $A = \{\omega_1\}$, $B = \{\omega_1, \omega_2\}$. Let $P(A) = \frac{1}{3}$ and $P(B) = \frac{1}{2}$.

   (a) Find $P(\{\omega_2\})$; TIP: Think at $B - A$
   (b) Find $P(\{\omega_3\})$; TIP: Think at $\Omega - B$
   (c) Find $P(\{\omega_1, \omega_3\})$.

3. Consider $\Omega = \{\omega_1, \omega_2, \omega_3, \omega_4\}$ and the events $A = \{\omega_1\}$, $B = \{\omega_1, \omega_2\}$, $C = \{\omega_1, \omega_2, \omega_4\}$. Let $P(A) = \frac{1}{4}$, $P(B) = \frac{9}{20}$, $P(C) = \frac{7}{15}$.

   (a) Find $P(\{\omega_2\})$, $P(\{\omega_3\})$, $P(\{\omega_4\})$; TIP: Think at $B - A$, $\Omega - B$, $C - B$.
   (b) Find $P(\{\omega_1, \omega_2, \omega_3\})$.

3.2 Combinatorial Probabilities with EXCEL®

In many problems, all the elements of Ω are equiprobable, id est, $p_i = \frac{1}{n}$. In this case, the evaluation of probabilities is brought to counting the number of elements and determining its ratio to the total number of elements: for instance, $P(E = \{\omega_{i_1}, \ldots, \omega_{i_k}\}) = \frac{k}{n}$. This is the situation in classical problems of probability involving coins, balls, cards, dices etc. These problems are mainly combinatorial, id est, they use combinatorics to evaluate the number of elements of E and Ω.

EXCEL® offers functions for solving Under EXCEL®, the built-in functions `PERMUT` and `COMBIN` furnish values of permutations and combinations ***without repetition***: the elements are drawn simultaneously – or an element drawn is not returned to the draw: in a draw, each element may appear at most once. We have

$$PERMUT(n; k) = k!\binom{n}{k} = \frac{n!}{(n-k)!}; \quad COMBIN(n; k) = \binom{n}{k} = \frac{n!}{k!(n-k)!}$$

For permutations and combinations ***with repetition***, the functions are COMBINA and PERMUTATIONA :

$$PERMUTATIONA(n; k) = n^k; \quad COMBINA(n; k) = \binom{n+k-1}{n-1}.$$

In addition, EXCEL proposes the function `FACT(`n`)` which returns $n!$.

Example

In a 52-card deck of playing cards, you draw simultaneously two at random. What is the probability that they are both face cards? To evaluate the probability, we start by evaluating n: there are 52 possibilities for the first card and 51 for the second one, so that $n = 52 \times 51 = PERMUT(52;\ 2)$. The deck has 12 face cards, so that there are 12 possibilities for the first face and 11 for the second: we have $k = 12 \times 11 = PERMUT(12;\ 2)$. We may evaluate the probability by entering "`= PERMUT(12;2)/PERMUT(52;2)`" in a cell of a workbook: the result is `0,049773576`.

Now, assume that you take a card at random, put it back, reshuffle and take another card at random. Then, repetition is possible: $n = 52 \times 52 = PERMUTATIONA\ (52;\ 2)$, while $k = 12 \times 12 = PERMUTATIONA(12;\ 2)$. Entering "`= PERMUTATIONA(12;2)/PERMUTATIONA(52;2)`" in a cell furnishes the result `0,053254`. ∎

Example

In a 52-card deck of playing cards, you draw simultaneously two at random. What is the probability that you get one ace? Here, it is convenient to consider the complementary event: no ace is draw. For this event, $k = 48 \times 47 = PERMUT(48;\ 2)$. Thus, its probability is `0,850679`. So, the probability of getting one ace is `0,149321`. With repetition, these probabilities are, respectively, `0,852071` and `0,147929`. ∎

Exercises

1. In a 52-card deck of playing cards, you draw simultaneously 5 at random. What is the probability of getting

   (a) 5 hearts; TIP: there are 13 hearts, so *PERMUT*(13; 5) manners of receiving 5 hearts.
   (b) A full (5 cards of the same color); TIP: there are 4 colors, thus 4 × *PERMUT*(13; 5) manners of receiving 5 cards of the same color.
   (c) Four aces; TIP: there are 48 × 5 × *PERMUT*(4; 4) possible draws of the 4 aces.
   (d) 4 cards of same numbers (or faces); TIP: there are 13 different cards for each color. Thus, 13 × 48 × 5 × *PERMUT*(4; 4) possibilities of receiving 4 identical numbers (or faces).

2. In a 52-card deck of playing cards, you draw sequentially 5 at random. At each draw, the card is put back in the deck, which is reshuffled. What is the probability of getting

   (a) 5 hearts; TIP: there are 13 hearts, so *PERMUTATIONA*(13; 5) manners of receiving 5 hearts.
   (b) 5 cards of the same color; TIP: there are 4 colors, thus 4 × *PERMUTATIONA*(13; 5) manners of receiving 5 cards of the same color.
   (c) Exactly four aces; TIP: there are 48 × 5 × *PERMUTATIONA*(4; 4) possible draws of exactly 4 aces.
   (d) At least 4 aces; TIP: there are 52 × 5 × *PERMUTATIONA*(4; 4) possible draws with at least 4 aces.

3. We roll three dice numbered from 1 to 6. All the faces are equiprobable. What is the probability of getting

   (a) Three aces; TIP: there are 6 × 6 × 6 possibilities for the result and a single one for three aces
   (b) A sum of points equal to 9; TIP: there are 25 ways to get 9 points
   (c) A sum equal to 10. TIP: there are 27 ways to get 10 points.

4. An urn contains 20 numbered balls of three colors: 6 red, 12 green, 2 blue. We draw at random 2 balls simultaneously. What is the probability of getting

   (a) Two green balls; TIP: there are *PERMUT*(20; 2) possible draws and *PERMUT*(12; 2) draws involving only green balls
   (b) Two balls of same color; TIP: there are *PERMUT*(12; 2) + *PERMUT*(6; 2) *PERMUT* (2; 2) draws involving a single color
   (c) Two balls of different colors. TIP: this is the complementary event to the previous one. Otherwise, you have 6 × 14 + 12 × 8 + 2 × 18 possible draws of different colors.

3.3 Conditional Probability, Bayes' Formula, and Independence

The probability of an event A conditionally to an event B such that $P(B) > 0$ is

$$P(A|B) = \frac{P(A \cap B)}{P(B)}. \tag{3.1}$$

Notice that, if $P(A)\, P(B) > 0$:

$$P(A \cap B) = P(B) \times P(A|B) = P(A) \times P(B|A).$$

Thus, if $P(A)\, P(B) > 0$:

$$P(B|A) = \frac{P(B) \times P(A|B)}{P(A)}. \tag{3.2}$$

This equation is usually referred as ***Bayes' formula***. It is often used when Ω is cut in k disjoint subsets: $\Omega = E_1 \cup \ldots \cup E_k$, $E_i \cap E_j = \varnothing$, *if* $i \neq j$; $P(E_i)$, $P(A \mid E_i)$ are known for $1 \leq i \leq k$ and we desire to determine $P(E_j \mid A)$. In this case,

$$P(A) = P\left(\bigcup_{i=1}^{k} (A \cap E_i)\right) = \sum_{i=1}^{k} P(A \cap E_i) = \sum_{i=1}^{k} P(E_i)\, P(A|E_i)$$

and, if $\forall j$: $P(E_j) > 0$, then

$$P(E_j|A) = \frac{P(E_j)\, P(A|E_j)}{P(A)} = \frac{P(E_j)\, P(A|E_j)}{\sum_{i=1}^{k} P(E_i)\, P(A|E_i)}. \tag{3.3}$$

We say that

$A, B \subset \Omega$ are independent

if and only if

$P(A \cap B) = P(A) \times P(B)$. ∎

This means that A and B do not affect each other, id est, if $P(A)\, P(B) > 0$:

$$P(A|B) = P(A) \quad \text{and} \quad P(B|A) = P(B)$$

It is possible to extend the definition of conditional probability to include negligible events – see, for instance (Coletti & Scozzafava, 2000).

Example

A fair coin has head and tail results equiprobable. It is tossed twice. One of the results is head, what is the probability that the other is tail? Let us denote H a head and T a tail. Here,

$\Omega = \{HH, HT, TH, TT\}$. Let us consider the events $B = \{HH, HT, TH\}$ (one of them is a head) and the event $B = \{HT, TH, TT\}$ (one of them is a tail). We have $A \cap B = \{HT, TH\}$ and

$$P(A|B) = \frac{P(A \cap B)}{P(B)} = \frac{2/4}{3/4} = \frac{2}{3}$$ ∎

Example

A sample contains 50% of green pins. Globally, 20% of the pins are damaged, but there is a difference between the colors: 30% of the green pins are damaged. What is the probability that a damaged pin is green? Let E_1 = *the pin is green*; E_2 = *the pin is not green*. We have $P(E_1) = 0{,}5$, $P(E_2) = 0{,}5$. Let A = *pin is damaged*. We have $P(A|E_1) = 0{,}3$, $P(A) = 0{,}2$. Then

$$P(E_1|A) = \frac{P(E_1) \times P(A|E_1)}{P(A)} = \frac{0{,}5 \times 0{,}3}{0{,}2} = 0{,}75.$$ ∎

Exercises

1. Overall, it is estimated that 1% of the parts manufactured in a factory are defective. The quality control makes mistakes by accepting 1% of non-compliant parts and rejecting 2% of compliant parts. What is the probability that an accepted part is compliant?

   TIP: Let E_1 = *the part is accepted*, A = *the part is compliant*

2. Two boxes contain the same number of pins, divided between two colors: red and green. In box 1, 30% are red and 70% are green; in box 2, 40% are red pins and 60% are green. The boxes are equiprobable.

   (a) We take a red pin: what is the probability that it came from box 1?
   (b) We take a green pin: what is the probability that it came from box 2?
   (c) How are these values modified if box 1 is much larger than box 2, so that its probability is $\frac{3}{4}$?

   TIP: let E_i = *box i*, then either A = *red pin* or A = *green pin*.

3. An urn contains 20 numbered balls of three colors: 1–6 in red, 7–18 in green, 19–20 in blue. We draw a ball at random.

   (a) The number is multiple of 3: what is the probability that the color is green?
   (b) The number has two digits: what is the probability that the color is blue?
   (c) The number is pair: what is the probability that the color is red?

   TIP: let E_i = *color i*, A = *property of the number*.

4. Three urns contain 10 numbered balls in red and blue each, but the number of red balls is different in each one: 5 in urn 1, 4 in urn 2, 3 in urn 3. The urns are equiprobable. A red ball is drawn.

 (a) What is the probability that it came from urn 1?
 (b) What is the probability that it did not come from urn 2?
 (c) What is the probability that it came from urn 3?
 TIP: let $E_i = urn\ i$, $A = property\ of\ the\ ball$.

5. Determine if the events A and B are independent or not, when

 (a) $P(A) = \frac{1}{3}, P(B) = \frac{2}{3}, P(A \cap B) = \frac{1}{2}$.
 (b) $P(A) = \frac{1}{2}, P(B) = \frac{2}{3}, P(A \cap B) = \frac{1}{3}$.
 (c) $P(A) = \frac{3}{5}, P(B) = \frac{5}{6}, P(A \cap B) = \frac{1}{2}$.
 (d) $P(A) = \frac{3}{5}, P(B) = \frac{5}{6}, P(A \cap B) = \frac{2}{5}$.
 TIP: verify the equality defining independence.

3.4 Random Variables

A random variable X is an application $X : \Omega \to \mathbb{R}$, *id est*, X is a numerical characteristic of the elements of Ω. In statistics, we are interested in general information on X. Namely, we are **not** interested in individual values $X(\omega)$, but in the global behavior of X on Ω. For instance, we are not interested in the case $X(\omega)$ of the individual ω, but in the proportion of individuals for which X takes a given value. Probabilities on the values of X derive from the probabilities on Ω: indeed, given a subset $A \subset \mathbb{R}$, we may consider $X^{-1}(A) = \{\omega \in \Omega : X(\omega) \in A\}$. Then, $P(X \in A) = P(X^{-1}(A))$.

The main tool for the manipulation of random variables is the **cumulative distribution function** (***CDF***):

$$F(x) = P(X < x) = P(I(x)), I(x) = X^{-1}((-\infty, x))$$

The **probability density function** (***PDF***) is $f = F'$ (if it exists). Then, $P(X \in A) = \int_A f(x)dx$. F is usually called the **distribution** of X. ∎

Notice that ***different variables may have the same distribution*** (see examples below). Cumulative functions have many useful properties. For instance,

$$F(-\infty) = 0, \quad F(+\infty) = 1,$$

$$P(a \leq X < b) = F(b) - F(a),$$

$$P(X = a) = [F(a)] = F(a+) - F(a-),$$

$$P(X \geq x) = 1 - F(x),$$

$$F \text{ is monotonous non decreasing, continuous at left.}$$ ∎

Remark

If the probability P on Ω is defined by a mass function μ, then

$$F(x) = \sum_{\omega \in I(x)} \mu(\omega) = \sum_{\substack{\omega \text{ such that} \\ X(\omega) < x}} \mu(\omega).$$

If the probability P on Ω is defined by a mass density μ, then

$$F(x) = \int_{I(x)} \mu(\omega) d\omega = \int_{\substack{\omega \text{ such that} \\ X(\omega) < x}} \mu(\omega) d\omega.$$ ∎

Exercises

1. Consider $\Omega = (0, 1)$ and the mass density $\mu(\omega) = 2\omega$. Find the distribution of $X(\omega) = e^{\omega}$. TIP: If $I(x) \neq \emptyset$, then $I(x) = (0, \ln(x))$.
2. Consider $\Omega = \{-2, -1, 1, 2\}$ and the mass function $\mu(i) = \frac{|i|}{6}$. Find the distribution of $X(\omega) = \omega^2$.
3. Consider $\Omega = (0, 1)$ and the mass density $\mu(\omega) = 1$. Find the distribution of $X(\omega) = \omega^2$. TIP: If $I(x) \neq \emptyset$, then $I(x) = (0, \sqrt{x})$.
4. Consider $\Omega = \{-2, -1, 0, 1, 2\}$ and the mass function $\mu(i) = \frac{1}{5}$. Find the distribution of $X(\omega) = \omega^2$.

3.4.1 Statistics of a Random Variable

Let us introduce some quantities concerning a random variable:

The **mean** $E(X)$ of X is the constant that best approximates the variable X in the least-squares sense.

The **variance** $V(X)$ is the mean squared error in the approximation.

The **standard deviation** of X is $\sigma(X) = \sqrt{V(X)}$.

The **moment of order *k*** of X, denoted $M_k(X)$, is the mean of X^k.
A **median** of X is a value m such that $P(X \leq m) \geq \frac{1}{2}$ and $P(X \geq m) \geq \frac{1}{2}$. ∎

The median is characterized by

$$F(m+) \geq \frac{1}{2}, \quad F(m-) \leq \frac{1}{2}. \tag{3.4}$$

Formally, these quantities are evaluated by Riemman-Stieltjes integrals: for any regular function ϕ

$$E(\phi(X)) = \int_{-\infty}^{+\infty} \phi(x) dF(x). \tag{3.5}$$

These expressions are evaluated with

- $\phi(X) = X$ for the determination of the mean $E(X)$;
- $\phi(X) = (X - E(X))^2$ for the determination of the variance $V(X)$;
- $\phi(X) = X^k$ for the determination of the moment $M_k(X)$;
- $\phi(X) = e^{itX}$ for the determination of the characteristic function $\varphi(t)$.

Notice that

$$V(X) = E\left(X^2\right) - (E(X))^2 = M_2(X) - M_1(X). \tag{3.6}$$

We have

$$V(X) = 0 \Leftrightarrow X = E(X) \text{ almost surely (a.s.)} \tag{3.7}$$

Remark
If X has a PDF f, then

$$E(\phi(X)) = \int_{-\infty}^{+\infty} \phi(x) f(x) dx.$$

If Ω has a mass function μ, then

$$E(\phi(X)) = \sum_{\omega \in \Omega} \mu(\omega) \phi(X(\omega)).$$

If Ω has a mass density μ, then

$$E(\phi(X)) = \int_{\Omega} \mu(\omega) \phi(X(\omega)) d\omega.$$ ∎

Example

Let us consider $\Omega = (0, 1)$ and the probability $P((a, b)) = b - a$ ($\mu(\omega) = 1$). Let $X(\omega) = \omega^2$. Then $X(\omega) < x \Leftrightarrow \omega^2 < x$. Thus

$$I(x) = \emptyset, if\ x \leq 0; I(x) = \left(0, \sqrt{x}\right), if\ 0 < x \leq 1; I(x) = \Omega, if\ x > 1$$

Thus,

$$F(x) = 0, if\ x \leq 0; F(x) = \sqrt{x}, if\ 0 < x \leq 1; F(x) = 1, if\ x > 1.$$

The PDF is

$$f(x) = 0, if\ x \leq 0; f(x) = \frac{1}{2\sqrt{x}}, if\ 0 < x \leq 1; f(x) = 0, if\ x > 1.$$

The mean of X is

$$E(X) = \int_0^1 xf(x)dx = \int_0^1 \frac{\sqrt{x}}{2}dx = \frac{1}{3}.$$

We have

$$E\left(X^2\right) = \int_0^1 x^2 f(x)dx = \int_0^1 \frac{\sqrt{x^3}}{2}dx = \frac{1}{5}.$$

Thus,

$$V(X) = \frac{1}{5} - \frac{1}{9} = \frac{4}{45}.$$

Let us consider

$$J(\alpha) = \int_0^1 (x - \alpha)^2 f(x)dx = \alpha^2 - \frac{2}{3}\alpha + \frac{1}{5}.$$

The minimum of J is attained at $\alpha = E(X)$ and $J(E(X)) = V(X)$. ∎

Example

Let us consider $\Omega = (0, 1)$ and the probability $P((a, b)) = b - a$. Let $X_1(\omega) = \sin(2\pi\omega)$; $X_2(\omega) = \cos(2\pi\omega)$. Let F_i be the CDF of X_i, $i = 1, 2$. Then $F_i(x) = 0$, *if* $x \leq -1$; 1; $F_i(x) = 1$, *if* $x > 1$ and

$$F_i(x) = 1 - \frac{1}{\pi}\arccos(x) = \frac{1}{2} + \frac{1}{\pi}\arcsin(x), if\ -1 < x \leq 1.$$

Both the variables have the same distribution, although they are different. We have

$$f(x) = \frac{1}{\pi\sqrt{1-x^2}}, \quad if -1 < x \le 1$$

Since f is symmetrical with respect to the origin, we have $E(X_i) = 0$, while $E(X_i^2) = \frac{1}{2}$. Thus, $V(X_i) = ½$. ∎

Exercises

1. Let $\Omega = \{0, 1, \ldots, n\}$ and $X(\omega) = \omega \bmod 2$ (remainder after division by 2). Assume the elements of Ω as equiprobable.

   (a) Show that $X(\Omega) = \{0, 1\}$.
   (b) Determine $p = P(X = 0)$ when $n = 2k$.
   (c) Determine $p = P(X = 0)$ when $n = 2k + 1$.
   (d) Determine the CDF F as a function of p.

2. Let X be a continuous random variable having the CDF F. Assume that $F(x) = x^3$ for $0 < x < 1$.

   (a) Show that $P(X < 0) = 0$.
   (b) Show that $P(X > 0) = 1$.
   (c) Find the PDF f.
   (d) Determine the mean and the variance of X.

3. Let $\Omega = \{1, 2, 3, 4, 5\}$ and $X(\omega) = \frac{1}{\omega+1}$. Assume the elements of Ω as equiprobable.

   (a) Determine $X(\Omega)$.
   (b) Determine the mean and the variance of X.

4. Let $\Omega = (0, 1)$ and $X(\omega) = \frac{1}{\omega+1}$, with $P((a, b)) = b - a$ $(\mu(\omega) = 1)$.

   (a) Show that $X(\Omega) = \left(\frac{1}{2}, 1\right)$.
   (b) Show that the CDF of X verifies $F(x) = 2 - \frac{1}{x}$, for $½ < x < 1$.
   (c) Determine the mean and the variance of X.

5. Let $\Omega = (-1, 1)$ and $X(\omega) = \omega^3$, with $P((a,b)) = \frac{b-a}{2}\left(\mu(\omega) = \frac{1}{2}\right)$. Find the CDF, the PDF, the mean and the variance of X.

3.4.2 Numerical Evaluation of Statistics

As shown in Sect. 3.8, a simple way to the estimation of the statistics of a random variable consists in the use of samples: for instance, we can generate a sample from the variable and use it to estimate the statistics of the variable using the statistics of the sample.

Such a point of view is developed in Sect. 3.8. Here, we consider the numerical evaluation of statistics using Eq. (3.5), which involves numerical integration. The use of this approach is possible only in the situations where the distribution of the variable is known: Eq. (3.5) requests the knowledge of the CDF F – or, according to the Remark page 99, the knowledge of the CDF or of the mass density. A first approximation consists in evaluating

$$E(\phi(X)) \approx \int_{-A}^{A} \phi(x)dF(x) \tag{3.8}$$

and using numerical integration to evaluate the left-hand side. For instance, we can consider $-A = x_0 < x_1 < \ldots < x_n = A$ and

$$E(\phi(X)) \approx \sum_{i=0}^{n} \phi\left(x_{i+\frac{1}{2}}\right)(F(x_{i+1}) - F(x_i)), x_{i+\frac{1}{2}} = \frac{x_{i+1} + x_i}{2}. \tag{3.9}$$

If the PDF f is known, we can use the approximation

$$E(\phi(X)) \approx \int_{-A}^{A} \phi(x)f(x)dx. \tag{3.10}$$

Again, the left-hand side can be evaluated by numerical integration. In this case, the methods presented in Sect. 2.5 can be used. Of course, a simple estimate may be generated by

$$E(\phi(X)) \approx \sum_{i=0}^{n} p_{i+\frac{1}{2}}\phi\left(x_{i+\frac{1}{2}}\right), p_{i+\frac{1}{2}} = f\left(x_{i+\frac{1}{2}}\right)(x_{i+1} - x_i). \tag{3.11}$$

A mass density μ can be used into an analogous way with a partition of the domain of the values of ω:

$$E(\phi(X)) \approx \sum_{i=0}^{n} p_{i+\frac{1}{2}}\phi\left(X\left(\omega_{i+\frac{1}{2}}\right)\right), p_{i+\frac{1}{2}} = \mu\left(\omega_{i+\frac{1}{2}}\right)(\omega_{i+1} - \omega_i). \tag{3.12}$$

Example

Let us consider $\Omega = (0, 1)$ and the probability $P((a, b)) = b - a$ $(\mu(\omega) = 1)$. Let $X(\omega) = \omega^2$. Consider a partition of Ω in n equally spaced subintervals: $\omega_i = i/n$. The results furnished by Eq. (3.12) for three values of n are shown in Table 3.1.

Table 3.1 Numerical evaluation of the mean and variance using the mass function

n	100	500	1000	*exact*
$E(X)$	0,333325	0,333333	0,33333325	1/3
$E(X^2)$	0,199983334	0,199999333	0,199999833	1/5
$V(X)$	0,088877778	0,088888444	0,088888778	4/45

We can also use the PDF $f(x) = 1/2\sqrt{x}$ and a partition of (0,1) to evaluate $\int_0^1 x^k f(x)dx$. The results furnished by Eq. (3.11) are shown in Table 3.2.

Table 3.2 Numerical evaluation of the mean and variance using the PDF

n	100	500	1000	*exact*
$E(X)$	0,333362736	0,333336015	0,333334286	1/3
$E(X^2)$	0,199996957	0,199999876	0,199999969	1/5
$V(X)$	0,088866244	0,088886978	0,088888223	4/45

Gauss points can be used if we provide VBA functions evaluating $xf(x)$ and $x^2f(x)$: the results appear in Table 3.3.

Table 3.3 Numerical evaluation of the mean and variance using Gauss points

n	10	20	50	*exact*
$E(X)$	0,333378021	0,333339316	0,333333733	1/3
$E(X^2)$	0,199999671	0,199999989	0,2	1/5
$V(X)$	0,088858766	0,088884889	0,088888622	4/45

Finally, we evaluate the same statistics using Eq. (3.9) – in this case, we must provide a VBA function evaluating $F(x)$: the results appear in Table 3.4. ∎

Table 3.4 Numerical evaluation of the mean and variance using the CDF

n	10	20	50	*exact*
$E(X)$	0,333537053	0,333351761	0,333339866	1/3
$E(X^2)$	0,20000051	0,200000009	0,200000002	1/5
$V(X)$	0,088753544	0,088876613	0,088884536	4/45

Exercises

1. Let X be a continuous random variable having the CDF F. Assume that $F(x) = x^3$ for $0 < x < 1$.

   (a) Determine the PDF $f(x)$.
   (b) Consider a partition of (0, 1) in $n = 100$ equally spaced subintervals: $x_i = i/n$. Use this partition and $f(x)$ to estimate $E(X)$, $E(X^2)$, $E(X^3)$, $V(X)$.
   (c) Use $n = 20$ Gauss points and $f(x)$ to estimate $E(X)$, $E(X^2)$, $E(X^3)$, $V(X)$.
   (d) Determine the exact values of $E(X)$, $E(X^2)$, $E(X^3)$, $V(X)$.
   (e) Determine the error in the approximation.

2. Let X be a continuous random variable on $\Omega = (0, 1)$ such that $X(\omega) = \frac{1}{\omega+1}$, $\mu(\omega) = 1$,

   (a) Consider a partition of (0, 1) in $n = 250$ equally spaced subintervals: $\omega_i = i/n$. Use this partition and $\mu(\omega)$ to estimate $E(X)$, $E(X^2)$, $V(X)$.
   (b) Use $n = 20$ Gauss points and $f(x)$ to estimate $E(X)$, $E(X^2)$, $V(X)$.
   (c) Determine the exact values of $E(X)$, $E(X^2)$, $V(X)$.
   (d) Determine the error in the approximation.

3. Let X be a continuous random variable on $\Omega = (-1, 1)$ such that $X(\omega) = \Omega^3$, $\mu(\omega) = 1/2$,

   (a) Consider a partition of (0, 1) in $n = 1000$ equally spaced subintervals: $\omega_i = i/n$. Use this partition and $\mu(\omega)$ to estimate $E(X)$, $E(X^2)$, $V(X)$.
   (b) Use $n = 50$ Gauss points and $f(x)$ to estimate $E(X)$, $E(X^2)$, $V(X)$.
   (c) Determine the exact values of $E(X)$, $E(X^2)$, $V(X)$.
   (d) Determine the error in the approximation.

3.4.3 Classical Inequalities

Two classical inequalities are useful when dealing with random variables. The first one is Markov's inequality: for any increasing positive function $g : \mathbb{R} \to \mathbb{R}$ such that $g(x) > 0$

$$P(X \geq x) \leq \frac{E(g(X))}{g(x)}. \tag{3.13}$$

As a consequence, if $X \geq 0$ is a positive random variable such that $E(X) > 0$, then

$$P(X \geq \lambda E(X)) \leq \frac{1}{\lambda}. \tag{3.14}$$

The second is Tchebichev's inequality

$$P(|X - E(X)| \geq \varepsilon) \leq \frac{V(X)}{\varepsilon^2}. \tag{3.15}$$

Finally, we have the Jensen's inequality: for any convex function $g : \mathbb{R} \to \mathbb{R}$, bounded from below,

$$g(E(X)) \leq E(g(X)). \tag{3.16}$$

Exercises

1. A company has a mean daily production of 1000 units of a given product.

   (a) Use Markov's inequality to find an upper bound for the probabilities of productions of 1010, 1100, 2000 unities.
   (b) Assume that the standard deviation of the daily production is 10. Use Tchebichev's inequality to find an upper bound for these probabilities. TIP: Consider $P(|X - 1000| \geq x - 1000)$.

2. When a fair coin is tossed n times, the number of Heads X has a mean $n/2$ and a variance $n/4$.

   (a) Determine an upper bound for the probability of obtaining a deviation of 10% of the mean. TIP: Consider $P\left(\left|X - \frac{n}{2}\right| > \alpha \frac{n}{2}\right)$.

(b) Find a lower bound for the probability of obtaining at least 95% of the expected Heads. TIP: $P(|X - m| > a) = P(X < m - a) + P(X > m + a)$

(c) Determine a number of tosses for which the observed frequency of Heads deviates from 1/2 by at most 1% with a probability 99%. TIP: Find n such that $P\left(\left|X - \frac{n}{2}\right| > 0.01\frac{n}{2}\right) \le 0.01$

3.4.4 Characteristic Function and Moments

The **characteristic function** of X is $\varphi(t) = E(e^{itX})$. ∎

In the sequel, we shall use the expansion of the characteristic function in terms of the moments:

$$\varphi(t) = \sum_{n=0}^{+\infty} \frac{(it)^k}{k!} M_k(X). \tag{3.17}$$

Recall that

$$e^z = \sum_{k=0}^{+\infty} \frac{z^k}{k!} = 1 + z + \frac{z^2}{2} + \frac{z^3}{6} + \ldots$$

Taking $z = itX$, we have

$$e^{itX} = \sum_{k=0}^{+\infty} \frac{(it)^k}{k!} X^k = 1 + itX + \frac{(it)^2}{2} X^2 + \frac{(it)^3}{6} X^3 + \ldots$$

Thus, the characteristic function of X, $\varphi(t) = E(e^{itX})$,expands as

$$\varphi(t) = \sum_{k=0}^{+\infty} \frac{(it)^k}{k!} M_k(X) = 1 + itM_1(X) + \frac{(it)^2}{2} M_2(X) + \frac{(it)^3}{6} M_3(X) + \ldots$$

as indicated.

Exercises

1. Determine the characteristic functions of the following random variables

(a) X is a random variable such that $P(X = 0) = 1$.
(b) X is a Bernoulli variable such that $P(X = 1) = p$, $P(X = 0) = 1 - p$.
(c) X is Poisson distributed with parameter $\lambda : P(X = k) = e^{-\lambda}\lambda^k/k!$.
(d) X is uniformly distributed on (a, b) – *its PDF is* : $f(x) = 1/(b - a)$ on (a, b)
(e) X is exponentially distributed –*its PDF is* : $f(x) = \lambda e^{-\lambda x}$ on $(0, +\infty)$.

3.5 Random Vectors and Pairs of Random Variables

A random vector $\boldsymbol{X} = (X_1, \ldots, X_n)$ of dimension n is an application $\boldsymbol{X} : \Omega \to \mathbb{R}^n$, *id est*, $\boldsymbol{X}$ is a vector of numerical characteristics of the elements of Ω. Analogously to random variables, probabilities on the values of $\boldsymbol{X}$ are generated by the probabilities on Ω: for given a subset $\boldsymbol{A} \subset \mathbb{R}^n$, we may consider $\boldsymbol{X}^{-1}(\boldsymbol{A}) = \{\omega \in \Omega : \boldsymbol{X}(\omega) \in \boldsymbol{A}\}$. Then, $P(\boldsymbol{X} \in \boldsymbol{A}) = P(\boldsymbol{X}^{-1}(\boldsymbol{A}))$. The cumulative function is defined analogously to the one-dimensional situation: for $\boldsymbol{X} = (X_1, \ldots, X_n)$ and $\boldsymbol{x} = (x_1, \ldots, x_n)$, we denote "$\boldsymbol{X} < x$" the region "$X_1 < x_1, X_2 < x_2, \ldots, X_n < x_n$", id est, "$\boldsymbol{X} < x$" $= \prod_{i=1}^{n}(-\infty, x_i)$. Then $\boldsymbol{I}(\boldsymbol{x}) = \boldsymbol{X}^{-1}\left(\prod_{i=1}^{n}(-\infty, x_i)\right)$ and $F(\boldsymbol{x}) = P(\boldsymbol{I}(\boldsymbol{x}))$. The PDF is defined as $f = \frac{\partial^n F}{\partial x_1 \partial x_2 \ldots \partial x_n}$.

For a couple $\boldsymbol{X} = (X_1, X_2)$ of random variables, we have

The **cumulative distribution function** (***CDF***) of $\boldsymbol{X} = (X_1, X_2)$ is:

$$F(x_1, x_2) = P(X_1 < x_1, X_2 < x_2)$$

The **probability density function** (***PDF***) is $f = \frac{\partial^2 F}{\partial x_1 \partial x_2}$ (if it exists). The CDF F is the **distribution** of X.

The **marginal distribution** of X_i is $F_i(x_i) = P(X_i < x_i)$. We have

$$F_1(x_1) = F(x_1, +\infty), F_2(x_2) = F(+\infty, x_2)$$

The **marginal density** of X_i is $f_i = \frac{\partial F_i}{\partial x_i} = \int_{-\infty}^{+\infty} f(x_1, x_2) dx_i$

The **copula** associated to $\boldsymbol{X} = (X_1, X_2)$ is, for $0 < c_1, c_2 < 1$:

$$C(c_1, c_2) = P(F_1 < c_1, F_2 < c_2)$$

The **covariance** between X_1 and X_2 is

$$cov(X_1, X_2) = E((X_1 - E(X_1))(X_2 - E(X_2))) = E(X_1 X_2) - E(X_1)E(X_2).$$

The **linear correlation coefficient** between X_1 and X_2 is

$$\rho(X_1, X_2) = \frac{cov(X_1, X_2)}{\sqrt{V(X_1)V(X_2)}}.$$

The best approximation of X_i by an affine function of X_j in the least squares sense is $\ell(X_j) = \alpha X_j + \beta$, with

$$\alpha = \frac{cov(X_i, X_j)}{V(X_j)}, \beta = E(X_i) - \alpha E(X_j).$$

We have $E((X_i - \ell(X_j))^2) = V(X_i)(1 - |\rho(X_i, X_j)|^2)$. Thus,

$$X_i = \ell(X_j) \ a.s. \Leftrightarrow |\rho(X_i, X_j)| = 1.$$

The covariance matrix $\boldsymbol{C}(\boldsymbol{X})$ and the correlation matrix $\boldsymbol{\rho}(\boldsymbol{X})$ of $\boldsymbol{X}$ are given by

$$C_{ij}(\boldsymbol{X}) = Cov(X_i, X_j), \quad \rho_{ij}(\boldsymbol{X}) = Cov(X_i, X_j).$$

The **conditional mean** of X_i with respect to X_j is the best approximation of X_i by a function $\psi_j(X_j)$ in the least squares' sense. We have

$$E(X_2|X_1) = \psi_1(x_1) = \int x_2 f(x_2|x_1)\, dx_2, \quad f(x_2|x_1) = \frac{f(x_1,x_2)}{f_1(x_1)}.$$

$$E(X_1|X_2) = \psi_2(x_2) = \int x_1 f(x_1|x_2)\, dx_1, \quad f(x_1|x_2) = \frac{f(x_1,x_2)}{f_2(x_2)}.$$ ∎

Remark

If X has a PDF f, then

$$E(\phi(X_1, X_2)) = \int\int_{-\infty}^{+\infty} \phi(x_1, x_2) f(x_1, x_2) dx_1 dx_2.$$

If Ω has a mass function μ, then

$$E(\phi(X_1, X_2)) = \sum_{\omega \in \Omega} \mu(\omega)\phi(X_1(\omega), X_2(\omega)).$$

If Ω has a mass density μ, then

$$E(\phi(X_1, X_2)) = \int_{\Omega} \mu(\omega)\phi(X_1(\omega), X_2(\omega)) d\omega.$$ ∎

Example

Let us consider $\Omega = (0, 1)$ and the probability $P((a, b)) = b - a$. Let $X_1(\omega) = \omega^2$; $X_2(\omega) = \omega$. We have, for $0 < x_1, x_2 < 1$:

$$X_1 < x_1, X_2 < x_2 \Leftrightarrow \omega < \min\{\sqrt{x_1}, x_2\}$$

Thus, for $0 < x_1, x_2 < 1$:

$$F(x_1, x_2) = \min\{\sqrt{x_1}, x_2\} = \begin{cases} \sqrt{x_1}, & \text{if } x_1^2 < x_2 \\ x_2, & \text{if } x_1^2 > x_2 \end{cases}$$

This function cannot be derived in the usual sense, due to the change of definition across the curve $x_2 = x_1^2$. The density is null in the subregions $x_2 < x_1^2$ and $x_2 > x_1^2$, but corresponds to a Dirac mass – see, for instance, (Souza de Cursi, 2015) – on the curve separating these regions. The marginal distributions are, for $0 < x_1, x_2 < 1$:

$$F_1(x_1) = \sqrt{x_1}, \quad F_2(x_2) = x_2.$$

Although we cannot derive F, marginal densities may be determined:

$$f_1(x_1) = \frac{1}{2\sqrt{x_1}}, \quad f_2(x_2) = 1.$$

The copula associated to $\boldsymbol{X}$ is $C(c_1, c_2) = \min\{c_1, c_2\}$. We have

$$E(X_1^k) = \frac{1}{2}\int_0^1 x_1^{k-\frac{1}{2}} dx_1 = \frac{1}{2k+1},$$

$$E(X_2^k) = \int_0^1 x_2^k dx_2 = \frac{1}{k+1}.$$

Thus,

$$E(X_1) = \frac{1}{3}, V(X_1) = \frac{4}{45}; \quad E(X_2) = \frac{1}{2}, V(X_2) = \frac{1}{12}.$$

In addition,

$$E(X_1X_2) = \frac{1}{\int_0^1 d\omega}\int_0^1 X_1(\omega)X_2(\omega)d\omega = \int_0^1 \omega^3 d\omega = \frac{1}{4},$$

$$cov(X_1, X_2) = \frac{1}{12}, \quad \rho(X_1, X_2) = \frac{\sqrt{15}}{4}.$$

The best least squares approximation by linear functions are

$$X_1 \approx X_2 - \frac{1}{6}; \quad X_2 \approx \frac{15}{16}X_1 + \frac{1}{16}.$$

The conditional mean $E(X_i| X_j)$ is determined by finding $\psi(X_j)$ that minimizes

$$J(\psi) = E\Big((X_i - \psi(X_j))^2\Big).$$

Then

$$E(X_1|X_2) = X_2^2, E(X_2|X_1) = \sqrt{X_1}.$$ ∎

Example

Let us consider $\Omega = \{-2, -1, 0, 1, 2\}$. Assume that all the elements are equiprobable. Let $X_1(\omega) = \omega^2$; $X_2(\omega) = |\omega|$. We have

$$\boldsymbol{X}(\Omega) = \{(4,2), (1,1), (0,0), (1,1), (4,2)\}$$

Thus,

$$P(\boldsymbol{X} = (4,2)) = P(\boldsymbol{X} = (1,1)) = \frac{2}{5}; P(\boldsymbol{X} = (0,0)) = \frac{1}{5}.$$

The marginal distributions are

$$P(X_1 = 4) = P(X_1 = 1) = \frac{2}{5}; P(X_1 = 0) = \frac{1}{5}.$$

$$P(X_2 = 2) = P(X_2 = 1) = \frac{2}{5}; P(X_2 = 0) = \frac{1}{5}.$$

We have

$$E(X_1) = 2, V(X_1) = \frac{14}{5}; \quad E(X_2) = \frac{6}{5}, V(X_2) = \frac{14}{5}$$

In addition, $E(X_1X_2) = \frac{18}{5}$, so that

$$cov(X_1,X_2) = \frac{6}{5}, \quad \rho(X_1,X_2) = \frac{3\sqrt{5}}{7}.$$

The best least squares approximation by linear functions are

$$X_1 \approx \frac{15}{7}X_2 - \frac{4}{7}; \quad X_2 \approx \frac{3}{7}X_1 + \frac{12}{35}.$$

The conditional mean $E(X_i | X_j)$ is determined by finding $\psi(X_j)$ that minimizes

$$J(\psi) = E\left(\left(X_i - \psi(X_j)\right)^2\right).$$

For instance, we evaluate $E(X_1 | X_2)$ by minimizing

$$J(\psi) = \frac{2}{5}(4 - \psi(2))^2 + \frac{2}{5}(1 - \psi(1))^2 + \frac{1}{5}(0 - \psi(0))^2,$$

so that $\psi(2) = 4$, $\psi(1) = 1$, $\psi(0) = 0$: we may take $\psi(X_2) = X_2^2$. In addition

$$P(X_1 = i|X_2 = j) = \frac{P(\boldsymbol{X} = (i,j))}{P(X_2 = j)}.$$

Then

$$P(X_1 = 4|X_2 = 2) = P(X_1 = 1|X_2 = 1) = P(X_1 = 0|X_2 = 0) = 1.$$

The other probabilities are null. To evaluate $E(X_2 | X_1)$, we minimize

$$J(\psi) = \frac{2}{5}(2 - \psi(4))^2 + \frac{2}{5}(1 - \psi(1))^2 + \frac{1}{5}(0 - \psi(0))^2,$$

so that $\psi(4) = 2$, $\psi(1) = 1$, $\psi(0) = 0$: we may take $\psi(X_1) = \sqrt{X_1}$. Here

$$P(X_2 = 2|X_1 = 4) = P(X_2 = 1|X_1 = 1) = P(X_2 = 0|X_1 = 0) = 1$$

and the other probabilities are null. ∎

Let us introduce the independence between variables:

$\boldsymbol{X} = (X_1, X_2)$ is a couple of **independent** variables if and only if:

$$F(x_1, x_2) = F_1(x_1)F_2(x_2)$$

In such a case, we say that X_1 and X_2 are **independent**. It is **equivalent to** any of the following conditions:

- The density of the pair is the product of the individual densities: $f(x_1, x_2) = f_1(x_1)f_2(x_2)$;
- the **copula** associated to $\boldsymbol{X} = (X_1, X_2)$ is, for $0 < c_1, c_2 < 1$:

$$C(c_1, c_2) = P(F_1 < c_1)P(F_2 < c_2) = c_1c_2;$$

- the characteristic function of the pair is the product of the of the individual characteristic functions: $\phi(t_1, t_2) = E(e^{it_1X_1 + it_2X_2}) = \phi_1(t_1)\phi_2(t_2)$, $\phi_j(s) = E(e^{isX_j})$. ∎

Indeed, notice that

$$P(F_i < c_i) = P\left(X_i < F_i^{-1}(c_i)\right) = F_i\left(F_i^{-1}(c_i)\right) = c_i.$$

Thus, independence of (X_1, X_2) is equivalent to $C(c_1, c_2) = c_1c_2$. Furthermore, if X_1 and X_2 are independent, then $E(e^{it_1X_1 + it_2X_2}) = E(e^{it_1X_1})E(e^{it_2X_2})$, so that $\phi(t_1, t_2) = \phi_1(t_1)\phi_2(t_2)$. Furthermore:

If X_1 and X_2 are independent, **then** $cov\ (X_1, X_2) = \rho(X_1, X_2) = 0$, $E(X_1| X_2) = E(X_1)$, $E(X_2| X_1) = E(X_2)$. **The converse is false, except for Normally distributed variables** (see below). ∎

Example

Let us consider a couple $\boldsymbol{X} = (X_1, X_2)$ having as PDF

$$f(x_1, x_2) = \begin{cases} 1, & \text{if } 0 < x_1, x_2 < 1 \\ 0, & \text{otherwise.} \end{cases}$$

The marginal densities verify

$$f_i(x_i) = \begin{cases} 1, & \text{if } 0 < x_i < 1 \\ 0, & \text{otherwise.} \end{cases}$$

Thus, $f(x_1, x_2) = f_1(x_1)f_2(x_2)$ and the variables are independent. Notice that, for $0 < x_i < 1$: $F_i(x_i) = x_i$ and $F(x_1, x_2) = x_1x_2 = F_1(x_1)F_2(x_2)$.

Analogously, $P(F_1 < c_1, F_2 < c_2) = P(X_1 < c_1, X_2 < c_2) = c_1c_2$. Notice that $f(x_1| x_2) = f(x_2| x_1) = 1$, for $0 < x_1, x_2 < 1$. Thus, $E(X_1|X_2) = E(X_2|X_1) = \frac{1}{2} = E(X_1) = E(X_2)$. ∎

Example
Let us consider a couple $\boldsymbol{X} = (X_1, X_2)$ having as PDF

$$P(\boldsymbol{X} = (0,0)) = P(\boldsymbol{X} = (0,1)) = P(\boldsymbol{X} = (1,0)) = P(\boldsymbol{X} = (1,1)) = \frac{1}{4}$$

The marginal distributions are

$$P(X_i = 0) = P(X_i = 1) = \frac{1}{2}.$$

Thus,

$$P(\boldsymbol{X} = (i,j)) = P(X_1 = i, X_2 = j) = P(X_1 = i)P(X_2 = j)$$

and the variables are independent. Notice that $P(X_1 = i| X_2 = j) = P(X_1 = i)$ and $P(X_2 = j|X_1 = i) = P(X_2 = j)$, so that $E(X_1|X_2) = E(X_2|X_1) = \frac{1}{2} = E(X_1) = E(X_2)$. ∎

Exercises

1. Let $\Omega = \{0, 1, \ldots, 6k\}$ and $=(X_1, X_2)$, $X_1(\omega) = \omega \; mod \; 2$, $X_2(\omega) = \omega \; mod \; 3$. Assume the elements of Ω as equiprobable ($\mu(\omega) = 1/6k$).

   (a) Show that $\boldsymbol{X}(\Omega) = \{(0, 0), (0, 1), (0, 2), (1, 0), (1, 1), (1, 2)\}$.
   (b) Determine the probabilities of the elements of $\boldsymbol{X}(\Omega)$.
   (c) Are the variables independent?

2. Let $\Omega = (0, 1)$, with $P((a, b)) = b - a$ ($\mu(\omega) = 1$) and $\boldsymbol{X} = (X_1, X_2)$, $X_1(\omega) = \omega^2$, $X_2(\omega) = \omega^4$.

   (a) Find the CDF F of $\boldsymbol{X}$.
   (b) Determine the PDF f of $\boldsymbol{X}$.
   (c) Determine the marginal distributions and densities.
   (d) Are the variables independent?
   (e) Find $E(X_2| X_1)$.

3. Let $\boldsymbol{X} = (X_1, X_2)$ be a pair of random variables having as density

$$f(x_1, x_2) = \begin{cases} x_1 + x_2, & \textit{if } 0 < x_1, x_2 < 1; \\ 0, & \textit{otherwise}. \end{cases}$$

   (a) Find the CDF F of $\boldsymbol{X}$.
   (b) Determine the marginal distributions and densities.
   (c) Are the variables independent?
   (d) Find $E(X_2| X_1)$.

3.6 Discrete and Continuous Random Variables

It is usual to consider **discrete** and **continuous** random variables, according to the range of X: if the image $X(\Omega)$ is a finite or enumerable set, then we say that X is discrete. If the image $X(\Omega)$ is an interval of real numbers (eventually infinite or semi-infinite), we say that X is continuous.

3.6.1 Discrete Variables

As previously remarked, we say that X is discrete if and only if

$$X(\Omega) = \{X_1, \ldots, X_n\} \; (finite) \; or \; X(\Omega) = \{X_i : i \in \mathbb{N}\} (enumerable).$$

Assume that $X_1 < X_2 < \ldots < X_i < X_{i+1} < \ldots$. Let $p_i = P(X = X_i)$. The CDF of X is

$$F(x) = \begin{cases} 0, & if\ x \leq \min\{X_i\}; \\ \sum_{k=1}^{i} p_k, & if\ X_i < x \leq X_{i+1}; \\ 1, & if\ x > \sup\{X_i\}. \end{cases} \tag{3.18}$$

In addition,

$$E(\phi(X)) = \sum_i p_i \phi(X_i). \tag{3.19}$$

Thus,

$$M_k(X) = \sum_i p_i X_i^k, E(X) = \sum_i p_i X_i, E(X^2) = \sum_i p_i X_i^2, \varphi(t) = \sum_j p_j e^{itX_j}.$$

Examples of discrete variables are:

- The uniform distribution on $\{1, \ldots, n\}$: $P(X = i) = \frac{1}{n}$, for $1 \leq i \leq n$. We have

$$E(X) = \frac{n+1}{2}, V(X) = \frac{n^2 - 1}{12}, \sigma(X) = \sqrt{\frac{n^2 - 1}{12}}.$$

- Bernoulli $\mathcal{B}(p)$: describes success (value 1) or failure (value 0) in a trial for a probability of success p. $X(\Omega) = \{0, 1\}$, $P(X = 1) = p$, $P(X = 0) = q = 1 - p$. We have

$$E(X) = p, V(X) = pq, \quad \sigma(X) = \sqrt{pq}.$$

- Binomial $\mathcal{B}(n,p)$: gives the number of successful trials among n when the probability of success is p. It may be interpreted as the law of a sum of n independent Bernoulli laws. $X(\Omega) = \{0, 1, \ldots, n\}$, $P(X = k) = \binom{n}{k} p^k q^{n-k}$, $q = 1 - p$. We have

$$E(X) = np, V(X) = npq, \quad \sigma(X) = \sqrt{npq}.$$

  Notice that $\mathcal{B}(1,p) = \mathcal{B}(p)$.

EXCEL® offers some built-in functions to deal with Bernoulli and Binomial. `BINOM.DIST(n;k;p;FALSE)` returns the value of $p(X = k)$ for $\mathcal{B}(n,p)$. `BINOM.DIST(n;k;p;TRUE)` returns the value of $p(X \leq k)$ for $\mathcal{B}(n,p)$. `BINOM.INV(n;p;a)` returns the smallest value of k such that $p(X \leq k) \geq a$ for $\mathcal{B}(n,p)$. `BINOM.DIST.RANGE(n;p;k1;k2)` returns the value of $p(k1 \leq X \leq k2)$ for $\mathcal{B}(n,p)$.

- Negative Binomial $\mathcal{NB}(n,p)$: gives the number of failures necessary to get n successes when the probability of success is p. $X(\Omega) = \{0, 1, \ldots\}$, $P(X = k) = \binom{n+k-1}{n-1} p^n q^k$, $q = 1 - p$. We have

$$E(X) = n\frac{q}{p}, V(X) = n\frac{q}{p^2}, \quad \sigma(X) = \frac{1}{p}\sqrt{nq}.$$

For the Negative Binomial, `NEGBINOM.DIST(k;n;p;flag)` returns, for $\mathcal{NB}(n,p)$, either the value of $p(X = k)$ or the value of $p(X \leq k)$, according to the value of flag (`FALSE` or `TRUE`).

- Poisson $\mathcal{P}(\lambda)$: $X(\Omega) = \mathbb{N} = \{0, 1, \ldots, n, \ldots\}, P(X = k) = \frac{\lambda^k}{k!} e^{-\lambda}$. We have

$$E(X) = \lambda, V(X) = \lambda, \quad \sigma(X) = \sqrt{\lambda}.$$

A Poisson's law approximates the sum of a large number of Bernoulli laws having small parameter p $(\lambda = np)$.

The EXCEL® built-in function to deal with Poisson's variables is `POISSON.DIST(k;λ;flag)`, which returns, for $\mathcal{P}(\lambda)$, either the value of $p(X = k)$ or $p(X \leq k)$, according to the value of flag (`FALSE` or `TRUE`).

- Multinomial distribution $\mathcal{M}(n,p)$: it is a generalization of the binomial distribution to the situation where we consider a population which is subdivided in k distinct subpopulations of probabilities $p_1, \ldots, p_k$, respectively – we have $p_1, + \ldots + p_k = 1$. A trial takes simultaneously n elements from the population and we count the number elements X_i of each subpopulation i. For $n_1 + \ldots + n_k = n$, we have

$$P(X_1 = n_1, \ldots, X_k = n_k) = \frac{n!}{n_1! \ldots n_k!} p_1^{n_1} p_2^{n_2} \ldots p_k^{n_k} = \frac{\left(\sum_{i=1}^k n_i\right)!}{\prod_{i=1}^k (n_i!)} \prod_{i=1}^k p_i^{n_i}.$$

Notice that each X_i is a binomial variable $\mathcal{B}(n, p_i)$. The variables $X_1, \ldots, X_k$ are not independent: $cov\ (X_i, X_j) = -np_ip_j, for\ i \neq j$; $cov\ (X_i, X_i) = np_iq_i$.

EXCEL® does not provide the multinomial distribution. The function `MULTINOMIAL(`n_1; ...; n_k`)` furnishes $\left(\sum_{i=1}^{k} n_i\right)!/\prod_{i=1}^{k}(n_i!)$.

3.6.2 Continuous Variables Having a PDF

Let X be a continuous random variable having f as PDF. Then,

$$E(\phi(X)) = \int_{-\infty}^{+\infty} \phi(x)f(x)dx. \tag{3.20}$$

Thus

$$M_k(X) = \int_{-\infty}^{+\infty} x^k f(x)dx, E(X) = \int_{-\infty}^{+\infty} xf(x)dx. \tag{3.21}$$

$$E(X^2) = \int_{-\infty}^{+\infty} x^2 f(x)dx, \varphi(t) = \int_{-\infty}^{+\infty} e^{itx} f(x)dx. \tag{3.22}$$

Examples of continuous variables are:

- Uniform distribution on (a, b):

$$f(x) = \begin{cases} \frac{1}{b-a}, & if\ x \in (a,b) \\ 0, & if\ x \notin (a,b) \end{cases}$$

We have

$$E(X) = \frac{a+b}{2}, V(X) = \frac{(b-a)^2}{12}, \sigma(X) = \frac{b-a}{\sqrt{12}}.$$

- Standard Gaussian $N(0, 1)$

$$f(x) = \frac{1}{\sqrt{2\pi}} e^{-\frac{x^2}{2}}$$

We have

$$E(X) = 0, V(X) = 1, \sigma(X) = 1.$$

- Normal $N(m, \sigma)$

$$f(x) = \frac{1}{\sigma\sqrt{2\pi}} e^{-\frac{1}{2}\left(\frac{x-m}{\sigma}\right)^2}$$

We have

$$E(X) = m, V(X) = \sigma^2, \sigma(X) = \sigma.$$

The built-in functions for Normal variables are `NORM.DIST`, `NORM.INV`. For Standard Gaussian, use `NORM.S.DIST`, `NORM.S.INV`. `NORM.DIST(x,m,σ,flag)` and `NORM.S.DIST(x,flag)` return either $f(x)$ or $p(X < x)$, according to flag (`FALSE` or `TRUE`). `NORM.INV(a,m,σ)` and `NORM.S.INV(a)` return x such that $p(X < x) = a$.

If X_1 and X_2 are both normal and $cov(X_1, X_2) = \rho(X_1, X_2) = 0$ **then** X_1 and X_2 are independent. ∎

- Chi-squared (Pearson's distribution) χ^2: the distribution of Z^2, where Z is Standard Gaussian.

$$f(x) = \frac{1}{\sqrt{2\pi x}} e^{-\frac{x}{2}}.$$

We have $E(X) = 1$, $V(X) = 2$.

- Chi-squared with n degrees of freedom $\chi^2(n)$: the distribution of $Z_1^2 + \ldots + Z_n^2$, where $Z_1, \ldots, Z_n$ are independent Standard Gaussian variables.

$$f(x) = \frac{1}{2^{\frac{n}{2}}\Gamma\left(\frac{n}{2}\right)} e^{-\frac{x}{2}} x^{\frac{n}{2}-1}.$$

We have $E(X) = n$, $V(X) = 2n$.

The built-in functions for Chi-squared variables are `CHISQ.DIST`, `CHISQ.INV`.

- Student-Fisher with n degrees of freedom $SF(n)$: the distribution of $\frac{Y}{\sqrt{\frac{Q}{n}}}$, where Y, Q are independent, Y is Standard Gaussian, Q is $\chi^2(n)$.

$$f(x) = \frac{\Gamma\left(\frac{n+1}{2}\right)}{\sqrt{n\pi}\Gamma\left(\frac{n}{2}\right)}\left(1 + \frac{x^2}{n}\right)^{\frac{n+1}{2}}.$$

We have $E(X) = 0$, $V(X) = \frac{n}{n-2}$ (for $n > 2$).

The built-in functions for Student-Fisher variables are `T.DIST`, `T.INV`.

- Behrens-Fisher-Snedecor (BFS, or simply Fisher) with (n_1, n_2) degrees of freedom: the distribution of $\frac{Q_1/n_1}{Q_2/n_2}$, where Q_i is $\chi^2(n_i)$, $Q_1$and Q_2 independent

> The built-in functions for BFS variables are F.DIST, F.INV.

- Log-normal: the distribution of X such that $\ln(X)$ is Normal $N(m, \sigma)$.

> The built-in functions for log-normal variables are LOGNORM.DIST, LOGNORM.INV.

Exercises

1. Let X_1 and X_2 be two independent variables having Bernoulli distributions of same parameter $p > 0$. Determine the range and the probabilities of $Z_1 = X_1 - X_2$, $Z_2 = X_1X_2$.
2. Let X be a continuous random variable uniformly distributed on $(-1, 1)$. Evaluate

   (a) $P\left(-\frac{1}{2} < X < \frac{1}{3}\right)$;
   (b) $P\left(X > \frac{1}{3}\right)$;
   (c) $M_3(X)$.

3. An urn contains 10 red balls and 30 blue balls. 5 balls are drawn sequentially, with replacement (the ball is put back in the urn after drawing). Determine the probabilities of getting

   (a) Exactly three red balls.
   (b) At least two red balls.
   (c) Less than three red balls. TIP: look at binomial

4. An urn contains 10 red balls and 30 blue balls. A ball is drawn. If it is blue, it is put back in the urn and we make another draw. The game stops when a red ball is drawn. Let n be the number of draws before stopping. Find the probabilities of

   (a) $n = 5$;
   (b) $n \geq 5$;
   (c) $n \leq 10$. TIP: look at negative binomial

5. Let X be a random variable which is normally distributed $N(1, 2)$. Find the probabilities

   (a) $P(-1 < X < 1)$;
   (b) $P\ (X < 3)$;
   (c) $P\ (X > 1)$.

6. Let X be a random variable which is normally distributed $N(0, 2)$ and $Y = e^{-X+1}$. Find the probabilities

   (a) $P(1 < Y < 2)$;
   (b) $P(Y < 3)$;
   (c) $P(Y > 2)$; TIP: $\ln(Y)$ is $N(-1, 2)$.

7. Let X be a random variable which is normally distributed $N(0, 2)$ and $Y = X^2$. Find the probabilities

   (a) $P(1 < Y < 2)$;
   (b) $P(Y < 3)$;
   (c) $P(Y > 2)$; TIP: $\frac{X}{2}$ is Standard Gaussian

8. Let X_1, X_2 be two independent random variables normally distributed $N(0, 2)$ and $Y = \frac{X_1}{\sqrt{X_2^2}}$. Find the probabilities

   (a) $P(1 < Y < 2)$;
   (b) $P(Y < 3)$;
   (c) $P(Y > 2)$; TIP: $\frac{X_i}{2}$ is Standard Gaussian.

9. Let X_1, X_2 be two independent random variables normally distributed $N(0, 2)$ and $Y = \frac{X_1^2}{X_2^2}$. Find the probabilities

   (a) $P(1 < Y < 2)$;
   (b) $P(Y < 3)$;
   (c) $P(Y > 2)$; TIP: $\frac{X_i}{2}$ is Standard Gaussian.

Supplementary Exercises

1. The statistics of car crashes show the following connection with the local maximum speed regulation (source NHTSA, data for 2017):

Max speed (mph)	30	40	50	55	≥ 60	No limit
Accidents (thousands)	418	311	248	327	233	60

   Given a crash, what is the probability of that the max speed was 40 mph ? inferior or equal to 40? at least 55?

2. A candidate for a game must choose between 10 boxes containing prizes whose value is respectively 0, 10, 20, 25, 50, 100, 200, 500, 1000, 10,000 dollars. What is the probability of him making more than a thousand dollars? Less than a hundred dollars? What is the mean value of the winnings?

3. A Sphynx waits for travelers to pass by and asks them a question which they answer with a probability of hitting the right answer equal to $p > 0$. In case of error, the traveler is devoured by the Sphynx. The number of travelers passing each week near the Sphynx follows a Poisson's law of parameter $\lambda > 0$. What is the probability that the Sphynx will not eat anyone in a week?

   TIP: consider the events "number of travelers passing is n" and "all n answers are good", the product of probabilities, group the product λp and sum over all the values of n.

4. A model is destined to predict an event. When the prediction is positive, the success rate is 85%. When the answer is negative, the success rate is of 60%. The global probability of the event is 10%.

   (a) The model furnishes a positive answer. What is the probability that the event arises in reality?
   (b) The model furnishes a negative answer. What is the probability that the event does not occur in reality?

5. You have in your pocket two dices. One of the dices is regular, numbered from 1 to 6, but the other one is unfair: the face numbered one was replaced by a second face numbered 6. In both the dices, all the faces are equiprobable. You take a dice at random, equiprobably and roll it. The face up is 6.

   (a) What is the probability of this result?
   (b) What is the probability that the dice is not the unfair one?

6. Two models M_1 and M_2 are destined to predict an event. When the prediction is positive, the success rate is 85% for M_1 and 70% for M_2. When the answer is negative, the success rate is of 60% for M_1 and 80% for M_2. The global probability of the event is 10%.

   (a) Both the models furnish a positive answer. What is the probability that the event arises in reality?
   (b) Both the models furnish a negative answer. What is the probability that the event does not occur in reality?
   (c) M_1 furnishes a positive answer and M_2 furnishes a negative answer. What is the probability that the event arises in reality?
   (d) M_1 furnishes a negative answer and M_2 furnishes a positive answer. What is the probability that the event does not occur in reality?

7. Let $\boldsymbol{X} = (X_1, \ldots, X_n)^t$ be a vector of independent variables. Let $\boldsymbol{A}$ be a $m \times n$ matrix of real numbers and $\boldsymbol{Y} = \boldsymbol{AX}$.

   (a) Verify that $\boldsymbol{Y}^t\boldsymbol{Y} = \boldsymbol{X}^t\boldsymbol{A}^t\boldsymbol{AX}$.
   (b) Verify that $E(\boldsymbol{Y}) = \boldsymbol{A}E(\boldsymbol{X})$.
   (c) Verify that $\mathbf{C}(\boldsymbol{Y}) = E(\boldsymbol{X}^t\boldsymbol{A}^t\boldsymbol{AX}) - E(\boldsymbol{X})^t\boldsymbol{A}^t\boldsymbol{A}E(\boldsymbol{X})$.

8. Let $\boldsymbol{X} = (X_1, \ldots, X_n)$ be a vector of **independent standard gaussian variables.** Let $\boldsymbol{A}$ be a $m \times n$ matrix of real numbers and $\boldsymbol{Y} = \boldsymbol{AX}$.

   (a) Show that $E(\boldsymbol{Y}) = \boldsymbol{0}$.
   (b) Both the models furnish a negative answer. What is the probability that the event does not occur in reality?
   (c) M_1 furnishes a positive answer and M_2 furnishes a negative answer. What is the probability that the event arises in reality?
   (d) M_1 furnishes a negative answer and M_2 furnishes a positive answer. What is the probability that the event does not occur in reality?

9. Two models M_1 and M_2 are destined to predict an event. When the prediction is positive, the success rate is 85% for M_1 and 70% for M_2. When the answer is negative, the success rate is of 60% for M_1 and 80% for M_2. The global probability of the event is 10%.

   (a) Both the models furnish a positive answer. What is the probability that the event arises in reality?
   (b) Both the models furnish a negative answer. What is the probability that the event does not occur in reality?
   (c) M_1 furnishes a positive answer and M_2 furnishes a negative answer. What is the probability that the event arises in reality?
   (d) M_1 furnishes a negative answer and M_2 furnishes a positive answer. What is the probability that the event does not occur in reality?

3.7 Sequences of Random Variables

One of the central points in Uncertainty Quantification is the approximation of random variables: we look for representations of the observed variability of a system as a function of the known or assumed variability of certain parameters and inputs. Such a representations are generally approximations based on the construction of a sequence $\{X_n : n \geq 0\}$ of random variables that converge to the exact variable X: $X_n \longrightarrow X$. In probability, it is usual to manipulate different definitions of convergence:

1. Convergence in the quadratic mean: in quadratic mean if and only if $E((X_n - X)^2) \longrightarrow 0$, for $n \longrightarrow +\infty$;
2. Almost sure convergence: the event $E = \{X_n \longrightarrow X\}$ is almost sure, id est, $P(E) = 1$;
3. Convergence in probability: for any $\varepsilon > 0$, the event $E_n(\varepsilon) = \{\|X_n - X\| \geq \varepsilon\}$ verifies $P(E_n(\varepsilon)) \longrightarrow 0$, for $n \longrightarrow +\infty$;
4. Convergence in distribution: let F_n be the CDF of X_n and F be the CDF of X. $X_n \longrightarrow X$ in distribution if and only if $F_n(x) \longrightarrow F(x)$ at any point x where F is continuous.

One of the fundamental results concerning the convergence of sequences of random variables is **Levy's theorem** (Lévy, 1922):

Theorem (Lévy):
Let $\{X_n : n \in \mathbb{N}\}$ be a sequence of random variables such that the characteristic function of X_n is φ_n. Let X be a random variable having as characteristic function φ. Then $X_n \longrightarrow X$ in distribution if and only if $\varphi_n(t) \longrightarrow \varphi(t)$ almost everywhere. ■

Levy's theorem furnishes a practical method for the approximation of a random variable X by approximating its characteristic function φ – as previously observed, $\varphi(t) = E(e^{itX})$,expands as

$$\varphi(t) = \sum_{k=0}^{+\infty} \frac{(it)^k}{k!} M_k(X) = 1 + itM_1(X) + \frac{(it)^2}{2} M_2(X) + \frac{(it)^3}{2} M_3(X) + \ldots$$

We may consider X_n as the variable having as characteristic function

$$\varphi_n(t) = \sum_{k=0}^{n} \frac{(it)^k}{k!} M_k(X) \approx E\left(e^{itX_n}\right).$$

Then, from Levy's theorem, $X_n \longrightarrow X$ in distribution. In practice, we look for a variable verifying the equations:

$$M_k(X_n) = M_k(X), \quad for\ 1 \leq k \leq n. \tag{3.23}$$

Example
Let us consider a gaussian variable $X \sim N(0, 1)$. Since

$$M_1(X) = 0, M_2(X) = 1, M_3(X) = 0,$$

We may consider a variable X_3 such that

$$M_k(X_3) = M_k(X), 1 \leq k \leq 3.$$

For instance, a discrete variable such that

$$P(X_3 = -1) = P(X_3 = 1) = \frac{1}{2}.$$ ■

A second theorem concerns gaussian variables

Theorem:
Let $\{X_n : n \in \mathbb{N}\}$ be a sequence of normally distributed random variables. If $X_n \longrightarrow X$ in distribution, then X is normally distributed. In addition, $X_n \longrightarrow X$ in distribution if and only if $E(X_n) \longrightarrow E(X)$, $V(X_n) \longrightarrow V(X)$. ■

Exercises

1. Let X be a gaussian variable $X \sim N(0, 1)$. Find a discrete variable X_5 such that

$$M_k(X_5) = M_k(X), 1 \leq k \leq 5.$$

TIP: $M_4(X) = 3$, $M_5(X) = 0$. A solution is to look for a discrete variable taking the values $-\sqrt{a}, -1, 0, 1, \sqrt{a}$, with $P(X_5 = -1) = P(X_5 = 1)$ and $P(X_5 = -\sqrt{a}) = P(X_5 = \sqrt{a})$. The equations $M_2(X_5) = M_2(X)$ and $M_4(X_5) = M_4(X)$ form a linear system which furnishes $P(X_5 = \pm 1)$ and $P(X_5 = \pm\sqrt{a})$.

2. Let X be a gaussian variable $X \sim N(0, 1)$. Find a discrete variable X_7 such that

$$M_k(X_5) = M_k(X), 1 \leq k \leq 7.$$

TIP: $M_6(X) = 15$, $M_7(X) = 0$. A solution is to look for a discrete variable taking the values $-\sqrt{b}, -\sqrt{a}, -1, 0, 1, \sqrt{b}, \sqrt{a}$, with $P(X_7 = -1) = P(X_7 = 1)$, $P(X_7 = -\sqrt{a}) = P(X_7 = \sqrt{a})$ and $P(X_7 = -\sqrt{b}) = P(X_7 = \sqrt{b})$. The equality of the moments of even order form a linear system which furnishes $P(X_5 = \pm 1)$, $P(X_5 = \pm\sqrt{a})$ and $P(X_5 = \pm\sqrt{b})$.

3. Let X be a gaussian variable $X \sim N(0, 1)$ and Y_1, Y_2 be two independent discrete variables such that

$$P(Y_i = -a) = P(Y_i = a) = p; P(Y_i = 0) = 1 - 2p.$$

Let $X_5 = Y_1 + Y_2$ be a random variable. Determine p and a such that

$$M_k(X_5) = M_k(X), 1 \leq k \leq 5.$$

TIP: Use the equations $M_2(X_5) = M_2(X)$ and $M_4(X_5) = M_4(X)$.

4. Consider $\Omega = (0, 1)$ and the mass density $\mu(\omega) = 1$. Let $X_n, X : \Omega \to \mathbb{R}$ be random variables such that $X_n(\omega) = \frac{\omega}{n}$ and $X(\omega) = 0$.

(a) Determine the CDF F_n of X_n. TIP: X_n is uniform on $\left(0, \frac{1}{n}\right)$.
(b) Find the CDF F of X. TIP: $X = \delta_0$ (i.e., $P(X = 0) = 1$).
(c) Show that $X_n \longrightarrow X$ in distribution.
TIP: Show that, for $x > 0$ and n large enough $F_n(x) = 1$.
(d) Show that $X_n \longrightarrow X$ in probability.
TIP: Show that, for $\varepsilon > 0$ and n large enough $P(X_n \geq \varepsilon) = 0$.
(e) Show that $X_n \to X$ almost surely.
TIP: Consider the set $A = \{\omega \in \Omega : X_n(\omega) \to 0\}$
(f) Show that $X_n \to X$ in the quadratic mean. TIP: Evaluate $E\left(X_n^2\right)$

5. Let $S_n = \frac{X_n}{n}$, where $\{X_n : n \in \mathbb{N}\}$ is a sequence of independent discrete random variables such that $(0 < p < 1)$:

$$P(X_n = k) = \frac{p}{n}\left(1 - \frac{p}{n}\right)^k, k \geq 0.$$

(a) Let $t \in \mathbb{R}$ and $n(t) \in \mathbb{N}$ verify $n(t) \leq nt < n(t) + 1$. Verify that

$$P(X_n \leq nt) = P(X_n \leq n(t)) = 1 - \left(1 - \frac{p}{n}\right)^{n(t)+1}$$

(b) Verify that

$$\left(1 - \frac{p}{n}\right)^{nt+2} \leq \left(1 - \frac{p}{n}\right)^{n(t)+1} \leq \left(1 - \frac{p}{n}\right)^{nt+1}$$

(c) Conclude that $P(S_n \leq t) \longrightarrow 1 - e^{-pt}$.
(d) Show that S_n converges in distribution to an exponential law.

3.8 Samples

A **sample** from X is a set of independent observations of X, *id est*, a set $\mathcal{X} = \{X_1, \ldots, X_n\}$, where each X_i has the same distribution as X and is independent from X_j, $\forall j \neq i$. The **empirical mean** of the sample is

$$\overline{X_n} = \frac{1}{n}\sum_{i=1}^{n} X_i. \tag{3.24}$$

A **median** of the sample is a middle value that cuts the sample in two equal parts. If the sample is ordered in an increasing order, the median is $X_{\frac{n}{2}}$ for n pair; while for n odd, the median is the arithmetic mean of $X_{\frac{n-1}{2}}, X_{\frac{n+1}{2}}$.

Population's variance of the sample is

$$V_p = \frac{1}{n}\sum_{i=1}^{n}\left(X_i - \overline{X_n}\right)^2 = \frac{1}{n}\sum_{i=1}^{n} X_i^2 - \left(\overline{X_n}\right)^2. \tag{3.25}$$

We have

$$V_p = \frac{D_n^2}{n}, \quad D_n^2 = \sum_{i=1}^{n}\left(X_i - \overline{X_n}\right)^2. \tag{3.26}$$

Population's standard deviation is $s_p = \sqrt{V_p}$. This Pearson's approach considers the values $\mathcal{X}$ as being a population of equiprobable individuals and apply the formulae.

Since the elements of the sample are variates from X, all these quantities are random variables – thus, we may evaluate their means, and variances. For instance,

$$E\left(\overline{X_n}\right) = E(X), \quad V\left(\overline{X_n}\right) = \frac{1}{n}V(X), \quad \sigma\left(\overline{X_n}\right) = \frac{\sigma(X)}{\sqrt{n}}. \tag{3.27}$$

For Normal variables, $E(V_p) \neq V(X)$, so that it may be preferable to use the **sample's variance** and the associated **sample's standard deviation**:

$$V_n = \frac{D_n^2}{n-1}, \quad s_n = \sqrt{V_n}. \tag{3.28}$$

Observe that, from Tchebichev's inequality (Eq. 3.15):

$$P\left(\left|\overline{X_n} - E(X)\right| \geq \varepsilon\right) \leq \frac{V(X)}{n\varepsilon^2},$$

so that $\overline{X_n} \longrightarrow E(X)$ in probability – this result is the **weak law of large numbers**. The reader may find in the literature extensions of this basic result, such as the **strong law of large numbers**: for X regular enough, $\overline{X_n} \longrightarrow E(X)$ almost surely.

In addition, the values of these quantities change with the sample: a **confidence interval** must be associated to the values of the mean and the variance of the sample, to take into account the error margins. To define a confidence interval to the estimation $\widehat{\xi}$ of quantity ξ, we must choose a **confidence level** $1 - \alpha$ (α is the **risk**, id est, the probability of rejection of the real value) and determine an interval (ξ_{min}, ξ_{max})that $P\left(\widehat{\xi} \in (\xi_{min}, \xi_{max})\right) = 1 - \alpha$. Confidence intervals are often based on the **Central Limit Theorem**

Theorem (Central Limit):
Let $\{X_n : n \in \mathbb{N}\}$ be a sequence of independent random variables having the same distribution of finite mean m and variance σ^2. Let

$$Z_n = \frac{\overline{X_n} - m}{\frac{\sigma}{\sqrt{n}}}.$$

Then Z_n converges in distribution to a Standard Gaussian $N(0, 1)$. ∎

Indeed, we use $\widehat{m} = \overline{X_n}$ as estimation of the mean: by determining from $N(0, 1)$ a number z_α such that $P(|Z| \leq z_\alpha) = 1 - \alpha$, we evaluate an error margin $\Delta_\alpha = \frac{z_\alpha \sigma}{\sqrt{n}}$ and we generate a confidence interval $\left(\overline{X_n} - \Delta_\alpha, \overline{X_n} + \Delta_\alpha\right)$. When σ is unknown, we may use the **Cochran's Theorem**

Theorem (Cochran):
Let $(X_1, \ldots, X_n)$ be a vector of independent random variables having the same distribution $N(m, \sigma)$. Let

$$Q_n^2 = \frac{D_n^2}{\sigma^2}, \quad T_n = \frac{\overline{X_n} - m}{\frac{s_n}{\sqrt{n}}}.$$

Then Q_n^2 is chi-squared $\chi^2(n - 1)$ and T_n is Student-Fisher $SF(n - 1)$ ∎

Using this result, we may determine from $SF(n-1)$ a number t_α such that $P(|T_n| \leq t_\alpha) = 1 - \alpha$, and evaluate an error margin $\Delta_\alpha = \frac{t_\alpha s_n}{\sqrt{n}}$, what generates a confidence interval $\left(\overline{X_n} - \Delta_\alpha, \overline{X_n} + \Delta_\alpha\right)$. Cochran's theorem furnishes also a confidence interval for the variance: we determine from the distribution $\chi^2(n-1)$ two numbers A_1, A_2 such that $P\left(Q_n^2 \leq A_1\right) = P\left(Q_n^2 \geq A_2\right) = \frac{\alpha}{2}$. Then, $\left(\frac{D_n^2}{A_2}, \frac{D_n^2}{A_1}\right)$ is a confidence interval for $\widehat{\sigma^2}$.

The data may also be used to test hypothesis made about the mean, such as $H_0 : m = m_0$, $H_0 : m > m_0$, or $H_0 : m < m_0$. A test verifies if the data is incompatible with the hypothesis at a risk level α. If there is an incompatibility, the hypothesis must be rejected. Otherwise, the data cannot reject the hypothesis: the user may consider that the hypothesis remains tentatively valid. Tests are based on non-rejection intervals, analogous to confidence intervals: for instance, we may use the non-rejection regions given in Table 3.5.

Table 3.5 Non-rejection intervals for tests on the mean

H_0	Non-rejection interval	t_α
$m = m_0$	$\overline{X_n} \in (m_0 - \Delta_\alpha, m_0 + \Delta_\alpha)$	$P(\lvert T_n\rvert \leq t_\alpha) = 1 - \alpha$
$m < m_0$	$\overline{X_n} \in (-\infty, m_0 + \Delta_\alpha)$	$P(T_n \leq t_\alpha) = 1 - \alpha$
$m > m_0$	$\overline{X_n} \in (m_0 - \Delta_\alpha, +\infty)$	$P(T_n \leq t_\alpha) = \alpha$

Analogously, hypothesis on the variance may be tested, such as H_0: $\sigma = \sigma_0$, H_0: $\sigma > \sigma_0$, or H_0: $\sigma < \sigma_0$. Examples of non-rejection regions are given in Table 3.6.

Table 3.6 Non-rejection intervals for tests on the variance σ^2

H_0	Non-rejection interval	A_1, A_2
$\sigma^2 = \sigma_0^2$	$D_n^2 \in (A_1\sigma_0^2, A_2\sigma_0^2)$ $s_n^2 \in \left(\frac{A_1\sigma_0^2}{n-1}, \frac{A_2\sigma_0^2}{n-1}\right)$	$P\left(\chi_{n-1}^2 \leq A_1\right) = \frac{\alpha}{2}, P\left(\chi_{n-1}^2 \geq A_2\right) = \frac{\alpha}{2}$
$\sigma^2 < \sigma_0^2$	$D_n^2 \in (0, A_2\sigma_0^2)$ $s_n^2 \in \left(0, \frac{A_2\sigma_0^2}{n-1}\right)$	$P\left(\chi_{n-1}^2 \geq A_2\right) = \alpha$
$\sigma^2 > \sigma_0^2$	$D_n^2 \in (A_1\sigma_0^2, +\infty)$ $s_n^2 \in \left(\frac{A_1\sigma_0^2}{n-1}, +\infty\right)$	$P\left(\chi_{n-1}^2 \leq A_1\right) = \alpha$

EXCEL® has built-in functions to calculate these quantities: `AVERAGE` evaluates $\overline{X_n}$; `VAR.P` evaluates V_p, `VAR.S` evaluates V_n, `STDEV.P` evaluates s_p, `STDEV.S` evaluates s_n. `MEDIAN` calculates the median. `T.INV` and `T.INV.2T` gives the values of t_α. `CONFIDENCE.NORM` and `CONFIDENCE .T` furnish the values of Δ_α. `CHISQ.INV` furnishes A_1, `CHISQ.INV.RT` furnishes A_2. The `ANALYSIS TOOLPACK` also furnishes tools for the evaluation of statistics of samples.

Example

Open a blank workbook an enter "`=SEQUENCE(9;1;-1;0,25)`" in cell `A1`. This will create 9 numbers starting at `-1`, with a step `0,25` in cells `A1` to `A9`. Give the name `A` to cells `A1:A9`. Select cell `C1` and enter "`=AVERAGE(A)`". You will get the value `0`, which is the mean of the sample. Select `C2` and enter "`=VAR.P(A)`": the value is `0,41667`. Select `D2` and enter "`=STDEV.P(A)`": the value is `0,645497`. Select `C3` and enter "`=VAR.S(A)`": the value is `0,46875`. Select `D3` and enter "`=STDEV.S(A)`": the value is `0,684653`. Select `C4` and enter "`=MEDIAN(A)`": the value is `0` – the middle point of the sample. Select `C5` and enter "`=CONFIDENCE.T(0,05;D3;9)`": the value is `0,526271`, which corresponds to Δ_α evaluated using $SF(8)$ and $\alpha = 0.05$ (risk of 5%, confidence of 95%). Thus, the 95% confidence interval for the mean is (−0,53, 0,53). To determine t_α, select cell `C6` and enter "`=T.INV.2T(0,05;8)`": the value is `2,306004`. To recover the value of Δ_α, select cell `D6` and enter "`=C6*D3/SQRT(9)`": the value is `0,526271` again. Select cell `C7` and enter "`=CHISQ.INV(0,025;8)`": the value is `2,179731`, which corresponds to A_1. Select cell `C8` and enter "`=CHISQ.INV.RT(0,025;8)`": the value is `17,53455`, which corresponds to A_2. Select `D7` and enter "`=8*D3^2/C7`": the value is `1,720396`, which corresponds to $\frac{D_n^2}{A_1}$. Select `D8` and enter "`=8*D3^2/C8`": the value is `0,213864`, which corresponds to $\frac{D_n^2}{A_2}$. The confidence interval for the variance is (0.21, 1.72). The confidence interval for the standard deviation is obtaining by taking the square roots (Excel built-in function `SQRT`): we obtain the interval (0.46, 1.31). To test $H_0 : m < m_0 = -0.3$, select `F1` and enter "`=T.INV(0,95;8)`": the value is `1,859548` – it corresponds to t_α. Then, select `F2` and enter "`=F1*D3/SQRT(9)`": the value is `0,424382` and corresponds to Δ_α. Since $\overline{X_n} < m_0 + \Delta_\alpha = 0{,}12$, the hypothesis cannot be rejected. To test $H_0 : m > m_0 = 0.3$, select `G1` and enter "`=T.INV(0,05;8)`": the value is `-1,85955` – it corresponds to t_α. Then, select `G2` and enter "`=G1*D3/SQRT(9)`": the value is `-0,42438` and corresponds to Δ_α. Since $\overline{X_n} > m_0 + \Delta_\alpha = -0{,}12$, the hypothesis cannot be rejected. To test the hypothesis $H_0 : m = m_0 = 0.1$, select `H1` and enter "`=0,1-C5`": the value is `-0,26031`. Then, select `H2` and enter "`=0,1+C5`": the value is `0,460313`: the interval of non-rejection is (−0.26,0.46). Since $\overline{X_n} = 0$, the hypothesis cannot be rejected. To test $H_0 : \sigma^2 < \sigma_0^2 = 0.5$, select `F4` and enter "`=CHISQ.INV.RT(0,05;8)`": the value is `15,50731`, which corresponds to A_2. Then, select `F5` and enter "`=F4*0,5/8`": the value is `0,969207`: the interval of non-rejection for s_n^2 is (0, 0.97). Since $s_n^2 = 0.42$, the hypothesis cannot be rejected. To test $H_0 : \sigma^2 > \sigma_0^2 = 1.2$, select `G4` and enter "`=CHISQ.INV(0,05;8)`": the value is `2,732237`, which corresponds to A_1. Then, select `G5` and enter "`=G4*1,2/8`": the value is `0,409896`: the interval of non-rejection is (0.41, +∞). Again, the hypothesis cannot be rejected. To test $H_0 : \sigma^2 = \sigma_0^2 = 0.8$, select `H4` and enter "`=CHISQ.INV(0,025;8)`": the value is `2,179731`, which corresponds to A_1; select `I4` and enter "`=CHISQ.INV.RT(0,025;8)`": the value is `17,53455`, which corresponds to A_2. Then, select `H5` and enter "`=H4*0,8/8`": the value is `0,217793`; select `I5` and enter "`=I4*0,8/8`": the value is `1,753455`: the interval of non-rejection is (0.22, 1.75) and the hypothesis cannot be rejected. ∎

Remark

As shown in the preceding example, the data may be compatible with different hypothesis – eventually contradictory. Indeed, the information provided by the sample corresponds to the confidence intervals generated using the risk α chosen. A test cannot distinguish among values belonging to the confidence intervals generated. ∎

Example

Now, let us consider the use of the ANALYSIS TOOLPACK to make the analysis performed in the preceding example: go to *Data>Data Analysis*

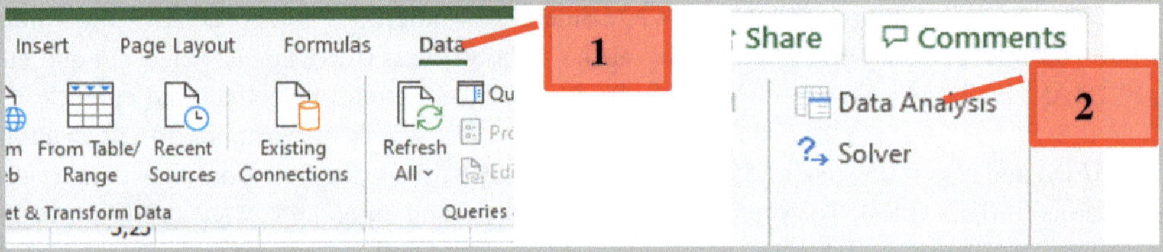

This will open the window *Data Analysis*. Select *Descriptive Statistics*

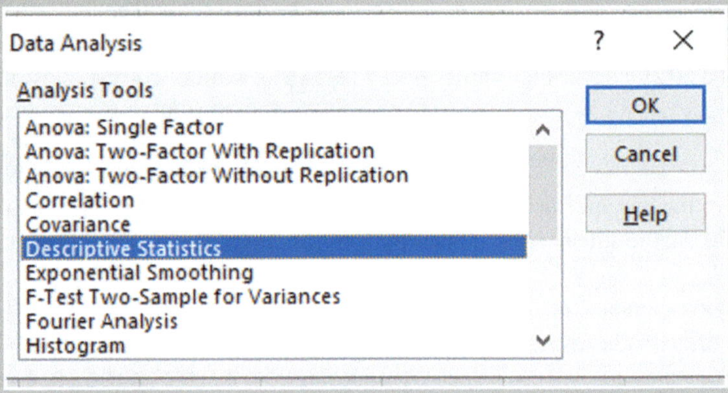

Enter A in the input range: EXCEL® will recognize the cells. Select a cell for the beginning of the output (here, E1). Check *Summary Statistics* and give a confidence level for the mean. Press *OK*

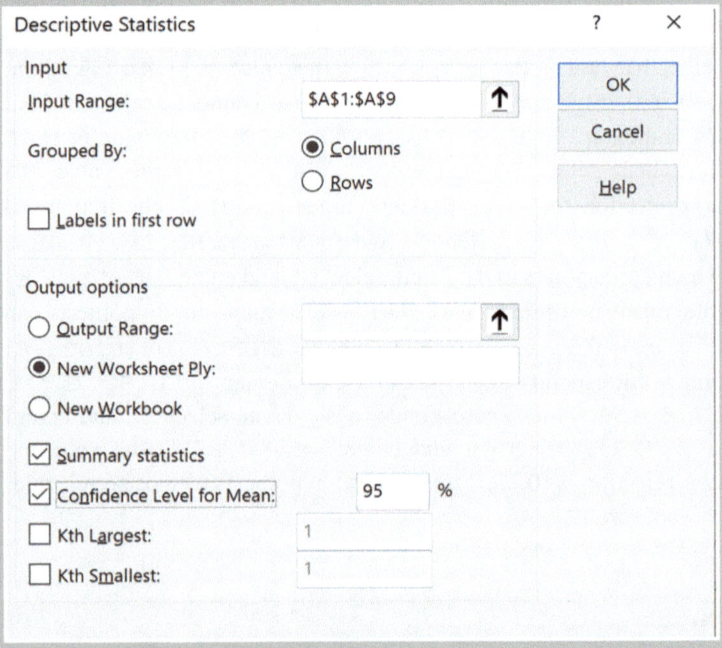

The result will be a complete set of statistics of the sample, including the value of $\Delta = 0{,}53$ for the confidence interval of the mean. ■

	A	B
1	*Column1*	
2		
3	Mean	0
4	Standard Error	0,22821773
5	Median	0
6	Mode	#N/A
7	Standard Deviation	0,6846532
8	Sample Variance	0,46875
9	Kurtosis	-1,2
10	Skewness	0
11	Range	2
12	Minimum	-1
13	Maximum	1
14	Sum	0
15	Count	9
16	Confidence Level(95,0%)	0,52627103

The `ANALYSIS TOOLPACK` does not include the tests of the preceding example. It includes tests for the comparisons of two samples. ■

Analogously, EXCEL® has tools for the comparison of statistics of two samples from variables $\boldsymbol{X}$ and $\boldsymbol{Y}$. For instance, we may test the hypothesis $\boldsymbol{H_0}$: $\boldsymbol{m_X = m_Y}$, $\boldsymbol{H_0}$: $\boldsymbol{m_X < m_Y}$, $\boldsymbol{H_0}$: $\boldsymbol{m_X > m_Y}$, $\boldsymbol{H_0}$: $\boldsymbol{\sigma_X^2 = \lambda\sigma_Y^2}$, $\boldsymbol{H_0} : \boldsymbol{\sigma_X^2} < \lambda\boldsymbol{\sigma_Y^2}$, $\boldsymbol{H_0} : \boldsymbol{\sigma_X^2} > \lambda\boldsymbol{\sigma_Y^2}$. Such a tests are based in the same principles as the preceding ones: a region of non-rejection is defined for a risk $\boldsymbol{\alpha}$ and the hypothesis is rejected if the observed values are not in the region. Otherwise, the hypothesis is considered as compatible with the data.

For two samples, the tests for the comparison of the means are reduced to those for a single variable $\boldsymbol{\bar{Z} = \overline{X_{n_X}} - \overline{Y_{n_Y}}}$, which is generally assumed to be normally distributed $N(\boldsymbol{m_Z}, \boldsymbol{\sigma_{\bar{Z}}})$. We have $\boldsymbol{m_Z = m_X - m_Y}$, but a supplementary assumption is needed for the normality of the variable $\boldsymbol{\overline{Z}}$ – for instance, one among the following:

- A large amount of data in each sample: sizes n_X, n_Y of the samples large enough to use gaussian approximations for each empirical mean. Let s_X, s_Y be the in this case, is approximately $\sigma_{\bar{Z}} \approx s_{\bar{Z}} = \sqrt{\frac{s_X^2}{n_X} + \frac{s_Y^2}{n_Y}}$;
- Coupled observations: the data is formed by couples (X_i, Y_i): in this situation, $n_X = n_Y$ and we have a sample $Z_i = X_i - Y_i$ from $Z = X - Y$, which has as mean $m_Z = m_X - m_Y$ and tests are performed on m_Z using $\sigma_{\bar{Z}} \approx s_Z/\sqrt{n_X}$, where s_Z is evaluated by the formula for the sample's standard deviation;

- The variance is the same for both the samples: $\sigma_X = \sigma_Y = \sigma$. In this case, $\sigma_{\overline{Z}} = \sigma\sqrt{\frac{1}{n_X} + \frac{1}{n_Y}}$ and $\frac{D^2}{\sigma^2} = \frac{(n_X-1)s_X^2+(n_Y-1)s_Y^2}{\sigma^2}$ is approximately $\chi^2(n_X + n_Y - 2)$. Then, the tests are performed using these variables.

The first approach is often used even for small datasets, but its use cannot be rigorously justified in such a situation. Under one of these assumptions,

$$m_X = m_Y + a \Leftrightarrow m_Z = a, m_X < m_Y + a \Leftrightarrow m_Z < a, m_X > m_Y + a \Leftrightarrow m_Z > a.$$

Thus, the comparison reduces to a test on the value of $\overline{Z}$. Examples of non-rejection intervals for the first assumption are given in Table 3.7, where Z is a standard Gaussian variable.

Table 3.7 Non-rejection intervals for comparisons of means, assuming normality of $\overline{Z} = \overline{X_{n_X}} - \overline{Y_{n_Y}}$

H_0	Non-rejection interval	z_α
$m_X = m_Y + a$	$\overline{Z} \in (a - z_\alpha\sigma_{\overline{Z}}, a + z_\alpha\sigma_{\overline{Z}})$	$P(Z \leq z_\alpha) = 1 - \frac{\alpha}{2}$
$m_X < m_Y + a$	$\overline{Z} \in (-\infty, a + z_\alpha\sigma_{\overline{Z}})$	$P(Z \leq z_\alpha) = 1 - \alpha$
$m_X > m_Y + a$	$\overline{Z} \in (a + z_\alpha\sigma_{\overline{Z}}, +\infty)$	$P(Z \leq z_\alpha) = \alpha$

Z is a standard gaussian variable $N(0, 1)$

For the comparison of variances, we observe that

$$F = \frac{\sigma_Y^2}{\sigma_X^2}\frac{s_X^2}{s_Y^2} = \frac{s_X^2/\sigma_X^2}{s_Y^2/\sigma_Y^2} \sim BFS(n_X - 1, n_Y - 1).$$

The tests are based on this parameter. Examples of non-rejection regions are given in Table 3.8.

Table 3.8 Non-rejection intervals for comparisons of variances

H_0	Non-rejection interval	$F_1,\ F_2$
$\sigma_X^2 = \lambda\sigma_Y^2$	$\frac{s_X^2}{s_Y^2} \in (\lambda F_1, \lambda F_2)$	$P(F \leq F_1) = \frac{\alpha}{2}, P(F \geq F_2) = \frac{\alpha}{2}$
$\sigma_X^2 < \lambda\sigma_Y^2$	$\frac{s_X^2}{s_Y^2} \in (0, \lambda F_2)$	$P(F \geq F_2) = \alpha$
$\sigma_X^2 > \lambda\sigma_Y^2$	$\frac{s_X^2}{s_Y^2} \in (\lambda F_1, +\infty)$	$P(F \leq F_1) = \alpha$

Example

Open a blank workbook an enter "=SEQUENCE(9;1;-1;0,25)" in cell A1. Then, select cell B1 and enter "=SEQUENCE(25;1;-1,2;0,1)". This will create two sequences of numbers: 9 numbers starting at -1, with a step 0,25 in cells A1 to A9 and 25 numbers starting at -1,2, with a step 0,1 in cells B1 to B25. Give the name X to cells A1:A9 and name Y to cells B1:B25. Select cell D1 and enter "=AVERAGE(X)". Then, select cell D2 and enter "=AVERAGE(Y)". You will get the value 0 in both the cells – these are the empirical means $\overline{X_{n_X}}$ and $\overline{Y_{n_Y}}$, with $n_X = 9$, $n_Y = 25$. Select E1 and enter "=STDEV.S(X)": the value is 0,684653. Select E2 and enter "=STDEV.S(Y)": the value is 0,73598. Select D3 and enter "=D1-D2", which is the value of $\overline{X_{n_X}} - \overline{Y_{n_Y}}$. Select E3 and enter "=SQRT(E1^2/9 + E2^2/25)", which is the value of $\sqrt{\frac{s_X^2}{n_X} + \frac{s_Y^2}{n_Y}}$: 0,27157.

To test $H_0 : m_X > m_Y + 0.4$ under the first assumption, with a risk $\alpha = 0{,}05$ (5%), select `G1` and enter "`=NORM.S.INV(0,05)`": the value is `-1,64485` – it corresponds to z_α. Then, select `G2` and enter "`=0,4+G1*E3`": the value is `-0,04669`. Since $\overline{Z} = 0 > -0.04$, the hypothesis cannot be rejected. To test the hypothesis $H_0 : m_X < m_Y - 0{,}4$, with the same risk, select `H1` and enter "`=NORM.S.INV(0,95)`": the value is `1,64485`. Then, select `H2` and enter "`=-0,4+H1*E3`": the value is `0,04669`: since $\overline{Z} = 0 < 0.04$, the hypothesis cannot be rejected. To test $H_0 : m_X = m_Y$, with a risk 5%, select `I1` and enter "`=NORM.S.INV(0,975)`": the value is `1,959963985`. Then, select `I2` and enter "`=I1*E3`": the value is `0,532266463`: the interval of non-rejection is $(-0.53, 0.53)$ since $\bar{Z} = 0$, the hypothesis cannot be rejected.

To test $H_0 : \sigma_X^2 < 0.5\,\sigma_Y^2$, enter "`=F.INV.RT(0,05;8;24)`" in `G4`: the value is `2,355081`, which corresponds to F_2. Then, select `G5` and enter "`=0,5*G4`": the value is `1,177541`: the interval of non-rejection for s_X^2/s_Y^2 is (0, 1.18). Since $s_X^2/s_Y^2 = 0.87$, the hypothesis cannot be rejected. To test $H_0 : \sigma_X^2 > 1.5\,\sigma_Y^2$, select `H4` and enter "`=F.INV(0,05;8;24)`": the value is `0,321002576`, which corresponds to F_1. Then, select `H5` and enter "`=H4*1,5`": the value is `0,481503864`: the interval of non-rejection is $(0.48, +\infty)$. Again, the hypothesis cannot be rejected. To test $H_0 : \sigma_X^2 = \sigma_Y^2$, select `I4` and enter "`=F.INV(0,025;8;24)`": the value is `0,253342837`, which corresponds to F_1; select `I5` and enter "`=F.INV.RT(0,025;8;24)`": the value is `2,779134581`, which corresponds to F_2: the interval of non-rejection is (0.25, 2.78). Again, the hypothesis cannot be rejected. ∎

Exercises

1. In an poll, a sample of 1000 voters gives 52% to a candidate. Give a confidence interval for his or her score, with a 5% risk. What is the interval for a 1% risk? What is the size of the sample to get a confidence interval of length $\pm 1\%$?

   TIP: The score of the candidate in the sample is $\overline{X_n}$, where X is a Bernoulli variable taking the value 1 if the voter votes for the candidate; 0 if not. The mean p of X is the score of the candidate. Recall that the variance of X is $p(1-p)$ and may be estimated by s_n on the sample.

2. At the end of each day, an agency compiles statistic on the average waiting time (AWT) of customers. Over 30 days, we have the following results (in minutes):

AWT	0–5	5–10	10–15	15–20	20–25	25–30
Number of days	2	7	11	6	3	1

   Use these data to estimate the mean and the variance of the mean of the AWT. Give confidence intervals for these quantities with $\alpha = 5\%$.
   TIP: Each AWT is an empirical mean, so that it is approximately gaussian. Use the center of each class to evaluate the mean and the variance.

3. A quality department must check whether the impurity content of a product complies with the legislation, which requires a mean of 5 and a maximum variance of 2, assuming that the distribution of X is Normal,

(a) A sample of 20 elements is analyzed and leads to an empirical mean $\overline{X_{20}} = 3$ and an empirical standard deviation $s_{20} = 1.1$. Determine confidence intervals with risk 5% for the men and the variance.
(b) A new sample of 400 elements is analyzed and leads to an empirical mean $\overline{X_{400}} = 3.2$ and an empirical standard deviation $s_{400} = 1.3$. Determine confidence intervals with risk 5% for the mean and the variance.
(c) Can the quality department confirm that the variance is inferior to 2 with a risk of 5%?
(d) Can it confirm that the mean is inferior to 5, with a risk of 5%?

4. Two methods of measurement of a quantity furnish the results below:

Method 1	44,45,45,48,48,46,46,46,46,47,47,47,47,47,47,47,49,49,49,49,50,50,51,51,52
Method 2	45,45,48,46,46,46,46,46,46,47,47,47, 47,47,47, 47,47,47,49,49,49,50,50,50

(a) Determine the empirical mean and the standard variance of each method. Give confidence intervals for these quantities, with a risk of 5%.
(b) May we consider the means of the samples as equal, with a risk of 5%?
(c) May we consider that the variances of both the methods are equal, with a risk of 5%?

5. The results of the students of two groups are given below:

Group 1	11,8,4,12,7,8,6,6,3,5
Group 2	8,12,9,14,10,11,3,5,9,9

(a) Determine the empirical mean and the standard variance of each method. Give confidence intervals for these quantities, with a risk of 5%.
(b) May we consider the group 2 as better, with a risk of 5%?
(c) May we consider that the variances of both the methods are equal, with a risk of 5%?

6. The results of the students of two groups are given below:

Group 1	13,14,19,7,9,13,11,1,12,7,17,17,11,6,16,7,11,20,13
Group 2	7,17,4,12,15,9,7,12,8,14,14,9,9,9,10,8,11,17,14

(a) Determine the empirical mean and the standard variance of each method. Give confidence intervals for these quantities, with a risk of 5%.
(b) May we consider the group 2 as better, with a risk of 5%?
(c) May we consider that the variances of both the methods are equal, with a risk of 5%?

3.8.1 Maximum Likelihood Estimators

Samples are often used to estimate parameters from distributions. For instance, to make a prediction about the results of a poll, we need to estimate the parameter $\boldsymbol{p} = (p_1, \ldots, p_k)$ of a multinomial distribution. Analogously, to determine if the proportion of non-compliant elements in a product satisfies a given condition, we must estimate the parameter p of a Bernoulli distribution.

In general, we have a model distribution $f(x, \boldsymbol{\theta})$, where $\boldsymbol{\theta} = (\theta_1, \ldots, \theta_r)$ is a vector of unknown parameters – to be determined from the sample. For a discrete distribution, f is a probability: $f(x, \boldsymbol{\theta}) = P(X = x)$. For a continuous distribution, f is the density of probability.

The objective is to find an estimator $\widehat{\boldsymbol{\theta}}$ of $\boldsymbol{\theta}$. A classical method to estimate the values of $\boldsymbol{\theta}$ is the Maximum of the Likelihood: let $\mathcal{X} = \{X_1, \ldots, X_n\}$ be a sample from X. The Likelihood is defined as

$$L(\mathcal{X},\boldsymbol{\theta}) = \prod_{i=1}^{n} f(X_i,\boldsymbol{\theta}). \tag{3.29}$$

It is often convenient to consider the log-Likelihood, given by

$$\log L(\mathcal{X},\boldsymbol{\theta}) = \sum_{i=1}^{n} \log\left(f(X_i,\boldsymbol{\theta})\right). \tag{3.30}$$

The Maximum Likelihood Estimator (MLE) of $\boldsymbol{\theta}$ is the element which maximizes L or, equivalently, $\log L$. ∎

Example
Let us consider the estimation of the parameter p of a Bernoulli distribution. In this case, the sample $\mathcal{X}$ verifies

$$L(\mathcal{X},p) = \prod_{i=1}^{n} p^{X_i}(1-p)^{1-X_i}$$

and

$$\log L(\mathcal{X},p) = \log(p)\sum_{i=1}^{n} X_i + \left(n - \sum_{i=1}^{n} X_i\right)\log(1-p).$$

The maximum of likelihood corresponds to

$$\widehat{p} = \frac{1}{n}\sum_{i=1}^{n} X_i,$$

id est, the empirical mean of the sample. ∎

Exercises

1. Let X be a discrete variable exponentially distributed: $P(X = k) = \lambda^k e^{-\lambda}/k!$. Let $\mathcal{X} = \{X_1, \ldots, X_n\}$ be a sample from X.

   (a) Determine the likelihood function.
   (b) Find the log-likelihood.
   (c) Determine the MLE for λ.

   TIP: Solve the equation $\partial \log L/\partial \widehat{\lambda} = 0$.

2. Let X be a discrete variable, binomially distributed $\mathcal{B}(n,p)$, where n is given. Let $\mathcal{X} = \{X_1, \ldots, X_n\}$ be a sample from X.

   (a) Determine the likelihood function.
   (b) Find the log-likelihood.
   (c) Determine the MLE for p.
   TIP: Solve the equation $\partial \log L/\partial \widehat{p} = 0$.

3. Let X be a discrete variable, geometrically distributed: $P(X = k) = p(1 - p)^k$. Let $\mathcal{X} = \{X_1, \ldots, X_n\}$ be a sample from X.

   (a) Determine the likelihood function.
   (b) Find the log-likelihood.
   (c) Determine the MLE for p.
   TIP: Solve the equation $\partial \log L/\partial \widehat{p} = 0$.

4. Let X be a continuous variable exponentially distributed: its density is $f(x) = \lambda e^{-\lambda x}$ $(x > 0)$. Let $\mathcal{X} = \{X_1, \ldots, X_n\}$ be a sample from X.

   (a) Determine the likelihood function.
   (b) Find the log-likelihood.
   (c) Determine the MLE for λ.
   TIP: Solve the equation $\partial \log L/\partial \widehat{\lambda} = 0$.

5. Let X be a continuous variable normally distributed: its density is $f(x) = \frac{1}{\sigma\sqrt{2\pi}} e^{-\frac{1}{2}\left(\frac{x-m}{\sigma}\right)^2}$. Let $\mathcal{X} = \{X_1, \ldots, X_n\}$ be a sample from X.

   (a) Determine the likelihood function.
   (b) Find the log-likelihood.
   (c) Determine the MLE for $\boldsymbol{\theta} = (m, \sigma)$.
   TIP: Solve the equations $\partial \log L/\partial \widehat{m} = \partial \log L/\partial \widehat{\sigma} = 0$

3.8.2 Samples from Random Vectors

When $\boldsymbol{X} = (X_1, \ldots, X_k)$ is a random vector of dimension k, a sample is a set of vectors. Then, the empirical mean and the variance may be determined for each component X_i, $1 \leq i \leq k$. Analogously, tests on the mean and on the variance may be performed for each component. However, other analyses can be performed, such as the calculation of covariances, correlations, affine approximations, and conditional expectations. For instance, we evaluate the Pearson's covariance as

$$cov_P(X_{k_n}, X_{\ell_n}) = \frac{1}{n}\sum_{i=1}^{n}\left(X_{ki} - \overline{X_{k_n}}\right)\left(X_{\ell i} - \overline{X_{\ell_n}}\right) = \overline{X_{kn}X_{\ell n}} - \overline{X_{kn}} \times \overline{X_{\ell n}}.$$

where

$$\overline{X_{kn}X_{\ell n}} = \frac{1}{n}\sum_{i=1}^{n} X_{ki}X_{\ell i}.$$

The standard covariance of the sample is

$$cov_n(X_{k_n}, X_{\ell_n}) = \frac{1}{n-1}\sum_{i=1}^{n}\left(X_{ki} - \overline{X_{k_n}}\right)\left(X_{\ell i} - \overline{X_{\ell_n}}\right).$$

The empirical correlation is

$$r(X_{k_n}, X_{\ell_n}) = \frac{cov_P(X_{k_n}, X_{\ell_n})}{\sqrt{V_P(X_{k_n})V_P(X_{\ell_n})}} = \frac{cov_n(X_{k_n}, X_{\ell_n})}{\sqrt{V_n(X_{k_n})V_n(X_{\ell_n})}}.$$

r is independent of the choice between Pearson's and standard's definitions.

EXCEL® proposes built-in functions for the analysis of samples of vectors. `TREND` (Y, X) determines the linear approximation $Y \approx aX + b$. `FORECAST.LINEAR`$(z;\ Y;\ X)$ evaluates the linear approximation $az + b$ at z. `SLOPE` (Y, X) gives a; `INTERCEPT` (Y, X) gives b. As an alternative, we may generate a scatter plot of the pairs (X, Y) and use the tools to add a trendline and its equation. Finally, the ANALYSIS TOOLPACK may be used. `COVARIANCE.P` (Y, X) and `COVARIANCE.S` (Y, X) furnish their Pearson's and standard covariance, respectively. `CORREL` (Y, X) furnishes their correlation

Example

Open a blank Workbook, enter the values below:

	A	B	C	D	E
1	1	2	3	4	5
2	4,5	6,8	9,6	10,8	12,5
3	4,84	6,84	8,84	10,84	12,84

Give the name "X" to the cells `A1:E1` and "Y" to the cells `A2:E2`. Select cell `A3` and enter "`=TREND(Y;X)`". The results are the following

3	4,84	6,84	8,84	10,84	12,84

Row 3 contains the approximations $Y \approx aX + b$. Select cell `A5` and enter "`=SLOPE(Y;X)`". Select Cell `B5` and enter "`=INTERCEPT(Y;X)`". The value in cell `A5` is $a = 2$; the value in cell `B5` is $b = 2.84$.

Select cell `A1:E2`. Go to *Insert* and select *Scatter*

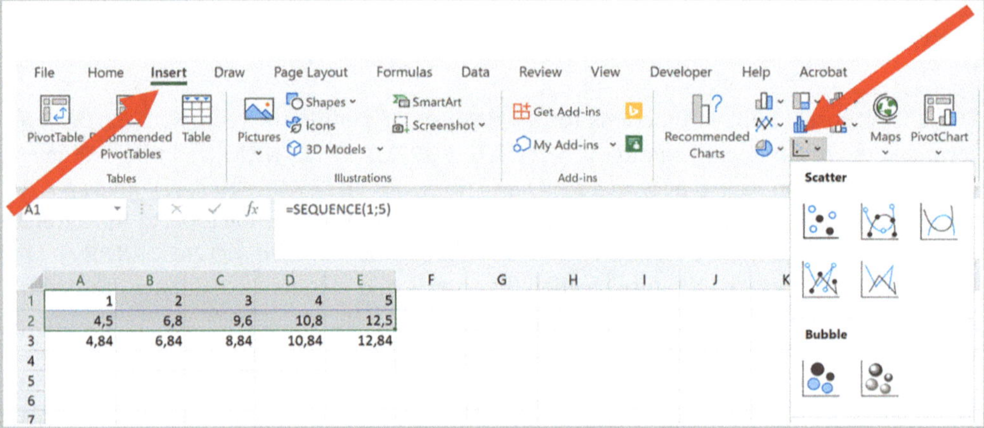

Select the first graph

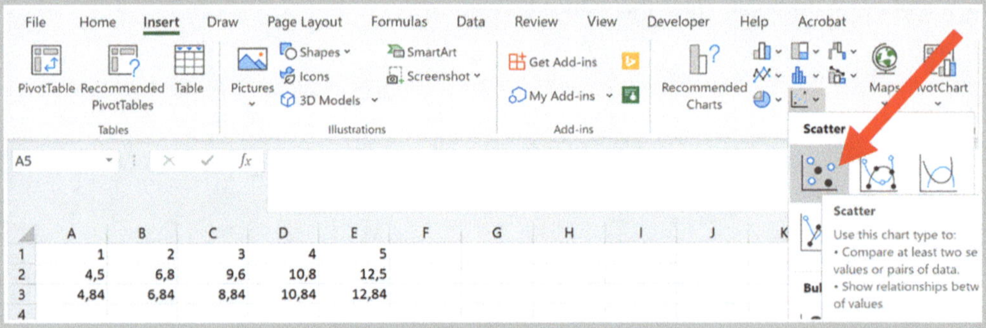

Then, click on the icon "+" at the upper right side and select "*Trendline>More Options*"

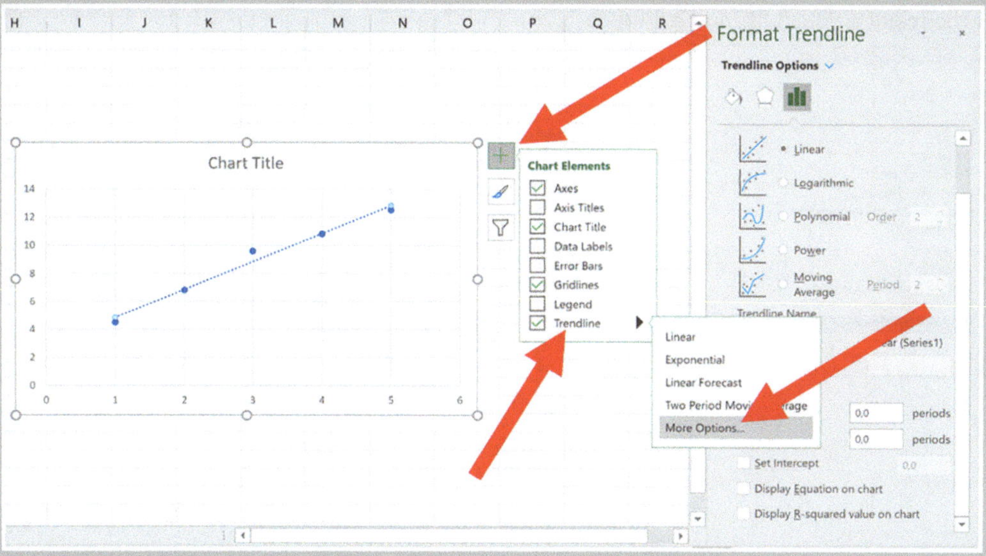

Click on "Display Equation on chart". The equation "$Y = aX + b$" appears on the chart. Here, $a = 2$, $b = 2.84$.

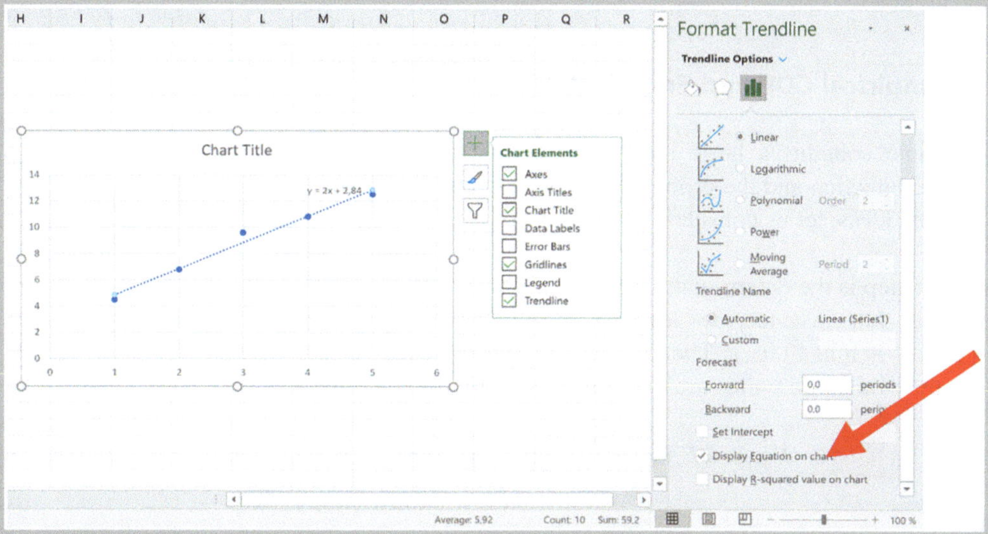

To use the ANALYSIS TOOLPACK, the variables must be in columns: select cell A7 and enter "=TRANSPOSE(X)"; select cell B7 and enter = "TRANSPOSE(Y)". Then, go to *Data>Data Analysis*:

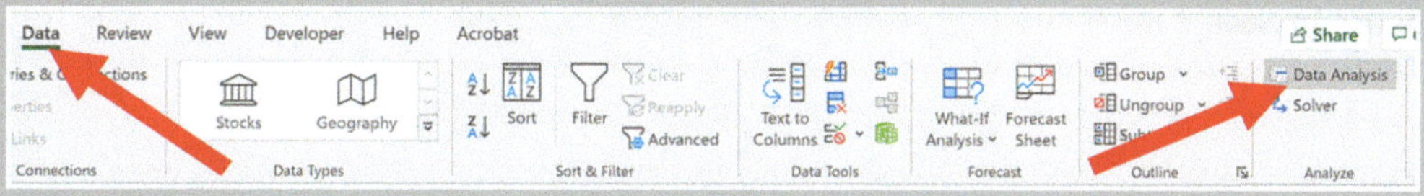

Select "Regression" and furnish the requested information. Recall that data concerning X and Y was copied to `A7:A11` and `B7:B11`, respectively. Choose "New Worksheet Ply" for the results. Select "Line Fit Plots".

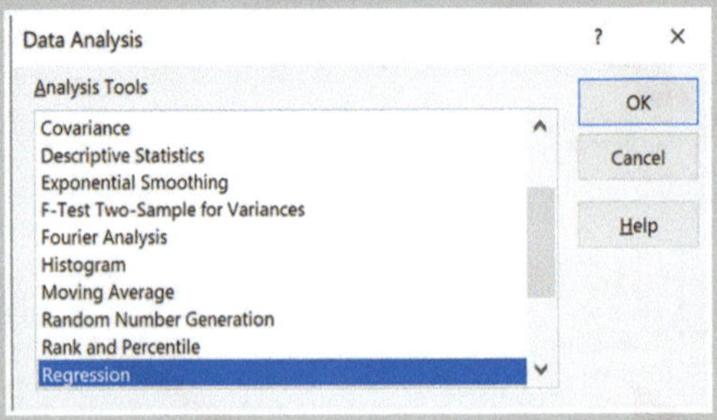

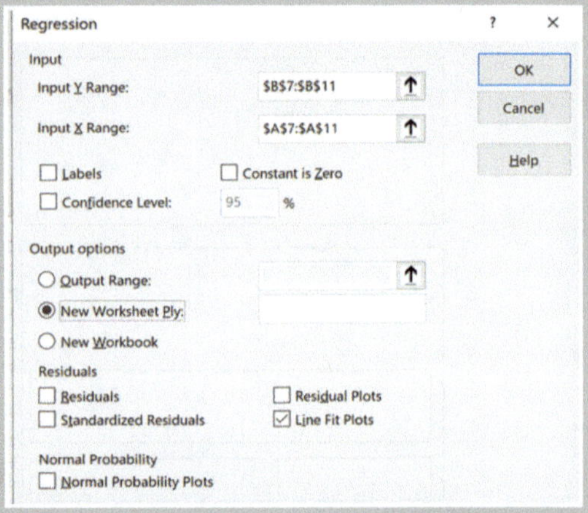

The results appear in a new sheet and include the empirical correlation coefficient, the regression coefficients and their confidence intervals, the plot of the data and of the approximated values, the errors at each point and cutting values for test comparing the means and variances. ∎

3.8.3 Empirical CDF and Empirical PDF

If the sample contains a large number of elements, we may determine approximations of the Cumulative Function and the probability density of the variable.

To fix the ideas, let us consider a sample $\mathcal{X} = \{X_1, \ldots, X_n\}$ of a random variable X, with n large enough.

The first step is the determination of the CDF: we start by choosing a set of ne evaluation points $x_1, \ldots, x_{ne}$. For instance, we can fix a number of subintervals n_{sub} and determine a step $\delta x = (\max \mathcal{X} - \min \mathcal{X})/n_{sub}$. Then we set $x_i = \min \mathcal{X} + (i-1)\delta x$.

Once the evaluation points are determined, we approximate the CDF as

$$F(x_i) \approx F_e(x_i) = \frac{card(\mathcal{X}_i)}{n}, \mathcal{X}_i = \{X_j \in \mathcal{X} : X_j < x_i\},$$

id est,

$$F_e(x_i) = \frac{number\ of\ elements\ from\ X\ that\ are\ inferior\ to\ x_i}{n}.$$

Then, the PDF is approximated as the numerical derivative of F_e, generated by one of the methods introduced in Sect. 2.8.

Example

Let us exemplify the procedure: let us consider $X \sim \chi^2(9)$. We generate a sample of 1000 variates from X:

Open a blank workbook enter the value 9 in `B2` and the instruction =`CHISQ.INV(RAND();$B$2)` in `A2`.

A	B	C
X	nddl	min
=CHISQ.INV(rand();$B$2)		

Copy `A2` and paste until `A1002`

996	4,175467
997	6,831262
998	9,154369
999	13,52316
1000	9,763955
1001	6,824807
1002	3,631053

Enter the formula =`MIN(A2:A1002)` in `D1`; the formula =`MAX(A2:A2002)` in `D2`; the number of subintervals in `D3`; the formula =`(D2-D1)/D3` in `D4`; the formula =`D1` in `F1`; the formula =`F1+$D$4` in `F2`. Copy `F2` and paste in `F3:F(nsub+1)` – here, until `F21`.

C	D	E	F
min	1,281267	Xe	1,281267
max	25,27261		2,480834
nsub	20		3,680401
h	1,199567		4,879968
			6,079535
			7,279102
			8,478669
			9,678236

Enter the formula = `COUNTIF(A2:A1002;"<" & F1)/1001` in `H1`. Copy `H1` and paste it in `H2:H(nsub+1)` – here, until `H21`. The values in `H1: H(nsub+1)` are the values of *Fe*.

H
0
0,014985
0,061938
0,146853
0,261738
0,397602
0,513487
0,615385

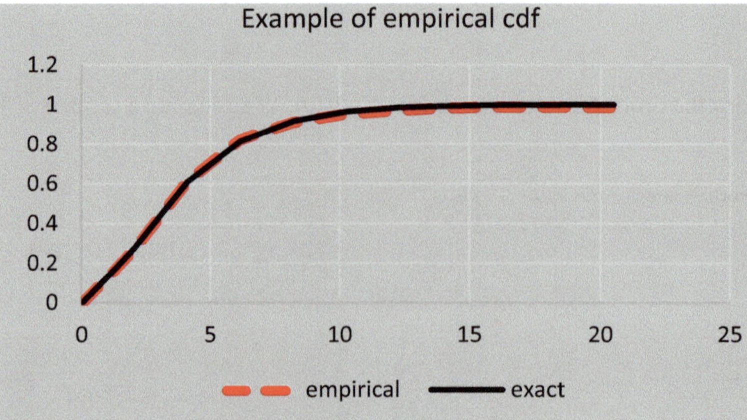

Fig. 3.1 Comparison between the empirical and the exact cdf

To determine the PDF, we must calculate the derivatives of the empirical CDF *Fe* (Fig. 3.1). For instance, we may use a variational derivative (see Sect. 2.8) (Fig. 3.2)

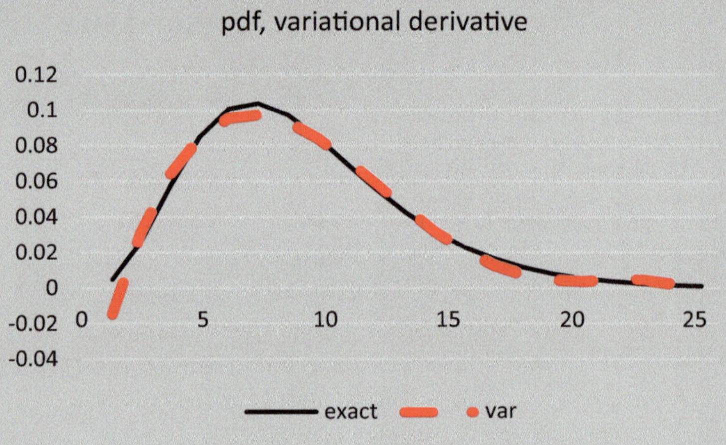

Fig. 3.2 Comparison between the empirical and the exact pdf ∎

We may define a VBA module for the generation of the CDF:

```
Function empirical_repartition(x, n)
Dim resu() As Variant
ReDim resu(1 To 2)
a = WorksheetFunction.Min(x)
b = WorksheetFunction.Max(x)
h = (b - a) / n
Dim xabs() As Double
ReDim xabs(1 To n + 1)
Dim Fx() As Double
ReDim Fx(1 To n + 1)
m = UBound(x, 1)
For i = 1 To n + 1
  xabs(i) = a + (i - 1) * h
  s = 0
  For j = 1 To m
    If x(j) < xabs(i) Then
      s = s + 1
    End If
  Next j
  Fx(i) = s / m
Next i
resu(1) = xabs
resu(2) = Fx
empirical_repartition = resu
End Function
```

We may call this function in a sub. For instance:

```
Sub test2()
 nddl = Cells(2, 2).value
 x = readcol(1, 2)
 a = WorksheetFunction.Min(x)
 b = WorksheetFunction.Max(x)
 nsub = Cells(3, 4).value
 resu = empirical_repartition(x, nsub)
 t = resu(1)
 y = resu(2)
End Sub
```

Function `readcol(jcol, il)` reads the contents of column `jcol`, starting at line `il`. As an option, you may indicate the last column `ns` to be read. An analogous function `readline` was previously introduced in Sect. 1.14.

```
Function readcol(col, il, Optional ns)
Dim x() As Double
If IsMissing(ns) Then
   ns = Cells(Rows.Count, col).End(xlUp).Row
End If
ReDim x(1 To ns - il + 1)
For i = il To ns
   x(i - il + 1) = Cells(i, col).value
Next i
readcol = x
End Function
```

Example

Let us exemplify the use of VBA: let us consider again $X \sim \chi^2(9)$. We generate a sample of 1000 variates from X by the same way as in the preceding example. Then, we run the Sub test2 and we add to the workbook a chart containing two series of data – the exact values of the CDF and the values generated by the function `empirical_repartition`. The results are shown in Fig. 3.3.

The PDF is generated by a numerical derivative of the empirical CDF *Fe*. Here, we use a variational derivative (see Sect. 2.8). The results are shown in Fig. 3.4.

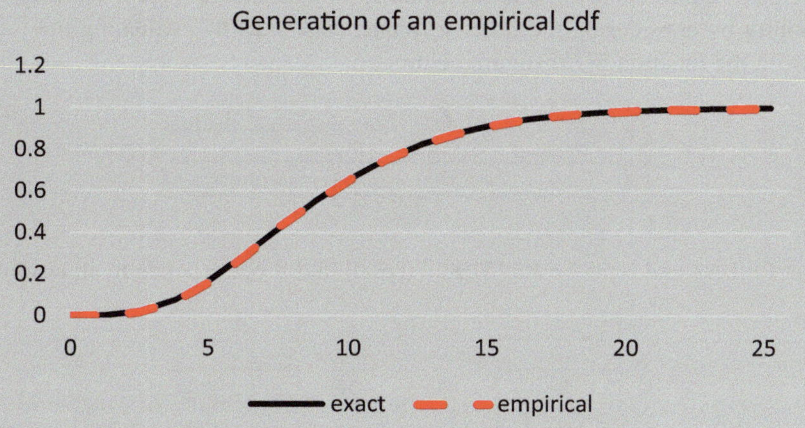

Fig. 3.3 Comparison between the empirical and the exact cdf

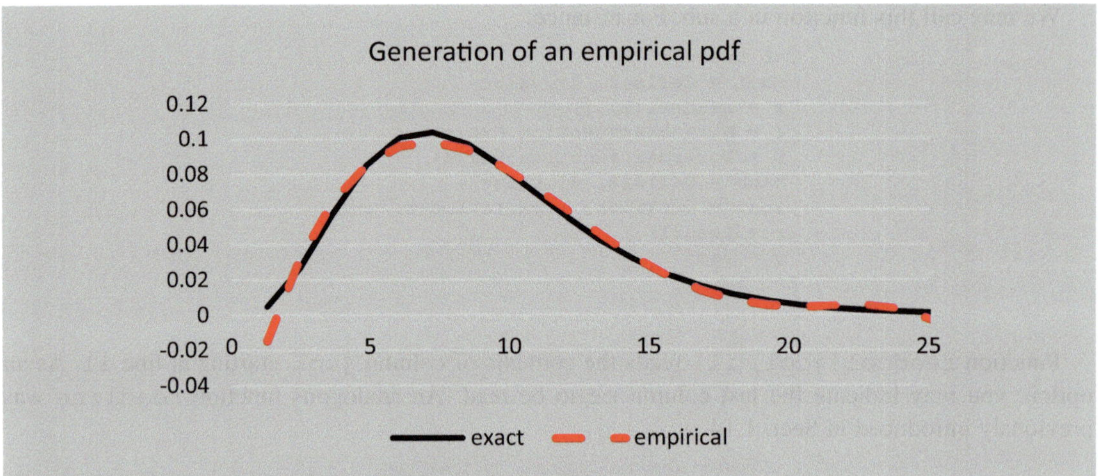

Fig. 3.4 Comparison between the empirical and the exact pdf ∎

3.9 Frequentist Probabilities with EXCEL®

In practice, probabilities are associated with relative frequencies. A table analogous to Table 3.9 below is transformed in probabilities by evaluating the relative frequencies of each possible element $p_i = \frac{o_i}{O}$, $O = \sum_i o_i$. O is the total number of observations.

Table 3.9 Example of a table of observed values: $p_i = \frac{o_i}{O}$, $O = \sum o_i$

Element	ω_1	ω_2	...	ω_n
Number of observations	o_1	o_2	...	o_n
Probability	p_1	p_2	...	p_n

For instance, let us consider the individual choices of a sample of city's citizen for their evening out: 54 people go to a music show, 98 to the cinema, 34 to a theatre's play. We may transform this data in probability by considering the relative frequencies: start by evaluating the total number of observations with the function SUM (figure below)

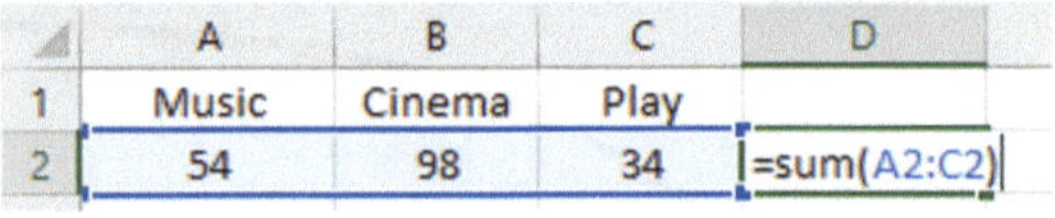

Then, select Cell A3 and type "= A2/$D$2" (recall that $ corresponds to an absolute address)

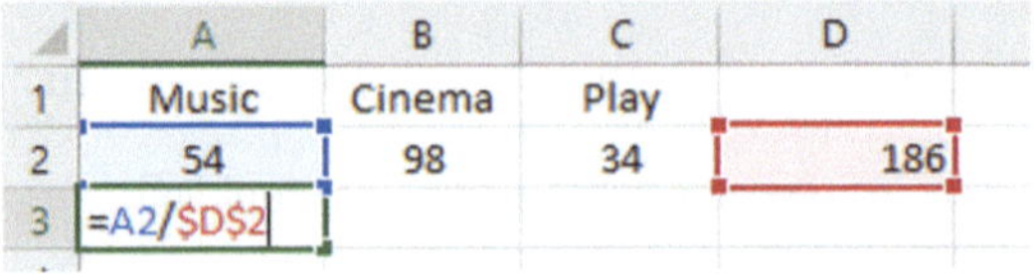

Copy the formula and paste it in cells B3 and C3: you will obtain the probabilities in line 3:

	A	B	C	D
1	Music	Cinema	Play	
2	54	98	34	186
3	0,29032258	0,526882	0,182796	

As an alternative, you may assign a name to the group of cells `A2 : C2` – for instance, let us use the name `X`. You will obtain the same result by selecting the cell `A3` and entering "= `X/SUM(X)`".

3.9.1 Testing Adequacy to a Distribution

Data in tables of relative frequencies, analogous to Table 3.9 may be compared to theoretical distributions, to test if prior assumptions on the distribution are coherent with the observations. One of the popular tests is the χ^2 – *test*, based on the following theorem:

Theorem (Fisher-Pearson):
Let $\boldsymbol{Y}$ be a variable having a multinomial distribution $\mathcal{M}(N,\boldsymbol{p})$. Let $\boldsymbol{f}: \mathbb{R}^r \to \mathbb{R}^n$ and $\boldsymbol{\theta} \in \mathbb{R}^r$ be such that $\boldsymbol{p} = \boldsymbol{f}(\boldsymbol{\theta})$. Let $\widehat{\boldsymbol{\theta}}$ be the ***maximum likelihood estimation*** of $\boldsymbol{\theta}$ and $\widehat{\boldsymbol{p}} = \boldsymbol{f}\left(\widehat{\boldsymbol{\theta}}\right)$. Let us define

$$Z_i = \frac{Y_i - N\widehat{p}_i}{\sqrt{N\widehat{p}_i}}.$$

Then $\|\mathbf{Z}\|^2$ is approximatively $\chi^2(n - r - 1)$ for large values of n:

$$\|\mathbf{Z}\|^2 \longrightarrow \chi^2(\mathrm{k} - \mathrm{r} - 1) \quad \text{for } \mathrm{n} \longrightarrow +\infty.$$ ∎

Assume that the data in Table 3.9 corresponds to a variable X. According to this theorem, we may test the hypothesis H_0 : *X has a density f*$(x, \boldsymbol{\theta})$ as follows: use the data in Table 3.9 to estimate the r parameters $\boldsymbol{\theta}$ and estimate the value $\widehat{p}_i = P(o_i)$. Then, evaluate

$$Q^2 = \sum_{i=1}^{n} \frac{(o_i - T_i)^2}{T_i} \quad (T_i = O\widehat{p}_i). \tag{3.31}$$

T_i is the expected value if the hypothesis is true. According to the theorem, Q^2 is approximatively $\chi^2(n - r - 1)$: we may determine a critical value χ^2_α such that $P(Q^2 \geq \chi^2_\alpha) = \alpha$. The non-rejection region is $(0, \chi^2_\alpha)$.

EXCEL® offers a built-in function `CHISQ.TEST`$(o_1, \ldots, o_n; T_1, \ldots, T_n)$ to evaluate $a = P(\chi^2 \geq Q^2)$ from $\chi^2(n - 1)$: for $r = 0$, H_0 is rejected if $a < \alpha$. For $r > 0$, `CHISQ.INV.RT`$(\alpha; n - r - 1)$ furnishes χ^2_α.

Example

A coin is tossed 100 times. We observe 60 Heads and 40 tails.

To test the hypothesis that the coin is fair, we consider $T_H = T_T = 50$, which are the expected values for a fair coin. Then:

$$Q^2 = \frac{(40-50)^2}{50} + \frac{(60-50)^2}{50} = 4$$

Since no parameter was estimated, $r = 0$. Since $n = 2$, we look for χ^2_α at the distribution $\chi^2(1)$: under EXCEL®, "=CHISQ.INV.RT(0,05;1)" furnishes the value 3.84. Thus, the hypothesis is rejected with a risk of 5%. Analogously, `CHISQ.TEST` furnishes 0.0455 < 0.05 and the hypothesis is rejected with a risk 5%. ∎

Example

Three products are proposed to 100 consumers. 32 choose product 1, 40 choose product 2, 28 choose product 3. To test the hypothesis that the distribution is uniform, we evaluate

$$Q^2 = \frac{\left(32-\frac{100}{3}\right)^2}{\frac{100}{3}} + \frac{\left(40-\frac{100}{3}\right)^2}{\frac{100}{3}} + \frac{\left(60-\frac{100}{3}\right)^2}{\frac{100}{3}} = 2.24$$

Since no parameter was estimated, $r = 0$. Since $n = 3$, we look for χ^2_α at the distribution $\chi^2(2)$: under EXCEL®, "=CHISQ.INV.RT(0,05;2)" furnishes the value 5.99. Thus, the hypothesis is not rejected with a risk of 5%. Analogously, `CHISQ.TEST` furnishes 0.33 > 0.05 and the hypothesis is not rejected with a risk 5%. ∎

Example

The number of non-compliant parts in 100 samples of 5 elements is the following.

Number	0	1	2	3	4	5
Occurrences	81	10	6	2	1	0

To test the hypothesis that the distribution of the non-compliant parts is binomial $\mathcal{B}(5, p)$ we must start by estimating p. The MLE estimator for p is

$$\widehat{p} = \frac{1}{O}\sum_{i=0}^{4} o_i X_i = \frac{81 \times 0 + 10 \times 1 + 6 \times 2 + 3 \times 2 + 4 \times 1}{5 \times 100}$$

Thus, $\widehat{p} = 0.064$. Then, $Q^2 \approx 151$. One parameter was estimated: $r = 1$. We look for χ^2_α at the distribution $\chi^2(4)$: under EXCEL®, "=CHISQ.INV.RT(0,05;4)" furnishes the value 9.5. Thus, the hypothesis is rejected with a risk of 5%. ∎

Exercises

1. A dice rolls 100 times. The results are the following:

Face	1	2	3	4	5	6
Observations	17	20	25	14	21	13

Can we assume that the dice is fair?

2. A sample of a variable furnishes the results below:

Value	0	1	2	3	4
Observations	25	10	30	30	5

Can we assume that X is uniformly distributed?

3. A sample of a variable furnishes the results below:

Value	0	1	2	3	4	5	6
Observations	11	9	7	9	5	10	8

Can we assume that X is uniformly distributed?

4. A sample of a variable furnishes the results below:

Value	0	1	2	3
Observations	40	42	13	5

Can we assume that X is Poisson distributed?
TIP: Estimate the parameter λ, then evaluate T_i. Here, $r = 1$.

5. A sample of a variable furnishes the results below:

Value	0	1	2	3	4	5	6
Observations	5	15	42	35	30	10	8

Can we assume that X is Poisson distributed?
TIP: Estimate the parameter λ, then evaluate T_i. Here, $r = 1$.

6. A sample of a variable furnishes the results below:

Value	0	1	2	3	4
Observations	16	50	120	90	42

Can we assume that X is Binomially distributed $\mathcal{B}(4, p)$?
TIP: Estimate the parameter p, then evaluate T_i. Here, $r = 1$.

7. A sample of a variable furnishes the results below:

Value	0	1	2	3
Observations	200	100	10	5

Can we assume that X is Binomially distributed $\mathcal{B}(3, p)$?
TIP: Estimate the parameter p, then evaluate T_i. Here, $r = 1$.

8. A sample of a variable furnishes the results below:

Value	8–10	10–12	12–14	14–16	16–18
Observations	5	67	510	370	60

Can we assume that X is normally distributed N(m, σ)?
TIP: Use the values of f at the centers of the classes to estimate the parameters m, σ, then evaluate T_i. Here, $r = 2$.

9. A sample of a variable furnishes the results below:

Value	0–10	10–20	20–30	30–40	40–50	50–60
Observations	10	210	560	180	20	5

Can we assume that X is normally distributed N(m, σ)?
TIP: Use the values of f at the centers of the classes to estimate the parameters m, σ, then evaluate T_i. Here, $r = 2$.

3.9.2 Testing Independence

A second situation often found concerns *contingency tables*, id est, bidimensional tables, containing observations of a couple (X, Y), the model situation is Table 3.10. The margin of X is $o_{i\blacksquare} = \sum_j o_{ij}$ and the margin of Y is $o_{\blacksquare j} = \sum_i o_{ij}$. The total number of observations is $O = \sum_{i,j} o_{ij}$

Table 3.10 Example of a table of observed values of (X, Y)

Variables	Y_1	Y_2	...	Y_n	Margin X
X_1	o_{11}	o_{12}	...	o_{1n}	$o_{1\blacksquare}$
X_2	o_{21}	o_{22}	...	o_{2n}	$o_{2\blacksquare}$
⋮	⋮	⋮	⋱	⋮	⋮
X_m	o_{m1}	o_{m2}	...	o_{mn}	$o_{m\blacksquare}$
Margin Y	$o_{\blacksquare 1}$	$o_{\blacksquare 2}$	...	$o_{\blacksquare n}$	O

Such a table can be transformed in probabilities by using the transformations: $p_{ij} = \frac{o_{ij}}{O}$, $p_{i\blacksquare} = \frac{o_{i\blacksquare}}{O}$, $p_{\blacksquare j} = \frac{o_{\blacksquare j}}{O}$. Let us illustrate this situation by introducing another variable: the outing is made in a group or not. We consider the data below:

	A	B	C	D	E
1		Music	Cinema	Play	margin
2	group	40	38	9	87
3	not group	14	60	25	99
4	margin	54	98	34	186

Select the cell G2 and type "= `B2/$E$4`" (again, $ fixes the cell). Press ENTER, then copy the cell G2 and paste it on the range G2:J4. You will get the results below:

F	G	H	I	J
	Music	Cinema	Play	margin
group	0,215054	0,204301	0,048387	0,467742
not group	0,075269	0,322581	0,134409	0,532258
margin	0,290323	0,526882	0,182796	1

As previously mentioned, an alternative consists in assigning a name to the group of cells `B2:D4` – for instance, let us use the name `A`. You will obtain the same result by selecting the cell `G2` and entering "= `A/SUM(A)`".

When considering contingency tables, we are often interested in evaluating if the variables are independent or not. We may test the hypothesis H_0 : *X and Y are independent* by using the χ^2test: if the variables are independent, then $P(X = X_i, Y = Y_j) = P(X = X_i)P(Y = Y_j)$. We start by estimating $P(X = X_i) \approx p_{i\blacksquare}$, $P(Y = Y_j) \approx p_{\blacksquare j}$, so that $\widehat{p_{ij}} = p_{i\blacksquare}\, p_{\blacksquare j}$. Then, we evaluate

$$Q^2 = \sum_{i=1}^{m}\sum_{j=1}^{n}\frac{\left(o_{ij} - T_{ij}\right)^2}{T_{ij}} \quad \left(T_{ij} = O\widehat{p_{ij}}\right). \tag{3.32}$$

Q^2 is approximatively $\chi^2((m - 1)(n - 1))$: we may determine a critical value χ_α^2 such that $P\left(Q^2 \geq \chi_\alpha^2\right) = \alpha$. The non-rejection region is $\left(0, \chi_\alpha^2\right)$.

Example

Let us consider 100 observations of a couple (*X*, *Y*) leading to the contingency table at below left. To test the independence, we must determine p_{ij} (Table below at right).

X \ Y	-1	0	1
-1	20	0	10
0	0	40	0
1	10	0	20

X \ Y	-1	0	1
-1	0.2	0	0.1
0	0	0.4	0
1	0.1	0	0.2

Thus, $p_{i\blacksquare} = (0.3, 0.4, 0.3)$, $p_{\blacksquare j} = (0.3, 0.4, 0.3)$. Thus, $Q^2 \approx 111$. The number of degrees of freedom is $(3 - 1)\,(3 - 1) = 4$: for $\alpha = 0.05$, $\chi_\alpha^2 = 9.5$. Since $Q^2 > \chi_\alpha^2$, the hypothesis is rejected.

Example

Let us consider 100 observations of a couple (*X*, *Y*) leading to the contingency table at below left. To test the independence, we must determine p_{ij} (Table below at right).

X \ Y	0	1
0	28	22
1	21	29

X \ Y	0	1
0	0.28	0.22
1	0.21	0.29

Thus, $p_{i\blacksquare} = (0.5, 0.5)$, $p_{\blacksquare j} = (0.49, 0.51)$. So, the theoretical values are

X \ Y	0	1
0	28.5	26.5
1	34.2	31.8

X \ Y	0	1
0	0.285	0.265
1	0.342	0.318

and $Q^2 \approx 1.58$ The number of degrees of freedom is (2 – 1) (2 – 1) = 1: for $\alpha = 0.05$, $\chi^2_\alpha = 3.8$. Since $Q^2 < \chi^2_\alpha$, the hypothesis is accepted.

Exercises

1. A sample of two variables furnishes the results below:

X \ Y	1	2	3	4
1	50	15	6	50
2	47	13	9	35
3	35	33	12	34

Can we assume that *X* and Y are independent?

2. A sample of two variables furnishes the results below:

X \ Y	1	2	3
1	200	190	110
2	300	150	50

Can we assume that *X* and Y are independent?

3. A sample of two variables furnishes the results below:

X \ Y	1	2	3	4
1	150	250	80	10
2	75	110	40	10
3	40	100	130	10

Can we assume that *X* and Y are independent?

3.10 Generating Uniform Random numbers

In this section, we examine three different ways to generate uniform random numbers: using built-in functions of EXCEL®, calling the ANALYSIS TOOLPACK or using VBA programming. The use of VBA becomes interesting if you want to collect your subprograms in modules or extensions which may be reused in other workbooks.

3.10.1 Using Built-In Functions

EXCEL® proposes built-in functions for the generation of random numbers uniformly distributed: RAND and RANDARRAY. As their name indicates, RAND generates a single random number, uniformly distributed on (0, 1). RANDARRAY is more flexible: it takes optional arguments allowing to define the interval RANDARRAY ([rows], [columns], [min], [max], [integer]). For instance, RA NDARRAY(m; n) generates a $m \times n$ array of uniformly distributed numbers on (0, 1); RANDARR AY(m; n; a; b) generates a $m \times n$ array of uniformly distributed numbers on (a, b);. RANDARRAY (m; n; a; b; *true*) generates a $m \times n$ array of integers uniformly distributed numbers on (a, b).

As observed in the preceding, random numbers generated by built-in functions are volatile, except if you fix the volatile expressions as indicated in Sect. 1.8. If not, the values will change at ant modification of the workbook.

To use RAND, open a EXCEL® workbook, select cell A1, type = *rand* () and ENTER. The cell contains a random variate from the uniform distribution on (0, 1), analogous to the result below

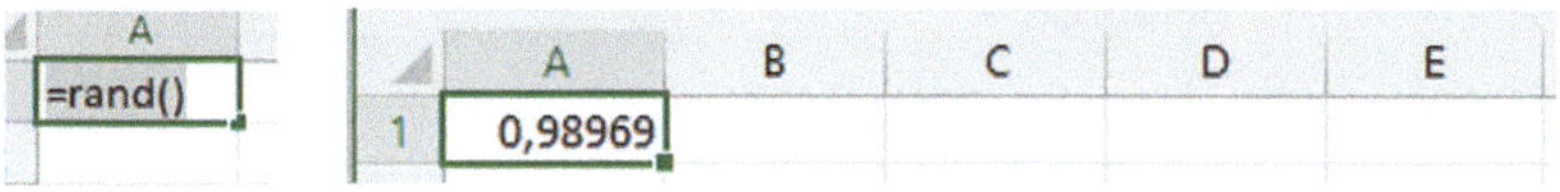

To generate a table of random numbers, the formula must be replicated on the cells defining the table.

To use RANDARRAY, reselect A1 and type =*randarray*(2; 3; −1; 4), then ENTER. You will get a 2 × 3 table of random values on (−1, 4).

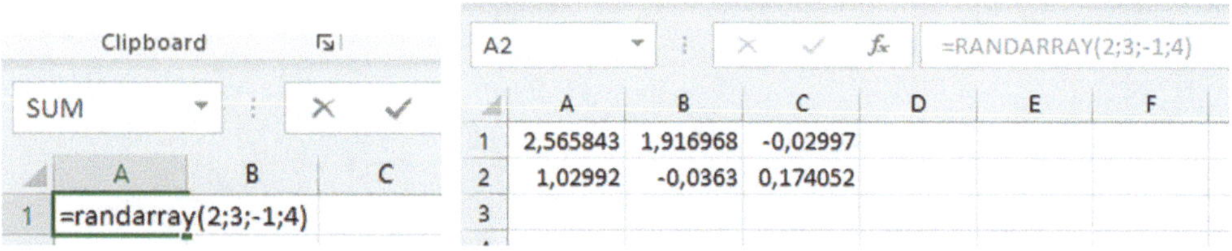

Select E1 and type =*randarray*(2; 3; −5; 10; *true*). You will get a 2 × 3 table of random integer values on (−5, 10).

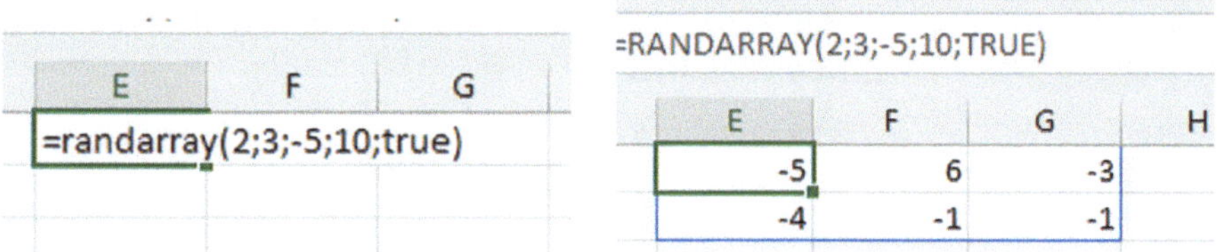

3.10.2 Using the Data Analysis Tool

It is also possible to use the Data Analysis Tool. Go to the *Data>Data Analysis*

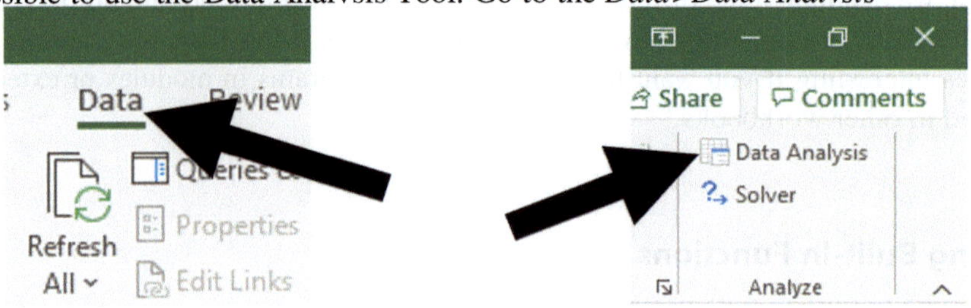

Choose "Random Number Generation"	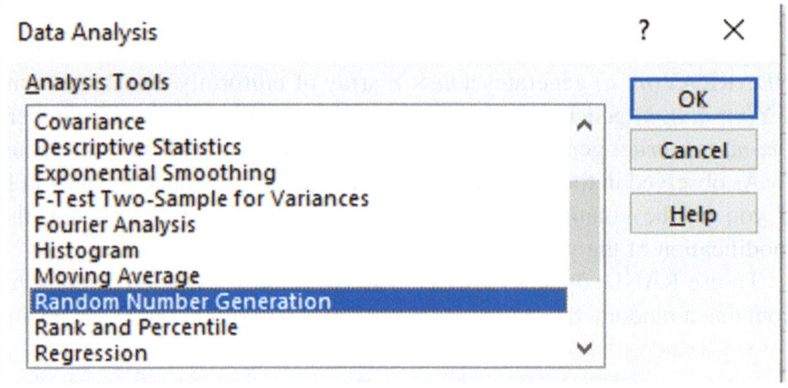
Select "Uniform Distribution" and define the parameters: here, we generate a 2 × 3 table of random values on (−2, 3), starting at cell `I1`.	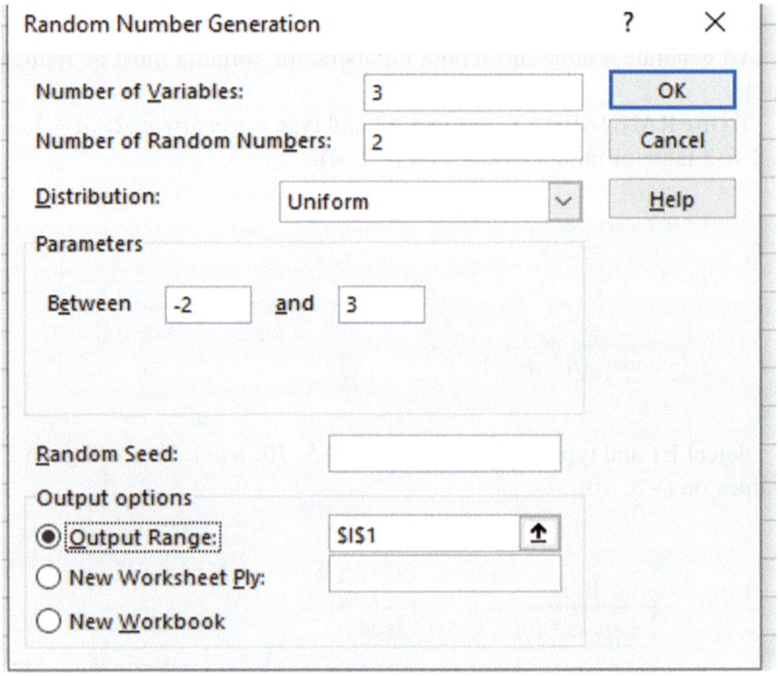

An example of result is:

I	J	K
-0,80688	2,088565	-1,23948
0,050691	2,62035	2,014557

3.10.3 Using VBA

Open a blank workbook named *Book1* (of course, you may use a different name) and insert a VBA module as indicated in Sect. 1.5.

Type the text:

(General)

```
Public Function unigen(a, b, ms, Optional ns = 0)
Randomize
Dim x()
If ns > 0 Then
   ReDim x(1 To ms, 1 To ns)
   For i = 1 To ms
       For j = 1 To ns
           x(i, j) = a + (b - a) * Rnd()
       Next j
   Next i
Else
   ReDim x(1 To ms)
   For i = 1 To ms
       x(i) = a + (b - a) * Rnd()
   Next i
End If
unigen = x
End Function
```

Notice the structure of the code: *"Public"* means that all the workbook can use the function *unigen*. Otherwise, you may declare the function as *"Private"* – the, only the module where it is defined may use it. *"Optional"* indicates that the parameter *ns* may be omitted and its default value is zero. *"Randomize"* restarts the generator of random numbers, so that you will have different results at each run; *"Dim x()"* defines a dynamic array (variable size); *"If ns > 0"* tests if the user defined a value for *ns* or not and takes the adequate redefinition of *x*; *"ReDim x(...)"* gives to *x* the adequate size (normally, arrays start from 0 in VBA. Here, they start at 1); *"unigen = x"* returns the values.

Save the workbook with extension *.xlsm*, close the window for VBA and come back to the worksheet.

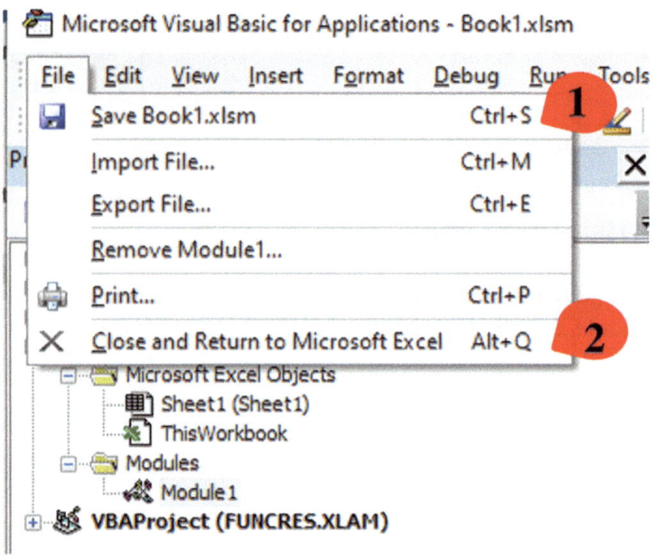

Select the cell A1 and type *"= unigen(−1;2;2;4)"*. Press ENTER. You will get a result similar to the one given at the right side

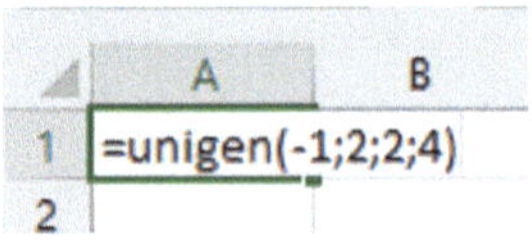

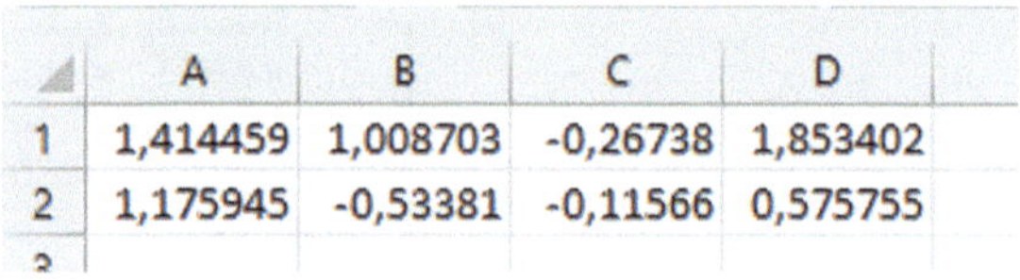

Now, select cell F1 and type *"= unigen(0;3;4)"*. Press ENTER. The result will be similar to the one given below:

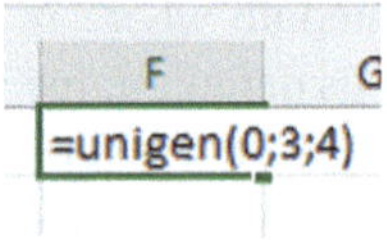

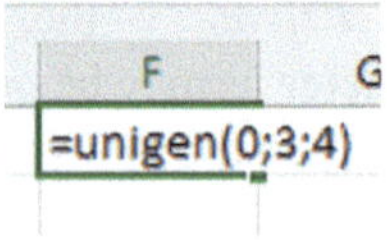

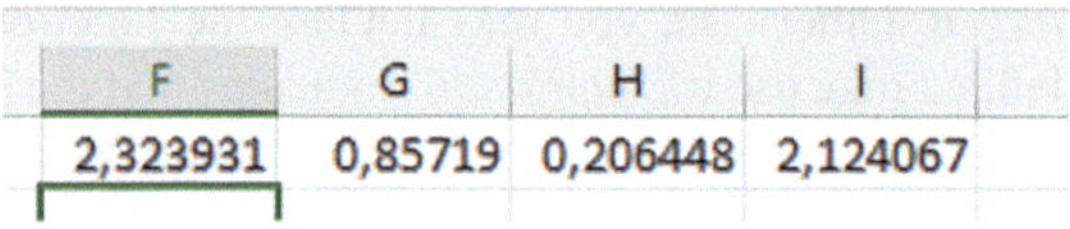

This subprogram may be found in the module *RandomNumbers.bas*. As an alternative, you may load the module to your workbook, as indicated in Sect. 1.6. It is also available in the add-in *RandomNumbers.xlam*, which may be loaded to the workbook as indicated in Sect. 1.2. To test this method, we can generate $ns = 10{,}000$ variates from the uniform distribution on $(-1, 2)$ and compare the empirical distribution of the sample generated by `unigen` and the exact one (Fig. 3.5).

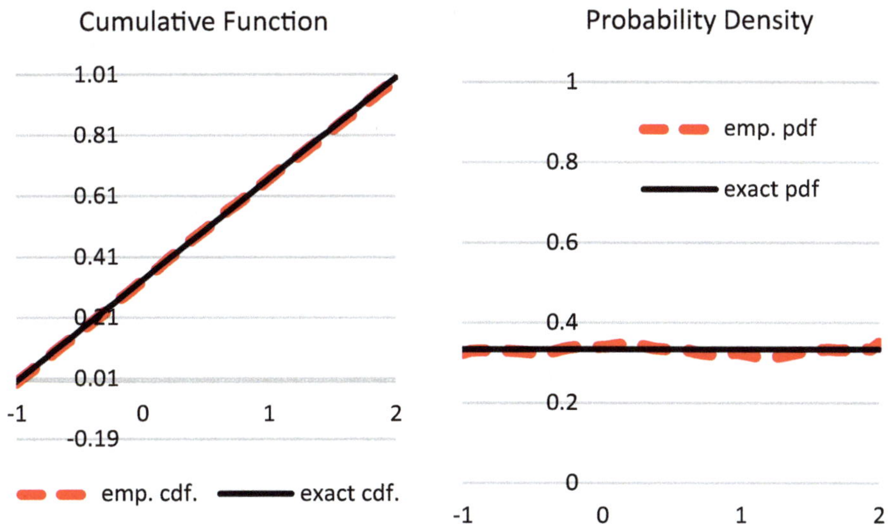

Fig. 3.5 Results furnished by the function `unigen`. The empirical sample is formed by 10,000 variates from the uniform distribution on $(-1, 2)$ generated by `unigen`. The empirical probability density is the SPH derivative of the cumulative function

3.11 Generating Normal Random Numbers

The generation of normally distributed $N(m, \sigma)$ random numbers may use any of the three preceding options: built-in functions, ANALYSIS TOOLPACK or VBA.

3.11.1 Using Built-In Functions

The simplest way to generate a table of $m \times n$ variates from $N(m, \sigma)$ is to generate a random $m \times n$ array of probabilities and using the built-in function NORMINV to find the variates. You can generate the probabilities using the uniform distribution on (0, 1). Let us illustrate the procedure: open a blank workbook and enter the value of m in A2 and the value of σ in B2. Then select the cell D1 and type = *norminv(randarray(2;4);A2;B2)*.

	A	B	C	D	E	F
1	m	sigma		=norminv(randarray(2;4);A2;B2)		
2	3	2				

Press ENTER: you will get a 2×4 array of variates from $N(m, \sigma)$.

	A	B	C	D	E	F	G
1	m	sigma		-0,22037	0,862683	1,50753	4,540549
2	3	2		4,441271	5,201174	1,852958	1,007882

If you desire to control the values of the probabilities, you may split the procedure in two sub steps: firstly, generate the 2 × 4 array of probabilities by typing = *randarray(2;4)* in the cell A4. Pressing ENTER will bring a table of probabilities.

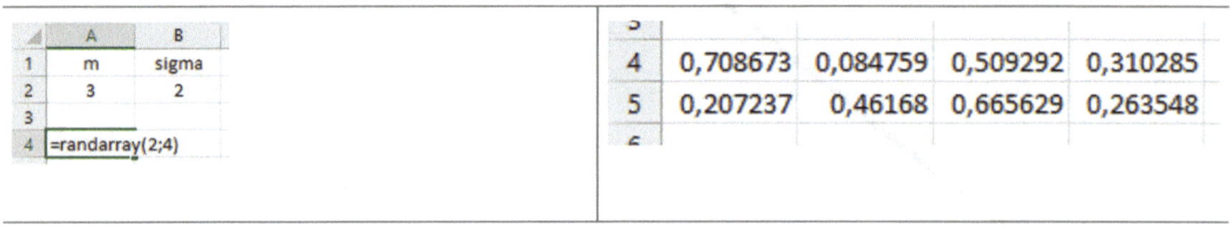

Then, select D1 and type =*norminv(A4#;A2;B2)*. Again, you get a 2 × 4 array of variates from $N(m, \sigma)$.

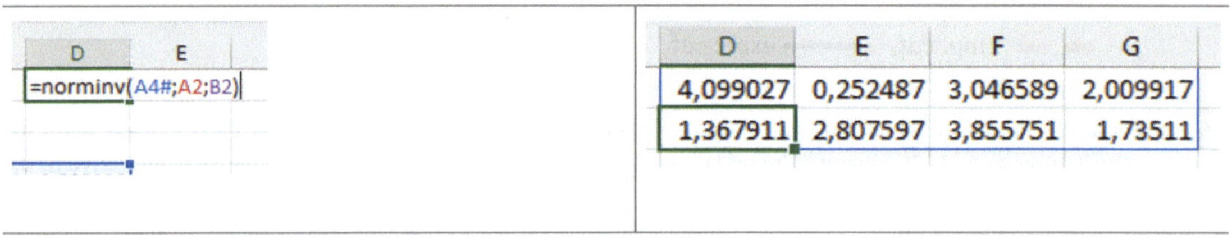

3.11.2 Using the Analysis ToolPack

The Data Analysis Tool provides Normally distributed Random Numbers. To use it, you must perform the steps previously described: go to the *Data>Data Analysis*, choose "Random Number Generation", but now select "Normal". Enter the dimensions, the mean and the variance (you must enter numbers: addresses of cells are not accepted), the chice for the exhibition of the results and click OK. You must get results similar to those of the figure below.

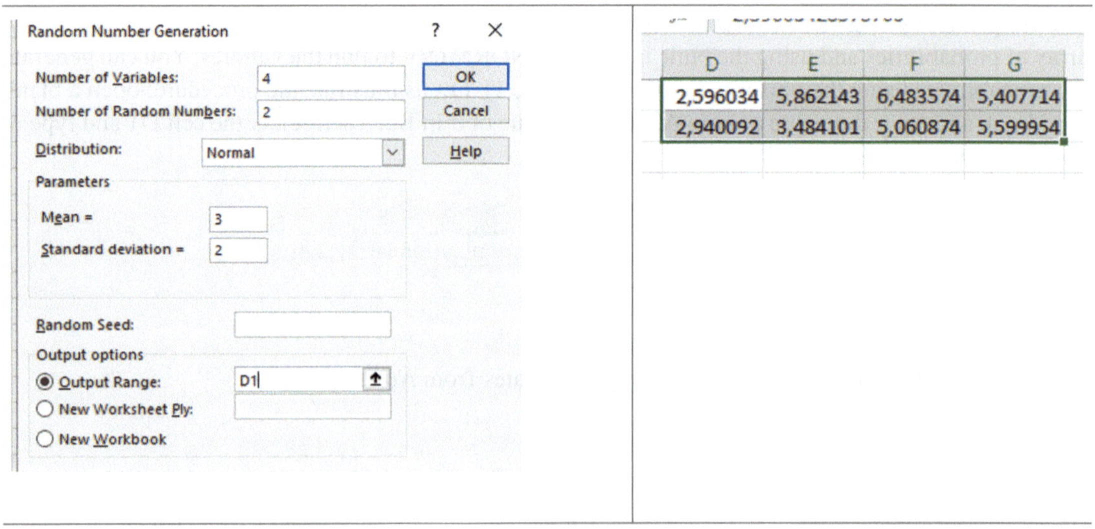

3.11.3 Using VBA

There are many methods to generate variates from a normal distribution using VBA. For instance, we may

1. Use the built-in functions inside VBA
2. Use the log-trig generator, originally proposed by Box and Muller (Box, 1958).

The log-trig generator uses the following result:

> Let $U_1,\ U_2$ be two independent random variables, both uniformly distributed on (0, 1); $Z_1 = \sqrt{-2\ln(U_1)}\ \sin(2\pi U_2)$, $Z_2 = \sqrt{-2\ln(U_1)}\ \cos(2\pi U_2)$. Then $Z = (Z_1, Z_2)$ is a standard gaussian vector, id est, Z_1, Z_2 are two independent random variables, having the same distribution $N(0,\ 1)$.

Thus, we may transform samples from uniform distributions into standard gaussian samples. To get a sample from $N(m,\ \sigma)$, we use $X_i = m + \sigma Z_i$: then, X_1, X_2 are two independent random variables, both having the same distribution $N(m,\ \sigma)$. The log-trig generator is useful for the generation of random gaussian vectors: we may generate $2k$ independent normal variables $X_1, \ldots, X_{2k}$ by using $2k$ independent uniform variables $U_1, \ldots, U_{2k}$ and $Z_1, \ldots, Z_{2k}$ such that (Z_{2i-1}, Z_{2i}) is generated by transformation of (U_{2i-1}, U_{2i}). As an alternative, we may use the polar form of the generator: $Z_1 = V_1\sqrt{-2\ln(V)/V}$, $Z_2 = V_2\sqrt{-2\ln(V)/V}$, with $V_i = 2U_i - 1\ (i = 1, 2)$; $V = V_1^2 + V_2^2$.

In any case, you must perform the steps previously defined: go to *Developer>Visual Basic>VBAProject> Insert>Module* and type the code that implements the method. For instance, let us consider the use of built-in functions: an example of code is

```
Public Function normgen_builtin(m, sigma, ms, Optional ns = 0)
Randomize
Dim x()
If ns > 0 Then
   ReDim x(1 To ms, 1 To ns)
   For i = 1 To ms
       For j = 1 To ns
           x(i, j) = Application.WorksheetFunction.NormInv(Rnd(), m, sigma)
       Next j
   Next i
Else
   ReDim x(1 To ms)
   For i = 1 To ms
       x(i) = Application.WorksheetFunction.NormInv(Rnd(), m, sigma)
   Next i
End If
normgen_builtin = x
End Function
```

If this module is inserted in an EXCEL® workbook (see Sect. 1.5), you may execute it in the workbook. For instance, enter the code at left in cell `A1`: you will get a 2 × 5 sample from $N(2,\ 1)$ (example at right).

	A	B	C
1	=normgen_builtin(2;1;2;5)		

	A	B	C	D	E
1	3,802419	1,821489	3,158566	1,861551	2,510594
2	-0,12216	-0,01646	2,085389	0,740719	-0,24173

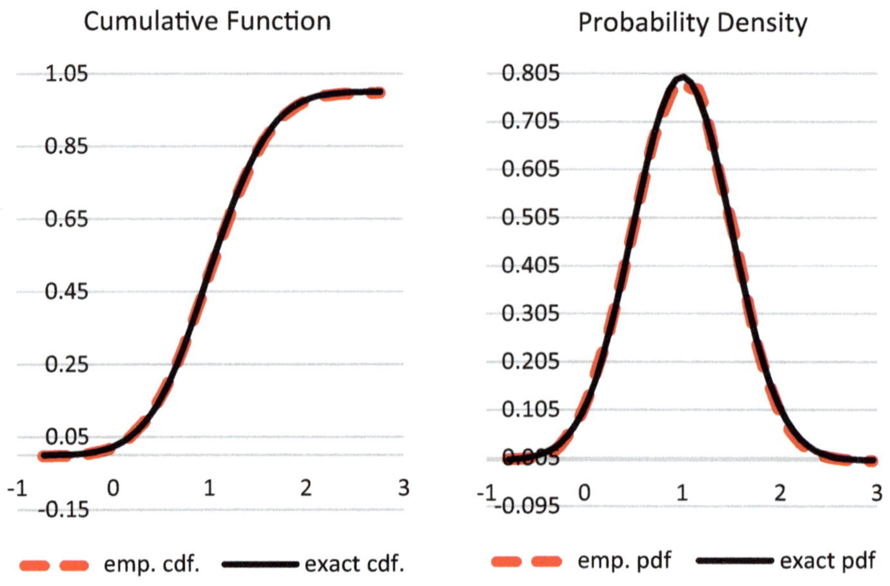

Fig. 3.6 Results furnished by the function `normgen_builtin`. The empirical sample is formed by 10,000 variates from $N(1, 0.5)$ generated by `normgen_builtin`

To test this method, we may generate $ns = 10{,}000$ variates from $N(1, 0.5)$ using `normgen_builtin` and we compare the empirical distribution of the variates with the theoretical one. The results are shown in Fig. 3.6 above.

The classical Box-Muller generator may be implemented as follows:

```
Public Function normgen_bm(m, sigma, ms, Optional ns = 0)
Randomize
Dim x() As Double
If ns > 0 Then
    ntot = ns * ms
Else
    ntot = ms
End If
If ntot Mod 2 = 0 Then
    kmax = ntot / 2
Else
    kmax = (ntot + 1) / 2
End If
Dim aux()
ReDim aux(1 To 2 * kmax)
For k = 1 To kmax
     a1 = Sqr(-2 * Application.WorksheetFunction.Ln(Rnd()))
     a2 = 2 * Application.WorksheetFunction.Pi * Rnd()
     aux(2 * k - 1) = a1 * Sin(a2)
     aux(2 * k) = a1 * Cos(a2)
Next k

If ns > 0 Then
    ReDim x(1 To ms, 1 To ns)
    k = 0
    For i = 1 To ms
        For j = 1 To ns
            k = k + 1
            x(i, j) = m + sigma * aux(k)
        Next j
    Next i
Else
    ReDim x(1 To ms)
    k = 0
    For i = 1 To ms
        k = k + 1
        x(i) = m + sigma * aux(k)
    Next i
End If
normgen_bm = x
End Function
```

An example of use and results are given below:

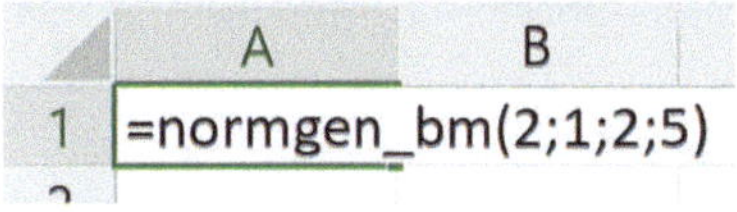

	A	B
1	=normgen_bm(2;1;2;5)	

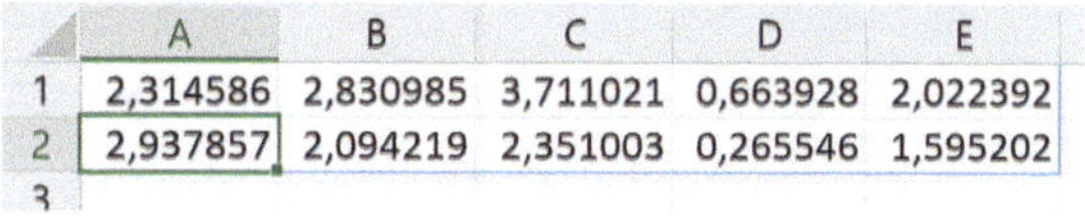

	A	B	C	D	E
1	2,314586	2,830985	3,711021	0,663928	2,022392
2	2,937857	2,094219	2,351003	0,265546	1,595202

We can test this method analogously to the preceding one: we generate $ns = 10{,}000$ variates from $N(1, 0.5)$ using `normgen_bm` and we compare the empirical distribution of the variates with the theoretical one. The results are shown in Fig. 3.7.

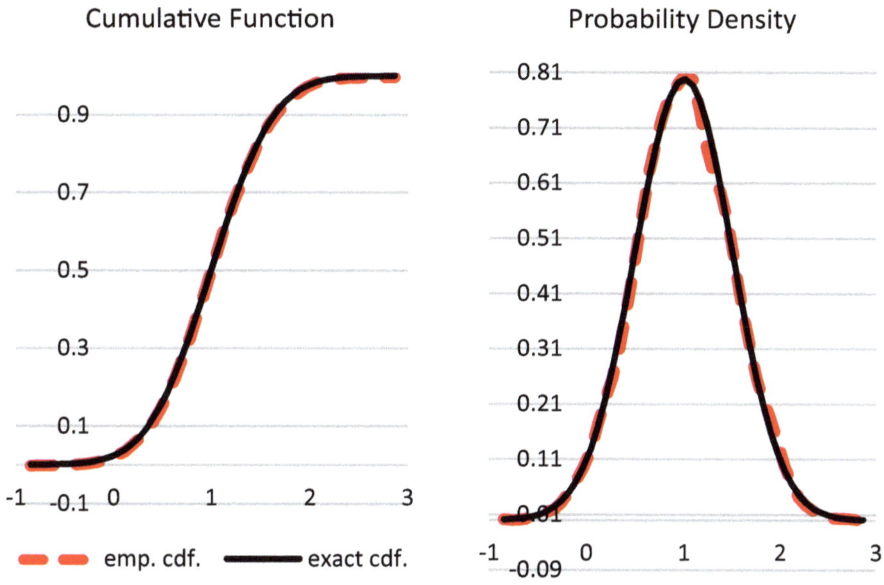

Fig. 3.7 Results furnished by the function `normgen_bm`. The empirical sample is formed by 10,000 variates from $N(1, 0.5)$ generated by `normgen_bm`

3.12 Generating Triangular Random Numbers

Triangular distributions are mainly used when the information available on the random variables is limited to a minimal value x_{min}, a maximal value x_{max} and a probable value x_p. A triangular distribution with these parameters is referred as $T(x_{min}, x_{max}, x_p)$. The PDF associated to a triangular distribution forms a triangle and its equation is

$$f(x) = \begin{cases} \dfrac{2(x - x_{min})}{(x_{max} - x_{min})(x_p - x_{min})}, & \text{if } x_{min} \le x \le x_p; \\ \dfrac{2(x_{max} - x)}{(x_{max} - x_{min})(x_{max} - x_p)}, & \text{if } x_p \le x \le x_{max}; \\ 0, & \text{otherwise}. \end{cases}$$

Its CDF is

$$F(x) = \begin{cases} 0, & \text{if } x \leq x_{min}; \\ \dfrac{(x - x_{min})^2}{(x_{max} - x_{min})(x_p - x_{min})}, & \text{if } x_{min} \leq x \leq x_p; \\ 1 - \dfrac{(x_{max} - x)^2}{(x_{max} - x_{min})(x_{max} - x_p)}, & \text{if } x_p \leq x \leq x_{max}; \\ 1, & \text{otherwise.} \end{cases}$$

The generation of samples from a triangular distribution uses the uniform generator. Indeed, let U be uniformly distributed on (0,1) and consider

$$X = \begin{cases} x_{min} + \sqrt{U(x_{max} - x_{min})(x_p - x_{min})}, & \text{if } U < \dfrac{x_p - x_{min}}{x_{max} - x_{min}} \\ x_m - \sqrt{(1 - U)(x_{max} - x_{min})(x_{max} - x_p)}, & \text{otherwise.} \end{cases}$$

Then, X is $T(x_{min}, x_{max}, x_p)$. This method of generation is implemented in *RandomNumbers.xlam* and *RandomNumbers*.bas – the corresponding function is `trigen` (x_{min}, x_{max}, x_p, *ms*). As an example, let us generate a sample formed by 10,000 variates from $T(-1, 3, 0)$ and determine the empirical CDF and PDF (the PDF is generated by SPH numerical derivation of the cumulative function. The results are shown in Fig. 3.8.

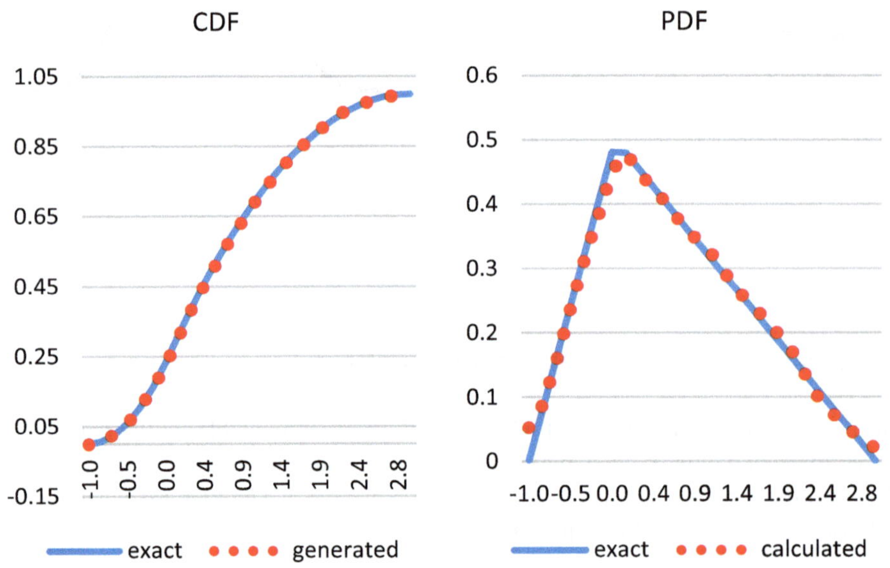

Fig. 3.8 Results furnished by the function `trigen`. The empirical sample is formed by 10,000 variates from $T(-1, 3, 0)$ furnished by `trigen`

3.13 Generating Random Numbers by Inversion

Let X be a random variable having as cumulative function F. Let $U = F(X)$. Then, for $u \in (0, 1)$

$$P(U < u) = P(F(X) < u) = P\left(X < F^{-1}(u)\right) = F\left(F^{-1}(u)\right) = u.$$

Consequently, U is uniformly distributed on (0, 1). Thus, samples from X may be obtained from samples from the uniform distribution, if F^{-1} is available: we generate $u_1, \ldots, u_{ns}$ from the uniform distribution and we set $x_i = F^{-1}(u_i)$.

This approach is very general and allows the generation of samples from arbitrary distributions for which F^{-1} can be determined.

As an example, let us consider the arcsine distribution, for which

$$F(x) = \frac{2}{\pi} \arcsin \sqrt{\frac{x-a}{b-a}}, a < x < b.$$

Here,

$$F^{-1}(u) = a + (b-a) \sin^2\left(\frac{\pi}{2}u\right)$$

Let us generate a sample of 10,000 variates using $a = -1, b = 2$, and use it to estimate the CDF and the PDF of X. The results are shown in Fig. 3.9.

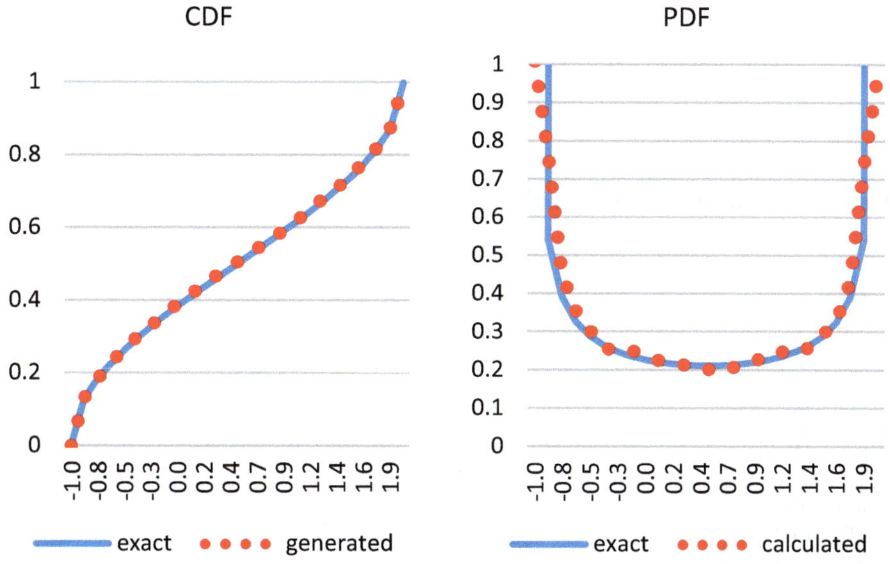

Fig. 3.9 Results furnished by the method of inversion. The empirical sample is formed by 10,000 variates from the arcsin distribution, generated by inversion

3.14 Generating Discrete Random Numbers

A discrete random number is a random variable taking its values in a finite set $\{X_1, \ldots, X_k\}$ with probabilities $p_i = P(X = X_i)$. To generate discrete random numbers, you must furnish the values X_i and the probabilities p_i. You may use a workbook, ANALYSIS TOOLPACK or VBA.

3.14.1 Using Built-In Functions

Open a blank workbook, enter the values of X_i in line 1, the values of p_i in line 2 and evaluate the cumulative sum of probabilities F_i in line 3.

1	X(i)	3	5	6	9
2	p(i)	0,2	0,3	0,4	0,1
3	F(i)	0,2	0,5	0,9	1

Enter the number of variates *ns* to be generated in `B5`. Generate *ns* uniform random numbers in line 6. Then, compare each uniform number u_i to the cumulative sum in line 3. Choose X_i corresponding to the minimal value F_k such that $u_i \leq F_k$. For instance, enter in `B7` the formula "`=IF(B6<$B$3;$B$1;IF(B6<$C$3; $C$1;IF(B6<$D$3;$D$1;IF(B6<$E$3;$E$1))))`". Then copy this formula to the other cells in line 7. The results appear:

5	ns	10									
6	u	0,365673	0,410191	0,554695	0,320738	0,163707	0,513733	0,506083	0,699044	0,051486	0,068013
7	X	5	5	6	5	3	6	6	6	3	3

3.14.2 Using the Analysis Toolpack

The ANALYSIS TOOLPACK proposes a tool for the generation of discrete random numbers. As in the preceding method, open a blank workbook, enter the values of X_i in column 1, the values of p_i in column 2 – to use the ANALYSIS TOOLPACK, the data must be in columns.

	A	B
1	X(i)	p(i)
2	3	0,2
3	5	0,3
4	6	0,4
5	9	0,1
6		
7	X	
8		

Go to *Data>Data Analysis* and choose the option "*Random Number Generation*" and l'option "*Discrete*".

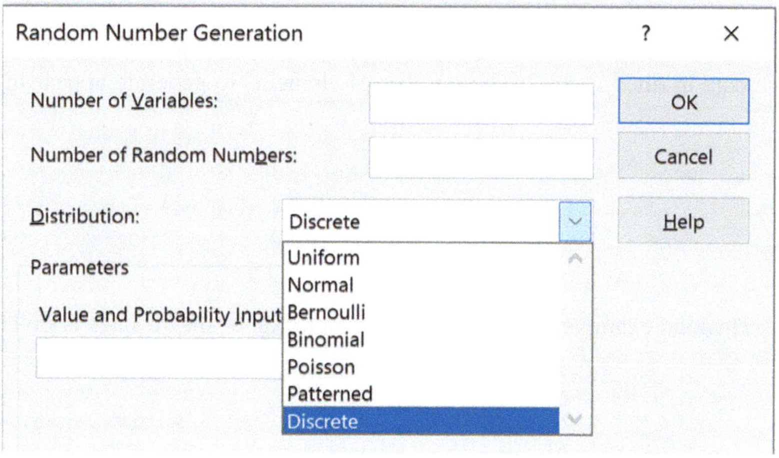

Enter the data: number of columns in "*Number of Variables*", number of rows in "*Number of Random Numbers*", the range containing the values and the probabilities, the output range. Click on "*OK*"

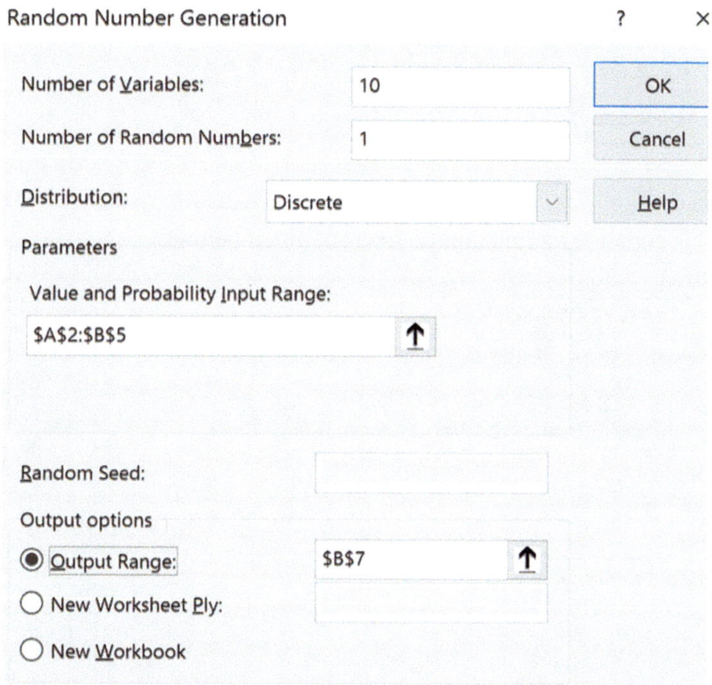

The results appear in line 7

7	X	6	5	9	9	6	3	6	6	6	6

3.14.3 Using VBA

To use VBA, we assume that the workbook has two names "X" and "p" corresponding respectively to the data in lines 1 and 2. The number of elements to generate appear in B3.

	A	B	C	D	E
1	X(i)	3	5	6	9
2	p(i)	0,2	0,3	0,4	0,1
3	ns	10			

Then, an example of macro for the generation of the variates is given below:

```
Sub gendiscrete()
Dim x As Variant
nX = "[" & ActiveWorkbook.Name & "]" & ActiveSheet.Name & "!" & "X"
x = Range(nX)
Dim prob As Variant
nP = "[" & ActiveWorkbook.Name & "]" & ActiveSheet.Name & "!" & "p"
prob = Range(nP)
ns = Cells(3, 2).Value
xs = discgen(x, prob, ns)
Cells(5, 2).Value = "X"
For i = 1 To ns
   Cells(5, i + 1).Value = xs(i)
Next i
End Sub
```

This code uses a function discgen, given below:

```
Function cumprob(prob)
n = UBound(prob, 2)
Dim aux() As Double
ReDim aux(1 To n)
aux(1) = prob(1, 1)
For i = 2 To n
     aux(i) = aux(i - 1) + prob(1, i)
Next i
cumprob = aux
End Function
```

```
Function discgen(x, prob, ns)
F = cumprob(prob)
Dim aux() As Double
ReDim aux(1 To ns)
Randomize
For i = 1 To ns
    u = Rnd()
    j = 1
    p = F(j)
    Do Until p > u
      j = j + 1
      p = F(j)
    Loop
    aux(i) = x(1, j)
Next i
gendiscrete = aux
End Function
```

Running the macro produces the results in line 5:

5	X	6	5	5	3	3	9	3	3	3	6

To test the quality of the random numbers generated, the code runs with $ns = 10{,}000$ and we evaluate the relative frequency of each value X_i. The results are exhibited in Table 3.11.

Table 3.11 Relative frequencies of each state for two simulations

	State			
nsim	3	5	6	9
1000	0,189	0,3	0,413	0,098
10,000	0,1958	0,3024	0,4014	0,1004
exact	0,2	0,3	0,4	0,1

3.15 Generating Regular Random Functions

The main difference between random numbers and random functions is a matter of dimensionality. For instance, let us consider a function $y : (0, 1) \to \mathbb{R}$: to define such a function, we must give the value $y(t)$ for any $t \in (0, 1)$, so that the definition of y requests the definition of infinitely many values.

Indeed, functions are infinite dimensional objects, id est, functions are elements from infinite dimensional vector spaces: contrarily to vectors from $\mathbb{R}^n$-which are fully characterized by a finite number of coefficients (n real values) – functions needs an infinite number of parameters to be completely defined.

Some classes of functions can be characterized using Hilbert basis – fortunately, these classes are general enough for practical purposes. Recall that a Hilbert basis for a real function space V is a family $\{\varphi_i : i \geq 1, i \in \mathbb{N}\} \subset V$ such that

$$\forall y \in V : y = \sum_{i=1}^{+\infty} y_i \varphi_i, y_i \in \mathbb{R}, \forall i \geq 1. \tag{3.33}$$

Thus, we may generate an approximation by considering a finite sum, such as, for instance

$$\forall y \in V : y \approx P_n y = \sum_{i=1}^{n} y_i \varphi_i, y_i \in \mathbb{R}, \quad 1 \leq i \leq n. \tag{3.34}$$

More generally, we may consider

$$\forall y \in V : y \approx P_n y = \sum_{i=1}^{n} y_{j_i} \varphi_{j_i}, y_i \in \mathbb{R}, \quad j_i \geq 1, j_i \in \mathbb{N}. \tag{3.35}$$

These expressions generalize to the case where $\boldsymbol{y} = (y_1, \ldots, y_d)$ is a vector of real functions. Typical Hilbert basis for functions defined on an interval (a, b) are furnished by polynomials:

$$\varphi_i(t) = \left(\frac{t-a}{b-a}\right)^{i-1}, a < b;$$

or trigonometrical functions:

$$\varphi_1(t) = 1; \varphi_{2k+1}(t) = \sin\left(k\left(\frac{t-a}{b-a}\right)\right), \varphi_{2k}(t) = \cos\left(k\left(\frac{t-a}{b-a}\right)\right) \ (k \geq 1).$$

We may also consider periodical functions

$$\varphi_1(t) = 1; \varphi_{2k+1}(t) = \sin\left(2\pi k\left(\frac{t-a}{b-a}\right)\right),$$

$$\varphi_{2k}(t) = \cos\left(2\pi k\left(\frac{t-a}{b-a}\right)\right)\ (k \geq 1).$$

To generate approximations of a given y, we may use the standard scalar product of square summable functions

$$(y, z) = \int_a^b y(t)z(t)dt$$

with an orthogonal projection approach. For instance, when using the approximation (3.34), we have

$$(y - P_n y, \varphi_i) = 0, 1 \leq i \leq n.$$

Thus, the coefficients $\mathbf{y} = (y_1, \ldots, y_k)^t$ are the solutions of a linear system

$$Ay = B, A_{ij} = (\varphi_i, \varphi_j), B_i = (y, \varphi_i).$$

If the family $\{\varphi_i : i \geq 1,\ i \in \mathbb{N}\}$ is formed by orthogonal elements, then $(\varphi_i, \varphi_j) = 0$, if $i \neq j$. In this case,

$$y_i = \frac{B_i}{A_{ii}}.$$

For an orthonormal family, in addition, $A_{ii} = 1$, so that $y_i = B_i$. The reader will find in the literature examples of orthonormal families, such as, for instance, Legendre Polynomials. An arbitrary family $\{\phi_i : i \geq 1,\ i \in \mathbb{N}\}$ may be transformed into an orthonormal family $\{\varphi_i : i \geq 1,\ i \in \mathbb{N}\}$ by the Gram-Schmidt's procedure:

$$\varphi_i = \frac{r_i}{\sqrt{(r_i, r_i)}}; r_1 = \phi_1, r_i = \phi_i - \sum_{j=1}^{i-1} (\phi_i, \varphi_j)\varphi_j\ (i > 1).$$

Example

Let us consider the function $y : (-1, 1) \to \mathbb{R}$ given by $y(t) = e^t$. Let us look for a polynomial approximation of order 5. We have

$$y_1 = \frac{5(-539 + 73e^2)}{2e} \approx 0.368887,$$

$$y_2 = -\frac{45(-569 + 77e^2)}{e} \approx 0.706556,$$

$$y_3 = \frac{840(-133 + 18e^2)}{e} \approx 0.930079,$$

$$y_4 = -\frac{70(-2431 + 329e^2)}{e} \approx 0.013995,$$

$$y_5 = \frac{630(-133 + 18e^2)}{e} \approx 0.697559.$$

Approximated values may be obtained by the methods presented in 2.5 (Fig. 3.10).

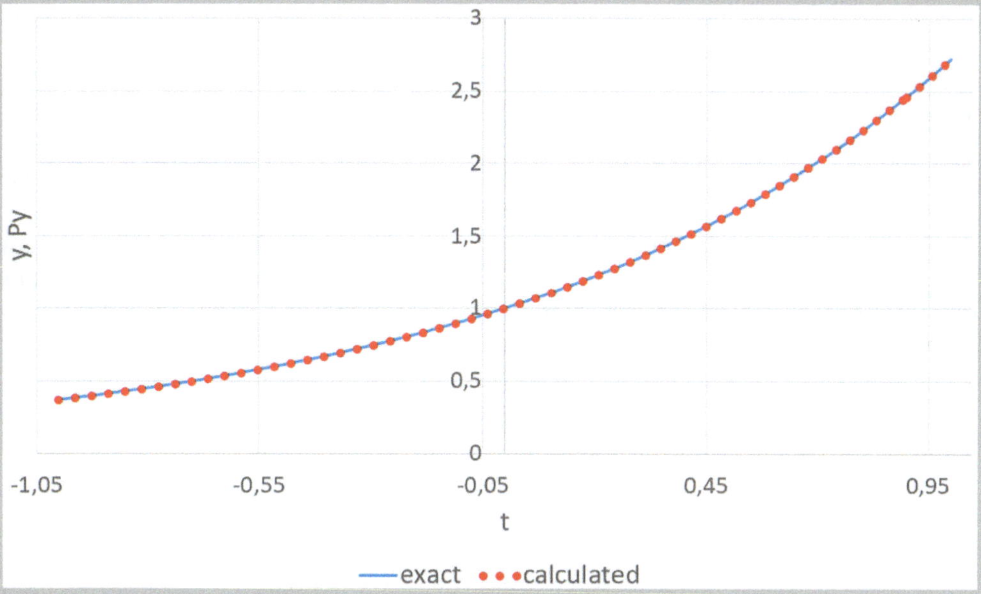

Fig. 3.10 Approximation of e^t at order 5 by a polynomial family ∎

Hilbert expansions may be used to generate regular random functions – the functions generated will have the same regularity as the Hilbert basis considered. Different methods may be used to achieve the generation. Whichever you choose, you must previously define the Hilbert basis $\{\varphi_i : i \geq 1, i \in \mathbb{N}\}$

1. Method 1: use Eq. (3.34) with a fixed k. If you choose this option, you must select k before starting the generation. Now, let the value of k be given: you must generate the coefficients $y_1, \ldots, y_k$. To do this, you must choose a distribution taking values on $\mathbb{R}^k$: for instance, you can generate each coefficient as a variate from an independent gaussian distribution $N(0, \sigma)$.
2. Method 2: use Eq. (3.34) with a variable k having as minimal value k_0. If you choose this option, you must determine k at each generation. For instance, generate a random integer k_1 from a discrete distribution taking values from 0 to $+\infty$ (for instance Poisson's distribution). Then, take $k = k_0 + k_1$ and apply method 1. As an alternative, you can use the built-in function `RANDBETWEEN` of EXCEL®: `RANDBETWEEN(k0,kmax)` generates an integer k such that `k0`$\leq k \leq$`kmax`.
3. Method 3: use Eq. (3.35) with a fixed k. Again, you must select k before starting the generation. Now, let the value of k be given: you must generate the indexes $j_1, \ldots j_k$. For instance, take $j_\ell = 1 + \delta_\ell$, where δ_ℓ is a variable having an independent Poisson distribution. Once the indexes are defined, you must generate the coefficients $y_1, \ldots, y_k$. Analogously to the preceding methods, a

distribution taking values on $\mathbb{R}^k$ is needed: for instance, you can use an independent gaussian distributions $N(0, \sigma)$.

4. Method 4: use Eq. (3.35) with a variable k having as minimal value k_0. In this case, select k as in method 2. Then use method 3.

Example

Let us consider the polynomial family on $(-1, 1)$. Let us apply the Method 1: open a blank workbook. In cell B1, enter the formula "=`SEQUENCE(1;k;1)`". You will get the values $-1, -0.9, -0.8, \ldots, 1$ in cells `B1:V1`. Give the name t to these cells.

Enter the values of k and σ for the generation of the coefficients. Give the names `k` to `B3` and `sigma` to `D3`. In cell `B4`, enter the formula "=`SEQUENCE(1;k;1)`". Import *RandomNumbers.bas*. Select `B5` and enter the formula "=`normgen_builtin(0;sigma;k)`". An example of result is given below:

2			method 1			
3	k	5	sigma	5		
4	i	1	2	3	4	5
5	y_i	-0,60545	-5,86539	6,520787	-1,6184	7,935759

To evaluate the values of the random function, select B6 and enter "=`(t+1)/2`". Give the name `s` to `B6:V6`. Select `B7` and enter the formula of the polynomial: "=`B5+C5*s+D5*s^2+E5*s^3+F5*s^4`". You will get the values of the random function in line 7.

6	s	0	0,05	0,1	0,15	0,2	0,25	0,
7	py	-0,60545	-0,88257	-1,12761	-1,33999	-1,51795	-1,65854	-1,7576

To use Method 4: select cell `B9` and enter the formula "=`RANDBETWEEN(3,10)`" to get an integer between 3 and 10; In line 11, generate the coefficients y_{j_i} with "=`normgen_builtin(0;sigma;B9)`" in `B11`. To get the indexes j_i, select `B12` and enter the formula "=`RANDBETWEEN(1,20)`" to get an index between 1 and 20. The indexes in line 10 were generated by "=`SEQUENCE(1;B9;1)`" in `B10`. An example of results is given below:

8			method 4								
9	k	10									
10	i	1	2	3	4	5	6	7	8	9	10
11	y_i	0,151529	10,64455	-7,83833	-8,04581	-2,01558	4,876366	2,838525	2,594825	4,575005	-0,91248
12	j_i	19	15	8	14	2	3	13	20	4	17

Then you may generate the values of the random function analogously to Method 1.

13	py	0	-0,08802	-0,14822	-0,17721	-0,17166	-0,12859	-0,0457	0,077979	0,2411

The curves generated are exhibited in the Fig. 3.11 below.

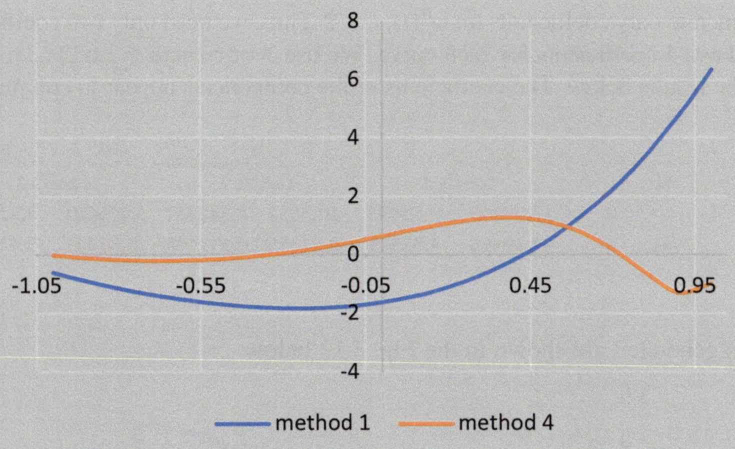

Fig. 3.11 Random functions generated by a polynomial family ∎

3.16 Generating Regular Random Curves

A random regular curve in $\mathbb{R}^2$ may be generated by using its parametric form: $\boldsymbol{x} = (x_1(t), x_2(t))$, $t \in (0, T)$. The initial and the final values may be prescribed by considering families $\{\varphi_i : i \geq 1, i \in \mathbb{N}\}$ having properties of periodicity or such that their values at the boundary points are equal to zero. For instance, to generate a curve such that $\boldsymbol{x}(0) = \boldsymbol{x_0}, \boldsymbol{x}(T) = \boldsymbol{x_T}$, we may consider

$$x_i(t) = x_i(0) + \frac{x_i(T) - x_i(0)}{T} t + \sum_{j=1}^{k} x_{i,j}\, \varphi_j^0(t), \tag{3.36}$$

where $\varphi_j^0(0) = \varphi_j^0(T) = 0$. For instance,

$$\varphi_j^0(t) = \varphi_j(t) - \left(\varphi_j(T) - \varphi_j(0)\right) \frac{t}{T} - \varphi_j(0). \tag{3.37}$$

To generate periodic curves, we may consider the family

$$\varphi_{2k-1}^{per}(t) = \sin\left(2k\pi\left(\frac{t-a}{b-a}\right)\right), \varphi_{2k}^{per}(t) = \cos\left(2k\pi\left(\frac{t-a}{b-a}\right)\right) \ (k \geq 1) \tag{3.38}$$

and

$$x_i(t) = \sum_{j=1}^{k} x_{i,j}\, \varphi_j^{per}(t), \tag{3.39}$$

Example

Let us generate three random curves in $\mathbb{R}^2$ such that $\boldsymbol{x}(0) = (1, 0)$,$\boldsymbol{x}(1) = (0, 1)$. Here, $T = 1$ and we use Eqs. (3.36) and (3.37) with a polynomial family of order 4. Since

$$\varphi_1^0(t) = 1 - (1-1)\mathrm{t} - 1 = 0,$$

$$\varphi_2^0(t) = \mathrm{t} - (1-0)\mathrm{t} - 0 = 0,$$

We need to generate only coefficients for $\varphi_i^0(t), i > 2$. Thus, we need only two coefficients for each component, id est, 4 coefficients for each curve. We use "`=normgen_builtin(0;5;2;6)`" in `C2` to get the results below. The coefficients of the components appear in columns.

	A	B	C	D	E	F	G	H
1	x0		coeffs 1		coeffs 2		coeffs 3	
2	1	0	-7,70749	-3,55425	-10,3657	2,348552	5,726601	-4,25745
3	x1		-4,10466	-4,53511	-0,82815	0,912069	-4,91607	-2,41362
4	0	1						

The curves generated are shown in the Fig. 3.12 below:

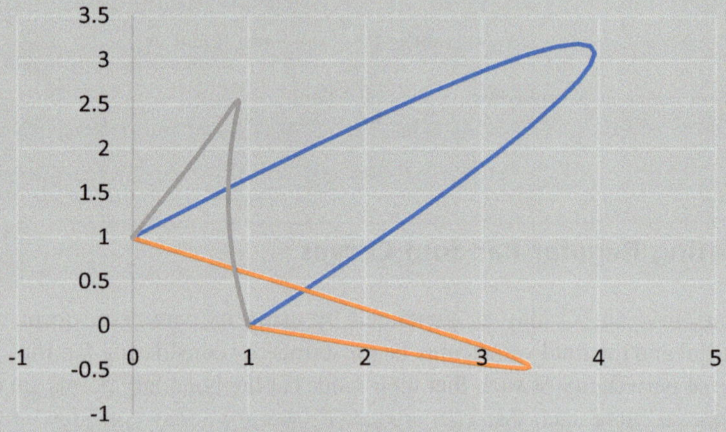

Fig. 3.12 Random curves generated by a polynomial family (order 4)

Other expression may be used for φ_i^0. For instance, we can use $\varphi_i^0(t) = \sin(i\pi t)$. An example of order 3 is shown below - the coefficients of the components appear in columns (Fig. 3.13):

	A	B	C	D	E	F	G	H
1	x0		coeffs 1		coeffs 2		coeffs 3	
2	1	0	-2,18854	1,605489	6,329064	-8,18363	-1,59271	10,80754
3	x1		1,223097	5,76869	-1,51174	3,7615	-7,99716	-3,03788
4	0	1	-5,61256	8,868552	-0,87016	0,938819	-4,57884	-4,33623

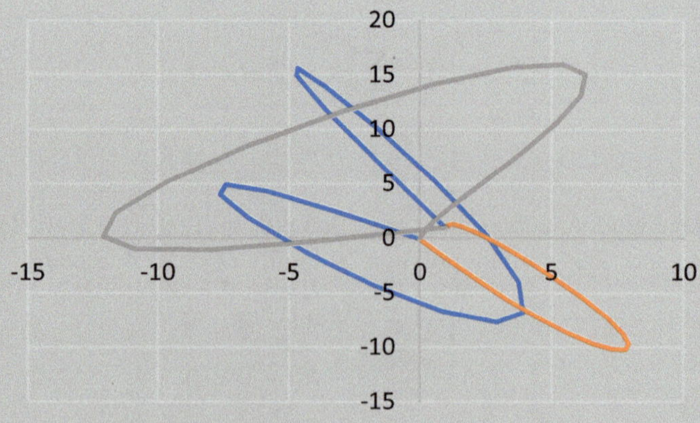

Fig. 3.13 Random curves generated by a sinusoidal family (order 3)

The curves generated are less or more complex according to the degree used in the generation. For instance, the polynomial family of order 3 furnishes the results below (Fig. 3.14):

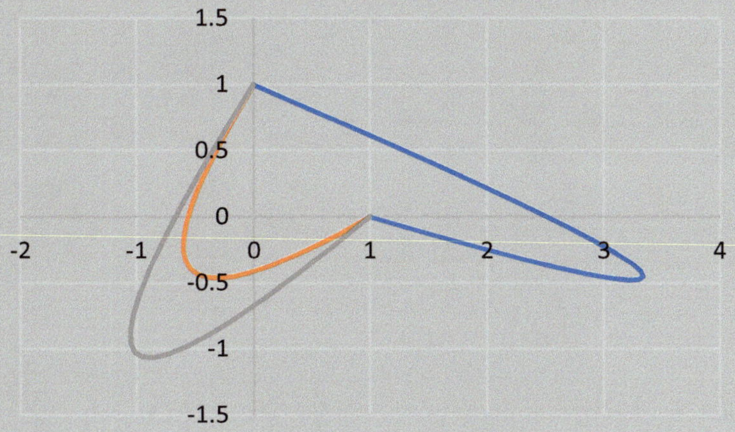

Fig. 3.14 Random curves generated by a polynomial family (order 3) ∎

Example
Let us generate three random curves in $\mathbb{R}^2$ such that $\boldsymbol{x}(0) = \boldsymbol{x}(1)$. Here, $T = 1$ and we use Eqs. (3.36) and (3.37) with a polynomial family of order 4. To generate periodic curves, we take $\boldsymbol{x}_0 = \boldsymbol{x}_T$, chosen at random.

	A	B	C	D	E	F	G	H
1	x0_1		coeffs 1		coeffs 2		coeffs 3	
2	0,411041	0,270255	-4,82149	-1,02352	1,23711	3,051058	0,247859	3,723362
3	x0_2		-2,30942	2,114307	1,40858	1,010077	-2,19491	0,055435
4	0,074845	-0,01191						
5	x0_2							
6	-0,46736	-0,83404						

The curves generated are shown in the Figs. 3.15 and 3.16 below:

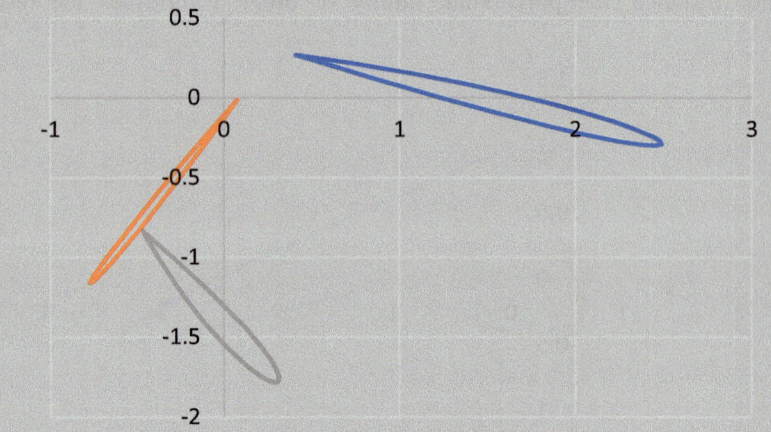

Fig. 3.15 Random periodic curves generated by a polynomial family (order 4)

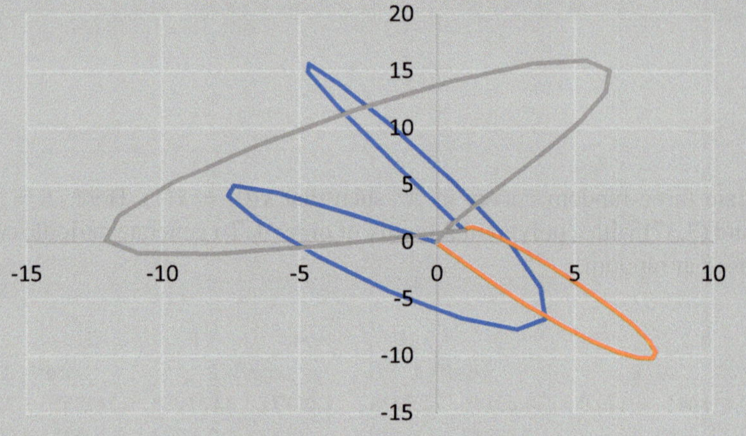

Fig. 3.16 Random curves generated by a sinusoidal family (order 3)

An example using the family $\varphi_i^0(t) = \sin(i\pi t)$ of order 3 is shown below - the coefficients of the components appear in columns:

	A	B	C	D	E	F	G	H
1	x0		coeffs 1		coeffs 2		coeffs 3	
2	1	0	-2,18854	1,605489	6,329064	-8,18363	-1,59271	10,80754
3	x1		1,223097	5,76869	-1,51174	3,7615	-7,99716	-3,03788
4	0	1	-5,61256	8,868552	-0,87016	0,938819	-4,57884	-4,33623
5								

The curves are exhibited in Fig. 3.16.

An example using the family $\varphi_i^{per}(t)$ (Eq. 3.38) of order 2 is shown below (Fig. 3.17):

	A	B	C	D	E	F	G	H
1	x0_1		coeffs 1		coeffs 2		coeffs 3	
2	0,057377	-0,36811	2,607841	-3,04282	4,879009	-3,57481	7,66344	0,725704
3	x0_2		-3,81257	1,08542	1,377428	1,831946	7,040804	5,952539
4	-0,86922	0,010724						
5	x0_2							
6	0,139357	-0,03603						

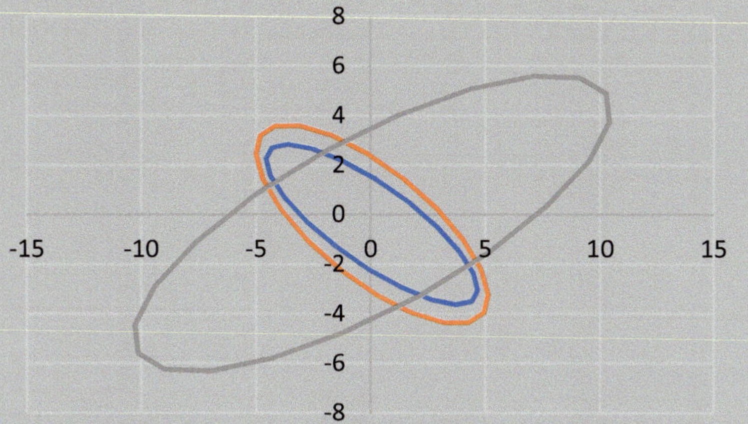

Fig. 3.17 Random periodic curves generated by the periodic family ∎

Stochastic Processes

4

Eduardo Souza de Cursi

The formal study of stochastic processes is not in the scope of this book, which focus on the use of EXCEL® to solve practical problems. In our context, stochastic processes may be considered as random variables indexed by the time, id est, families $\{X(t) = X_t : t \in \mathcal{T} \subset \mathbb{R}\}$, such that $X(t) = X_t$ is a random variable, for any $t \in \mathcal{T}$. Recalling that a random variable X_t is an application $\Omega \ni \omega \to X_t(\omega) \in \mathbb{R}$, we may consider a stochastic process as an application $X : \Omega \times \mathcal{T} \to \mathbb{R}$ and $X_t(\omega) = X(\omega, t)$. In the theory of stochastic processes, it is usual to make a distinction between discrete time (when $\mathcal{T}$ is a discrete set) and continuous time (when $\mathcal{T}$ is an interval). In many applications, $\mathcal{T} = \mathbb{N}$ or $\mathcal{T} = \mathbb{Z}$.

Analogously to random variables, we may define statistics of a stochastic process:

The **mean** of X is $m_X(t) = E(X_t)$.
The **variance** of X is $V_X(t) = V(X_t) = E(X_t^2) - (E(X_t))^2$
The **standard deviation** of X is $\sigma_X(t) = \sqrt{V_X(t)}$.
The **moment of order *k*** of X is $M_{X,k}(t) = E(X_t^k)$.
X is a **second-order process** if and only if $E(X_t^2) < \infty, \forall t$. ■

Notice that $E(X_t^2) < \infty \Longrightarrow E(X_t) < \infty$, so that the mean of a second order process is also finite. Let us introduce other definitions:

The **autocorrelation** of X is $R_{XX}(t, s) = E(X_t X_s)$.
The **autocovariance** of X is $C_{XX}(t, s) = E(X_t X_s) - E(X_t)E(X_s)$.
Notice that $R_{XX}(t, s) = R_{XX}(s, t)$ and $C_{XX}(t, s) = C_{XX}(s, t)$.
The **crosscorrelation** of X and Y is $R_{XY}(t, s) = E(X_t Y_s)$.
The **crosscovariance** of X and Y is $C_{XY}(t, s) = E(X_t Y_s) - E(X_t)E(Y_s)$ ■

For $\mathcal{T} = \mathbb{N}$ or $\mathcal{T} = \mathbb{Z}$, t and s are integers and the stochastic process is said to be discrete. Other interesting properties concern stationarity and ergodicity

E. Souza de Cursi, *Uncertainty Quantification and Stochastic Modelling with EXCEL*, Springer Texts in Business and Economics, https://doi.org/10.1007/978-3-030-77757-9_4

4.1 Stationarity and Ergodicity

X is **weakly stationary** if and only if $V_X(t)$ is finite for any t; m_X is independent from t: $m_X(t) = m_0$, $\forall\, t$ and $R_{XX}(s, t) = R_X(|s - t|)$, $\forall\, s, t$. R_X is extended by symmetry to negative values: $R_X(a) = R_X(-a)$, so that $R_{XX}(s, t) = R_X(s - t) = R_X(t - s)$.

X is **stationary** if and only if the distribution of $(X(t_1), \ldots, X(t_n))$ coincides with the distribution of $(X(t_1 + \tau), \ldots, X(t_n + \tau))$ for any n, any $(t_1, .., t_n)$ and any τ. If X is stationary, the X is weakly stationary. Let X be weakly stationary. X is **ergodic in the mean** if and only if $m_0 = \lim\limits_{T\to+\infty} \frac{1}{T}\int_0^T X(s)ds$. It is **ergodic in correlation** if and only if $R_X(\tau) = \lim\limits_{T\to+\infty} \frac{1}{T} \times \int_0^T X(s)X(s+\tau)ds$. ■

Notice that, for a weakly stationary process,

$$C_{XX}(t,s) = R_{XX}(t,s) - m_X^2 = R_X(|s-t|) - m_0^2 = C_X(|s-t|)$$

For ergodic processes, we may evaluate the mean and/or the autocorrelation from a single realization, by approximating the limits by the evaluation at a large T. For discrete processes (for instance, $\mathcal{T} = \mathbb{N}$ or $=\mathbb{Z}$), the integrals are replaced by sums:

$$m_0 = \lim_{n\to\infty} \frac{1}{n}\sum_{i=0}^{n} X_i; \qquad R_X(s) = \lim_{n\to\infty} \frac{1}{n}\sum_{i=0}^{n} X_i X_{i+s}.$$

In practice, these quantities are estimated by finite sums. For instance, we may use VBA to evaluate these values. An example of code is the following:

```
Function read_signal(line, col, ns)
Dim x() As Double
ReDim x(1 To ns)
For i = 1 To ns
    x(i) = Cells(line, col + i - 1).Value
Next i
read_signal = x
End Function
```

```
Function mean(x)
mean = WorksheetFunction.Average(x)
End Function
```

```
Function variance(x)
variance = WorksheetFunction.Var_S(x)
End Function
```

```
Function autocorr(k, x)
ns = UBound(x, 1)
s = 0
For i = 1 To ns - k
    s = s + x(i) * x(i + k)
Next i
autocorr = s / (ns - k)
End Function
```

```
Function vautocorr(kmax, x)
Dim resu() As Double
ReDim resu(0 To kmax)
For k = 0 To kmax
    resu(k) = autocorr(k, x)
Next k
vautocorr = resu
End Function
```

```
Function covar(k, x)
aux = mean(x)
covar = autocorr(k, x) - aux * aux
End Function
```

```
Function vcovar(kmax, x)
Dim resu() As Double
ReDim resu(0 To kmax)
For k = 0 To kmax
    resu(k) = covariance(k, x)
Next k
vecovar = resu
End Function
```

The function `read_signal` reads the data corresponding to the values $X_1, \ldots, X_{ns}$ in the line specified by `line`, starting at column `col`. Functions `mean` and `variance` evaluate the mean and the variance of the signal `x(1),..., x(ns)`. Function `vautocorr` furnishes a vector containing the autocorrelations $R_X(s)$, $0 \leq s \leq kmax$. Function `vcovariance` furnishes a vector containing the autocovariances $C_X(s)$, $0 \leq s \leq kmax$. To use these functions together, we need a sub which calls them. For instance, open a blank workbook and entre the data below:

	A	B	C	D	E	F	G	H	I	J	K
1	X	0,120762	0,351975	0,430462	0,340533	0,576352	0,902625	0,714455	0,741438	0,398821	0,986865
2	mean										
3	variance										
4	autocorr										
5	covar										
6											
7											

Then, add the module at left and execute it as a macro: the results appear as below

```
Sub test()
x = read_signal(1, 2, 10)
m = mean(x)
Cells(2, 2).Value = m
s2 = variance(x)
Cells(3, 2).Value = s2
kmax = 4
rx = vautocorr(kmax, x)
For j = 0 To kmax
  Cells(4, j + 2).Value = rx(j)
Next j
cx = vcovar(kmax, x)
For j = 0 To kmax
  Cells(5, j + 2).Value = cx(j)
Next j
End Sub
```

A	B	C	D	E	F
X	0,264056	0,87051	0,609802	0,270287	0,548867
mean	0,556429				
variance	0,075961				
autocorr	0,377978	0,324555	0,353121	0,338316	0,34466
covar	0,068365	0,014942	0,043508	0,028703	0,035047

In the analysis of weakly stationary process, we may use the partial autocorrelations, defined below:

The **partial autocorrelation** $\rho(s)$ of a discrete process X is

$$\rho(s) = E((X_{t+s} - P_s X_{t+s})(X_t - P_s X_t)),$$

Where $P_s X_t$ is the best approximation of X_t by an affine function of $X_{s,t} = (X_{t+1}, \ldots, X_{t+s-1})$, id est,

$$P_s X_t = \arg\min \left\{ E\left((Y - X_t)^2\right) : Y \text{ is an affine function of } X_{s,t} \right\}$$

Let

$$A = (C_X(i-j), \quad 1 \le i, \quad j \le s). \tag{4.1}$$

and

$$B = \left(B_{ij}, \ 1 \le i, \ j \le s\right), \ B_{ij} = C_X(i-j), \quad if\, j < s; \quad B_{is} = C_X(i). \tag{4.2}$$

Then

$$\rho(s) = \frac{\det(A)}{\det(B)}, \tag{4.3}$$

In addition, $\rho(s)$ coincides with the coefficient of X_t for the best approximation of X_{t+s} by an affine function of $X_t, \ldots, X_{t+s-1}$. ∎

The reader will find in the literature algorithms for the evaluation of partial autocorrelations, such as the algorithm of Durbin-Levinson or the algorithm of Schur. The Durbin-Levinson's algorithm evaluates recursively the coefficients $a_1^s, \ldots, a_s^s$ such that the best approximation of X_{t+s} by an affine function of $X_t, \ldots, X_{t+s-1}$ is $X_{t+s} = \sum_{i=1}^{s} a_i^s X_{t+s-i}$: $\rho(s) = a_s^s$. Schur's algorithm evaluates $\rho(s)$ (see, for instance, (Pourahmadi, 2001)). To evaluate partial autocorrelations using Eq. (4.3) we may use the following code:

```
Function vpartcorr(kmax, x)
aux = vcovar(kmax, x)
Dim mat1() As Double
Dim mat2() As Double
Dim resu() As Double
ReDim resu(0 To kmax)
resu(0) = 1
resu(1) = aux(1) / aux(0)
For k = 2 To ncor
   ReDim mat1(1 To k, 1 To k)
   For i = 1 To k
       mat1(i, i) = aux(0)
       For j = i + 1 To k
           mat1(i, j) = aux(j - i)
           mat1(j, i) = aux(j - i)
       Next j
   Next i
   ReDim mat2(1 To k, 1 To k)
   For i = 1 To k
       For j = 1 To k - 1
           mat2(i, j) = mat1(i, j)
       Next j
       mat2(i, k) = aux(i + 1)
   Next i
   aux1 = WorksheetFunction.MDeterm(mat2)
   aux2 = WorksheetFunction.MDeterm(mat1)
   resu(k) = aux1 / aux2
Next k
vpartcorr = resu
End Function
```

For instance, add the lines at left in the Sub test previously introduced and run the sub as a macro:

```
rhox = vpartcorr(kmax, x)
For j = 0 To kmax
  Cells(6, j + 2).Value = rhox(j)
Next j
End Sub
```

4	autocorr	0,152825	0,137676	0,117214	0,133059	0,143379
5	covar	0,029415	0,014266	-0,0062	0,009648	0,019969
6	part corr	1	0,484981	0	0	0

Example
Let X_t be a discrete variable Poisson distributed with parameter λt, id est,

$$P(X_t = n) = \frac{(\lambda t)^n}{n!} e^{-\lambda t}.$$

Then,

$$m_X(t) = \lambda t, \quad V_X(t) = \lambda t, \quad R_{XX}(s,t) = \lambda \min(s,t) + \lambda^2 st.$$

So,

$$C_{XX}(s,t) = \lambda \min(s,t).$$

This process, usually called Poisson Process, is not weakly stationary. Thus, it is not ergodic.

∎

Example
Let $X_t = A_1 \cos(\omega t) + A_2 \sin(\omega t)$, where $\omega \in \mathbb{R}$, while A_1, A_2 are independent random variables such that $E(A_i) = m_i, V(A_i) = \sigma_i^2$. Then,

$$m_X(t) = m_1 \cos(\omega t) + m_2 \sin(\omega t),$$

$$V_X(t) = \sigma_1^2 \cos^2(\omega t) + \sigma_2^2 \sin^2(\omega t),$$

$$C_{XX}(s,t) = \sigma_1^2 \cos(\omega t)\cos(\omega s) + \sigma_2^2 \sin(\omega t)\sin(\omega s)$$

$$R_{XX}(s,t) = C_{XX}(s,t) + m_X(s)m_X(t)$$

This process, usually called Harmonic Process, is not weakly stationary for $m_i \neq 0$. Thus, it is not ergodic for $m_i \neq 0$. If $m_1 = m_2 = 0$ and $\sigma_1^2 = \sigma_2^2 = \sigma^2$, then

$$m_X(t) = 0, \quad V_X(t) = \sigma^2, \quad R_{XX}(s,t) = \sigma^2 \cos(\omega(s-t)).$$

In this case, the process is ergodic in the mean and in the correlation. ∎

We are often interested in the use of the past values for the prediction of future values of a process. As previously observed, the best prediction for the value of X_{n+1} knowing $\mathbf{X}_n = (X_1, .. X_n)$ is $E(X_{n+1} | \mathbf{X}_n)$. The difference $e_{n+1} = X_{n+1} - E(X_{n+1} | \mathbf{X}_n)$ is the **innovation process**. We have

$$E(e_n) = 0; \qquad E(e_m e_n) = 0 \ if \ m \neq n.$$

We say that the process is **deterministic** if and only if $e_n = 0, \forall n$. Notice that a deterministic process remains random: it *is not* deterministic in the sense where X_n is not a random variable: in the context of stochastic processes, the expression deterministic means that the innovation is null.

It is usual to separate the processes into two parts: the trend and the fluctuations. In general, the trend is first determined and then the fluctuations are studied. Trend may be deterministic or stochastic.

To achieve such a decomposition, we write $X_t = f(t) + \delta X_t$, where f is a deterministic function (for instance, $f(t) = m_X(t)$). If there exists f such that δX_t is stationary, then we say that X_t is **stationary in trend.**

We say that the trend is **seasonal** when the trend is a periodic function.

As an alternative, we may write $X_{t+1} = X_t + \delta X_t$. If $\delta X_t = X_{t+1} - X_t$ is stationary, we say that the process is **stationary in difference** or **integrated.**

More generally, we may consider $\delta^2 X_t = \delta X_{t+1} - \delta X_t, \ldots, \delta^k X_t$ processes that are **stationary in difference of order k or integrated of order k** ■

Checking the stationarity of a process can be an arduous task. For AR or ARMA processes (see Sect. 4.5), stationarity tests exist in the literature, referred as "unit root tests", such as the Dickey-Fuller test (Dickey & Fuller, 1979), the Philips-Perron test (Phillips & Perron, 1988), the LMC test (Leybourne & McCabe, 1994), or the KPSS test (Kwiatkowski et al., 1992). As mentioned, these tests assume that the process is AR (see Sect. 4.5). Sometimes, they also use a trend model: in practice, the trend model is either assumed or estimated from data – in the first case, arguments are furnished to justify the choice; in the second one, a procedure of estimation is used (for instance, least squares).

Example

Let $X_{n+1} = X_n + \varepsilon_n$, where $X_0 = 0$ and $\{\varepsilon_n : n \in \mathbb{N}\}$ is a family of independent gaussian variables of same distribution $N(0, 1)$. Then $\delta X_n = X_{n+1} - X_n = \varepsilon_n$. Thus, the distribution of δX_n is independent from t: we have $m_{\delta X}(t) = 0$, $V_{\delta X}(t) = 1$, $R_{\delta X \delta X}(s, t) = 0$, *if* $s \neq t$, $R_{\delta X \delta X}(t, t) = 1$. Thus, the process is stationary in difference.

We have $m_X(t) = 0$, $V_X(t) = t$, $R_{XX}(t, s) = \min\{t, s\}$, $C_{XX}(t, s) = \min\{t, s\}$.

Let us consider the partial autocorrelations: the projection of X_{n+1} onto the subspace generated by $\mathcal{X}_{1,n} = \{X_n\}$ is $P_n X_{n+1} = X_n$ (since their coefficient of linear correlation is $\rho(X_{n+1}, X_n) = 1$ – see page 106). Thus, $\rho(1) = 1$. For $k > 1$,the projection of X_{n+k} onto the subspace generated by $\mathcal{X}_{k,n} = \{X_n, \ldots, X_{n+k-1}\}$ is $P_n X_{n+k} = X_{n+k-1}$, so that $\rho(k) = 0$. ■

Example

Let $X_{n+1} = h(n) + \frac{1}{2} X_n + \varepsilon_n$, where $X_0 = 0$ and $\{\varepsilon_n : n \in \mathbb{N}\}$ is a family of independent gaussian variables of same distribution $N(0, 1)$. Then $X_{n+1} = \sum_{k=0}^{n} \frac{h(n-k)}{2^k} + \sum_{k=0}^{n} \frac{\varepsilon_{n-k}}{2^k}$. Thus, the trend is $f(n) = \sum_{k=0}^{n} \frac{h(n-k)}{2^k}$ and $\delta X_n = \sum_{k=0}^{n} \frac{\varepsilon_{n-k}}{2^k}$. We have $m_{\delta X}(n) = 0$, $V_{\delta X}(n) = \frac{4}{3}\left(1 - \frac{1}{4^{n+1}}\right)$. Thus, the process is not stationary in trend. ■

Example

Let $X_{n+1} = h(n) + \frac{1}{2}X_n + \varepsilon_n$, where $\{\varepsilon_n : n \in \mathbb{Z}\}$ is a family of independent gaussian variables of same distribution $N(0, 1)$. There is a trend-stationary version of the process, by taking $X_n = \sum_{k=0}^{+\infty} \frac{h(n-k)}{2^k} + \sum_{k=0}^{+\infty} \frac{\varepsilon_{n-k}}{2^k}$. Then, the trend is $f(n) = \sum_{k=0}^{+\infty} \frac{h(n-k)}{2^k}$ and $\delta X_n = \sum_{k=0}^{+\infty} \frac{\varepsilon_{n-k}}{2^k}$. We have $m_{\delta X}(n) = 0$, $V_{\delta X}(t) = \frac{4}{3}$. The distribution of δX_n is $\left(0, \frac{2}{\sqrt{3}}\right), \forall n$. ∎

Exercises

1. Let $X_{n+1} = a + X_n + \varepsilon_n$, where $X_0 = 0$ and $\{\varepsilon_n : n \in \mathbb{N}\}$ is a family of independent gaussian variables of same distribution $N(0, 1)$.

   (a) Show that $X_{n+1} = (n+1)a + \sum_{k=0}^{n} \varepsilon_k$
   (b) Show that $m_X(n) = na$.
   (c) Show that $V_X(n) = n$.
   (d) Show that $R_{XX}(m, n) = mna^2 + \min\{m, n\}$
   (e) Analyze the process $Y_n = X_{n+1} - X_n$.
   (f) Is this process stationary of some kind?

2. Let $\{\varepsilon_n : n \in \mathbb{N}\}$ be a family of independent gaussian variables of same distribution $N(0, 1)$ and $X_{2k} = 0$, $X_{2k+1} = 2\varepsilon_{2k-1}$

   (a) Show that $m_X(n) = 0$.
   (b) Show that $V_X(n) = 0$, *if n pair*; $V_X(n) = 4$, *ifn odd*.
   (c) Show that $R_{XX}(m, n) = 0$ *if* $m \neq n$;
   (d) Determine $R_{XX}(m, m)$
   (e) Analyze the process $Y_n = X_n - X_{n-1}$.
   (f) Is this process stationary of some kind?

3. Let $\{\varepsilon_n : n \in \mathbb{N}\}$ be a family of independent gaussian variables of same distribution $N(0, 1)$ and $X_n = \varepsilon_n \varepsilon_{n+1} \varepsilon_{n+2}$

   (a) Show that $m_X(n) = 0$.
   (b) Show that $V_X(n) = 1$.
   (c) Show that $R_{XX}(m, n) = 0$ *if* $m \neq n$
   (d) Determine $R_{XX}(m, m)$
   (e) Is this process stationary of some kind?

4. Let $\{\varepsilon_n : n \in \mathbb{N}\}$ be a family of independent gaussian variables of same distribution $N(0, 1)$ and $X_n = \varepsilon_n + \varepsilon_{n+1} + \varepsilon_{n+2}$

   (a) Show that $m_X(n) = 0$.
   (b) Show that $V_X(n) = 3$.
   (c) Show that $R_{XX}(m, n) = 0$ *if* $|m - n| \geq 3$

(d) Determine $R_{XX}(m, m)$, $R_{XX}(m, m+1)$, $R_{XX}(m, m+2)$
(e) Is this process stationary of some kind?

5. Let $\{\varepsilon_n : n \in \mathbb{N}\}$ be a family of independent gaussian variables of same distribution $N(0, 1)$ and $X_n = \varepsilon_n + a\varepsilon_{n+1}$.

   (a) Show that $m_X(n) = 0$.
   (b) Show that $V_X(n) = 1 + a^2$.
   (c) Show that $R_{XX}(m, n) = 0$ *if* $|m - n| \geq 1$
   (d) Determine $R_{XX}(m, m)$, $R_{XX}(m, m+1)$
   (e) Is this process stationary of some kind?

6. Let $\{\varepsilon_n : n \in \mathbb{N}\}$ be a family of independent gaussian variables of same distribution $N(0, 1)$ and $X_n = an + b + a_0\varepsilon_n$.

   (a) Determine $m_X(n)$.
   (b) Determine $V_X(n)$.
   (c) Is this process stationary?
   TIP: Analyze the process $\delta X_n = X_n - X_{n-1}$

7. Let $\{\varepsilon_n : n \in \mathbb{N}\}$ be a family of independent gaussian variables of same distribution $N(0, 1)$ and $X_n = an + b + a_0\varepsilon_n + a_1\varepsilon_{n-1}$.

   (a) Determine $m_X(n)$.
   (b) Determine $V_X(n)$.
   (c) Is this process stationary?
   TIP: Analyze the process $Y_n = X_{n+1} - 2X_n + X_{n-1}$

8. Let $\{\varepsilon_n : n \in \mathbb{N}\}$ be a family of independent gaussian variables of same distribution $N(0, 1)$ and $X_n = h(n) + a_0\varepsilon_n + a_1\varepsilon_{n-1}$ and $Y_n = X_n - h(n)$

   (a) Determine $m_Y(n)$. Conclude $m_X(n)$
   (b) Determine $V_Y(n)$. Infer the value of $V_X(n)$.
   (c) Let $c_0, c_1, \ldots, c_k$ be such that $E\left(\sum_t (X_n - p(n))^2\right)$ is minimal, with $p(n) = c_0 + c_1 n + \ldots + c_k n^k$. Show that $\sum_t (h(n) - p(n))^2$ is minimal.

   TIP: Show that $E\left(\sum_t (X_t - p(t))^2\right) = \sum_t E\left((h(t) - p(t))^2\right)$

4.2 Determination of the Distribution of a Stationary Process

If X is stationary, then the probability distribution of X_n is the same for all the values of n. In such a situation, we may use a trajectory to determine the common distribution, by considering the points in the trajectory as a sample from this common distribution. For instance, we may determine a representation of X by one of the methods presented in Chap. 5. Once the representation is determined, if furnishes the CDF and the PDF of X, as shown in Chap. 5.

Example

Let us consider the situation where the available data consists in 250 points of $X_n = 1 + \varepsilon_n - 2\varepsilon_{n-1} + 0.5\varepsilon_{n-2}$, where $\{\varepsilon_n : n \in \mathbb{Z}\}$ is a family of independent gaussian variables of same distribution $N(0, 1)$ – it is a gaussian white noise (see Sect. 4.3). We use the data to generate a polynomial representation $X \approx c_0 + c_1U + c_2U^2 + \ldots c_kU^k$.

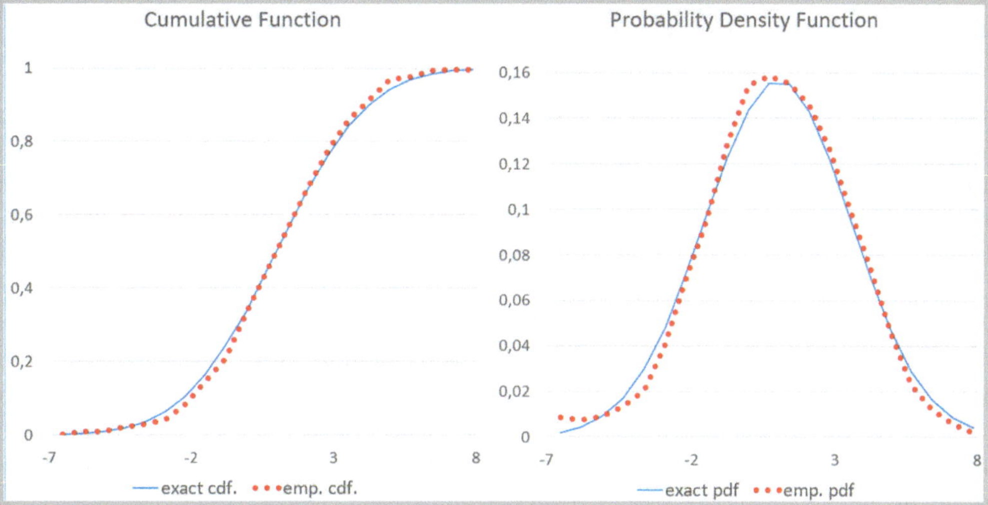

Fig. 4.1 Cumulative function and probability density of process $X_n = 1 + \varepsilon_n - 2\varepsilon_{n-1} + 0.5\varepsilon_{n-2}$, corresponding to the approximation by a polynomial of degree 9 of an uniform variable U. The PDF is generated by SPH numerical derivative, with a gaussian kernel.. Dashed line is the approximation. Continuous line is the exact value

Since we do not know the exact nature of X, we use U uniformly distributed on (0, 1) and $k = 9$. The results obtained for the CDF and the PDF of X are shown in Fig. 4.1.

Of course, it is also possible to use a gaussian variable $U \sim N(0, 1)$. The corresponding results are exhibited in Fig. 4.2:

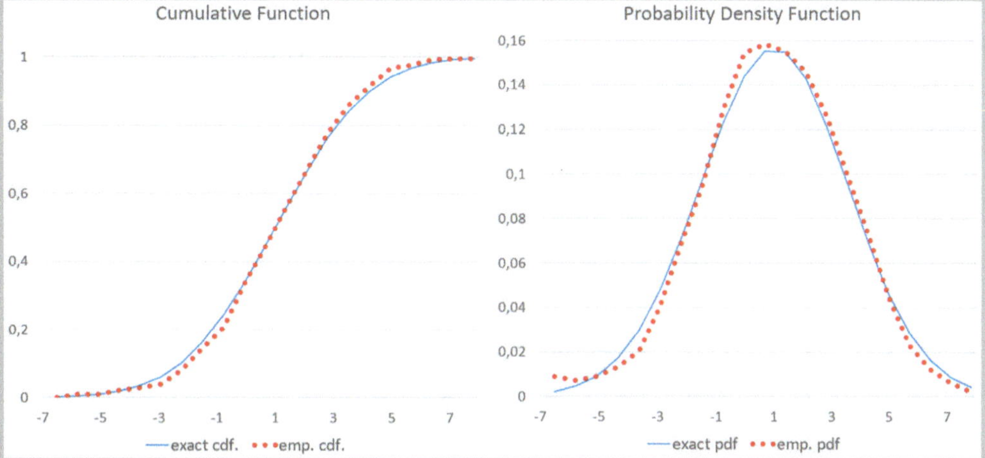

Fig. 4.2 Cumulative function and probability density of process $X_n = 1 + \varepsilon_n - 2\varepsilon_{n-1} + 0.5\varepsilon_{n-2}$, corresponding to the approximation by a polynomial of degree 9 of a gaussian variable U. The PDF is generated by SPH numerical derivative, with a gaussian kernel. Dashed line is the approximation. Continuous line is the exact value ∎

4.3 White Noise

X is a **white noise process** if and only if X is ergodic and

$$m_X(t) = 0 \text{ and } R_X(\tau) = \sigma^2\delta(\tau),$$

where δ is the Dirac's mass and $\sigma > 0$. ■

Notice that the definition does not mention the distribution of X_t: indeed, a white noise may be generated by different distributions. For instance, a white noise generated by an uniform distribution is an *uniform white noise*. Analogously, a white noise generated by a gaussian distribution is a *gaussian white noise*. To generate a white noise, we must generate independent variables for each t. For instance, we may consider X_t as uniformly distributed on $\left(-\sigma\sqrt{3},\ \sigma\sqrt{3}\right)$ or normally distributed $N(0, \sigma)$.

As an example, we may generate 10^4 samples of a single variate from the uniform distribution on $\left(-\sqrt{3}, \sqrt{3}\right)$ – we use the ANALYSIS TOOLPACK with a number of variables equal to 10000 and number of Random numbers equal to 1. The random numbers generated simulate a white noise on the time interval (0, 10), with a time step $\Delta t = 0{,}001$: the element number i of the sample corresponds to $X(t_i)$, $t_i = (i - 1)\Delta t$. We use the data to evaluate the autocorrelation for $\tau \in (-1, 1)$. The results are shown in Fig. 4.3.

Into an analogous way, we may generate a sample of a single value of 10000 gaussian variables $N(0, 1)$ and simulate a white noise as indicated above. The corresponding autocorrelation $R_X(\tau)$ for $\tau \in (-1, 1)$ is shown in Fig. 4.4.

Remark
One of the properties of white noise processes concerns innovation: if $\{e_n : n \in \mathbb{N}\}$ is an innovation process and $E\left(e_n^2\right) = \sigma^2, \forall n$ then e_n is a white noise. ■

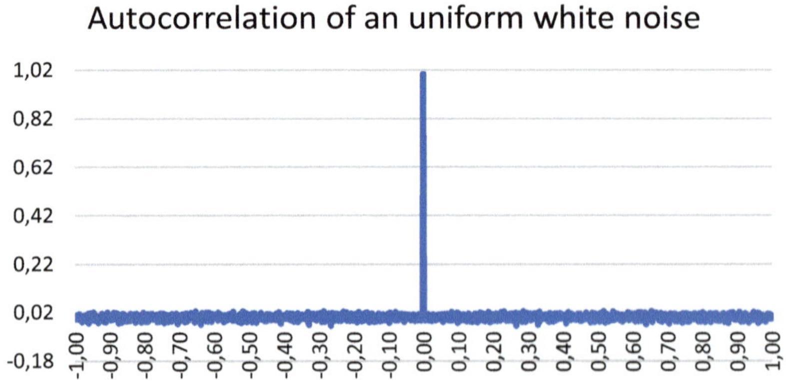

Fig. 4.3 Autocorrelation from a uniform white noise generated by the ANALYSIS TOOLPACK

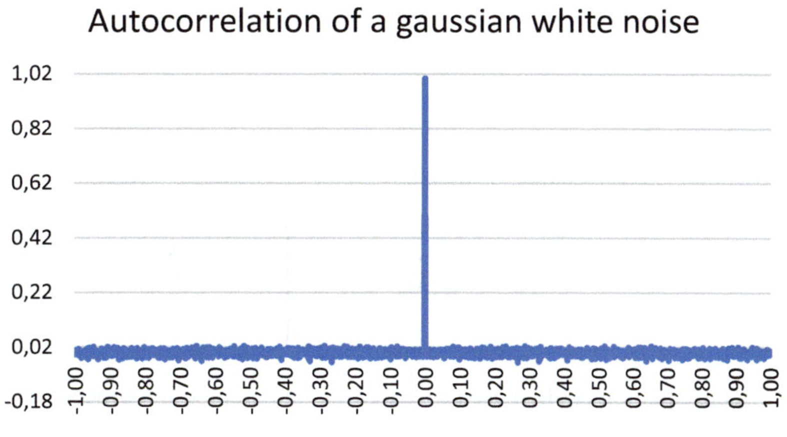

Fig. 4.4 Autocorrelation from a gaussian white noise generated by the ANALYSIS TOOLPACK

An example of code for the generation of a white noise using VBA is the following

```
Function wngen(ns)
aux = RandomNumbers.normgen_bm(0, 1, ns)
wngen = aux
End Function
Sub wnsheet()
ns = Cells(1, 2).Value
e = wngen(ns)
Cells(2, 1).Value = "white noise"
For i = 1 To ns
   Cells(2, i + 1).Value = e(i)
Next i
End Sub
```

This code takes the value of ns in cell `A2` end generates the sample of the white noise in line 2, starting at cell `B2`. It uses the module *RandomNumbers.bas* (see page 10 to import a module into the workbook), which contains `normgen_bm` (see page 154) – as an alternative, you can include the code of `normgen_bm` as a subprogram in your workbook (see page 7 to include a module in the workbook).

4.4 Moving Average Processes

X is a **moving average of order k**,denoted ***MA*(*k*)**, if and only if there exists a discrete white noise $\varepsilon = \{\varepsilon_n : n \in \mathbb{Z}\}$, a number $c \in \mathbb{R}$ and a vector of coefficients $\boldsymbol{a} = (a_1, \ldots, a_k)$ such that

$$\forall n \in \mathbb{Z} : X_n = c + \varepsilon_n + \sum_{i=1}^{k} a_i \varepsilon_{n-i} = c + \sum_{i=0}^{k} a_i \varepsilon_{n-i} \quad (a_0 = 1) \qquad \blacksquare$$

A moving average process is weakly stationary. Then,

$$E(X_n) = c, \;\; V(X_n) = \sigma^2 \sum_{i=0}^{k} a_i^2,$$

where σ^2 is the variance of the white noise ε. The autocorrelation of $MA(k)$ is

$$R_X(p) = E\big(X_n X_{n+p}\big) = \begin{cases} c^2 + \sigma^2 \sum_{j=0}^{k-|p|} a_j a_{j+|p|}, & if\ |p| < k \\ c^2, & otherwise. \end{cases}$$

This equation is often used to determine the order k of a process (k): indeed, $R_X(p)$ is constant for $p \geq k$.

The simulation of a MA process for given coefficients c, $a_1, \ldots, a_k$ is performed by generating the white noise (see Sect. 4.3) and using the expression of X_n. An example of code for the simulation is given below at left. To run the code, we may use a sub analogous to the one given at the right side:

```
Function magen(c, a, ns)
k = UBound(a, 1)
e = wngen(ns + k + 1)
Dim x() As Double
ReDim x(1 To ns)
For i = 1 To ns
    n = k + i
    s = e(n)
    For j = 1 To k
        s = s + a(j) * e(n - j)
    Next j
    x(i) = s + c
Next i
magen = x
End Function
```

```
Sub masheet()
ns = Cells(1, 2).Value
c = Cells(2, 2).Value
lastcol = Cells(3, Cells.Columns.Count).End(xlToLeft).Column
k = lastcol - 1
Dim a() As Double
ReDim a(1 To k)
For i = 1 To k
   a(i) = Cells(3, i + 1).Value
Next i
x = magen(c, a, ns)
Cells(4, 1).Value = "MA(" & CStr(k) & ")"
For i = 1 To ns
   Cells(4, i + 1).Value = x(i)
Next i
End Sub
```

This `Sub` reads the size of the sample *ns* in cell `A2`, *c* in cell `B2` and the coefficients $a_1, \ldots, a_k$ in line 3, starting in cell `C2`. The results appear in line 4, starting at cell `D2`. The code uses the function wngen previously introduced (page 181).

A central result concerning Moving Average processes is Wold's theorem:

Theorem (Wold)
Let $\{X_n : n \in \mathbb{Z}\}$ be a second-order stationary process. Then, X_n corresponds to a process *MA* (∞): there exists a deterministic process $\{D_n : n \in \mathbb{N}\}$ and a sequence of real coefficients $\{a_n : n \in \mathbb{N}\}$ such that

$$\forall n \in \mathbb{Z} : X_n = D_n + \sum_{i=0}^{+\infty} a_i \varepsilon_{n-i}$$

with $a_0 = 1$, $\sum_{i=0}^{+\infty} a_i^2 < \infty$, $cov(D, \varepsilon_i) = 0$, $\forall\, i$. ■

Wold's theorem suggests that second order stationary processes may be approximated by considering truncations $MA(k)$ of the infinite series $MA(\infty)$:

$$X_n \approx D_n + \sum_{i=0}^{k} a_i \varepsilon_{n-i}. \tag{4.4}$$

The simulation of a stationary process by using Wold's representation is made by using the approximation given in Eq. (4.4): the function magen previously introduced furnishes the simulation.

The coefficients of the approximation may be determined from a trajectory of X_n: indeed, we can determine m_X and $R_X$from the trajectory. Since

$$E(X_n) \approx \mathrm{E}\left(D_n + \sum_{i=0}^{k} a_i \varepsilon_{n-i}\right) = E(D_n), \tag{4.5}$$

we may take the deterministic process as being the mean of X_n: $D_n = m_X$. In addition,

$$E(X_n X_{n+p}) \approx E(D_n D_{n+p}) + E\left(\sum_{i,j=0}^{k} a_i a_j \varepsilon_{n-i} \varepsilon_{n+p-j}\right),$$

so that, for $0 \leq p \leq k$:

$$R_X(p) \approx m_X^2 + \sigma_\varepsilon^2 \sum_{i=0}^{k-p} a_i a_{i+p},$$

Taking a white noise of variance 1, we have:

$$\sum_{i=0}^{k-p} a_i a_{i+p} + \left(m_X^2 - R_X(p)\right) = 0. \tag{4.6}$$

These equations connect the mean, the coefficients, and the autocorrelation: they may be used to determine unknowns among these data. For instance, if the mean m_X and the coefficients $a_1, \ldots, a_k$ are known, then we may determine the values of R_X. Conversely, if the values of R_X are given, we may determine the mean m_X and the coefficients $a_1, \ldots, a_k$.

Example

As an example, let us consider the process $X_n = 1 + \varepsilon_n - 2\varepsilon_{n-1} + 0.5\varepsilon_{n-2}$.

Assume that the coefficients are known: we may use Eq. (4.6) to determine R_X: $R_X(0) = 6.25$, $R_X(1) = R_X(-1) = -2$, $R_X(2) = R_X(-2) = 1.5$, $R_X(p) = 1$, if $|p| > 2$.

Conversely, assume that the values of $R_X(p)$ are known for $p = 0, \ldots, 10$. We determine m_X, $a_1, \ldots, a_k$ (recall that $a_0 = 1$): Eq. (4.66) forms a system of nonlinear equations to be solved by the methods presented in Sect. 2.3. For instance, we use the SOLVER to minimize the RMS error $\sqrt{\frac{\sum_{i=0}^{10}\left(\widehat{R_i}-R_i\right)^2}{11}}$ between the predicted vector $\widehat{R} = \left(\widehat{R_X}(0), \ldots, \widehat{R_X}(10)\right)$ and the real values $R = (R_X(0), \ldots, R_X(10))$. We adopt a strategy of increasing k: a first solution $m_X^{(0)}$ is obtained for $k = 0$. This solution is extended by $a_1 = 0$ and $\left(m_X^{(0)}, 0\right)$ used as starting point to solve the nonlinear system with $k = 1$. We repeat the process until $k = 10$: each solution $\left(m_X^{(k)}, a_1^{(k)}, \ldots, a_k^{(k)}\right)$ is extended by a zero and $\left(m_X^{(k)}, a_1^{(k)}, \ldots, a_k^{(k)}, 0\right)$ is used as starting point to determine the solution $\left(m_X^{(k+1)}, a_1^{(k+1)}, \ldots, a_k^{(k+1)}, a_{k+1}^{(k+1)}\right)$. An example of workbook is given below (Fig. 4.5)

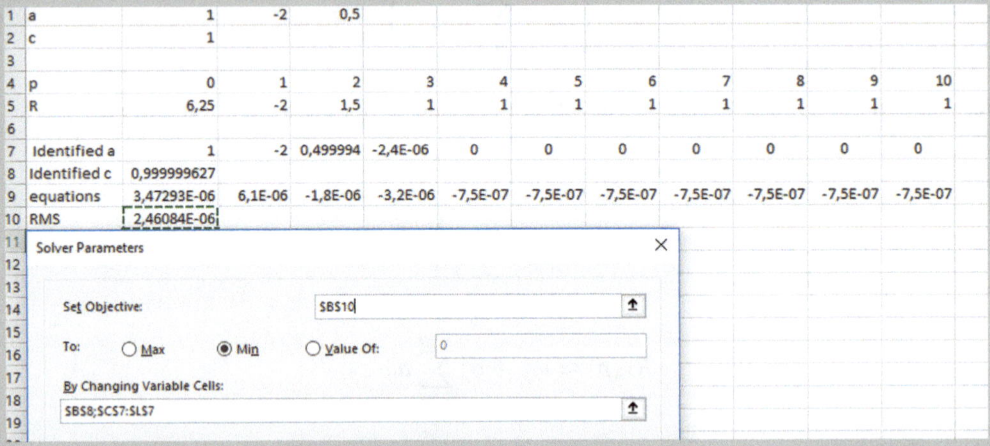

	A	B	C	D	E	F	G	H	I	J	K	L
1	a	1	-2	0,5								
2	c	1										
3												
4	p	0	1	2	3	4	5	6	7	8	9	10
5	R	6,25	-2	1,5	1	1	1	1	1	1	1	1
6												
7	Identified a	1	-2	0,499994	-2,4E-06	0	0	0	0	0	0	0
8	Identified c	0,999999627										
9	equations	3,47293E-06	6,1E-06	-1,8E-06	-3,2E-06	-7,5E-07	-7,5E-07	-7,5E-07	-7,5E-07	-7,5E-07	-7,5E-07	-7,5E-07
10	RMS	2,46084E-06										

Solver Parameters

Set Objective: $B$10

To: Max Min Value Of: 0

By Changing Variable Cells:

$B$8;$C$7:$L$7

Fig. 4.5 Example of workbook with solution for k = 10

The results for $k \leq 5$ are exhibited in Table 4.1 below. We observe that the numerical solution is close to the exact one.

Table 4.1 Results obtained for different values of k. The results do not improve for $k > 2$

k	$(m_X;\ a_1, \ldots, a_k)$	RMS error
0	(1, 08)	1.57
1	(0, 97; −2, 12)	0,31
2	(1.00; −2, 00; 0, 50)	2.9E−6
3	(1.00; −2, 00; 0, 50; −2.4E − 6)	2.5E−6
4	(1.00; −2, 00; 0, 50; −2.4E − 6; 0)	2.5E−6
5	(1.00; −2, 00; 0, 50; −2.4E − 6; 0; 0)	2.5E−6

Since $m_X = 1$, we have $C_X(0) = 5.25$, $C_X(1) = C_X(-1) = -3$, $C_X(2) = C_X(-2) = -0.5$ For $|p| > 2$, $C_X(p) = 0$. Thus, $\rho(1) = -0, 57$, $\rho(2) = -0, 21$, $\rho(3) = -0, 12$, $\rho(4) = -0, 07$, $\rho(5) = -0, 04$, $\rho(6) = -0, 02$, etc. (Fig. 4.6).

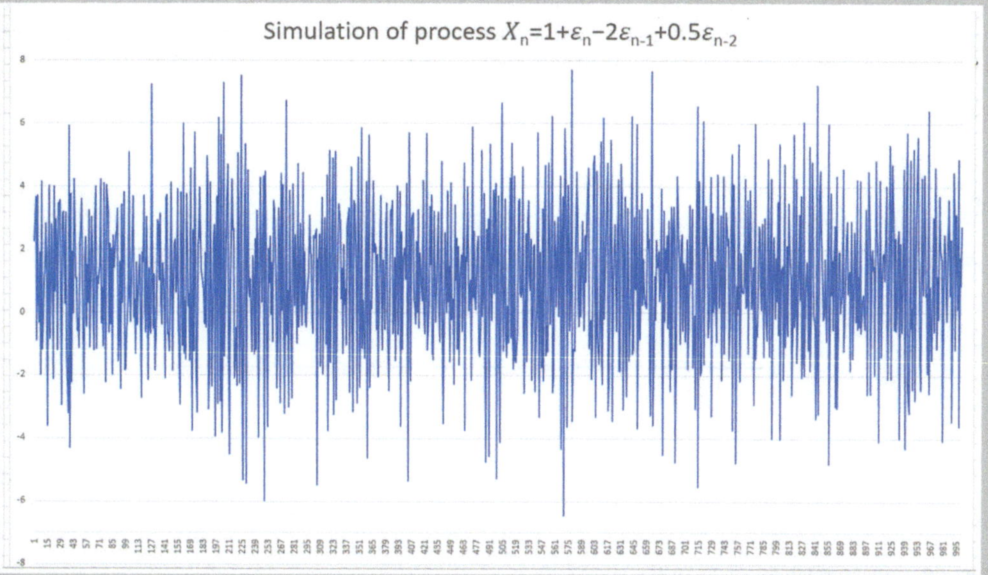

Fig. 4.6 Simulation of X_n by function magen

In general, the exact values of R_X are unknown: we have only estimations from a finite number of observations. Let us illustrate this situation by considering that the available data is a trajectory of $n = 1000$ points $X_1, \ldots, X_n$.

We may generate these data by using the function `magen` previously introduced, but we may also use a workbook (see example below) where the white noise is furnished by the ANALYSIS TOOLPACK in line 1. We generate 10003 values from the white noise and use them to evaluate X_n in line 3. The mean of the process is evaluated as the mean of line 3 and is placed in `B5`. The RMS error is evaluated in cell `B11` (Fig. 4.7).

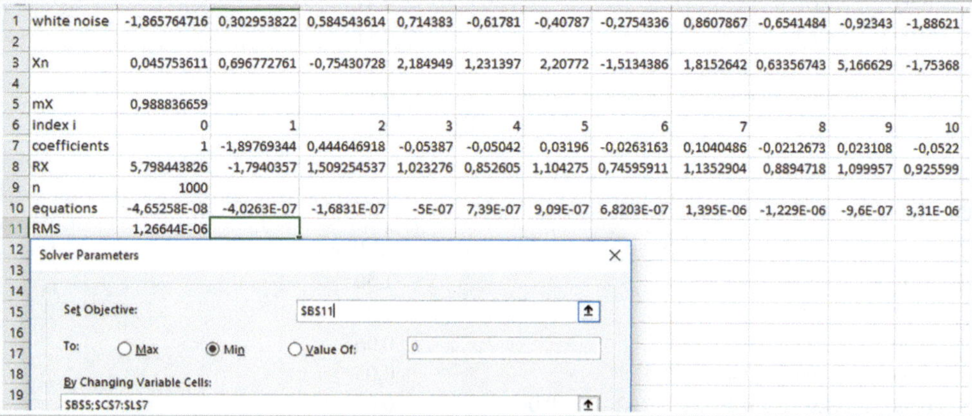

1	white noise	-1,865764716	0,302953822	0,584543614	0,714383	-0,61781	-0,40787	-0,2754336	0,8607867	-0,6541484	-0,92343	-1,88621
2												
3	Xn	0,045753611	0,696772761	-0,75430728	2,184949	1,231397	2,20772	-1,5134386	1,8152642	0,63356743	5,166629	-1,75368
4												
5	mX	0,988836659										
6	index i	0	1	2	3	4	5	6	7	8	9	10
7	coefficients	1	-1,89769344	0,444646918	-0,05387	-0,05042	0,03196	-0,0263163	0,1040486	-0,0212673	0,023108	-0,0522
8	RX	5,798443826	-1,7940357	1,509254537	1,023276	0,852605	1,104275	0,74595911	1,1352904	0,8894718	1,099957	0,925599
9	n	1000										
10	equations	-4,65258E-08	-4,0263E-07	-1,6831E-07	-5E-07	7,39E-07	9,09E-07	6,8203E-07	1,395E-06	-1,229E-06	-9,6E-07	3,31E-06
11	RMS	1,26644E-06										

Fig. 4.7 Example of workbook with solution for k = 10

In a first analysis, only the first 1000 values of X_n are used. The autocorrelations are shown in Fig. 4.8: as expected, they become almost constant for $s > 2$, which corresponds to the order of the process.

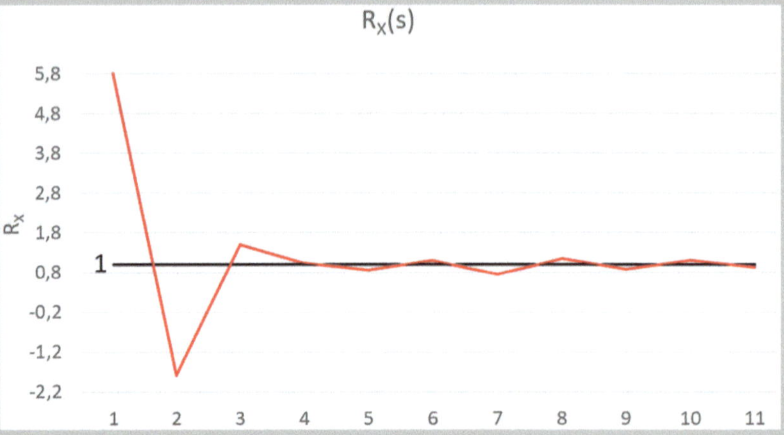

Fig. 4.8 Autocorrelations of X_n in Fig. 4.6

Examples of determination of the coefficients by minimization of the RMS error are shown in Tables 4.2 and 4.3 below, as in Fig. 4.9. The coefficients found are close to the exact ones: the results are good for both the samples (sizes 1000 and 10,000, respectively). The errors are mainly due to the numerical evaluation of R_X, which does not furnish the exact results, but only an approximation – the deviations in the values of R_X generate errors on the values of the coefficients – but the agreement is good. ∎

Table 4.2 Results obtained for different values of k (*sample size* 1000)

k	c	a_1	a_2	a_3	a_4	a_5	RMS error
0	1,06						1,43
1	0,97	−2,01					0,31
2	0,99	−1,89	0,48				0,11
3	0,99	−1,89	0,47	−0,01			0,11
4	0,99	−1,91	0,43	−0,08	−0,08		0,11
5	0,99	−1,90	0,46	−0,03	−0,01	0,08	0,10

Table 4.3 Results obtained for different values of k(*sample size* 10000)

k	c	a_1	a_2	a_3	a_4	a_5	RMS error
0	1,07						1,50
1	0,97	−2,08					0,29
2	0,99	−1,96	0,46				0,06
3	1,00	−1,96	0,48	0,02			0,05
4	0,99	−1,97	0,46	−0,02	−0,04		0,05
5	1,00	−1,96	0,47	0,01	−0,01	0,04	0,05

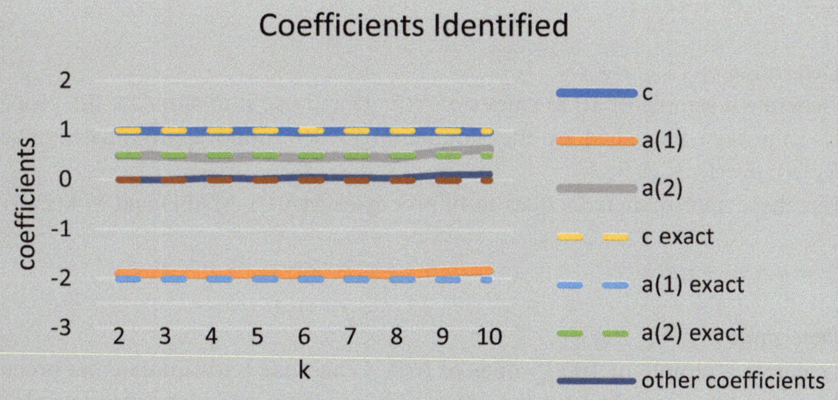

Fig. 4.9 Coefficients identified for each *k* (***sample size* 1000**)

Remark

The analysis of MA processes is often made using ***delay (or lag) operators***. The basic delay operator is $L(X_t) = X_{t-1}$. Thus, $X_{n-k} = L^k X_n$ and we have

$$X_n = c + \varepsilon_n + \sum_{i=1}^{k} a_i \varepsilon_{n-i} \Leftrightarrow X_n - c = B(L)(\varepsilon_n), \quad B(x) = 1 - \sum_{i=1}^{k} a_i x^i.$$

The polynomial B and its roots play an essential rule in the theory of MA and AR processes (see Sect. 4.5 for AR). Indeed, let x_i, $i = 1, \ldots, n$ be the roots of p. Then, $\mathrm{B}(L) = a_k(x_1 - L)\ldots(x_k - L)$. Recalling that product of the roots is equal to $\frac{1}{a_k}$, we have $\mathrm{B}(L) = \left(Id - \frac{L}{x_1}\right)\ldots\left(Id - \frac{L}{x_k}\right)$, where Id is the identity operator. Thus, $\mathrm{B}(L)^{-1}$ exists if and only if not any root of $\mathrm{B}(L)$ has a modulus equal to 1: $|x_i| \neq 1$, $i = 1, \ldots, n$. If $|x_i| > 1$, we have (see, for instance, (Dautray & Lions, 2012))

$$\left(Id - \frac{L}{x_i}\right)^{-1} = \sum_{n=0}^{+\infty} \frac{1}{x_i^{n+1}} L^n$$

Thus, if $|x_i| > 1$, $i = 1, \ldots, n$, then $\varepsilon_n = B(L)^{-1}(X_n - c)$. ∎

Exercises

1. Let $X_n = 1 + \varepsilon_n$.

   (a) Determine m_X, V_X, R_X, C_X.
   (b) Generate a sample of 1000 values of $N(0, 1)$ and use it to simulate the process.
   (c) Use the data furnished by the simulation to determine approximate values of m_X, $R_X(p)$, $p = 0, 1, 2, 3, 4$.
   (d) Use these approximated values to fit processes MA(0), MA(1), MA(2) to the data.

2. Let $X_n = 1 + \varepsilon_n + \frac{1}{2}\varepsilon_{n-1}$.

   (a) Determine m_X, V_X, R_X, C_X.
   (b) Generate a sample of 1000 values of $N(0, 1)$ and use it to simulate the process.
   (c) Use the data furnished by the simulation to determine approximate values of m_X, $R_X(p), p = 0, 1, 2, 3, 4$.
   (d) Use these approximated values to fit processes MA(0), MA(1) and MA(2) to the data.

3. Let $X_n = \mu + \varepsilon_n + \alpha\varepsilon_{n-1}$.

   (a) Determine m_X, V_X, R_X, C_X.
   (b) Generate a sample of 1000 values of $N(0, 1)$ and use it to simulate the process.
   (c) Use the data furnished by the simulation to determine approximate values of m_X, $R_X(p), p = 0, 1, 2, 3, 4$.
   (d) Use these approximated values to fit processes MA(0), MA(1) and MA(2) to the data.

4. Let $X_n = \varepsilon_n - \varepsilon_n - \alpha\varepsilon_{n-2}$.

   (a) Determine m_X, V_X, R_X, C_X.
   (b) Generate a sample of 1000 values of $N(0, 1)$ and use it to simulate the process.
   (c) Use the data furnished by the simulation to determine approximate values of m_X, $R_X(p), p = 0, 1, 2, 3, 4$.
   (d) Use these approximated values to fit processes MA(0), MA(1) and MA(2) to the data.

5. Let $X_n = \varepsilon_n - \alpha\varepsilon_{n-2}$.

   (a) Determine m_X, V_X, R_X, C_X.
   (b) Generate a sample of 1000 values of $N(0, 1)$ and use it to simulate the process.
   (c) Use the data furnished by the simulation to determine approximate values of m_X, $R_X(p), p = 0, 1, 2, 3, 4$.
   (d) Use these approximated values to fit processes MA(0), MA(1) and MA(2) to the data.

4.5 Autoregressive Processes

X is **auto regressive of order k**, denoted $\boldsymbol{AR(k)}$, if and only if X is weakly stationary and there exists a discrete white noise $\varepsilon = \{\varepsilon_n : n \in \mathbb{Z}\}$, a number $c \in \mathbb{R}$ and a vector of coefficients $\boldsymbol{a} = (a_1, \ldots, a_k)$ such that

$$\forall n \in \mathbb{Z} : X_n = c + \varepsilon_n + \sum_{i=1}^{k} a_i X_{n-i}. \qquad \blacksquare$$

A condition to verify if the process is stationary is furnished by the result below (see also the remark page 187):

Theorem

The process $AR(k)$ given by $X_n = c + \varepsilon_n + \sum_{i=1}^{k} a_i X_{n-i}$ is stationary if and only if the polynomial $A(x) = 1 - \sum_{i=1}^{k} a_i x^i$ has not any root of modulus inferior or equal to 1: $A(x) = 0 \Longrightarrow |x| > 1$. ■

Let us denote $E(X_n) = \mu$. We have

$$\mu = c + \sum_{i=1}^{k} a_i \mu,$$

So that, for $1 - \sum_{i=1}^{k} a_i \neq 0$,

$$\mu = \frac{c}{1 - \sum_{i=1}^{k} a_i}. \tag{4.7}$$

Assume that the white noise has a variance $\sigma^2 = 1$. Then,

$$E(X_n X_n) = c\mu + 1 + \sum_{i=1}^{k} a_i E(X_n X_{n-i})$$

and, for $p > 0$,

$$E(X_n X_{n+p}) = c\mu + \sum_{i=1}^{k} a_i E(X_n X_{n+p-i}).$$

Thus,

$$R_X(p) = c\mu + \delta_{p0} + \sum_{i=1}^{k} a_i R_X(i - p) \tag{4.8}$$

Equations (4.7) and (4.8) are usually referred as the Yule-Walker equations. Taking the values =0, 1, . . ., k, we obtain $k + 1$ linear equations for the determination of the quantities $c\mu, a_1, \ldots, a_k$:

$$\begin{pmatrix} R_X(1) & \cdots & R_X(k) & 1 \\ R_X(0) & \cdots & R_X(k-1) & 1 \\ \vdots & \vdots & \vdots & \vdots \\ R_X(2-k) & \cdots & R_X(1) & 1 \\ R_X(1-k) & \cdots & R_X(0) & 1 \end{pmatrix} \begin{pmatrix} a_1 \\ \vdots \\ a_k \\ c\mu \end{pmatrix} = \begin{pmatrix} R_X(0) - 1 \\ R_X(1) \\ \vdots \\ R_X(k-1) \\ R_X(k) \end{pmatrix}$$

Once $c\mu$ is known, c is determined by Eq. (4.7). We have

$$c = \sqrt{c\mu \times \left(1 - \sum_{i=1}^{k} a_i\right)}$$

The values of the coefficients may also be determined by a least squares procedure. Indeed, we have

$$\begin{pmatrix} 1 & X_k & \cdots & X_1 \\ 1 & X_{k+1} & \cdots & X_2 \\ \vdots & \vdots & \vdots & \vdots \\ 1 & X_{k+n} & \cdots & X_{n+1} \end{pmatrix} \begin{pmatrix} c \\ a_1 \\ \vdots \\ a_k \end{pmatrix} = \begin{pmatrix} X_{k+1} \\ X_{k+2} \\ \vdots \\ X_{k+n+1} \end{pmatrix} - \begin{pmatrix} \varepsilon_{k+1} \\ \varepsilon_{k+2} \\ \vdots \\ \varepsilon_{k+n+1} \end{pmatrix}.$$

Thus, the vector of coefficients $(c, a_1, \ldots, a_k)^t$ is the solution of a noisy linear system. Since the noise is gaussian, the Maximum Likelihood Estimator of the coefficients coincides with the least squares solution of the linear system where the vector of noise is neglected, id est, the solution of

$$\begin{pmatrix} 1 & X_k & \cdots & X_1 \\ 1 & X_{k+1} & \cdots & X_2 \\ \vdots & \vdots & \vdots & \vdots \\ 1 & X_{k+n} & \cdots & X_{n+1} \end{pmatrix} \begin{pmatrix} c \\ a_1 \\ \vdots \\ a_k \end{pmatrix} = \begin{pmatrix} X_{k+1} \\ X_{k+2} \\ \vdots \\ X_{k+n+1} \end{pmatrix}. \tag{4.9}$$

Notice that the linear system is overdetermined for $n > k + 1$.

Since the process is weakly stationary, $R_X(p - i) = R_X(i - p)$ and we may use the Yule-Walker equations to determine the values of μ, $R_X(p)$, when the values of $c, a_1, \ldots, a_k$ are known. Indeed, we start by the determination of μ by using Eq. (4.7). The second step is the use of Eq. (4.8): for a stationary process, so that Eq. (4.8) form a linear system for $R_X(p), p = 0, \ldots k$. For instance, for $k = 1$:

$$\begin{pmatrix} 1 & -a_1 \\ -a_1 & 1 \end{pmatrix} \begin{pmatrix} R_X(0) \\ R_X(1) \end{pmatrix} = \begin{pmatrix} c\mu + 1 \\ c\mu \end{pmatrix}$$

For $k = 2$:

$$\begin{pmatrix} 1 & -a_1 & -a_2 \\ -a_1 & 1 - a_2 & 0 \\ -a_2 & -a_1 & 1 \end{pmatrix} \begin{pmatrix} R_X(0) \\ R_X(1) \\ R_X(2) \end{pmatrix} = \begin{pmatrix} c\mu + 1 \\ c\mu \\ c\mu \end{pmatrix}$$

For $k = 3$:

$$\begin{pmatrix} 1 & -a_1 & -a_2 & -a_3 \\ -a_1 & 1 - a_2 & -a_3 & 0 \\ -a_2 & -a_1 - a_3 & 1 & 0 \\ -a_3 & -a_2 & -a_1 & 1 \end{pmatrix} \begin{pmatrix} R_X(0) \\ R_X(1) \\ R_X(2) \\ R_X(3) \end{pmatrix} = \begin{pmatrix} c\mu + 1 \\ c\mu \\ c\mu \\ c\mu \end{pmatrix}$$

For $k = 4$:

$$\begin{pmatrix} 1 & -a_1 & -a_2 & -a_3 & -a_4 \\ -a_1 & 1-a_2 & -a_3 & -a_4 & 0 \\ -a_2 & -a_1-a_3 & 1-a_4 & 0 & 0 \\ -a_3 & -a_2-a_4 & -a_1 & 1 & 0 \\ -a_4 & -a_3 & -a_2 & -a_1 & 1 \end{pmatrix} \begin{pmatrix} R_X(0) \\ R_X(1) \\ R_X(2) \\ R_X(3) \\ R_X(4) \end{pmatrix} = \begin{pmatrix} c\mu+1 \\ c\mu \\ c\mu \\ c\mu \\ c\mu \end{pmatrix}$$

Finally, we may determine $R_X(p)$, for $p > k$, by using Eq. (4.8) again:

$$R_X(k+n) = c\mu + \sum_{i=\mathrm{n}}^{k+n-1} a_{k+n-i} R_X(i). \tag{4.10}$$

For a process $AR(k)$, $\rho(s)$ is evaluated by using the orthogonal projection property: indeed,

$$X_{t+s} = c + \varepsilon_{t+s} + \sum_{i=1}^{k} a_i X_{t+s-i}.$$

Thus, only $X_{t+s-k}, \ldots, X_{t+s-1}$ appear in the expression X_{t+s} as an affine function of $x_{s,t} = (X_t, \ldots, X_{t+s-1})$. Therefore, on the one hand, $\rho(s) = 0$, for $s > k$ (in this case, X_t does not appear in the affine expression of X_{t+s}); on the other hand, $\rho(k) = a_k$ (since $t + k - i = t$ for $i = k$).

Simulations of a process $AR(k)$ may be generated by two ways:

1. If k values $X_0, \ldots, X_{k-1}$ are given, we may generate a sample from the white noise ε and use the recurrence in the definition of the $AR(k)$ to determine $X_k, X_{k+1}, \ldots$;
2. Otherwise, it is necessary to determine the inverse $A(L)^{-1}$ of $A(L) = Id - \sum_{i=1}^{k} a_i L^i$. We have (see the remark page 187):

$$A(L)^{-1} = \sum_{i=0}^{+\infty} p_i L^i. \tag{4.11}$$

The unknown coefficients are determined by using that $A(L)^{-1}A(L) = Id$: we have

$$p_0 = 1; \qquad p_n = \sum_{j=1}^{\min\{k,n\}} p_{n-j} a_j, \qquad n > 0; \tag{4.12}$$

what determines the coefficients p_i. Then,

$$X_n = A(L)^{-1}(c + \varepsilon_n) \tag{4.13}$$

Then, we may generate a sample from the white noise and use this equation to generate X. An example of code for the simulation is shown below:

```
Function invcp(a, kmax)
Dim coef() As Double
ReDim coef(0 To kmax)
k = UBound(a, 1)
coef(0) = 1
coef(1) = coef(0) * a(1)
For n = 2 To kmax
   mmax = WorksheetFunction.Min(k, n)
   s = 0
   For j = 1 To mmax
      s = s + coef(n - j) * a(j)
   Next j
   coef(n) = s
Next n
invcp = coef
End Function
```

```
Function argen(c, a, kmax, ns)
p = invcp(a, kmax)
e = wngen(ns + kmax + 1)
Dim x() As Double
ReDim x(1 To ns)
For i = 1 To ns
    n = k + i
    s = e(n) + c
    For j = 1 To k
        s = s + p(j) * (e(n - j) + c)
    Next j
    x(i) = s
Next i
argen = x
End Function
```

Function `invcp` determines the coefficients p_n, $0 \leq n \leq kmax$ of $p(L)^{-1}$ and function `argen` generates a sample $X_1, \ldots, X_{ns}$ from the process. Values of c and $a_1, \ldots, a_k$ must be supplied. The subprogram `argen` uses `wngen` previously introduced (page 181). To run this code, you need a `Sub` calling `argen`. An example is given at the below: Sub `arsheet` reads c in cell `B1`, the values of a in line 2, starting at `B2`, the maximal value of n is read in cell `B3`, the size of the sample in cell `B4`. The values of $X_1, \ldots, X_{ns}$ are written in line 5, starting at `B5`.

```
Sub arsheet()
c = Cells(1, 2).Value
lca = Cells(2, Cells.Columns.Count).End(xlToLeft).Column
k = lca - 1
Dim a() As Double
ReDim a(1 To k)
For i = 1 To k
   a(i) = Cells(2, i + 1).Value
Next i
kmax = Cells(3, 2).Value
ns = Cells(4, 2).Value
x = argen(c, a, kmax, ns)
For i = 1 To ns
    Cells(5, i + 1).Value = x(i)
Next i
End Sub
```

Remark

An alternative approach for the identification and the simulation of an AR process consists in the elimination of the mean c. Indeed, let $Y_n = X_n - \mu$. Then,

$$Y_n = \varepsilon_n + \sum_{i=1}^{k} a_i Y_{n-i}$$

Thus, Y_n verifies the Yule-Walker equations with $c = 0$:

$$R_Y(p) = \delta_{p0} + \sum_{i=1}^{k} a_i R_Y(i-p).$$

For instance,

$$\begin{pmatrix} R_Y(1) & \cdots & R_Y(k) \\ R_Y(0) & \ldots & R_Y(k-1) \\ \vdots & \vdots & \vdots \\ R_Y(2-k) & \ldots & R_Y(1) \end{pmatrix} \begin{pmatrix} a_1 \\ \vdots \\ a_k \end{pmatrix} = \begin{pmatrix} R_Y(0)-1 \\ R_Y(1) \\ \vdots \\ R_Y(k-1) \end{pmatrix}$$

Analogously, the Least Squares approach leads to the linear system:

$$\begin{pmatrix} Y_k & \cdots & Y_1 \\ \vdots & \cdots & \vdots \\ Y_{k+n} & \cdots & Y_{n+1} \end{pmatrix} \begin{pmatrix} a_1 \\ \vdots \\ a_k \end{pmatrix} = \begin{pmatrix} Y_{k+1} \\ \vdots \\ Y_{k+n+1} \end{pmatrix}.$$

The characteristic polynomial A is the same for X and Y, so that we have $Y_n = A(L)^{-1}(\varepsilon_n)$. ∎

Example

Let us consider the process $X_n = \varepsilon_n + \frac{5}{6}X_{n-1} - \frac{1}{6}X_{n-2}$. We have $c = 0, k = 2, a_1 = \frac{5}{6}, a_2 = -\frac{1}{6}$. Thus, $\mu = 0$. Since the roots of the polynomial $A(x) = 6 - 5x + x^2$ are $x_1 = 2,\ x_2 = 3$, the process is stationary. We have:

$$\begin{pmatrix} 1 & -\frac{5}{6} & \frac{1}{6} \\ -\frac{5}{6} & \frac{7}{6} & 0 \\ \frac{1}{6} & -\frac{5}{6} & 1 \end{pmatrix} \begin{pmatrix} R_X(0) \\ R_X(1) \\ R_X(2) \end{pmatrix} = \begin{pmatrix} 1 \\ 0 \\ 0 \end{pmatrix}$$

Thus, $R_X(0) = \frac{21}{10}, R_X(1) = \frac{3}{2}, R_X\,(2) = \frac{9}{10}$. Now, we use Eq. (4.10) to determine the other values of R_X:

$$R_X(3) = R_X(2+1) = a_2R_X(1) + a_1R_X(2) = \frac{1}{2};$$

$$R_X(4) = R_X(2+2) = a_2R_X(2) + a_1R_X(3) = \frac{4}{15};$$

$$R_X(5) = R_X(2+3) = a_2R_X(3) + a_1R_X(4) = \frac{5}{4}\ldots$$

$$R_X(6) = \frac{5}{36},\quad \mathrm{R_X}(7) = \frac{77}{1080},\quad R_X(8) = \frac{47}{1296},\ldots$$

Now, assume that our data is the values of $R_X(4)$, …, $R_X(8)$. We use the Yule-Walker equations (4.8) to determine $c\mu$, a_1, a_2:

$$\begin{pmatrix} R_X(5) & R_X(4) & 1 \\ R_X(6) & R_X(5) & 1 \\ R_X(7) & R_X(6) & 1 \end{pmatrix}\begin{pmatrix} a_1 \\ a_2 \\ c\mu \end{pmatrix} = \begin{pmatrix} R_X(6) \\ R_X(7) \\ R_X(8) \end{pmatrix}$$

The results furnished by EXCEL® are exhibited below

matrix	0,138889	0,266667	1		second	0,071296
	0,071296	0,138889	1		member	0,036265
	0,036265	0,071296	1			0,018338
a1	0,833333	exact	0,833333			
a2	-0,16667		-0,16667			
c*mu	-1,4E-17		0			

Of course, we may consider also

$$\begin{pmatrix} R_X(-1) & R_X(-2) & 1 \\ R_X(0) & R_X(-1) & 1 \\ R_X(1) & R_X(0) & 1 \end{pmatrix}\begin{pmatrix} a_1 \\ a_2 \\ c\mu \end{pmatrix} = \begin{pmatrix} R_X(0) - 1 \\ R_X(1) \\ R_X(2) \end{pmatrix}$$

The results are almost the same:

matrix	1,5	0,9	1		second	1,1
	2,1	1,5	1		member	1,5
	1,5	2,1	1			0,9
a1	0,833333	exact	0,833333			
a2	-0,16667		-0,16667			
c*mu	1,11E-16		0			

A simulation of $X_n$may be generated by Eqs. (4.11), (4.12), and (4.13). We start by finding the values of p_n: Eq. (4.12) furnishes the results in Table 4.4. For $n \geq 15$, the coefficients have a modulus inferior to $1E - 4$.

Table 4.4 Values of p_n furnished by Eq. (4.12)

n	0	1	2	3	4	5	6	7
p_n	1	0,83	0,53	0,30	0,16	0,09	0,04	0,02
n	8	9	10	11	12	13	14	15
p_n	0,01	6E−3	3E−3	1E−3	7E−4	4E−4	2E−4	9E−5

We use the ANALYSIS TOOLPACK to generate a sample of a gaussian white noise of variance $\sigma^2 = 1$, formed by 10015 elements and the coefficients for $n \leq 14$ to generate a simulation of $X_1, \ldots, X_{10000}$. The simulation is presented in Fig. 4.10.

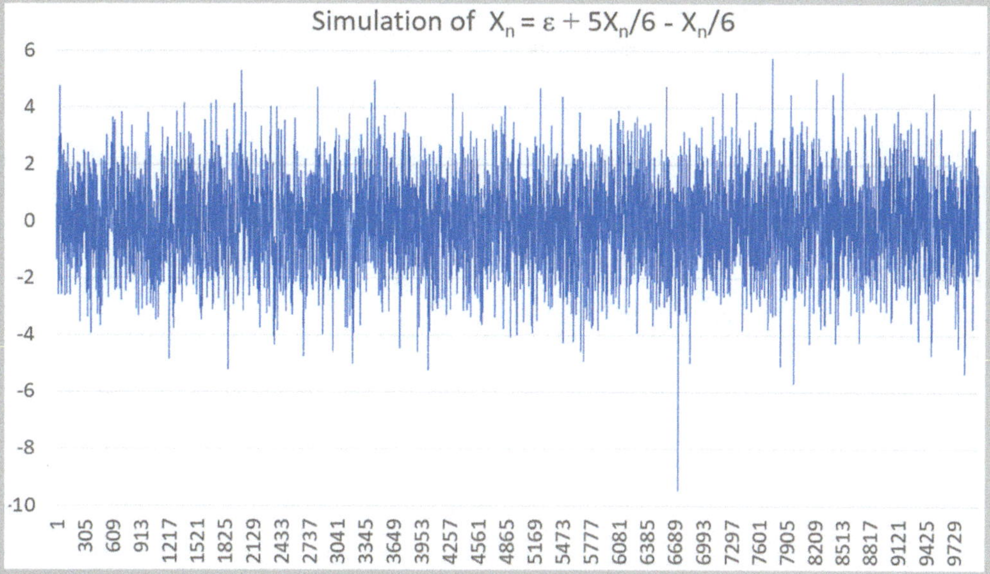

Fig. 4.10 Simulation of X_n using the ANALYSIS TOOLPACK

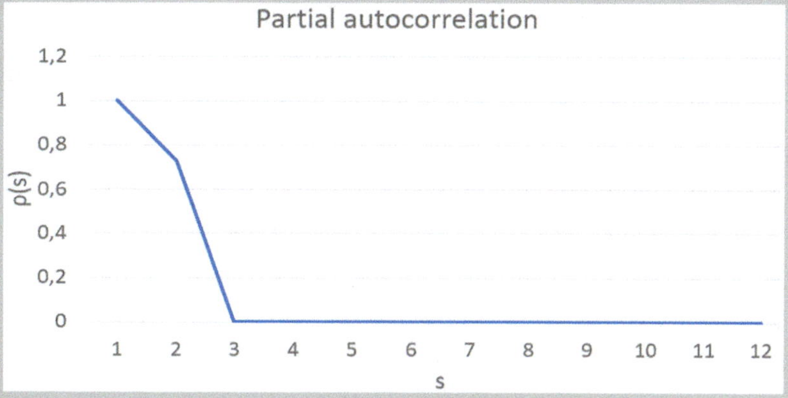

Fig. 4.11 Partial autocorrelation of the trajectory generated by simulation. As expected, the partial autocorrelation becomes zero for $s > 2$

In practice, we have not a perfect knowledge of the process, since only a sample is available. We illustrate such a situation by considering that the available data is the first np points from the simulation above. The partial autocorrelation for $np = 1000$ is exhibited in Fig. 4.11: it falls to zero at $s = 3$, so that the order of the process is identified as $k = 2$.

We used the trajectory with $np = 1000$ to estimate the values of $R_X(p), p \leq 10$. The results are shown in Table 4.5.

Table 4.5 RMS error in the estimation of $R_X(p)$, $p \leq 10$

np	250	1000	2500	5000	10,000
RMS error	0,24	0,097	0,032	0,021	0,026

Then, we used these values to estimate the coefficients by the Yule-Walker equations (4.10), with $p = 0$, 1, 2 results are shown in Table 4.6. Results using $p = 1$, 2, 3 were analogous (Figs. 4.12, 4.13, 4.14, and 4.15).

Table 4.6 RMS error in the estimation of c, a_1, a_2 by Yule-Walker

np	250	1000	2500	5000	10,000
RMS error	0,19	0,14	0,12	0,08	0,05

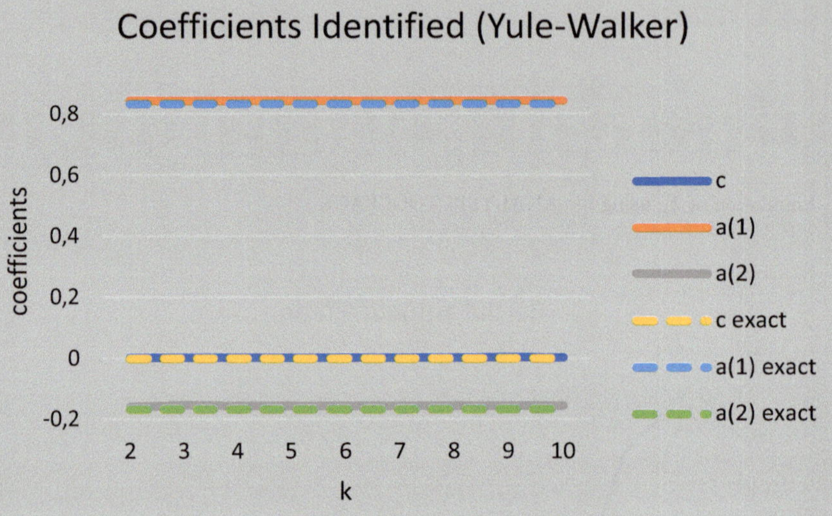

Fig. 4.12 Values of the coefficients c,a_1,a_2. For $k \geq 2$, the values obtained are close to the exact ones

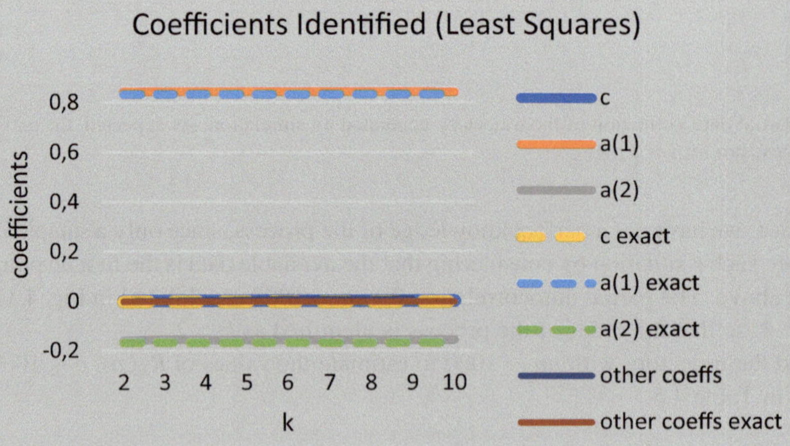

Fig. 4.13 Least Squares identification for a trajectory of $np = 1000$ points and values of $k \geq 2$

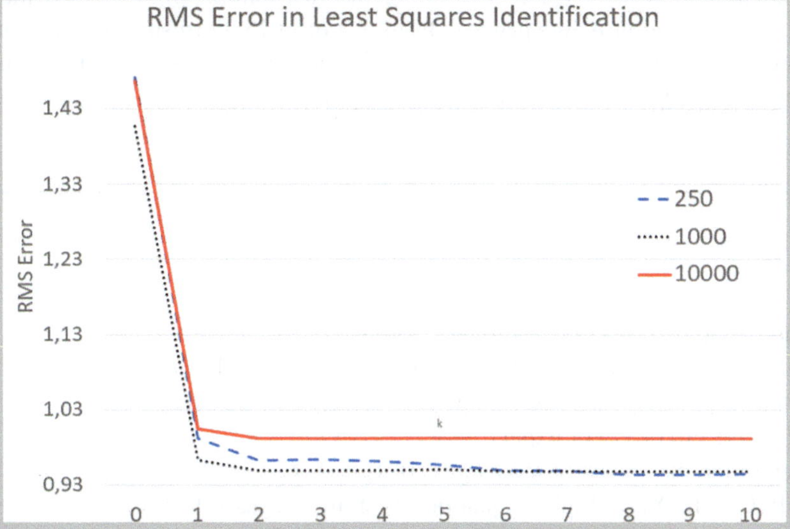

Fig. 4.14 RMS error in the Least Squares identification for different numbers of points ***np* = 250,1000,10000**. For $k \geq 2$, the RMS error is stable and the corresponding coefficients are near zero

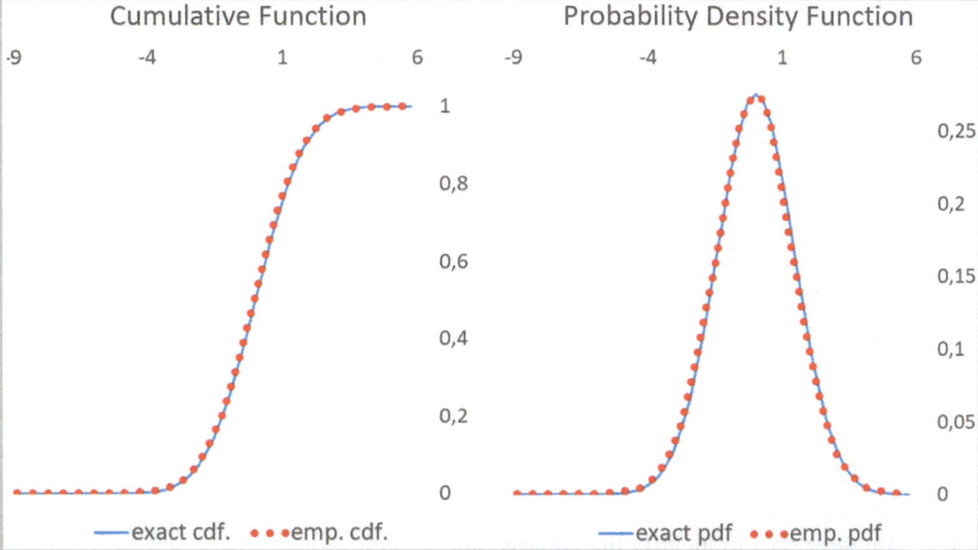

Fig. 4.15 Cumulative function and probability density of process $X_n = \varepsilon_n + \frac{5}{6}X_{n-1} - \frac{1}{6}X_{n-2}$, corresponding to a simulation of 10,000 variates. The PDF was obtained by numerical derivative of the CDF, evaluated by the SPH approach, with a gaussian kernel. Dashed line is the approximation. Continuous line is the exact value

As an alternative, we may use the least squares approach. We solved the linear system of Eq. (4.9) for $np = 250$, 1000, 10000 and different values of k. The results are shown in Fig. 4.15. We observe that the results are almost stable for $k \geq 2$. In addition, the coefficients a_i, $i > k$ are almost zero.

The simulation may be used to determine the distribution of X_n- since the process is stationary, the distribution is independent from n. Eq. (4.13) furnishes a representation of X_n as a MA process, which may be used to determine the exact distribution of X_n. The empirical distribution and the exact one are compared in Fig. 4.15. ∎

Exercises

1. Let $X_n = \alpha X_{n-1} + \varepsilon_n$.

   (a) Show that the process is stationary if $|\alpha| < 1$.
   TIP: Verify that the characteristic equation is $A(x) = 1 - \alpha x$.
   (b) Assuming that $|\alpha| < 1$, find the coefficients p_n defining $A(L)^{-1}$.
   TIP: Use Eq. (4.12) or expand $\frac{1}{1-\alpha x}$.
   (c) Determine the values of $R_X(p)$, $p = 0, 1, 2, 3, 4$.
   TIP: Use the Yule-Walker equations as shown pages 189–190.
   (d) Determine $\rho_X(p)$, $p = 0, 1, 2, 3, 4$.
   TIP: Use Eq. (4.3) and the properties shown page 190.
   (e) Generate a sample of 1000 values of $N(0, 1)$ and use it to simulate the process. TIP: Eq. (4.13)
   (f) Use these data to identify the process.
   TIP: Use the Yule-Walker equations (4.7) and (4.8).

2. Let $X_n = \varepsilon_n + \frac{5}{6}X_{n-1} - \frac{1}{6}X_{n-2}$.

   (a) Show that the process is stationary.
   TIP: Find the roots of $A(x) = 6 - 5x + x^2$.
   (b) Find the coefficients p_n defining $A(L)^{-1}$.
   TIP: Use Eq. (4.12) or expand $\frac{-2}{1-\frac{x}{3}} + \frac{3}{1-\frac{x}{2}}$.
   (c) Determine the values of $R_X(p)$, $p = 0, 1, 2, 3, 4$.
   TIP: Use the Yule-Walker equations as shown pages 189–190.
   (d) Determine $p_X(p)$, $p = 0, 1, 2, 3, 4$.
   TIP: Use Eq. (4.3) and the properties shown page 190.
   (e) Generate a sample of 1000 values of $N(0, 1)$ and use it to simulate the process.
   TIP: Eq. (4.13)
   (f) Use these data to identify the process.
   TIP: Use the Yule-Walker equations (4.7) and (4.8).

3. Let $X_n = \varepsilon_n + \frac{7X_{n-1}}{6} - \frac{X_{n-2}}{3}$.

   (a) Show that the process is stationary.
   TIP: Find the roots of $A(x) = 6 - 7x + 2x^2$.
   (b) Find the coefficients p_n defining $A(L)^{-1}$.
   TIP: Use Eq. (4.12) or expand $\frac{3}{1-\frac{x}{3}} - \frac{4}{1-\frac{2x}{3}}$.

(c) Determine the values of $R_X(p), p = 0, 1, 2, 3, 4$.
TIP: Use the Yule-Walker equations as shown pages 189–190.
(d) Determine $\rho_X(p), p = 0, 1, 2, 3, 4$.
TIP: Use Eq. (4.3) and the properties shown page 190.
(e) Generate a sample of 1000 values of $N(0, 1)$ and use it to simulate the process.
TIP: Eq. (4.13)
(f) Use these data to identify the process.
TIP: Use the Yule-Walker equations (4.7) and (4.8).

4. Let $X_n = \varepsilon_n + \frac{13X_{n-1}}{12} - \frac{3X_{n-2}}{8} + \frac{X_{n-3}}{24}$.

(a) Show that the process is stationary.
TIP: Verify that the roots of the characteristic polynomial are 2,3,4.
(b) Find the coefficients p_n defining $A(L)^{-1}$.
TIP: Use Eq. (4.12) or expand $\frac{6}{1-\frac{x}{2}} - \frac{8}{1-\frac{x}{3}} + \frac{3}{1-\frac{x}{4}}$.
(c) Determine the values of $R_X(p), p = 0, 1, 2, 3, 4$.
TIP: Use the Yule-Walker equations as shown pages 189–190.
(d) Determine $\rho_X(p), p = 0, 1, 2, 3, 4$.
TIP: Use Eq. (4.3) and the properties shown page 190.
(e) Generate a sample of 1000 values of $N(0, 1)$ and use it to simulate the process.
TIP: Eq. (4.13)
(f) Use these data to identify the process.
TIP: Use the Yule-Walker equations (4.7) and (4.8).

4.6 ARMA Processes

A process ARMA is the combination between an autoregressive and a moving average processes:

X is $\boldsymbol{ARMA(r, s)}$ if and only if X is weakly stationary and there exists a discrete white noise $\varepsilon = \{\varepsilon_n : n \in \mathbb{Z}\}$, a number $c \in \mathbb{R}$ and two vectors of coefficients $\boldsymbol{a} = (a_1, \ldots, a_r)$ and $\boldsymbol{b} = (b_1, \ldots, b_s)$ such that

$$\forall n \in \mathbb{Z} : X_n = c + \varepsilon_n + \sum_{i=1}^{r} a_i X_{n-i} + \sum_{i=1}^{s} b_i \varepsilon_{n-i}.$$ ■

Let L be the delay operator (see remark page 187) and let us denote

$$A(L) = Id - \sum_{i=1}^{r} a_i L^i, B(L) = Id + \sum_{i=1}^{s} b_i L^i. \quad (4.14)$$

Then,

$$A(L)X_n = c + B(L)\varepsilon_n$$

and

$$X_n = A(L)^{-1}c + A(L)^{-1}B(L)\varepsilon_n \tag{4.15}$$

The determination of $A(L)^{-1}$ is made by Eqs. (4.11) and (4.12). In addition, let us set

$$A(L)^{-1}B(L) = \sum_{n=0}^{+\infty} q_n L^n.$$

Let us set $b_0 = 1$. Then

$$A(x)^{-1}B(x) = \sum_{i=0}^{+\infty} p_i x^i \times \sum_{j=0}^{s} b_j x^j.$$

Thus,

$$A(x)^{-1}B(x) = \sum_{i=0}^{+\infty} \sum_{j=0}^{s} p_i b_j x^{i+j},$$

id est,

$$A(x)^{-1}B(x) = \sum_{n=1}^{+\infty} x^n \left(\sum_{j=0}^{\min(s,n)} p_{n-j} b_j \right).$$

Hence,

$$q_n = \sum_{j=0}^{\min(s,n)} p_{n-j} b_j. \tag{4.16}$$

This equation is the equivalent for $A(L)^{-1}B(L)$ of Eq. (4.12) for $A(L)^{-1}$. Thus, the simulation of a process ARMA analogous to the simulation of a process AR:

1. If r values $X_0, \ldots, X_{r-1}$ are given, we may generate a sample from the white noise ε and use the recurrence in the definition of the $ARMA(r, s)$ to determine $X_r, X_{r+1}, \ldots$;
2. Otherwise, it is necessary to determine $A(L)^{-1}$ and $A(L)^{-1}B(L)$ by using Eqs. (4.12) and (4.16) (see also the remark page 187). Then, we may generate a sample from the white noise and use Eq. (4.15) to generate X.

ARMA processes satisfy a family of Yule-Walker equations:

$$\mu\left(1 - \sum_{i=1}^{r} a_i\right) = c; \tag{4.17}$$

$$R_X(p) = c\mu + \sum_{i=1}^{r} a_i R_X(i-p), \textit{if } p > s. \tag{4.18}$$

Analogously to *AR* processes, these equations may be used to determine the quantities $c\mu$, $a_1, \ldots,$ a_r. Taking the values $=s + 1, \ldots, s + r$, we obtain $k + 1$ linear equations for the determination of the quantities $c\mu$, $a_1, \ldots, a_k$:

$$\begin{pmatrix} R_X(s) & \cdots & R_X(s-r+1) & 1 \\ R_X(s+1) & \ldots & R_X(s-r+2) & 1 \\ \vdots & \vdots & \vdots & \vdots \\ R_X(s+r-1) & \ldots & R_X(s) & 1 \\ R_X(s+r) & \cdots & R_X(s+1) & 1 \end{pmatrix} \begin{pmatrix} a_1 \\ \vdots \\ a_r \\ c\mu \end{pmatrix} = \begin{pmatrix} R_X(s+1) \\ R_X(s+2) \\ \vdots \\ R_X(s+r) \\ R_X(s+r+1) \end{pmatrix}$$

Once the system is solved, we may determine c, analogously to an AR process. As an alternative, the identification of the AR part of the process may also be made by the Least Squares approach, by solving the linear system (4.9).

Finally, we observe that

$$Y_n = X_n - c - \sum_{i=1}^{r} a_i X_{n-i} = \varepsilon_n + \sum_{i=1}^{s} b_i \varepsilon_{n-i}. \tag{4.19}$$

Thus, Y_n is a process *MA*(s) having a mean equal to zero. Thus, the coefficients b_i, $i = 1, \ldots, s$ verify Eqs. analogous to Eq. (4.4):

$$\sum_{i=0}^{s-p} b_i b_{i+p} = R_Y(p). \tag{4.20}$$

Taking $p = 0, \ldots, s - 1$, we obtain a system of nonlinear equations which determines the coefficients $b_1, \ldots, b_s$.

The main difficult in the identification of ARMA processes is the determination of the orders (r, s). If these orders are known, the methods for the identification of AR and MA may be applied. But the selection a convenient couple (r, s) may involve difficulties. One of the methods to choose the values of (r, s) is the use of model selection tools, such as AIC (Akaike, 1974), BIC (Schwarz, 1978) or HQ (Hannan & Quinn, 1979). The reader will find in the literature other methods for the identification of ARMA processes (see, for instance, (Choi, 1992)).

Example

Let us consider the process $X_n = \frac{16X_{n-1}}{15} - \frac{4X_{n-2}}{15} + \varepsilon_n + \frac{5\varepsilon_{n-1}}{6} + \frac{\varepsilon_{n-2}}{6}$. We have $c = 0$, $r = s = 2$, $a_1 = \frac{16}{15}, a_2 = -\frac{4}{15}, b_1 = \frac{5}{6}, b_2 = \frac{1}{6}$. Thus, $\mu = 0$. We have also

$$A(x) = 1 - \frac{16x}{15} + \frac{4x^2}{15}, \quad B(x) = 1 + \frac{5x}{6} + \frac{x^2}{6}$$

The roots of the polynomial A are $x_1 = \frac{3}{2}, x_2 = \frac{5}{2}$, so that the process is stationary. To simulate the process, we need to determine the coefficients p_n: Eqs. (4.12) and (4.16) furnish the data below (Tables 4.7 and 4.8 and Figs. 4.16 and 4.17)

Table 4.7 Coefficients of $A(L)^{-1}$: values of p_n furnished by Eq. (4.16)

n	0	1	2	3	4	5	6	7
p_n	10,000	10,667	0,8711	0,6447	0,4554	0,3139	0,2133	0,1439
n	8	9	10	11	12	13	14	15
p_n	0,0966	0,0646	0,0432	0,0288	0,0192	0,0128	0,0086	0,0057
n	16	17	18	19	20	21	22	23
p_n	0,0038	0,0025	0,0017	0,0011	0,0008	0,0005	0,0003	0,0002
n	24	25	26	27	28	29	30	31
p_n	0,0001	0,0001	0,0001	0,0000	0,0000	0,0000	0,0000	0,0000

Table 4.8 Coefficients of $A(L)^{-1}B(L)$: values of q_n furnished by Eq. (4.16)

n	0	1	2	3	4	5	6	7
p_n	10,000	19,000	19,267	15,484	11,379	0,8008	0,5508	0,3740
n	8	9	10	11	12	13	14	15
p_n	0,2520	0,1691	0,1132	0,0756	0,0505	0,0337	0,0225	0,0150
n	16	17	18	19	20	21	22	23
p_n	0,0100	0,0067	0,0044	0,0030	0,0020	0,0013	0,0009	0,0006
n	24	25	26	27	28	29	30	31
p_n	0,0004	0,0003	0,0002	0,0001	0,0001	0,0001	0,0000	0,0000

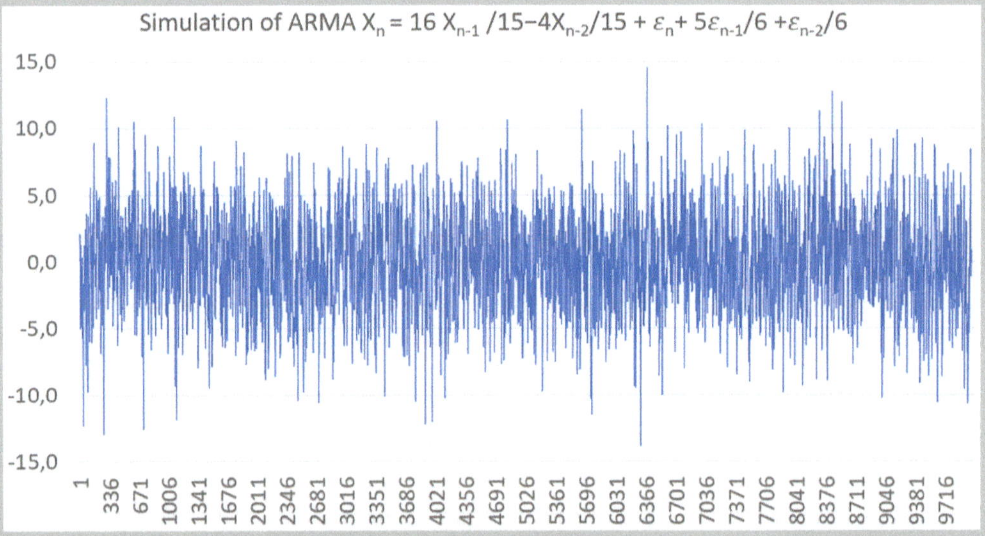

Fig. 4.16 Simulation of an ARMA process

We choose to simulate the process using the values of p_n for $n \le 30$. We generate 10030 variates from a gaussian white noise, and we use Eq. (4.18) to identify the coefficients of the AR part of the model: for $r = 2$, we obtain $c = 0, 1$; $a_1 = 1, 09$; $a_2 = -0.29$ (to be compared with the exact values $c = 0$; $a_1 = 1, 07$; $a_2 = -0.27$). Then, we consider the process Y_n given by Eq. (4.19) and we use Eq. (4.20) to identify the MA part of the process. We obtain $b_1 = 0.83$, $b_2 = 0.14$ (to be compared with the exact values $b_1 = 0.83$; $b_2 = 0.17$). The empirical distribution and the exact one are compared in Fig. 4.10. ∎

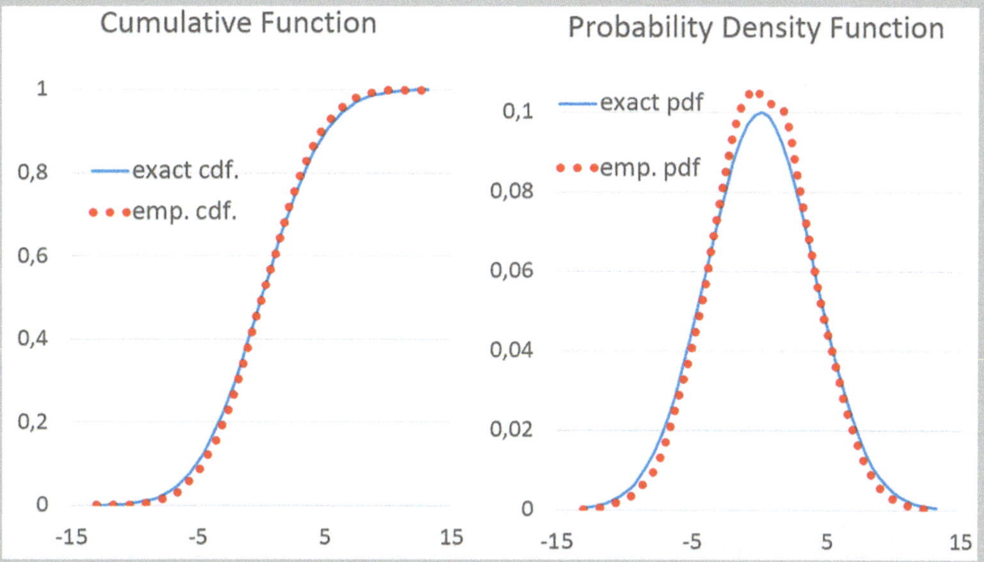

Fig. 4.17 Cumulative function and probability density of process $X_n = \frac{16}{15}X_{n-1} - \frac{4}{15}X_{n-2} + \varepsilon_n + \frac{5}{6}\varepsilon_{n-1} + \frac{1}{6}\varepsilon_{n-2}$, corresponding to a simulation of 10,000 variates

Exercises

1. Let $X_n = \frac{X_{n-1}}{2} + \varepsilon_n + \frac{\varepsilon_{n-1}}{3}$.

   (a) Determine $A(x)$, $B(x)$.
   (b) Verify that the process is stationary.
   (c) Determine the coefficients p_n of the expansion of $A(x)^{-1}$
   (d) Determine the coefficients q_n of the expansion of $A(x)^{-1}B(x)$
   (e) Generate a sample of 1000 values of $N(0, 1)$ and use it to simulate X.
   (f) Use the data furnished by the simulation to determine approximate values of m_X, $R_X(p), p = 0, 1, 2, 3, 4$.
   (g) Use these approximated values to fit a process $ARMA(1, 1)$ to the data.

2. Let $X_n = \frac{X_{n-1}}{3} + \varepsilon_n + \frac{\varepsilon_{n-1}}{2}$.

   (a) Determine $A(x)$, $B(x)$.
   (b) Verify that the process is stationary.
   (c) Determine the coefficients p_n of the expansion of $A(x)^{-1}$
   (d) Determine the coefficients q_n of the expansion of $A(x)^{-1}B(x)$
   (e) Generate a sample of 1000 values of N(0,1) and use it to simulate X.
   (f) Use the data furnished by the simulation to determine approximate values of m_X, $R_X(p), p = 0, 1, 2, 3, 4$.
   (g) Use these approximated values to fit a process $ARMA(1, 1)$ to the data.

3. Let $Y_n = 2 + \frac{3Y_{n-1}}{4} - \frac{Y_{n-2}}{8} + \varepsilon_n - \frac{\varepsilon_{n-1}}{16}$.

   (a) Let $X_n = Y_n - m_X$. Show that $X_n = \frac{3}{4}X_{n-1} - \frac{1}{8}X_{n-2} + \varepsilon_n - \frac{1}{16}\varepsilon_{n-1}$
   (b) Show that $m_X = \frac{16}{3}$.
   (c) Determine $A(x)$, $B(x)$ for the process X_n
   (d) Verify that the process X_n is stationary
   (e) Determine the coefficients p_n of the expansion of $A(x)^{-1}$
   (f) Determine the coefficients q_n of the expansion of $A(x)^{-1}B(x)$
   (g) Generate a sample of 1000 values of N(0,1) and use it to simulate X.
   (h) Use the data furnished by the simulation to determine approximate values of m_X, $R_X(p), p = 0, 1, 2, 3, 4$.
   (i) Use these approximated values to fit a process $ARMA(2, 1)$ to the data.

4.7 Markov Processes

Markov processes – also called Markov chains – are processes without memory: the future value X_{n+1} depends only on the actual state X_n. In other words, the conditional probability of X_{n+1} knowing the past values $X_0, \ldots, X_n$ depends only upon X_n:

$$P(X_{n+1} < x_{n+1} | X_i < x_i, \quad 0 \le i \le n) = P(X_{n+1} < x_{n+1} | X_n < n).$$

Therefore, a Markov process is fully characterized by its probability of transition:

$$P(t,x,s,y) = P(X_t < x | X_s < y).$$

The associated probability density is $p(t,x,s,y) = \frac{\partial^2 P(t,x,s,y)}{\partial x \partial y}$. The process is stationary if and only if $P(t, x, s, y) = P(t - s, x, y)$ and $p(t, x, s, y) = p(t - s, x, y)$.

Let us consider a stationary Markov chain where X can take a finite number of values $\{x_1, \ldots, x_k\}$. In the language of Markov chains, x_i is referred as a possible state of the chain. Then,

$$p(n, x_i, x_j) = \pi_{ij}(n).$$

We have

$$\pi_{ij}(n) \ge 0, \quad \sum_{i=1}^{k} \pi_{ij}(n) = 1$$

These probabilities may be collected into a matrix $\Pi(n) = (\pi_{ij}(n) : 1 \le i, j \le k)$. In the framework of Markov processes, the matrix $\mathbf{\Pi(1)}$ is simply denoted $\mathbf{\Pi}$ and is called transition matrix. The chain may be represented by a graph where the arcs represent the transitions. As an example, we represent the case where $k = 4$ in Fig. 4.18.

We have

$$P(X_{n+1} = x_i) = \sum P(\mathrm{X}_{n+1} = x_i | X_n = x_j) P(X_n = x_j)$$

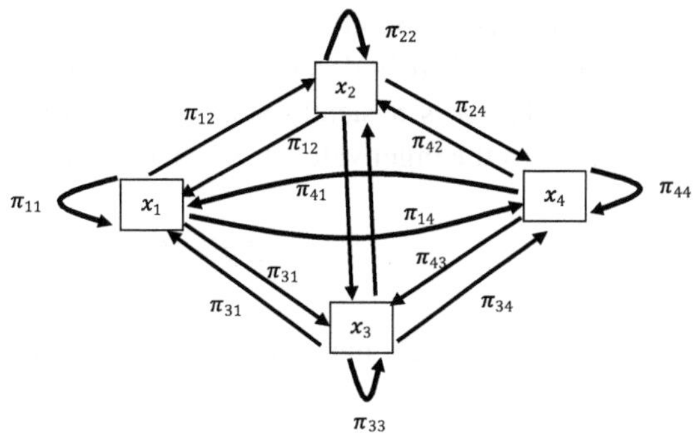

Fig. 4.18 Graphical representation of a Markov Chain with 4 states

Let us introduce

$$\boldsymbol{p}^{(s)} = \left(\boldsymbol{p}_1^{(s)}, \ldots, \boldsymbol{p}_k^{(s)}\right)^t, \ P(X_s = x_i) = p_i^{(s)}.$$

Then,

$$\boldsymbol{p}^{(1)} = \boldsymbol{\Pi}\boldsymbol{p}^{(0)}.$$

Thus, by recurrence

$$\boldsymbol{p}^{(n)} = \boldsymbol{\Pi}^n\boldsymbol{p}^{(0)} \Leftrightarrow \boldsymbol{\Pi}(\mathrm{n}) = \boldsymbol{\Pi}^n.$$

The states of a Markov chain are usually classified in recurrent, transient, and absorbing.

A state is recurrent when the probability of visiting at some time is equal to 1: $P(\exists n \in \mathbb{N} : X_n = x_i) = 1$. Markov chains with a finite number of states have at least one recurrent state.

A state is transient when such a probability is strictly inferior to 1: $P(\exists n \in \mathbb{N} : X_n = x_i) < 1$.

A state is absorbing when it cannot be leaved, id est, $\exists n \in \mathbb{N} : P(X_{n+s} = x_i | X_n = x_i) = 1, \forall s \geq 0$. Not all the Markov chains have absorbing states. An absorbing state is characterized by the fact that $\pi_{ii}(n) = 1$, so that the transition matrix $\Pi(n)$ contains columns of an identity matrix. A Markov chain that has an absorbing state is said an absorbing chain. For such a chain, it is usual to renumber the states to write the transition matrix under the canonical form

$$\boldsymbol{\Pi} = \begin{pmatrix} \boldsymbol{B} & \boldsymbol{0} \\ \boldsymbol{A} & \boldsymbol{Id} \end{pmatrix}$$

Under the canonical form, $\boldsymbol{A}$ gives the probability of transition from a non-absorbing to state to an absorbing state, while $\boldsymbol{B}$ gives the probability of transitions between non-absorbing states. Matrix $\boldsymbol{M} = (\boldsymbol{Id} - \boldsymbol{B})^{-1}$ plays an essential rule in the theory of Markov processes. For instance, M_{ij} gives the expected number of visits to state i when starting at state j. Analogously, $M_{\blacksquare \mathrm{j}} = \sum_{i=1}^{k} M_{ij}$ gives the expected number of steps before absorption when starting at state j. The matrix $\boldsymbol{N} = \boldsymbol{BM}$ gives the probability N_{ij} of an absorption by the *ith.* absorbing state when the initial state is the *jth* non-absorbing state.

We say that $\boldsymbol{p}$ is an stationary probability if and only if

$$\boldsymbol{p} = \boldsymbol{\Pi}\boldsymbol{p}.$$

A stationary probability is a fixed point of the chain: if the chain reaches such a probability, then the probability distribution remains constant for all the posterior times. If $\boldsymbol{p}^{(n)} \to$p for $n \to +\infty$ then $= \boldsymbol{\Pi p}$, so that p is a steady state probability. Notice that $\boldsymbol{p}$ corresponds to an eigenvector of $\boldsymbol{\Pi}$, associated to the eigenvalue 1, so that it is a solution of the overdetermined linear system

$$\boldsymbol{Ap} = \boldsymbol{B},\ \boldsymbol{A} = \begin{pmatrix} \boldsymbol{Id} - \boldsymbol{\Pi} \\ \boldsymbol{1} \end{pmatrix},\ \boldsymbol{B} = \begin{pmatrix} \boldsymbol{0} \\ 1 \end{pmatrix}.$$

This linear system may solved by Least Squares:

$$\boldsymbol{A}^t\boldsymbol{Ap} = \boldsymbol{A}^t\boldsymbol{B}.$$

To verify if the Least Squares solution is a stationary probability, we must evaluate the error $e = \|\boldsymbol{Ap} - \boldsymbol{B}\|$. If $e = 0$, the solution fond is a stationary probability. Otherwise, not. An example of macro to find a stationary probability by this method is the given below (Fig. 4.19):

```
Sub statprob()
Dim PI As Variant
nPI = "[" & ActiveWorkbook.name & "]" & ActiveSheet.name & "!" & "PI"
PI = Range(nPI)
k = UBound(PI, 1)
If UBound(PI, 2) <> k Then
   MsgBox "PI is not square !"
   Exit Sub
End If
Dim a() As Double
ReDim a(1 To k + 1, 1 To k)
Dim b() As Double
ReDim b(1 To k + 1, 1 To 1)
For j = 1 To k
   b(j, 1) = 0
   For i = 1 To k
       a(i, j) = PI(i, j)
   Next i
   a(k + 1, j) = 1
Next j
b(k + 1, 1) = 1
at = WorksheetFunction.Transpose(a)
aux1 = WorksheetFunction.MMult(at, a)
aux2 = WorksheetFunction.MMult(at, b)
aux3 = WorksheetFunction.MInverse(aux1)
p = WorksheetFunction.MMult(aux3, aux2)
aux4 = WorksheetFunction.MMult(a, p)
Dim aux5() As Double
ReDim aux5(1 To k)
For i = 1 To k
   aux5(i) = aux4(i, 1) - b(i, 1)
Next i
e = Sqr(WorksheetFunction.SumSq(aux5) / k)
For j = 1 To k
   Cells(j, k + 4).Value = p(j, 1)
Next j
Cells(k + 2, 2).Value = e
End Sub
```

Fig. 4.19 Example of VBA module for the determination of the stationary probability

From Perron-Frobenius theorem, it is expected that $\mathbf{\Pi}^n$ converges to a matrix having all the columns equal to $\boldsymbol{p}$, i. e.,

$$\mathbf{\Pi}^n \to (\boldsymbol{p}, \ldots, \boldsymbol{p}).$$

Thus, a second way to find a stationary probability consists in calculating powers of $\mathbf{\Pi}$ until convergence to a matrix having the form above. An example of VBA function evaluating $\mathbf{\Pi}^n$ is shown in Fig. 4.20.

A Markov chain may be periodical, id est, be such that there exists a number k_T such that $\boldsymbol{p}^{(k_T)} = \boldsymbol{p}^{(0)}, \forall \boldsymbol{p}^{(0)}$, id est, such that $\mathbf{\Pi}^{k_T} = \boldsymbol{Id}$. In such a case, $\boldsymbol{p}^{(n+k_T)} = \boldsymbol{p}^{(n)}, \forall n \in \mathbb{N}$. Indeed, a second classification of the states involves periodicity: a state is said to be periodic with period k_T if and only if there is $k_T > 1$ such that $p_{ii}^{(n)} = 0$ when n is not a multiple of k_T and $p_{ii}^{(n)} > 0$ otherwise. In such a case, the chain may return to the state x_i at each k_T steps.

To simulate the chain, we must generate a sequence $X_1, \ldots, X_n$ from an initial value $X_0 = x_{i_0}$. To do this, we choose an initial vector $\bar{\boldsymbol{p}}^{(0)}$ having a single component equal to 1 at position i_0 and the

```
Function matrixpower(PI, n)
k = UBound(PI, 1)
Dim aux() As Double
ReDim aux(1 To k, 1 To k)
For i = 1 To k
   For j = 1 To k
       aux(i, j) = PI(i, j)
   Next j
Next i
For kk = 1 To n - 1
     auxnew = WorksheetFunction.MMult(aux, PI)
     For i = 1 To k
        For j = 1 To k
           aux(i, j) = auxnew(i, j)
        Next j
     Next i
Next kk
matrixpower = aux
End Function
```

```
Sub matrixp()
Dim PI As Variant
nPI = "[" & ActiveWorkbook.name & "]" & ActiveSheet.name & "!" & "PI"
PI = Range(nPI)
k = UBound(PI, 1)
If UBound(PI, 2) <> k Then
   MsgBox "PI is not square !"
   Exit Sub
End If
ii = ActiveCell.Row
jj = ActiveCell.Column
n = Cells(ii, jj).Value
aux = matrixpower(PI, n)
For i = 1 To k
   For j = 1 To k
      Cells(ii + i, jj + j).Value = aux(i, j)
   Next j
Next i
End Sub
```

Fig. 4.20 Example of VBA module for the evaluation of $\mathbf{\Pi}^n$. The user must run the Macro matrixp(), which calls the function matrixpower. The Macro reads the value $\boldsymbol{n}$ in the active cell and evaluates $\mathbf{\Pi}^n$

other components null. Then, we evaluate $\boldsymbol{p}^{(1)} = \boldsymbol{\Pi}\bar{\boldsymbol{p}}^{(0)}$ and we choose the value x_{i_1} at random, among the values x_1, …, x_k, according to the probabilities $p_1^{(1)}, \ldots, p_1^{(k)}$. For instance, we may generate a variate u_1 from the uniform distribution on (0,1) and find its position in the cumulative distribution associated to $\boldsymbol{p}^{(1)}$: we take the value x_{i_1} such that $p_1^{(1)} + \ldots + p_1^{(i_1-1)} < u_1 < p_1^{(1)} + \ldots + p_1^{(i_1)}$ Then, we set $\bar{\boldsymbol{p}}^{(1)}$as the vector having a single component equal to 1 at position i_1 and the other components null and we make the same operations again. An example of Macro for the simulation is shown in Fig. 4.21.

```
Sub markovsim()
Dim PI As Variant
nPI = "[" & ActiveWorkbook.name & "]" & ActiveSheet.name & "!" & "PI"
PI = Range(nPI)
k = UBound(PI, 1)
If UBound(PI, 2) <> k Then
   MsgBox "PI is not square !"
   Exit Sub
End If
i_n = Cells(k + 2, 2).Value
Dim x() As Long
nsim = Cells(k + 4, 2).Value
ReDim x(0 To nsim)
x(0) = i_n
Dim p_n() As Variant
ReDim p_n(1 To k, 1 To 1)
For i = 1 To k
   p_n(i, 1) = 0
Next i
p_n(i_n, 1) = 1
Dim c_pn() As Double
ReDim c_pn(1 To k + 1)
c_pn(1) = 0
Randomize
For n = 1 To nsim
     p_new = WorksheetFunction.MMult(PI, p_n)
     For i = 1 To k
        p_n(i, 1) = p_new(i, 1)
     Next i
     For i = 1 To k
         c_pn(i + 1) = c_pn(i) + p_n(i, 1)
     Next i
     u = Rnd()
     For i = 2 To k + 1
        If c_pn(i) > u Then
            aux = i - 1
            Exit For
        End If
     Next i
     For i = 1 To k
        p_n(i, 1) = 0
     Next i
     i_n = aux
     p_n(i_n, 1) = 1
     x(n) = i_n
Next n
Cells(k + 6, 1).Value = "states"
For i = 0 To nsim
   Cells(k + 6, i + 2).Value = x(i)
Next i
End Sub
```

Fig. 4.21 Example of VBA module for the simulation of a Markov chain

To use this macro, you must enter the matrix $\mathbf{\Pi}$ in a workbook and give it the name `PI`. You must enter the initial state in column B, two lines after the last line of PI. Th number of states to simulate is given in column B, two lines after the initial state. The results appear two lines after the number of simulations (see example below)

Example

Let us consider a stationary Markov chain such that

$$\Pi = \begin{pmatrix} \frac{1}{2} & 0 & \frac{1}{2} \\ 0 & \frac{1}{2} & \frac{1}{2} \\ \frac{1}{2} & \frac{1}{2} & 0 \end{pmatrix}$$

Let us open a blank workbook, enter the data at right and give the name `PI` to cells `B1:D3`:

	A	B	C	D
1	PI	0,5	0	0,5
2		0	0,5	0,5
3		0,5	0,5	0
4				

In cell `B6`, enter "`=MUNIT(3)-PI`". In cells `B9:D9`, enter "`1`". Give the name "`A`" to cells `B6:D9`. In cells `F6:F9`, enter the values `0, 0, 0, 1`. Give them the name "`B`"

	A	B	C	D	E	F
6	A	0,5	0	-0,5	B	0
7		0	0,5	-0,5		0
8		-0,5	-0,5	1		0
9		1	1	1		1

Run the macro `statprob()`. The results appear in `G1:G4`. You may also enter the formula for the Least Squares solution of the linear system in `H6`: "`=MMULT(MINVERSE(MMULT(TRANSPOSE(A);A));MMULT(TRANSPOSE(A);B))`". Give the name `p` to cells `H6:H9`. In `H10`, enter the formula "`=SQRT(SUMSQ(MMULT(A;p)-B)/4)`". If the candidate p is a stationary probability, the value in `H10` is zero. Here, $p = \left(\frac{1}{3}, \frac{1}{3}, \frac{1}{3}\right)$ is a stationary probability.

G	H
0,333333	
0,333333	
0,333333	
1,36E-16	
p	0,333333
	0,333333
	0,333333
e	1,18E-16

Select cell I1 and enter the value 100. Execute the macro matrixp(). The results appear in J2:L4: we see that all the columns are identical and coincide with the value of $\boldsymbol{p}$.

	J	K	L
100			
	0,333333	0,333333	0,333333
	0,333333	0,333333	0,333333
	0,333333	0,333333	0,333333

To simulate this Markov chain, we use the macro given above. The workbook appears as:

	A	B	C	D
1	PI	0,5	0	0,5
2		0	0,5	0,5
3		0,5	0,5	0
4				
5	i0	3		
6				
7	nsim	1000		
8				

When the macro is executed, the results appear in line 9. Typical results are shown in Table 4.9. ∎

Table 4.9 Relative frequencies of each state for two simulations

	State		
nsim	1	2	3
1000	0,319	0,344	0,337
10,000	0,3299	0,3384	0,3317

Example

Let us consider a stationary Markov chain such that

$$\boldsymbol{\Pi} = \begin{pmatrix} 0 & 1 & 0 \\ 0 & 0 & 1 \\ 1 & 0 & 0 \end{pmatrix}.$$

Then, $\boldsymbol{\Pi}^3 = \boldsymbol{Id}$. Thus, the chain is periodical, with $k_T = 3$. Let use EXCEL® verify that these results. Let us repeat the operations to find a stationary probability, as indicated in the preceding example. The result appears as below:

	A	B	C	D	E	F	G	H	I	J	K	L
1	PI	0	1	0		p	0,333333		100			
2		0	0	1			0,333333			0,333333	0,333333	0,333333
3		1	0	0			0,333333			0,333333	0,333333	0,333333
4						e	1,36E-16			0,333333	0,333333	0,333333
5												
6	A	1	-1	0	B	0	p	0,333333				
7		0	1	-1		0		0,333333				
8		-1	0	1		0		0,333333				
9		1	1	1		1	e	0				
10												

In this case, the stationary probability is $\boldsymbol{p} = \left(\frac{1}{3}, \frac{1}{3}, \frac{1}{3}\right)$.

To verify periodicity, select cell `N1`, enter the value 3 and execute the macro matrixp(). The results appear in O2:Q4: we see that $\boldsymbol{\Pi}^3 = \boldsymbol{Id}$

N	O	P	Q
3			
	1	0	0
	0	1	0
	0	0	1

To simulate the chain, we use the workbook below:

	A	B	C	D
1	PI	0	1	0
2		0	0	1
3		1	0	0
4				
5	i0	3		
6				
7	nsim	10000		
8				

When the macro `markovsim()` is executed, the results of the simulation appear in line 9. Typical results are shown in Table 4.10. ∎

Table 4.10 Relative frequencies of each state for two simulations

	State		
nsim	1	2	3
1000	0,333	0,334	0,334
10,000	0,3333	0,3334	0,3334

Example

Let us consider a stationary Markov chain such that

$$\boldsymbol{\Pi} = \begin{pmatrix} 0,2 & 0 & 0,4 & 0 \\ 0 & 1 & 0,2 & 0 \\ 0,8 & 0 & 0,1 & 0 \\ 0 & 0 & 0,3 & 1 \end{pmatrix}.$$

This Markov chain has 2 absorbing states: 2 and 4. By renumbering the states: $1 \rightarrow 1$, $2 \rightarrow 3$, $3 \rightarrow 2$, $4 \rightarrow 4$, we obtain the canonical form

$$\boldsymbol{\Pi} = \begin{pmatrix} 0,2 & 0,4 & 0 & 0 \\ 0,8 & 0,1 & 0 & 0 \\ 0 & 0,2 & 1 & 0 \\ 0 & 0,4 & 0 & 1 \end{pmatrix}.$$

Thus,

$$\boldsymbol{A} = \begin{pmatrix} 0 & 0,2 \\ 0 & 0,1 \end{pmatrix}, \ \boldsymbol{B} = \begin{pmatrix} 0,2 & 0,4 \\ 0,8 & 0,1 \end{pmatrix}.$$

A typical workbook for this chain is shown below:

	A	B	C	D	E	F	G	H	I	J	K	L	M	N	O
1	B	0,2	0,4		M	2,25	1		j	1	2		N	0,4	0,4
2		0,8	0,1			2	2		$M_{\blacksquare j}$	4,25	3			0,6	0,6
3															
4	PI	0,2	0,4	0	0										
5		0,8	0,1	0	0										
6		0	0,2	1	0										
7		0	0,3	0	1										

Here,

$$\boldsymbol{M} = (\boldsymbol{Id} - \boldsymbol{B})^{-1} = \begin{pmatrix} 2,25 & 1 \\ 2 & 2 \end{pmatrix},$$

$$M_{\blacksquare 1} = M_{11} + M_{21} = 4,25, \ M_{\blacksquare 2} = M_{12} + M_{22} = 3,$$

$$\boldsymbol{N} = \boldsymbol{AM} = \begin{pmatrix} 0,4 & 0,4 \\ 0,6 & 0,6 \end{pmatrix}.$$

We use the Macro `markovsim` to simulate 30 steps $X_1, \ldots, X_{30}$ of the chain, starting from $X_0 \in \{1, 2\}$ of the canonical form ($\{1,3\}$ in the original chain). The simulation is repeated *nrep* times and we use the results to evaluate the mean number of steps before absorption and the mean number of visits at the non-absorbing states. The results are shown in Tables 4.11 and 4.12. They are close to the theoretical values. ∎

Table 4.11 Results for starting state $X_0 = 1$

	visits			Absorption at	
nrep	State 1	State 2	Steps before absorption	State 3	State 4
1000	2245	1992	4237	0,406	0,594
10,000	2249	1998	4247	0,405	0,595

Table 4.12 Results for starting state $X_0 = 3$

	visits			Absorption at	
nrep	State 1	State 2	Steps before absorption	State 3	State 4
1000	1073	2034	3107	0,400	0,600
10,000	1023	2021	3045	0,401	0,599

Exercises

1. Let us consider a Markov chain having states $\{1, 2\}$ and transition matrix $(0 \leq a, b \leq 1)$

$$\mathbf{\Pi} = \begin{pmatrix} a & b \\ 1-a & 1-b \end{pmatrix}$$

(a) Draw the graphical representation of the chain.
(b) Determine the values of $P(X_{n+1} = 1 | X_n = 2)$ and $P(X_{n+1} = 2 | X_n = 1)$
(c) Determine the values of a, b such that the chain is absorbing.
TIP: the diagonal values must be equal to 1.
(d) Determine the values of a, b such that the chain is periodical.
TIP: Determine the values such that $\mathbf{\Pi}^2 = \boldsymbol{Id}$.
(e) Find a general expression of the eigenvector $\boldsymbol{v}$ of $\mathbf{\Pi}$ associated to the eigenvalue 1.
TIP: set $\boldsymbol{v} = (\alpha, \beta)$ and solve the linear system $\mathbf{\Pi}\boldsymbol{v} = \boldsymbol{v}$.
(f) Determine the conditions on $\boldsymbol{v}$ such that $\boldsymbol{v} = \boldsymbol{p}$.
TIP: $p_1, p_2 \geq 0$; $p_1 + p_2 = 1$.
(g) Simulate the chain for convenient values of a, b.

2. Let us consider a Markov chain having states $\{1, 2, 3, 4\}$ and transition matrix

$$\mathbf{\Pi} = \begin{pmatrix} 0,02 & 0,12 & 0 & 0,7 \\ 0,91 & 0,4 & 0,6 & 0 \\ 0 & 0,48 & 0,4 & 0 \\ 0,07 & 0 & 0 & 0,3 \end{pmatrix}$$

(a) Draw a graphical representation of the chain
(b) Determine the values of $P(X_{n+1} = 2 | X_n = 3)$ and $P(X_{n+1} = 3 | X_n = 2)$

(c) Determine the stationary probability.
TIP: use the macro matrixp() to evaluate $\mathbf{\Pi}^n$ for n = 100.
(d) Simulate the chain.

3. This exercise comes from (Sinclair, 2005), under Creative Commons Attribution License 1.0. Consider a machine that can be in the states idle, busy, waiting, broken, in repair – numbered respectively 1 to 5. The transition matrix is

$$\mathbf{\Pi} = \begin{pmatrix} 0,05 & 0,1 & 0 & 0 & 0,75 \\ 0,93 & 0,43 & 0,6 & 0 & 0 \\ 0 & 0,43 & 0,35 & 0 & 0 \\ 0,02 & 0,04 & 0,05 & 0,8 & 0 \\ 0 & 0 & 0 & 0,2 & 0,25 \end{pmatrix}$$

(a) Draw a graphical representation of the chain
(b) Determine the values of $P(X_{n+1} = 2 | X_n = 4)$ and $P(X_{n+1} = 4 | X_n = 2)$
(c) Determine the stationary probability.
(d) Simulate the chain.

4. This exercise comes from (Chen, 2008), under Creative Commons Attribution License 1.0 license. Consider a Markov chain to generate music: the states are the notes A, A#, B, C,D, E,F,G,G# – numbered respectively 1 to 9. The transition matrix is

$$\mathbf{\Pi} = \begin{pmatrix} \frac{4}{19} & 1 & \frac{7}{15} & 0 & 0 & \frac{4}{19} & 0 & \frac{1}{5} & 0 \\ 0 & 0 & 0 & 0 & 0 & \frac{1}{19} & 0 & 0 & 0 \\ \frac{3}{19} & 0 & \frac{1}{15} & \frac{6}{15} & 0 & 0 & \frac{1}{5} & \frac{1}{5} & \frac{3}{4} \\ 0 & 0 & \frac{4}{15} & \frac{3}{15} & \frac{3}{11} & \frac{3}{19} & 0 & \frac{2}{5} & 0 \\ \frac{2}{19} & 0 & 0 & \frac{6}{15} & \frac{3}{11} & 0 & 0 & 0 & 0 \\ \frac{1}{19} & 0 & \frac{3}{15} & 0 & \frac{5}{11} & \frac{5}{19} & \frac{1}{5} & 0 & \frac{1}{4} \\ 0 & 0 & 0 & 0 & 0 & \frac{4}{19} & 0 & 0 & 0 \\ \frac{6}{19} & 0 & 0 & 0 & 0 & \frac{1}{19} & \frac{3}{5} & \frac{1}{5} & 0 \\ \frac{3}{19} & 0 & 0 & 0 & 0 & \frac{1}{19} & 0 & 0 & 0 \end{pmatrix}$$

(a) Determine the values of $P(X_{n+1} = 2 | X_n = 6)$ and $P(X_{n+1} = 6 | X_n = 2)$
(b) Determine the stationary probability.
(c) Simulate the chain.

4.8 Diffusion Processes

Stochastic Differential Equations (SDE) are a common tool in Mathematical Finance, namely the famous model of Fischer Black and Myron Scholes (Black & Scholes, 1973) for the price of options, initially introduced as a partial differential equation and reinterpreted by Robert Cox Merton as a SDE (Merton, 1973). Previous works tending to introduce probability in the analysis of finance were made by (Regnault, 1863; Cowles, 1933) and namely by (Bachelier, 1900), which firstly introduced Brownian motion in finance.

Since these works, SDE has become a central tool in finance, but also in other fields such as, for instance, stochastic optimization and reliability. The mathematical foundations are connected to Itô's Calculus (Itô, 1944; Itô, 1946; Itô, 1950; Itô, 1951).

The solutions of SDE are usually referred as diffusion processes. In the sequel, we examine how to simulate and use diffusion processes with EXCEL®.

4.8.1 Time Integral and Derivative of a Process

Standard integrals are defined as limits of finite sums. For instance, to define $\int_a^b f(t)dt$, we may consider a partition

$$\mathcal{T} = \{a = t_0 < t_1 < \ldots < t_{n-1} < t_n = b\} \tag{4.21}$$

of the interval (a, b) and the finite sum

$$S(\mathcal{T},f) = \sum_{i=1}^{n} (t_i - t_{i-1})f(t_{i-1}). \tag{4.22}$$

For a regular function f,

$$S(\mathcal{T},f) \to \int_a^b f(t)dt$$

when $n \to +\infty$ and $\max\{t_i - t_{i-1}\} \to 0$.

We may define the integral of a stochastic process with respect to time into an analogous way. For instance, let us consider a stochastic process $\{X(t),\ t \in (a, b)\}$. We may consider the finite sum

$$I(\mathcal{T},\mathrm{X}) = \sum_{i=1}^{n} (t_i - t_{i-1})X(t_{i-1}). \tag{4.23}$$

If the process has integrable variances, id est, if

$$\int_a^b E\left(X^2(t)\right)dt < \infty,$$

then the finite sum converges for $n \to +\infty$ and $\max\{t_i - t_{i-1}\} \to 0$. In this case, we have, analogously the standard functions (see, for instance, (Föllmer, 1981)):

$$I(\mathcal{T},\mathrm{X}) \to I(X) = \int_a^b X(t)dt. \tag{4.24}$$

I(*X*) is called the integral of the process *X* with respect to the time. It is linear:

$$I(\alpha X + \beta Y) = \alpha I(X) + \beta I(Y), \ \forall \alpha, \beta \in \mathbb{R}.$$

and

$$\int_a^c X_t dt + \int_c^b X_t dt = \int_a^b X_t dt. \tag{4.25}$$

From Eq. (4.23), we have

$$E(I(X)) = \int_a^b m_X(t)dt, \quad E\left(I(X)^2\right) = \int\int_a^b R_{XX}(s,t)dsdt. \tag{4.26}$$

Let us define

$$dY_t = X_t dt \Leftrightarrow Y_t = Y_0 + \int_0^t X_s ds, \qquad \forall t > 0. \tag{4.27}$$

Then, X_t is the derivative of Y_t with respect to time. We have

$$m_Y(t) = m_Y(0) + \int_0^t m_X(s)ds, \qquad \forall t > 0 \Rightarrow m_Y = \frac{d}{dt} m_X. \tag{4.28}$$

Assuming that Y_0 and X are independent, we have

$$\begin{aligned} R_{YY}(s,t) = E\left(Y_0^2\right) + m_Y(0)\left(\int_0^s m_X(\tau_1)d\tau_1 + \int_0^t m_X(\tau_2)d\tau_2\right) \\ + \int_{\tau_1=0}^{\tau_1=s} \int_{\tau_2=0}^{\tau_2=t} R_{XX}(\tau_1,\tau_2)d\tau_1 d\tau_2, \end{aligned} \tag{4.29}$$

so that

$$R_{XX}(s,t) = \frac{\partial^2}{\partial s \partial t} R_{YY}(s,t). \tag{4.30}$$

Into an analogous way, we have

$$R_{YX}(s,t) = \frac{\partial}{\partial t} R_{YY}(s,t). \tag{4.31}$$

Example

Let us consider the process $Y_t = A_1 \cos(\omega t) + A_2 \sin(\omega t)$, where $\omega \in \mathbb{R}$, while A_1, A_2 are independent random variables such that $E(A_i) = 0$, $V(A_i) = \sigma^2$, $i = 1, 2$. We have

$$m_Y(t) = 0, \;\; V_Y(t) = \sigma^2, \;\; R_{YY}(s,t) = \sigma^2 \cos(\omega(s-t)).$$

Let $dY_t = X_t dt, \; Y_0 = A_1$. We have

$$R_{XX}(s,t) = \frac{\partial^2}{\partial s \partial t} R_{YY}(s,t) = \omega^2 \sigma^2 \cos(\omega(s-t)).$$

$$R_{YX}(s,t) = \frac{\partial}{\partial t} R_{YY}(s,t) = \omega \sigma^2 \sin(\omega(s-t)).$$

Indeed,

$$Y_t = A_1 - A_1 \omega \int_0^t \sin(\omega s) ds + A_2 \omega \int_0^t \cos(\omega s) ds,$$

So that, we have

$$X_t = -A_1 \omega \sin(\omega t) + A_2 \omega \cos(\omega t).$$ ∎

Example

Let us consider the process $X_t = A_1 \cos(\omega t) + A_2 \sin(\omega t)$, where $\omega \in \mathbb{R}$, while A_1, A_2 are independent random variables such that $E(A_i) = 0$, $V(A_i) = \sigma^2$, $i = 1, 2$. We have

$$m_X(t) = 0, \;\; V_X(t) = \sigma^2, \;\; R_{XX}(s,t) = \sigma^2 \cos(\omega(s-t)).$$

Let $dY_t = X_t dt, \; Y_0 = 0$. We have

$$m_Y(t) = 0$$

and

$$R_{YY}(s,t) = \int_{\tau_1=0}^{\tau_1=s} \int_{\tau_2=0}^{\tau_2=t} R_{XX}(\tau_1, \tau_2) d\tau_1 d\tau_2,$$

so that

$$R_{YY}(s,t) = \sigma^2 \frac{\cos(\omega(s-t)) - \cos(\omega s) - \cos(\omega t) + 1}{\omega^2},$$

Indeed, let

$$Y_t = A_1 \int_0^t \sin(\omega s)ds + A_2 \int_0^t \cos(\omega s)ds.$$

Then,

$$Y_t = \frac{A_1}{\omega}(1 - \cos(\omega t)) + \frac{A_2}{\omega} \sin(\omega t)$$

and we have

$$E(Y_t Y_s) = \sigma^2 \frac{\cos(\omega(s - t)) - \cos(\omega s) - \cos(\omega t) + 1}{\omega^2}.$$

∎

Exercises

1. Consider the process $X_t = A_1 + A_2 t$, where A_1, A_2 are independent random variables such that $E(A_i) = 0, V(A_i) = 1, i = 1, 2$.

   (a) Determine $R_{XX}(s, t)$.
   (b) Let $Y_t = X_t dt,\ Y_0 = 0$. Determine $R_{YY}(s, t)$.
   (c) Let $dX_t = Y_t dt$. Determine $R_{YY}(s, t)$.

2. Consider the process $X_t = A_1 + A_2 \exp(-t)$, where A_1, A_2 are independent random variables such that $E(A_i) = 0, V(A_i) = 1, i = 1, 2$.

   (a) Determine $R_{XX}(s, t)$.
   (b) Let $Y_t = X_t dt,\ Y_0 = 0$. Determine $R_{YY}(s, t)$.
   (c) Let $dX_t = Y_t dt$. Determine $R_{YY}(s, t)$.

3. Consider a process X_t such that $R_{XX}(s, t) = \exp(-|s - t|)$.

   (a) Let $Y_t = X_t dt,\ Y_0 = 0$. Determine $R_{YY}(s, t)$.
   (b) Let $dX_t = Y_t dt$. Determine $R_{YY}(s, t)$.

4. Consider a process X_t such that $R_{XX}(s, t) = 1 - |s - t|$, *if* $|s - t| < 1$; $R_{XX}(s, t) = 0$, *otherwise*.

   (a) Let $Y_t = X_t dt,\ Y_0 = 0$. Determine $R_{YY}(s, t)$.
   (b) Let $dX_t = Y_t dt$. Determine $R_{YY}(s, t)$.

4.8.2 Simulation of the Time Integral of a White Noise

The simulation of the process defined by the time integral (4.24) or (4.27) can request, on the one hand, special simulation features and, on the other hand, special quadrature formulae.

Indeed, a simulation is made by choosing a final time of simulation $T > 0$ and a partition of the interval $(0, T)$ in a maximal number of steps >0: $0 = t_0 < t_1 < \ldots < t_{ns-1} < t_{ns} = T$. For instance, we may take $t_i = i\Delta t$, $\Delta t = \frac{T}{ns}$, $0 \le i \le ns$. Then, we generate a sequence of values $X_1, \ldots, X_{ns}$ such that $X_i = X(t_i)$ and use them to evaluate

$$Y_{n+1} = Y_n + \int_{t_n}^{t_{n+1}} X_s ds, \quad 0 \le n \le ns - 1. \tag{4.32}$$

By this way, we generate $Y_1, \ldots, Y_{ns}$ where $Y(t_j) = Y_j$. Such a procedure requests some precautions. On the one hand, the generation of $X_1, \ldots, X_n, \ldots$ must be consistent with the statistical properties of X_t – namely $E(X_n) = m_X(t_n)$ and $E(X_iX_j) = R_{XX}(t_i, t_j)$. On the other hand, the evaluation of the integral at the right hand side must be consistent with Eqs. (4.26) and (4.29). For instance, let us consider the approximation

$$\int_{t_n}^{t_n+\Delta t} X_s ds \approx \Delta t(\alpha X_n + \beta). \tag{4.33}$$

The conditions (4.26) yield that

$$\alpha m_X(t_n) + \beta = \frac{1}{\Delta t} \int_{t_n}^{t_n+\Delta t} m_X(s) ds$$

$$\alpha^2 R_{XX}(t_n, t_n) + 2\alpha\beta m_X(t_n) + \beta^2 = \frac{1}{\Delta t^2} \int_{t_n}^{t_n+\Delta t} \int_{t_n}^{t_n+\Delta t} R_{XX}(s, t) ds dt$$

If the process is weakly stationary, these equations read as:

$$\alpha m_X + \beta = m_X$$

$$\alpha^2 R_X(0) + 2\alpha\beta m_X + \beta^2 = \frac{1}{\Delta t^2} \int_{t_n}^{t_n+\Delta t} \int_{t_n}^{t_n+\Delta t} R_X(|s - t|) ds dt$$

id est,

$$\beta = (1 - \alpha) m_X$$

$$\alpha^2 C_X(0) + m_X^2 = \frac{1}{\Delta t^2} \int_{t_n}^{t_n+\Delta t} \int_{t_n}^{t_n+\Delta t} R_X(s - t) ds dt$$

If R_X is a regular function, we can take as approximation

$$\int_{t_n}^{t_n+\Delta t} \int_{t_n}^{t_n+\Delta t} R_X(s - t) ds dt \approx \Delta t^2 R_X(0),$$

so that $\alpha \approx 1$, $\beta \approx 0$ are convenient values. But these values may be not convenient if R_X is not a regular function. For instance, if $R_X(s) = \sigma^2 \delta(s)$, we obtain

$$\alpha = \sqrt{\frac{1}{\Delta t} - \frac{m_X^2}{\sigma^2}}, \quad \beta = (1 - \alpha) m_X.$$

If $X_t = \varepsilon_t$ is a white noise, we have $m_X = 0$, so that

$$\alpha = \frac{1}{\sqrt{\Delta t}}, \quad \beta = 0. \tag{4.34}$$

In the sequel, we shall consider the time integral of a white noise:

$$E_t = E_0 + \int_0^t \varepsilon_s ds. \tag{4.35}$$

According to Eq. (4.34), the simulations use the approximation

$$\int_{t_n}^{t_n+\Delta t} \varepsilon_s ds \approx \varepsilon_n \sqrt{\Delta t}. \tag{4.36}$$

We have $m_E(t) = 0$ and

$$R_{EE}(s,t) = \sigma^2 \int_{\tau_1=0}^{\tau_1=s} \int_{\tau_2=0}^{\tau_2=t} \delta(\tau_1 - \tau_2) d\tau_1 d\tau_2 = \sigma^2 \min\{s,t\}. \tag{4.37}$$

Taking into account Eqs. (4.23) and (4.24) and the theorem enounced page 120, E_t is gaussian. In addition,

$$0 \leq r \leq s \leq t \Rightarrow E((E_t - E_s)E_r) = \sigma^2 r - \sigma^2 r = 0. \tag{4.38}$$

Thus, on the one hand, $E_t \sim N(0, \sigma\sqrt{t})$ and, on the other hand, $(E_t - E_s)$ is independent from E_r if $0 \leq r \leq s \leq t$. We have, for $0 \leq s \leq t$:

$$E\left((E_t - E_s)^2\right) = \sigma^2(t - s).$$

Thus, $E_t - E_s \sim N(0, \sigma\sqrt{t-s})$ for $0 \leq s \leq t$.

Let us consider the process $X_t = \varepsilon_{t\,+\,\delta t}$ and

$$Y_t = \int_0^t \varepsilon_{s+\delta t} ds = \int_{\delta t}^{t+\delta t} \varepsilon_s ds = \int_0^{t+\delta t} \varepsilon_s ds - \int_0^{\delta t} \varepsilon_s ds = E_{t+\delta t} - E_{\delta t} = \delta E_t.$$

Let us introduce $\delta_k E_t = E_{t+\delta t_k} - E_{\delta t_k}$. Then, $R_{ij}(s, t) = E(\delta_i E_t \times \delta_j E_s)$ verifies

$$\begin{aligned} R_{ij}(s,t) = \sigma^2 (& \min\{t + \delta t_i, s + \delta t_j\} - \min\{\delta t_i, s + \delta t_j\} \\ & - \min\{t + \delta t_i, \delta t_j\} + \min\{\delta t_i, \delta t_j\}). \end{aligned} \tag{4.39}$$

Thus, for a process

$$X_t = \sum_{i=0}^{k} a_i \varepsilon_{t+\delta ti},$$

$Y_t = \int_0^t X_t ds$ verifies

$$Y_t = \sum_{i=0}^{k} a_i \delta_i E_t,\ \delta_i E_t = E_{t+\delta t_i} - E_{\delta t_i}$$

and

$$R_{YY}(s,t) = E(Y_t Y_s) = \sum_{i=0}^{k} \sum_{j=0}^{k} a_i a_j R_{ij}(s,t). \tag{4.40}$$

A process

$$X_t = \sum_{i=0}^{k} a_i \varepsilon_{t-\delta ti}$$

may be analyzed into an analogous way: notice that

$$\int_{\delta t}^{t+\delta t} \varepsilon_s ds \sim \int_{-\delta t}^{t-\delta t} \varepsilon_s, \tag{4.41}$$

So that Eqs. (4.39) and (4.40) apply yet.

Example

Let us consider a white noise $\{\varepsilon_t : t \geq 0\}$ of variance 1 and the process $X_t = \varepsilon_t + \varepsilon_{t+1}$. Let $Y_t = \int_0^t X_s ds$. We have $k = 1$, $a_0 = a_1 = 1$, $\delta_0 t = 0$, $\delta_1 t = 1$. Thus,

$$Y_t = E_t + E_{t+1} - E_1.$$

$$R_{00}(s,t) = \min\{t,s\} - \min\{0,s\} - \min\{t,0\} + \min\{0,0\} = \min\{t,s\};$$

$$\begin{aligned} R_{01}(s,t) &= \min\{t,s+1\} - \min\{0,s+1\} - \min\{t,1\} + \min\{0,1\} \\ &= \min\{t,s+1\} - \min\{t,1\}; \end{aligned}$$

$$\begin{aligned} R_{11}(s,t) &= \min\{t+1,s+1\} - \min\{1,s+1\} - \min\{t+1,1\} + \min\{1,1\} \\ &= \min\{t+1,s+1\} - 1 = \min\{t,s\}; \end{aligned}$$

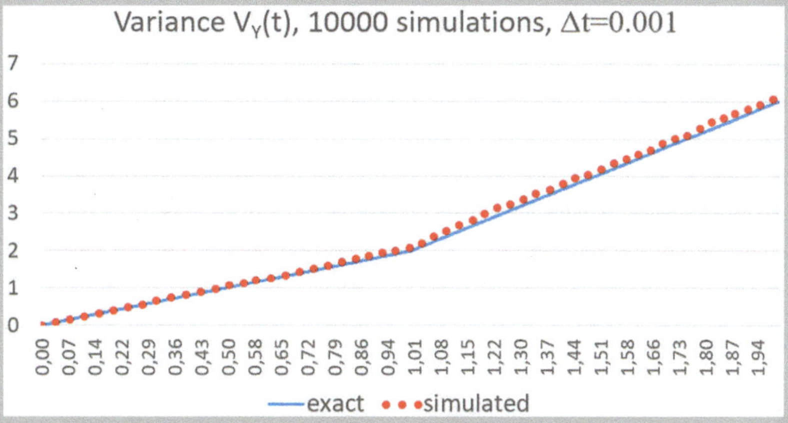

Fig. 4.22 Results obtained by simulation of the process Y_t

For $t = s$,

$$R_{00}(t,t) = t, R_{01}(t,t) = t - \min\{t,1\},\ R_{11}(t) = t$$

Thus,

$$V_Y(t) = 2t,\ \text{if } t < 1;\ V_Y(t) = 4t - 2,\ \text{if } t > 1.$$

Figure 4.22 shows the results of a numerical evaluation of the variances for $t \in (0, 3)$ by using 10000 simulations of Y_t with a step $\Delta t = 0.01$. ∎

Example

Let us consider a white noise $\{\varepsilon_t : t \geq 0\}$ of variance 1 and the process $X_t = \varepsilon_t - \varepsilon_{t-1}$. Let $Y_t = \int_0^t X_s ds$. We have $k = 1, a_0 = 1, a_1 = -1, \delta_0 t = 0, \delta_1 t = -1$. We apply the analogy given in Eq. (4.41)

$$E_{t-1} - E_{-1} \sim E_{t+1} - E_1 \Rightarrow Y_t \sim E_t - (E_{t+1} - E_1).$$

Thus,

$$V_Y(t) = 2t, \text{if } t < 1; V_Y(t) = 2, \text{if } t > 1.$$

Figure 4.23 shows the results of a numerical evaluation of the variances for $t \in (0, 2.4)$ by using 10000 simulations of Y_t with a step $\Delta t = 0.01$.

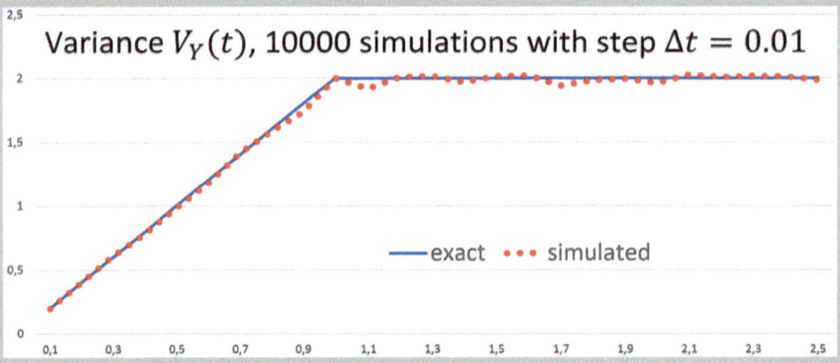

Fig. 4.23 Results obtained by simulation of the process Y_t ∎

Exercises

1. Consider a a white noise $\{\varepsilon_t : t \geq 0\}$ of variance 1 and the process $X_t = \varepsilon_t + \varepsilon_{t-2}$.

   (a) Identify the coefficients a_0, a_1, a_2.
   (b) Determine $R_{ij}(t, t)$, $0 \leq i, j \leq 2$.
   (c) Evaluate the variance of Y_2 by using simulations of the process Y_t.

2. Consider a a white noise $\{\varepsilon_t : t \geq 0\}$ of variance 1 and the process $X_t = \varepsilon_t - \varepsilon_{t+2}$.

   (a) Identify the coefficients a_0, a_1, a_2.
   (b) Determine $R_{ij}(t, t)$, $0 \leq i, j \leq 2$.
   (c) Evaluate the variance of Y_2 by using simulations of the process Y_t.

4.8.3 Brownian Motion

Brownian motion takes his name from Robert Brown, a botanist which observed chaotic motions of particles in suspension (Brown, 1828). The mathematical theory of Brownian motion was established much later (Paley et al., 1933), after its use by (Bachelier, 1900), the questions by (Einstein, 1905) and (Pearson, 1905) and the model by (Langevin, 1908). Due to the formalism introduced by Wiener, Brownian motion is also known as Wiener Process.

$\{W_t = W(t) : t \geq 0\}$ is a **Wiener process** (or **Brownian motion**) if and only if $W(0) = 0$ and $dW_t = \varepsilon_t dt$, where $\{\varepsilon_t = \varepsilon(t) : t \in \mathbb{R}\}$ is a gaussian white noise of variance $\sigma^2 = 1$.

$\{\mathbf{W}_t = \mathbf{W}(t) : t > 0\}$ is a ***n*-dimensional Wiener process** (or ***n*-dimensional** Brownian motion) if and only if $\mathbf{W}(t) = (W_1(t), \ldots, W_n(t))$, where each W_i is a Wiener process and W_i is independent from W_j for $i \neq j$. ■

Thus, a Brownian motion corresponds to an integral of a white noise. As shown in Sect. 4.8.2, we have

$$m_W(t) = 0,\ R_{WW}(s,t) = \min\{s,t\},\ E\left(W_t^2\right) = t. \tag{4.42}$$

$$0 \leq r \leq s \leq t \Rightarrow E((W_t - W_s)W_r) = r - r = 0. \tag{4.43}$$

$$W_t - W_s \sim N\left(0, \sqrt{t-s}\right),\ E\left((W_t - W_s)^2\right) = t - s,\ \mathit{for}\ 0 \leq s \leq t. \tag{4.44}$$

Let $\varphi : \mathbb{R} \to \mathbb{R}$ be a regular function and

$$I_t(\varphi) = \int_0^t \varphi(W_s)ds. \tag{4.45}$$

For a regular φ:

$$E(I_t(\varphi)) = \int_0^t E(\varphi(W_s))ds. \tag{4.46}$$

The simulation of a Brownian motion can be made by the method previously presented (Sect. 4.8.2):

$$W_{n+1} = W_n + \sqrt{\Delta t}\, Z_n, 0 \le n \le ns - 1, \;\; Z_n \sim N(0, 1). \tag{4.47}$$

For instance, we may use one of the VBA functions below:

```
Function wienersim(ns, dt)
Dim W() As Double
ReDim W(0 To ns)
h = Sqr(dt)
W(0) = 0
Randomize
For n = 1 To ns
    aux = Rnd()
    If Abs(aux) < 0.000001 Then
     Randomize
     aux = Rnd()
    End If
    Z = WorksheetFunction.Norm_S_Inv(aux)
    W(n) = W(n - 1) + h * Z
Next n
wienersim = W
End Function
```

```
Function wienersim2(T, dt)
ns = WorksheetFunction.RoundUp(T / dt, 0) + 1
Dim W() As Double
ReDim W(0 To ns)
h = Sqr(dt)
W(0) = 0
n = 0
tt = 0
Randomize
Do While tt < T
    n = n + 1
    tt = tt + dt
    aux = Rnd()
    If Abs(aux) < 0.000001 Then
     Randomize
     aux = Rnd()
    End If
    Z = WorksheetFunction.Norm_S_Inv(aux)
    W(n) = W(n - 1) + h * Z
Loop
wienersim2 = W
End Function
```

Function `wiernersim` simulates ns steps of the Brownian motion. Function `wienersim2` simulates the motion until time `T`. Both use a time step Δt=`dt`.

Example

Let us consider $T = 4$. We simulate a bidimensional Brownian motion with $ns = 4000$, so that $\Delta t = 1E - 3$. The results are shown in Fig. 4.24.

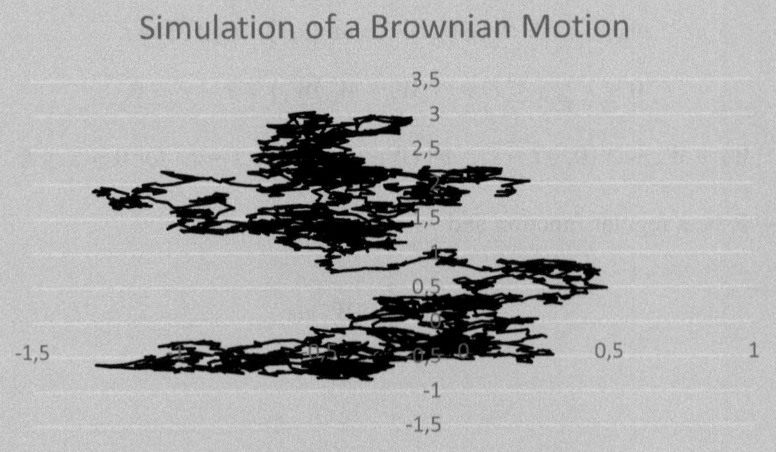

Fig. 4.24 Simulation of a Brownian Motion until $T = 4$, with $ns = 4000$ steps

We evaluate $E(\|X(T)\|^2)$ using a sample of 10000 realizations with different values of ns. The theoretical value is $E(\|X(T)\|^2) = 8$. The results are shown in Table 4.13. ∎

Table 4.13 Results furnished by the simulation of a Brownian motion until $T = 4$

ns	400	4000	40,000
$E(\|X(T)\|^2)$	8,0	8,1	8,0
$\sigma(\|X(T)\|^2)$	7,9	8,0	8,0

The simulation of a time integral (4.45) is made by generating a Brownian motion and then, using it to evaluate the integral. For instance, we may use the code at right.

```
Function intbrownsim(T, dt)
W = wienersim2(T, dt)
s = 0
tt = 0
i = 0
Do While tt < T
    s = s + phi(W(i))
    tt = tt + dt
    i = i + 1
Loop
s = s * dt
intbrownsim = s
End Function
```

Example

Let us consider $\varphi(W_s) = W_s^2$. Then, $E(I_t(\varphi)) = \frac{t^2}{2}$. We generate a sample of 10,000 variates from $I_t(\varphi)$ with $\Delta t = 1E - 3$ and we use the sample to estimate the mean $E(I_t(\varphi))$. The results are shown in Fig. 4.25. ∎

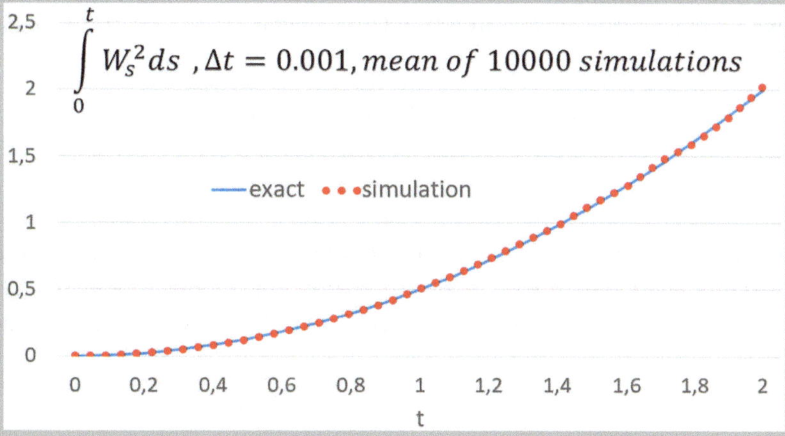

Fig. 4.25 Empirical mean of the simulation of the time integral of $\varphi(W_t) = W_t^2$, with $\Delta t = 1E - 3$

Exercises

1. For each function below, generate a sample of 1000 variates from $I_t(\varphi)$ for $t = 1$ and use the sample to estimate the mean value of $I_t(\varphi)$.

(a) $\varphi(W_s) = \exp(W_s)$.
(b) $\varphi(W_s) = \sin(2\pi W_s)$.
(c) $\varphi(W_s) = W_s^4$.

4.8.4 Random Walks

Random walks generalize Brownian motion: a random walk is a Markov process which generates trajectories in the space $\mathbb{R}^n$. A discrete random walk generates points $X_1, \ldots, X_n, \ldots$ in $\mathbb{R}^n$ and the trajectory connects the points generated. A continuous random walk generates a trajectory $\{X_t = X(t) : t \geq 0;\ X(0) = X_0\}$.

In general, a discrete random walk is defined by the distribution of the displacements $\Delta X_n = X_n - X_{n-1}$ and the starting point X_0. The displacements ΔX_n are often multiples of a random vector having a known distribution: $\Delta X_n = hZ_n$, $h > 0$, Z_n known. In practice, the distribution of Z_n is often taken as independent of n, what is equivalent to take $Z_n \sim Z$. For instance, a discrete Brownian motion is a random walk where Z is a gaussian variable $N(0, 1)$ – parameter h is connected to the time discretization: $h = \sqrt{\Delta t}$.

Let us assume that $E(Z) = 0$, $V(Z) = \sigma^2$ and that Z_i, Z_j are independent for $i \neq j$. Noticing that

$$X_n = X_0 + \sum_{i=1}^{n} \Delta X_i, \tag{4.48}$$

We have

$$E(X_n) = X_0, \quad V(X_n) = nh^2\sigma^2. \tag{4.49}$$

To simulate a discrete random walk, we need to know the values of X_0, h and the distribution of Z. Then, we choose a finite number of steps $ns > 0$ and we calculate $X_1, \ldots, X_{ns}$ by the recurrence (Z_n is a variate from Z).

$$X_{n+1} = X_n + hZ_n, \quad 0 \leq n \leq ns - 1. \tag{4.50}$$

Example

Let us consider a bidimensional random walk starting at $X_0 = (0, 0)$ defined by $\Delta X_n = hZ_n$, $Z_n \sim Z$, $Z = (\cos\theta, \sin\theta)$, $P(\theta = 0) = P\left(\theta = \frac{\pi}{2}\right) = P(\theta = \pi) = P\left(\theta = \frac{3\pi}{2}\right) = \frac{1}{4}$. We generate a sample of $ns = 10000$ variates of Z (see Sect. 3.14). The resulting path is shown below (Fig. 4.26):

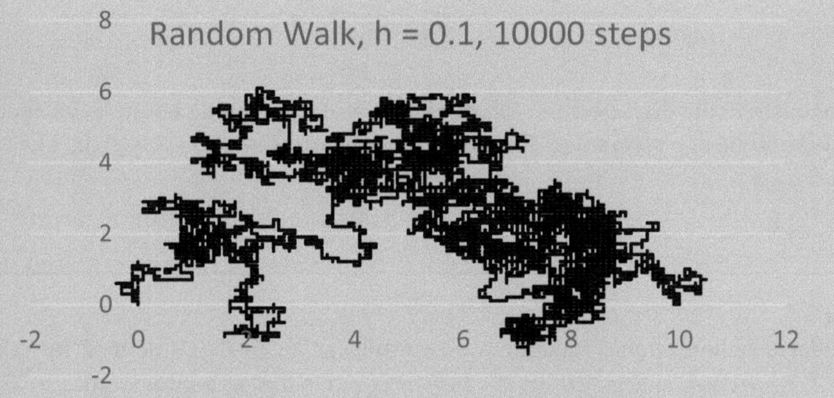

Fig. 4.26 Simulation of a random walk with 4 possible displacements at each step

Example

Let us consider a bidimensional random walk starting at $X_0 = (0, 0)$ defined by $dX_t = Z_t$, $Z_t \sim Z, Z = (\cos\theta, \sin\theta)$, θ uniformly distributed on $(0, 2\pi)$. We simulate $ns = 10000$ steps with $\Delta t = 0.1$. The trajectory simulated is shown below (Fig. 4.27):

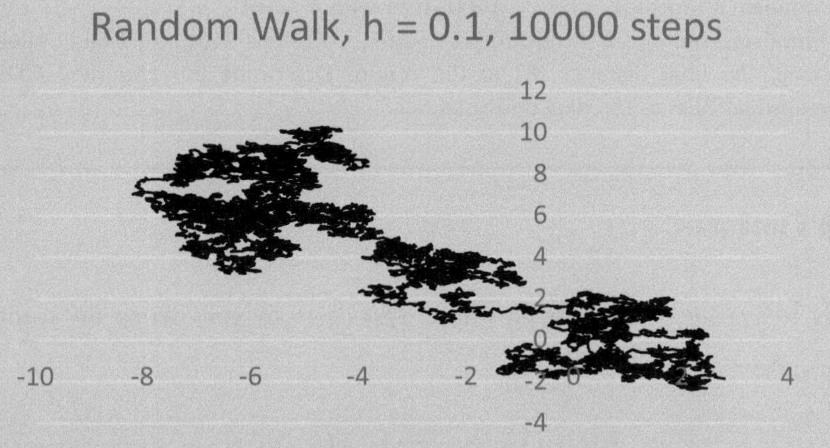

Fig. 4.27 Simulation of a random walk with possible displacements uniformly distributed

Remark

Continuous random walks are generally defined by their initial point X_0 and their derivatives:

$$dX_t = Z_t dt, \quad X(0) = X_0. \tag{4.51}$$

Thus (see Sect. 4.8.1):

$$X_t - X_0 = \int_0^t Z_s ds, \quad \forall t > 0. \tag{4.52}$$

Analogously to the discrete case, the distribution of Z_t is often chosen as independent of t: $Z_n \sim Z$. For instance, a continuous Brownian motion corresponds to $Z \sim N(0, 1)$. ∎

Exercises

1. Consider a bidimensional random walk starting at $X_0 = (0, 0)$ defined by $\Delta X_n = hZ_n$, $Z_n \sim Z$, $Z = (\cos\theta, \sin\theta)$, $P\left(\theta = (i-1)\frac{2\pi}{n}\right) = \frac{1}{n}$, $1 \le i \le n$, with $n = 10$.

   (a) Simulate a trajectory of $ns = 1000$ steps with $h = 0.1$.
   (b) Simulate $nsim = 1000$ trajectories of $ns = 100$ steps with $h = 1$ and collect the data about the final distance d_{ns} to the origin. Determine the empirical CDF and the empirical PDF of the data collected.

2. Consider a bidimensional random walk starting at $X_0 = (0, 0)$ defined by $\Delta X_n = hZ_n$, $Z_n \sim Z$, $Z = (\cos\theta, \sin\theta)$, where θ is normally distributed with mean 0 and standard deviation π.

   (a) Simulate a trajectory of $ns = 1000$ steps with $h = 0.1$.
   (b) Simulate $nsim = 1000$ trajectories of $ns = 100$ steps with $h = 1$ and collect the data about the final distance d_{ns} to the origin. Determine the empirical CDF and the empirical PDF of the data collected.

4.8.5 Itô's Integrals

Analogously to $\int_a^b f(t)dt$, we may approximate $\int_a^b f(t)dg(t)$ by considering the partition given in Eq. (4.21) and the finite sum

$$S(\mathcal{T}, f, g) = \sum_{i=1}^{n} (g(t_i) - g(t_{i-1}))f(t_{i-1}). \tag{4.53}$$

For regular functions f and g,

$$S(\mathcal{T}, f, g) \to \int_a^b f(t)dg(t)$$

when $n \to +\infty$ and $\max\{t_i - t_{i-1}\} \to 0$.

Let us consider two stochastic processes $\{X(t), t \in (a, b)\}$, $\{Y(t), t \in (a, b)\}$ and the finite sum

$$I(\mathcal{T}, \mathrm{X}, \mathrm{Y}) = \sum_{i=1}^{n} (Y(t_i) - \mathrm{Y}(t_{i-1}))X(t_{i-1}). \tag{4.54}$$

If both processes have integrable variances, id est, if

$$\int_a^b E\left(X^2(t)\right)dt < \infty \qquad and \qquad \int_a^b E\left(Y^2(t)\right)dt < \infty,$$

then this finite sum converges for $n \rightarrow +\infty$ and $\max\{t_i - t_{i-1}\} \rightarrow 0$. We have, analogously the standard functions (see, for instance, (Föllmer, 1981)):

$$I(\mathcal{T}, \mathrm{X}, \mathrm{Y}) \rightarrow I(X, Y) = \int_a^b X(t)dY(t). \tag{4.55}$$

$I(X, Y)$ is referred as Itô's integral of the process X with respect to the process Y. Itô's integral is linear:

$$I(\alpha X + \beta Y, Z) = \alpha I(X, Z) + \beta I(Y, Z),$$

$$I(X, \alpha Y + \beta Z) = \alpha I(X, Y) + \beta I(X, Z).$$

Remark

For standard integrals, we may use different forms of discretization to evaluate $\int_a^b f(t)dg(t)$. For instance, we may consider

$$S_3(\mathcal{T}, f, g) = \sum_{i=1}^{n} (g(t_i) - g(t_{i-1}))\frac{f(t_{i+1}) + f(t_i)}{2}$$

For regular functions, such a finite sum converges to the same value $\int_a^b f(t)dt$. However, such a property is not verified for stochastic integrals. Indeed,

$$I_3(\mathcal{T}, X, Y) = \sum_{i=1}^{n} (Y(t_i) - \mathrm{Y}(t_{i-1}))\frac{X(t_{i+1}) + X(t_i)}{2}$$

leads to the Stratonovich's integral (Stratonovich, 1966; Fisk, 1963), which is a different stochastic integral, with distinct properties. ∎

When $Y = W$ and $X = \varphi(W)$, supplementary properties may be established. Let us introduce

$$I_W(\varphi) = I(\varphi(W), W) = \int_a^b \varphi(W(t))dW(t). \tag{4.56}$$

Then,

$$E(I_W(\varphi)) = 0, \tag{4.57}$$

$$E(I_W(\varphi)I_W(\psi)) = \int_a^b E(\varphi(W(t))\psi(W(t)))dt, \tag{4.58}$$

$$E\left(I_W(\varphi)^2\right) = \int_a^b E\left(\varphi(W(t))^2\right)dt. \tag{4.59}$$

The reader will find in the literature methods for the numerical evaluation of $I_W(\varphi)$ – for instance, (Chorin, 1973; Blankenship & Baras, 1981).

A supplementary property of $I_W(\varphi)$ is Itô's formula: for a regular function $u : \mathbb{R} \to \mathbb{R}$ (see, for instance, (Föllmer, 1981))

$$I_W(u') = (W(b)) - u(W(a)) - \frac{1}{2}I_t(u''),$$

id est,

$$\int_a^b u'(W(t))dW(t) = u(W(b)) - u(W(a)) - \frac{1}{2}\int_a^b u''(W(t))dt.$$

Itô's formula extends to multidimensional situations where $u : \mathbb{R}^p \to \mathbb{R}$ as

$$\int_a^b \nabla u(\mathbf{W}(t))d\mathbf{W}(t) = u(\mathbf{W}(b)) - u(\mathbf{W}(a)) - \frac{1}{2}\int_a^b \Delta u(\mathbf{W}(t))dt.$$

4.8.6 Itô's Calculus

Itô's calculus manipulates stochastic differential equations (SDE) having the general form

$$dX_t = a(t, W_t)dW_t + b(t, W_t)dt. \tag{4.60}$$

Equation (4.60) defines a **Itô's diffusion**. Its interpretation is an equality involving Itô's integrals:

$$X(t) - X(0) = \int_0^t a(s, W(s))dW(s) + \int_0^t b(s, W(s))dt, \quad \forall s > 0.$$

For instance, Itô's formula may be rewritten as a SDE:

$$du(W_t) = u'(W_t)dW_t + \frac{1}{2}u''(W_t)dt. \tag{4.61}$$

SDE follow some standard rules. For instance, let $dX_t = a(t, W_t)dW_t + b(t, W_t)dt$, $dY_t = c(t, W_t)dW_t + d(t, W_t)dt$ and $Z_t = \alpha X_t + \beta Y_t$. Then, for $\alpha, \beta \in \mathbb{R}$:

$$dZ_t = (\alpha a(t, W_t) + \beta c(t, W_t))dW_t + (\alpha b(t, W_t) + \beta d(t, W_t))dt.$$

If $Z_t = X_t Y_t$, then

$$dZ_t = X_t dY_t + Y_t dX_t + (a(t, W_t)c(t, W_t))dt. \tag{4.62}$$

Let $X_t = x(W_t)$, $Y_t = y(W_t)$ and $Z_t = X_t Y_t$. Then,

$$dZ_t = X_t dY_t + Y_t dX_t + x'(W_t)y'(W_t)dt. \tag{4.63}$$

If $Z_t = y(X_t)$, then

$$dZ_t = y'(X_t)dX_t + \frac{1}{2}y''(X_t)(x'(W_t))^2 dt. \tag{4.64}$$

If $Z_t = F(t, X_t)$ and $dX_t = a(t, W_t)dW_t + b(t, W_t)dt$, then

$$dZ_t = \frac{\partial F}{\partial t}(t, X_t)dt + \frac{\partial F}{\partial x}(t, X_t)dX_t + \frac{1}{2}\frac{\partial^2 F}{\partial x^2}(t, X_t)(a(t, W_t))^2 dt. \tag{4.65}$$

In the multidimensional situation, $\boldsymbol{Y}_t = \boldsymbol{F}(t, \boldsymbol{X}_t)$, $\boldsymbol{F} : \mathbb{R} \times \mathbb{R}^n \to \mathbb{R}^m$. Then, we have

$$d(Y_i)_t = \frac{\partial F_i}{\partial t}(t, \boldsymbol{X}_t) + (\boldsymbol{A}.\boldsymbol{dX}_t)_i + \frac{1}{2}\boldsymbol{dX}_t^t \boldsymbol{B}_i \boldsymbol{dX}_t, \tag{4.66}$$

(4.66)

with

$$A_{ij} = \frac{\partial F_i}{\partial x_j}, \quad B_{ijk} = \frac{\partial^2 F_i}{\partial x_j \partial x_k}.$$

If $\boldsymbol{dX}_t = \boldsymbol{a}(t, \boldsymbol{W}_t)d\boldsymbol{W}_t + \boldsymbol{b}(t, \boldsymbol{W}_t)dt$, then

$$d(Y_i)_t = \frac{\partial F_i}{\partial t}(t, \boldsymbol{X}_t) + (\boldsymbol{A}.\boldsymbol{dX}_t)_i + \frac{1}{2}\boldsymbol{a}_t^t \boldsymbol{B}_i \boldsymbol{a}_t, \tag{4.67}$$

(4.67)

These rules are often used to manipulate SDE to get explicit solutions or to generate representations of the solutions of differential equations, such as Feynman-Kac formulae. The reader will find in the literature many representations of the solutions of Differential Equations and Partial Differential Equations: see, for instance, (Souza de Cursi, 1994; Zhou & Cai, 2016; Kharroubi & Pham, 2015; Morillon, 1997; Milstein & Tretyakov, 2012; Milstein & Tretyakov, 2013; Milstein & Tretyakov, 2004; Milstein & Tretyakov, 2020).

Example

The classical Black-Scholes model for the evolution of the price of an option reads as

$$dS_t = \sigma S_t dW_t + \mu S_t dt.$$

Let

$$F(t,x) = F_0 \exp\left(\mu t - \frac{\sigma^2}{2}t + \sigma x\right).$$

We have

$$\frac{\partial F}{\partial t} = \left(\mu - \frac{\sigma^2}{2}\right)F(t,x),\ \frac{\partial F}{\partial x} = \sigma F(t,x),\ \frac{\partial^2 F}{\partial x^2} = \sigma^2 F(t,x).$$

Thus, if $Y_t = F(t, X_t)$ and $dX_t = a(t, W_t)dW_t + b(t, W_t)dt$, we have

$$dY_t = \left(\mu - \frac{\sigma^2}{2}\right)Y_t dt + \sigma Y_t dX_t + \frac{1}{2}\sigma^2 Y_t (a(t, W_t))^2 dt$$

Let us choose $X_t = W_t$. Then, $dX_t = dW_t$ and $a(t, W_t) = 1$, $b(t, W_t) = 0$. Consequently,

$$dY_t = \mu Y_t dt + \sigma Y_t dW_t$$

Thus, the solution is

$$S_t = S_0 \exp\left(\mu t - \frac{\sigma^2}{2}t + \sigma W_t\right).$$

Notice that $S_t = A_t U_t$, $A_t = S_0 \exp\left(\mu t - \frac{\sigma^2}{2}t\right)$, $U_t = \exp(\sigma W_t)$. V_t is deterministic and U_t is a lognormal variable (recall that $W_t \sim N(0, \sqrt{t})$). We have

$$E(S_t) = A_t E(U_t),\ \ V(S_t) = A_t^2 V(U_t),$$

with

$$E(U_t) = \exp(\sigma^2 t),\ \ V(U_t) = \exp(2\sigma^2 t) - \exp(\sigma^2 t).$$

Thus,

$$E(S_t) = S_0 \exp(\mu t),\ \ V(S_t) = S_0^2 \exp(2\mu t)\left(\exp(\sigma^2 t) - 1\right).$$ ∎

Example

The Vacisek's model for the evolution of the interest rates reads as

$$dX_t = \sigma dW_t + \theta(\mu - X_t)dt.$$

Let

$$F(t,x) = xe^{\theta t}.$$

We have

$$\frac{\partial F}{\partial t} = \theta F, \quad \frac{\partial F}{\partial x} = e^{\theta t}, \frac{\partial^2 F}{\partial x^2} = 0.$$

Thus, if $Y_t = F(t, X_t)$ and $dX_t = a(t, W_t)dW_t + b(t, W_t)dt$, we have

$$dY_t = \theta X_t e^{\theta t} dt + e^{\theta t} dX_t = \sigma e^{\theta t} dW_t + \theta \mu e^{\theta t} dt$$

Thus,

$$Y_t = \mathrm{Y}_0 + \sigma \int_0^t e^{\theta s} dW_s + \theta \mu \int_0^t e^{\theta s} ds$$

Thus, the solution is

$$X_t = X_0 e^{-\theta t} + \mu\left(1 - e^{-\theta t}\right) + \sigma e^{-\theta t} \int_0^t e^{\theta s} dW_s.$$

We have

$$E(S_t) = X_0 e^{-\theta t} + \mu\left(1 - e^{-\theta t}\right), \quad \mathrm{V}(S_t) = \frac{\sigma^2}{2\theta}\left(1 - e^{-2\theta t}\right).$$

Vacisek's model is an example of Ornstein-Uhlenbeck process. ∎

Example

Let us consider the partial differential equation

$$\frac{\partial u}{\partial t} - \frac{a^2}{2} \Delta u = f \; on \; Q = \Omega \times (0, T),$$

$$u(x, 0) = u_0(x), x \in \Omega, \quad u(x, t) = u_0(x, t), x \in \partial\Omega, t \in (0, T)$$

Let us consider

$$d\mathbf{X}_t = a d\mathbf{W}_t, \quad dT_t = -dt, \quad \mathbf{X}_0 = \mathbf{x}, T_0 = t, \quad Y_t = u(\mathbf{X}_t, T_t).$$

Let us introduce $\mathbf{Z}_t = (\ \mathbf{X}_t,\ T_t,\ )$. Then,

$$d\mathbf{Z}_t = \begin{pmatrix} a & 0 & \cdots & 0 \\ 0 & \ddots & \cdots & \vdots \\ \vdots & \vdots & a & 0 \\ 0 & \cdots & \cdots & 0 \end{pmatrix} d\mathbf{W}_t + \begin{pmatrix} 0 \\ 0 \\ \vdots \\ -1 \end{pmatrix} dt.$$

Let $F(\mathbf{Z}_t) = u(\mathbf{X}_t, T_t)$. We have

$$A_{1i} = \frac{\partial F}{\partial Z_i} = \frac{\partial u}{\partial x_i}, A_{1,n+1} = \frac{\partial F}{\partial Z_{i+1}} = \frac{\partial u}{\partial t},$$

$$B_{1ij} = \frac{\partial^2 F}{\partial Z_i \partial Z_j} = \frac{\partial^2 u}{\partial x_i \partial x_j},$$

$$B_{1i,n+1} = \frac{\partial^2 F}{\partial Z_i \partial Z_{n+1}} = \frac{\partial^2 u}{\partial x_i \partial t},$$

$$B_{1,n+1,n+1} = \frac{\partial^2 F}{\partial Z_{n+1}^2} = \frac{\partial^2 u}{\partial t^2}.$$

$$dY_t = \frac{\partial u}{\partial t}\, dT_t + \sum \frac{\partial u}{\partial x_i}\, d(X_i)_t + \frac{1}{2}\, a^2 \times \Delta u\, dt,$$

id est,

$$dY_t = \left(\frac{a^2}{2}\Delta u - \frac{\partial u}{\partial t}\right) dt + \sum \frac{\partial u}{\partial x_i}\, a\, dW_i.$$

Thus

$$dY_t = -f(\boldsymbol{X}_t, T_t)dT_t + a\boldsymbol{\nabla u}.d\boldsymbol{W}$$

and

$$Y_t - Y_0 = -\int_0^t f(\boldsymbol{X}_s, T_s)ds + a \int_0^t \boldsymbol{\nabla u}(\boldsymbol{X}_s, T_s).d\boldsymbol{W}$$

Let

$$\tau(x) = \inf\{t : \boldsymbol{Z}_t \notin Q\}.$$

Then, $Y_{\tau(x)} = u_0(X_{\tau(x)}, T_{\tau(x)})$, $Y_0 = u(t, x)$ and

$$u(x,t) = u_0\left(X_{\tau(x)}, T_{\tau(x)}\right) + \int_0^{\tau(x)} f(\boldsymbol{X}_s, T_s)ds - a \int_0^t \boldsymbol{\nabla u}(\boldsymbol{X}_s, T_s).d\boldsymbol{W}.$$

Since $E\left(\int_0^t \boldsymbol{\nabla u}(\boldsymbol{X}_s, T_s).d\boldsymbol{W}\right) = 0$, we have

$$u(t,x) = E\left(u_0\left(X_{\tau(x)}, T_{\tau(x)}\right) + \int_0^{\tau(x)} f(\boldsymbol{X}_s, T_s)ds \right).$$

This equality is an example of Feynman-Kac formula. ∎

Example

Let us consider the partial differential equation

$$-\frac{a^2}{2}\Delta u = f \text{ on } \Omega,$$

$$u(x) = u_0(x), \quad x \in \partial\Omega$$

Let us consider

$$d\boldsymbol{X}_t = ad\boldsymbol{W}_t, \quad \boldsymbol{X}_0 = \boldsymbol{x}, \quad Y_t = u(\boldsymbol{X}_t).$$

Let $F(Z_t) = u(X_t)$. We have

$$dY_t = \sum \frac{\partial u}{\partial x_i} d(X_i)_t + \frac{1}{2} a^2 \times \Delta u \, dt,$$

id est,

$$dY_t = dt\frac{a^2}{2}\Delta u + \sum \frac{\partial u}{\partial x_i} a \, dW_i.$$

Thus

$$dY_t = -f(\boldsymbol{X}_t)dX_t + a\boldsymbol{\nabla u}.d\boldsymbol{W}$$

and

$$Y_t - Y_0 = -\int_0^t f(\boldsymbol{X}_s)ds + a\int_0^t \boldsymbol{\nabla u}(\boldsymbol{X}_s).d\boldsymbol{W}$$

Let

$$\tau(x) = \inf\{t : \boldsymbol{Z}_t \notin \Omega\}.$$

Then, $Y_{\tau(x)} = u_0(X_{\tau(x)})$, $Y_0 = u(x)$ and

$$u(x) = u_0\left(X_{\tau(x)}\right) + \int_0^{\tau(x)} f(\boldsymbol{X}_s)ds - a\int_0^t \boldsymbol{\nabla u}(\boldsymbol{X}_s).d\boldsymbol{W}.$$

Since $E\left(\int_0^t \boldsymbol{\nabla u}(\boldsymbol{X}_s).d\boldsymbol{W}\right) = 0$, we have

$$u(x) = E\left(u_0\left(X_{\tau(x)}\right) + \int_0^{\tau(x)} f(\boldsymbol{X}_s)ds\right).$$

This equality is another example of Feynman-Kac formula. ∎

4.8.7 Numerical Simulation of Stochastic Differential Equations

Let us consider the following SDE;

$$dX_t = a(t, X_t)dW_t + b(t, X_t)dt, X_0 = x_0. \tag{4.68}$$

The reader will find in the literature many works concerning methods for the simulation of SDE analogous to (4.68). For instance, (Milstein, 1973; Pardoux & Talay, 1985; Milstein & Tretyakov, 2004; Graham & Talay, 2013; Kloeden & Platen, 1989; Talay, 2015; (Boukhetala & Guidoum, 2011). One of the most popular ones is Euler's method: let us introduce a maximum time of simulation $T > 0$, a maximum number of time steps $ns > 0$, $\Delta t = T/ns$, $t_i = i\Delta t$, $i \geq 0$, $W_i = W(t_i)$, $X_i = X(t_i)$. Then, Eq. (4.60) is discretized as

$$\begin{gathered} X_{n+1} - X_n = a_n \Delta W_n + b_n \Delta t, . \\ a_n = a(t_n, X_n), \;\; b_n = b(t_n, X_n), \;\; \Delta W_n = (W_{n+1} - W_n) \end{gathered} \tag{4.69}$$

A simple way to generate ΔW_n consists in using that

$$\Delta W_n \sim N\left(0, \sqrt{\Delta t}\right). \tag{4.70}$$

Thus,

$$\begin{gathered} X_{n+1} = X_n + a_n \sqrt{\Delta t} Z_n + b_n \Delta t \\ W_{n+1} = W_n + \sqrt{\Delta t} Z_n, Z_n \sim N(0, 1) \end{gathered} \tag{4.71}$$

Examples of VBA functions for the simulation by Euler's method are the following ones:

```
Function sde_euler_evol(T, dt, X0) As Variant
ns = WorksheetFunction.RoundUp(T / dt, 0)
Dim x() As Double
ReDim x(0 To ns)
dw = normgen_bm(0, Sqr(dt), ns + 1)
x(0) = X0
tt = 0
For n = 1 To ns
    an = a(tt, x(n - 1))
    bn = b(tt, x(i - 1))
    x(n) = x(n - 1) + an * dw(n) + bn * dt
    tt = tt + dt
Next n
sde_euler_evol = x
End Function
```

```
Function sde_euler_sim(T, dt, X0) As Variant
ns = WorksheetFunction.RoundUp(T / dt, 0)
Dim x As Double
x = X0
dw = normgen_bm(0, Sqr(dt), ns + 1)
tt = 0
w = 0
For n = 1 To ns
    an = a(tt, x)
    bn = b(tt, x)
    x = x + an * dw(n) + bn * dt
    tt = tt + dt
Next n
sde_euler_sim = x
End Function
```

Function `sde_euler_evol` returns a vector x containing all the values $X_0, X_1, \ldots, X_{ns}$ such that `ns*dt` = T. Function `sde_euler_sim` returns a the single value of X_{ns}. To use these functions, the reader must provide functions $a(t, X)$ and $b(t, X)$ (see examples)

A second popular method of discretization is Milstein's method:

$$\begin{gathered} X_{n+1} - X_n = a_n \Delta W_n + \left(b_n \Delta t + c_n\left(\Delta W_n^2 - \Delta t\right)\right), \\ c_n = \frac{1}{2} a_n \frac{\partial a_n}{\partial x}, \;\; \frac{\partial a_n}{\partial x} = \frac{\partial a}{\partial x}(t_n, X_n). \end{gathered} \tag{4.72}$$

Analogously to Euler's method, we may generate $W_{n+1} - W_n$ by (4.70), so that

$$\begin{gathered} X_{n+1} - X_n = a_n \sqrt{\Delta t} Z_n + a_n \Delta W_n + \Delta t\left(b_n + c_n\left(Z_n^2 - 1\right)\right) \\ W_{n+1} = W_n + \sqrt{\Delta t} Z_n, Z_n \sim N(0, 1) \end{gathered} \tag{4.73}$$

Examples of code for the simulation by Milstein's method are given below:

```
Function sde_mils_evol(T, dt, X0) As Variant
ns = WorksheetFunction.RoundUp(T / dt, 0)
Dim x() As Double
ReDim x(0 To ns)
dw = normgen_bm(0, Sqr(dt), ns + 1)
x(0) = X0
tt = 0
For n = 1 To ns
   an = a(tt, x(n - 1))
   bn = b(tt, x(n - 1))
   cn = c(tt, x(n - 1))
   x(n) = x(n - 1) + an * dw(n) + bn * dt + cn * (dw(n) * dw(n) - dt)
   tt = tt + dt
Next n
sde_mils_evol = x
End Function
```

```
Function sde_mils_sim(T, dt, X0) As Variant
ns = WorksheetFunction.RoundUp(T / dt, 0)
Dim x As Double
x = X0
dw = normgen_bm(0, Sqr(dt), ns + 1)
tt = 0
w = 0
For n = 1 To ns
   an = a(tt, x)
   bn = b(tt, x)
   cn = c(tt, x)
   x = x + an * dw(n) + bn * dt + cn * (dw(n) * dw(n) - dt)
   tt = tt + dt
Next n
sde_mils_sim = x
End Function
```

Example

The classical Black-Scholes model for the evolution of the price of an option reads as

$$dS_t = \sigma S_t dW_t + \mu S_t dt.$$

We simulated this SDE with $\mu = 0.1$, $\sigma = 0.3$, $S_0 = 1$. We generated a sample of *nsim* variates from S_T, which was used to evaluate the mean $E(S_T)$ and $V(S_T)$ for several times T. Figure 4.28 exhibits an example of result. A comparison of the results furnished by Milstein's and Euler's methods for $T = 1$ is shown in Tables 4.14 and 4.15.

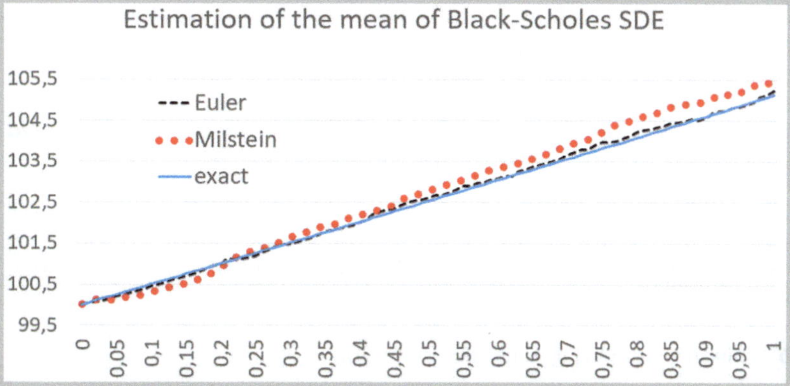

Fig. 4.28 Simulation of Black-Scholes SDE with $\mathbf{\Delta t = 0.001}$.Mean of ***nsim*** $\mathbf{= 10000}$ realizations

Table 4.14 Estimation of $E(S_T)$ by simulation of the SDE

	$\Delta t = 0.01$			$\Delta t = 0.001$		
nsim	1000	5000	10,000	1000	5000	10,000
Euler	106,54%	104,88%	105,50%	104,15%	104,75%	105,29%
Milstein	105,98%	105,16%	105,46%	103,65%	105,96%	105,00%
exact	105,13%	105,13%	105,13%	105,13%	105,13%	105,13%

Table 4.15 Estimation of $V(S_T)$ by simulation of the SDE

	$\Delta t = 0.01$			$\Delta t = 0.001$		
nsim	1000	5000	10,000	1000	5000	10,000
Euler	0,11	0,10	0,11	0,10	0,10	0,11
Milstein	0,10	0,10	0,11	0,10	0,11	0,10
exact	0,10	0,10	0,10	0,10	0,10	0,10

We observe that both the approaches furnish results close to the exact ones. ∎

Example

The Vacisek's model (Vasicek, 1977) for the evolution of the interest rates reads as

$$dX_t = \sigma dW_t + \theta(\mu - X_t)dt.$$

We simulated this SDE with $\mu = 0.1$, $\sigma = 0.3$, $\theta = 2$, $X_0 = 0.1$. Analogously to the preceding example, a sample of *nsim* variates from X_T was generated and used to evaluate the mean $E(X_T)$ and $V(X_T)$ for several times T. Figure 4.29 shows an example of result. A comparison of the results furnished by Milstein's and Euler's methods for $T = 2$ is shown in Tables 4.16 and 4.17. In this case, $\frac{\partial a_n}{\partial x} = 0$, so that both the methods furnish close results.

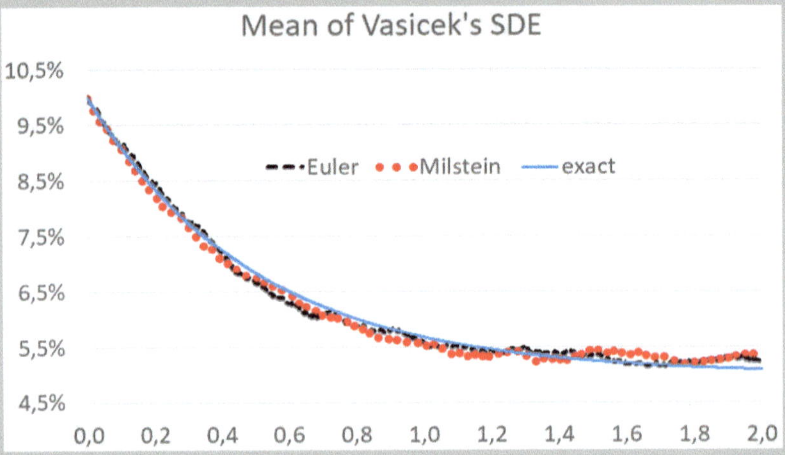

Fig. 4.29 Simulation of Vacisek's SDE with $\mathbf{\Delta t = 0.001}$.Mean of ***nsim*** **= 10000** realizations

Table 4.16 Estimation of $E(X_T)$ by simulation of the SDE

	$\Delta t = 0.01$			$\Delta t = 0.001$		
nsim	1000	5000	10,000	1000	5000	10,000
Euler	5,66%	5,31%	4,77%	5,21%	5,14%	5,18%
Milstein	5,12%	5,43%	4,99%	5,78%	5,21%	5,25%
exact	5,09%	5,09%	5,09%	5,09%	5,09%	5,09%

Table 4.17 Estimation of $V(X_T)$ by simulation of the SDE

	$\Delta t = 0.01$			$\Delta t = 0.001$		
nsim	1000	5000	10,000	1000	5000	10,000
Euler	0,02	0,02	0,02	0,02	0,02	0,02
Milstein	0,02	0,02	0,02	0,02	0,02	0,02
exact	0,02	0,02	0,02	0,02	0,02	0,02

Again, both the approaches furnish results close to the exact ones. ∎

Example

The Cox-Ingersoll-Ross model (Cox et al., 1985) for the instantaneous interest rates is the solution of the SDE:

$$dX_t = \sigma\sqrt{X_t}dW_t + \theta(\mu - X_t)dt.$$

Then (Jafari & Abbasian, 2017)

$$E(X_t) = X_0 e^{-\theta t} + \mu\left(1 - e^{-\theta t}\right),$$

$$V(X_t) = \frac{\sigma^2}{\theta}X_0\left(e^{-\theta t} - e^{-2\theta t}\right) + \frac{\mu\sigma^2}{2\theta}\left(1 - e^{-\theta t}\right)^2.$$

We simulated this SDE with $\mu = 0.1$, $\sigma = 0.3$, $\theta = 2$, $X_0 = 1$. Figure 4.30 exhibits an example of result. Results furnished by Milstein's and Euler's methods for $T = 2$ are shown in Fig. 4.30 and Tables 4.18 and 4.19.

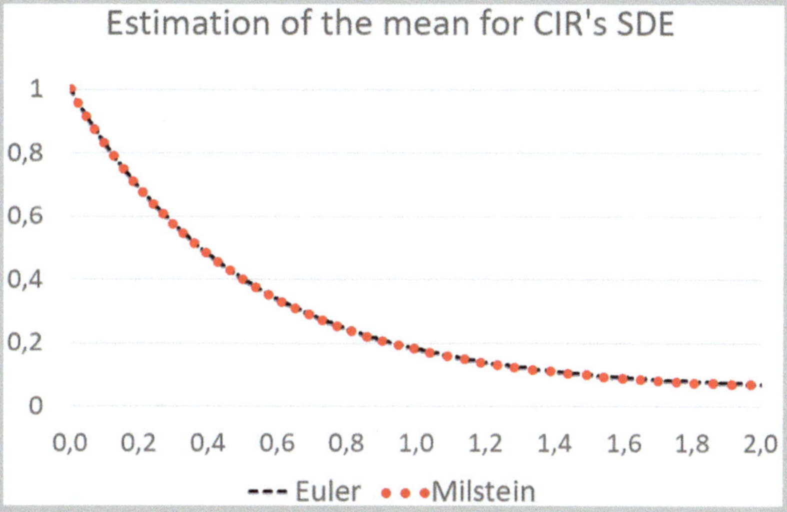

Fig. 4.30 Simulation of CIR's SDE with $\Delta t = 0.001$. Mean of ***nsim*** = **10000** realizations

Table 4.18 Estimation of $E(X_T)$ by simulation of the SDE

	$\Delta t = 0.01$			$\Delta t = 0.001$		
nsim	1000	5000	10,000	1000	5000	10,000
Euler	6,83%	6,68%	6,64%	6,73%	6,73%	6,77%
Milstein	6,61%	6,67%	6,61%	6,68%	6,75%	6,74%
exact	6,74%	6,74%	6,74%	6,74%	6,74%	6,74%

Table 4.19 Estimation of $V(X_T)$ by simulation of the SDE

	$\Delta t = 0.01$			$\Delta t = 0.001$		
nsim	1000	5000	10,000	1000	5000	10,000
Euler	1,98E−3	1,84E−3	1,88E−3	2,02E−3	1,91E−3	1,90E−3
Milstein	1,99E−3	1,92E−3	1,85E−3	1,86E−3	1,85E−3	1,87E−3
exact	1,89E−3	1,89E−3	1,89E−3	1,89E−3	1,89E−3	1,89E−3

Again, both the approaches furnish results close to the exact ones. ∎

Example
Consider the Partial Differential Equation

$$-\Delta u = f \ on \ \Omega, \ \ u(x) = u_0(x) \ \ on \ \ \partial\Omega,$$

with

$$\Omega = \{(x_1, x_2, x_3) : x_1 + x_2 + x_3 \leq 1, x_1 \geq 0, x_2 0, x_3 \geq 0\},$$

$$f(x) = -6, u_0(x) = x_1^2 + x_2^2 + x_3^2.$$

Let us consider

$$d\boldsymbol{X}_t = \sqrt{2} d\boldsymbol{W}_t, \ \ \boldsymbol{X}_0 = x = \left(\frac{1}{3}, 0, \frac{1}{2}\right).$$

We evaluate

$$u(x) \approx \frac{1}{nsim} \sum_{i=1}^{nsim} Y_i, \ \ Y_i = u_0\left(\boldsymbol{X}_{\tau_i(x)}\right) + \int_0^{\tau_i(x)} f(\boldsymbol{X}_s) ds$$

The exact value is $u(\boldsymbol{X}_0) = \frac{13}{36} \approx 0,173611111$. The results obtained with Euler's method and $nsim = 1000$ are shown in Table 4.20. The results for $\Delta t = 1E - 8$ and different values of $nsim$ are shown in Table 4.21. We observe that the results are close to the exact value. ∎

Table 4.20 Error in the estimation of $u(x)$ by simulation of the SDE. (***nsim* = 1000**)

Δt	1E−2	1E−3	1E−4	1E−5	1E−6	1E−7	1E−8
error	3E−3	−2E−3	1E−4	−1E−5	−1E−5	2E−5	−3E−6

Table 4.21 Error in the estimation of $u(x)$ by simulation of the SDE. (**Δt = 1*E* − 8**)

nsim	1000	5000	10,000	25,000
error	−3E−6	2E−6	2E−6	7E−7

Example

Consider the Partial Differential Equation

$$\frac{\partial u}{\partial t} - \Delta u = f \ on\ \Omega \times (0, T),$$

$$u(x,t) = u_0(x,t) \ on\ \partial\Omega \times (0,T), \quad u(x,0) = u_0(x,0) \ on\ \Omega$$

With

$$\Omega = \{(x_1, x_2, x_3) : x_1 + x_2 + x_3 \le 1, x_1 \ge 0, x_2 0, x_3 \ge 0\},$$

$$f(x) = (x_1^2 + x_2^2 + x_3^2 - 6)e^t, \quad u_0(x) = x_1^2 + x_2^2 + x_3^2.$$

Let us consider

$$d\boldsymbol{X}_t = \sqrt{2} d\boldsymbol{W}_t, dT_t = -dt, \boldsymbol{X}_0 = x = \left(\frac{1}{3}, 0, \frac{1}{2}\right), T_0 = 1.$$

We evaluate

$$u(x,t) \approx \frac{1}{nsim} \sum_{i=1}^{nsim} Y_i, \quad Y_i = u_0\left(\boldsymbol{X}_{\tau_i(x)}, T_{\tau_i(x)}\right) + \int_0^{\tau_i(x)} f(\boldsymbol{X}_s, T_s) ds$$

The exact value is $u(\boldsymbol{X_0}, T_0) = \frac{13}{36} e \approx 0,471923929$. The results obtained with $nsim = 1000$ are shown in Table 4.22. In this situation, the influence of Δt is preponderant: increasing $nsim$ does not improve the result – such a behavior is caused by the presence of T_t in the SDE. ∎

Table 4.22 Error in the estimation of $\boldsymbol{u}(\boldsymbol{x}, \boldsymbol{t})$ by simulation of the SDE. (***nsim*** **= 1000**)

Δt	1E−2	1E−3	1E−4	1E−5	1E−6	1E−7	1E−8
error	1E−2	8E−4	−2E−3	3E−4	5E−5	4E−5	3E−6

Exercises

1. Consider the Constant Elasticity of Variance (CEV) model:

$$dX_t = \mu X_t dt + \sigma X_t^\gamma dW_t.$$

(a) Take $\mu = 0.2$, $\sigma = 0.8$, $\gamma = 1.3$, $X_0 = 1$. Simulate a trajectory of $ns = 100$ steps with $\Delta t = 0.01$. Generate a chart showing the values of a call with strike K: $Y_t = \max\{0, X_t - K\}$, $K = 0.5$.

(b) Simulate $nsim = 1000$ trajectories of $ns = 100$ steps with $\Delta t = 0.01$. Use this sample to estimate the mean and the variance of X_1. Generate a chart showing the values of the empirical mean of the value of the call $Y_t = \max\{0, X_t - K\}$, $K = 0.5$.

2. Consider the Chan-Karloyi-Logstaff-Sanders (CKLS) model

$$dX_t = (\mu + \theta X_t)dt + \sigma X_t^{\gamma} dW_t.$$

(a) Take $\mu = 0.2$, $\sigma = 0.8$, $\gamma = 0.3$, $\theta = 0.01$, $X_0 = 1$. Simulate a trajectory of $ns = 1000$ steps with $\Delta t = 0.01$. Generate a chart showing the values of a call with strike K: $Y_t = \max\{0, X_t - K\}$, $K = 0.5$.
(b) Simulate $nsim = 1000$ trajectories of $ns = 100$ steps with $\Delta t = 0.01$. Use this sample to estimate the mean and the variance of X_1. Generate a chart showing the values of the empirical mean of the call $Y_t = \max\{0, X_t - K\}$, $K = 0.5$.

3. Consider the Black-Scholes model with $\mu = 0.05$, $\sigma = 0.2$, $S_0 = 1$.

(a) Simulate a trajectory of $ns = 1000$ steps with $\Delta t = 0.01$. Generate a chart showing the values of a put with strike K: $Y_t = \max\{0, K - X_t\}$, $K = 1.5$.Simulate $nsim = 1000$ trajectories of $ns = 100$ steps with $\Delta t = 0.01$. Use this sample to estimate the mean and the variance of X_1. Generate a chart showing the values of the empirical mean of a put with strike K: $Y_t = \max\{0, K - X_t\}$, $K = 1.5$.

4. Consider the Vacisek's model with $\mu = 0.05$, $\sigma = 0.2$,$\theta = 3$, $X_0 = 1$.

(a) Simulate a trajectory of $ns = 1000$ steps with $\Delta t = 0.01$. Generate a chart showing the values of a call with strike K: $Y_t = \max\{0, X_t - K\}$, $K = 0.5$.
(b) Simulate $nsim = 1000$ trajectories of $ns = 100$ steps with $\Delta t = 0.01$. Use this sample to estimate the mean and the variance of X_1. Generate a chart showing the values of the empirical mean of a call with strike K: $Y_t = \max\{0, X_t - K\}$, $K = 0.5$.

5 Representation of Random Variables

As previously observed, random variables are a comfortable tool to model variability, whenever statistical data is available. Nevertheless, the complete knowledge of their distribution is rarely available, since only samples may be obtained. In some situations, even the variable itself cannot be observed – it is a hidden variable – and only its effects may be observed: variability of the response of the system is observed, but the cause of this variability remains unidentified. To use random variables in models, it is necessary to have information such as its distribution (for instance, its cumulative distribution or its density) and some of its statistical properties (for instance, mean, variance, mode). Classical approaches start by assumptions on the distribution – for instance, assuming that the variable under consideration is Gaussian or a particular transformation of a Gaussian variable. In the classical approaches, severe errors of model are difficulty to correct, even in the Bayesian approaches, since the conjugate distributions are predetermined, so that the choice of the prior distribution seriously constrains the result for the posterior distribution. Statistical Learning approaches are more flexible but request a large amount of data and face difficulties when the cause of the heterogeneity is unobserved. Uncertainty Quantification (UQ) proposes an alternative approach tending to introduce both more flexibility and economic use of data: on the one hand, the connection of the basic variable chosen for the representation (which may be interpreted as a "prior") and the variable to be represented (which may be interpreted as a "posterior") is relaxed, so that severe errors may be corrected. In addition, variability generated by hidden variables may be represented by explicit ones. On the other hand, reasonable quantities of data are enough to the determination of the representations.

In this chapter, we present some UQ methods for the representation of an observed variability. The methods may be implemented in workbooks by operations on cells that do not involve VBA or using VBA – this last option is more efficient. In the sequel, we consider both the approaches, but the reader will be able to verify by himself the power of the VBA approach.

5.1 The UQ Approach for the Representation of Random Variables

Let us consider the following situation: assume that U is a random variable and that $X = X(U)$ is other random variable which depends upon U- for instance, X is the solution of a linear system or an

Supplementary Information The online version of this chapter (https://doi.org/10.1007/978-3-030-77757-9_5) contains supplementary material, which is available to authorized users.

E. Souza de Cursi, *Uncertainty Quantification and Stochastic Modelling with EXCEL*,
Springer Texts in Business and Economics, https://doi.org/10.1007/978-3-030-77757-9_5

optimization problem which depends on U but the expression of the variable X as a function of U is not known: only a few data is available – a reduced sample of the couple (U, X) is given, with a number of elements insufficient to produce a statistical distribution. In other words, we are interested in finding the solution of the model problem below:

Problem
Let

$$\mathcal{S} = \{(U_1, X_1), \ldots, (U_{ns}, X_{ns})\} \tag{5.1}$$

be a sample of a couple of random variables (X, U) such that $X = X(U)$. Find the probability distribution of $X(U)$ – *id est*, its Cumulative Distribution Function (CDF) and, if possible, its Probability Density Function (PDF). ∎

As indicated, we are mainly interested in situations where the available data is limited: in such a context, UQ methods may be used to determine the distribution of X.

The UQ approach may be summarized as follows: to represent $X = X(U)$, we may consider a family of functions (for instance, a Hilbert basis – see 3.15, page 161. See also Eqs. (5.33) and (5.34)).

$$\mathcal{F} = \{\varphi_j : 1 \leq j \leq k\} \tag{5.2}$$

and approximate

$$X \approx PX(U) = \sum_{j=1}^{k} x_j \varphi_j(U) \ . \tag{5.3}$$

For instance, we may consider a polynomial family generated by the functions $\varphi_i(U) = U^{i-1}$ or a normalized polynomial family $\varphi_i(U) = ((U - a)/(b - a))^{i-1}$, $a = \min U$, $b = \max U$. Available data must be used to determine the coefficients $\boldsymbol{x} = (x_1, \ldots, x_k)$, which are the unknowns to be found: UQ furnishes a collection of methods for the determination of $\boldsymbol{x}$, including complex situations, where the distribution of U is unknown or the values of U are missing. Once $\boldsymbol{x}$ is determined, the expression (5.3) may be used to determine approximations of the CDF and the PDF of X, to evaluate probabilities of events connected to X or furnish evaluations of statistics of X. Equation (5.3) may be written under matrix form as

$$X \approx PX(U) = \boldsymbol{\Phi}(U)\boldsymbol{x}, \boldsymbol{x} = (x_1, \ldots, x_{k+1})^t, \boldsymbol{\Phi}(\mathrm{U}) = (\varphi_1(U), \ldots, \varphi_k(U)) \ . \tag{5.4}$$

If the joint distribution of the couple (U, X) is known, the coefficients $\boldsymbol{x}$ can be determined by solving the linear equations (see Eqs. (5.23), (5.24), and (5.25) in the sequel) that characterize the orthogonal projection (see Eqs. (5.21) and (5.22) in the following). For instance, let us consider U as uniformly distributed on $(-1, 1)$ and $X = e^U$. In such a situation, we can determine the coefficients by solving the linear system shown in Eqs. (5.23), (5.24), and (5.25), page 266. Then, for $k = 6$, $a = -1$, $b = 1$, we find the coefficients shown in Table 5.1. As shown in Fig. 5.1, the approximation is exceptionally good.

Table 5.1 Coefficients of the expansion of e^U (polynomial family, $U \sim \mathcal{U}(-1, 1)$)

i	1	2	3	4	5	6	*Normalized*
$\boldsymbol{x}$	1,00	1,00	0,50	0,17	0,04	0,01	*NO*
	0,37	0,74	0,70	0,63	0,05	0,28	*YES*

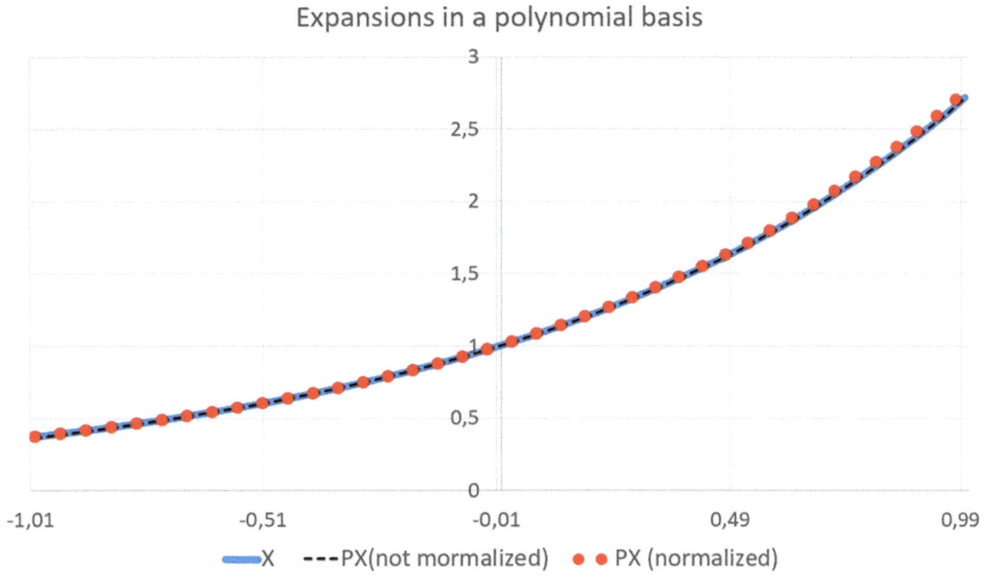

Fig. 5.1 Expansion of X as a polynomial function of U. When the joint distribution is known, the coefficients can be determined by orthogonal projection – but such a situation is rare in practice. UQ considers situations where a limited number of observations is available

Table 5.2 An example of data for the UQ solution of the model problem

U	0,110263	0,818171	0,218055	−0,428266	−0,521897
X	1,11657	2,26635	1,24366	0,65164	0,59339
U	0,972777	0,498947	−0,063814	0,439192	0,899228
X	2,64528	1,64699	0,93818	1,55145	2,45770

Table 5.3 Coefficients obtained by collocation from data in Table 5.2 (polynomial family)

i	1	2	3	4	5	6	*Normalized*
x	1,00	1,00	0,50	0,17	0,04	0,01	*NO*
	0,594	0,886	0,667	0,325	0,112	0,061	*YES*

Unfortunately, such a situation is rare in practice. In most of real situations, the available information reduces to limited samples of observations of the couple. UQ consider this last kind of situations: only a limited amount of data is available and/or situations where the variable U is missing.

Let us illustrate the UQ approach by considering the situation where the available information reduces to the type of law of U and a sample of 10 values of the couple (U, X) (see Table 5.2).

As observed, we can consider a polynomial family and the data from Table 5.2 is used to determine the coefficients of the approximation (5.3). The simplest method for such a determination is collocation (see Sect. 5.2 in the sequel), which furnishes the results shown below for $k = 6$ (Table 5.3):

Knowing the coefficients, we may generate a large sample from X, by using a large sample from U: we generate 10^5 variates from $U(-0.521897, 0.972777)$ – notice that no information is available outside this interval – and use them to generate a sample from X by using the approximation – the computational cost is negligible and EXCEL® furnishes the result in a fraction of second. The, we may use this simulated data to approximate the Cumulative Distribution (CDF) of X by using the

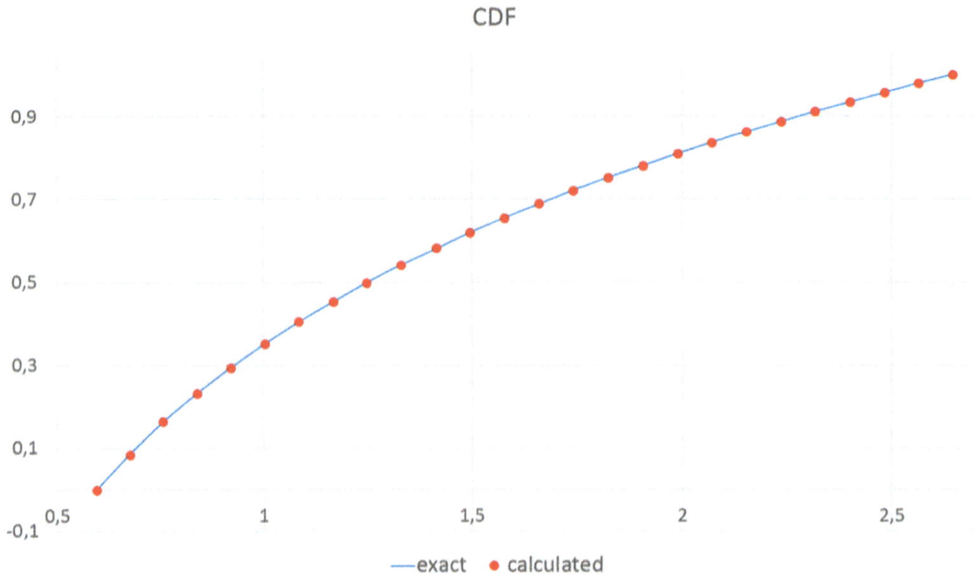

Fig. 5.2 Approximation of the CDF of X by using a large sample of simulated data generated by UQ. The approximated CDF corresponds to the red points and the exact one to the blue curve. The normalized approximation was used

empirical CDF from the simulated data – the evaluation of the empirical CDF is made by the method previously introduced in Sect. 3.8.3. The results are shown in Fig. 5.2.

As we can observe, the agreement between the calculated CDF and the exact one is excellent. Once the CDF is determined, we may seek for the Probability Density (PDF), which is the derivative of the CDF. Since the CDF is given only at a discrete set of points, we must use a numerical method to approximate its derivative. Here, we use the particle method (see Sect. 2.8. See also (Souza de Cursi, 2015)). The results are exhibited in Fig. 5.3: as we see, the agreement is particularly good.

The determination of such a representation allows us to evaluate probabilities of events concerning the variable X. It furnishes also better evaluations of the statistics of X, such as the mean and the variance.

For instance, assume that we are interested in evaluating $P(X > 2, 4)$ and $P(X < 0, 7)$: we can generate a large sample of variates from X – as previously observed, the cost of the generation is simply the cost of the evaluation of the approximation PX and EXCEL® performs it in a fraction of second – and evaluate the probability on this sample. Using a sample of 10^5 variates from $U(-0.521897, 0.972777)$, we obtain the results in Table 5.4.

Notice that, on the one hand, the probabilities cannot be evaluated from the sample in Table 5.2 and, on the other hand, the estimations of the mean and the variance issued from Table 5.2 are respectively 1,511 and 0,491: the use of the UQ representation provides a significant improvement.

If the distribution of U is unknown, you can be interested in finding it too. In such a situation, an adapted method must be used – such as, for instance, Moments Matching (see Sect. 5.4). In general, it is necessary to introduce an artificial variable A, having a known distribution: for instance, let us consider the data for U and X in Table 5.5, where U is uniformly distributed on $(-1, 1)$ and $X = e^U$. Consider an artificial variable $A \sim N(0, 1)$ – notice that the distribution of A is different from the distribution of U.

We may generate a sample$(A_1, \ldots, A_{ns})$ of $ns = 10$ variates from $N(0, 1)$ using, for instance the function `normgen_builtin` from *RandomNumbers.bas*. To use this sample to find a

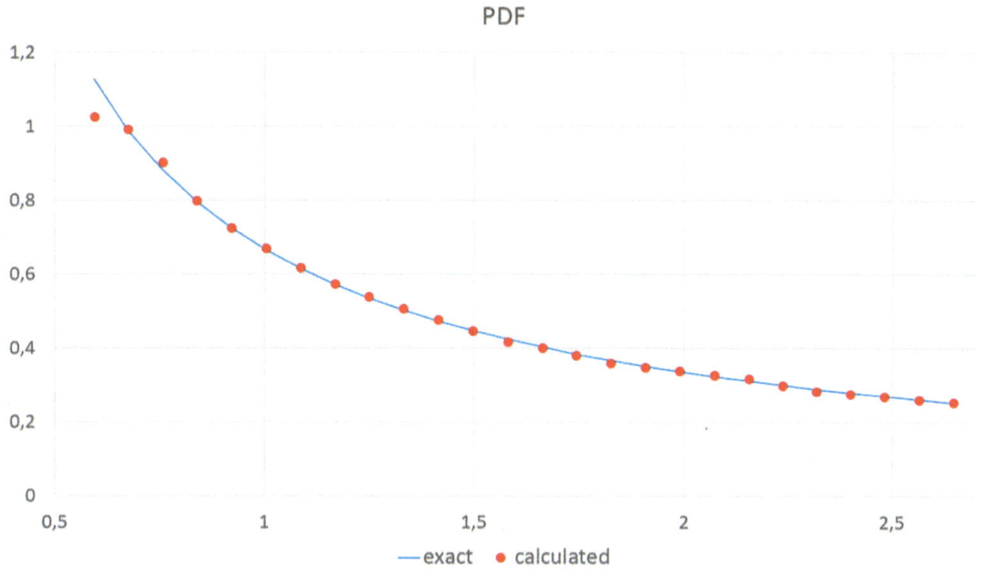

Fig. 5.3 Approximation of the PDF of X by using a large sample of simulated data generated by UQ. The approximated PDF corresponds to the red points and the exact one to the blue curve. The curve is obtained by SPH derivation of the CDF

Table 5.4 Evaluation of probabilities and statistics of the variable X by UQ

	$P(X > 2,4)$	$P(X < 0,7)$	Mean	Variance
Exact	0,0651	0,111	1,373	0,338
Evaluated	0,0651	0,110	1,372	0,337

Data in Table 5.2 is insufficient to evaluate the probabilities and leads to larger errors for the statistics

Table 5.5 An example of data including an artificial Gaussian variable

A	−2,07681	−1,17285	−0,86457	−0,52899	−0,11835
U	−0,9762	−0,74019	−0,60681	−0,42832	−0,15648
X	0,376742	0,477024	0,545087	0,651602	0,855151

A	0,082604	0,264137	0,592105	0,908595	1,257556
U	0,061595	0,310196	0,502534	0,600561	0,867986
X	1,063532	1,363692	1,652905	1,823141	2,38211

representation $PU = \sum_{i=1}^{k} u_i \varphi_i(A)$ of U as a function of A, it is necessary to create an artificial link between A and U – for instance, we can reorder the data into a way that both A and U are increasingly ordered: $U_1 < U_2 < \ldots < U_{ns}$ and $A_1 < A_2 < \ldots < A_{ns}$ – this generates a positive correlation between U and A. Now, our data takes the form in Table 5.5.

We consider an artificial sample formed by the pairs (A_i, U_i) and collocation can furnish coefficients $\boldsymbol{u} = (u_1, \ldots, u_k)$ defining an expansion of U as a function of A: we can solve the linear system

$$(PU)_\ell = \sum_{i=1}^{k} u_i \varphi_i(A_\ell) = U_\ell \ , \ell = 1, \ldots, ns \ .$$

The results furnished by collocation can be improved by determining the vector $\boldsymbol{u}$ that minimizes a measure of the deviation between the empirical moments of U and the empirical moments of PU – for instance, we may consider the sum of relative errors:

$$J(\boldsymbol{u}) = \sum_{i=1}^{m} \left| \frac{M_i(PU) - M_i(U)}{M_i(U)} \right|, M_i(Z) = \frac{1}{ns} \sum_{\ell=1}^{ns} Z_\ell^i$$

Minimization may be achieved by the SOLVER and collocation can furnish a starting point. For instance, the data in Table 5.5 with a normalized polynomial family furnishes the coefficients in Table 5.6.

The expansion in Table 5.6 can be used to generate a larger sample from U, by generating a large sample from A and using the expansion to obtain values of U.

However, a precaution is necessary: indeed, we must respect the interpolation interval and to generate only values of A on on $(\min A, \max A) = (-2, 07681, 1, 257556)$. Values outside this interval correspond to an extrapolation, for which collocation furnishes bad results: these values must be eliminated from the sample – from our original sample of 10^4 variates, 321 were eliminated by this reason. The remaining 9679 values of A were used to generate new values from U. The results are shown in Fig. 5.4.

Table 5.6 Coefficients obtained by collocation (C) and Moments Matching (M3) from data in Table 5.4 (normalized polynomial)

i	1	2	3	4	5
C	−0,97324	0,500184	1,133785	8,489041	−8,88642
M3	−0,97557	0,523057	1,394686	7,195578	−7,6199

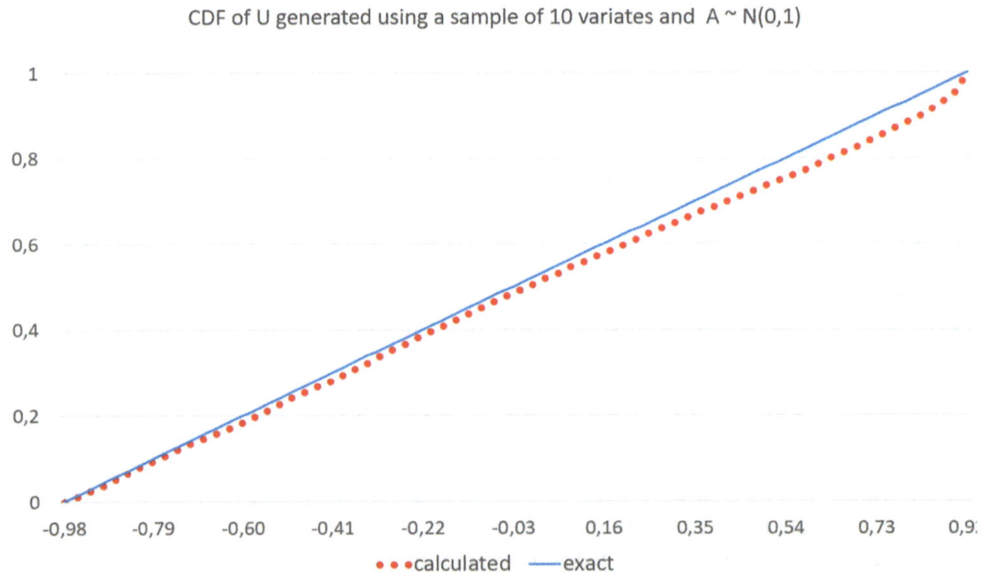

Fig. 5.4 Approximation of the CDF of U by using a large sample of simulated data generated by UQ expansion on an artificial variable A. The basis is formed by normalized polynomials and the degree of approximation is $k = 5$. The approximated CDF corresponds to the red points and the exact one to the blue curve. The variable A is gaussian and the original samples contain 10 elements. The expansion was generated by Moments Matching using the collocation solution as starting point

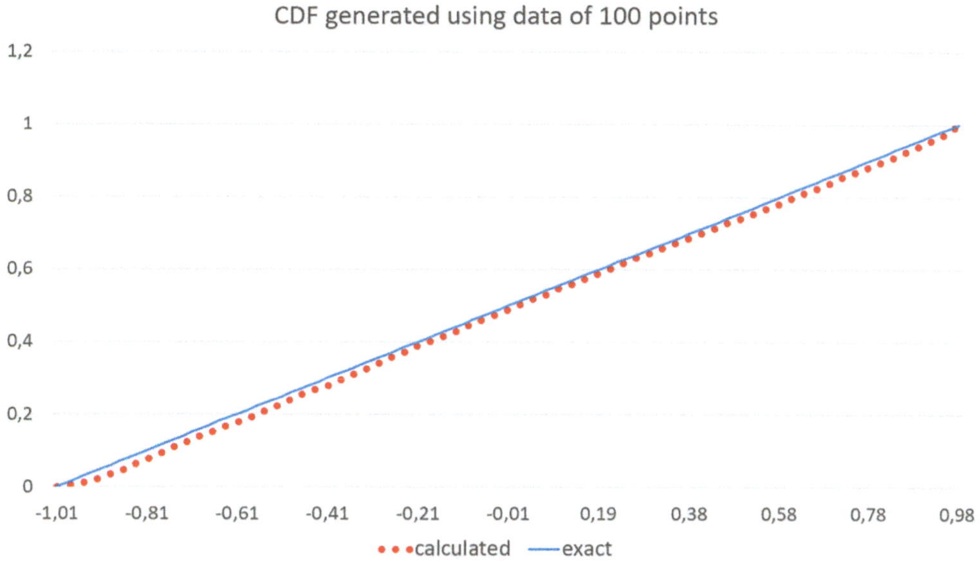

Fig. 5.5 Approximation of the CDF of U by using a large sample of simulated data generated by UQ expansion on an artificial variable A. The variable A is gaussian and we used a normalized polynomial with $k = 6$

Table 5.7 Expansion of X as a function of U: Coefficients obtained by collocation (C) from data in Table 5.5 (normalized polynomial)

i	1	2	3	4	5	6
C	0,376738	0,696332	0,622194	0,46988	0,044316	0,172647

The results may be improved by using more data, other basis or a higher degree of approximation. For instance, results obtained using a dataset of 100 points and a normalized polynomial of degree 6 are shown in Fig. 5.5.

Now, the distribution of U is known and we can determine the distribution of X. If we are not interested in the connection between U and X, we can determine an expansion of X as a function of the artificial variable. Otherwise, we must determine an expansion of X as function of U.

For instance, collocation furnishes the coefficients in Table 5.7 for an expansion of X as function of U. Then, we can generate a large sample from X by using the large sample from U previously determined. The results are shown in Figs. 5.6 and 5.7. The evaluation of the probabilities and statistics shown in Table 5.4 leads to the results in Table 5.8.

Of course, better results are obtained when the distribution of U is known: in this case, the expansion in Table 5.7 furnishes the results in Figs. 5.8 and 5.9.

As previously observed, different basis can be used. For instance, we can use a trigonometric basis

$$\varphi_1(u) = 1; \varphi_{2i}(u) = \cos\left(i\frac{u-a}{b-a}\right), \quad \varphi_{2i+1}(u) = \sin\left(i\frac{u-a}{b-a}\right) \; (i = 1, 2, \ldots) \tag{5.5}$$

Using the trigonometric basis and the data in Table 5.2 and $k = 13$, the coefficients found are shown in Table 5.9. The results for the CDF and the PDF are similar to those found by the polynomial family (see, for instance, Figs. 5.10 and 5.11)

In the rest of this Chapter, we shall present a panel of the UQ methods with illustrations of their use.

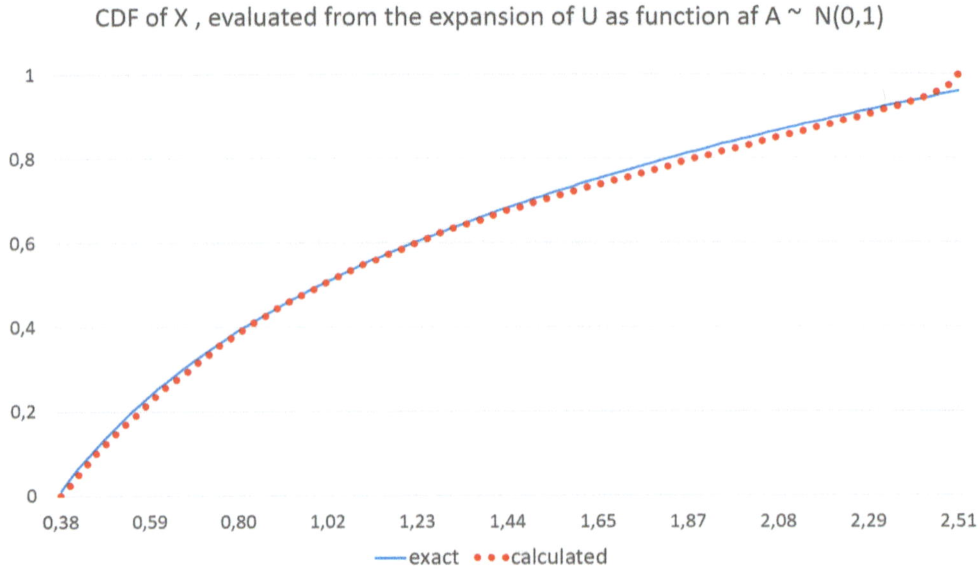

Fig. 5.6 Approximation of the CDF of U in the case where the distribution of U is unknown. Here, the artificial variable is $A \sim N(0, 1)$

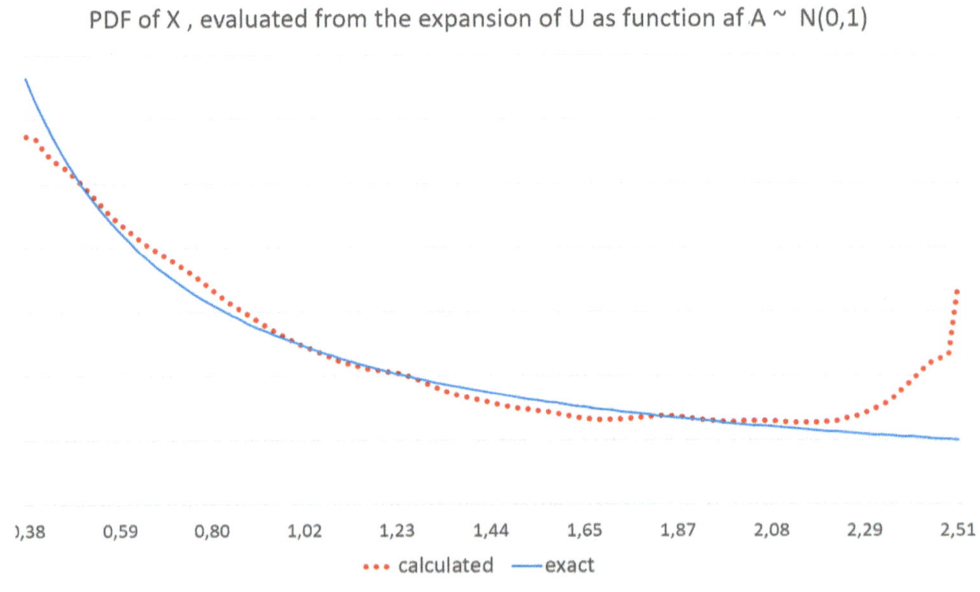

Fig. 5.7 Approximation of the PDF of X in the case where the distribution of U is unknown. Here, the artificial variable is $A \sim N(0, 1)$

5.2 Collocation

In the framework of UQ, collocation may be considered as the simplest way to determine a representation of random variables – but this approach faces difficulties when the variable U is missing or the data contains significant errors.

Table 5.8 Evaluation of probabilities and statistics of the variable X by UQ

	$P(X > 2)$	$P(X < 0, 7)$	Mean	Variance
Exact	0,153	0,322	1,175	0,432
Evaluated	0,166	0,313	1,184	0,424

Data in Table 5.5 is insufficient to evaluate the probabilities and leads to larger errors for the statistics

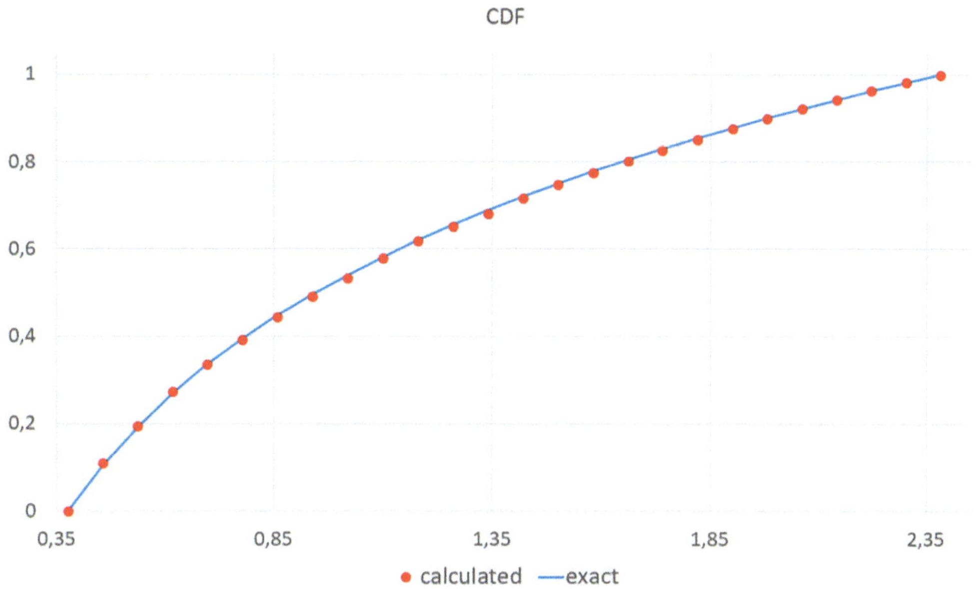

Fig. 5.8 Approximation of the CDF of X in the case where the distribution of U is known

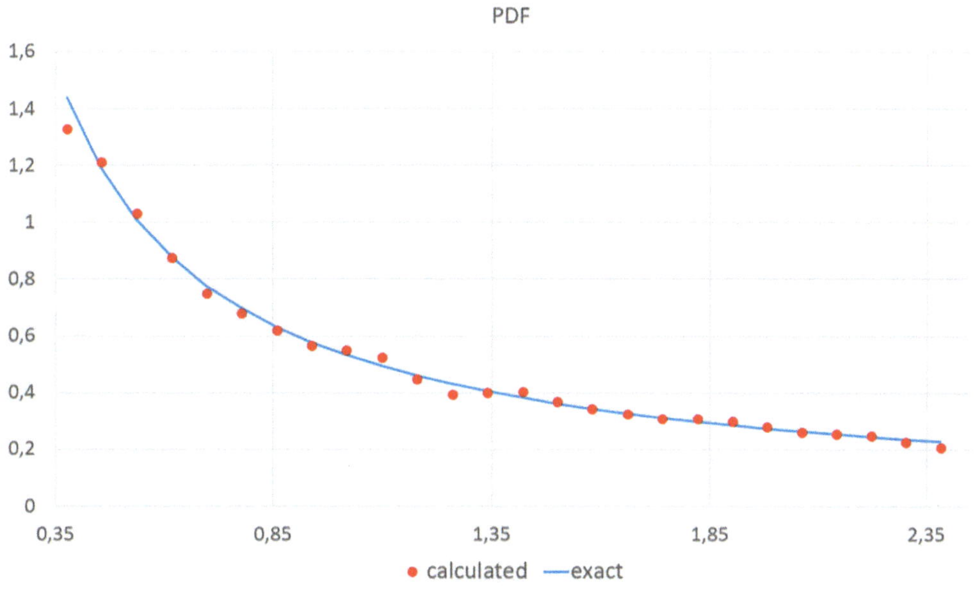

Fig. 5.9 Approximation of the CDF of X in the case where the distribution of U is known

Table 5.9 Expansion of X as a function of U: Coefficients obtained by collocation (C) from data in Table 5.5 (normalized trigonometrical family)

i	1	2	3	4	5	6	7
C	0,9396	0,8290	0,2724	0,4107	−0,6999	−0,4741	−0,5176
i	8	9	10	11	12	13	
C	−0,24314	0,300917	0,588993	0,12089	−0,24567	−0,03981	

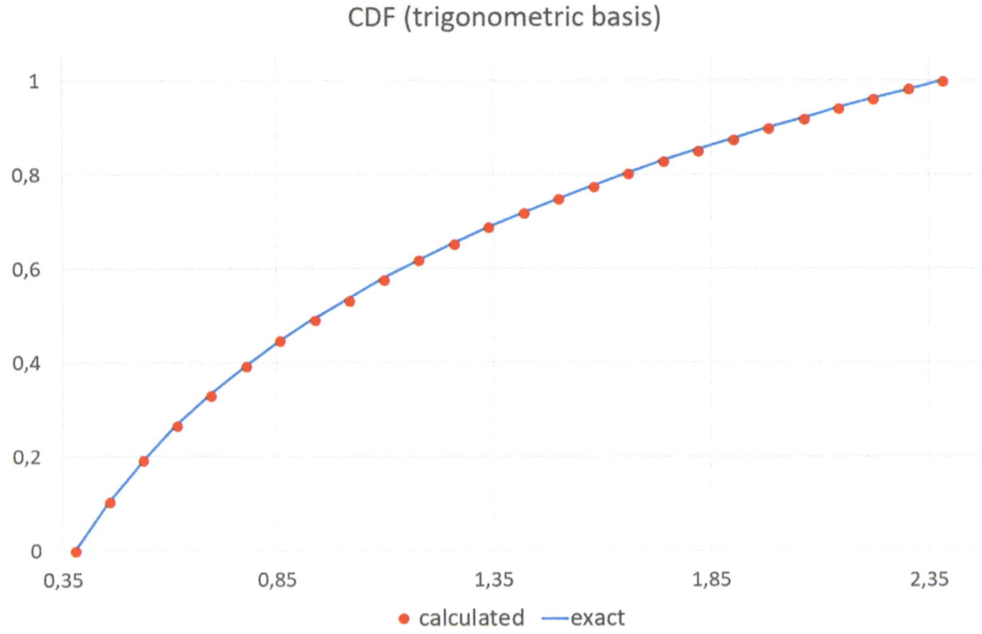

Fig. 5.10 Approximation of the CDF of X using a trigonometrical family: the results are similar to those furnished by the polynomial family

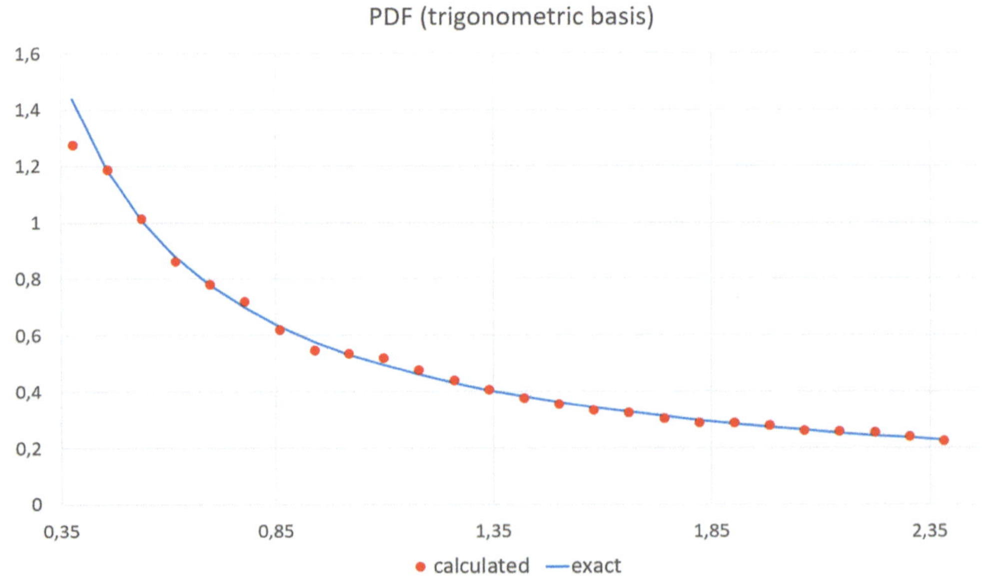

Fig. 5.11 Approximation of the PDF of X using a trigonometrical family: the results are similar to those furnished by the polynomial family

Collocation summarizes as follows: since $X \approx PX(U)$, we must have

$$X_i = X(U_i) \approx PX(U_i), \quad 1 \leq i \leq ns \tag{5.6}$$

id est, we consider that the Eq. (5.3) is satisfied at each point of the sample. Thus,

$$\sum_{j=1}^{k} x_j \varphi_j(u_i) = X_i \tag{5.7}$$

Range the data in vectors $\boldsymbol{X} = (X_1, \ldots, X_{ns})^t$, $\boldsymbol{U} = (X_1, \ldots, X_{ns})^t$. Then,

$$\boldsymbol{Ax} = \boldsymbol{B}, \tag{5.8}$$

with $\boldsymbol{A} = (\boldsymbol{\Phi}(U_1), \ldots, \boldsymbol{\Phi}(U_{ns}))^t$ (line i of $\boldsymbol{A}$ contains $\boldsymbol{\Phi}(U_i)$) and $\boldsymbol{B} = \boldsymbol{X}$, id est,

$$\boldsymbol{A} = \left(A_{ij} : 1 \leq i \leq ns, 1 \leq j \leq k\right), \quad A_{ij} = \varphi_j(U_i); \tag{5.9}$$

$$B = (B_i : 1 \leq i \leq ns), \quad B_i = X_i \tag{5.10}$$

Thus, $\boldsymbol{x}$ is the solution of the linear system (5.8), (5.9), and (5.10). Notice that Matrix $\boldsymbol{A}$ is formed by ns rows and k columns: in general, $ns > k$, so that the system is overdetermined. We have (see Sect. 2.3.4)

$$(\boldsymbol{A}^t\boldsymbol{A})\boldsymbol{x} = \boldsymbol{A}^t\boldsymbol{B}. \tag{5.11}$$

To obtain a more stable result, we may consider the Levenberg-Marquardt approximation

$$\boldsymbol{x} \approx \boldsymbol{x}_{\boldsymbol{\varepsilon}} \quad \textit{with} \ (\boldsymbol{A}^t\boldsymbol{A} + \boldsymbol{\varepsilon Id})\boldsymbol{x}_{\boldsymbol{\varepsilon}} = \boldsymbol{A}^t\boldsymbol{B} \ . \tag{5.12}$$

Here, $\varepsilon > 0$ is a parameter of regularization – Tikhonov's regularization – to be chosen as small enough (often inferior to $1E - 3$). For $\varepsilon = 0$, Eq. (5.12) reduces to Eq. (5.11) – thus, Eq. (5.11) corresponds to a particular value of ε. We may write $\boldsymbol{x} = (\boldsymbol{A}^t\boldsymbol{A})^{-1}\boldsymbol{A}^t\boldsymbol{B}$ and $\boldsymbol{x}_{\boldsymbol{\varepsilon}} = (\boldsymbol{A}^t\boldsymbol{A} + \boldsymbol{\varepsilon Id})^{-1}\boldsymbol{A}^t\boldsymbol{B}$ – these expressions may be directly written in EXCEL®. For instance, we may use the EXCEL® matrix formula MMULT(MINVERSE(MMULT(TRANSPOSE(A),A)),MMULT(TRANSPOSE(A), B)) to determine $\boldsymbol{x}$. EXCEL® evaluates efficiently these expressions whenever the size of the problems remains small. For large or medium problems, we may obtain better results by using numerical techniques of solution of the linear system – such as those presented in Sect. 2.1, which are more efficient and stable for the numerical determination of $\boldsymbol{x}$.

As an alternative, we may look for the minimum of

$$J(\boldsymbol{x}) = \frac{1}{2}\|\boldsymbol{Ax} - \boldsymbol{B}\|^2, \tag{5.13}$$

or its regularized form

$$J(\boldsymbol{x}) = \frac{1}{2}\|\boldsymbol{Ax} - \boldsymbol{B}\|^2 + \frac{\varepsilon}{2}\|\boldsymbol{x}\|^2, \tag{5.14}$$

Notice that Eq. (5.14) corresponds to Eq. (5.13) when $\varepsilon = 0$: Eq. (5.13) is a particular case of Eq. (5.14). In this case, we need to use an optimization procedure: EXCEL® offers the SOLVER to this task (see Sects. 2.2.1 and 2.2.2)

5.2.1 Finding the Coefficients of the Expansion in a Worksheet

For small problems, the determination of the coefficients of the expansion can be made directly in a worksheet.

For instance, open a blank workbook and enter the data given in Table 5.1 in columns `A` and `B` as shown at right. Give the name "`U`" to the data in column `A`, starting at `A2` (here, until `A11`). Give the name "`X`" to the data in column `B`, starting at `B2` (here, until `B11`). Enter the formula "`=MIN(U)`" in `C2` and the formula "`=MAX(U)`" in `C4`. Give the name `lb` to `C2` and the name `ub` to `C4`.

	A	B	C
1	U	X	lb
2	0,110263	1,116572	-0,5219
3	0,818171	2,26635	ub
4	0,218055	1,243655	0,972777
5	-0,42827	0,651638	
6	-0,5219	0,593394	
7	0,972777	2,645282	
8	0,498947	1,646986	
9	-0,06381	0,938179	
10	0,439192	1,551453	
11	0,899228	2,457705	

To make a polynomial approximation using $\varphi_i(u) = \left(\frac{u-lb}{ub-lb}\right)^{i-1}$, enter the formula "`=(U-lb)/(ub-lb)`" in `D2`. Give the name `s` to the data in column `D` (here `D2:D11`). Then, enter "`=MAX(SIGN(s);1)`" in `E2` (EXCEL® gives a `#NUM!` when evaluating 0^0, so that you cannot enter "`=s^0`"), "`=s^1`" in `F2`, "`=s^2`" in `G2` etc. Go up to the order of approximation k you choose. Give the name "A" to the matrix generated.

D	E	F
s	A	
0,422942	1	=s^1
0,896562	1	
0,495059	1	
0,062643	1	
0	1	
1	1	
0,682988	1	
0,306477	1	
0,643009	1	
0,950792	1	

For instance, for $k = 6$, we have the results at right. Here, `A` is the name of `E2:J11`.

E	F	G	H	I	J
A					
1	0,422942	0,17888	0,075656	0,031997986	0,013533
1	0,896562	0,803823	0,720677	0,646130883	0,579296
1	0,495059	0,245083	0,121331	0,060065785	0,029736
1	0,062643	0,003924	0,000246	1,53988E-05	9,65E-07
1	0	0	0	0	0
1	1	1	1	1	1
1	0,682988	0,466472	0,318595	0,217596166	0,148615
1	0,306477	0,093928	0,028787	0,008822456	0,002704
1	0,643009	0,41346	0,265859	0,170949456	0,109922
1	0,950792	0,904006	0,859522	0,817226585	0,777013

Enter the value of ε in `L1`. Give it the name `epsilon`. Enter the value of k used to create `A` in `L2`. Give it the name `k`.

K	L
epsilon	1,00E-03
k	6

In cell `M2`, enter the formula "`=MMULT(MINVERSE(MMULT(TRANSPOSE(A);A)+epsilon*MUNIT(k));MMULT(TRANSPOSE(A);X)`")
You get the coefficients of the polynomial approximation of degree $k-1$, starting from `M2`.

M
x
0,593291
0,902007
0,615147
0,349043
0,157741
0,028212

To generate a sample from X, we must generate a sample from U and evaluate the go to `B13` and en- ter the formula `=RANDARRAY(1;10;lb.;ub)`. Give the name `us` to the data generated. If you desire more points, change "`10`" to the desired number of variates.

13	us	=randarray(1;10;lb;ub)

Then, go to cell `B14` and enter the formula "`=(us-lb)/(ub-lb)`". Give the name `sn` to the data in line 14. Go to cell B15 and enter the formula "`= M2 + M3*sn + M4*sn^2 + M5*sn^3 + M6*sn^4 + M7*sn^5`". The sample appears in line 15.

13	us	-0,41054	-0,4978	-0,12455	0,244483	-0,50491	0,33201	0,303752	-0,4409
14	sn	0,074505	0,01612	0,265839	0,51274	0,011362	0,5713	0,552394	0,05413
15	PX(us)	0,664059	0,607993	0,883935	1,276464	0,60362	1,392985	1,354232	0,64398

If a large amount of data is generated, the new sample may be used to find the CDF and the PDF of X, as shown in Sect. 3.8.3.

If you prefer to solve the problem using the formulation (5.14), you must start by creating a cell containing the value of J.

For instance, give the name `coef` to the cells containing the coefficients (`M2:M7`).
Select cell M2 and enter the formula "`=RANDARRAY(k,1)`".
Then, select `L3` and enter the formula "`=SUMSQ(MMULT(A;coef)-X)+epsilon*SUMSQ(coef)/2`". Give the name "J" to `L3`.

K	L	M
epsilon	1,00E-03	x
k	6	0,525439
J	1,496694	0,604684
		0,703458
		0,075981
		0,782843
		0,848977

Now, run the SOLVER as shown in Sect. 2.2.1. The objective cell is J (cell L3) and the variables are coef (cells M2:M7).
If the value of J is small at the end of the calculations, the coefficients found may be considered as correct.

K	L	M
epsilon	1,00E-03	x
k	5	0,593196
J	0,000848	0,899025
		0,62452
		0,346004
		0,152563
		0,029872

5.2.2 Solution Using VBA

The use of VBA provides more flexibility in the analysis, namely for the use of other basis than the polynomial one, the automatic calculation of the coefficients and the automatic production of charts. The VBA solution uses some of the elements previously introduced: for the determination of the coefficients, a basis $\{\varphi_i : i \geq 1\}$ and a method of solution of linear system are used. The determination of the CDF and the PDF requests the generation of samples, the evaluation of expansions and of empirical CDFs and their derivatives. All these elements may be found in the preceding chapters and in the modules and add-ins furnished by the book. For instance, LinSys.bas and LinSys.xlam contain the methods of solution of Linear Systems presented in Sect. 2.1. You can either import the module or include the add-in in your workbook.

For instance, we can get the coefficients by using the VBA code below. Here, we use LUSolution - which is part of LinSys - to solve the linear system (5.12).

```
Function collocation(x, u, k, basis, epsilon)
nx = UBound(x, 1)
Dim A() As Double
ReDim A(1 To nx, 1 To k)
Dim B() As Double
ReDim B(1 To nx, 1 To 1)
For i = 1 To nx
    B(i, 1) = x(i)
    For j = 1 To k
       A(i, j) = basis.value(j - 1, u(i))
    Next j
Next i
at = WorksheetFunction.Transpose(A)
ata = WorksheetFunction.MMult(at, A)
atb = WorksheetFunction.MMult(at, B)
For i = 1 To k
   ata(i, i) = ata(i, i) + epsilon
Next i
collocation = LUSolution(ata, squeeze(atb))
End Function
```

Function squeeze transforms a $m \times n$ matrix having two indexes in a vector having a single index.
It maps $(i, j) \rightarrow k = (i-1)n + j$.
This function may be found in VectorTools

```
Function squeeze(vx) As Variant
m = UBound(vx, 1)
n = UBound(vx, 2)
Dim x() As Variant
ReDim x(1 To m * n)
ind = 0
For i = 1 To m
  For j = 1 To n
    ind = ind + 1
    x(ind) = vx(i, j)
  Next j
Next i
squeeze = x
End Function
```

To use the code, you need a basis (defined by a class, such as `poly`, see Sect. 1.17) and a Sub that calls this function. For instance, you can use the code at right.

```
Sub solve_collocation()
u = readcol(1, 2)
x = readcol(2, 2)
Dim p As New poly
p.lb = WorksheetFunction.Min(u)
p.ub = WorksheetFunction.Max(u)
p.zero = Cells(2, 3).value
k = Cells(4, 3).value
epsilon = Cells(6, 3).value
c = collocation(x, u, k, p, epsilon)
Cells(1, 4) = "coeffs"
For i = 1 To k
Cells(i + 1, 4).value = c(i)
Next i
End Sub
```

Let us exemplify: open a blank workbook, add these modules, import `poly.cls`, `LinSys.bas`, `expansion.cls` (see Sects. 1.6 and 1.17).

Enter the data in columns `A`, `B`, `C` as shown at right.
Run the macro `solve_collocation`.
You will get the results in column `D`.

	A	B	C	D
1	U	X	zero	coeffs
2	0,110263	1,116572	1,00E-03	0,593291
3	0,818171	2,26635	k	0,902007
4	0,218055	1,243655	6	0,615147
5	-0,42827	0,651638	epsilon	0,349043
6	-0,5219	0,593394	1,00E-03	0,157741
7	0,972777	2,645282	ns	0,028212
8	0,498947	1,646986	1,00E+05	
9	-0,06381	0,938179	ne	
10	0,439192	1,551453	25	
11	0,899228	2,457705		

Add `newsample` below to your module. It generates a sample of `ns` variates from *X*. For `unigen`, see Sect. 3.10.3 (or import `RandomNumbers.bas`). Add the lines at right to `solve_collocation`. For `graph1`, see Sect. 1.14. `empirical_repartition` is defined in Sect. 3.8.3.

```
Function newsample(c, basis, ns)
us = unigen(basis.lb, basis.ub, ns)
Dim ap As New expansion
Set ap.phi = basis
newsample = ap.proj(c, us)
End Function
```

```
Next i
ns = Cells(8, 3).value
xs = newsample(c, p, ns)
ne = Cells(10, 3).value
resu = empirical_repartition(xs, ne)
xabs = resu(1)
Fe = resu(2)
Dim Fex() As Double
ReDim Fex(1 To UBound(Fe, 1))
For i = 1 To UBound(Fe, 1)
    Fex(i) = WorksheetFunction.Ln(xabs(i))
Next i
ReDim gdata(1 To 2, 3)
gdata(1, 1) = xabs
gdata(1, 2) = Fe
gdata(1, 3) = "empirical"
gdata(2, 1) = xabs
gdata(2, 2) = Fex
gdata(2, 3) = "exact"
Dim ttl As String
ttl = "Empirical CDF"
Call graph1(gdata, ttl)
End Sub
```

Run the macro `solve_collocation` again. You will get a figure with the estimation of the CDF of X and the exact one for comparison (Fig. 5.12).

If you desire to evaluate the PDF, you must add supplementary code at the end of the macro `solve_collocation`:

```
Call graph1(gdata, ttl)
Dim dFex() As Double
ReDim dFex(1 To UBound(Fex, 1))
For i = 1 To UBound(Fex, 1)
  dFex(i) = 1 / xabs(i)
Next i
d = (xabs(2) - xabs(1))
sig = 0.75 * d
kernel = "gaussian"
der_kernel = "der_gaussian"
dFe = sph_derivation(xabs, Fe, xabs, Fe, kernel, der_kernel, sig)
gdata(1, 2) = dFe
gdata(2, 1) = xabs
gdata(2, 2) = dFex
ttl = "Empirical PDF"
Call graph1(gdata, ttl)
End Sub
```

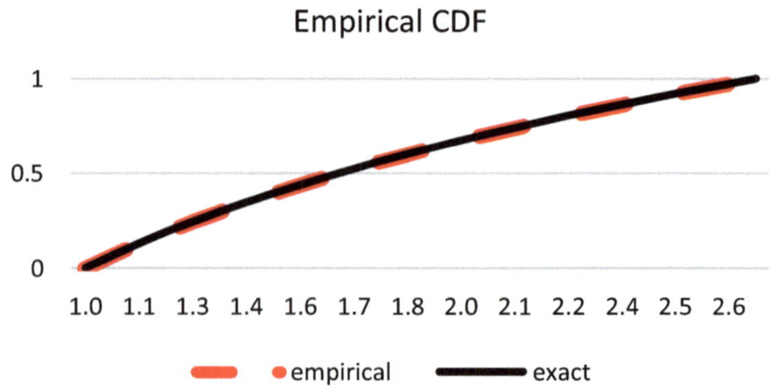

Fig. 5.12 Empirical cumulative function (VBA solution)

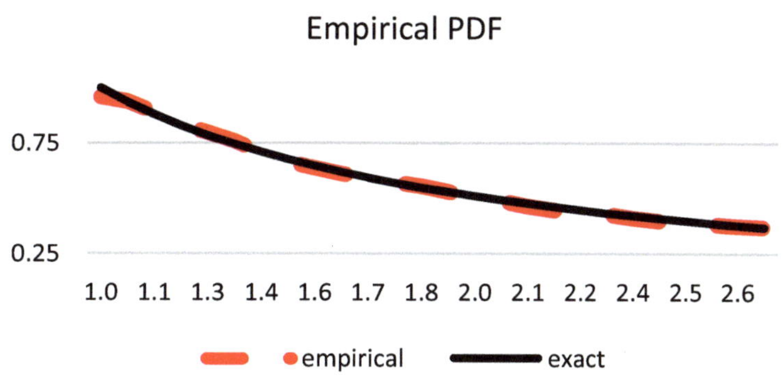

Fig. 5.13 Empirical probability density (VBA solution)

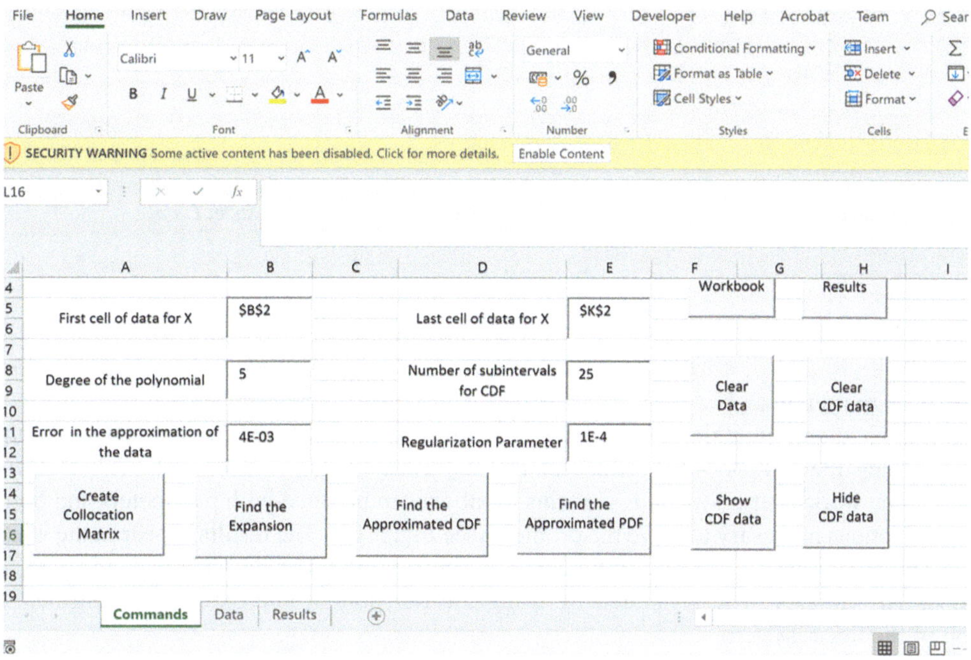

Fig. 5.14 The workbooks contain macros: the user must authorize their execution by clicking on *"Enable Content"*. The workbooks contain four (4) worksheets: *Commands*, *Data* and *Results* are visible. *CDFdata* is hidden

Add this code, import the module `InterpDerivInteg.bas` and run `solve_collocation again`. You will get a new chart, with the estimation of the PDF of X and the exact one (Fig. 5.13)

5.2.3 Solution Using an Adapted Workbook

The solutions of Eqs. (5.11) and (5.12) are implemented in the EXCEL® workbooks *collocation_LinSys.xlsm* and *collocation_minimize.xlsm*, respectively.

Both these workbooks contain macros, so that the user must authorize their execution as previously shown (see Fig. 5.14): *collocation_LinSys.xlsm* uses the formula `MMULT(MINVERSE(MMULT`

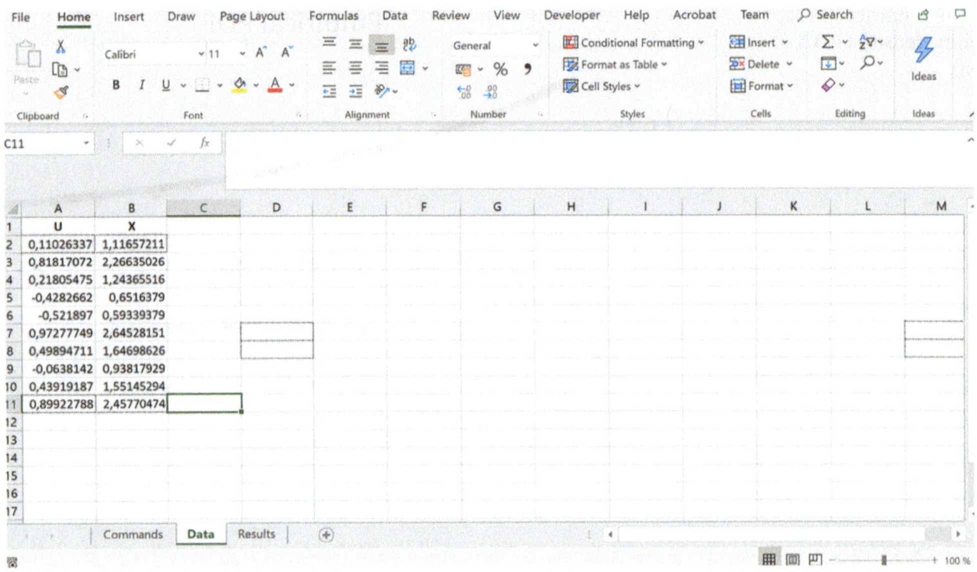

Fig. 5.15 The values of (U_i, X_i) must be given in the columns "*A*" and "*B*" of the sheet *Data*

`(TRANSPOSE(A),A)),MMULT(TRANS POSE(A),B))`, while *collocation_minimize.xlsm* calls the SOLVER to minimize J.

Both the workbooks contain 4 Worksheets, named *Commands, Data, Results* and *CDFdata*. The data concerning the sample – the values of (U_i, X_i) must be given in the spreadsheet *Data* (see Fig. 5.15). The worksheet *Commands* contains Textboxes to be filled with parameters and buttons to execute the actions necessary to solve the problem (see Fig. 5.16). The results appear in the worksheet *Results*. The spreadsheet *CDFdata* is hidden – it contains working data corresponding to a large sample of 10^5 variates of X generated by using the polynomial approximation furnished by the solution of the linear system.

The steps to solve the problem with these workbooks are the following:

- Insert the data corresponding to the pairs (U_i, X_i) – for instance, in the columns A and B of the Worksheet *Data*, with U_1 at *A2*, X_1 at *B2*, U_2 at *A3*, X_2 at *B3*, and so on.
- Fill the data in the Worksheet *Commands*: First and Last Cells of U and X; degree of the polynomial; regularization parameter ε(recall that $\varepsilon = 0$ corresponds to an unregularized problem); number of subintervals for the CDF. The selection of the cells may be performed by DoubleClicks:
  - Go to the Worksheet *Data* and DoubleClick on the cell A2: it becomes red

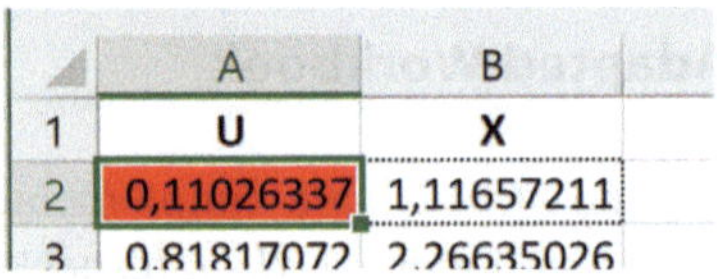

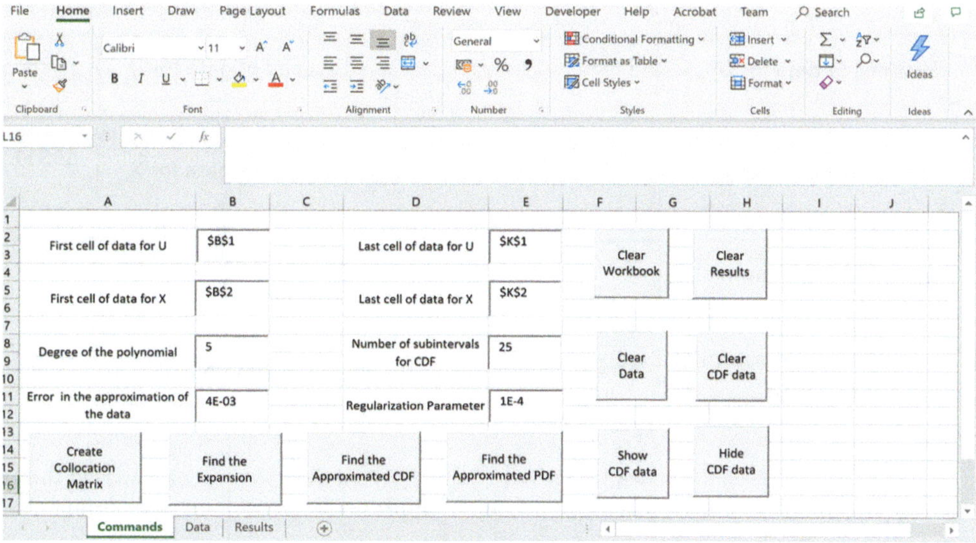

Fig. 5.16 *Commands* defines the parameters of the approximation and executes the actions for the determination of x

– Go to the Worksheet *Commands* and DoubleClick in the TextBox for the first cell of U: the content becomes *$A$2* (*$* is unnecessary, but it is automatically included by EXCEL® when the DoubleClick is used)

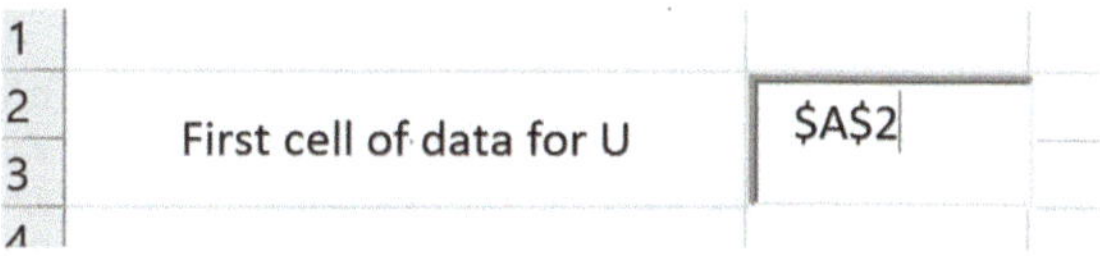

– Come back to the Worksheet *Data* and make the same procedure to select all the cells corresponding to First and Last Cells of U and X – the cells in Data turn white again after a double click on a TextBox: after the previous action, we have

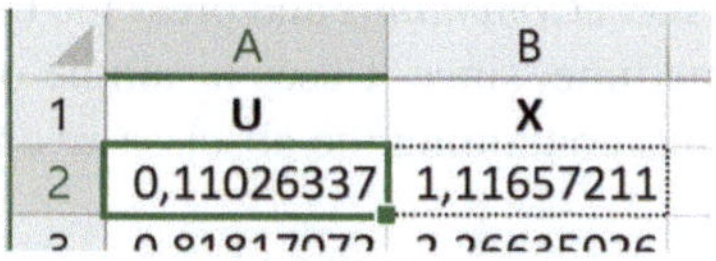

- After entering all the data, the fields in the Worksheet *Commands* are filled. For instance, entering the data in Fig. 5.15 causes the Worksheet *Commands* to appear as follows:

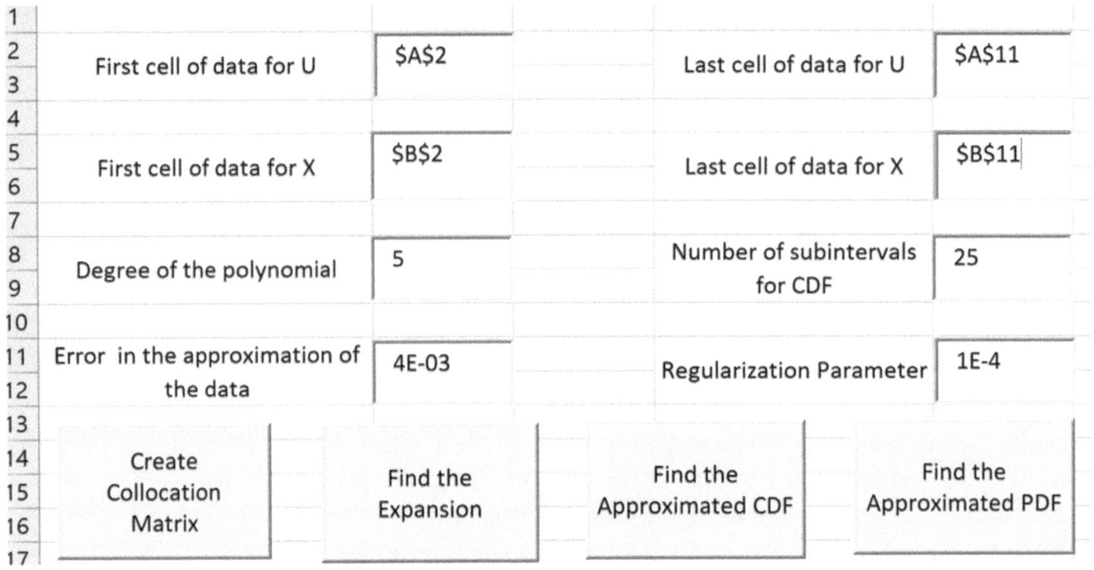

- Now, click on the button "Find the expansion" to get the results: this button generates $\boldsymbol{A}$, $\boldsymbol{B}$ and $\boldsymbol{x}$ (the button "Create Collocation Matrix" generates only $\boldsymbol{A}$ and $\boldsymbol{B}$). Then, go to the Worksheet *Results*. The results appear as below:

	A	B	C	D	E	F	G	H	I	J
1	A							B		x
2	1	0,42294185	0,17887981	0,07565576	0,03199799	0,01353329		1,11657211		0,59326222
3	1	0,89656158	0,80382267	0,72067652	0,64613088	0,57929613		2,26635026		0,89228406
4	1	0,4950588	0,24508322	0,12133061	0,06006578	0,0297361		1,24365516		0,6472676
5	1	0,06264293	0,00392414	0,00024582	1,5399E-05	9,6463E-07		0,6516379		0,33374578
6	1	0	0	0	0	0		0,59339379		0,13524735
7	1	1	1	1	1	1		2,64528151		0,0433195
8	1	0,68298759	0,46647204	0,31859461	0,21759617	0,14861548		1,64698626		
9	1	0,30647664	0,09392793	0,02878672	0,00882246	0,00270388		0,93817929		
10	1	0,64300882	0,41346034	0,26585865	0,17094946	0,10992201		1,55145294		
11	1	0,95079222	0,90400585	0,85952174	0,81722659	0,77701268		2,45770474		
12										
13	error	0,00376839								

- The coefficients of the expansion appear in the column entitled $\boldsymbol{x}$. Error indicates the mean square error. Come back to the Worksheet *Commands* and press the button "Find the Approximated CDF" . The evaluation requires some time, since it uses a sample of 10^5 variates. Now, the Worksheet *Results* contains 2 new columns, containing the abscissa *xe* and the values of the CDF *Fe* at these abscissae (Fig. 5.17).
- Press the button "Find the Approximated PDF": one new column is added to the Worksheet *Results*, containing the values of the CDF *dFe* at the same abscissa (Fig. 5.17).

The procedure is the same for both the Workbooks *collocation_minimize.xlsm* and *collocation_LinSys.xlsm*. As an example, let us consider the situation presented in Table 5.2: the results shown in Figs. 5.2 and 5.3 are generated by *collocation_LinSys.xlsm* with $\varepsilon = 0$ and the

a)

	L	M
	xe	CDF Fe
	0,36444416	0
	0,45857141	0,11247
	0,55269865	0,20405
	0,64682589	0,28304
	0,74095313	0,35087
	0,83508037	0,41167
	0,92920761	0,46476
	1,02333486	0,5129
	1,1174621	0,55713
	1,21158934	0,59706
	1,30571658	0,63447
	1,39984382	0,66894
	1,49397107	0,7019

b)

	L	M	N
	xe	CDF Fe	PDF dFe
	0,36444416	0	1,18569539
	0,45857141	0,11247	1,09064471
	0,55269865	0,20405	0,91055162
	0,64682589	0,28304	0,78117607
	0,74095313	0,35087	0,68404081
	0,83508037	0,41167	0,60543932
	0,92920761	0,46476	0,53857062
	1,02333486	0,5129	0,49080947
	1,1174621	0,55713	0,44736208
	1,21158934	0,59706	0,41112788
	1,30571658	0,63447	0,38204712
	1,39984382	0,66894	0,35822077
	1,49397107	0,7019	0,33549444

Fig. 5.17 Contents of the worksheet Results after clicking. (**a**) After clicking "Find the Approximated CDF". (**b**) After clicking "Find the Approximated PDF"

Table 5.10 Coefficients obtained by *collocation_minimize* from data in Table 5.2

i	1	2	3	4	5	6
x	0,593	0,901	0,615	0,357	0,161	0,017

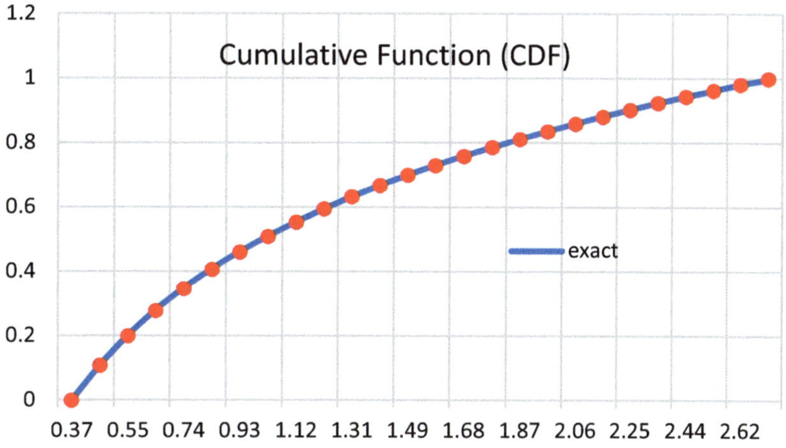

Fig. 5.18 Results furnished by *collocation_minimize* from data in Table 5.2

polynomial basis $\varphi_i(U) = U^{i-1}$. The results furnished by *collocation_minimize.xlsm* with $\varepsilon = 0$, $k = 6$ and the normalized polynomial basis

$$\varphi_i(U) = \left(\frac{U-a}{b-a}\right)^{i-1}, \quad \begin{aligned} a &= \min\{U_i : 1 \le i \le n\} \\ b &= \max\{U_i : 1 \le i \le n\} \end{aligned} \tag{5.15}$$

are shown in Table 5.10 and Figs. 5.18 and 5.19.

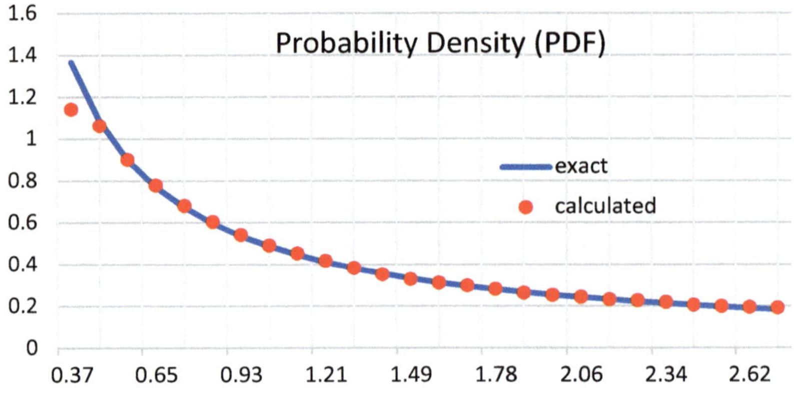

Fig. 5.19 Results furnished by *collocation_minimize* from data in Table 5.2

Exercises

1. Consider the Walras model, where the excess demand X is a function of the price deviation U. The following data was observed:

U	1,000	0,819	0,670	0,549	0,449	0,368
X	−1,033	−0,578	−0,296	−0,169	−0,097	−0,049

U	0,301	0,247	0,202	0,165	0,135	0,111
X	−0,029	−0,014	−0,009	−0,004	−0,002	−0,001

   (a) Use the data above to determine by collocation a representation of the excess demand as a function of the price deviation.
   (b) Assume that $U \sim N(0.4, 0.3)$. Use the representation found to determine the CDF and the PDF of X.
   TIP: generate a large sample from N(0.4,0.3) and use it to determine the CDF. The use the methods of Sect. 2.8 to find the derivative of the CDF.
   (c) Assume that the distribution of U is unknown. Introduce an artificial uniformly distributed variable A and use it to determine a representation of U as function of A by collocation. Use the representation found to determine the CDF and the PDF of X.

2. Consider a Household stocks model, where the households X are a function of the propensity to consume U. The following data was observed:

U	0,0580	0,0609	0,2197	0,3883	0,4677	0,5439
X	0,0000	0,0021	0,0060	0,0078	0,0123	0,0126

U	0,7550	0,8709	0,8761	0,9045	0,9133	0,9778
X	0,0050	0,0022	0,0023	0,0019	0,0020	0,0001

   (a) Use the data above to determine a representation of the Households as a function of the propensity to consume by collocation.
   (b) Assume that $U \sim N(0.5, 0.1)$. Use the representation found to determine the CDF and the PDF of X.

TIP: generate a large sample from N(0.5,0.1) and use it to determine the CDF. The use the methods of Sect. 2.8 to find the derivative of the CDF.

(c) Assume that the distribution of U is unknown. Introduce an artificial uniformly distributed variable A and use it to determine a representation of U as function of A by collocation. Use the representation found to determine the CDF and the PDF of X.

3. Consider a Keynesian model of consumption, where the consumption X is a function of the propensity to consume U. The following data was observed:

U	0,2138	0,2483	0,2947	0,2951	0,3354	0,3425
X	8,7482	8,7470	7,0831	6,1580	6,8340	5,7317

U	0,3627	0,3654	0,4223	0,5067	0,5422	0,5549
X	5,1731	5,1117	4,3880	4,6680	4,9407	4,6589

(a) Use the data above to determine a representation of the consumption as a function of the propensity to consume by collocation.

(b) Assume that $U \sim N(0.5, 0.1)$. Use the representation found to determine the CDF and the PDF of X.

TIP: generate a large sample from N(0.5,0.1) and use it to determine the CDF. The use the methods of Sect. 2.8 to find the derivative of the CDF.

(c) Assume that the distribution of U is unknown. Introduce an artificial uniformly distributed variable A and use it to determine a representation of U as function of A by collocation. Use the representation found to determine the CDF and the PDF of X.

5.3 Variational Approximation

The variational approach is based on the notion of *orthogonal projection*. Recall that, given a vector $\boldsymbol{v}$ from a vector space V and a linear subspace $S \subset V$, $\boldsymbol{Pv}$ – the orthogonal projection of $\boldsymbol{v}$ onto S, is the element of S which minimizes the distance to $\boldsymbol{v}$:

$$\boldsymbol{Pv} \in S \textit{ and } \|\boldsymbol{v} - \boldsymbol{Pv}\| = \min\{\|\boldsymbol{v} - \boldsymbol{w}\| : \boldsymbol{w} \in S\} \tag{5.16}$$

If the scalar product in V is denoted by $(\blacksquare,\blacksquare)$, we have (see Fig. 5.20)

$$\boldsymbol{Pv} \in S \textit{ and } (\boldsymbol{v} - \boldsymbol{Pv}, \boldsymbol{w} - \boldsymbol{Pv}) = 0, \ \forall \boldsymbol{w} \in S. \tag{5.17}$$

Finally, by taking $\widetilde{\boldsymbol{w}} = \boldsymbol{Pv} + \boldsymbol{w} \in S$ in Eq. (5.16), we have

$$\boldsymbol{Pv} \in S \textit{ and } (\boldsymbol{v} - \boldsymbol{Pv}, \boldsymbol{w}) = 0, \ \forall \boldsymbol{w} \in S \ . \tag{5.18}$$

In the context of UQ, we consider $S = [\mathcal{F}]$ as the subspace of the random variables generated by all the linear combinations of the elements of $\mathcal{F}$:

$$S = [\mathcal{F}] = \left\{ \sum_{j=1}^{k} a_j \varphi_j(U) : \boldsymbol{a} = (a_1, \ldots, a_k) \in \mathbb{R}^k \right\} \tag{5.19}$$

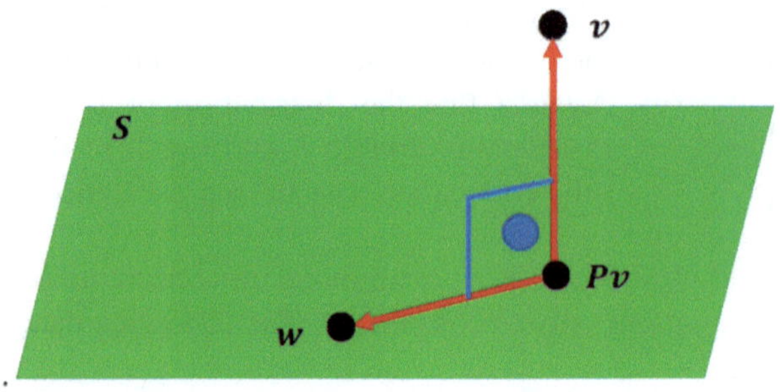

Fig. 5.20 Orthogonal projection: the vector $v - Pv$ is orthogonal to any element from V

Since S is formed by random variables, a convenient scalar product is

$$(Y, Z) = \mathrm{E}(YZ) \tag{5.20}$$

Then, we look for PX as the orthogonal projection of X onto S: from Eqs. (5.18), (5.19), and (5.20), we obtain

$$PX = \sum_{j=1}^{k} x_j \varphi_j(U) \text{ and } E\left(X - \sum_{j=1}^{k} x_j \varphi_j(U), w\right) = 0, \ \forall w \in S \ . \tag{5.21}$$

Taking $w = \varphi_i$, we have

$$\sum_{j=1}^{k} x_j E\big(\varphi_j(U)\varphi_j(U)\big) = E(X, \varphi_i(U)), \ 1 \le i \le k \ . \tag{5.22}$$

what corresponds to the linear system

$$\boldsymbol{Ax} = \boldsymbol{B}, \tag{5.23}$$

with $\boldsymbol{A} = E(\boldsymbol{\Phi}(\mathrm{U})^t \boldsymbol{\Phi}(\mathrm{U}))$ and $\boldsymbol{B} = E(\boldsymbol{\Phi}(\mathrm{U})^t X)$, id est,

$$\boldsymbol{A} = \left(A_{ij} : 1 \le i \le k, 1 \le j \le k\right), \quad A_{ij} = E\big(\varphi_i(U)\varphi_j(U)\big); \tag{5.24}$$

$$B = (B_i : 1 \le i \le k), \ B_i = E(X\varphi_i(U)) \tag{5.25}$$

Again, $\boldsymbol{x}$ is the solution of a linear system, now given by (5.23), (5.24), and (5.25). Here, the Matrix $\boldsymbol{A}$ is formed by k rows and k columns, so that the system is, general, determined. Here, we use the data to generate the linear system: indeed, we may use the data to approximate

$$A_{ij} \approx \frac{1}{ns} \sum_{r=1}^{ns} \varphi_i(U_r)\varphi_j(U_r); \ B_i \approx \frac{1}{ns} \sum_{r=1}^{ns} \varphi_j(U_r)\varphi_j(U_r) \tag{5.26}$$

5.3.1 Finding the Coefficients of the Expansion in a Worksheet

Analogously to the approach by collocation, the linear system (5.23) may be solved by calling the SOLVER or implementing a convenient EXCEL® formula.

The first steps are analogous to those made for the solution by collocation: open a blank workbook, enter the data `U`, `X`; determine `lb`, `ub` and generate `s`. Remember that names `U`, `X`, `s`, `lb`, `ub` must be defined. The worksheet will be similar to this one.

	A	B	C	D
1	U	X	lb	s
2	0,110263	1,116572	-0,5219	0,422942
3	0,818171	2,26635	ub	0,896562
4	0,218055	1,243655	0,972777	0,495059
5	-0,42827	0,651638		0,062643
6	-0,5219	0,593394		0
7	0,972777	2,645282		1
8	0,498947	1,646986		0,682988
9	-0,06381	0,938179		0,306477
10	0,439192	1,551453		0,643009
11	0,899228	2,457705		0,950792

Define the place for the matrix A by entering the values of the index i from φ_i at the sides of the matrix. A_{ij} will be placed at the intersection between the line with i and the column with j.

E	F	G	H	I	J	K
A	1	2	3	4	5	6
1						
2						
3						
4						
5						
6						

Enter "=1" in cell `F2` (corresponding to (1,1)). Go to cell `G2` (corresponding to (1,2)) and enter the formula "`=AVERAGE(s^(G$1+$E2-2))`". Copy the formula in `G2` and paste it on all the other empty cells of the matrix `A`. Give the name "A" to `F2:K7`.

E	F	G	H	I	J	K
A	1	2	3	4	5	6
1	1	0,546047	0,410958	0,339067042	0,29528	0,266082
2	0,546047	0,410958	0,339067	0,295280472	0,266082	0,245161
3	0,410958	0,339067	0,29528	0,266082052	0,245161	0,229281
4	0,339067	0,29528	0,266082	0,245160926	0,229281	0,216663
5	0,29528	0,266082	0,245161	0,229281135	0,216663	0,206267
6	0,266082	0,245161	0,229281	0,216662582	0,206267	0,197458

Go to an empty column and define the place for B. For instance, choose column "M". In M2, enter the formula "=AVERAGE(X)". In M3, enter the formula "=SUMPRODUCT(X;s^($E3-1))/COUNT(A:A)". Copy the formula in M3 and paste it in the rest of the empty cells of B. Give the name "B" to M2:M7.

M
B
1,511121
1,055272
0,869375
0,759076
0,686046
0,633777

Now, you have the matrices *A* and *B* and you need only to solve the linear system by one of the methods available. For instance, go to cell P2 and enter the formula "=MMULT(MINVERSE(A+epsilon* MUNIT(k));B)". The values of the coefficients appear in P2:P7.

N	O	P
epsilon	1,00E-06	**x**
k	6	0,593396
		0,886463
		0,66684
		0,324719
		0,11238
		0,061434

5.3.2 Solution Using VBA

The VBA solution for the variational approximation is analogous to the one previously presented for the collocation approach: you need just to modify the calculations determining A and B, as shown here:

```
Function variational(x, u, k, basis, epsilon)
nx = UBound(x, 1)
Dim A() As Double
ReDim A(1 To k, 1 To k)
Dim B() As Double
ReDim B(1 To k)
Dim ap As New expansion
Set ap.phi = basis
For i = 1 To k
    y_i = ap.value(i - 1, u)
    B(i) = WorksheetFunction.SumProduct(x, y_i) / nx
    A(i, i) = WorksheetFunction.SumProduct(y_i, y_i) / nx
    For j = i + 1 To k
       y_j = ap.value(j - 1, u)
       aux = WorksheetFunction.SumProduct(y_i, y_j) / nx
       A(i, j) = aux
       A(j, i) = aux
    Next j
Next i
For i = 1 To k
   A(i, i) = A(i, i) + epsilon
Next i
variational = LUSolution(A, B)
End Function
```

Then, modify two lines of the Sub `solve_collocation`:

The name and the evaluation of c.	`Sub solve_variational()`
	`c = variational(x, u, k, p, epsilon)`

Thus, the new Sub is identical to `solve_collocation`, excepted its name and a line : now the name is `solve_variational` and the line evaluation c is `c = variational( ... ).`
Run the sub. You will get results analogous to those at right and the charts below (Figs. 5.21 and 5.22).

C	D
zero	coeffs
1,00E-03	0,596991
k	0,905811
6	0,581681
epsilon	0,338912
1,00E-03	0,170914
ns	0,054456
1,00E+05	
ne	
25	

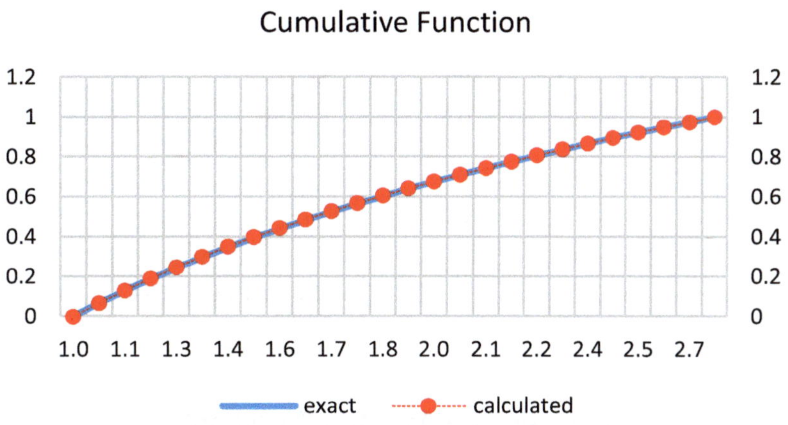

Fig. 5.21 Empirical cumulative function (VBA solution)

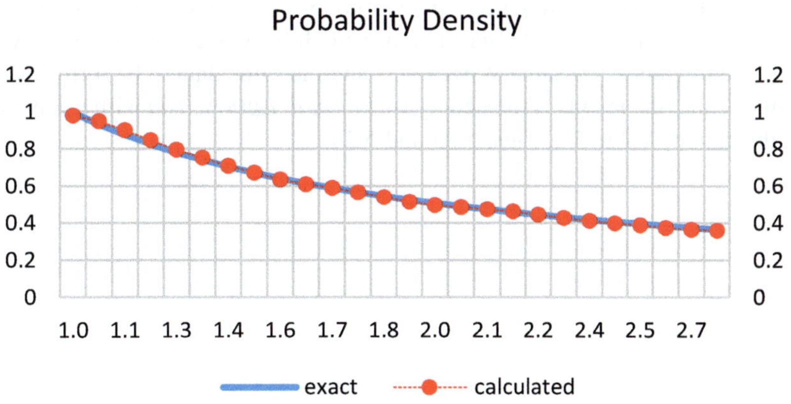

Fig. 5.22 Empirical probability density (VBA solution)

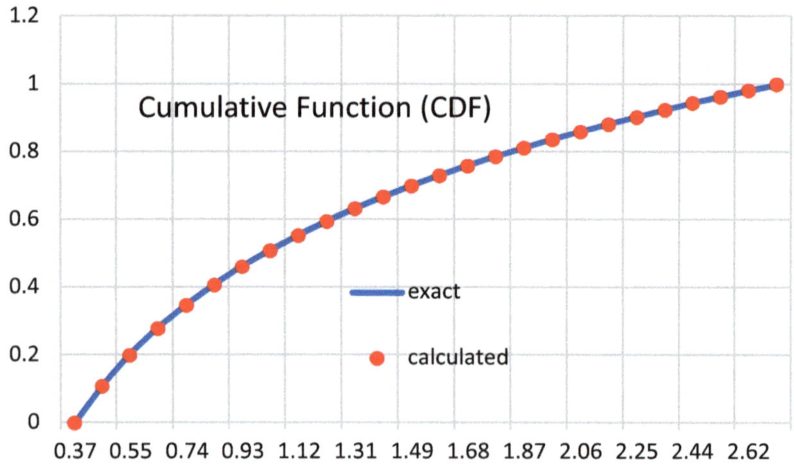

Fig. 5.23 Results furnished by the variational approximation from data in Table 5.2

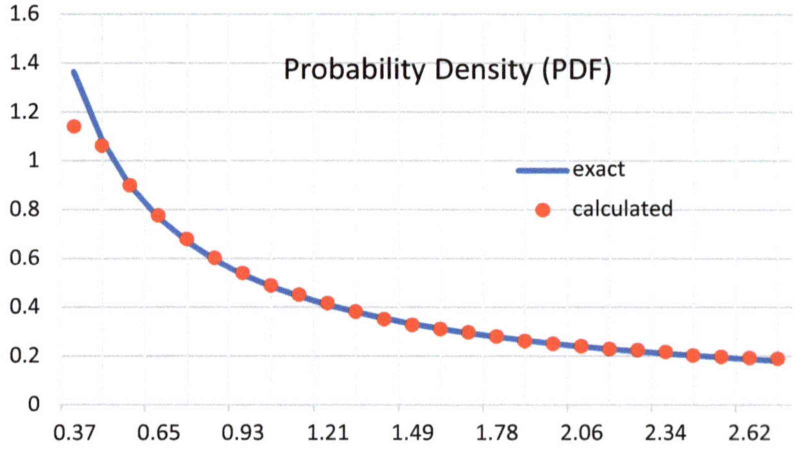

Fig. 5.24 Results furnished by the variational approximation from data in Table 5.2

5.3.3 Solution Using an Adapted Workbook

As in the preceding situation, two workbooks implement these solutions: *variational_minimize.xlsm* and *variational_LinSys.xlsm*. The use of these workbooks is entirely analogous to the use of the preceding ones: the same Worksheets and the same data to be furnished by the same way. We give in the sequel the results furnished by *variational_LinSys.xlsm* with φ_i given by Eq. (5.15) with $k = 6$ and the data of Table 5.2, Figs. 5.23 and 5.24.

Exercises

1. Consider the Walras model, where the excess demand X is a function of the price deviation U. The following data was observed:

U	1,000	0,819	0,670	0,549	0,449	0,368
X	−1,033	−0,578	−0,296	−0,169	−0,097	−0,049

U	0,301	0,247	0,202	0,165	0,135	0,111
X	−0,029	−0,014	−0,009	−0,004	−0,002	−0,001

(a) Use the data above to determine a variational representation of the excess demand as a function of the price deviation.

(b) Assume that $U \sim N(0.4, 0.3)$. Use the representation found to determine the CDF and the PDF of X.

TIP: generate a large sample from N(0.4,0.3) and use it to determine the CDF. The use the methods of Sect. 2.8 to find the derivative of the CDF.

(c) Assume that the distribution of U is unknown. Introduce an artificial uniformly distributed variable A and use it to determine a representation of U as function of A by the variational approach. Use the representation found to determine the CDF and the PDF of X.

2. Consider a Household stocks model, where the households X are a function of the propensity to consume U. The following data was observed:

U	0,0580	0,0609	0,2197	0,3883	0,4677	0,5439
X	0,0000	0,0021	0,0060	0,0078	0,0123	0,0126

U	0,7550	0,8709	0,8761	0,9045	0,9133	0,9778
X	0,0050	0,0022	0,0023	0,0019	0,0020	0,0001

(a) Use the data above to determine a variational representation of the Households as a function of the propensity to consume.

(b) Assume that $U \sim N(0.5, 0.1)$. Use the representation found to determine the CDF and the PDF of X.

TIP: generate a large sample from N(0.5,0.1) and use it to determine the CDF. The use the methods of Sect. 2.8 to find the derivative of the CDF.

(c) Assume that the distribution of U is unknown. Introduce an artificial uniformly distributed variable A and use it to determine a variational representation of U as function of A. Use the representation found to determine the CDF and the PDF of X.

3. Consider a Keynesian model of consumption, where the consumption X is a function of the propensity to consume U. The following data was observed:

U	0,2138	0,2483	0,2947	0,2951	0,3354	0,3425
X	8,7482	8,7470	7,0831	6,1580	6,8340	5,7317

U	0,3627	0,3654	0,4223	0,5067	0,5422	0,5549
X	5,1731	5,1117	4,3880	4,6680	4,9407	4,6589

(a) Use the data above to determine a variational representation of the consumption as a function of the propensity to consume.

(b) Assume that $U \sim N(0.5, 0.1)$. Use the representation found to determine the CDF and the PDF of X.

TIP: generate a large sample from N(0.5,0.1) and use it to determine the CDF. The use the methods of Sect. 2.8 to find the derivative of the CDF.

(c) Assume that the distribution of U is unknown. Introduce an artificial uniformly distributed variable A and use it to determine a representation of U as function of A by the variational approach. Use the representation found to determine the CDF and the PDF of X.

5.4 Moments Matching Method

In the preceding situations, the data was composed of pairs (U_i, X_i), so that the connection between the variables X and U was empirically contained in the data. We may also be interested in situations where such an information is missing. For instance, when we have only information about the values of X: a variability is observed, but we cannot find the cause U of the variability – or it is simply not observed. If the information about U is missing, the methods above cannot be used directly. An alternative for this situation is the *Moments Matching Method* (*M3*), which deals with situations where U is unknown. This method usually requests more data (id est, more values of X) than the preceding ones. It is also considered as difficulty to use, since it involves hard optimization problems, often requiring global optimization methods. In addition, *M3* starts by breaking the connection between the observations of X and the values of PX, since it concentrates on the connection of their probability distributions: in principle, PX is not an approximation of X: their distributions may be close, but the variables may differ significantly. Nevertheless, when *M3* is correctly used, it may solve situations where U is unknown, what justifies our interest.

From the mathematical standpoint, *M3* is based on a consequence of Levy's theorem (see page 120), which shows that is possible to approach the distribution of a random variable X by approximating its moments (see page 120). Thus, we may find an approximation of the distribution of X by using the distribution of Z such that

$$M_i(Z) = M_i(X), \quad 1 \le i \le m. \tag{5.27}$$

To find such a distribution, an arbitrary regular random variable Z may be considered – the unique limitation is the existence of m moments. Taking $Z = PX$ defined as in the preceding (see Eq. (5.3)) furnishes a non-linear system of equations

$$F_p(\boldsymbol{x}) = \mu_p(\boldsymbol{x}) - M_p(X) = 0, \quad 1 \le i \le m, \tag{5.28}$$

where

$$\mu_p(\boldsymbol{x}) = M_p(PX) = E((PX)^p) = E\left(\left(\sum_{j=1}^{k} x_j \varphi_j(U)\right)^p\right) = E((\boldsymbol{\Phi}(U)\boldsymbol{x})^p) \tag{5.29}$$

One of the interesting features of this approach is the following one: here, U appears as arbitrary – the unique limitation remains the one formulated above: the existence of the first m moments of

$Z = \mathbf{\Phi}(U)\boldsymbol{x} = \sum_{j=1}^{k} x_i \varphi_j(U)$. By this approach, the connection between U and X is generated by Eq. (5.31) and Levy's theorem, so that we may, in principle, use a variable that has no connection with X. In practice, the use of a variable which is unconnected to X (so, independent from X) often requests more information on X – *id est*, a larger sample – to obtain good results. Of course, if the information about the couple (U, X) is available – as in Table 5.2 – the data for U may be used to the fitting, with best results with smaller sets of data.

Errors in the determination of $\boldsymbol{x}$ may be interpreted as the introduction of noise: indeed, the characteristic function ϕ_X of X may be interpreted as the Fourier transform of f_X – the probability density of X. An error in the evaluation of the coefficients $\boldsymbol{x}$ generates an error in the approximation of ϕ_X by ϕ_{PX}, what is equivalent to adding a noise to f_X – observe that a small error on the PDF f_X may correspond to a large error on the CDF $F_X(x)$, which is the integral of f_X on $(-\infty, x)$. This is one of the main difficulties when using the approach by *M3*: on the one hand, an excellent precision on $\boldsymbol{x}$ is necessary; on the other hand, we must solve a hard numerical problem, which requests adapted methods. Nevertheless, as previously remarked, when these problems are correctly solved, the approach *M3* furnishes results for situations where the other approaches cannot be applied, such as, for instance, situations where only observations of X are given and U is unknown.

The approach by Moments Matching may be implemented according to different formulations, which are exposed in the sequel. We shall illustrate the method by considering the situation where U is a random variable having a uniform distribution on the interval $(-1, 1)$ and $X = e^U$. The available information is a sample of 25 values of the couple (U, X) is available (see Table 5.11).

5.4.1 The Standard Formulation of M3

If $m = k$, Eqs. (5.28) and (5.29) may be solved by methods of resolution of nonlinear equations, such as, for instance, Newton-Raphson or fixed-point methods. When $m > k$, we must look for a generalized solution, such as, for instance, the minimum of the distance between the vectors of moments

$$J(\boldsymbol{x}) = d(\boldsymbol{\mu}(\boldsymbol{x}), \boldsymbol{M}(X)), \tag{5.30}$$

Table 5.11 Data for the method M3

U	0,110263	0,818171	0,218055	−0,428266	−0,521897
X	1,11657	2,26635	1,24366	0,65164	0,59339
U	0,972777	0,498947	−0,063814	0,439192	0,899228
X	2,64528	1,64699	0,93818	1,55145	2,45770
U	0,948851	0,825434	−0,158727	−0,874752	0,472945
X	2,58274	2,282872	0,853229	0,416965	1,60471
U	−0,752068	0,217750	−0,823237	0,435408	0,028596
X	0,471391	1,24328	0,439008	1,54559	1,02901
U	0,812861	−0,2440565	0,077486	−0,453719	−0,817560
X	2,25435	0,783443	1,08057	0,635261	0,441507

where

$$\boldsymbol{\mu}(\boldsymbol{x}) = (\mu_1(\boldsymbol{x}), \ldots, \mu_m(\boldsymbol{x})) \ , \ \boldsymbol{M}(X) = (M_1(X), \ldots, M_m(X)) \tag{5.31}$$

For instance, we may use the distance

$$d_1(\boldsymbol{A}, \boldsymbol{B}) = \|\boldsymbol{A} - \boldsymbol{B}\| \ (absolute\ error). \tag{5.32}$$

In practice, we may use a pseudo-norm instead of a norm. For instance, we may consider

$$d_2(\boldsymbol{A}, \boldsymbol{B}) = \sum_{p=1}^{m} \frac{|A_i - B_i|}{|B_i|} \ (sum\ of\ relative\ errors) \tag{5.33}$$

When using EXCEL®, the SOLVER handles both the situations defined above: solution of the system of nonlinear equations (5.28) and (5.29) or minimization of J given by (5.32) and (5.33). The second one corresponds to a standard minimization procedure – it is implemented in the workbook *m3_minimize.xlsm*.

The main difficulty in this approach is to obtain a good approximation of the minimizers of J defined by (5.30). Indeed, this function is characterized by its strong nonconvexity, so that it does not satisfy one of the main assumptions in the convergence results of the standard methods of optimization – the convexity of the objective function – here, of J. In the lack of convexity, the standard methods of deterministic optimization often fail unless for initial points carefully chosen – in fact, for initial points near the solution.

To overcome the difficulty of convergence, global optimization methods were proposed by many authors. The SOLVER offers some Global optimization methods: MultiStart and Evolutionary. In the Workbook *m3_minimize.xlsm*, a combination of those methods is used: initially, GRG multistart is used and, then, evolutionary method is applied to the result furnished by GRG. This procedure is iterated according to the choice of the user. Between two iterations, stochastic perturbations are used to improve the result (see, for instance, Pogu & Souza de Cursi, 1994; Autrique & Souza de Cursi, 1997; El Mouatasim et al., 2006; Bouhadi et al., 2004a, b). If good initial points are furnished, perturbations may be ignored in the minimization procedure.

Let us start by the use of a basic worksheet – analogously to the preceding approaches, Moment Matching may be implemented in a worksheet without using VBA.

The first steps are the same as in the preceding: open a blank workbook, enter the data `X`, `U`; determine `lb`, `ub` and generate `s`. Remember that names `U`, `X`, `s`, `lb`, `ub` must be defined. The worksheet will be like this one.

	A	B	C	D
1	X	U	lb	s
2	1,116572	0,110263	-0,87475	0,533153
3	2,26635	0,818171	ub	0,916317
4	1,243655	0,218055	0,972777	0,591496
5	0,651638	-0,42827		0,241666
6	0,593394	-0,5219		0,190987
7	2,645282	0,972777		1
8	1,646986	0,498947		0,743533
9	0,938179	-0,06381		0,438931
10	1,551453	0,439192		0,71119
11	2,457705	0,899228		0,96019
12	2,58274	0,948851		0,987049

Now, enter the values of k and m. In column `F`, enter the initial guess for the coefficients. For instance, you can use a random starting point by entering in `F2` the formula "`=RANDARRAY(k;1)`". Give them the name "`c`".
Then, evaluate `PX`: go to cell `G2` and enter the formula "`=$F$2+$F$3*s +$F$4*s^2+$F$5*s^3+$F$6*s^4+$F$7*s^5`".
Give the name "`PX`" to the data in column `G`.

E	F	G
k	c	PX
6	0,877679	1,502141
m	0,636496	2,782315
8	0,558827	1,630476
	0,485125	1,073141
	0,615786	1,023843
	0,069391	3,243305
		2,063265
		1,329733
		1,957662
		3,013585
		3,166414
		2,802146
		1,251034
		0,877679

In the next step, evaluate the moments: in column `H`, enter the indexes of the moments to evaluate. In `I2`, enter the formula "`=AVERAGE(X^H2)`". In `J2`, enter "`=AVERAGE(PX^H2)`". In `K2`, enter "`=I2-J2`" (absolute error). In `L2`, enter "`=K2/I2`" (relative error).
Copy `I2` and paste on the other cells of columns `I` up to the last moment to be evaluated.
Repeat the operation in columns `J, K, L`: copy the cell in line 2 and paste on the rest of the column.

H	I	J	K	L
index	M_i(X)	M_i(PX)	absolute	relative
1	1,311006	0,892385	0,418621	0,319313
2	2,238939	1,073621	1,165318	0,520478
3	4,48846	1,524562	2,963898	0,660337
4	9,837174	2,365088	7,472086	0,759577
5	22,67403	3,854609	18,81942	0,829999
6	53,85514	6,468971	47,38617	0,879882
7	130,3715	11,05974	119,3118	0,915168
8	319,669	19,147	300,522	0,940104
9	791,0416	33,44902	757,5926	0,957715
10	1971,122	58,83968	1912,283	0,970149
11	4938,886	104,0819	4834,804	0,978926
12	12431,88	184,9716	12246,91	0,985121

Evaluate the RMS and the maximum of columns `K` and `L`. For instance, the formula in `M2` is "`=SQRT(SUMSQ(K2:K9)/m)`". In `M4`, the formula is "`=MAX(ABS(K2:K9))`" (only the 8 first moments are considered).
Define the objective, which may be one of the cells `M2, M4, N2, N4` or a function combining the values of these cells. Here, we chose `N4` as objective.

M	N	O
RMS abs	RMS rel	objective
115,7662	0,755834	0,940104
max abs	max rel	
300,522	0,940104	

Run the SOLVER to minimize the objective (cell `O2`) by changing the values of `c`. (cells `F2:F7`). As observed, it is a hard optimization problem: we use the method evolutionary.

F
c
0,197553
0,363094
6,914528
-0,28643
-0,15721
-7,27257

M	N	O
RMS abs	RMS rel	objective
6,26364	0,041827	0,054494
max abs	max rel	
17,41995	0,054494	

The results may be improved by using a better starting point. For instance, we may use as initial point the solution furnished by collocation (calculate it elsewhere in the worksheet)

cini	0,41692
	0,773144
	0,683958
	0,541577
	0,028757
	0,200897

Using the starting point furnished by collocation, we obtain the results at right.

F
c
0,41691
0,773143
0,683933
0,541629
0,028902
0,200734

M	N	O
RMS abs	RMS rel	objective
0,000336	2,17E-06	2,94E-06
max abs	max rel	
0,000939	2,94E-06	

We may also change the objective. For instance, we considered `M4` as objective.

F
c
0,416881
0,773112
0,685088
0,539679
0,028977
0,201533

M	N	O
RMS abs	RMS rel	objective
3,13E-06	7,86E-07	3,9E-06
max abs	max rel	
3,9E-06	1,95E-06	

The distribution obtained is exhibited in the Figs. 5.25 and 5.26:

The use of VBA for the determination of the solution by Moment Matching allows to make automatic the operations described in the previous section. We may also generate automatically the

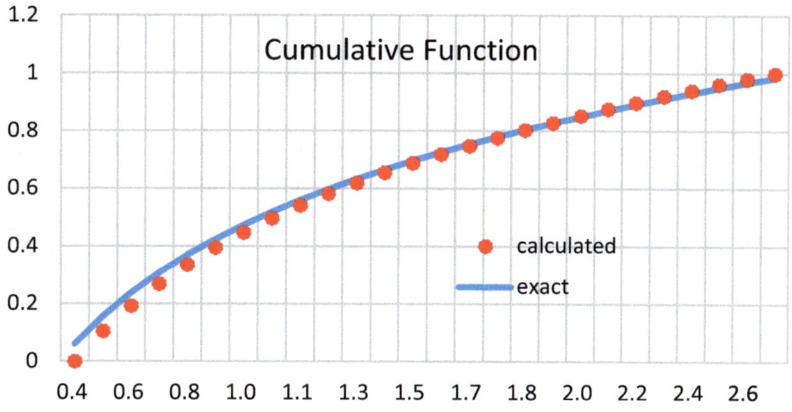

Fig. 5.25 Results furnished by the Moment Matching Method for data in Table 5.11

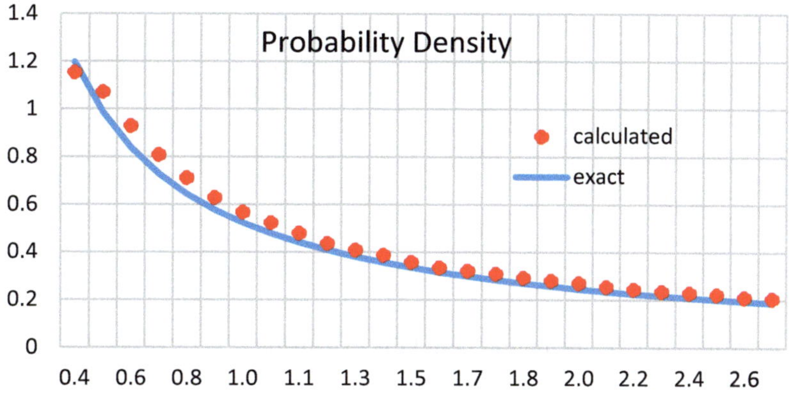

Fig. 5.26 Results furnished by the Moment Matching Method for data in Table 5.11

charts containing the CDF and the PDF calculated. For instance, we may create a workbook particularly adapted to this situation, as the workbook *m3_minimize.xlsm* – its use is analogous to the use of the preceding ones:

1. Insert the data corresponding to X_i – for instance, in the columns A and B of the Worksheet *Data*, with X_1 at A2, U_1 at B2, X_2 at A3, U_2 at B3, and so on. If no data is inserted in column B, you must choose a distribution for U in the corresponding *ComboBox*. If no data is inserted in the column C, the workbook will generate an initial point.

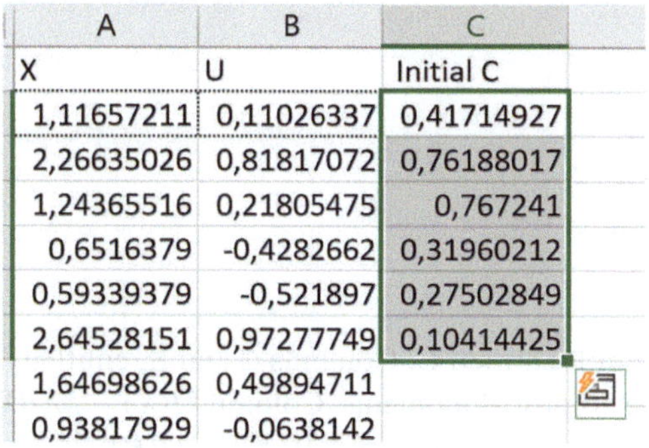

A	B	C
X	U	Initial C
1,11657211	0,11026337	0,41714927
2,26635026	0,81817072	0,76188017
1,24365516	0,21805475	0,767241
0,6516379	-0,4282662	0,31960212
0,59339379	-0,521897	0,27502849
2,64528151	0,97277749	0,10414425
1,64698626	0,49894711	
0,93817929	-0,0638142	

Worksheet *Data* containing information about X, U and $\boldsymbol{x}$ ($\boldsymbol{C}$ corresponds to $\boldsymbol{x}$)

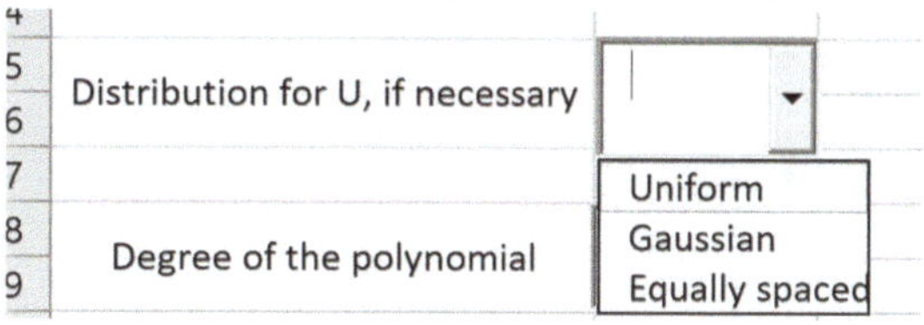

Combo Box in Worksheet *Commands* for the choice of the distribution of U – to be used when U is unknown

2. Fill the other data in the Worksheet *Commands*: degree of the polynomial k; number of subintervals for the CDF, number of moments m. The ranges corresponding to X, U and $\boldsymbol{x}$ will be automatically selected by the Workbook.

3. After entering all the data, the fields in the Worksheet *Commands* are filled. A typical situation is the following – it corresponds to the preceding data:

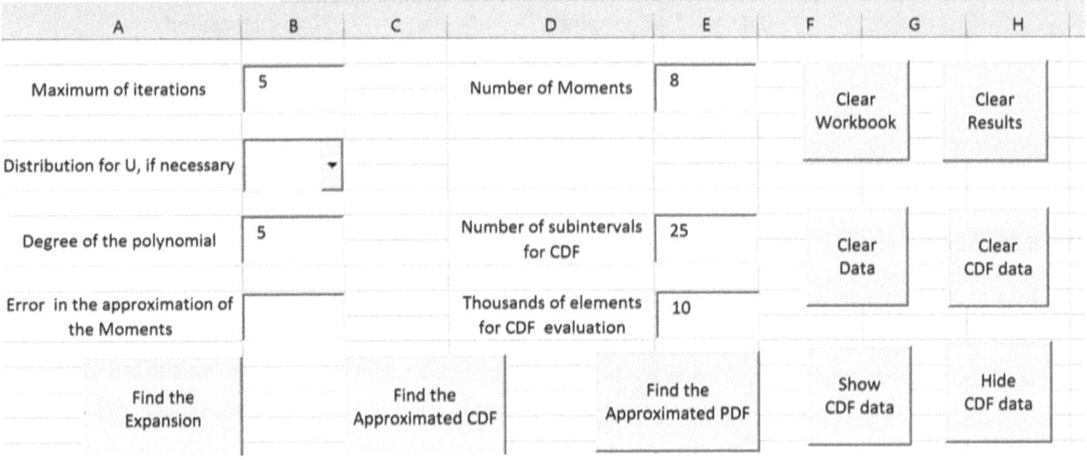

4. Now, click on the button "Find the expansion" 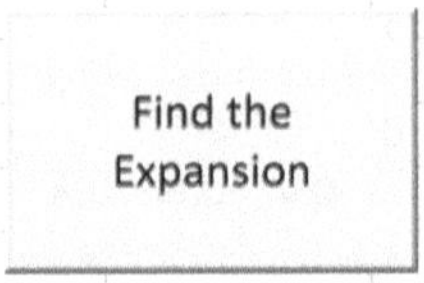

to get the results: this button generates $\boldsymbol{x}$. Then, go to the Worksheet *Results*. The results appear as below:

	A	B	C	D	E	F	G	H	I	J
1	X		U		C		M(X)	M(PX)	M(X)-M(PX)	Rel err
2	1,11657211		0,11026337		0,41810054		1,31100552	1,31121582	-0,0002103	0,00016041
3	2,26635026		0,81817072		0,76026945		2,23893853	2,23910984	-0,0001713	7,6516E-05
4	1,24365516		0,21805475		0,76620612		4,48845983	4,48850178	-4,194E-05	9,3448E-06
5	0,6516379		-0,4282662		0,32051669		9,8371736	9,83706117	0,00011243	1,1429E-05
6	0,59339379		-0,521897		0,27577962		22,6740309	22,673817	0,00021393	9,4348E-06
7	2,64528151		0,97277749		0,10428688		53,8551403	53,855015	0,00012535	2,3275E-06
8	1,64698626		0,49894711				130,371509	130,371735	-0,0002265	1,7371E-06
9	0,93817929		-0,0638142				319,668986	319,668932	5,4367E-05	1,7007E-07
0	1,55145294		0,43919187							
1	2,45770474		0,89922788						2,0369E-07	3,1902E-08
2	2,58274033		0,94885098							
3	2,2828716		0,82543413							
4	0,85322946		-0,1587268							
5	0,4169654		-0,874752							
6	1,60471367		0,47294534							
7	0,47139088		-0,7520676							

5. The coefficients of the expansion appear in the column E entitled $\boldsymbol{C}$. Columns G and H, entitled $M(X)$ and $M(PX)$respectively, contains the values of the empirical moments. Column $M(X) - M(PX)$ shows the difference and the last cell in this column shows the value of $\|\boldsymbol{M}(\boldsymbol{X}) - \boldsymbol{M}(\boldsymbol{PX})\|^2$.

6. Come back to the Worksheet *Commands* and press the button "Find the Approximated CDF"

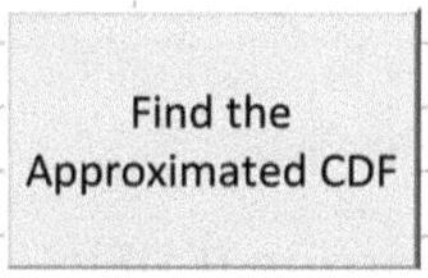

. The evaluation may require some time, depending on the number of variates selected in the corresponding TextBox of Worksheet *Commands*. Now, the Worksheet *Results* contains new data in columns K and L, containing the abscissa *xe* and the values of the CDF *Fe* at these abscissa:

K	L
xe	CDF Fe
0,3700903	0
0,46400477	0,1124
0,55791925	0,2018
0,65183373	0,2784
0,74574821	0,3466
0,83966268	0,4061
0,93357716	0,4616
1,02749164	0,5127
1,12140612	0,5521
1,21532059	0,5937
1,30923507	0,6327
1,40314955	0,6659
1,49706403	0,7019
1,5909785	0,7336

7. Press the button "Find the Approximated PDF"

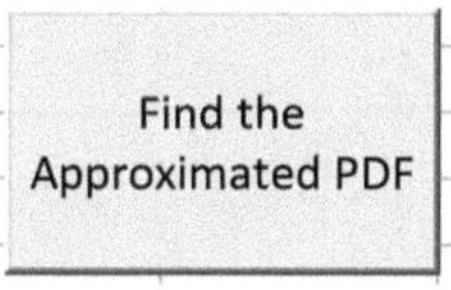

: one new column is added to the Worksheet *Results*, containing the values of the CDF *dFe* at the same abscissa:

K	L	M
xe	CDF Fe	PDF dFe
0,3700903	0	1,18670726
0,46400477	0,1124	1,08203466
0,55791925	0,2018	0,88949851
0,65183373	0,2784	0,77186286
0,74574821	0,3466	0,68084994
0,83966268	0,4061	0,61321305
0,93357716	0,4616	0,56582952
1,02749164	0,5127	0,48328545
1,12140612	0,5521	0,4332621
1,21532059	0,5937	0,42733915
1,30923507	0,6327	0,38559018
1,40314955	0,6659	0,36875305
1,49706403	0,7019	0,36014593
1,5909785	0,7336	0,34479508

The worksheet *Results* contain other information, such as the values of $PX(U_i)$ and the quadratic error between the vector $(PX(U_1), \ldots, PX(U_n))$ and the vector $(X_1, \ldots, X_n)$.

We use $k = 6$, $m = 8$. A first run of the method M3 generates a starting point and, then, we run the method again. The results are Shown in Table 5.12 and Figs. 5.27 and 5.28 – they are comparable to those furnished by the other approaches.

Now, let us suppose that the information about U is missing: only the values of X are available. In this situation, we must create an artificial variable U. For instance, we may generate a sample from an arbitrary random variable – for instance, uniformly distributed or normally distributed and use these values to fit the distribution of X. To create an artificial connection between U and X, both vectors are ordered into an increasing order and the method is applied to the resulting data. As an example, the

Table 5.12 Coefficients obtained by the Moments Matching from data in Table 5.11

i	1	2	3	4	5	6
x	0.418	0.753	0.799	0.293	0.261	0.122

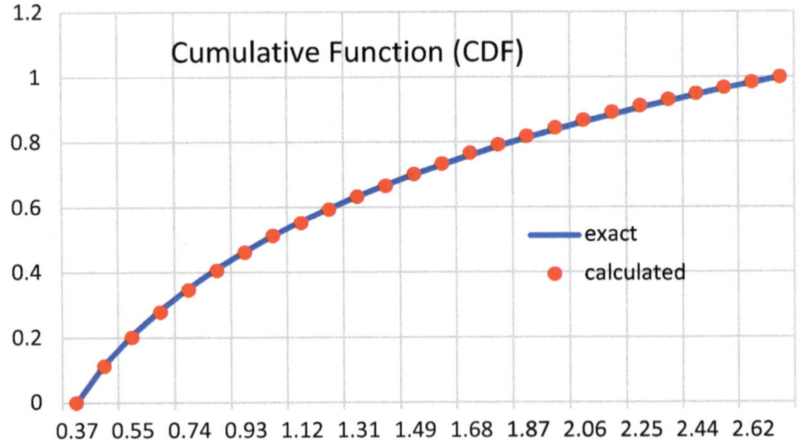

Fig. 5.27 CDF generated by Moments Matching for the data of Table 5.11

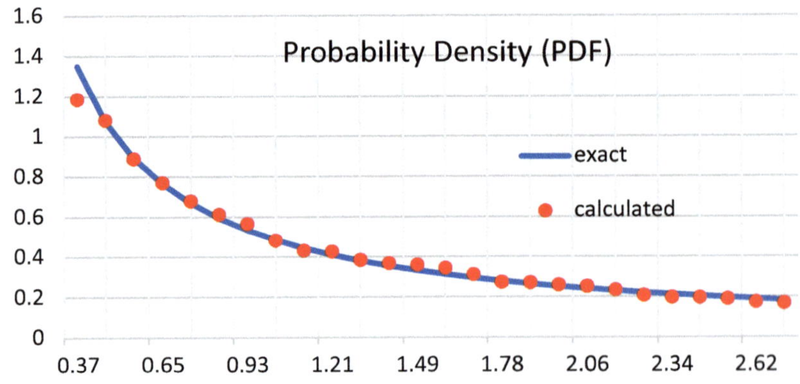

Fig. 5.28 PDF generated by Moments Matching for the data of Table 5.11

results obtained furnished by the simple workbook, using a gaussian variable of distribution $N(0, 1)$ are shown in the Figs. 5.29 and 5.30:

May the lack of information about U be compensated by more information about X? Let us consider a sample of $n = 50$ variates from X. The results furnished by the adapted workbook are shown in Figs. 5.31 and 5.32, Table 5.13.

As observed above, U is arbitrary: for instance, we may use a equally spaced grid: $U_i = (i - 1)/10$. The results are shown in Figs. 5.33 and 5.34, Table 5.14.

As we can see from these examples, Moment Matching may furnish information about the distribution of X, even in the situation where U is unknown. But these examples show also that the computational cost may be high: several iterations of the solver are necessary to obtain good results: *M3* is highly dependent of the quantity of data and of the quality of the optimization.

Fig. 5.29 CDF generated by Moments Matching (simple workbook) for the data of Table 5.11, with U unknown (a gaussian variable N(0,1) was used)

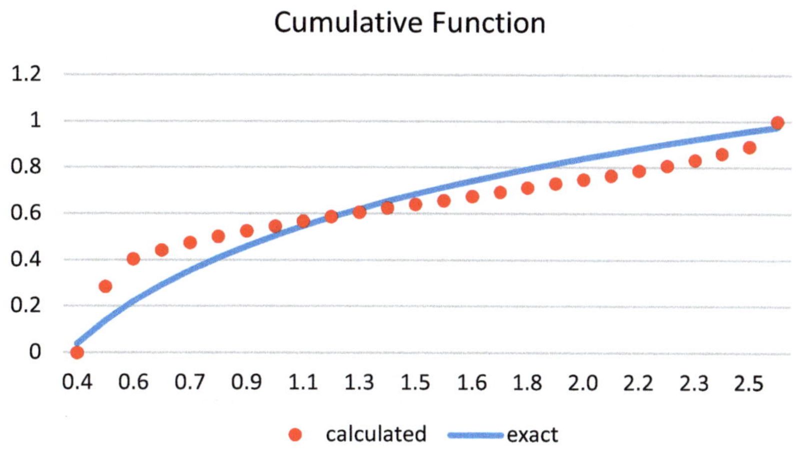

Fig. 5.30 PDF generated by Moments Matching (simple workbook) for the data of Table 5.11, with U unknown (a gaussian variable $N(0, 1)$ was used)

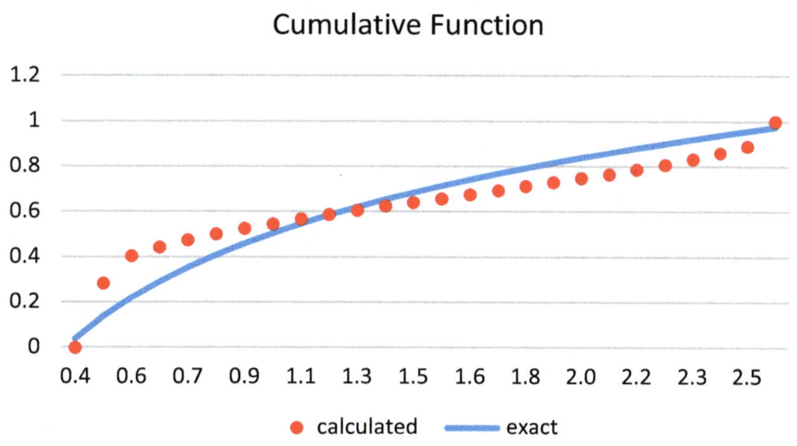

Fig. 5.31 CDF generated by Moments Matching for a sample of 50 values of X, with U unknown (a gaussian variable N(0,1) was used)

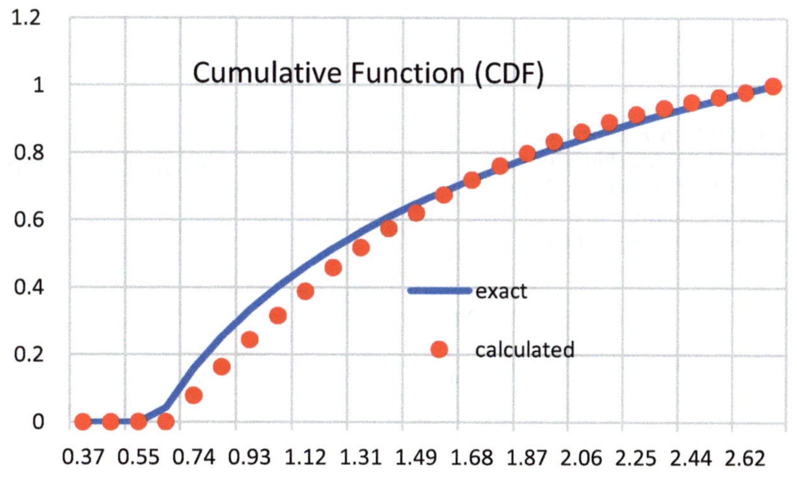

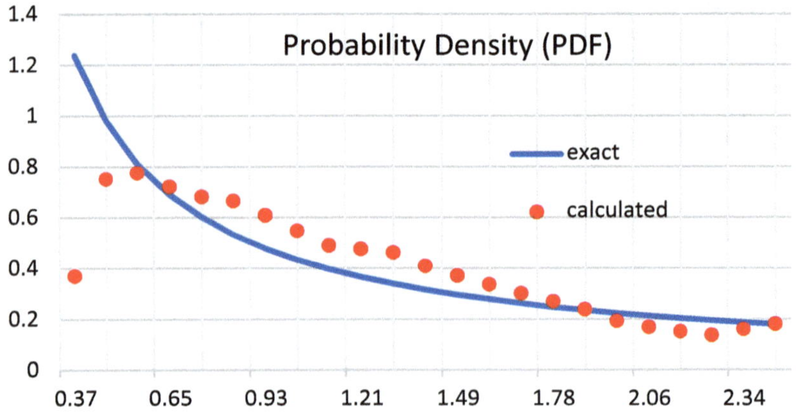

Fig. 5.32 PDF generated by Moments Matching for a sample of 50 values of X, with U unknown (a gaussian variable $N(0, 1)$ was used)

Table 5.13 Coefficients obtained by the Moments Matching, for a sample of 50 values of X, with U unknown (a gaussian variable $N(0, 1)$ was used)

i	1	2	3	4	5	6
x	0,450	0,128	1,325	13,94	10,24	0,378

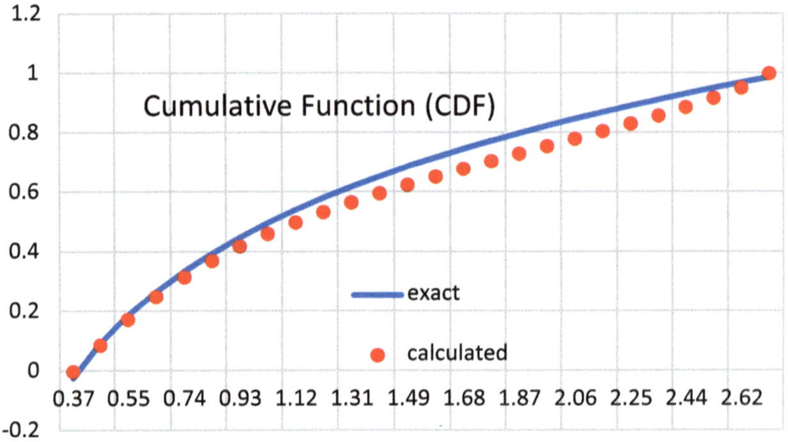

Fig. 5.33 CDF generated by Moments Matching, for a sample of 50 values of X, with U unknown (an uniform grid is used)

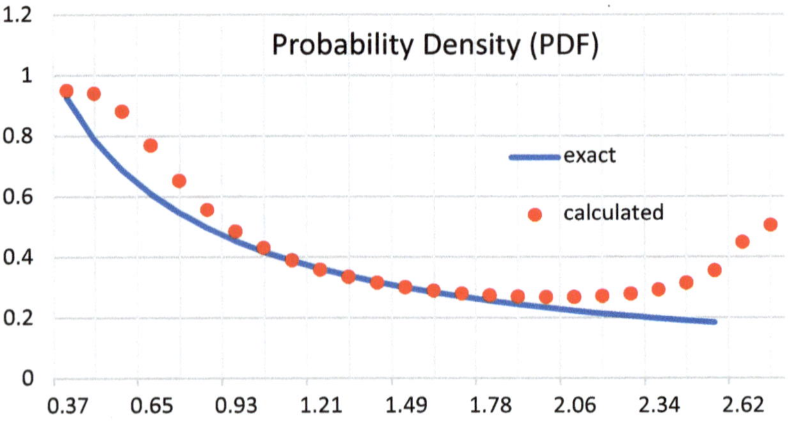

Fig. 5.34 PDF generated by Moments Matching, for a sample of 50 values of X, with U unknown (an uniform grid was used)

Table 5.14 Coefficients obtained by the Moments Matching, for a sample of 50 values of X, with U unknown (a gaussian variable N(0,1) was used)

i	1	2	3	4	5	6
x	0.418	0.753	0.799	0.293	0.261	0.122

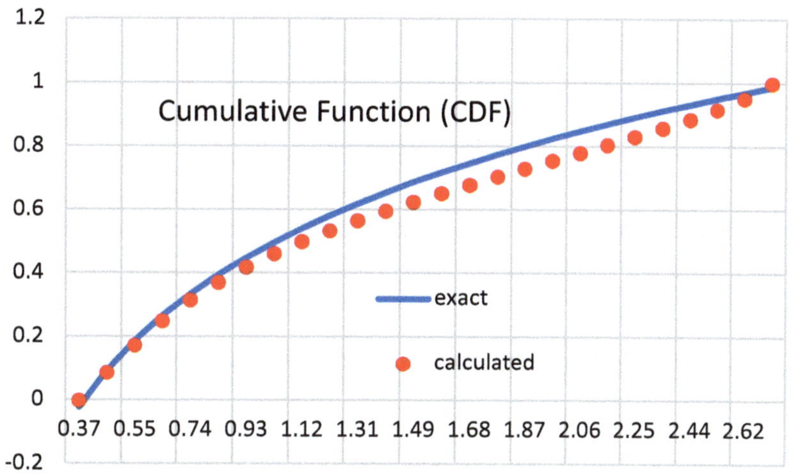

Fig. 5.35 CDF generated by Moments Matching (formulation by constrained optimization), with U unknown (an uniform grid is used)

5.4.2 Constrained Optimization Formulation of M3

As previously observed, *M3* may furnish solutions where only the distribution of *PX* is close to the distribution of *X*, but *PX* may differ significantly from *X*. An attempt to combine the approximation of *X* with the standard formulation of *M3* consists in considering Eq. (5.28) as constraints in the determination of the best approximation of *X*: let $\boldsymbol{A}$ and $\boldsymbol{B}$ be the collocation matrices given by Eqs. (5.8), (5.9), and (5.10): a good approximation minimizes the distance between the approximated values $\boldsymbol{Ax}$ and the observed values $\boldsymbol{B}$. By introducing a second distance (or pseudo-distance) δ, we may look for

$$\boldsymbol{x} = \arg\min\left\{\delta(\boldsymbol{Ay}, \boldsymbol{B}) : \quad F_p(\boldsymbol{y}) = 0, 1 \le p \le m, \boldsymbol{y} \in \mathbb{R}^k \right\}$$

Analogously to the standard formulation, a pseudo-distance may be used instead of a distance. In this formulation, the equality of the moments appears as a set of nonlinear constraints of optimization. As in the standard formulation, an alternative formulation is

$$\boldsymbol{x} = \arg\min\left\{\delta(\boldsymbol{Ay}, \boldsymbol{B}) : d(\boldsymbol{\mu}(\boldsymbol{y}), \boldsymbol{M}(X)) = 0, \boldsymbol{y} \in \mathbb{R}^k \right\}$$

In practice, it may be convenient to introduce some flexibility by considering $\varepsilon > 0$ and

$$\boldsymbol{x} = \arg\min\left\{\delta(\boldsymbol{Ay}, \boldsymbol{B}) : d(\boldsymbol{\mu}(\boldsymbol{y}), \boldsymbol{M}(X)) \le \varepsilon, \boldsymbol{y} \in \mathbb{R}^k\right\}$$

This approach is implemented in the Workbook *m3_cont.xlsm*, which is analogous to the preceding ones. In the sequel, we present the results furnished by this workbook with $d(\boldsymbol{A}, \boldsymbol{B}) = \|\boldsymbol{A} - \boldsymbol{B}\|$ (Figs. 5.35 and 5.36).

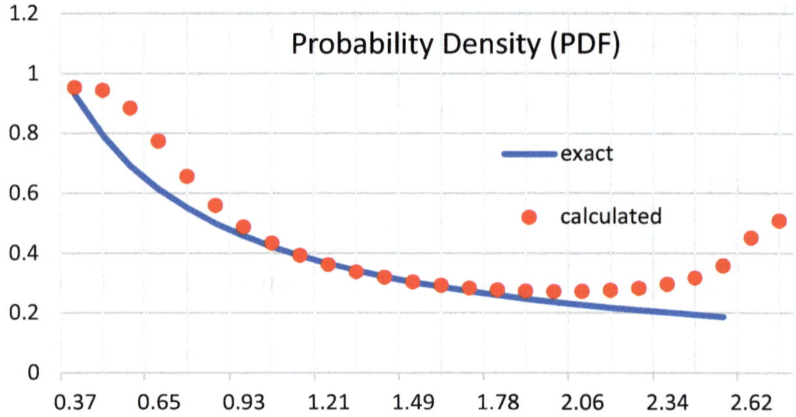

Fig. 5.36 PDF generated by Moments Matching (formulation by constrained optimization), with U unknown (an uniform grid is used)

5.4.3 Multiobjective Optimization Formulation of M3

The formulation of *M3* by constrained optimization may be reinterpreted in terms of multiobjective optimization: indeed, we may consider that we have two objectives – which may be eventually conflicting: on the one hand, to minimize $\delta(\mathbf{Ax}, \mathbf{B})$; on the other hand, to minimize $d(\boldsymbol{\mu}(\mathbf{x}), \mathbf{M}(X))$. Thus, we may consider the bi-objective optimization problem

$$\mathbf{x} = \arg\min \left\{ \begin{pmatrix} d(\mathbf{Ay}, \mathbf{B}) \\ \delta(\boldsymbol{\mu}(\mathbf{y}), \mathbf{M}(X)) \end{pmatrix} : \mathbf{y} \in \mathbb{R}^k \right\}$$

As usual in multiobjective optimization, the solution takes the form of a Pareto's front which must be constructed by an adapted method. When using the SOLVER from EXCEL®, no option allows the solution of multiobjective problems – the multiobjective problem must be reduced to a single-objective problem. This may be achieved by several ways: for instance, by using the method exposed in (Souza de Cursi, 2015; Zidani, 2013), which furnishes the complete Pareto's front. However, in the context of UQ, we are not interested in all the possible solutions, but in a single one, so that we may consider as single objective

$$J(\mathbf{x}) = \alpha_1 d(\mathbf{Ay}, \mathbf{B}) + \alpha_2 \delta(\boldsymbol{\mu}(\mathbf{y}), \mathbf{M}(X))$$

where α_1, α_2 are coefficients representing the weights of each objective. These coefficients may also be used to scale the objectives to give an equilibrated value of each term. This approach is implemented in the Workbook *m3_agreg.xlsm*, which is analogous to the preceding ones. In the sequel, we present the results furnished by this workbook with $\alpha_1 = 1$, $\alpha_2 = 100$, $\delta(\mathbf{A}, \mathbf{B}) = d(\mathbf{A}, \mathbf{B}) = \|\mathbf{A} - \mathbf{B}\|$. We use an uniform grid for U (Figs. 5.37 and 5.38).

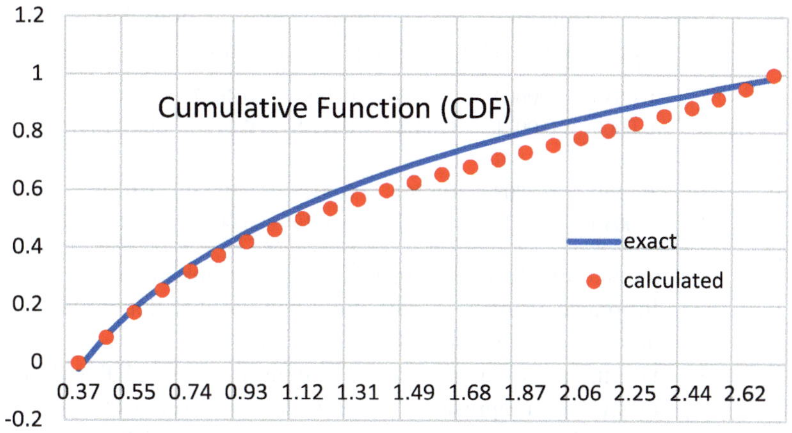

Fig. 5.37 CDF generated by Moments Matching (formulation by aggregation of two objectives), with U unknown (an uniform grid is used)

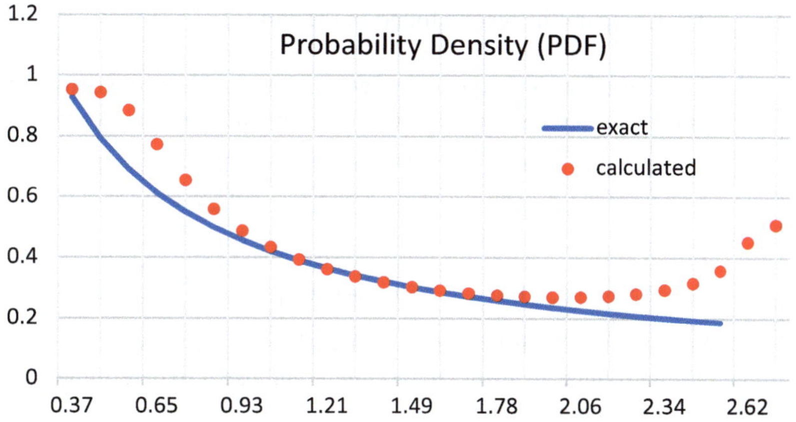

Fig. 5.38 PDF generated by Moments Matching (formulation by aggregation of two objectives), with U unknown (an uniform grid is used)

Exercises

1. Consider a Walras model, where the excess demand X is a function of the price deviation U. The following data was observed:

U	1,000	0,819	0,670	0,549	0,449	0,368
X	−1,033	−0,578	−0,296	−0,169	−0,097	−0,049

U	0,301	0,247	0,202	0,165	0,135	0,111
X	−0,029	−0,014	−0,009	−0,004	−0,002	−0,001

Assume that the distributions of X and U are unknown.

(a) Introduce an artificial variable A and use it to determine a representation of the excess demand as a function of A.
(b) Use the representation found to determine the CDF and the PDF of X.
(c) Determine a representation of X as a function of U.
(d) Determine a representation of U as function of A.
(e) Use the representations found in (c) and (d) to determine the CDF and the PDF of X
(f) Compare the results found in (b) and (e).

2. Consider a Household stocks model, where the households X are a function of the propensity to consume U. The following data was observed:

U	0,0580	0,0609	0,2197	0,3883	0,4677	0,5439
X	0,0000	0,0021	0,0060	0,0078	0,0123	0,0126

U	0,7550	0,8709	0,8761	0,9045	0,9133	0,9778
X	0,0050	0,0022	0,0023	0,0019	0,0020	0,0001

Assume that the distributions of X and U are unknown.

(a) Introduce an artificial variable A and use it to determine a representation of the excess demand as a function of A.
(b) Use the representation found to determine the CDF and the PDF of X.
(c) Determine a representation of X as a function of U.
(d) Determine a representation of U as function of A.
(e) Use the representations found in (c) and (d) to determine the CDF and the PDF of X
(f) Compare the results found in (b) and (e).

3. Consider a Keynesian model of consumption, where the consumption X is a function of the propensity to consume U. The following data was observed:

U	0,2138	0,2483	0,2947	0,2951	0,3354	0,3425
X	8,7482	8,7470	7,0831	6,1580	6,8340	5,7317

U	0,3627	0,3654	0,4223	0,5067	0,5422	0,5549
X	5,1731	5,1117	4,3880	4,6680	4,9407	4,6589

Assume that the distributions of X and U are unknown.

(a) Introduce an artificial variable A and use it to determine a representation of the excess demand as a function of A.
(b) Use the representation found to determine the CDF and the PDF of X.
(c) Determine a representation of X as a function of U.
(d) Determine a representation of U as function of A.
(e) Use the representations found in (c) and (d) to determine the CDF and the PDF of X
(f) Compare the results found in (b) and (e).

5.5 Multidimensional Expansions

The methods presented extend straightly to the situation where X or U are multidimensional.

5.5.1 Case Where U Is Multidimensional

In the situation where $\boldsymbol{U} = (U_1, \ldots, U_n)$ is a multidimensional vector, the family $\{\varphi_i : 1 \leq i \leq k\}$ is generated by functions $\varphi_{i,\,j} : \mathbb{R}^n \to \mathbb{R}$. For instance, we may consider a tensor product of basis $\{\varphi_{i,\,j} : 1 \leq i \leq k_j\}$:

$$\varphi_{\ell_n(i_1,\ldots,i_n)}(U) = \prod_{j=1}^{n} \varphi_{i_j,j}(U_j).$$

The map ℓ_n defines a renumbering of $(i_1, \ldots, i_n)$ which brings the vector to a single integer. For instance,

$$\ell_2(i_1, i_2) = i_1 + (i_2 - 1)k_1.$$

and

$$\ell_n(i_1, .., i_n) = \ell_{n-1}(i_1, \ldots, i_{n-1}) + (i_n - 1)\ell_{n-1}(k_1, \ldots, k_{n-1}).$$

For instance,

$$\ell_3(i_1, i_2, i_3) = \ell_2(i_1, i_2) + (i_3 - 1)\ell_2(k_1, k_2).$$

Then, the approximation reads as

$$X \approx PX = \boldsymbol{\Phi}(\boldsymbol{U})\boldsymbol{x} = \sum_{\ell_n=1}^{k} x_{\ell_n}\varphi_{\ell_n}(\boldsymbol{U}), \tag{5.34}$$

id est,

$$X \approx PX = \sum_{i_1=1}^{k_1}\sum_{i_2=1}^{k_2}\cdots\sum_{i_n=1}^{k_n} x_{\ell_n}\varphi_{1,i_1}(U_1)\varphi_{2,i_2}(U_2)\ldots\varphi_{n,i_n}(U_n).$$

Example

Let us consider the situation where the data is the following one:

X	4,45	2,32	2,36	3,28	3,32	2,54	4,45	2,02
U_1	0,95	0,57	0,15	1,00	0,18	0,38	0,95	0,32
U_2	−0,63	0,59	−0,18	0,56	−0,75	−0,07	−0,63	0,44

X	3,28	1,74	1,93	2,79	2,74	4,98	2,19	3,09
U_1	0,65	0,27	0,34	0,30	0,37	0,92	0,24	0,88
U_2	−0,31	0,84	0,64	−0,36	−0,26	−0,90	0,09	0,40

X	1,65	1,86	3,33	3,81	3,73	2,08	2,73	4,03
U_1	0,19	0,24	0,81	0,93	0,29	0,38	0,59	0,78
U_2	0,82	0,52	−0,07	−0,25	−0,87	0,49	0,08	−0,61

Let us consider an approximation by a polynomial family, with $k_1 = k_2 = 4$. Then, taking $a_i = \min(U_i)$, $b_i = \max(U_i)$.

$$\varphi_{1,i}(u_1) = \left(\frac{u_1 - a_1}{b_1 - a_1}\right)^{i-1}, \quad \varphi_{2,j}(u_2) = \left(\frac{u_2 - a_2}{b_2 - a_2}\right)^{j-1}.$$

and

$$\varphi_{\, j+4i-4}(u_1, u_2) = \varphi_{1,i}(u_1)\, \varphi_{2,j}(u_2).$$

Here, the numbering is

$$\ell_2(i_1, i_2) = i_2 + (i_1 - 1)k^2.$$

The collocation approach furnishes the coefficients

ℓ_2	1	2	3	4	5	6	7	8
i	1	1	1	1	2	2	2	2
j	1	2	3	4	1	2	3	4
X_{ℓ_2}	3,61	−4,01	2,69	−0,70	1,14	−1,30	3,02	−1,89

ℓ_2	9	10	11	12	13	14	15	16
i	3	3	3	3	4	4	4	4
j	1	2	3	4	1	2	3	4
X_{ℓ_2}	0,05	2,96	−6,89	4,27	0,38	−2,08	5,09	−3,27

The RMS error of the approximation is 2.5E-3. ∎

A simple way to implement the creation of a tensor basis for two-dimensional U is given in `tensorbasis.cls`:

```
Public phi1 As Variant
Public phi2 As Variant
Public indbase As Variant
Function value(k, x)
  aux = indbase(k + 1)
  i = aux(1)
  j = aux(2)
  value = phi1.value(i, x(1)) * phi2.value(j, x(2))
End Function
Function der1(k, x)
  aux = indbase(k + 1)
  i = aux(1)
  j = aux(2)
  der1 = phi1.d1(i, x(1)) * phi2.value(j, x(2))
End Function
Function der2(k, x)
  aux = indbase(k + 1)
  i = aux(1)
  j = aux(2)
  der2 = phi1.d1(i, x(1)) * phi2.d1(j, x(2))
End Function
```

```
Function create_indexes(k1, k2)
Dim v() As Variant
ReDim v(1 To k1 * k2)
k = 0
Dim vv(1 To 2)
For i = 0 To k1 - 1
  For j = 0 To k2 - 1
    k = k + 1
    vv(1) = i
    vv(2) = j
    v(k) = vv
  Next j
Next i
create_indexes = v
End Function
```

To use this class, you must start by importing it in your workbook (see 1.6). Then, you must add a sub that creates two basis `phi1` and `phi2`, creates the tensor product basis and generates the indexes `indbase`. For instance, for a polynomial family, the sub must contain the code below:

```
Dim p1 As New poly
p1.lb = lb1
p1.ub = ub1
p1.zero = zero
Dim p2 As New poly
p2.lb = lb2
p2.ub = ub2
p2.zero = zero
```

```
Dim p As New tensorbasis
Set p.phi1 = p1
Set p.phi2 = p2
indb = p.create_indexes(k1, k2)
p.indbase = indb
```

In this case, your code must previously define `lb1`, `ub1`, `lb2`, k1, k2 which corresponds to a_1, b_1, a_2, b_2, k_1, k_2, respectively. Since you use the polynomial basis, you need to import `poly.cls` too. An example of use is given in 6.1.

5.5.2 Case Where X Is Multidimensional

Let us assume that $\boldsymbol{X} = (X_1, \ldots, X_n)^t$. Then the approximation reads as

$$\boldsymbol{X} \approx \boldsymbol{PX} = \sum_{j=1}^{k} \boldsymbol{x_j}\varphi_j(U), \boldsymbol{x_j} = \left(x_{1,j}, \ldots, x_{n,j}\right)^t. \tag{5.35}$$

Thus, *each component* is approximated by an expansion analogous to (5.3) or (5.34):

$$X_i \approx PX_i = \sum_{j=1}^{k} x_{i,j}\varphi_j(U), i = 1, \ldots, n$$

Consequently, we can apply the methods presented to each component of the variable.

The coefficients $\boldsymbol{x}$ form a matrix $\boldsymbol{x} = (x_{i,\ j},\ 1 \leq i \leq n,\ 1 \leq j \leq k)$ – notice that each line of $\boldsymbol{x}$ contains the coefficients corresponding to the expansion of a component: column j of $\boldsymbol{x}$ contains the coefficients of the expansion of X_i:

$$\boldsymbol{x} = \begin{pmatrix} x_{1,1} & \cdots & x_{1,k} \\ \vdots & \cdots & \vdots \\ x_{n,1} & \cdots & x_{n,k} \end{pmatrix}. \tag{5.36}$$

We may also apply the methods to the whole matrix $\boldsymbol{x}$. For instance, let us consider the approach by collocation: assume that the data is a sample of values

$$\mathcal{S} = \{(\boldsymbol{U_1}, \boldsymbol{X_1}), \ldots, (\boldsymbol{U_{ns}}, \boldsymbol{X_{ns}})\} \tag{5.37}$$

Each $\boldsymbol{X_i}$ is a vector: $\boldsymbol{X_i} = (X_{i,\ 1}, \ldots, X_{i,\ n})^t$. Then, we may range the data in a matrix $\boldsymbol{\mathcal{X}} = \left(X_{i,j}, 1 \leq i \leq ns, 1 \leq j \leq n\right)$ – notice that each line of $\boldsymbol{\mathcal{X}}$ is an element of the sample: line i of $\boldsymbol{\mathcal{X}}$ contains the values of $\boldsymbol{X_i}$:

$$\boldsymbol{\mathcal{X}} = \begin{pmatrix} X_{1,1} & \cdots & X_{1,n} \\ \vdots & \cdots & \vdots \\ X_{ns,1} & \cdots & X_{ns,n} \end{pmatrix}. \tag{5.38}$$

Let $\boldsymbol{A}$ be the collocation matrix, defined in Eq. (5.9): the coefficients $\boldsymbol{x}$ verify

$$\boldsymbol{Ax^t} = \boldsymbol{\mathcal{X}}. \tag{5.39}$$

The situation is analogous for the variational approach: we consider the matrix $\boldsymbol{\mathcal{B}} = \left(\mathcal{B}_{i,j}, 1 \leq i \leq k, 1 \leq j \leq n\right)$ such that $\mathcal{B}_{ij} = E\left(X_j\varphi_i(U)\right)$, i.e.:

$$\mathcal{B} = \begin{pmatrix} E(X_1\varphi_1(U)) & \cdots & E(X_n\varphi_1(U)) \\ \vdots & \cdots & \vdots \\ E(X_1\varphi_k(U)) & \cdots & E(X_n\varphi_k(U)) \end{pmatrix}. \tag{5.40}$$

Now, let $\boldsymbol{A}$ be the variational matrix, defined in Eq. (5.24): the coefficients $\boldsymbol{x}$ verify Eq. (5.39) yet.

The extension of Moment Matching to random vectors requests some precaution: indeed, we must approach also the covariances, so that, we must consider the approximation of generalized moments having the form $E\left(\prod_{i=1}^{n} X_i^{m_i}\right)$ the vector of moments $\boldsymbol{M}(\boldsymbol{X})$ is replaced by a tensor of moments $\mathcal{M}(\boldsymbol{X}) = \left\{ \mathcal{M}_{p_1 \cdots p_n} = E\left(\prod_{i=1}^{n} X_i^{p_i}\right), 0 \leq p_i \leq m_i \right\}$ and we look for the minimum of

$$J(\boldsymbol{x}) = d(\boldsymbol{\mu}(\boldsymbol{x}), \mathcal{M}(\boldsymbol{X})), \tag{5.41}$$

where $\boldsymbol{\mu}(\boldsymbol{x})$ is the tensor of moments of the approximation: $\boldsymbol{\mu}(\boldsymbol{x}) = \left\{ \mu_{p_1 \cdots p_n} = E\left(\prod_{i=1}^{n} (PX_i)^{p_i}\right), 0 \leq p_i \leq m_i \right\}$.

Example

Let us consider the situation where the data is the following one:

U	−0,95	−0,92	−0,92	−0,86	−0,82	−0,81	−0,77	−0,72
X_1	−0,37	−0,37	−0,37	−0,36	−0,36	−0,36	−0,36	−0,35
X_2	−0,16	−0,25	−0,25	−0,42	−0,55	−0,57	−0,65	−0,77

U	−0,64	−1,59	−0,30	−0,26	−0,24	−0,22	0,03	0,05
X_1	−0,34	−0,32	−0,22	−0,20	−0,19	−0,17	0,03	0,05
X_2	−0,90	0,96	−0,82	−0,72	−0,69	−0,63	0,10	0,15

Let us consider an approximation by a polynomial family analogous to the preceding example, with $k = 8$. The collocation approach furnishes the coefficients

$x_{1,j}$	−0,32	−0,20	0,11	0,22	0,11	0,12	−0,02	0,02
$x_{2,j}$	0,96	0,90	−7,41	−29,06	79,29	−52,31	3,88	3,90

The RMS error of the approximation is 1E-5. By moments matching (M3), with $m_1 = m_2 = m$, we obtain coefficients and a RMS error that are close to these values, for $m = 5$ to 10. As observed, the main interest of M3 is in the situations where the variable U is unknown. For instance, we may use a sample from the uniform distribution or a sample from $N(0, 1)$. In this case, the coefficients change significatively and the RMS error increases to 3E-1. ∎

Exercises

1. Consider the capital variation X of a company in a given time period. Assume that X is a function of the depreciation rate U_1 and of the investment rate U_2. The following data was recorded:

U_2 \ U_1	0,05	0,075	0,1	0,15	0,175	0,2
0,1	8,11	8,79	9,14	9,37	9,53	9,66
0,15	6,99	7,92	8,36	8,60	8,75	8,84
0,2	5,96	7,18	7,74	8,04	8,21	8,31
0,25	4,96	6,48	7,19	7,56	7,77	7,90
0,3	3,97	5,82	6,67	7,12	7,39	7,54

   (a) Assume that the distributions of U_1 and U_2 are uniform. Then, determine a representation of X as a function of U ans use it to determine the CDF and the PDF of X.
   (b) Assume that the distribution of U_1 is triangular $T(0.05,0.2,0.1)$ and the distribution of U_2 is $T(0.1, 0.3, 0.2)$. How this change impacts the representation determine in (a)? determine the CDF and the PDF of X in this case and compare the results to those found in (a).

2. Consider the capital variation X of a national economy in a given time period. Assume that X is a function of two random variables U_1 and U_2. The following data was observed:

U_2 \ U_1	0,05	0,075	0,1	0,15	0,175	0,2
0,1	0,13	0,16375	0,21	0,26875	0,34	0,42375
0,15	0,1825	0,2175	0,265	0,325	0,3975	0,4825
0,2	0,235	0,27125	0,32	0,38125	0,455	0,54125
0,25	0,2875	0,325	0,375	0,4375	0,5125	0,6
0,3	0,34	0,37875	0,43	0,49375	0,57	0,65875

   (a) Assume that the distributions of U_1 and U_2 are uniform. Then, determine a representation of X as a function of U ans use it to determine the CDF and the PDF of X.
   (b) Assume that the distribution of U_1 is triangular $T(0.05,0.2,0.1)$ and the distribution of U_2 is $T(0.1, 0.3, 0.2)$. How this change impacts the representation determine in (a)? determine the CDF and the PDF of X in this case and compare the results to those found in (a).

3. Consider the economic function $Y = 10\, x_1^{\alpha} x_2^{\beta} - 2\, x_1 - 5x_2$

   (a) Assume that the distributions of α and β are uniformly distributed on (0.1, 0.5). Determine a representation of Y using a polynomial basis.
   (b) Assume $\alpha \sim T(0.1,0.5,0.2)$ and $\beta \sim T(0.1, 0.5, 0.4)$. Determine a representation of Y using a polynomial basis.

5.6 Random Functions

The representation of random functions is an extension of the methods previously presented. To fix the ideas, let us consider a function $X = X(t, U) : t \in (0, T)$, depending upon a random variable U. In such a situation, we look for an expansion having time-dependent coefficients:

$$X(t, U) \approx PX(t, U) = \sum_{j=1}^{k} x_j(t)\varphi_j(U). \tag{5.42}$$

Equation (5.42) for random functions is the equivalent of Eq. (5.3) for random variables. The unknowns to be determined are the functions $x_i(t)$, $t \in (0, T)$.

For instance, assume that a sample from X is available: $\{X_{i,\,j} = X(t_i, U_j) : 0 \leq i \leq nt,\ 1 \leq j \leq ns\}$, with $0 = t_0 < t_1 < \ldots < t_{nt} = T$. In such a situation, we may consider $\boldsymbol{X} = (X_0, \ldots, X_{nt})^t$, where $X_i = X(t_i)$. $\boldsymbol{X}$ is a random vector of dimension $nt + 1$, so that we can apply the approach introduced in Sect. 5.5.2, for each t_i

$$X(t_i, U) \approx PX(t_i, U) = \sum_{j=1}^{k} x_{i,j}\varphi_j(U).$$

The coefficients $x_{i,\,j}$ can be gathered in a matrix (5.36) and be determined by solving (5.39). Equivalently, we can consider a sample

$$\mathcal{S}_i = \{(U_1, X_{i,1}), \ldots, (U_{ns}, X_{i,ns})\}$$

and apply to $\mathcal{S}_i$ the methods presented in the preceding. We have

$$x_j(t_i) \approx x_{i,j}, 0 \leq i \leq nt, 1 \leq j \leq ns. \tag{5.43}$$

If necessary, the values of $x_j(t)$ for other values may be determined by interpolation (see Sect. 2.7, page 80).

Example
Let $X_t = \sin(\pi U t)$, where U is uniformly distributed on $(-1, 1)$.

Assume that the expression of X_t is unknown, but we have observations corresponding to $ns = 20$, $nt = 10$, $T = 1$, $t_i = i\Delta t$, $\Delta t = 0, 1$. We apply the procedure of collocation to determine a representation using a polynomial basis with $k = 6$. The RMS error on the approximation of the data is 1E-3. We generated a sample of 10000 variates from the uniform distribution on $(-1, 1)$ and evaluated the approximation and the exact value on this new sample. The RMS error on this new data was 1E-3. A comparison of the standard deviations is shown in the Fig. 5.39. ∎

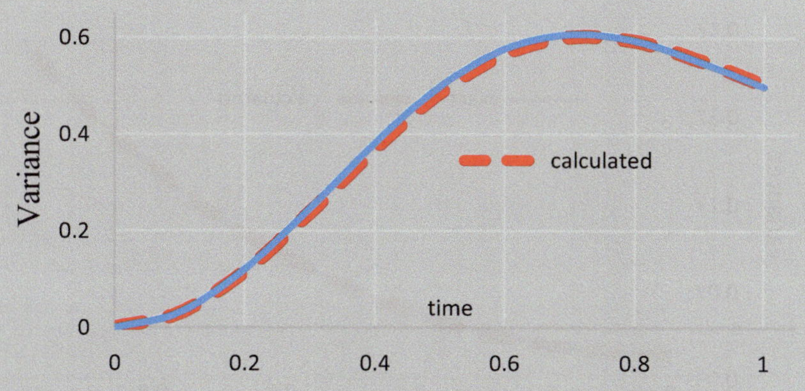

Fig. 5.39 Variance as function of time. The approximation furnishes values close to exact ones

Example
Let $X_t = e^{Ut}$, where U is uniformly distributed on $(-1, 1)$. Assume that we do not know the expression of X_t, but we have observations corresponding to $ns = 40$, $nt = 20$, $T = 1$, $t_i = i\Delta t$, $\Delta t = 0, 05$. We determine by collocation a representation using a polynomial basis with $k = 6$. The RMS error on the approximation of the data is 2E-4. We generated a sample of 10000 variates from the uniform distribution on $(-1, 1)$ and evaluated the approximation and the exact value on this new sample. The RMS error on this new data was 2E-4. A comparison of the means and variances is shown in the Figs. 5.40, 5.41, and 5.42.

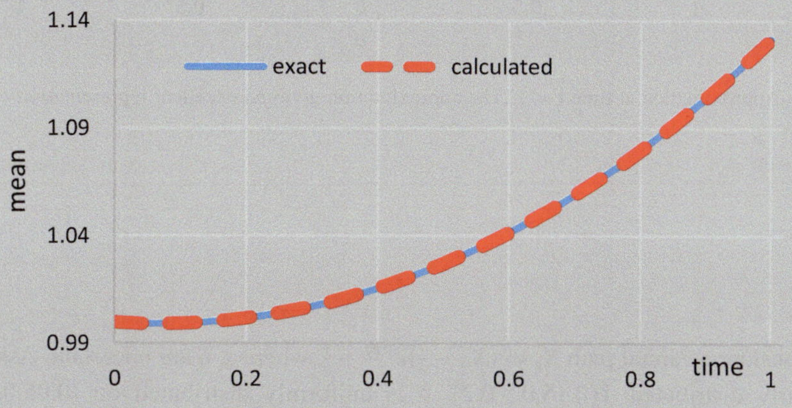

Fig. 5.40 Mean as function of time. The approximation is remarkably close to the exact values

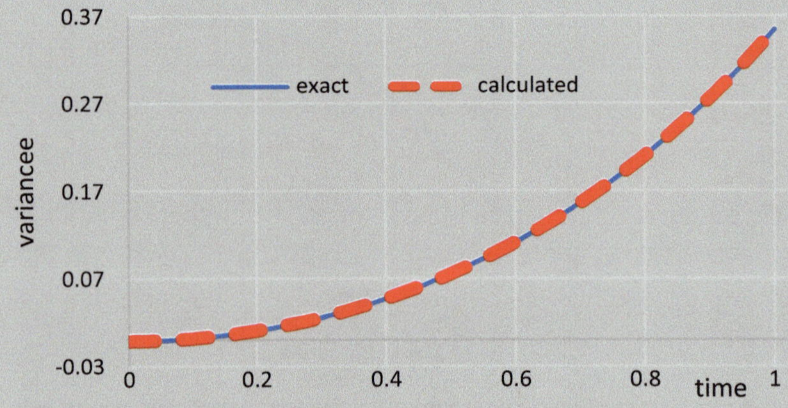

Fig. 5.41 Variance as function of time. The approximation is awfully close to the exact values

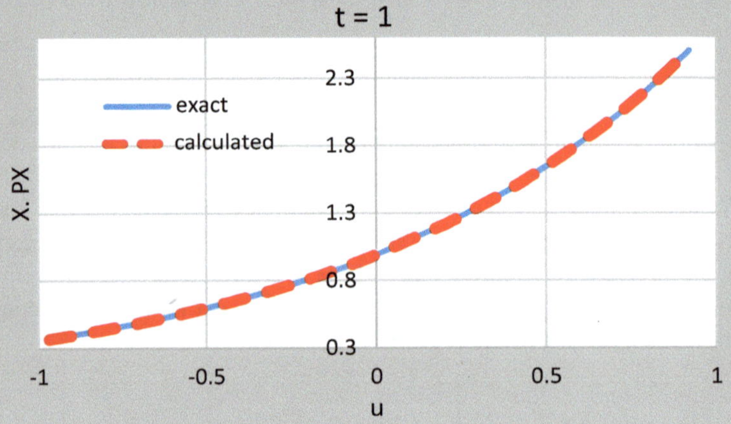

Fig. 5.42 Approximation at time $t = 1$. The approximation gives an excellent representation of the stochastic process. ∎

Example

Let us consider a capital path $X_t = \left(X_0 - \frac{s}{\delta}\right)e^{-\delta t} + \frac{s}{\delta}$, where s, δ are uncertain. Assume that s is triangularly distributed T(0.15,0.3,0.2), δ is uniformly distributed on (0.05,0.15) and the variables are independent.

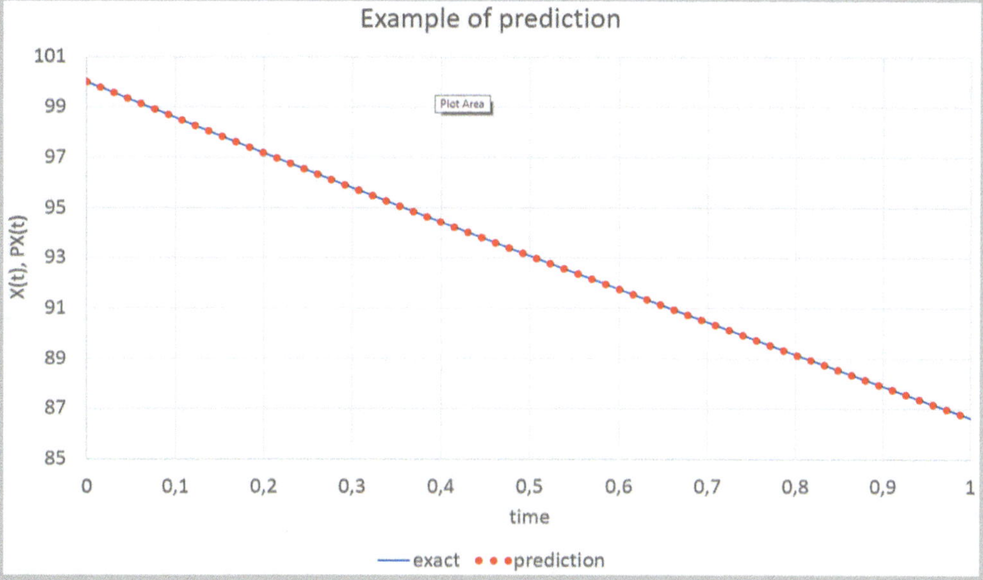

Fig. 5.43 Example of prediction ∎

Again, assume that we do not know the expression of X_t, but we have observations corresponding to the 25 combinations of the samples below:

s	0,171553736	0,179011042	0,223251145	0,249395849	0,249458937
δ	0,052162242	0,079594785	0,122753751	0,135923868	0,146172518

Assume that $nt = 20$, $T = 1$, $t_i = i\Delta t$, $\Delta t = 0, 05$. We apply the procedure of collocation to determine a representation using a polynomial basis with $k_1 = k_2 = 3$. An example of result is shown in the Fig. 5.43. ∎

Exercises

1. Consider a household's stocks time path given by

$$X_t = (1 - u)^2 \frac{t^2}{4}.$$

Assume that the marginal propensity to consume u is a random variable, uniformly distributed on (0.2, 0.8).
Determine a representation of X_t in the basis of trigonometric functions 1, sin (u), cos (u), sin (2u), cos (2u), . . .

2. Consider a capital path $X_t = \left(X_0 - \frac{s}{\delta}\right)e^{-\delta t} + \frac{s}{\delta}$, where s, δ are uncertain.
Assume that the distributions of s and δ are triangular: $s \sim T(0.1, 0.5, 0.25)$ and $\delta \sim T(0.05, 0.25, 0.15)$.

(a) Determine a representation of X_t in the polynomial basis $1, u_1, u_2, u_1^2, u_1 u_2, u_2^2, \ldots$
(b) Determine a representation of X_t in the tensor basis generated by the trigonometric functions 1, sin(u), cos (u), sin (u_1), cos (u), sin (2u), cos(2u), ...

3. Consider the family $X_t = u$, if $|t| < u$; $X_t = (|t| - u)^2 + u$, otherwise. Assume that the distribution of u is triangular: u~T(0, 1, 0.5)
(a) Determine a representation of X_t in the polynomial basis.
(b) Determine a representation of X_t in the trigonometric basis 1, sin(u), cos (u), sin (2u), cos (2u), ...

5.7 Random Curves

The analysis of the uncertainties in curves requests a particular approach, since it involves difficulties that remain unsuspected at the first glance.

From the formal point of view, a curve $\mathcal{C}$ is a set of points which may be represented by a parameterization or by an implicit equation. In the first case, the points of the curve are described by two functions:

$$\mathcal{C} = \{\boldsymbol{X} = (X_1, X_2) \in \mathbb{R}^2 : X_1 = X_1(t), X_2 = X_2(t), t \in (0, T)\}.$$

In the second approach, the curve is defined as the set of points that nullify an expression:

$$\mathcal{C} = \{\boldsymbol{X} = (X_1, X_2) \in \mathbb{R}^2 : \psi(X_1, X_2) = 0, with\ \boldsymbol{X} \in \Omega\}.$$

This simple consideration brings to us the first point to which we need to focus our attention – the distinction between the equations defining a curve and the set of points forming the curve. Indeed, the set of points is unique, but its representation by a parameterization or an equation is not unique.

For instance, let us consider a curve defined by the equation

$$\psi_1(\boldsymbol{X}, U) = \sqrt{X_1} + \sqrt{X_2} - U = 0, X_1 \geq 0, X_2 \geq 0.$$

We can rewrite this equation under the form

$$\psi_2(\boldsymbol{X}, U) = X_2 - \left(X_1 - 2U\sqrt{X_1} + U^2\right) = 0, X_1 \geq 0.$$

or, equivalently,

$$\psi_3(\boldsymbol{X}, U) = X_1 - \left(X_2 - 2U\sqrt{X_2} + U^2\right) = 0, X_2 \geq 0.$$

We can also generate a parametric representation of the curve:

$$X_1(t) = t^2 U^2, x_2(t) = (1 - t)^2 U^2, \mathrm{t} \in (0, 1)$$

The non-uniqueness of the parametric and implicit representation may introduce difficulties when uncertainty must be considered. For instance, let us suppose that U is uniformly distributed on (0, 1).

According to the description used, the statistical properties may differ. As an example, let us try to evaluate the mean of the family of curves.

If we consider the mean equation $\bar{\psi}_i(\boldsymbol{X}) = E(\psi_i(\boldsymbol{X}, U)) = 0$, we obtain:

$$\bar{\psi}_1(\boldsymbol{X}) = \sqrt{X_1} + \sqrt{X_2} - \frac{1}{2} = 0, X_1 \geq 0, X_2 \geq 0.$$

$$\bar{\psi}_2(\boldsymbol{X}) = X_2 - \left(X_1 - \sqrt{X_1} + \frac{1}{3}\right) = 0, X_1 \geq 0, X_2 \geq 0.$$

$$\bar{\psi}_3(\boldsymbol{X}) = X_1 - \left(X_2 - \sqrt{X_2} + \frac{1}{3}\right) = 0, X_1 \geq 0, X_2 \geq 0.$$

Analogously, the mean parameterization $\bar{X}_i(t) = E(X_i(t, U))$ leads to

$$\bar{X}_1(t) = \frac{t^2}{3}, \bar{X}_2(t) = \frac{1}{3}(1-t)^2, t \in (0, 1).$$

These curves are different each from other, as we can see in Fig. 5.44. Thus, the choice of the description of the curve may have a significant influence on the statistical properties. We find these same difficulties when we try to determine expansions. For instance, if we consider an expansion in a polynomial basis, using $k = 6$, we get

$$PX = \left(t^2U^2, (1-t)^2U^2\right) \quad (parametric)$$

$$PX = \left(t^2, t^2 - 2Ut + U^2\right) \quad (implicit\ \psi_1)$$

$$PX = \left(t, -U\sqrt{t} + \frac{1}{3}(1+3t)\right) \quad (implicit\ \psi_2)$$

$$PX = \left(-U\sqrt{t} + \frac{1}{3}(1+3t), t\right) \quad (implicit\ \psi_3)$$

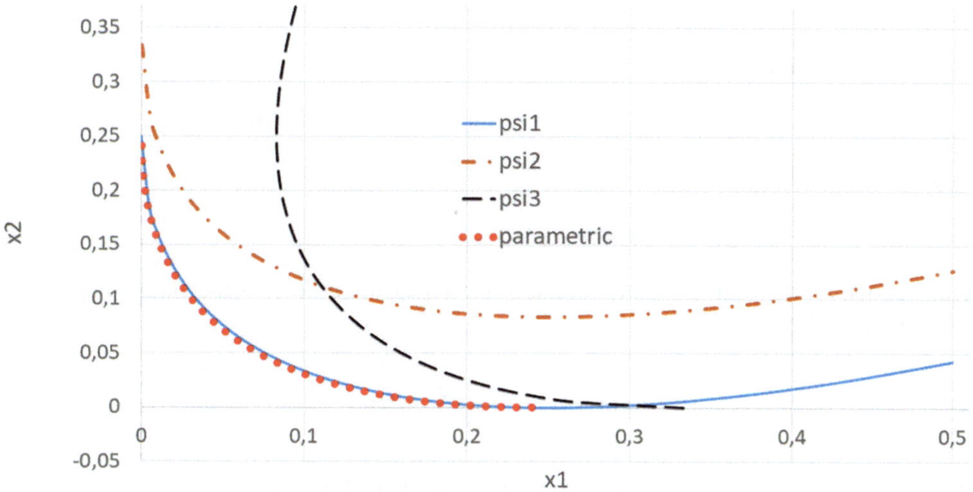

Fig. 5.44 Example of variability of the mean curve: according to the description used, the mean equation leads to a different curve

Considering the situations where the choice of a parameterization is dubious, it is necessary to set up a procedure which is independent of the parameterization or the implicit equation. To achieve it, we may consider the set of points $\mathcal{C}$ and build a method based on them: indeed, the set of points $\mathcal{C}$ is the same independently of the way the equations defining the curve are written. For instance, we may look for the sets of points generated by the representation

$$\boldsymbol{PX}(t,U) = \sum_{j=1}^{k} \boldsymbol{x}_j(t)\varphi_j(U)$$

and look for the representation that generates the best set of points, in the sense that they minimize the distance to the dataset. Formally, we consider a curve of the family as a set of points, whatever the choice of form to describe them: either

$$\mathcal{C}(U) = \{\boldsymbol{X} = (X_1, X_2) \in \mathbb{R}^2 : X_1 = X_1(t,U), X_2 = X_2(t,U), t \in (0,T)\} \tag{5.44}$$

or

$$\mathcal{C}(U) = \{\boldsymbol{X} = (X_1, X_2) \in \mathbb{R}^2 : \psi(X_1, X_2, U) = 0\}. \tag{5.45}$$

The data will be the sets of points $\mathcal{C}(u_1), \ldots, \mathcal{C}(u_{ns})$. Then, we consider the curves generated by the representations:

$$\mathcal{D}(\boldsymbol{x},u) = \left\{\boldsymbol{PX} = (PX_1, PX_2) \in \mathbb{R}^2 : PX_i = \sum_{j=1}^{k} x_{i,j}(s)\varphi_j(U) : s \in (s_{min}, s_{max})\right\}.$$

For each element from the sample, we may measure the distance between $\mathcal{D}(\boldsymbol{x},u_i)$ and $\mathcal{C}(u_i)$:

$$d_i(\boldsymbol{x}) = dist(\ \mathcal{D}(\boldsymbol{x},u_i), \mathcal{C}(u_i)). \tag{5.46}$$

Here *dist* is an adequate function to measure the distance between sets of points, such as, for instance, the distance from the points of $\mathcal{D}(\boldsymbol{x},u_i)$ to $\mathcal{C}(u)$:

$$dist(A,B) = d_{SH}(A,B) = \sup_{a \in A} \inf_{b \in B} \|a - b\| \tag{5.47}$$

or the Hausdorff distance:

$$dist(A,B) = d_H(A,B) = \max\{d_{SH}(A,B), d_{SH}(B,A)\} \tag{5.48}$$

The distances are collected in a vector

$$\boldsymbol{d}(\boldsymbol{x}) = (d_1(\boldsymbol{x}), \ldots, d_{ns}(x))^t.$$

Then, we look for the coefficients $\boldsymbol{x}$ that minimize a measure $g(\boldsymbol{d}(\boldsymbol{x}))$ of the magnitude of $\boldsymbol{d}(\boldsymbol{x})$, such as, for instance, its RMS norm:

$$g(\boldsymbol{d}(\boldsymbol{x})) = \sqrt{\frac{1}{ns}\sum_{i=1}^{ns} d_i^2(\boldsymbol{x})}\ . \tag{5.49}$$

Alternatively, we may consider other measures g, such as, for instance,

$$g(\boldsymbol{d}(\boldsymbol{x})) = \max\{d_i(\boldsymbol{x}) : 1 \le i \le ns\} \ .$$

Thus, we must consider two situations:

- Either the parameterization or implicit equation appears as clear and undoubtful. Then, we may use it to determine the representation: we consider the components as functions and we apply the methods previously presented for functions (Sect. 5.6).
- Or it is dubious, and we may use the approach by the Hausdorff distance.

An example of VBA code for the evaluation of the distances d_{SH} and d_H is given in Fig. 5.45. In this code: `distp2p` evaluates the distance between two points `p1`, `p2`- each point has two coordinates; `distp2c` evaluates the distance between a point `p1` and a curve `C` – the curve is an object of class `curve` as described in Sect. 1.17, page 33: `C` is a collection of two-dimensional vectors containing the coordinates of each point of the curve; `dsh` evaluates d_{SH}; `dh` evaluates the Hausdorff distance d_H between two curves `c1` and `C2`: each curve is an object of class `curve`; `distfam` generates the vector of distances between two families of curves `F1` and `F2` – both are objects of type `Collection`, and their elements of are curves: `F1(i)` is a curve. `F1` and `F2` must have the same number of elements; `distc2fam` generates the vector of distances between `c1` and the family `F2`.

As an example, let us consider two families of circles: `F1` is a family of 5 circles, centered at the origin and having radius 1,2,3,4,5. `F2` is a family of 5 identical circles of radius 5. We have `dsh(F1(i),F2(i))` = `dh(F1(i),F2(i))=5-i`, `dsh(F1(i),F1(j))` = `dh(F1(i),F1(j))` = `Abs(i-j)`.

Each circle is discretized in 21 points corresponding to the angles "`= SEQUENCE(1;21;0;PI()/10)`" in `B1`.

	A	B	C	D	E	F	G	H
1	theta	0	0,314159	0,628319	0,942478	1,256637	1,570796	1,884956	2,1

We generated the two families with these data, and we calculated the values of the Hausdorff distances between `F1(i)` and `F2(i)` as those between `F1(i)` and `F1(j)`. The results are shown in Table 5.15.

To determine a representation of the family, we may consider $k = 4$ and look for the coefficients $\boldsymbol{x} = (x_{i,j}(s_\ell) : 1 \le i \le 2,\ 1 \le j \le k,\ 0 \le \ell \le 20)$, $s_\ell = \ell/20$, which corresponds to $T = 1$, discretized in 20 subintervals of equal length. In this case, the parameterization is clear, and we may use the function approach presented in the previous section. However, we shall apply also the approach involving the Hausdorff distance for comparison purposes.

We apply initially the function approach, by minimizing the mean squared error for each s_ℓ. Examples of coefficients found are shown in Table 5.16. An example of curve predicted is shown in Fig. 5.46. In a second step, we use the SOLVER to minimize the RMS error (5.49). Examples of coefficients found are shown in Table 5.17. An example of curve predicted is shown in Fig. 5.47. The coefficients are almost equal in both the cases. The sample is generated by $U_i = i$, $1 \le i \le 5$. The Figures compare the prediction for $U = 3.5$ to the exact value. The accordance is particularly good.

```
Function distp2p(p1, p2)
distp2p = Sqr((p1(1) - p2(1)) ^ 2 + (p1(2) - p2(2)) ^ 2)
End Function
```

```
Function distp2c(p1, c)
npc = UBound(c, 1)
Dim d() As Double
ReDim d(1 To npc)
For i = 1 To npc
    d(i) = distp2p(p1, c(i))
Next i
distp2c = WorksheetFunction.Min(d)
End Function
```

```
Function dsh(c1, C2)
np1 = UBound(c1, 1)
Dim d() As Double
ReDim d(1 To np1)
For i = 1 To np1
  d(i) = distp2c(c1(i), C2)
Next i
dsh = WorksheetFunction.Max(d)
End Function
```

```
Function dh(c1, C2)
d12 = dsh(c1, C2)
d21 = dsh(C2, c1)
dh = WorksheetFunction.Max(d12, d21)
End Function
```

```
Function distfam(F1, F2, dist)
np1 = UBound(F1, 1)
Dim d() As Double
ReDim d(1 To np1)
For i = 1 To np1
  d(i) = Application.Run(dist, F1(i), F2(i))
Next i
distfam = d
End Function
```

```
Function distc2fam(c1, F2, dist)
np1 = F2.Count
Dim d() As Double
ReDim d(1 To np1)
For i = 1 To np1
  d(i) = Application.Run(dist, c1, F2(i))
Next i
distc2fam = d
End Function
```

Fig. 5.45 Example of VBA code for the implementation of the Hausdorff distance

Table 5.15 Calculated values of the Hausdorff distances for a family of circles. The results are exact

F1 to F1	$j = 1$	$j = 2$	$j = 3$	$j = 4$	$j = 5$	F1 to F2	
$i = 1$	0	1	2	3	4	$i = 1$	4
$i = 2$	1	0	1	2	3	$i = 2$	3
$i = 3$	2	1	0	1	2	$i = 3$	2
$i = 4$	3	2	1	0	1	$i = 4$	1
$i = 5$	4	3	2	1	0	$i = 5$	0

Table 5.16 Examples of coefficients found by the function approach for a family of circles

	$s_\ell = 0, 25$		$s_\ell = 0, 5$		$s_\ell = 1, 5$	
j	$x_{1,j}$	$x_{2,j}$	$x_{1,j}$	$x_{2,j}$	$x_{1,j}$	$x_{2,j}$
1	2,3E-3	−4,6E-5	7,2E-4	−4,6E-5	−8,1E-4	3,9E-4
2	−3,1E-3	1,0	−1,0	6,8E-5	1,1E-3	−1,0
3	1,2E-3	−2,1E-5	6,7E-4	−2,1E-5	−4,1E-4	1,9E-4
4	−1,3E-4	−2,7E-7	−8,3E-5	1,4E-6	4,6E-5	−2,0E-5

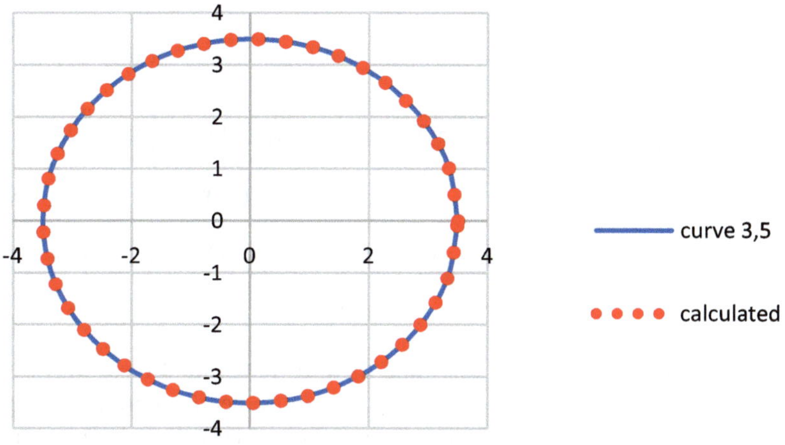

Fig. 5.46 Prediction by the function approach, with a polynomial basis, $k = 4$

Table 5.17 Examples of coefficients found by the Hausdorff approach, minimizing the RMS error (5.49), for a family of circles

	$s_\ell = 0, 25$		$s_\ell = 0, 5$		$s_\ell = 1, 5$	
j	$x_{1,j}$	$x_{2,j}$	$x_{1,j}$	$x_{2,j}$	$x_{1,j}$	$x_{2,j}$
1	2,3E-3	−4,6E-5	7,2E-4	−4,6E-5	−8,1E-4	3,9E-4
2	−3,1E-3	1,0	−1,0	6,8E-5	1,1E-3	−1,0
3	1,2E-3	−2,1E-5	6,7E-4	−2,1E-5	−4,1E-4	1,9E-4
4	−1,3E-4	−2,7E-7	−8,3E-5	1,4E-6	4,6E-5	−2,0E-5

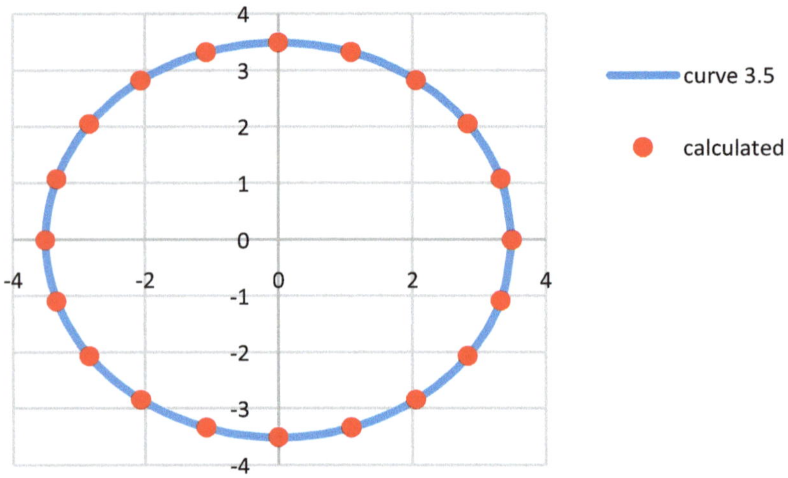

Fig. 5.47 Prediction by the Hausdorff approach, using a polynomial basis $k = 4$

Example

Let us consider the following Binh-Korn multiobjective optimization problem (see Sect. 2.6, page 79. See also (Bassi et al., 2018)):

$$f_1(\boldsymbol{x}) = 4\,(x_1 + u_1)^2 + 4(x_2 + u_2)^2,$$

$$f_2(\boldsymbol{x}) = \left(x_1 - \frac{u_1}{10} - 5\right)^2 + \left(x_2 - \frac{u_2}{10} - 5\right)^2$$

$$C = \{\boldsymbol{x} = (x_1, x_2) : \psi_i(\boldsymbol{x}) \le 0, 1 \le i \le 6\},$$

$$\psi_1(\boldsymbol{x}) = (x_1 - 5)^2 + x_2^2 - 25,\ \ \psi_2(\boldsymbol{x}) = 7.7 - (x_1 - 8)^2 - (x_2 - 3)^2,$$

$$\psi_3(\boldsymbol{x}) = -x_1, \psi_4(\boldsymbol{x}) = -x_2,\ \ \psi_5(\boldsymbol{x}) = x_1 - 5,\ \ \psi_6(\boldsymbol{x}) = x_2.$$

Assume that $u_1 \sim N(2, 0.1)$ and $u_2 \sim T(0, 1, 0.5)$ and these variables are independent. In this case, the Pareto front associated to the multiobjective optimization problem is a random curve. To determine a representation, we can consider a sample from the variables. Since they are independent, we may consider independent samples. For instance, let us consider the samples in Table 5.18.

Table 5.18 A small sample for the representation of Pareto fronts

u_1	1,793437	1,88848	1,999165	2,010893	2,01773
u_2	0,210664	0,407556	0,659623	0,787493	0,823312

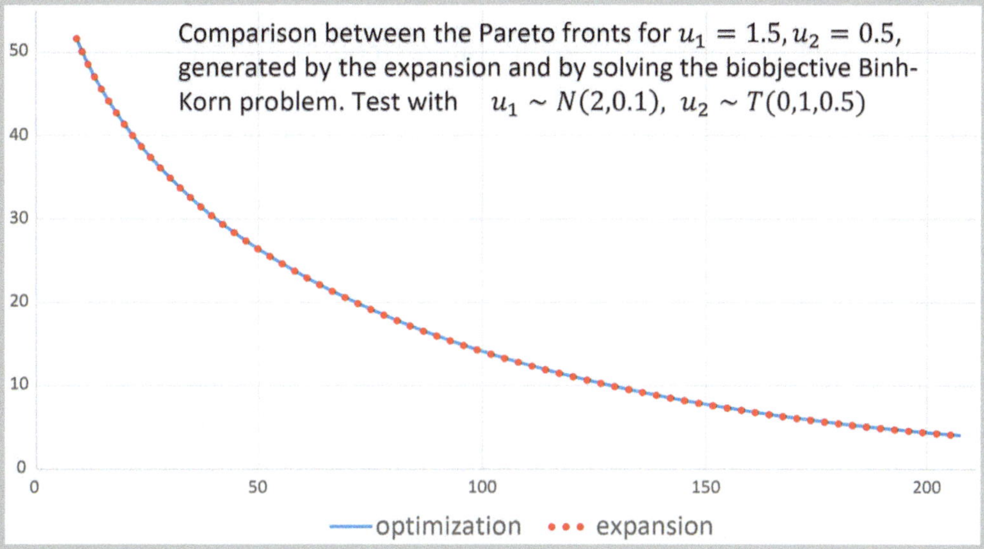

Fig. 5.48 Comparison between the Pareto front generated by the expansion and the solution of the multiobjective optimization problem (method of Sect. 2.6)

We have 5 values from each variable: independence allows us to generate a sample formed by 25 couples (u_{1_i}, u_{2_j}): we generate a sample of 25 Pareto fronts using these values – the bi-objective optimization problem is solved by the method presented in Sect. 2.6.

For the method introduced in Sect. 2.6, the parameterization involves a parameter $t \in (0, 1)$, which is used for all the curves. Thus, we may use this parametrization to determine the representation. Using discretized times $t_i = 0.01 * i, 0 \le i \le 100$, we determined the coefficients of an expansion with polynomial families and $k_1 = k_2 = 3$, for each time t_i, by collocation. The global RMS error was 1.3E-12.

The representation may be used to generate larger samples from the Pareto front. As an example, Fig. 5.48 compares the fronts generated by the expansion and by the solution of the multiobjective problem, for the values $u_1 = 1.5$; $u_2 = 0.1$.

Now, assume that u_1~T(1.5,2.5,2) and u_2~T(0, 1, 0.5):consider the samples in Table 5.19. In this case, the global RMS error was 7E-13. Figure 5.49 compares the fronts generated by the expansion and by the solution of the multiobjective problem, for the values $u_1 = 1.5$; $u_2 = 0.1$. ∎

Table 5.19 A small sample for the representation of Pareto fronts

u_1	1,755633	1,924332	2,021606	2,204238	2,405309
u_2	0,210664	0,407556	0,659623	0,787493	0,823312

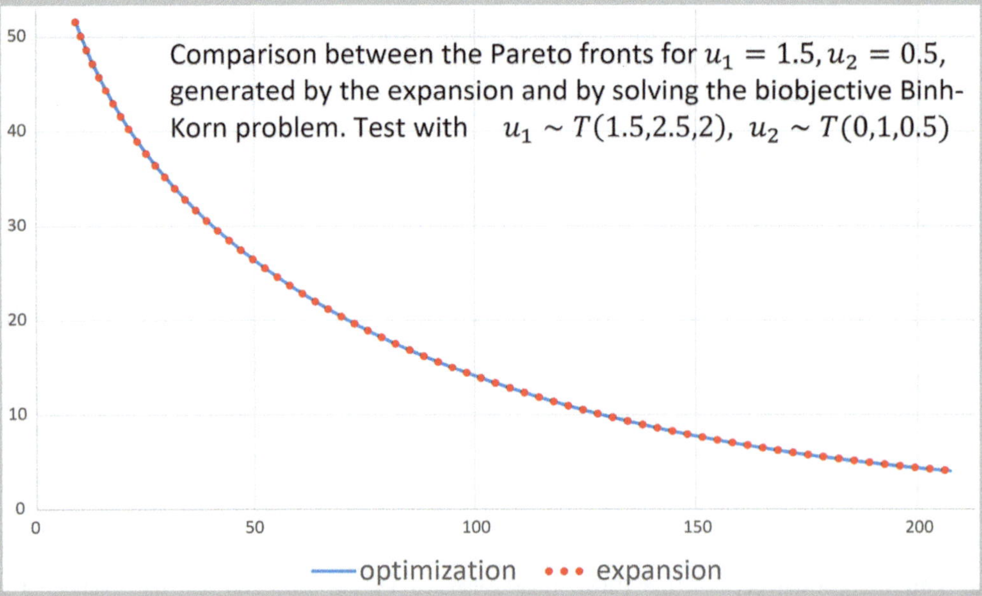

Fig. 5.49 Comparison between the Pareto front generated by the expansion and the solution of the multiobjective optimization problem (method of Sect. 2.6)

Example

Let us consider the classical portfolio selection problem: an investor has $C = 100$ units of capital to allocate by 1 year on three different options, based on a stock exchange having a mean annualized volatility of $\sigma = 20\%$. Option 1 has a mean return $r_1 = 10\%$ with a coefficient of volatility $\beta_1 = 1.05$. Option 2 has a mean return of $r_2 = 15\%$ with a coefficient $\beta_2 = 1.2$. Option 3 has a mean return of $r_3 = 5\%$ with a coefficient of volatility $\beta_3 = 0.8$. The analysis can be made by setting as unknowns the fraction of capital allocated to each option: we have three unknowns $0 \leq x_i \leq 1, 1 \leq i \leq 3$ may be reduced to two variables $\boldsymbol{x} = (x_1, x_2)$ by considering that $x_3 = 1 - x_1 - x_2$, with $x_1, x_2 \geq 0, x_1 + x_2 \leq 1$.

The classical analysis of the portfolio problem consists in evaluating the mean return by unit of capital

$$R(\boldsymbol{x}) = r_1x_1 + r_2x_2 + r_3(1 - x_1 - x_2)$$

and its variance:

$$V(\boldsymbol{x}) = \sigma^2 S(x), S(x) = \beta_1^2 r_1^2 x_1^2 + \beta_2^2 r_2^2 x_2^2 + (1 - x_1 - x_2)^2 r_3^2 \beta_3^2.$$

Then, we can look for the solution of maximal expected return, id est, a solution $\boldsymbol{x}^* = (x_1, x_2)$ that maximizes the value of R under the restriction of a maximal variance $V_{max} = \sigma^2 S_{max}$:

$$\boldsymbol{x}^* = \arg\max \{R(\boldsymbol{x}) : x_1, x_2 \geq 0, x_1 + x_2 \leq 1, S(\boldsymbol{x}) \leq S_{max}\}.$$

As an alternative, we can look for the solution of minimal volatility, id est, a solution $x^* = (x_1, x_2)$ that minimizes the value of S under the restriction of a minimal return R_{min}:

$$x^* = \arg\min \{S(x) : x_1, x_2 \geq 0, x_1 + x_2 \leq 1, R(x) \geq R_{min}\}.$$

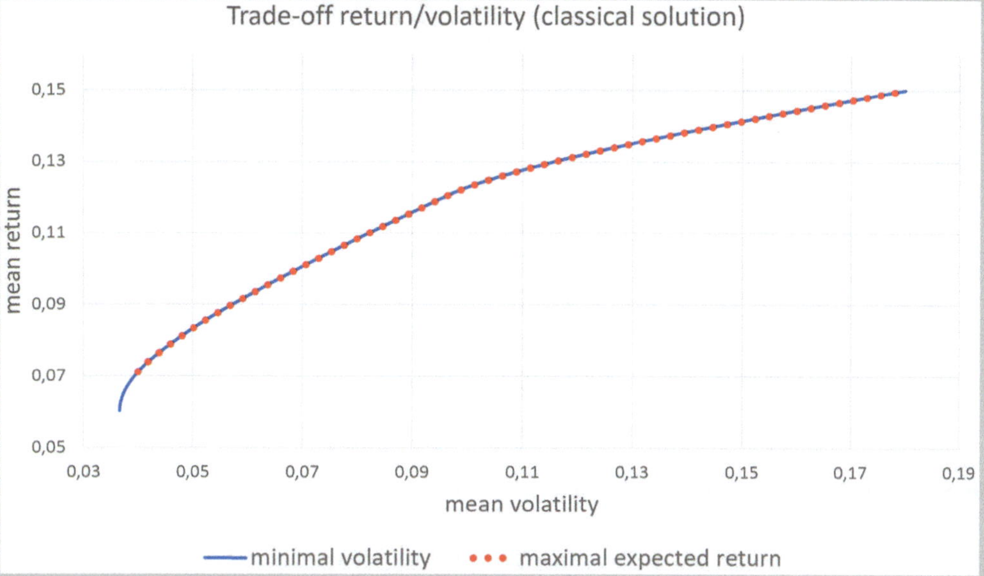

Fig. 5.50 Classical solution of the portfolio problem

Both the formulations can be solved by the SOLVER: varying the values of S_{max} and R_{min} generates a trade-off curve $(\sqrt{S(x^*)}, R(x^*))$ showing the possible compromises between Volatility and Return. The curves obtained are shown in Fig. 5.50: we observe that both the formulation produce the same trade-off curve.

Now, assume that the values of the coefficients β_1, β_2 are uncertain and each of these coefficients has a triangular distribution: $T(0.9,1.2,1.05)$ for β_1, $T(1.1,1.3,1.2)$ for β_2. We look for a representation of the trade-off curve. Analogously to the preceding situations, we consider a small sample, given in Table 5.20

Table 5.20 A small sample for the representation of the portfolio solution

β_1	0,984164403	0,996937932	1,017274687	1,062638505	1,091857263
β_2	1,12733492	1,187675572	1,230599464	1,236082147	1,272463665

Analogously to the preceding example, a natural parameterization is introduced by R_{min}, we may use this parametrization with a step 0.001 to determine the representation by collocation. The results appear in Fig. 5.51

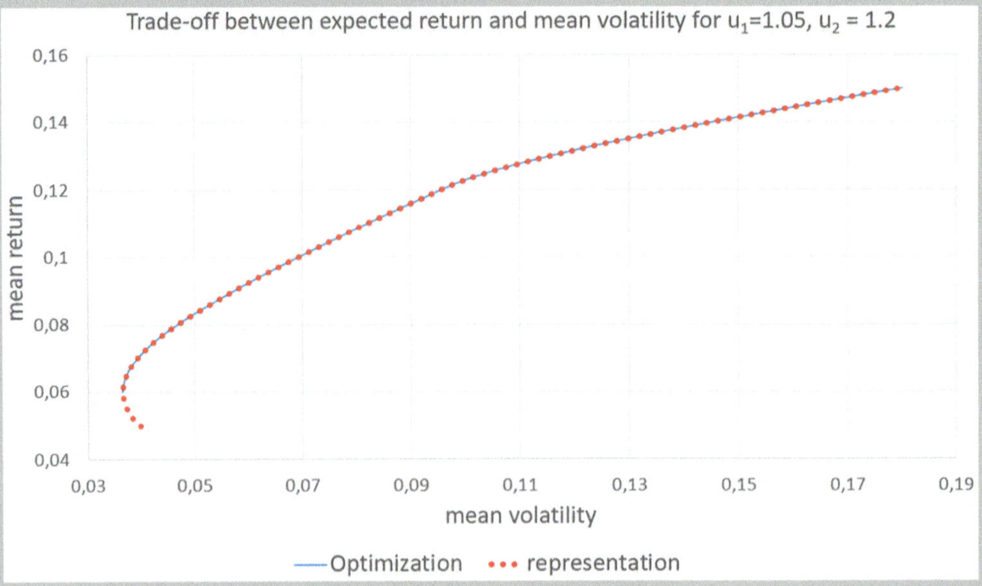

Fig. 5.51 Comparison between the prevision furnished by the representation and the solution of the optimization problem

Exercises

1. Consider the indifference curves associated to the economic function $X = 10\, x_1^{\alpha} x_2^{\beta} - 2\, x_1 - 5x_2$

   (a) Assume that the distributions of α and β are uniform. Then, determine an expansion of the indifference curves as a function of U in a polynomial basis.
   (b) Assume that the distribution of α is triangular $T(0.05{,}0.2{,}0.1)$ and the distribution of β is $T(0.1, 0.3, 0.2)$. How this change impacts the representation determine in (a)? Compare the results to those found in (a).

5.8 Mean, Variance and Confidence Intervals for Random Functions or Random Curves

In Sect. 3.8, we saw that we may use samples to obtain some statistics of random variables, such as the empirical mean, the variance and to determine confidence intervals that take into account the margin of error connected to the fact that the samples do not represent the whole population.

In the Sects. 5.2, 5.3, and 5.4 of this chapter we saw that we may use the samples to generate representations of random variables and then, use the representations to generate larger samples and use them to obtain the CDF and the PDF of the variables under consideration. Larger samples may also be used to evaluate the means and improve the evaluation of the variance and of the confidence interval.

In Sect. 5.5, these approaches were extended to multidimensional situations, so that analogous procedures may be used for vectors of random variables.

However, their extension to random functions and curves may face difficulties: as pointed in the preceding section, the determination of the mean of a family of random curves must take into account the non-uniqueness of the parametric and the implicit representations. These difficulties may concern also random functions, except for the situations where the parameterization to be used is undoubtful.

Indeed, analogously to Sect. 5.7, we may look for the mean of a family of curves either by considering a parameterization or the Hausdorff approach. Let us recall the definition of the family of curves (Eq. 5.44):

$$\mathcal{C}(u) = \left\{ \boldsymbol{X} = (X_1, X_2) \in \mathbb{R}^2 : X_1 = X_1(t, u), X_2 = X_2(t, u), t \in (0, T) \right\}$$

The mean of the family is a curve

$$E(\mathcal{C}) = \left\{ \overline{\boldsymbol{X}} = \left(\overline{X_1}, \overline{X_2}\right) \in \mathbb{R}^2 : \overline{X_1} = \overline{X_1}(t), \overline{X_2} = \overline{X_2}(t), t \in (0, T) \right\}$$

As shown in the preceding section, the solution consisting in taking the mean representation at each time $\overline{X_i}(t) = E(X_i(t, u))$ may lead to a result that depends on the parameterization used. As an alternative, we may use the variational formulation of the mean: if X is a random variable taking values in $\mathbb{R}$, then

$$E(X) = \arg\min \left\{ J(y) = E\left((X - y)^2 \right) : y \in \mathbb{R} \right\},$$

id est, $E(X)$ is the real number that minimizes the mean quadratic deviation from X. Analogously, the median is the value that minimizes the mean absolute deviation:

$$med(X) = \arg\min \{ I(y) = E(|x - y|) : y \in \mathbb{R} \},$$

The variance is the value of $J(E(X)) = E((X - E(X))^2)$. We may use these properties to determine the mean, the median and the variance of families of curves. Let

$$\mathcal{D} = \left\{ \boldsymbol{Y} = (Y_1, Y_2) \in \mathbb{R}^2 : Y_1 = Y_1(t), Y_2 = Y_2(t), t \in (0, T) \right\}$$

Then, we shall look for a curve $E(\mathcal{C})$ such that

$$E(\mathcal{C}) = \arg\min \left\{ J(\mathcal{D}) = E\left(dist^2(\mathcal{D}, \mathcal{C}(u)) \right) \right\},$$

where *dist* is a measure of the distance between the curves $\mathcal{D}$ and $\mathcal{C}(u)$. Analogously to Sect. 5.7, we may consider either a distance such as

$$dist^2(\mathcal{D}, \mathcal{C}(u)) = \int_0^T \| \boldsymbol{X}(t, u) - \bar{\boldsymbol{X}}(t) \|^2 dt \qquad (5.50)$$

or one of the set distances (see Eqs. (5.47) and (5.48), page 298).

In practice, only samples from X are available. Let $\mathfrak{C} = \{ \mathcal{C}_i = \mathcal{C}(u_i) : 1 \le i \le ns \}$ be a sample formed by ns variates from $\mathcal{C}(u)$: as an approximation, we may look for

$$E(C) \approx \overline{C} = \arg\min \left\{J(\mathcal{D}) = E\left(dist^2(\mathcal{D}, C_i)\right) : \quad C_i \in \mathfrak{C}\right\} . \tag{5.51}$$

For practical purposes, the distance is often not squared, and we look for

$$E(C) \approx \overline{C} = \arg\min \left\{I(\mathcal{D}) = E(dist(\mathcal{D}, C_i)) : \quad C_i \in \mathfrak{C}\right\} . \tag{5.52}$$

Notice that, formally, this definition corresponds to the median of the family, not to the mean. Then, we may proceed into a way analogous to the one used in Sect. 5.7: consider a family $\{\varphi_j : 1 \le j \le k\}$ and

$$\mathcal{D}(\boldsymbol{x}) = \left\{\boldsymbol{PX} = (PX_1, PX_2) \in \mathbb{R}^2 : PX_i = \sum_{j=1}^{k} x_{i,j}\varphi_j(s) : s \in (s_{min}, s_{max})\right\}.$$

Then, we may determine the coefficients $\boldsymbol{x}$ that minimize a measure $g(\boldsymbol{d}(\boldsymbol{x}))$ of the magnitude of $d(\boldsymbol{x})$, given by

$$\boldsymbol{d}(\boldsymbol{x}) = (d_1(\boldsymbol{x}), \ldots, d_{ns}(x))^t; d_i(\boldsymbol{x}) = dist(C(u_i), \mathcal{D}(\boldsymbol{x})).$$

For instance, we may take $g(\boldsymbol{d})$ as the norm of $\boldsymbol{d}$. If a discretization in s is used, such as $s_{min} = s_0 < s_1 < \ldots < s_{np} = s_{max}$, we may look for the values of $\boldsymbol{PX}(s_i)$, $1 \le i \le np$, i. e., for 2 np unknown values defining np points in $\mathbb{R}^2$.

However, these formulations present some difficulties yet: indeed, the optimization problems (5.51) and (5.52) may have multiple solutions and it may be necessary to impose supplementary conditions to select the convenient ones.

A simpler solution may be found, which may be employed if the sample under consideration is larger enough and all its elements have the same probability: we may look for the element of the sample which is the best representative of all the sample's population, id est:

$$E(C) \approx \overline{C} = \arg\min \left\{J(C_i) = E\left(dist\left(C_i, C_j\right)\right) : \quad C_j \in \mathfrak{C}\right\} . \tag{5.53}$$

In this case, we determine one of the curves of the family, corresponding to the minimal mean Hausdorff distance to the other members of the family – it may be considered as the best representative of the family – the element which is minimally off-centered. One of the interesting features of this approach is that it produces as result a member of the family, id est, a curve having the same general properties as the other members of the family.

In both the cases, the variance is estimated by $J(\overline{C})$. Confidence intervals with a risk α may be generated by considering quantiles: the confidence interval is formed by the curves of the family that generate a subset C_{conf} such that $\overline{C} \in C_{\text{conf}}$ and $P(\ C_{\text{conf}}) = 1 - \alpha$.

Example

Let us consider the family of circles

$$C(u) = \{X(t, u) = r(\sin(2\pi(t + u)), \cos(2\pi(t + u))), t \in (0, 1)\}$$

with u uniformly distributed on (1, 3). All the values of u generate the same circle, centered at the origin and having a radius r. Thus, we expect that the mean circle is the single curve of the family:

$$E(C) = \left\{\overline{X} = (\overline{X_1}, \overline{X_2}) \in \mathbb{R}^2 : \overline{X_1}^2(t) + \overline{X_2}^2(t) = \mathrm{r}^2, t \in (0, 1)\right\}$$

We have $E(X_1(t, u)) = E(X_2(t, u)) = 0$, so that the approach by the mean equation leads to the incorrect result $\overline{X_1}(t) = \overline{X_2}(t) = 0, \forall t \in (0,1)$. Using Eq. (5.50) leads to the same result. Indeed,

$$E\left(\int_0^1 \|\boldsymbol{X}(t,u) - \overline{\boldsymbol{X}}(t)\|^2 dt\right) = \int_0^1 \|\overline{\boldsymbol{X}}(t)\|^2 dt + 1,$$

so that the minimum corresponds to $\overline{\boldsymbol{X}}(t) = 0, \forall t$.

Let us use the Hausdorff distance with a discretization of the solution: we look for the points $\boldsymbol{PX}_i = (PX_{1,\,i}, PX_{2,\,i})$, $1 \leq i \leq np$ that minimize the mean Hausdorff distance to the family. To fix the ideas, let us consider the situation where we consider a single point: $np = 1, \boldsymbol{PX_1} = (a, b)$. The Hausdorff distance between the circle of radius r and the point is $r + \sqrt{a^2 + b^2}$, so that the minimum is attained for $\sqrt{a^2 + b^2} = 0$, id est, when $\boldsymbol{PX_1} = (0, 0)$. The result is similar for $np = 2$: let $\boldsymbol{PX_1} = (a, b), \boldsymbol{PX_2} = (-a, -b)$. The Hausdorff distance between the circle of radius r and the set of two points $\{\boldsymbol{PX_1}, \boldsymbol{PX_2}\}$ is $\sqrt{r^2 + a^2 + b^2}$, so that the minimum is attained when $a^2 + b^2 = 0$,id est, when the points coincide at the origin. This property generalizes to $np > 2$ points. Again, the results do not correspond to the expected one.

Let us use d_{SH}: again, consider two points (a, b), $(-a, -b)$. Now, the distance is $\left|r - \sqrt{a^2 + b^2}\right|$, which is minimal when $\sqrt{a^2 + b^2} = r$, id est, when the points belong to the circle. This property generalizes to $np > 2$ and the results correspond to the expected ones.

Finally, let us consider the approach (5.53): both d_{SH} and the Hausdorff distance d_H verify

$$dist\left(C_i, C_j\right) = 0, \forall i, j.$$

Thus, any member of the family is a solution $E(C) = C_i, \forall i$ – this result corresponds to the expected one (recall that all the C_i generate the same circle). In this case, the variance is equal to zero and the confidence intervals contain the single member of the family (Fig. 5.52).

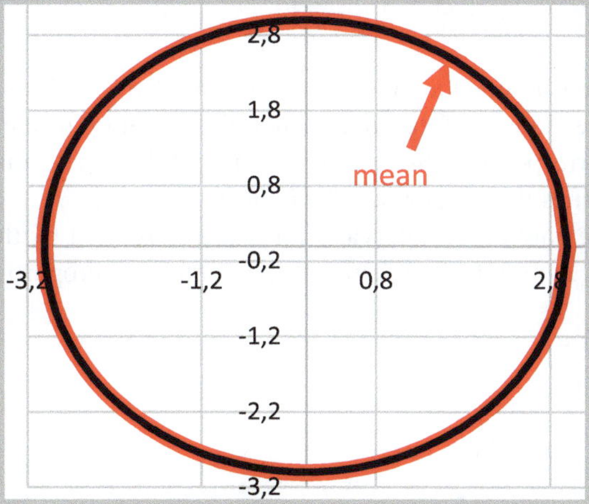

Fig. 5.52 A situation that frustrates several methods: the family is formed by a single circle of radius $r = 3$, with several different parameterizations. We expect the single curve of the family as its mean. However, many approaches furnish (0,0) as the mean

Example

Let us consider the family of circles

$$C(u) = \{\boldsymbol{X}(t,u) = u(\sin(2\pi t), \cos(2\pi t)), t \in (0,1)\}$$

with u uniformly distributed on (1, 5). The family is formed by concentric circles centered at the origin and having a radius $u \in (1, 5)$. Thus, we expect that the mean circle is the circle of mean radius:

$$E(C) = \left\{\overline{\boldsymbol{X}} = (\overline{X_1}, \overline{X_2}) \in \mathbb{R}^2 : \overline{X_1}^2(t) + \overline{X_2}^2(t) = 3^2, t \in (0,1)\right\}$$

We have

$$E(X_1(t,u)) = 3\sin(2\pi t), E(X_2(t,u)) = 3\cos(2\pi t)$$

so that the approach by the mean equation leads to the correct. Using Eq. (5.50) leads to the same result. Indeed,

$$E\left(\int_0^1 \|\boldsymbol{X}(t,u) - \overline{\boldsymbol{X}}(t)\|^2 dt\right) = J_1(\overline{X_1}) + J_2(\overline{X_2}),$$

where (c is a convenient constant)

$$J_1(\overline{X_1}) = \int_0^1 \left(\overline{X_1}^2(t) - 6\overline{X_1}(t)\sin(2\pi t)\right) dt + c,$$

$$J_2(\overline{X_2}) = \int_0^1 \left(\overline{X_2}^2(t) - 6\overline{X_2}(t)\cos(2\pi t)\right) dt + c.$$

Thus, the minimum is attained at $\overline{X_1}(t) = 3\sin(2\pi t), \overline{X_2}(t) = 3\cos(2\pi t)$.

The approach by the Hausdorff distance with a discretization of the solution leads again to results that do not correspond to the expected one, while d_{SH} furnishes the correct result. Approach (5.53) furnishes the exact result for both d_{SH} and d_H.

In this situation, the confidence interval with risk α is formed by the circles having radius from $3 - \Delta$ to $3 + \Delta$, $\Delta = 2(1 - \alpha)$. This region contains a proportion $1 - \alpha$ of the family (Fig. 5.53).

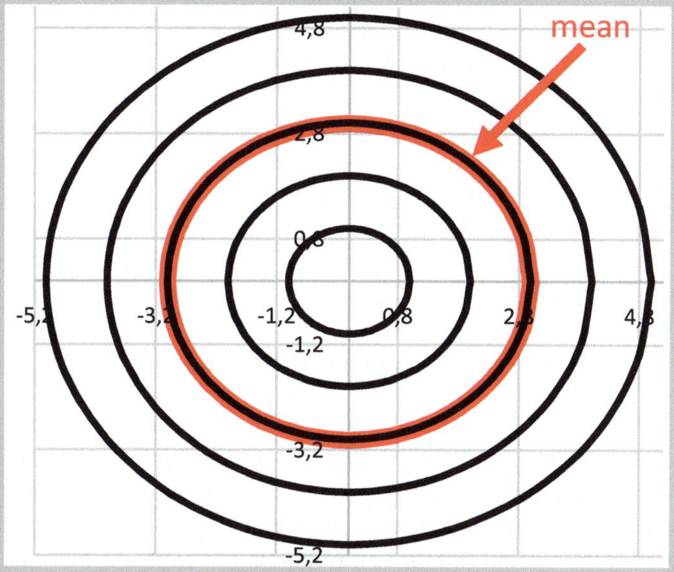

Fig. 5.53 A situation where all the approaches furnish the exact result: the family is formed by concentric circles of radius $u \in (1, 5)$, with the same parameterization

Example

Let us consider the indifference curves associated to the Leontief function ($\boldsymbol{u} = (u_1, u_2)$)

$$\mathcal{C}(\boldsymbol{u}) = \left\{ \boldsymbol{x} = (x_1, x_2) : \min \left\{ \frac{x_1}{u_1}, \frac{x_2}{u_2} \right\} = 1 \right\}$$

with $u_1, u_2 > 0$. In this case, the Hausdorff distance between two curves $\mathcal{C}(\boldsymbol{u})$ and $\mathcal{C}(\boldsymbol{v})$ can be computed analytically:

$$d_H(\mathcal{C}(\boldsymbol{u}), \mathcal{C}(\boldsymbol{v})) = \max \{ |u_1 - v_1|, |u_2 - v_2| \}$$

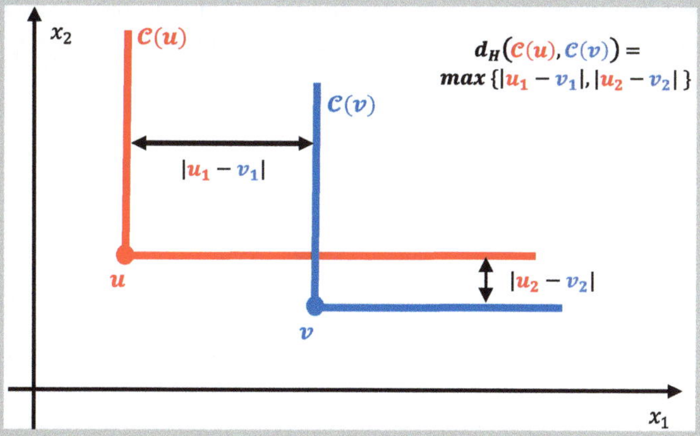

Fig. 5.54 Analytical evaluation of the Hausdorff distance between two Leontief's indifference curves

The mean distance between $C(\boldsymbol{u})$ and the family is the mean with respect to v:

$$\overline{d_H}(\boldsymbol{u}) = E_v(d_H(C(\boldsymbol{u}), C(\boldsymbol{v})))$$

The most representative element corresponds to the value

$$\boldsymbol{u}^* = \arg\min \left\{\overline{d_H}(\boldsymbol{u}) : \boldsymbol{u}\right\}.$$

In some situations, $\overline{d_H}(\boldsymbol{u})$ can be determined analytically, by integration. In addition, if the variables are independent and $\overline{\boldsymbol{u}} = (\overline{u_1}, \overline{u_2})$, where $\overline{u_i}$ is the median of u_i, then we can estimate

$$\overline{d_H}(\boldsymbol{u}^*) \geq E_v(d_H(C(\overline{\boldsymbol{u}}), C(\boldsymbol{v})))$$

Thus, if $\overline{\boldsymbol{u}}$ is a real value of $\boldsymbol{u}$, then $\boldsymbol{u}^* = \overline{\boldsymbol{u}}$. In general situations, it can be estimated from samples from (u_1, u_2). For instance, assume that u_1 and u_2 are independent and $u_1 \sim N(5, 0.2)$, $u_2 \sim T(4, 6, 5)$ – in this case, the exact solution is $u^* = (5, 5)$. We generate a sample of $\boldsymbol{u}$ by generating 200 values from each component and we use the sample to determine an approximation of $\boldsymbol{u}^*$ – the global sample from $\boldsymbol{u}$ is formed by 40000 points $(u_{1,\,i}, u_{2,\,j})$. We obtained $\boldsymbol{u}^* \approx (4.96, 5.00)$ – to be compared to the exact value (5, 5) (Fig. 5.54). ∎

Example

Let us consider the Cobb-Douglas function with two variables

$$cd(\boldsymbol{x}) = x_1^\alpha x_2^\beta$$

We are interested in the indifference curves $x_1^\alpha x_2^\beta = 1$, associated to this function, when α and β are uncertain: we look for the most representative curve and a confidence interval 80%.

Let us consider a first model where $\beta = 1 - \alpha$ and suppose that the experts agree that $\alpha \in (0.4, 0.6)$. We start by generating a family of curves: since no information about the distribution of α is given, let us assume a uniform distribution on the intervals. We generate a sample formed by a family of curves corresponding to na equally spaced subintervals for α and we apply the procedure previously introduced. The results obtained for $na = 100$ are shown in the Fig. 5.55:

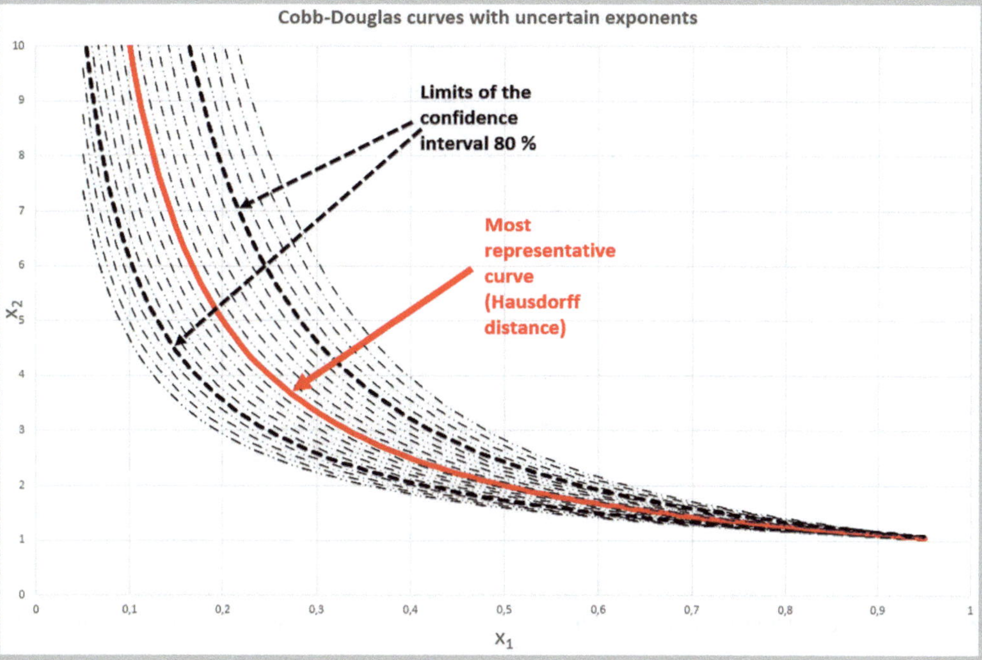

Fig. 5.55 First Cobb-Douglas model with uncertain exponents $\beta = 1 - \alpha$. α is uniformly distributed on (0.4, 0.6)

A second possibility when not any information is available about the distribution of α is the use of the triangular distribution – for instance, we can consider $T(0.4,0.6,0.5)$. Again, let us generate $na = 100$ variates from α and use the associated family of curves to find the most representative element. The result is shown in Fig. 5.56.

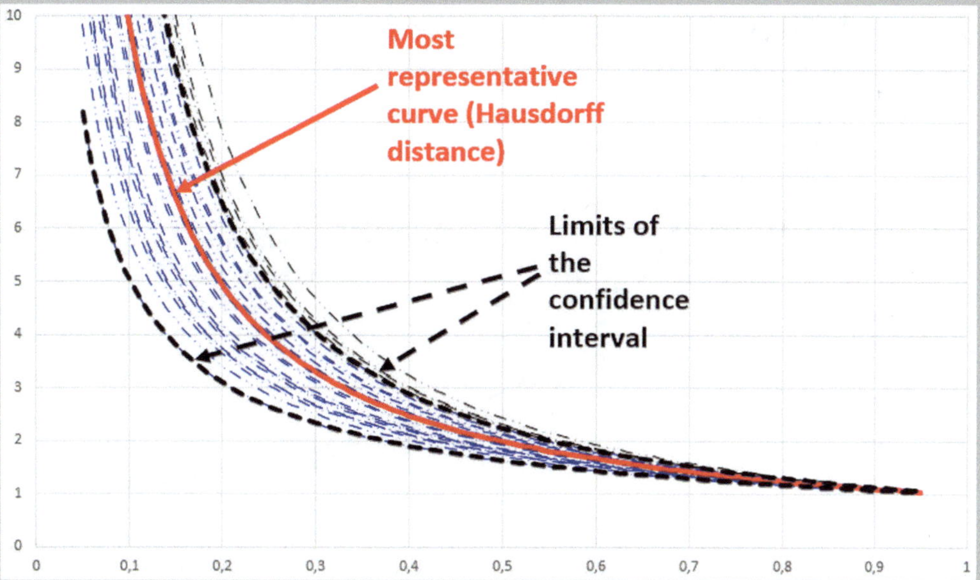

Fig. 5.56 First Cobb-Douglas model with uncertain exponents $\beta = 1 - \alpha$. α is triangular $T(0.4, 0.6,0.5)$

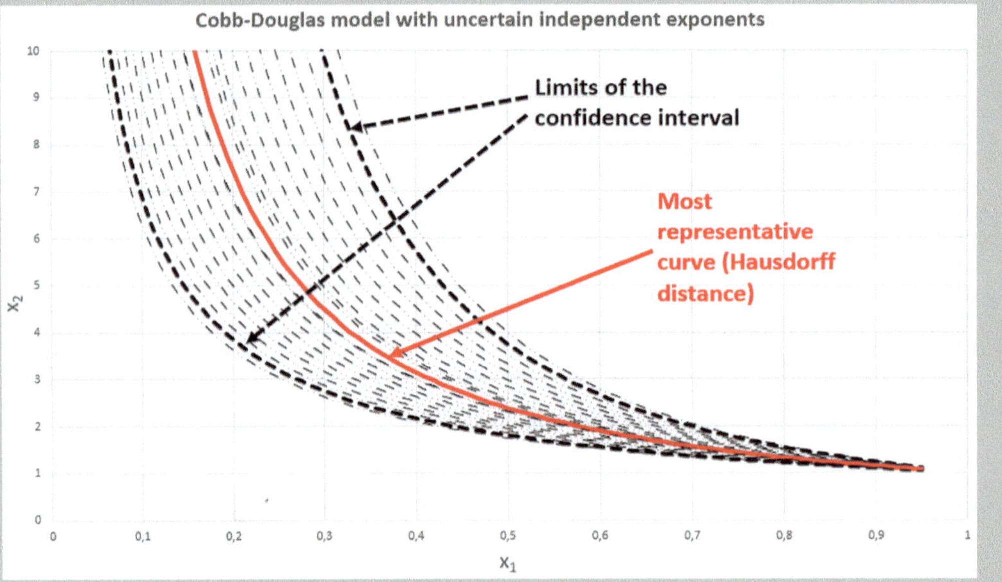

Fig. 5.57 Second Cobb-Douglas model with uncertain independent exponents α, β uniformly distributed

Consider a second model where α and β are independent and suppose that the experts agree that $\alpha \in (0.4, 0.6)$ and $\beta \in (0.3, 0.5)$. Again, we desire to determine the mean indifference curve and a confidence interval of level 80%. We start by generating a family of curves: since no information about the distribution of α and β is given, let us assume a uniform distribution on the intervals. We generate a sample formed by a family of curves corresponding to *na* subintervals for α and *nb* subintervals for β, all equally spaced. The results for $na = nb = 20$ are exhibited in Fig. 5.57. If the distributions are supposed triangular, the results correspond to those in Fig. 5.58

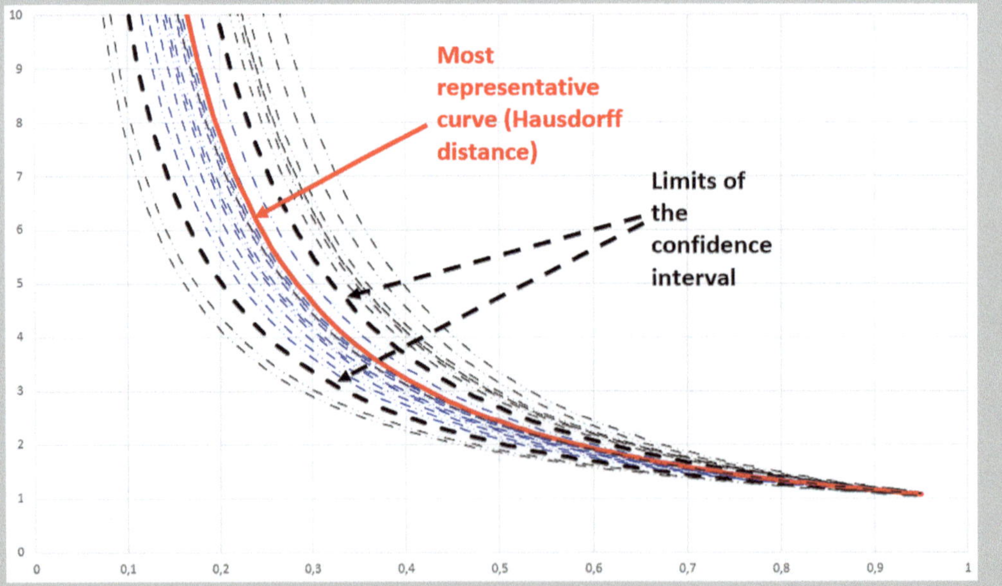

Fig. 5.58 Second Cobb-Douglas model with uncertain independent exponents α, β triangularly distributed $T(0.4, 0.6,0.5)$ and $T(0.3, 0.5,0.4)$

Example
Let us consider the Constant Elasticity Substitution function with two variables

$$ces(\boldsymbol{x}) = \left(\alpha x_1^{\beta} + (1-\alpha)x_2^{\beta}\right)^{\frac{1}{\beta}}.$$

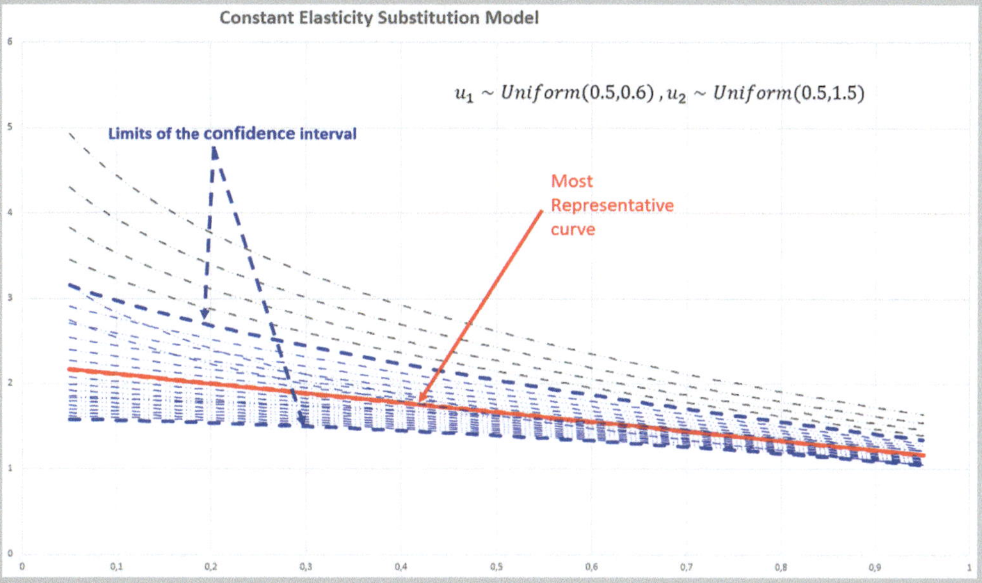

Fig. 5.59 Constant Elasticity Substitution model with uncertain independent exponents α, β both uniformly distributed

Let us analyze the indifference curves $ces(\boldsymbol{x}) = 1$, when α and β are uncertain: we look for the most representative curve and a confidence interval 80%.

Consider a model where α and β are independent and suppose that the experts agree that $\alpha \in (0.5, 0.6)$ and $\beta \in (0.5, 1.5)$. As in the preceding example, we start by generating a family of curves: since no information about the distribution of α and β is given, let us assume a uniform distribution on the intervals. We generate a sample formed by a family of curves corresponding to na subintervals for α and nb subintervals for β, all equally spaced. The results for $na = nb = 20$ are exhibited in Figs. 5.59 and 5.60. ∎

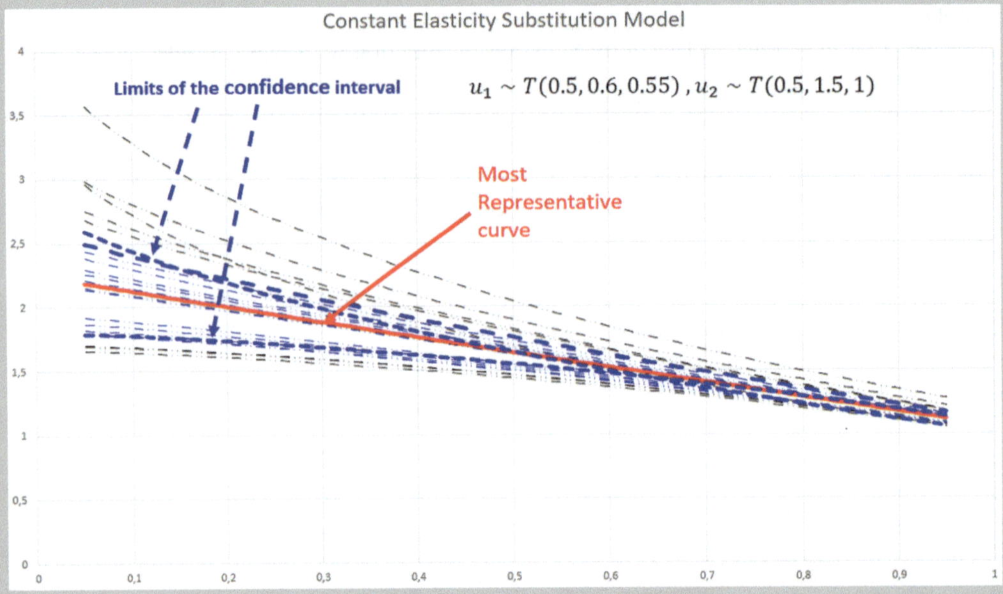

Fig. 5.60 Constant Elasticity Substitution model with uncertain independent exponents α, β., both triangular

Example

Let us consider the Pareto Front associated to the Binh-Korn problem involving two independent random variables $u_1 \sim N(2, 0.1)$ and $u_2 \sim T(0, 1, 0.5)$.

As shown in the example page 302, we may use a small sample and a discretization $t_i = 0.01 * i, 0 \leq i \leq 100$ to determine a representation of the family of the Pareto fronts.

Once the representation is determined, with the same data and parameters as in the example page 302, we can generate a large sample of 10,000 Pareto fronts using samples of 100 variates from each variable u_i, $i = 1, 2$. Then, we determine the mean point for each discretized time and determine a confidence interval 80%. The results are shown in Fig. 5.61.

Figure 5.62 compares the mean front generated by the expansions and the mean front corresponding to the most representative front among a sample of 2500 variates, generated by two samples of 50 values of each u_i, $i = 1, 2$. When using the expansion, the generation of a front involves the evaluation of a polynomial ate each t_i and the evaluation of the mean front correspond to the mean of the values determined at each t_i. When using the Hausdorff distance, the fronts are generated by solving the multiobjective optimization problem (method of Sect. 2.6) and then evaluating the Hausdorff distance of each member of the family to all the other members of the family. The computational cost of the evaluation by Hausdorff distance is much higher than the cost of the evaluation by using the expansion.

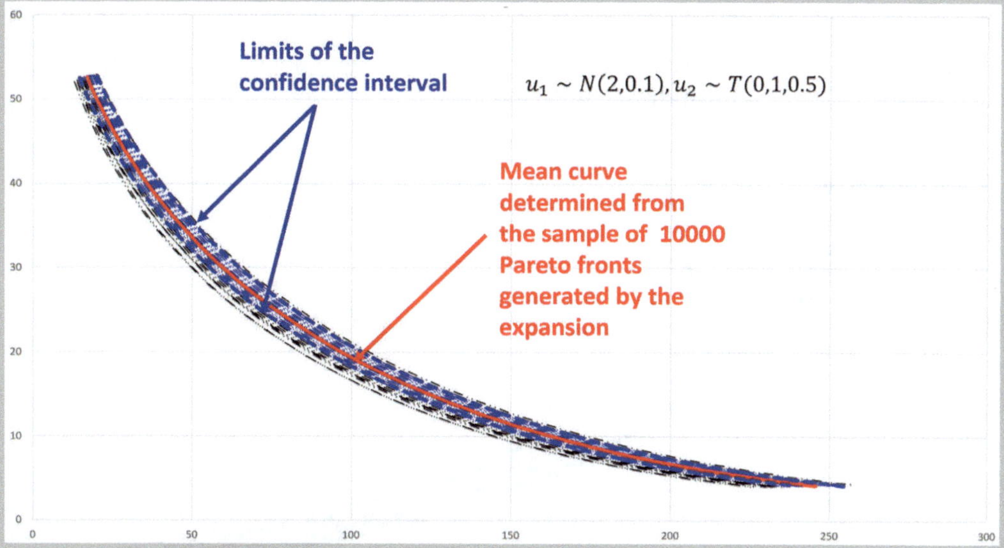

Fig. 5.61 Determination of the Mean Pareto Front and a confidence interval from the expansion, which is used to generate a sample of 10,000 Pareto fronts – the cost of the generation of a front is the evaluation of a polynomial at each value of t_i

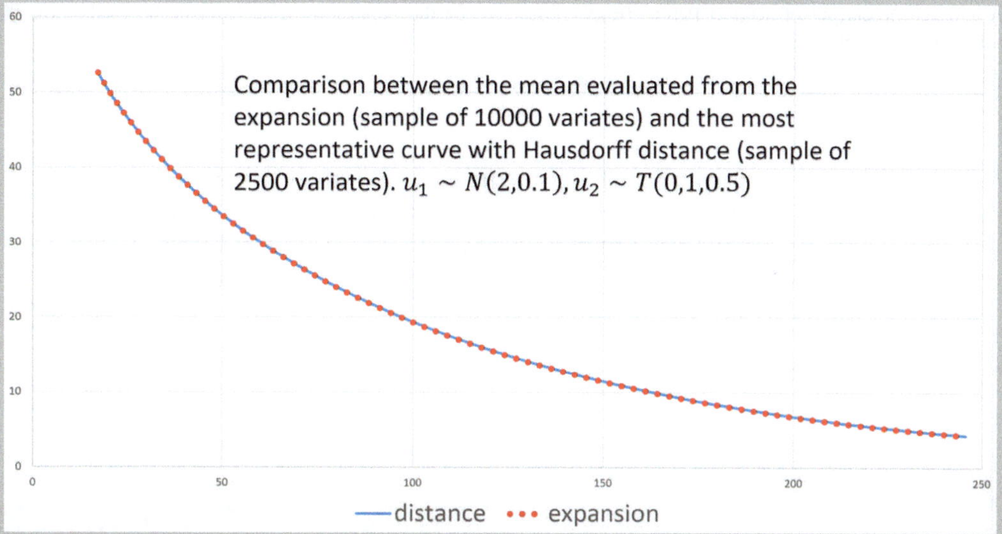

Fig. 5.62 Comparison between the mean Pareto Front generated by the expansion and the most representative Pareto front on a sample of 900 variates

The results for $u_1 \sim T(1.5{,}2.5,\ 2)$ and $u_2 \sim T(0,\ 1,\ 0.5)$ are shown in Figs. 5.63 and 5.64. ∎

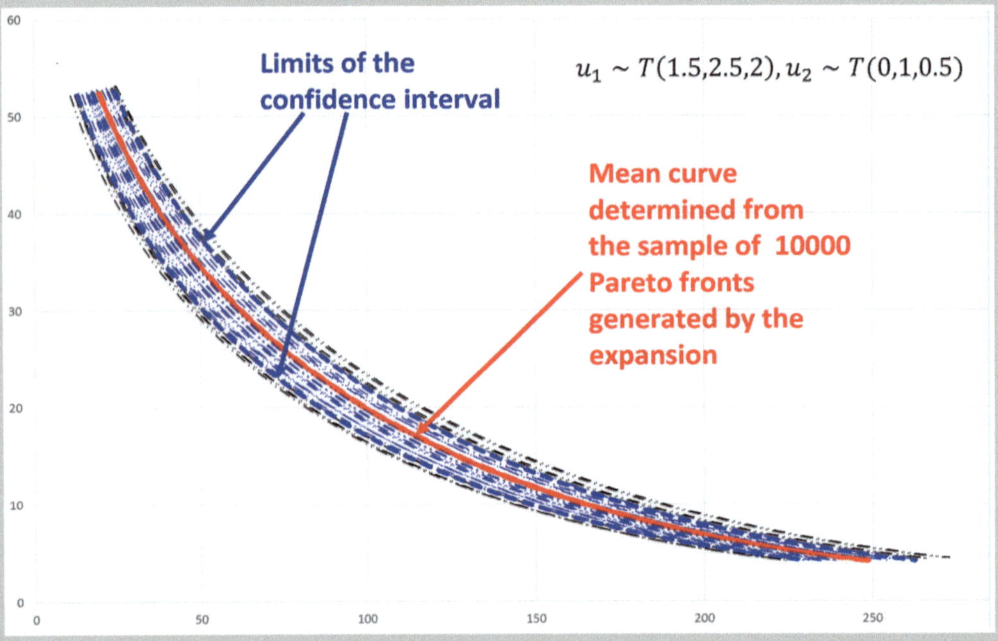

Fig. 5.63 Determination of the Mean Pareto Front and a confidence interval from the expansion, which is used to generate a sample of 10,000 Pareto fronts – the cost of the generation of a front is the evaluation of a polynomial at each value of t_i

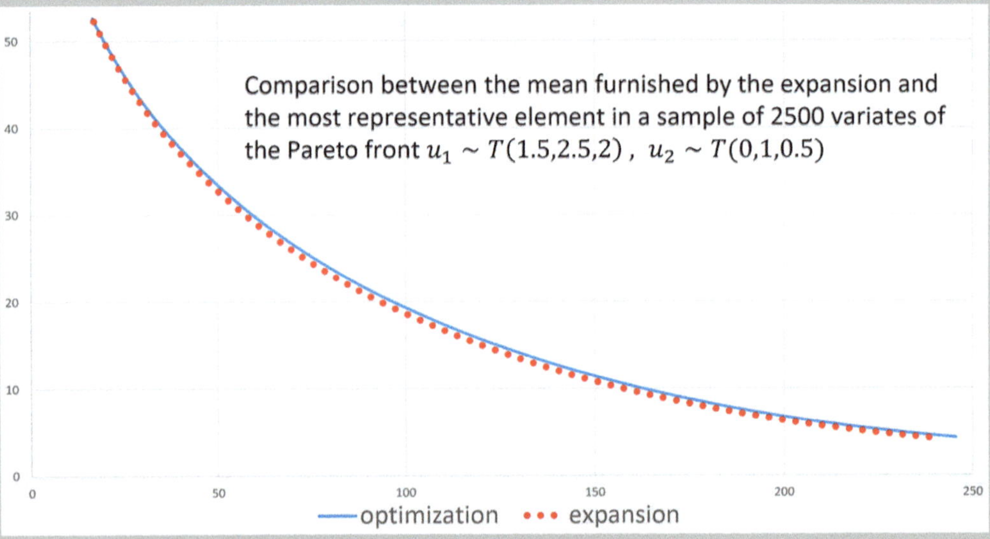

Fig. 5.64 Comparison between the mean Pareto Front generated by the expansion and the most representative Pareto front on a sample of 2500 variates

Example

Let us consider the portfolio selection problem introduced page 304: an investor desires to find the optimal allocation among three different options, in a stock exchange having a mean annualized volatility of $\sigma = 20\%$. Option 1 has a mean return $r_1 = 10\%$ with a coefficient of volatility $\beta_1 \sim T(0.9,1.2,1.05)$. Option 2 has a mean return of $r_2 = 15\%$ with a coefficient $\beta_2 = 1.2 \sim T(1.1,1.3,1.2)$. Option 3 has a mean return of $r_3 = 5\%$ with a coefficient of volatility $\beta_3 = 0.8$.

The unknowns are the fractions of capital $0 \le x_i \le 1$, $1 \le i \le 3$ and may be reduced to two variables $\boldsymbol{x} = (x_1, x_2)$ by considering that $x_3 = 1 - x_1 - x_2$, with $x_1, x_2 \ge 0, x_1 + x_2 \le 1$. Th mean return by unit of capital is

$$R(\boldsymbol{x}) = r_1 x_1 + r_2 x_2 + r_3(1 - x_1 - x_2)$$

and its variance is

$$V(\boldsymbol{x}) = \sigma^2 S(x), S(x) = \beta_1^2 r_1^2 x_1^2 + \beta_2^2 r_2^2 x_2^2 + (1 - x_1 - x_2)^2 r_3^2 \beta_3^2.$$

Page 304, we looked for the possible trade-offs between mean volatility $\sqrt{S(\boldsymbol{x}^*)}$ and mean return $R(\boldsymbol{x}^*)$, represented by a curve $(\sqrt{S(\boldsymbol{x}^*)}, R(\boldsymbol{x}^*))$ showing the possible compromises between Volatility and Return, with

$$\boldsymbol{x}^* = \arg\min \{S(\boldsymbol{x}) : x_1, x_2 \ge 0, x_1 + x_2 \le 1, R(\boldsymbol{x}) \ge R_{min}\}.$$

Then we found a representation of the trade-off curve. The representation can be used to generate a large sample of curves and the mean can be estimated from the sample generated. In Fig. 5.65, we compare the means furnished by the Hausdorff distance and the mean representation: the agreement between the curves is particularly good.

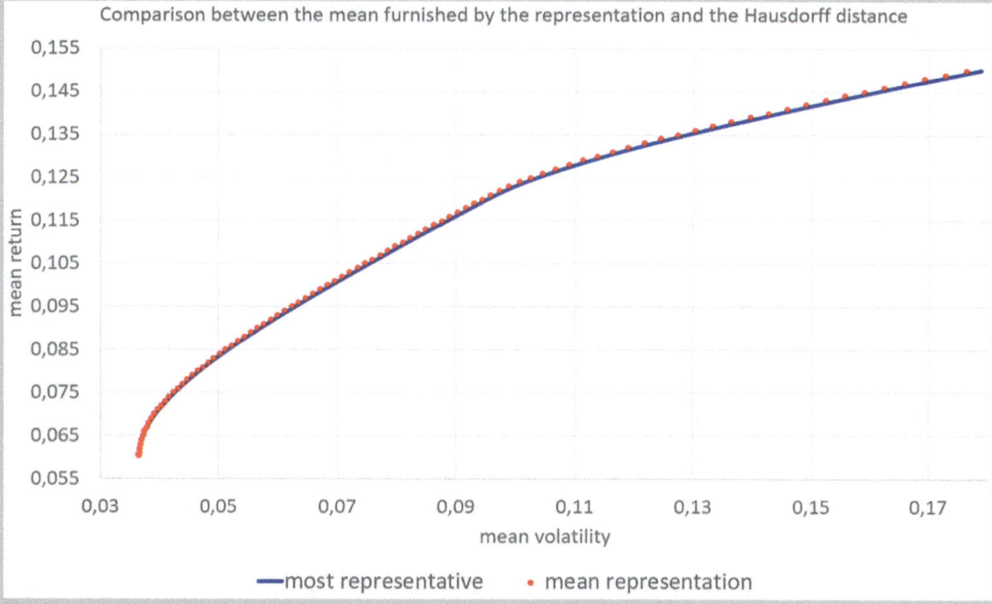

Fig. 5.65 Comparison between the mean evaluated from the representation and the most representative element evaluated using the Hausdorff distance

The sample generated by the representation can be used to determine a confidence interval 80%, using the Hausdorff distance. The results are shown in Fig. 5.66.

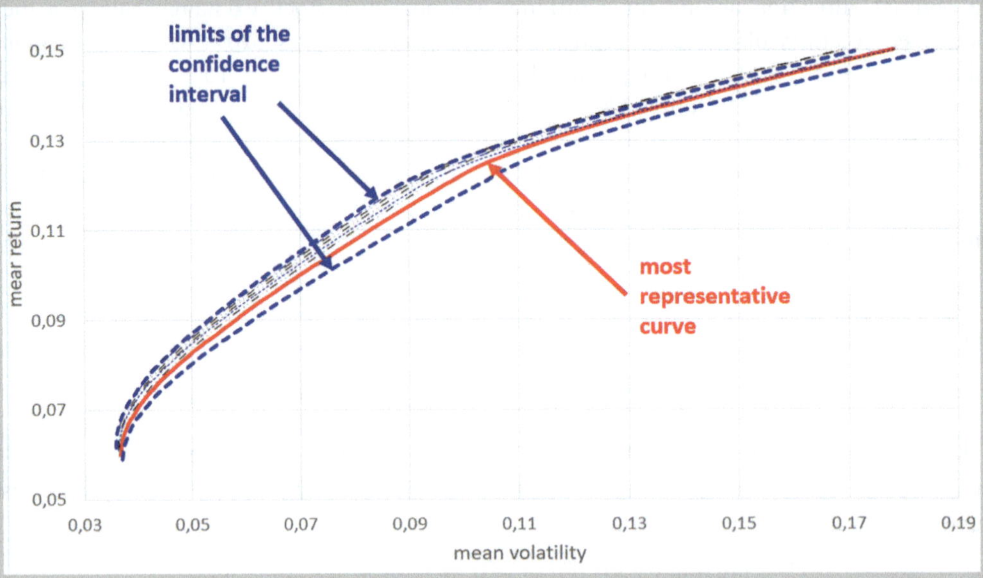

Fig. 5.66 Most representative curve and confidence interval 80% for the portfolio problem, both determined using the Hausdorff distance and the sample furnished by the expansion found page 304

When dealing with functions, the approach is analogous: we consider the graph of the function $t \longrightarrow X(t, u)$ as a curve:

$$\mathcal{C}(X) = \{(t, X(t, u)) : t \in (0, T)\}$$

and we apply the methods presented.

Example

Let us consider the family of random functions

$$X(t, u) = \cos(2\pi ut), t \in (0, 1).$$

with u uniformly distributed on $(-1, 1)$. Here

$$\mathcal{C}(u) = \{\boldsymbol{X}(t, u) = (t, \cos(2\pi ut)), t \in (0, 1)\}.$$

We have

$$E(X_1(t, u)) = t, E(X_2(t, u)) = \frac{\sin(2\pi t)}{2\pi t}$$

Analogously to the first example, this result may be considered as unsatisfactory, since the curve is not representative of the family. Let us look for the most representative element of the family. The result is shown in Fig. 5.67. ∎

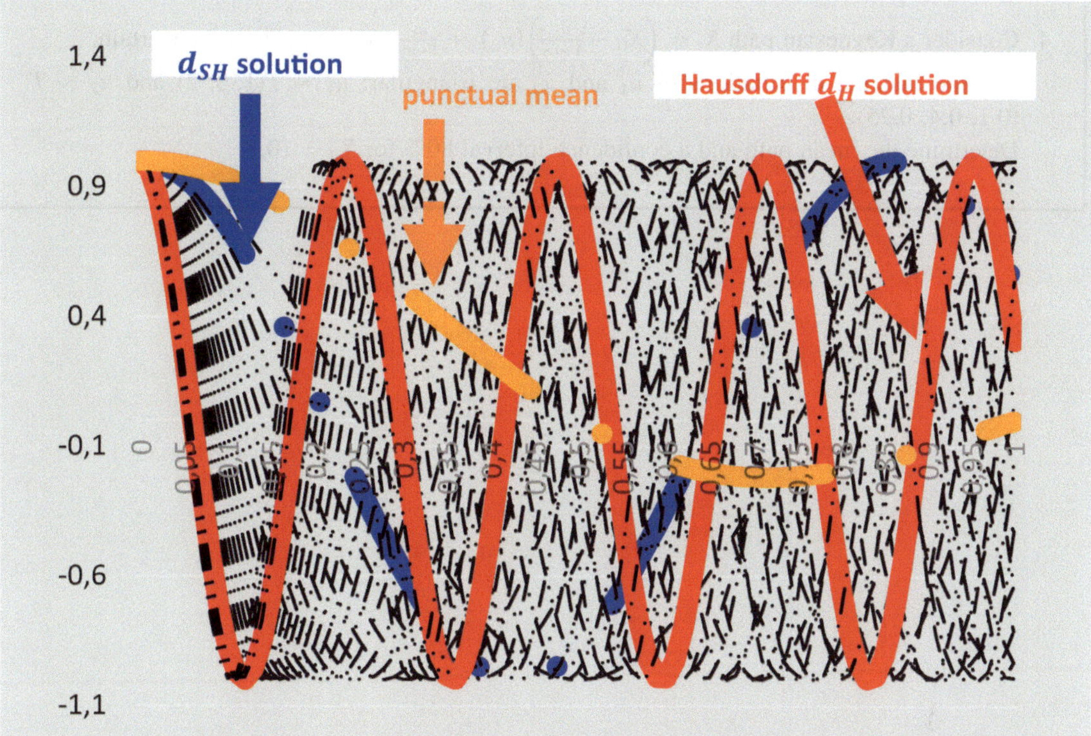

Fig. 5.67 Family of random curves. The mean evaluated by taking the punctual mean at fixed t furnishes a curve that is not representative of the family. We may look for an element that is the most representative in the sense of a distance. The results for two distances are shown in the Figure

Exercises

1. Consider a household's stocks time path given by

$$X_t = (1-u)^2 \frac{t^2}{4}.$$

   Assume that the marginal propensity to consume u is a random variable, uniformly distributed on (0.2, 0.8).
   Determine the mean path and a confidence interval 80%

2. Consider a capital path $X_t = \left(X_0 - \frac{s}{\delta}\right)e^{-\delta t} + \frac{s}{\delta}$, where s, δ are uncertain.
   Assume that the distributions of s and δ are triangular: $s \sim T(0.1, 0.5, 0.25)$ and $\delta \sim T(0.05, 0.25, 0.15)$.
   Determine the mean path and a confidence interval 80% for $X_0 = 10$.

3. Consider the family $X_t = u$, if $|t| < u$; $X_t = (|t| - u)^2 + u$, otherwise.
   Assume that the distribution of u is triangular: $u \sim T(0, 1, 0.5)$
   Determine the mean path and a confidence interval 80%.

4. Consider a Keynesian path $X_t = \left(X_0 - \frac{u_1}{1-u_2}\right)(u_2)^t + \frac{u_1}{1-u_2}$, where u_1, u_2 are uncertain. Assume that the distributions of u_1 and u_2 are triangular: $u_1 \sim T(1, 3, 2)$ and $u_2 \sim T(0.1, 0.4, 0.25)$.
Determine the mean path and a confidence interval 80% for $X_0 = 10$.

6 Uncertain Algebraic Equations

Solving equations is a basic activity in most fields of knowledge – for instance, Engineering, Management, Economics, … A simple example is furnished by the basic input-output analysis introduced by W. Leontieff in 1936 (Leontieff, 1936, 1937): the economy of a country may be empirically described by a matrix $\boldsymbol{T} = (T_{ij}, 1 \leq i, j \leq n)$ connecting inputs and outputs of the different economic sectors, such as the one presented in Fig. 6.1. Such a matrix synthetizes the empirical data about the connections and interdependencies between the economical fields under consideration. Nowadays, input-output matrices are used for statistical analysis and planning in Economics (Organisation for Economic Co-operation and Development, 2021). For instance, they may be used to furnish estimations of short-term impact of economic changes. The general form of $\boldsymbol{T}$ is shown in Table 6.1.

Table 6.1 Example of Input-Output Matrix

	Sector 1	....	Sector n
Sector 1	From 1 to 1	...	From n to 1
...	...	...	...
Sector n	From 1 to n	...	From n to n

$\boldsymbol{T}$ is referred as the flux matrix, productive matrix, technology matrix or Input-Output matrix. T_{ij} describes the part of consumption of goods produced in sector j for a unity production in sector i, *id est*, the fraction of the production of sector i absorbed by sector j. Line i of $\boldsymbol{T}$ gives the repartition of the production of sector i among all the sectors. Column j of $\boldsymbol{T}$ gives the fraction of the production of all the sectors necessary to the production of a unity in sector j. The matrix is usually normalized so that T_{ij} gives the fraction of the production of sector i consumed by sector j by unit of production of sector j – id est, by dividing column j by the total output of the sector j (id est, the total output of sector j is brought to a unity). The resulting matrix is called the structural matrix.

Let us exemplify the method of Leontieff by considering two sectors: agricultural and industry. Assume that the total production of the agricultural sector is of 250 units, while the production of the industry is of 160 units. From the 250 unities produced by the agricultural sector, 75 are consumed by itself, 50 by the industry and the rest is sold out on the markets (household, exportation). From the

Supplementary Information The online version of this chapter (https://doi.org/10.1007/978-3-030-77757-9_6) contains supplementary material, which is available to authorized users.

E. Souza de Cursi, *Uncertainty Quantification and Stochastic Modelling with EXCEL*, Springer Texts in Business and Economics, https://doi.org/10.1007/978-3-030-77757-9_6

160 units produced by the industry, 32 are destined to the agricultural sector, 16 to itself and the rest is sold out on markets. The data is synthetized in Table 6.2.

Table 6.2 An example of flux matrix

	Agricultural	Industry	Total Output
Agricultural	75	50	250
Industry	32	12	160

Table 6.2 may be transformed in a Table of values to consider financial aspects rather than the physical quantities produced. The normalization of Table 6.2 leads to the structural matrix in Table 6.3.

Table 6.3 Structural matrix ***T*** associated to Table 6.2

	Agricultural	Industry
Agricultural	0.3	0.3125
Industry	0.128	0.075

Matrix $\boldsymbol{T}$ is often completed by a vector $\boldsymbol{L}$ connecting the production to the Labor Market: L_i is the quantity of Labor requested in field i for the production of a unity. For instance, assume that the production requested 200 Labour units in the agricultural sector and 500 Labour units in industry. Then, vector $\boldsymbol{L}$ corresponds to the normalized line in Table 6.4.

Table 6.4 Structural matrix associated to Table 6.2

	Agricultural	Industry
Labour	200	500
Normalized (***L***)	0.8	3.125

From the empirical point of view, we may consider a simplified description of a local economy by a linear system involving the matrices $\boldsymbol{T}$ and $\boldsymbol{L}$. For instance, let the production be given by $\boldsymbol{X} = (X_1, \ldots, X_n)^t$, where X_i is the production of the sector i. Let the demand of each sector (id est, the goods available to consumers and exportation) be $\boldsymbol{B} = (B_1, \ldots, B_n)^t$ and the total Labor be $\boldsymbol{W} = (W_1, \ldots, W_n)^t$. These quantities verify the equations (see, for instance, (Arrous, 2000; Leontieff, 1986))

$$\boldsymbol{AX} = \boldsymbol{B}, \ \boldsymbol{A} = \boldsymbol{Id} - \boldsymbol{T}, \ \boldsymbol{W} = \boldsymbol{diag}(\boldsymbol{L})\boldsymbol{X} \tag{6.1}$$

Here, $\boldsymbol{diag}(\boldsymbol{L})$ is a diagonal matrix having $\boldsymbol{L}$ as diagonal.

Matrices $\boldsymbol{T}$ and $\boldsymbol{L}$ are deduced from empirical data and observations – thus, they contain errors, and their values are uncertain, due to the error margins. In addition, the demand $\boldsymbol{B}$ is also subjected to variability. Consequently, the analysis of Eq. (6.1) involves uncertainty.

As an example, assume that the demand is difficult to predict and is represented by independent random variables uniformly distributed: $B_1 = 125 + u_1$, with u_1 uniformly distributed on $(-10, 10)$, $B_2 = 110 + u_2$, with u_2 uniformly distributed on $(-5, 15)$. We are interested in finding the distribution of the Labor. In this situation, $\boldsymbol{X}$ is a random variable, which is a function of $\boldsymbol{B}$. Thus, $\boldsymbol{W}$ is also a random variable, which is yet a function of $\boldsymbol{B}$.

As a second example, we may consider innovation: a technical evolution provokes modifications in the structural matrix, but the changes are difficult to predict exactly. For instance, consider an agricultural evolution that produces a diminution of the self-consuming part and an increment of the part of industry consumed by the agricultural sector, but the analysts do not agree about the exact impact of these changes. Such an evolution leads to a new structural matrix

$$\boldsymbol{T} = \begin{pmatrix} 0.3 - u_1 & 0.3125 \\ 0.128 + u_2 & 0.075 \end{pmatrix}, \tag{6.2}$$

where u_1 is uniformly distributed on (0, 0.1) and u_2 is uniformly distributed on (0, 0.05).

In these simple situations, we can determine explicitly the connections. For instance, in the first situation,

$$\boldsymbol{X} = \boldsymbol{A}^{-1}\boldsymbol{B} \approx \begin{pmatrix} 246.914 + 1.523\, u_1 + 0.514\, u_2 \\ 153.086 + 0.211\, u_1 + 1.152\, u_2 \end{pmatrix}; \tag{6.3}$$

$$\boldsymbol{W} = \begin{pmatrix} L_1 X_1 \\ L_2 X_2 \end{pmatrix} \approx \begin{pmatrix} 222.222 + 1.37037\, u_1 + 0.462963\, u_2 \\ 478.395 + 0.658\, u_1 + 3.601\, u_2 \end{pmatrix}. \tag{6.4}$$

In the second situation,

$$\boldsymbol{X} = \boldsymbol{A}^{-1}\boldsymbol{B} \approx \begin{pmatrix} \dfrac{0.925\, B_1 + 0.3125\, B_2}{0.608 + 0.925\, u_1 - 0.3125\, u_2} \\ \dfrac{(0.128 + u_2)\, B_1 + (0.7 + u_1)\, B_2}{0.608 + 0.925\, u_1 - 0.3125\, u_2} \end{pmatrix} \tag{6.5}$$

$$\boldsymbol{W} = \begin{pmatrix} L_1 X_1 \\ L_2 X_2 \end{pmatrix} \approx \begin{pmatrix} \dfrac{0.9\, B_1 + 0.304\, B_2}{0.657 + u_1 - 0.338\, u_2} \\ \dfrac{0.432\, B_1 + 2.365\, B_2 + 3.378\, B_2\, u_1 + 3.378\, B_1\, u_2}{0.657 + u_1 - 0.338\, u_2} \end{pmatrix}. \tag{6.6}$$

These equalities allow the determination of the distribution of $\boldsymbol{W}$.

However, when considering more realistic situations, the number of sectors increase, so that the matrix $\boldsymbol{A}$ becomes larger. Analogously, when considering nonlinear situations, explicit solution may become difficult: it is necessary to provide other methods of solution.

A first approach consists in generating a sample from the variables and using this sample to determine an expansion by one of the methods shown in Chap. 5, but we shall examine a second approach.

Exercises

1. Consider the matrix

$$\boldsymbol{A} = \begin{pmatrix} 0.7 + u_1 & -0.5 - u_1 & -0.3 \\ -0.2 & 0.8 + u_2 & -0.3 - u_2 \\ -0.4 & -0.2 & 0.7 \end{pmatrix}$$

Assume that all the variables u_i are independent and triangularly distributed $T(0,0.5,0.2)$.

(a) Generate a sample of 10 variates from each random variable u_i.

(b) Use the values generate to generate a sample of 100 variates from $\boldsymbol{U} = (u_1, u_2)$.

(c) Use the sample from to determine a representation of the dominant eigenvalue of $\boldsymbol{A}$ as function of $\boldsymbol{U} = (u_1, u_2)$.

(d) Determine the Marginal CDF and PDF of the dominant eigenvalue.

2. Consider

$$\boldsymbol{A} = \begin{pmatrix} u_1 & 4 \\ 1 & \mathrm{u}_2 \end{pmatrix}$$

Assume that the distributions of u_1 and u_2 are triangular: $u_1 \sim T(1, 3, 2)$ and $u_2 \sim T(0,10,5)$.

(a) Generate a sample of 10 variates from each random variable u_i.

(b) Use these values to generate a sample of 100 variates from $\boldsymbol{U} = (u_1, u_2)$.

(c) Use the sample from to determine a representation of the eigenvalues of $\boldsymbol{A}$ as function of $\boldsymbol{U} = (u_1, u_2)$.

(d) Determine the Marginal CDF and PDF of the dominant eigenvalue.

3. Consider

$$A = \begin{pmatrix} 0.2 + \mathrm{u}_1 & \mathrm{u}_1 \\ \mathrm{u}_2 & -10 \end{pmatrix}$$

Assume that the distributions of u_1 and u_2 are triangular: $u_1 \sim T(1, 3, 2)$ and $u_2 \sim T(0, 2, 1)$.

(a) Generate a sample of 10 variates from each random variable u_i.

(b) Use these values to generate a sample of 100 variates from $\boldsymbol{U} = (u_1, u_2)$.

(c) Use the sample from to determine a representation of the eigenvalues of $\boldsymbol{A}$ as function of $\boldsymbol{U} = (u_1, u_2)$.

(d) Determine the Marginal CDF and PDF of the eigenvalues.

6.1 Uncertain Linear Systems

Let us consider a linear system

$$\boldsymbol{A}(\boldsymbol{U})\boldsymbol{X} = \boldsymbol{B}(\boldsymbol{U}), \tag{6.7}$$

where $\boldsymbol{U}$ is a random vector, $\boldsymbol{A}$ is a $n \times n$ matrix, $\boldsymbol{B}$ is a $n \times 1$ vector. Let us consider an expansion analogous to Eqs. (5.34) and (5.35) of Chap. 5:

$$\boldsymbol{X} \approx \boldsymbol{PX} = \sum_{j=1}^{k} \boldsymbol{x}_j \varphi_j(\boldsymbol{U}), \boldsymbol{x}_j = \left(x_{1,j}, \ldots, x_{n,j}\right)^{t}. \tag{6.8}$$

Then, using Eq. (6.7):

$$\boldsymbol{A}(\boldsymbol{U})\boldsymbol{PX} \approx \boldsymbol{B}(\boldsymbol{U}),$$

id est,

$$\sum_{j=1}^{k} \boldsymbol{A}(\boldsymbol{U})\boldsymbol{x}_j\varphi_j(\boldsymbol{U}) \approx \boldsymbol{B}(\boldsymbol{U}).$$

Thus, for $1 \leq r \leq n$

$$\sum_{j=1}^{k}\sum_{s=1}^{n} A_{rs}(\boldsymbol{U})x_{s,j}\varphi_j(\boldsymbol{U}) \approx B_r(\boldsymbol{U}),$$

so that, for $1 \leq i \leq k$

$$\sum_{j=1}^{k}\sum_{s=1}^{n} \varphi_i(\boldsymbol{U})A_{rs}(\boldsymbol{U})x_{s,j}\varphi_j(\boldsymbol{U}) \approx \varphi_i(\boldsymbol{U})B_r(\boldsymbol{U}),$$

Let us denote

$$\mathcal{A}_{risj} = E\big(\varphi_i(\boldsymbol{U})A_{rs}(\boldsymbol{U})\varphi_j(\boldsymbol{U})\big),\quad \mathcal{B}_{ri} = E(\varphi_i(\boldsymbol{U})B_r(\boldsymbol{U})) \tag{6.9}$$

Then, for $1 \leq r \leq n$, $1 \leq i \leq k$:

$$\sum_{j=1}^{k}\sum_{s=1}^{n} \mathcal{A}_{risj}x_{s,j} \approx \mathcal{B}_{ri}. \tag{6.10}$$

Equation (6.10) form a linear system for the unknown coefficients $(x_{s,j} : 1 \leq s \leq n, 1 \leq j \leq k)$. We may transform it in a classical linear system by using the map ℓ_2 introduced in Sect. 5.5.1. For instance, set

$$p = i + (r-1)k, q = j + (s-1)k. \tag{6.11}$$

Then, taking

$$\mathbb{A}_{pq} = \mathcal{A}_{risj}, \mathbb{B}_p = \mathcal{B}_{ri}, \mathbb{X}_q = x_{s,j}, \tag{6.12}$$

We have

$$\mathbb{A}\mathbb{X} = \mathbb{B}. \tag{6.13}$$

Thus, we may determine the coefficients of the expansion by solving the linear system (6.13). In practice, the information on $\boldsymbol{U}$ is frequently furnished by a sample $(\boldsymbol{U_m} : 1 \leq m \leq ns)$. Then,

$$\mathcal{A}_{risj} \approx \frac{1}{ns}\sum_{m=1}^{ns} \varphi_i(\boldsymbol{U_m})A_{rs}(\boldsymbol{U_m})\varphi_j(\boldsymbol{U_m}), \mathcal{B}_{ri} \approx \frac{1}{ns}\sum_{m=1}^{ns} \varphi_i(\boldsymbol{U_m})B_r(\boldsymbol{U_m}) \tag{6.14}$$

An example of VBA code for the generation of matrices $\mathbb{A}$ and $\mathbb{B}$ is given below:

```
Function varlinsys(u, basis, k, n)
ns = UBound(u, 1)
dimU = UBound(u, 2)
nn = k * n
Dim aa() As Double
ReDim aa(1 To nn, 1 To nn)
Dim bb() As Double
ReDim bb(1 To nn)
For i = 1 To nn
  bb(i) = 0
  For j = 1 To nn
    aa(i, j) = 0
  Next j
Next i
Dim uu()
ReDim uu(1 To dimU)
```

```
  For r = 1 To n
    For i = 1 To k
      p = i + (r - 1) * k
      aux1 = basis.value(i - 1, uu)
      bb(p) = bb(p) + Buu(r) * aux1
      For s = 1 To n
        For j = 1 To k
          q = j + (s - 1) * k
          aux2 = basis.value(j - 1, uu)
          aa(p, q) = aa(p, q) + aux1 * Auu(r, s) * aux2
        Next j
      Next s
    Next i
  Next r
Next m
For r = 1 To n
  For i = 1 To k
    p = i + (r - 1) * k
    bb(p) = bb(p) / ns
    For s = 1 To n
      For j = 1 To k
        q = j + (s - 1) * k
        aa(p, q) = aa(p, q) / ns
      Next j
    Next s
  Next i
Next r
Dim resu(1 To 2) As Variant
resu(1) = aa
resu(2) = bb
varlinsys = resu
End Function
```

To use `varlinsys`, you must define the sample u, the basis $\{\varphi_i : 1 \leq i \leq k\}$, the order of approximation k and the dimension n of the linear system. Let us illustrate its use by solving the Leontieff's examples previously presented.

Example

Let us consider the situation corresponding to Eq. (6.1), where $B_1 = 125 + u_1$, with u_1 uniformly distributed on $(-10, 10)$, $B_2 = 110 + u_2$, with u_2 uniformly distributed on $(-5, 15)$; while

$$T = \begin{pmatrix} 0.3 & 0.3125 \\ 0.128 & 0.075 \end{pmatrix}$$

We use the data furnished by a sample of 36 values from $\boldsymbol{U} = (u_1, u_2)$, formed by the pairs $(u_{1,\,i}, u_{2,\,j})$, $1 \leq i, j \leq 6$, given in Table 6.5.

Table 6.5 Data used for the generation of the sample

u_1	−936,334	−907,657	−805,736	−657,627	−446,154	4,120,922
u_2	−431,108	1,341,990	3,774,887	889,657	11,46,916	14,00444

In this case, $\boldsymbol{X}$ is given by Eq. (6.3). Since $\boldsymbol{U} = (u_1, u_2)$, we use the expansion (6.34), with two polynomial families of degree $k^1 = k^2 = 3$.

An example of VBA code for the creation of this family, using `tensorbasis.cls` (see page 288) is given below:

```
lb1 = -10
ub2 = 10
lb2 = -5
ub2 = 15

Dim p1 As New poly
p1.lb = lb1
p1.ub = ub1
p1.zero = zero
Dim p2 As New poly
p2.lb = lb2
p2.ub = ub2
p2.zero = zero
```

```
k1 = 3
k2 = 3
Dim p As New tensorbasis
Set p.phi1 = p1
Set p.phi2 = p2
indb = p.create_indexes(k1, k2)
p.indbase = indb
```

Then, we determine the coefficients of the expansion (6.34):

```
resu = varlinsys(u, p, k, n)
aa = resu(1)
bb = resu(2)
XX = LUSolution(aa, bb)
Dim x() As Double
ReDim x(1 To n, 1 To k)
For r = 1 To n
  For i = 1 To k
    pp = i + (r - 1) * k
    x(r, i) = XX(pp)
  Next i
Next r
```

We obtained the coefficients shown in Table 6.6. Figures 6.1 and 6.2 show the marginal distribution of each component of the solution: the accordance is particularly good. ∎.

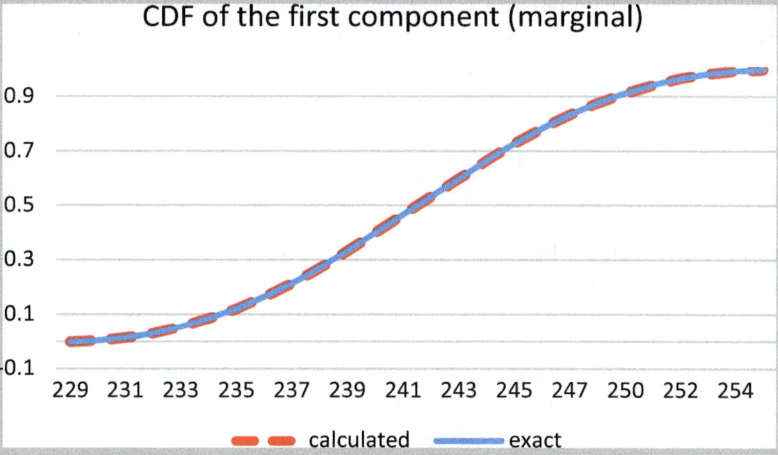

Fig. 6.1 CDF of the first component of the solution (marginal CDF of x_1)

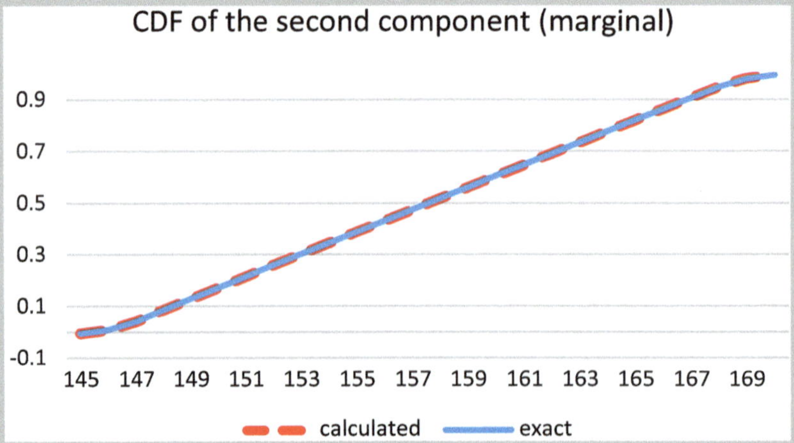

Fig. 6.2 CDF of the second component of the solution (marginal CDF of x_2)

Table 6.6 Coefficients obtained

$x_{1,j}$	229,12	10,29	0,00	15,23	0,00	0,00	0,00	0,00	0,00
$x_{2,j}$	145,22	23,05	0,00	2,11	0,00	0,00	0,00	0,00	0,00

Example

Let us consider the second situation corresponding to Eq. (6.1), where $B_1 = 125, B_2 = 110$;

$$\boldsymbol{T} = \begin{pmatrix} 0.3 - u_1 & 0.3125 \\ 0.128 + u_2 & 0.075 \end{pmatrix}$$

where u_1 is uniformly distributed on (0, 0.1) and u_2 is uniformly distributed on (0, 0.05).

We use the data furnished by a sample of 36 values from $\boldsymbol{U} = (u_1, u_2)$, formed by the pairs $(u_{1,i}, u_{2,j})$, $1 \le i, j \le 6$, given in Table 6.7.

Table 6.7 Data used for the generation of the sample

u_1	0,00086	0,040390	0,049284	0,072618	0,074695	0,096776
u_2	0,000932	0,030252	0,033205	0,038672	0,043191	0,046179

In this case, $\boldsymbol{X}$ is given by Eq. (6.5). As in the preceding example, we use the expansion (6.34), with two polynomial families of degree $k^1 = k^2 = 3$.

Table 6.8 Coefficients obtained

$x_{1,j}$	247,37	5,77	0,14	35,65	1,63	0,06	6,71	0,51	0,03
$x_{2,j}$	153,40	12,90	0,31	4,97	1,97	0,09	0,93	0,40	0,03

We obtained the coefficients shown in Table 6.8. Figures 6.3 and 6.4 show the marginal distribution of each component of the solution: again, the accordance is extremely good. ∎

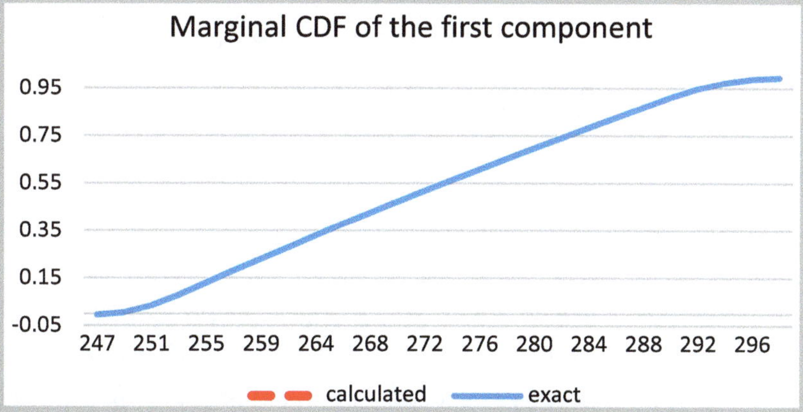

Fig. 6.3 CDF of the first component of the solution (marginal CDF of x_1)

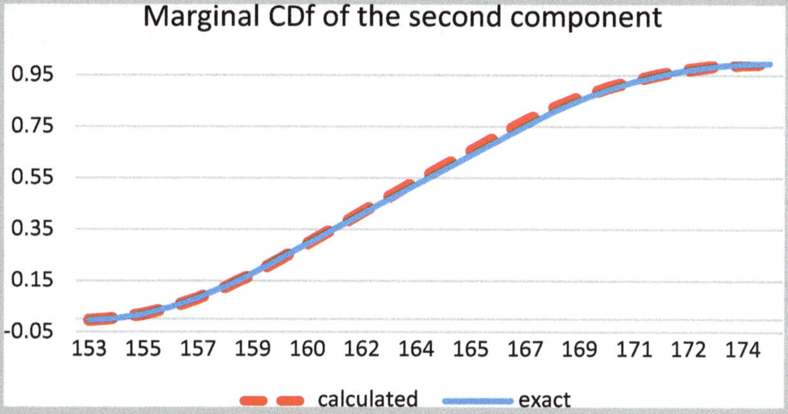

Fig. 6.4 CDF of the second component of the solution (marginal CDF of x_2)

6.1.1 Very Small Linear Systems

If your linear system has at just a few variables, you may adopt a simplified approach. Indeed, Cramer's rule may be used if you have just 2 or 3 variables. In his book (Cramer, 1750), Cramer gives the explicit solution of a linear system for 2 or 3 variables. Using Cramer's notation, a linear system

$$\begin{pmatrix} X^1 & Y^1 \\ X^2 & Y^2 \end{pmatrix} \begin{pmatrix} x \\ y \end{pmatrix} = \begin{pmatrix} B^1 \\ B^2 \end{pmatrix}$$

has as solution

$$x = \frac{-B^2Y^1 + B^1Y^2}{-X^2Y^1 + X^1Y^2} \; , \; y = \frac{B^2X^1 - B^1X^2}{-X^2Y^1 + X^1Y^2} \; .$$

A linear system

$$\begin{pmatrix} X^1 & Y^1 & Z^1 \\ X^2 & Y^2 & Z^2 \\ X^3 & Y^3 & Z^3 \end{pmatrix} \begin{pmatrix} x \\ y \\ z \end{pmatrix} = \begin{pmatrix} B^1 \\ B^2 \\ B^3 \end{pmatrix}$$

has as solution

$$x = \frac{B^3Y^2Z^1 - B^2Y^3Z^1 - B^3Y^1Z^2 + B^1\mathrm{Y}3Z^2 + B^2Y^1Z^3 - B^1Y^2Z^3}{X^3Y^2Z^1 - X^2Y^3Z^1 - X^3Y^1Z^2 + X^1Y^3Z^2 + X^2Y^1Z^3 - X^1Y^2Z^3},$$

$$y = \frac{B^3X^2Z^1 - B^2X^3Z^1 - B^3X^1Z^2 + B^1X^3Z^2 + B^2X^1Z^3 - B^1X^2Z^3}{-X^3Y^2\mathrm{Z}1 + X^2Y^3Z^1 + X^3Y^1Z^2 - X^1Y^3Z^2 - X^2Y^1Z^3 + X^1Y^2Z^3},$$

$$z = \frac{B^3X^2\mathrm{Y}1 - B^2X^3\mathrm{Y}1 - B^3X^1Y^2 + B^1X^3Y^2 + B^2X^1Y^3 - B^1X^2Y^3}{X^3Y^2Z^1 - X^2Y^3Z^1 - X^3Y^1Z^2 + X^1Y^3Z^2 + X^2Y^1Z^3 - X^1Y^2Z^3}.$$

Thus, for very small linear systems involving 2 or 3 unknowns, the explicit solution above may be used to generate samples. If the distributions of the terms X^i, Y^j, Z^k, B^ℓ are known, then large samples of the solutions may be generated. If small samples of these variables are given, then small samples of the solutions can be generated and a method of representation (Chap. 5) must be employed.

Example

Consider again the situation corresponding to Eq. (6.1), where $B_1 = 125 + u_1, u_1 \sim \mathcal{U}(-10, 10)$, $B_2 = 110 + u_2, u_2 \sim \mathcal{U}(-5, 15)$;

$$\boldsymbol{T} = \begin{pmatrix} 0.3 & 0.3125 \\ 0.128 & 0.075 \end{pmatrix}$$

In this case, $= \boldsymbol{Id} - \boldsymbol{T}$, so that

$$X^1 = 0.7, X^2 = -0.128, Y^1 = -0.3125, Y^2 = 0.925.$$

Thus,

$$-X^2\mathrm{Y}^1 + X^1Y^2 = 0.6075.$$

Since $B^1 = 125 + u_1$, $B^2 = 110 + u_2$, we have

$$-B^2Y^1 + B^1Y^2 = 150 + 0.925u_1 + 0.3125u_2,$$

$$B^2X^1 - B^1X^2 = 93 + 0.128u_1 + 0.7u_2.$$

So, we have

$$x_1 = \frac{150 + 0.925u_1 + 0.3125u_2}{0.6075},$$

$$x_2 = \frac{93 + 0.128u_1 + 0.7u_2}{0.6075}.$$

Finally

$$x_1 \approx 246.9136 + 1.522634\, u_1 + 0.514403\, u_2,$$

$$x_2 \approx 153,0864 + 0.2107\, u_1 + 1.152263\, u_2.$$

Thus, Cramer's rule furnishes the same result as Eq. (6.3). If data in Table 6.5 is given, we can generate a sample from x_1:

u_2 \ u_1	−9,36334	−9,07657	−8,05736	−6,57627	−4,46154	4,120922
−4,31108	230,4390	230,8757	232,4275	234,6827	237,9027	250,9706
1,34199	233,3470	233,7836	235,3355	237,5907	240,8106	253,8786
3,774887	234,5985	235,0351	236,5870	238,8421	242,0621	255,1300
8,89657	237,2331	237,6697	239,2216	241,4768	244,6967	257,7647
11,46916	238,5564	238,9931	240,5449	242,8001	246,0201	259,0880

This sample can be used to determine an expansion representing x_1 as a function of $\boldsymbol{U} = (u_1, u_2)$. An analogous table can be generated for x_2. Since the results are the same as in Eq. (6.3), we shall obtain the same results as in the Example page 328. ∎

Example

Consider again the situation corresponding to Eq. (6.1), where $B_1 = 125$, $B_2 = 110$, $u_1 \sim \mathcal{U}(0, 0.1)$, $u_2 \sim \mathcal{U}(0, 0.05)$;

$$\boldsymbol{T} = \begin{pmatrix} 0.3 - u_1 & 0.3125 \\ 0.128 + u_2 & 0.075 \end{pmatrix}$$

In this case, $\boldsymbol{A} = \boldsymbol{Id} - \boldsymbol{T}$ and

$$X^1 = 0.7 + u_1, X^2 = -0.128 - u_2, Y^1 = -0.3125, Y^2 = 0.925.$$

Thus,

$$-X^2Y^1 + X^1Y^2 = 0.6075 + 0.9025\, u_1 - 0.3125\, u_2.$$

Since $B^1 = 125$, $B^2 = 110$, we have

$$-B^2Y^1 + B^1Y^2 = 150,$$

$$B^2X^1 - B^1X^2 = 93 + 110\, u_1 + 125\, u_2.$$

So, we have

$$x_1 = \frac{150}{0.6075 + 0.9025\, u_1 - 0.3125\, u_2},$$

$$x_2 = \frac{93 + 110\, u_1 + 125\, u_2}{0.6075 + 0.9025\, u_1 - 0.3125\, u_2}.$$

If data in Table 6.7 is given, we can generate a sample from x_2:

u_2 \ u_1	0,00086	0,04039	0,049284	0,072618	0,074695	0,096776
0,000932	153,289	163,767	166,185	172,637	173,219	179,488
0,030252	152,673	162,693	165,002	171,161	171,717	177,694
0,033205	152,614	162,590	164,889	171,020	171,573	177,522
0,038672	152,505	162,401	164,681	170,761	171,309	177,208
0,043191	152,417	162,247	164,512	170,550	171,094	176,952

This sample can be used to determine an expansion representing x_2 as a function of $\boldsymbol{U} = (u_1, u_2)$. An analogous table can be generated for x_1. Since the results are the same as in Eq. (6.5), we shall obtain the same results as in the Example page 330. ∎

Exercises

1. Consider an input-output model where the structural matrix T and the Labor matrix are

$$\boldsymbol{T} = \begin{pmatrix} 0.3 + u_1 & 0.5 - u_1 & 0.3 \\ 0.2 & 0.2 & 0.3 \\ 0.4 & 0.2 & 0.3 \end{pmatrix}, \boldsymbol{L} = \begin{pmatrix} 300 \\ 150 + u_2 \\ 400 \end{pmatrix}$$

Assume that all the variables u_i are independent and triangularly distributed $T(-0.1,0.1,0)$.

(a) Generate a sample of 6 variates from each random variable u_i.
(b) Use the values generate to generate a sample of 36 variates from x_1.
(c) Use the sample from x_1 to determine a representation of x_1 as function of $\boldsymbol{U} = (u_1, u_2)$.
(d) Use `varlinsys` to determine a representation of x_1 as function of $\boldsymbol{U} = (u_1, u_2)$.
(e) Compare the results obtained in c) and d).
(f) Determine the Marginal distribution of x_1.

2. Consider a model for the equilibrium between two Markets is characterized by two variables (x_1, x_2) such that

$$\begin{pmatrix} 1 & 4 \\ 1 & -10 \end{pmatrix} \begin{pmatrix} x_1 \\ x_2 \end{pmatrix} = \begin{pmatrix} 200 - u_1 + u_2 \\ 300 + u_1 \end{pmatrix}$$

Assume that the distributions of u_1 and u_2 are triangular: $u_1 \sim T(0, 50, 25)$ and $u_2 \sim T(-50, 50, 0)$.

(a) Generate a sample of 6 variates from each random variable u_i.
(b) Use the values generate to generate a sample of 36 variates from x_1.
(c) Use the sample from x_1 to determine a representation of x_1 as function of $\boldsymbol{U} = (u_1, u_2)$.
(d) Use `varlinsys` to determine a representation of x_1 as function of $\boldsymbol{U} = (u_1, u_2)$.
(e) Compare the results obtained in c) and d).
(f) Determine the Marginal distribution of x_1.

3. Consider a model for the equilibrium between two Markets is characterized by two variables (x_1, x_2) such that

$$\begin{pmatrix} 0.2 & 1 \\ 1 & -10 \end{pmatrix} \begin{pmatrix} x_1 \\ x_2 \end{pmatrix} = \begin{pmatrix} 50 + u_1 \\ 300 + u_2 \end{pmatrix}$$

Assume that the distributions of u_1 and u_2 are triangular: $u_1 \sim T(0, 40, 20)$ and $u_2 \sim T(-50, 50, 0)$.
(a) Generate a sample of 6 variates from each random variable u_i.
(b) Use the values generate to generate a sample of 36 variates from x_1.
(c) Use the sample from x_1 to determine a representation of x_1 as function of $\boldsymbol{U} = (u_1, u_2)$.
(d) Use `varlinsys` to determine a representation of x_1 as function of $\boldsymbol{U} = (u_1, u_2)$.
(e) Compare the results obtained in c) and d).
(f) Determine the Marginal distribution of x_1.

6.2 Nonlinear Equations and Adaptation of an Iterative Code

The approach presented in the preceding section extends to nonlinear systems of equations:

$$\boldsymbol{f}(\boldsymbol{x}) = \boldsymbol{0} \Leftrightarrow \begin{cases} f_1(\boldsymbol{x}) = 0 \\ \vdots \\ f_n(\boldsymbol{x}) = 0 \end{cases} \tag{6.15}$$

Here, $\boldsymbol{f} : \mathbb{R}^n \longrightarrow \mathbb{R}$ defines the equations to be solved for the unknowns $\boldsymbol{x} = (x_1, \ldots, x_n)^t$. Equation (6.15) are generally solved by iterative methods, such as Newton-Raphson's method (see Sect. 2.3.3).

Codes for the numerical solution of algebraical equations are often available and one of the interesting possibilities offered by UQ is that of being able to study uncertainties using existing codes, without modifying them and without rewriting any of their parts.

As in the preceding situation, such an analysis may be made by generating a sample and then, using one of the preceding approaches to determine the probability distribution of the results furnished by the code.

Nevertheless, there is an alternative approach – called the *method of adaptation*, which allows to directly obtain the representation of the variability of the code output as a function of the parameters affected by the variability.

As an example, let us consider an iterative method which reads as

$$\boldsymbol{X}^{(r+1)} = \boldsymbol{\Psi}\left(\boldsymbol{X}^{(r)}, U\right) \tag{6.16}$$

$\boldsymbol{\Psi}$ is referred as the *iteration function*. Notice that a direct method may be considered as an iterative method with a single iteration.

If $\boldsymbol{X} = (X_1, \ldots, X_n)$, then $\boldsymbol{\Psi} = (\psi_1, \ldots, \psi_n)$ and we can describe the iterations component by component:

$$\boldsymbol{X}_i^{(r+1)} = \psi_i\left(\boldsymbol{X}^{(r)}, U\right), 1 \leq i \leq n. \tag{6.17}$$

To adapt this iterative method, we can consider approximations analogous to Eq. (6.8) (see Chap. 5 too):

$$\boldsymbol{X}^{(p)} \approx \boldsymbol{P}\boldsymbol{X}^{(p)} = \sum_{j=1}^{k} \boldsymbol{x}_j^{(p)} \varphi_j(U), \tag{6.18}$$

i. e.,

$$X_\ell^{(p)} \approx PX_\ell^{(p)} = \sum_{j=1}^{k} x_{j,\ell}^{(p)} \varphi_j(U), \ell = 1, \ldots, n \tag{6.19}$$

Then, we can write

$$\boldsymbol{P}\boldsymbol{X}^{(r+1)} \approx \boldsymbol{\Psi}\left(\boldsymbol{P}\boldsymbol{X}^{(r)}, U\right). \tag{6.20}$$

Thus,

$$\varphi_i(U)\boldsymbol{P}\boldsymbol{X}^{(r+1)} \approx \varphi_i(U)\boldsymbol{\Psi}\left(\boldsymbol{P}\boldsymbol{X}^{(r)}, U\right), 1 \leq i \leq k$$

and

$$E\left(\varphi_i(U)\boldsymbol{P}\boldsymbol{X}^{(r+1)}\right) \approx E\left(\varphi_i(U)\boldsymbol{\Psi}\left(\boldsymbol{P}\boldsymbol{X}^{(r)}, U\right)\right), 1 \leq i \leq k$$

We have

$$E\left(\varphi_j(U)\boldsymbol{P}\boldsymbol{X}^{(r+1)}\right) = \sum_{j=1}^{k} A_{i,j}\boldsymbol{x}_j^{(r+1)}, \ A_{i,j} = E(\varphi_i(U)\varphi_j(U)), \tag{6.21}$$

id est,

$$E\left(\varphi_j(U)PX_\ell^{(r+1)}\right) = \sum_{j=1}^{k} A_{i,j}x_{j,\ell}^{(r+1)}, \ A_{i,j} = E(\varphi_i(U)\varphi_j(U)), \tag{6.22}$$

Let

$$B_{i,\ell}^{(k)} = E\left(\varphi_i(U)\psi_\ell\left(\boldsymbol{P}\boldsymbol{X}^{(k)}, U\right)\right). \tag{6.23}$$

Then:

$$\sum_{j=1}^{k} A_{i,j}x_{j,\ell}^{(r+1)} = B_{i,\ell}^{(r)}, 1 \leq i \leq k \ \textit{and} \ 1 \leq \ell \leq n. \tag{6.24}$$

This is a linear system for the variables $x_{j,\ell}^{(r+1)}$. It can be seen as a sequence of n linear systems of the same matrix $\boldsymbol{A}$ and second member $\boldsymbol{B}_\ell^{(r)} = \left(B_{1,\ell}^{(r)}, \ldots, B_{k,\ell}^{(r)}\right)^t$: the solution of $\boldsymbol{A}\boldsymbol{Y}_\ell^{(r)} = \boldsymbol{B}_\ell^{(r)}$ furnishes $\boldsymbol{Y}_\ell^{(r)} = \left(x_{1,\ell}^{(r+1)}, \ldots, x_{k,\ell}^{(r+1)}\right)^t$.

This approach does not require any modification the code evaluating $\boldsymbol{\Psi}$: indeed, the calculation of $B_{i,\ell}^{(e)}$ requires only calls to the code that evaluates $\boldsymbol{\Psi}$ - it can be used "as it is", without any modification.

Such a method is called non-intrusive, since it does not request the user to plunge into the meanders of the code to modify it - it requests only calls to the code. An example of non-intrusive algorithm for the adaptation is given below:

Adaptation of an iterative method

Input: a sample from $U : (U_1, \ldots, U_{ns})$ is given, as a program which determines the value of $\boldsymbol{\Psi}(X, U)$. The basis φ_i and the degree of approximation k are given. Give the stopping conditions (for instance: maximum iteration number and minimal step at each iteration)

Output: the coefficients of the representation $\boldsymbol{PX}$ of $\boldsymbol{X}$.

Initialization: Determine the matrix $\boldsymbol{A}$ and give an initial guess $\left(x_{j,\ell}^{(0)} : 1 \le j \le k, 1 \le \ell \le n\right)$. Set the iteration number to zero: $r \leftarrow 0$.

1. Determine $\boldsymbol{Z_m} = \boldsymbol{PX}^{(r)}(U_m)$, for $1 \le m \le ns$.
2. Evaluate $\boldsymbol{T_m} = \boldsymbol{\Psi}(\boldsymbol{Z_m}, U_m)$, for $1 \le m \le ns$. Let $\boldsymbol{T_m} = (T_{1,m}, \ldots, T_{n,m})$
3. Determine $V_{m,i} = \varphi_i(U_m)$, $1 \le m \le ns$, $1 \le i \le k$.
4. For $1 \le \ell \le n$:
   (a) Evaluate $\boldsymbol{B}_\ell^{(r)} : B_{i,\ell}^{(r)} \approx \frac{1}{ns} \sum_{m=1}^{ns} V_{m,i} T_{\ell,m}$,
   (b) Solve the linear system $\boldsymbol{A}\boldsymbol{Y}_r^{(r)} = \boldsymbol{B}_r^{(r)}$,
   (c) Set $x_{i,\ell}^{(r+1)} = Y_{i,\ell}^{(r)}$, $1 \le i \le k$.
5. Increment the iteration number $r \longleftarrow r + 1$.
6. Test for the stopping conditions. If the iterations continue, then go to 1, else estimate $\boldsymbol{x} \approx \boldsymbol{x}^{(r)}$.

This approach is quite general and can be used for any situation where the code performing the iterations $\boldsymbol{X}^{(k+1)} = \boldsymbol{\Psi}(\boldsymbol{X}^{(k)}, U)$ can be executed. In many practical situations, $\boldsymbol{\Psi}(\boldsymbol{x}, u) = \boldsymbol{x} - \boldsymbol{\Phi}(\boldsymbol{x}, u)$, with $\boldsymbol{\Phi} = (\phi_1, \ldots, \phi_n)$. In such a situation, we may solve the linear systems

$$\sum_{j=1}^{k} A_{i,j} \delta x_{j,\ell}^{(r)} = \delta B_{i,\ell}^{(r)}, 1 \le i \le k \text{ and } 1 \le \ell \le n, \tag{6.25}$$

with

$$\delta B_{i,\ell}^{(k)} = E\left(\varphi_i(U)\phi_\ell\left(\boldsymbol{PX}^{(k)}, U\right)\right). \tag{6.26}$$

Then,

$$x_{j,\ell}^{(r+1)} = x_{j,\ell}^{(r)} - \delta x_{j,\ell}^{(r)}, 1 \le j \le k \text{ and } 1 \le \ell \le n. \tag{6.27}$$

As an example, let us consider relaxed Newton's iterations for the resolution of the equation $x^2 - 2x + u = 0$, where u is uniformly distributed on $(0, 1)$. The equation has two solutions for each u: $x_+ = 1 + \sqrt{1-u}$ and $x_- = 1 - \sqrt{1-u}$. We have

$$P(x_- < a) = 1 - (a-1)^2, 0 < a < 1; P(x_+ < a) = (a-1)^2, 1 < a < 2;$$

Relaxed Newton's iterations correspond to the iteration function

$$\Psi(x,u) = x - \Phi(x,u), \quad \Phi(x,u) = \omega \frac{x^2 - 2x + u}{2x - 2} \quad ,$$

Where $0 < \omega < 1$ is a relaxation parameter. Let us illustrate the procedure using a sample formed by 500 variates from u and the adaptation as shown in the algorithm above described, with a polynomial family and $k = 8$.

Open a blank workbook. Select cell A2 and enter the formula "= RANDARRAY(500;1;-1;1)". Give the name u to the data in column A. Enter the formula "=Min(u)" in B2. Give the name lb to B2. Enter the formula "=Max(u)" in B4. Give the name ub to B4. Enter the value of ω in B6, the size ns of the large sample to be generated at the end of the calculation in B8, the numerical zero in B10.
Select Cell C2 and enter the formula "=(u-lb)/(ub-lb)". Give the name s to the data in column C. Recall that RANDARRAY is volatile (see Sect. 1.8).
At the end of the operations, you will have a sheet like the one shown at right.

	A	B	C
1	u	lb	s
2	0,457475	-0,99452	0,729256
3	0,170339	ub	0,585044
4	-0,06757	0,996544	0,465557
5	0,792422	om	0,897481
6	0,59215	0,01	0,796896
7	-0,34108	ns	0,328188
8	-0,72202	10000	0,136861
9	0,388243		0,694484
10	-0,13005		0,434173
11	0,996301		0,999878
12	-0,1785		0,409842
13	0,631522		0,81667
14	-0,29486		0,3514
15	0,894399		0,948698
16	-0,99232		0,001107
17	-0,73867		0,128498
18	-0,48616		0,255323
19	0,012316		0,505678
20	0,720433		0,861325
21	0,006247		0,502629
22	0,296382		0,648348

If you desire to plot the results at the end of the calculations, you must sort the data in ascending order – otherwise the resulting chart will be unreadable. However, a column generated by RANDARRAY cannot be sorted: you must copy it *as values* to another column: For instance, select column A, copy it and paste as values on the same column A. Now you can sort the data in column A: select A2:A501, go to *Home>Sort & Filter* and sort the data in ascending order.

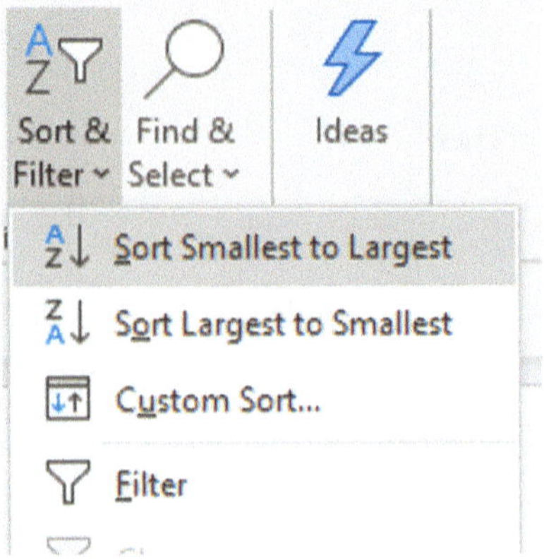

Select cell D2 and enter the value "1". Copy it up to D501. Enter the formulas "=s" in E2, "=s^2" in F2, "=s^3" in G2, etc. For an approximation with $k = 8$, the last colum is K: you must enter "=s^7" in K2. Give the names to these columns: "s_0" (to column s^0), "s_1" (to column s^1), "s_2" (to column s^2) etc. Give the name "mats" to D2:K501. If the data was ranged in ascending order, you will have a sheet analogous to the one below:

C	D	E	F	G	H	I	J	K
s	s^0	s^1	s^2	s^3	s^4	s^5	s^6	s^7
0	1	0	0	0	0	0	0	0
0,003547	1	0,003547	1,26E-05	4,46E-08	1,58E-10	5,61E-13	1,99E-15	7,06E-18
0,004235	1	0,004235	1,79E-05	7,6E-08	3,22E-10	1,36E-12	5,77E-15	2,44E-17
0,007394	1	0,007394	5,47E-05	4,04E-07	2,99E-09	2,21E-11	1,63E-13	1,21E-15
0,009537	1	0,009537	9,09E-05	8,67E-07	8,27E-09	7,89E-11	7,52E-13	7,17E-15

Select cell L1 and enter "= RANDARRAY(8;1;-5;5)". This will be the starting point $x^{(0)}$ of the iterations. Give the name "x0" to L2:L9. Go to M1 and enter the formula "=MMULT(mats;x0)". Give the name "PX" to M2:M501.
Enter "=(M2^2 -2*M2+A2)/(2*M2-2)" in N2. Copy this formula and paste it in M3:M501. Give the name "phi" to N2:N501".

L	M	N
x(0)	PX	phi
-4,24915	-4,24915	-2,43416
-0,83253	-4,25211	-2,43642
-0,28296	-4,25268	-2,43686
1,002258	-4,25532	-2,43888
-4,25911	-4,25712	-2,44025
-3,18778	-4,2583	-2,44115
0,048438	-4,25876	-2,4415
4,629598	-4,25933	-2,44194

Evaluate the matrix A: element A_{ij} is defined by the formula "=SUMPRODUCT(s_(i-1);s_(j-1))/500". For instance, the formula at S7 corresponds to "=SUMPRODUCT(s_(5-1);s_(4-1))/500" - it is shown at right. Give the name "A" to P3:W10.

	P	Q	R	S	T	U	V	W
	A							
	1	2	3	4	5	6	7	8
1	1	0,499858	0,332472	0,248756	0,198734	0,165525	0,141875	0,124176
2	0,499858	0,332472	0,248756	0,198734	0,165525	0,141875	0,124176	0,110433
3	0,332472	0,248756	0,198734	0,165525	0,141875	0,124176	0,110433	0,099455
4	0,248756	0,198734	0,165525	0,141875	0,124176	0,110433	0,099455	0,090486
5	0,198734	0,165525	0,141875	0,124176	0,110433	0,099455	0,090486	0,083022
6	0,165525	0,141875	0,124176	0,110433	0,099455	0,090486	0,083022	0,076715
7	0,141875	0,124176	0,110433	0,099455	0,090486	0,083022	0,076715	0,071317
8	0,124176	0,110433	0,099455	0,090486	0,083022	0,076715	0,071317	0,066644

=SUMPRODUCT(s_4;s_3)/500

Evaluate B: the formula for B_i is "=SUMPRODUCT(s_(i-1); phi)/500". For instance, the formula for B4 is shown below:

=SUMPRODUCT(s_3;phi)/500

Give the name "B" to Z2:Z9.
Determine the values of $x^{(1)}$. The formula in AB2 is "= x0-om*MMULT(MINVERSE(A);B)".

	Z	AB
B		x(1)
1	-3,06384	-1,56962
2	-1,67847	0,118205
3	-1,17541	1,810057
4	-0,90901	-1,315
5	-0,74283	-2,97446
6	-0,62879	-4,98019
7	-0,54546	-1,01656
8	-0,48181	-1,06142

Enable Iterative Calculations in the workbook (see Sect. 1.12).
Copy AB2:AB9 (x(1)) and paste it in L2:L9 (x(0)).

L	AB
x(0)	x(1)
-1,57864	-1,56962
0,115775	0,118205
1,81066	1,810057
-1,25677	-1,315
-3,19344	-2,97446
-4,72992	-4,98019
-1,1847	-1,01656
-1,03219	-1,06142

If the automatic calculation of the workbook is deactivated, the calculation must be started manually, as shown in Sect. 1.8. The iterations run until the stopping conditions are satisfied. Here, we used

☑ Enable iterative calculation

Maximum Iterations: 1 000

Maximum Change: 0,0000000001

L		Z	AB
x(0)	B		x(1)
-0,41777	1	-0,00114	-0,41778
0,96654	2	-0,00103	0,967156
-3,87626	3	-0,00096	-3,88576
25,70424	4	-0,00091	25,76346
-79,4481	5	-0,00086	-79,6293
128,6282	6	-0,00082	128,9165
-104,031	7	-0,00079	-104,26
33,35727	8	-0,00075	33,42903

You can continue the iterations if you desire: just recalculate the sheet (see Sect. 1.8): the results will change, with values of B expected to converge to zero. For instance, two calculations more of the sheet gives the result at right. You can generate different approximations by this way. Here, the results correspond to the branch x_- (see Fig. 6.5)

L		Z	AB
x(0)	B		x(1)
-0,41903	1	-6,9E-08	-0,41903
1,055286	2	-6,6E-08	1,055287
-5,23676	3	-6,3E-08	-5,23676
34,14491	4	-6,2E-08	34,14493
-105,126	5	-6E-08	-105,126
169,2602	6	-5,9E-08	169,2602
-136,114	7	-5,8E-08	-136,114
43,34273	8	-5,7E-08	43,34276

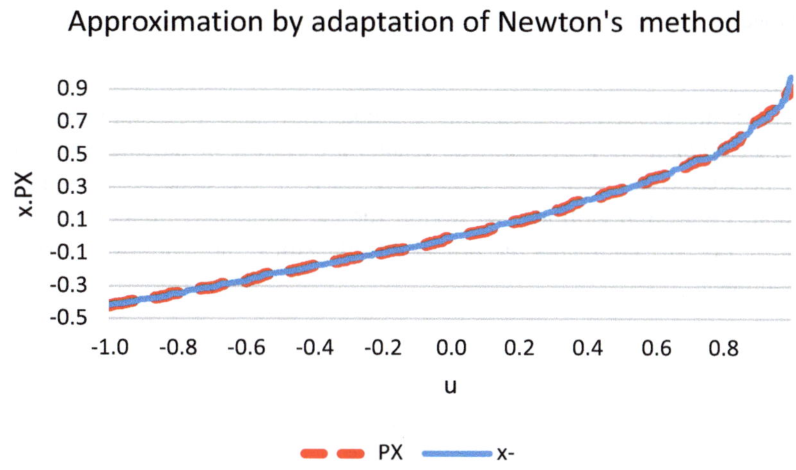

Fig. 6.5 Approximation of a root of a random equation by adaptation of Newton's method. In this case, the variable itself is approximated

According to the initial point and the sample of u, the method converges to x_- or x_+. For instance, a test with different data and a different initial point leads to the results in Fig. 6.6.

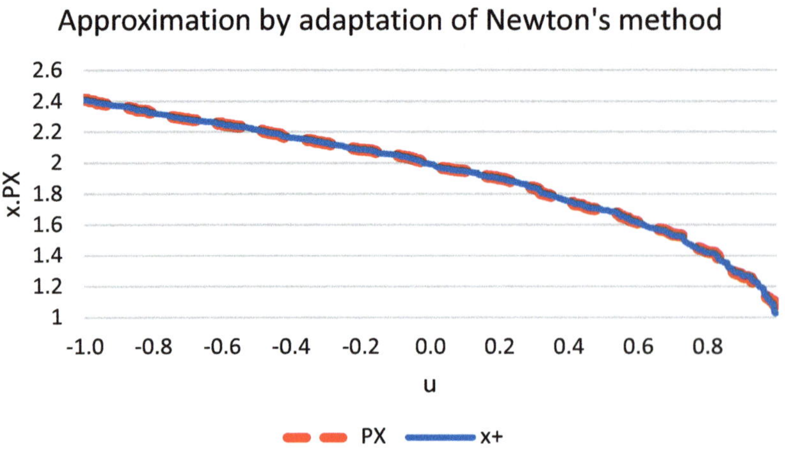

Fig. 6.6 Approximation of a root of a random equation by adaptation of Newton's method. In this case, the variable itself is approximated

Analogously to the preceding ones, this method may also be implemented in VBA. We implemented it in *adaptation_vba.xlsm*. The results are shown in the Figures below. Figure 6.7 compares the projection *PX* and the variable *X* itself – we observe that the accordance is particularly good. Figure 6.8 exhibits the results for the cumulative function – again, the results are good. Finally, Fig. 6.9 compares the probability densities: the numerical derivative was made by the SPH approach.

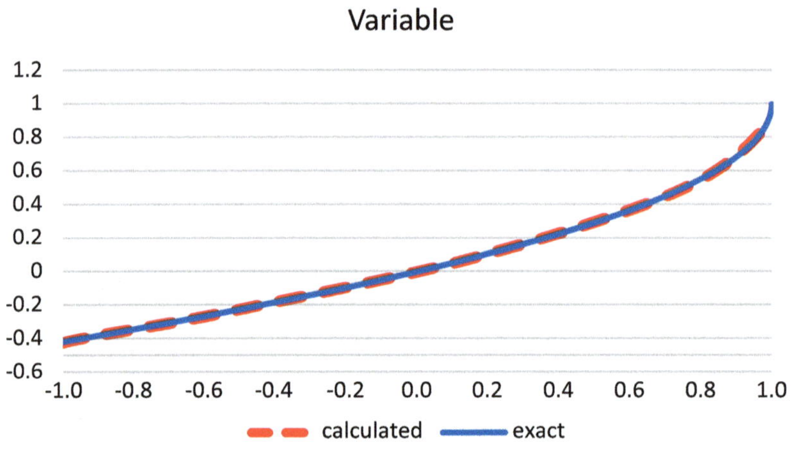

Fig. 6.7 Approximation of a root of a random equation by adaptation of Newton's method. In this case, the variable itself is approximated

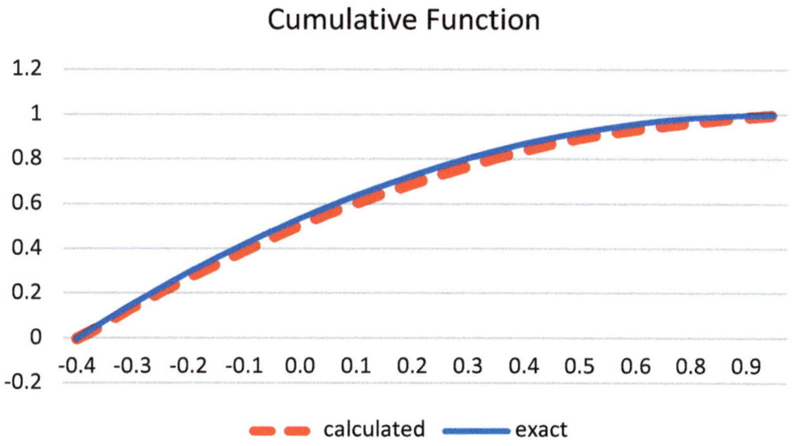

Fig. 6.8 Cumulative Function of a root of a random equation, determined by adaptation of Newton's method

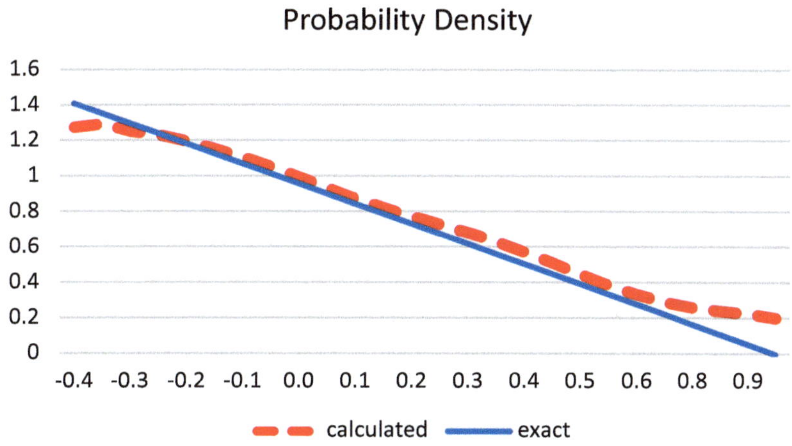

Fig. 6.9 Probability density of a root of a random equation, determined by adaptation of Newton's method. The calculated curve is obtained by the SPH derivative of the calculated CDF

An example of VBA code for the iterations is given in the sequel (Fig. 6.10). You must insert this code in a `Sub` of your workbook. The code assumes that you have also added an iteration function `psi(s, u, om)`. Other parameters that must be predefined are `nitmax` (maximal iteration number), `precmin` (minimal RMS difference between consecutive iterates), `matAinv` corresponds to $\boldsymbol{A}^{-1}$, `u` is the sample, `dimX` is the dimension of $\boldsymbol{X}$, `p` is the family $\{\varphi_i : 1 \leq i \leq k\}$ to be used, `us` is the sample from U and `ns` is its size. The coefficients `c` must be initialized before calling this code. This code supposes that U is unidimensional – it must be modified for multidimensional U.

```
nit = 0
prec = 1 + precmin
Dim continue As Boolean
continue = True
While continue
nit = nit + 1
  ReDim matB(1 To k, 1 To dimX)
  ReDim aux1(1 To dimX, 1 To ns)
  ReDim aux2(1 To k, 1 To ns)
  For j = 1 To ns
    uu = us(j)
    ReDim s(1 To dimX, 1 To 1)
    For kk = 1 To k
      auxx = p.value(kk - 1, uu)
      For jj = 1 To dimX
        s(jj, 1) = s(jj, 1) + c(jj, kk) * auxx
      Next jj
      aux2(kk, j) = auxx
    Next kk
    aux3 = psi(s, uu, om)
    For ii = 1 To dimX
      aux1(ii, j) = aux3(ii)
    Next ii
  Next j
  For kk = 1 To k
    For j = 1 To dimX
      ss = 0
      For ii = 1 To ns
        ss = ss + aux1(j, ii) * aux2(kk, ii)
      Next ii
      matB(kk, j) = ss / ns
    Next j
  Next kk
  If dimX = 1 Then
    aux = WorksheetFunction.MMult(matAinv, matB)
    For i = 1 To k
      cn(1, i) = aux(i, 1)
    Next i
  Else
    cn = WorksheetFunction.Transpose(WorksheetFunction.MMult(matAinv, matB))
  End If
  prec = 0
  For ii = 1 To dimX
   For jj = 1 To k
      auxx = cn(ii, jj) - c(ii, jj)
      prec = prec + auxx * auxx
      c(ii, jj) = cn(ii, jj)
   Next jj
  Next ii
  prec = Sqr(prec / (k * dimX))
  continue = nit < nitmax And prec > precmin
Wend
```

Fig. 6.10 Example of VBA code for the iterations of adaptation

It is interesting to notice that practically any iterative code may be adapted by this approach. For instance, Iterations from a descent method in optimization may be adapted. As an example, let us consider the iterations of the gradient descent with a fixed step $\mu > 0$.

$$\boldsymbol{X}^{(p+1)} = \boldsymbol{X}^{(p)} - \mu \nabla \boldsymbol{F}\left(\boldsymbol{X}^{(p)}, U\right) \tag{6.28}$$

These iterations correspond to

$$\Psi(\boldsymbol{x}, u) = \boldsymbol{x} - \mu \nabla \boldsymbol{F}(\boldsymbol{x}, u) \tag{6.29}$$

and may be adapted analogously to the Newton's iterations. As an example, let us consider the simple situation where $F(x, u) = (x - e^{-u})^2$, for which the exact solution is $x = e^{-u}$. Let us take u as uniformly distributed on $(-1,1)$. We run the iterations with $\mu = 0.1$ and we use the same sample and the same polynomial family used in the preceding example. The iterations of adaptation converge and furnish

$$x = (2.718, -5.428, 5.352, -3.222, 0.806, 0.648, -0.732, 0.228).$$

As we see in Fig. 6.11, the result is close to the exact solution.

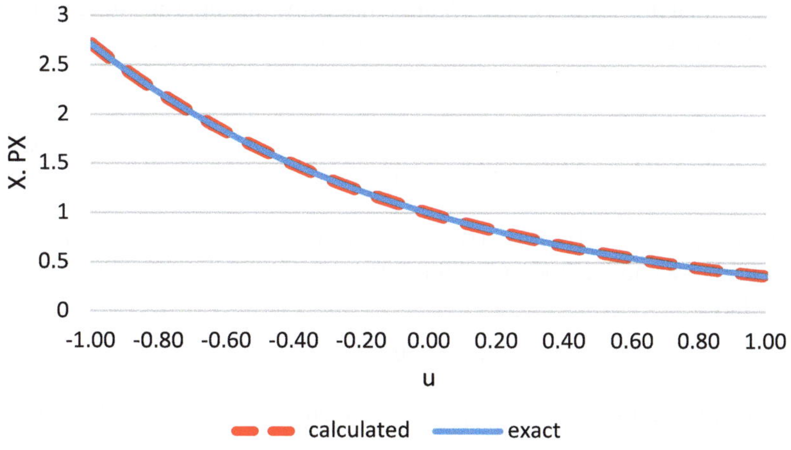

Fig. 6.11 Results for the adaptation of the gradient descent with a fixed step. U is uniform on $(-1, 1)$

Now, let us take u as normally distributed $N(0, 1)$. We run the iterations with the same parameters: $\mu = 0.1$, and same polynomial family used in the preceding example. The sample contains 500 variates from $N(0, 1)$. Adaptation iterations run and furnish

$$x = (2.720, -5.452, 5.430, -3.533, 1.780, -0.822, 0.298, -0.050).$$

As we see in Fig. 6.12, the result is close to the exact solution. Notice that, in this case, the exponential term leads to large values of X – but the numerical result nevertheless remains close to the exact one.

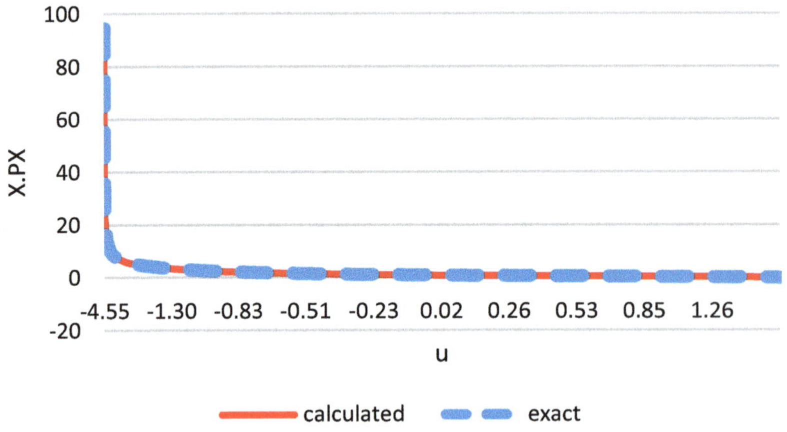

Fig. 6.12 Results for the adaptation of the gradient descent with a fixed step. U is gaussian $N(0, 1)$

Let us consider a final illustration of the adaptation method, to underline once again that practically any method based on recurrences can be studied by this approach: let us consider a differential equation:

$$\begin{pmatrix} x_1' \\ x_2' \end{pmatrix} = \mathbf{A}\begin{pmatrix} x_1 \\ x_2 \end{pmatrix}, \mathbf{A} = \begin{pmatrix} 0 & U \\ -U & 0 \end{pmatrix}, x_1(0) = 0, x_2(0) = 1, t \in (0, \mathrm{T}). \tag{6.30}$$

The solution is $x_1(t) = \sin(Ut)$, $x_2(t) = \cos(Ut)$. Assume that the Differential Equation is discretized by a Runge-Kutta method, using time steps $t_i = i\Delta t$, $0 \leq i \leq n$, $\Delta t = T/n$. Let $\boldsymbol{X_i} = (x_1(t_i), x_2(t_i))$. Then, we have

$$\boldsymbol{X_{i+1}} = \boldsymbol{\Psi}_i(\boldsymbol{X_i}, U), \tag{6.31}$$

where $\boldsymbol{\Psi_i}$ is the recurrence function associated to the method. For instance, for Euler's method,

$$\boldsymbol{\Psi}_i(\boldsymbol{X}_i, U) = \boldsymbol{X}_i + \Delta t\, \boldsymbol{A}(U)\boldsymbol{X}_i, \tag{6.32}$$

For other Runge-Kutta methods, the expression of Ψ_n is more complex, but we do not need to know the exact expression: as previously observed, the availability of a code performing the recurrence is enough to use the adaptation. For instance, the code `explicitRKstep` shown in Sect. 2.4, page 74 may be used.

Here, we do not look for the convergence of the iterations, but we generate a sequence of values $\boldsymbol{X_0}(\boldsymbol{U}), \boldsymbol{X_1}(\boldsymbol{U}), \ldots, \boldsymbol{X_n}(\boldsymbol{U})$. By the UQ approach, each $\boldsymbol{X_i}(\boldsymbol{U})$ is approximated by an expansion $\boldsymbol{PX_i}(U)$ as shown in Chap. 5. The expansion $\boldsymbol{PX_i}(U)$ has coefficients $\boldsymbol{x_i}$ to be determined (see Sect. 5.5.2) – the coefficients can be generated by adaptation of the recurrence (6.31)

For instance, let us assume that U is normally distributed, with a mean 2π and a standard deviation $\pi/6$. Assume that a sample formed by $ns = 20$ variates from U is available. Let us use RK methods of different orders and we evaluate the RMS error between the prevision generated by adaptation and the exact solution for all the values of U in the sample and all the values of $\boldsymbol{X_i}$, $0 \leq i \leq n$:

$$RMS = \left(\frac{1}{ns \times (n+1)} \sum_{j=1}^{ns} \sum_{i=0}^{n} \left\| \boldsymbol{PX}_i(U_j) - \boldsymbol{X}_i(U_j) \right\|^2 \right)^{\frac{1}{2}} \tag{6.33}$$

The results are shown in Table 6.9 and Fig. 6.13.

Table 6.9 RMS error in the adaptation of $RK(k)$ with n time steps

k	2	3	4	5	6	7	8
$n = 10$	2,9E-1	9,1E-2	8,7E-2	9,1E-2	6,2E-2	5,4E-2	5,0E-2
$n = 100$	2,2E-2	7,2E-3	7,2E-3	7,2E-3	5,1E-3	4,4E-3	3,9E-3
$n = 1000$	2,1E-3	7,1E-4	7,1E-4	7,1E-4	5,0E-4	4,3E-4	3,9E-4

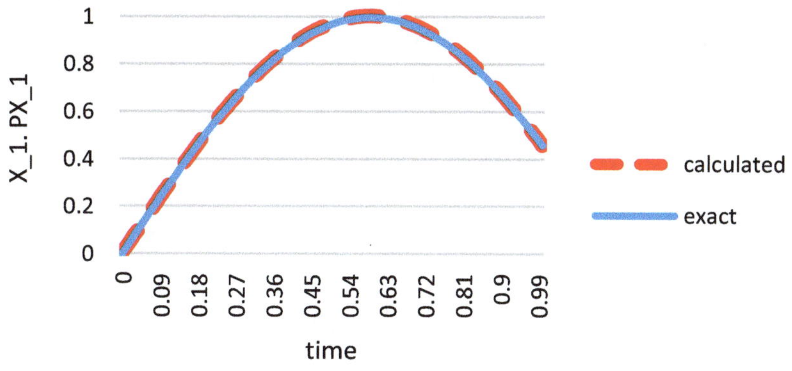

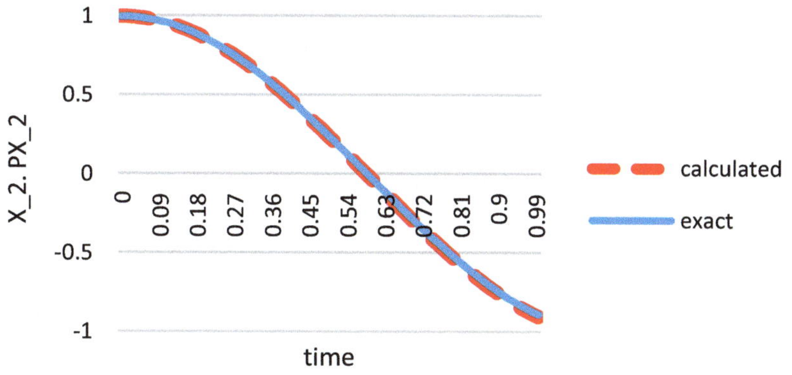

Fig. 6.13 Examples of results for the adaptation of the Runge-Kutta RK4. U is gaussian $N(0, 1)$

Exercises

1. Consider the adaptation of gradient descent iterations for the minimization of

$$f(\boldsymbol{x}) = \left(100 + u_1^2\right)x_1^2 + \left(50 + u_2^2\right)x_2^2 - (200 - u_1)x_1 + (50 + u_2)x_2.$$

Assume that $u_1 \sim N(0, 0.1)$, $u_2 \sim N(0,0.2)$.

(a) Generate a sample of 10 variates from each random variable u_i.

(b) Use these values to generate a sample of 100 variates from $\boldsymbol{U} = (u_1, u_2)$.

(c) Use the sample to determine a representation of $\boldsymbol{X} = \arg \min f$ as function of $\boldsymbol{U} = (u_1, u_2)$.

(d) Determine the CDF and the PDF of $\boldsymbol{X}$.

2. Consider the adaptation of the solution by RK of the differential equations

$$\frac{d}{dt}\begin{pmatrix} x_1 \\ x_2 \end{pmatrix} = \begin{pmatrix} -u_1 & 2 \\ 4 & -u_2 \end{pmatrix}\begin{pmatrix} x_1 \\ x_2 \end{pmatrix}, \begin{pmatrix} x_1(0) \\ x_2(0) \end{pmatrix} = \begin{pmatrix} 2 \\ 5 \end{pmatrix}$$

Assume that the distributions of u_1 and u_2 are triangular: $u_1 \sim T(0, 2, 1)$ and $u_2 \sim T(0, 8, 4)$.
(a) Generate a sample of 6 variates from each random variable u_i.
(b) Use these values to generate a sample of 36 variates from $\boldsymbol{U} = (u_1, u_2)$.
(c) Use the sample to determine a representation of $\boldsymbol{X}(1) = (x_1(1), x_2(1))$ as function of $\boldsymbol{U} = (u_1, u_2)$.
(d) Determine the CDF and the PDF of $\boldsymbol{X}$.

3. Consider the adaptation of gradient descent iterations for the minimization of

$$f(\boldsymbol{x}) = (1 + u_1^2)x_1^2 + (1 + u_2^2)x_2^2 + (10 - u_1)x_1 + (15 + u_2)x_2.$$

Assume that $u_1 \sim N(0, 0.2)$, $u_2 \sim T(0, 2, 1)$.
(a) Generate a sample of 10 variates from each random variable u_i.
(b) Use these values to generate a sample of 100 variates from $\boldsymbol{U} = (u_1, u_2)$.
(c) Use the sample to determine a representation of $\boldsymbol{X} = \arg\min f$ as function of $\boldsymbol{U} = (u_1, u_2)$.
(d) Determine the CDF and the PDF of $\boldsymbol{X}$.

4. Consider the adaptation of the solution by RK of the differential equations

$$\frac{d}{dt}\begin{pmatrix} x_1 \\ x_2 \end{pmatrix} = \begin{pmatrix} 0.2 - 5u_1 & -2.5 - 5u_1 \\ 1 & 0 \end{pmatrix}\begin{pmatrix} x_1 \\ x_2 \end{pmatrix} + \begin{pmatrix} 8 \\ 3 \end{pmatrix}, \begin{pmatrix} x_1(0) \\ x_2(0) \end{pmatrix} = \begin{pmatrix} 2 \\ 5 \end{pmatrix}$$

Assume that the distributions $u_1 \sim T(-0.1, 0.1,0)$ and $u_2 \sim N(0, 0.1)$.
(a) Generate a sample of 10 variates from each random variable u_i.
(b) Use these values to generate a sample of 100 variates from $\boldsymbol{U} = (u_1, u_2)$.
(c) Use the sample to determine a representation of $\boldsymbol{X}(1) = (x_1(1), x_2(1))$ as function of $\boldsymbol{U} = (u_1, u_2)$.
(d) Determine the CDF and the PDF of $\boldsymbol{X}$.

6.3 Iterative Evaluation of Eigenvalues

Eigenvalues and eigenvectors of matrices may be numerically determined by iterative methods. For instance, we can use Power Iterations, which read as

$$\boldsymbol{X}^{(p+1)} = \frac{\boldsymbol{Y}^{(p)}}{\|\boldsymbol{Y}^{(p)}\|}, \boldsymbol{Y}^{(p)} = \boldsymbol{A}(\boldsymbol{U})\boldsymbol{X}^{(p)}; \ \boldsymbol{X}^{(0)} \ given. \quad (6.34)$$

For convenient $\boldsymbol{X}^{(0)}$, the iterations converge to the eigenvalue of $\boldsymbol{A}$ having the largest modulus – which is said to be the dominant eigenvalue. To obtain such a convergence, the initial guess $\boldsymbol{X}^{(0)}$ must have a significant projection in the direction supported by the associated eigenvector. Analogously, inverse iteration produces the eigenvalue of $\boldsymbol{A}$ having the smallest modulus:

$$\boldsymbol{X}^{(p+1)} = \frac{\boldsymbol{Y}^{(p)}}{\left\| \boldsymbol{Y}^{(p)} \right\|} \quad , \ \boldsymbol{A}(\boldsymbol{U})\boldsymbol{Y}^{(p)} = \boldsymbol{X}^{(p)}; \ \boldsymbol{X}^{(0)} \ given. \tag{6.35}$$

Since these methods are iterative, they can be adapted as shown in the preceding section. For instance, iterations (6.34) can be adapted by using (6.8, 6.9, 6.10, 6.11, 6.12, and 6.13) setting

$$\boldsymbol{X}^{(p)} \approx \boldsymbol{P}\boldsymbol{X}^{(p)} = \sum_{j=1}^{k} \boldsymbol{x}_j^{(p)} \varphi_j(\boldsymbol{U}), \boldsymbol{x}_j^{(p)} = \left(x_{1,j}^{(p)}, \ldots, x_{n,j}^{(p)} \right)$$

$$\boldsymbol{Y}^{(p)} \approx \boldsymbol{P}\boldsymbol{Y}^{(p)} = \sum_{j=1}^{k} \boldsymbol{y}_j^{(p)} \varphi_j(\boldsymbol{U}), \boldsymbol{y}_j^{(p)} = \left(y_{1,j}^{(p)}, \ldots, y_{n,j}^{(p)} \right),$$

we have, for $1 \le i \le k$, $1 \le r \le n$ (see Eq. (6.9) for the definition of $\mathcal{A}$):

$$\sum_{j=1}^{k} \sum_{s=1}^{n} \mathcal{A}_{risj} x_{s,j}^{(p)} \approx c_{ri}^{(p)} = \sum_{j=1}^{k} \mathcal{M}_{ij} y_{rj}^{(p)}, \tag{6.36}$$

with

$$\mathcal{M}_{ij} = E\big(\varphi_i(\boldsymbol{U})\varphi_j(\boldsymbol{U})\big) \approx \frac{1}{ns} \sum_{m=1}^{ns} \varphi_i(\boldsymbol{U}_m)\varphi_j(\boldsymbol{U}_m)$$

Equation (6.36) may be interpreted as n independent linear systems of same matrix $\mathcal{M}$:

$$\mathcal{M}\mathcal{Y}_r^{(p)} = \mathcal{N}_r^{(p)} \quad , 1 \le r \le n$$

with $\mathcal{Y}_r^{(p)} = \left(y_{rj}^{(p)} : 1 \le j \le k \right)$ and

$$\mathcal{N}_r^{(p)} = \left(\sum_{j=1}^{k} \sum_{s=1}^{n} \mathcal{A}_{risj} x_{s,j}^{(p)}, 1 \le i \le k \right).$$

Solving it for a fixed r furnishes $\mathcal{Y}_r^{(p)}$and the solution for all the values of r furnishes $\boldsymbol{y}^{(p)}$. Then, we may generate $\boldsymbol{x}^{(p+1)}$ by normalizing $\boldsymbol{y}^{(p)}$: for instance, we can set $\boldsymbol{x}^{(p+1)} = \boldsymbol{y}^{(p)}/\|\boldsymbol{y}^{(p)}\|$. Alternatively, we may introduce

$$\mathcal{B}_{risj} = \delta_{rs}\mathcal{M}_{ij} \ ; \delta_{rs} = 0, if \ r \ne s, \delta_{rs} = 1, if \ r = s.$$

Then, using the map introduced in Eqs. (6.11) and (6.12), we have

$$\mathbb{B}\mathbb{Y}^{(p)} = \mathbb{A}\mathbb{X}^{(p)}.$$

```
Function a4powerit(u, basis, k, n)
ns = UBound(u, 1)
nn = k * n
Dim AA() As Double
ReDim AA(1 To nn, 1 To nn)
Dim BB() As Double
ReDim BB(1 To nn, 1 To nn)
For i = 1 To nn
  For j = 1 To nn
    AA(i, j) = 0
    BB(i, j) = 0
  Next j
Next i
For m = 1 To ns
  uuuu = u(m)
  Auuuu = a(uuuu)
  For i = 1 To k
    aux1 = basis.value(i - 1, uuuu)
    For j = 1 To k
      aux2 = basis.value(j - 1, uuuu)
      For r = 1 To n
        p = i + (r - 1) * k
        For s = 1 To n
          q = j + (s - 1) * k
          AA(p, q) = AA(p, q) + aux1 * Auuuu(r, s) * aux2
          If r = s Then
            BB(p, q) = BB(p, q) + aux1 * aux2
          End If
        Next s
      Next r
    Next j
  Next i
Next m
For i = 1 To k
  For j = 1 To k
    For r = 1 To n
      p = i + (r - 1) * k
      For s = 1 To n
        q = j + (s - 1) * k
        AA(p, q) = AA(p, q) / ns
        BB(p, q) = BB(p, q) / ns
      Next s
    Next r
  Next j
Next i
Dim resu(1 To 2) As Variant
resu(1) = AA
resu(2) = BB
a4powerit = resu
End Function
```

Fig. 6.14 Example of VBA code for the generation of matrices $\mathbb{A}$ and $\mathbb{B}$. $\mathbb{A}$ is denoted `AA` and $\mathbb{B}$ is denoted `BB` in the code. This code assumes that U is unidimensional: `u` is a vector containing a sample of `ns` variates from U

Inverse iterations may be adapted into an analogous way:

$$\sum_{j=1}^{k}\sum_{s=1}^{n}\mathcal{A}_{risj}y_{s,j}^{(p)}\approx C_{ri}^{(p)}=\sum_{j=1}^{k}\mathcal{M}_{ij}x_{rj}^{(p)},\qquad(6.37)$$

Again, we can use the map introduced in Eqs. (6.11) and (6.12) to rewrite the equations as a linear system

$$\mathbb{A}\mathbb{Y}^{(p)}=\mathbb{B}\mathbb{X}^{(p)}.$$

Here yet, we can use the normalization $x^{(p+1)}=y^{(p)}/\|y^{(p)}\|$.

Subspace Iteration may be adapted into a straight extension of these approaches: indeed, subspace iterations consider an initial subset of vectors $\{X^{(0,1)},\ldots,X^{(0,d)}\}$ and applies the method to each vector:

$$Y^{(p,i)}=A(U)X^{(p,i)}\ \left(\Leftrightarrow\mathbb{B}\mathbb{Y}^{(p,i)}=\mathbb{A}\mathbb{X}^{(p,i)}\right)$$

Then $\{X^{(p+1,1)},\ldots,X^{(p+1,d)}\}$ is the result of an orthonormalization procedure applied to $\{Y^{(p,1)},\ldots,Y^{(p,d)}\}$ (for instance, Gram-Schmidt). An example of VBA code for the generation of matrices $\mathbb{A}$ and $\mathbb{B}$ is given in Fig. 6.14. Examples of VBA code for Power Iterations and subspace iterations are given in Fig. 6.15.

```
Function powerit(XXini, AA, BB)
nn = UBound(XXini, 1)
Dim XX() As Double
ReDim XX(1 To nn, 1 To 1)
nXX = Sqr(WorksheetFunction.SumSq(XXini))
For i = 1 To nn
  XX(i, 1) = XXini(i, 1) / nXX
Next i
precmin = 0.0000000001
nitmax = 1000
nit = 0
prec = 1 + precmin
Dim continue As Boolean
continue = True
While continue
  nit = nit + 1
  ZZ = WorksheetFunction.MMult(AA, XX)
  YY = LUSolution(BB, squeeze(ZZ))
  nYY = Sqr(WorksheetFunction.SumSq(YY))
  s = 0
  For i = 1 To nn
    aux = YY(i) / nYY
    dx = XX(i, 1) - aux
    s = s + dx * dx
    XX(i, 1) = aux
  Next i
  prec = Sqr(s / nn)
  continue = nit < nitmax And prec > precmi
Wend
powerit = XX
End Function
```

```
Function subspaceit(XXini, AA, BB)
nvp = UBound(XXini, 1)
nn = UBound(XXini(1), 1)
XX = GramSchmidt(XXini)
precmin = 0.0000000001
nitmax = 1000
nit = 0
prec = 1 + precmin
Dim continue As Boolean
continue = True
Dim YY() As Variant
ReDim YY(1 To nvp)
While continue
  nit = nit + 1
  For n = 1 To nvp
    ZZ = WorksheetFunction.MMult(AA, XX(n))
    YY(n) = tocol(LUSolution(BB, squeeze(ZZ))
  Next n
  XXn = GramSchmidt(YY)
  prec = ndist(XXn, XX)
  For n = 1 To nvp
    XX(n) = XXn(n)
  Next n
  continue = nit < nitmax And prec > precmin
Wend
subspaceit = XX
End Function
```

Fig. 6.15 Example of VBA code for Power iterations and Subspace Iterations. $\mathbb{A}$ is denoted AA and $\mathbb{B}$ is denoted BB, $\mathbb{X}^{(p)}$ is denoted XX, $\mathbb{Y}^{(p)}$ is denoted YY. For Power iterations, the initial guess $\mathbb{X}^{(0)}$ is XXini. For Subspace Iterations, the initial guesses are XXini(i) $=\mathbb{X}^{(0,i)}, 1\leq i\leq d$. Function GramSchmidt performs orthonormalization and ndist evaluates a distance

Let us illustrate the approaches by using the matrix $A = P^t MP$, with

$$P = \begin{pmatrix} \cos(u) & -\sin(u) \\ -\sin(u) & \cos(u) \end{pmatrix}, M = \begin{pmatrix} \sin(u) & 0 \\ 0 & \cos(u) \end{pmatrix}$$

We have

$$A = \begin{pmatrix} \cos(u)\sin(u)(\cos(u) + \sin(u)) & \cos(u)\sin(u)(\cos(u) - \sin(u)) \\ \cos(u)\sin(u)(\cos(u) - \sin(u)) & \cos^3(u) + \sin^3(u) \end{pmatrix}$$

The exact eigenvalues are the elements of the diagonal of M: $\sin(u)$, $\cos(u)$. Here, we shall determine the eigenvalues as:

$$\begin{aligned} \lambda_1 &= \frac{1}{2}\left(A_{11} + A_{22} - \sqrt{(A_{11} - A_{12})^2 + 4A_{12}A_{21}}\right) \\ \lambda_2 &= \frac{1}{2}\left(A_{11} + A_{22} + \sqrt{(A_{11} - A_{12})^2 + 4A_{12}A_{21}}\right) \end{aligned} \tag{6.38}$$

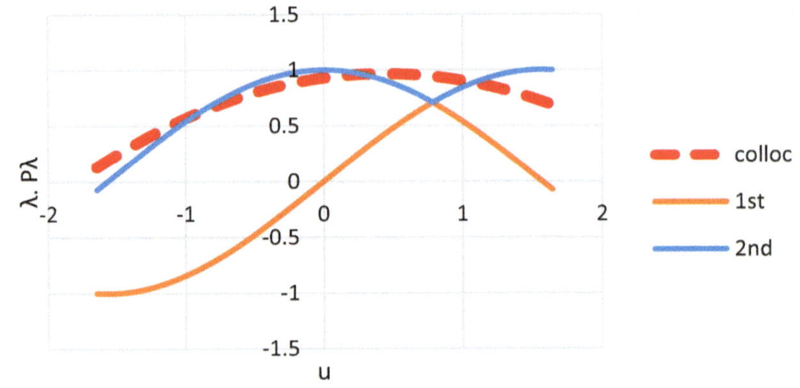

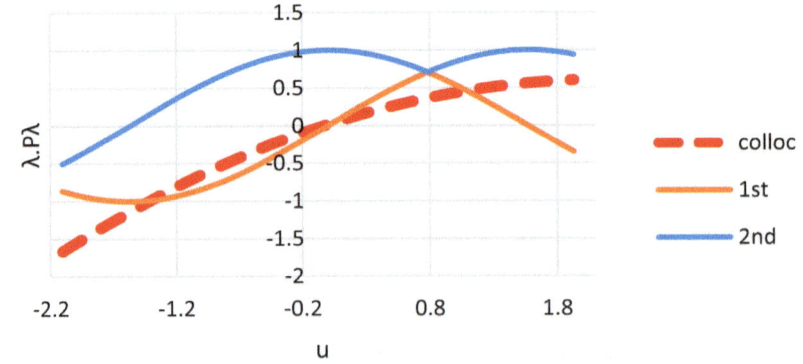

Fig. 6.16 Results obtained by collocation on a sample of 40 variates. U is $N(0, 1)$

Let us consider u as normally distributed $N(0, 1)$ and a sample of 40 variates from u. We may generate a sample of the eigenvalues using the sample and Eq. (6.38) and apply one of the methods presented in Chap. 5. For instance, the results obtained by collocation are shown in Fig. 6.16. We observe that the results are not so good as we can expect (Dashed lines correspond to the numerical results; continuous lines to the exact ones, furnished by Eq. (6.38)).

For comparison, the results furnished by Power Iteration are shown in Fig. 6.17 and the results furnished by Subspace iteration are shown in Fig. 6.18. We observe that the results have a better quality that those furnished by collocation.

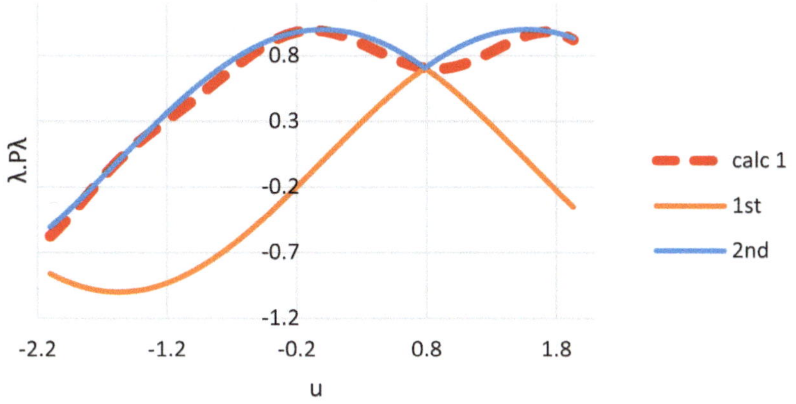

Fig. 6.17 Results obtained by Power Iteration on a sample of 40 variates. U is $N(0, 1)$. Initial Guess is randomly generated

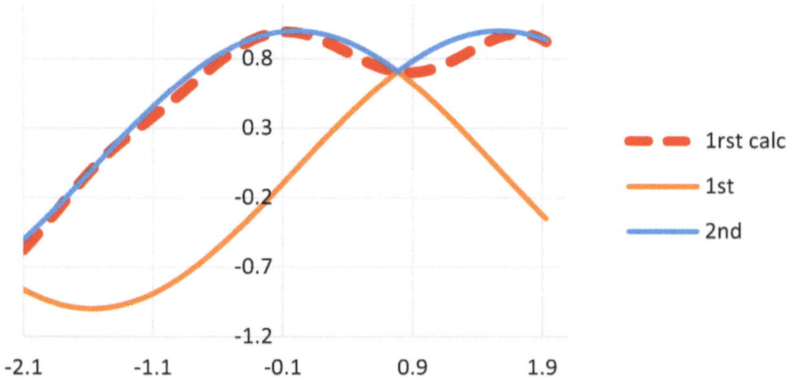

Fig. 6.18 Example of result obtained by Subspace Iteration on a sample of 40 variates. U is $N(0, 1)$. Initial Guess is randomly generated

Now, let us consider u as uniformly distributed on (0,2) and a sample of 40 variates from U. The results are exhibited in Figs. 6.19, 6.20, and 6.21. Again, the results furnished by the iterative methods are better than those furnished by collocation.

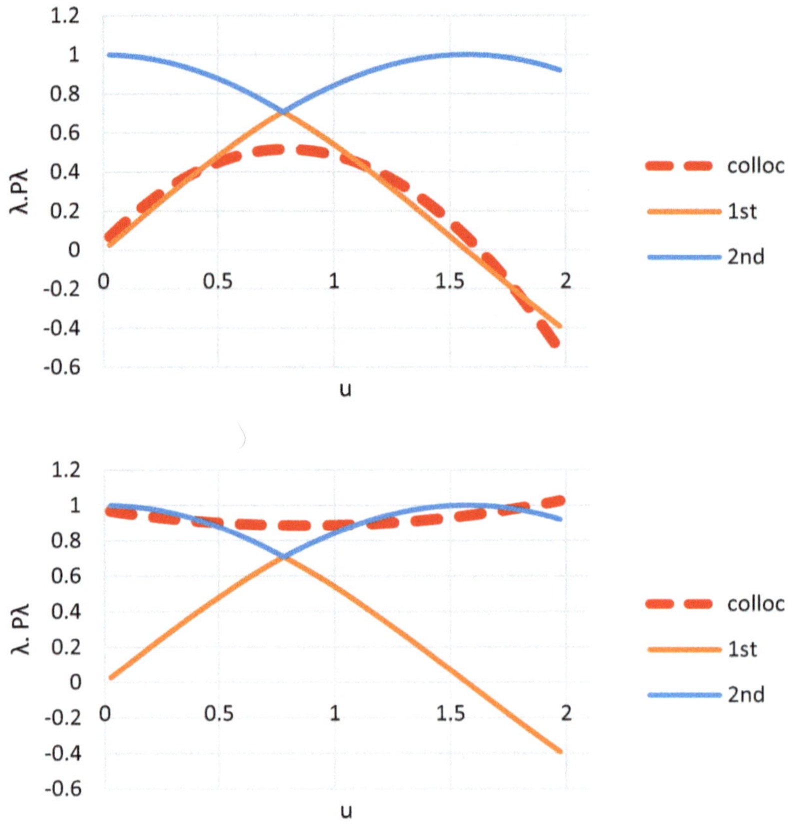

Fig. 6.19 Results obtained by collocation on a sample of 40 variates. U is $U(0, 2)$

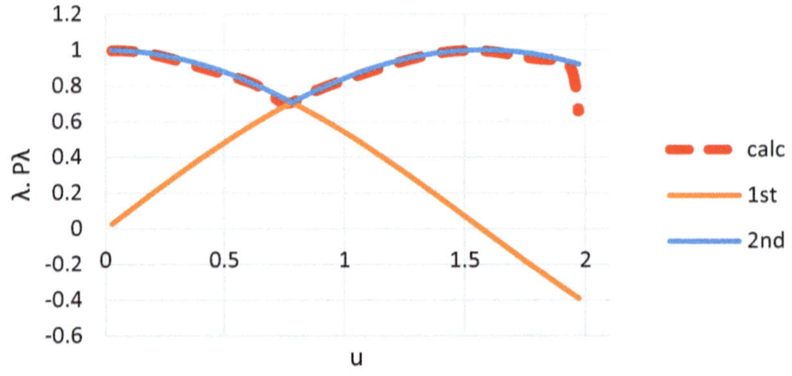

Fig. 6.20 Results obtained by Power Iteration on a sample of 40 variates. U is uniformly distributed on (0,2)

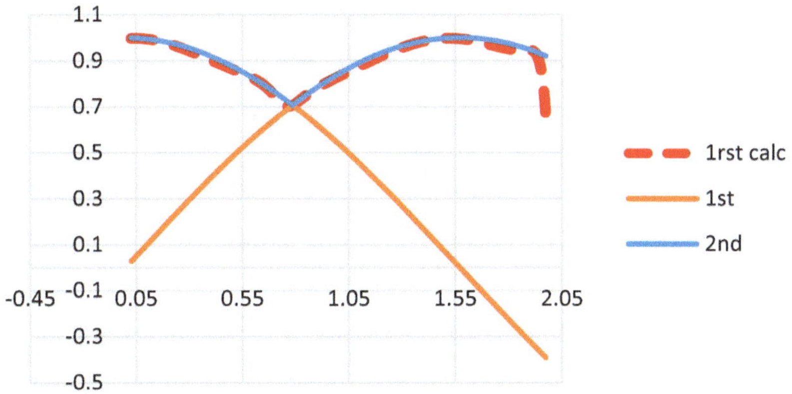

Fig. 6.21 Example of result obtained by Subspace Iteration on a sample of 40 variates. U is uniformly distributed on (0,2)

6.3.1 Very Small Matrices

If your Matrix has dimension 2, you can use a closed formula for the eigenvectors and eigenvalues: let

$$A = \begin{pmatrix} X^1 & Y^1 \\ X^2 & Y^2 \end{pmatrix}.$$

Then, the eigenvalues are

$$\lambda_1 = \frac{1}{2}\left(X^1 + Y^2 - \sqrt{\left(X^1\right)^2 + 4X^2Y^1 - 2X^1Y^2 + \left(Y^2\right)^2}\right),$$

$$\lambda_2 = \frac{1}{2}\left(X^1 + Y^2 + \sqrt{\left(X^1\right)^2 + 4X^2Y^1 - 2X^1Y^2 + \left(Y^2\right)^2}\right).$$

The corresponding eigenvectors are

$$v_1 = \begin{pmatrix} -\dfrac{-X^1 + Y^2 + \sqrt{\left(X^1\right)^2 + 4X^2Y^1 - 2X^1Y^2 + \left(Y^2\right)^2}}{2X^2} \\ 1 \end{pmatrix},$$

$$v_2 = \begin{pmatrix} -\dfrac{-X^1 + Y^2 - \sqrt{\left(X^1\right)^2 + 4X^2Y^1 - 2X^1Y^2 + \left(Y^2\right)^2}}{2X^2} \\ 1 \end{pmatrix}.$$

It is possible to give analogous formulae for matrices of order 3, but they are more complex: let

$$A = \begin{pmatrix} X^1 & Y^1 & Z^1 \\ X^2 & Y^2 & Z^2 \\ X^3 & Y^3 & Z^3 \end{pmatrix}.$$

Set

$$a = -X^1 - Y^2 - Z^3,$$

$$b = -X^2Y^1 + X^1Y^2 - X^3Z^1 - Y^3Z^2 + X^1Z^3 + Y^2Z^3,$$

$$c = X^3Y^2Z^1 - X^2Y^3Z^1 - X^3Y^1Z^2 + X^1Y^3Z^2 + X^2Y^1Z^3 - X^1Y^2Z^3,$$

$$p = b - \frac{a^3}{3}, q = c + \frac{2a^3}{27} - \frac{ab}{3},$$

$$\Delta = \frac{4}{27}p^3 + q^2, D_+ = \frac{1}{2}\left(-q + \sqrt{\Delta}\right), D_- = \frac{1}{2}\left(-q - \sqrt{\Delta}\right),$$

$$\bar{z} = -\frac{1}{2} + i\frac{\sqrt{3}}{2}.$$

Then, the eigenvalues are

$$\lambda_1 = D_+^{\frac{1}{3}} + D_-^{\frac{1}{3}}, \lambda_2 = \bar{z}D_+^{\frac{1}{3}} + \bar{z}^2D_-^{\frac{1}{3}}, \lambda_3 = \bar{z}^2D_+^{\frac{1}{3}} + \bar{z}D_-^{\frac{1}{3}}.$$

The eigenvector associated the eigenvalue λ is

$$v = \begin{pmatrix} -\lambda Z^1 + Y^2Z^1 - Y^1Z^2 \\ X^2Z^1 + \lambda Z^2 - X^1Z^2 \\ \lambda^2 - \lambda\left(X^1 + X^2\right) - X^2Y^1 + X^1Y^2 \end{pmatrix}.$$

This expression assumes that at least one of the lines of v is non null. Alternative expressions exist for the situation where this assumption is not verified: for instance, we can replace ("$\rightarrow$" means "replaced by") $X^1 \rightarrow X^1, Y^1 \rightarrow Z^1, X^2 \rightarrow X^3, Y^2 \rightarrow Z^3, Z^1 \rightarrow Y^1, Z^2 \rightarrow Y^3$ or $X^1 \rightarrow Y^2, Y^1 \rightarrow Z^2, X^2 \rightarrow Y^3$, $Y^2 \rightarrow Z^3, Z^1 \rightarrow X^2, Z^2 \rightarrow X^3$.

Exercises

1. Consider the matrix

$$\boldsymbol{A} = \begin{pmatrix} 0.7 + u_1 & -0.5 - u_1 & -0.3 \\ -0.2 & 0.8 + u_2 & -0.3 - u_2 \\ -0.4 & -0.2 & 0.7 \end{pmatrix}$$

Assume that all the variables u_i are independent and triangularly distributed $T(0,0.5,0.2)$.

(a) Generate a sample of 10 variates from each random variable u_i.

(b) Use the values generate to generate a sample of 100 variates from $\boldsymbol{U} = (u_1, u_2)$.

(c) Use the sample from to determine a representation of the dominant eigenvalue of $\boldsymbol{A}$ as function of $\boldsymbol{U} = (u_1, u_2)$.

(d) Determine the Marginal CDF and PDF of the dominant eigenvalue.

2. Consider

$$A = \begin{pmatrix} u_1 & 4 \\ 1 & \mathrm{u}_2 \end{pmatrix}$$

Assume that the distributions of u_1 and u_2 are triangular: $u_1 \sim T(1, 3, 2)$ and $u_2 \sim T(0,10,5)$.
(a) Generate a sample of 10 variates from each random variable u_i.
(b) Use these values to generate a sample of 100 variates from $U = (u_1, u_2)$.
(c) Use the sample from to determine a representation of the eigenvalues of A as function of $U = (u_1, u_2)$.
(d) Determine the Marginal CDF and PDF of the dominant eigenvalue.

3. Consider

$$A = \begin{pmatrix} 0.2 + \mathrm{u}_1 & \mathrm{u}_1 \\ \mathrm{u}_2 & -10 \end{pmatrix}$$

Assume that the distributions of u_1 and u_2 are triangular: $u_1 \sim T(1, 3, 2)$ and $u_2 \sim T(0, 2, 1)$.
(a) Generate a sample of 10 variates from each random variable u_i.
(b) Use these values to generate a sample of 100 variates from $U = (u_1, u_2)$.
(c) Use the sample from to determine a representation of the eigenvalues of A as function of $U = (u_1, u_2)$.
(d) Determine the Marginal CDF and PDF of the eigenvalues.

7 Random Differential Equations

Differential equations are first of all equations, so that the methods used for equations may be used – at least in principle.

Nevertheless, differential equations have an essential characteristic: the unknowns are fields – i. e. functions, which are elements belonging to infinitely dimensional vector spaces, contrarily to the equations manipulated in the preceding chapters, whose unknowns are vectors from $\mathbb{R}^n$, which is finite dimensional. We shall analyze the consequences of passing from finite dimension to infinite dimension in the sequel (Sect. 7.3).

As seen in Sect. 2.4, differential equations are generally solved by discretization – which brings them to a finite number of unknowns. Consequently, as underlined in Sect. 6.2, we may apply the UQ approach of adaptation to the discretization: the recurrence equations defining successive values generated may be adapted to generate the coefficients of the expansions of the discretized values of solution of the differential equation (see page 346). For instance, let us consider the following ordinary differential equation:

$$\frac{d}{dt}\begin{pmatrix} x_1 \\ x_2 \end{pmatrix} = \begin{pmatrix} u\sqrt{x_2} \\ 2ux_1^2 \end{pmatrix}, \quad x_1(0) = 1, \quad x_2(0) = 1, \quad \mathrm{t} \in (0, \mathrm{T}). \tag{7.1}$$

The solution is $x_1(t) = e^{ut}, x_2(t) = e^{2ut}$. Let us use the adaptation of Runge-Kutta of 4th. order, with $n = 100$, $T = 1$. Let us consider samples of 20 values from u and an expansion of order $k = 6$. We consider two distributions for u: uniform on (0,1) and $N(0.5,0.1)$. The results obtained are shown in Table 7.1. The RMS errors is evaluated as in Eq. (6.33) page 346. Examples of prevision is shown in Figs. 7.1 and 7.2.

E. Souza de Cursi, *Uncertainty Quantification and Stochastic Modelling with EXCEL*, Springer Texts in Business and Economics, https://doi.org/10.1007/978-3-030-77757-9_7

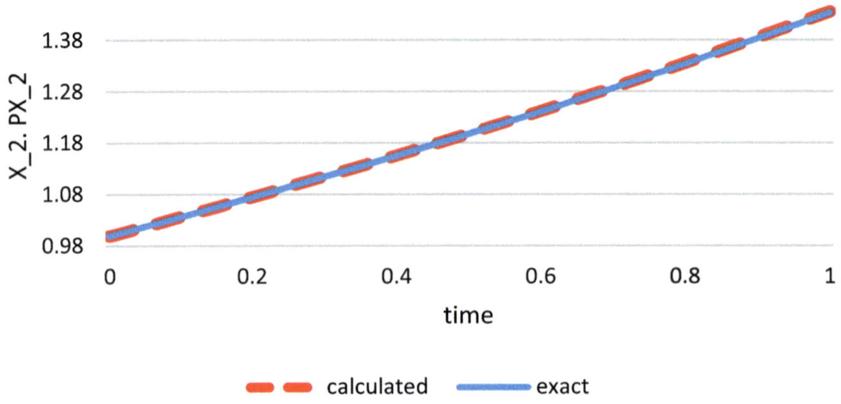

Fig. 7.1 Example of results for the adaptation of the Runge-Kutta RK4. U is uniformly distributed on (0, 1)

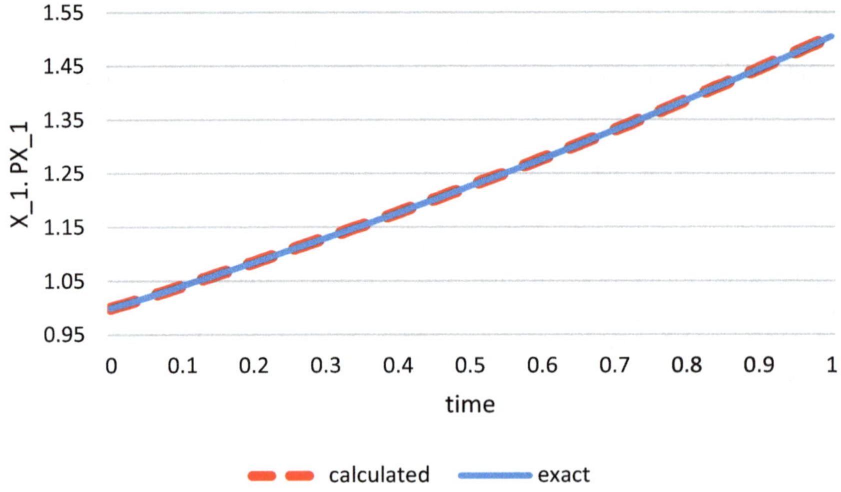

Fig. 7.2 Example of results for the adaptation of the Runge-Kutta RK4. U is normally distributed $N(0.5,0.1)$

Table 7.1 RMS error in the adaptation of RK(nrk) with 100 time steps

nrk	2	3	4	5	6	7	8
uniform	2,5E−3	8,3E−4	8,3E−4	8,2E−4	5,8E−4	5,1E−4	4,4E−4
$N(0.5,0.1)$	3,1E−3	1,0E−3	1,0E−3	1,0E−3	7,1E−4	6,2E−4	5,4E−4

However, there are also specific methods for differential equations. We shall examine some of them in this Chapter.

7.1 Linear Differential Equations

Let us consider the Ordinary Differential Equations (ODE):

$$\frac{d\boldsymbol{X}}{dt} = \boldsymbol{A}(t,\boldsymbol{U})\boldsymbol{X} + \boldsymbol{B}(t,\boldsymbol{U}), \quad \boldsymbol{X}(0) = \boldsymbol{X}_0(\boldsymbol{U}). \tag{7.2}$$

Here, $\boldsymbol{A}$ is a $n \times n$ matrix, $\boldsymbol{B}$ is a $n \times 1$ vector, $\boldsymbol{X_0}$ is a $n \times 1$ vector, $\boldsymbol{X}$ is a $n \times 1$ vector. Analogously to the approach adopted for the analysis of Linear Systems in Sect. 6.1, we consider an expansion analogous to Eqs. (5.34) and (5.35) of Chap. 5:

$$\boldsymbol{X} \approx \boldsymbol{PX} = \sum_{j=1}^{k} \boldsymbol{x_j}(t)\varphi_j(\boldsymbol{U}), \quad \boldsymbol{x_j} = \left(x_{1,j}, \ldots, x_{n,j}\right). \tag{7.3}$$

Thus, we look for an expansion such that

$$\frac{d}{dt}\boldsymbol{PX} = \boldsymbol{A}(t,\boldsymbol{U})\boldsymbol{PX} + \boldsymbol{B}(t,\boldsymbol{U}), \quad \boldsymbol{PX}(0) = \boldsymbol{X_0}(\boldsymbol{U}). \tag{7.4}$$

id est,

$$\begin{gathered}\sum_{j=1}^{k} \varphi_j(\boldsymbol{U})\frac{d}{dt}\boldsymbol{x_j}(t) = \sum_{j=1}^{k} \varphi_j(\boldsymbol{U})\boldsymbol{A}(t,\boldsymbol{U})\boldsymbol{x_j}(t) + \boldsymbol{B}(t,\boldsymbol{U}),\\ \sum_{j=1}^{k} \boldsymbol{x_j}(0)\varphi_j(\boldsymbol{U}) = \boldsymbol{X_0}(\boldsymbol{U})\end{gathered} \tag{7.5}$$

Thus, for $1 \le r \le n$

$$\begin{gathered}\sum_{j=1}^{k} \varphi_j(\boldsymbol{U})\frac{d}{dt}x_{r,j}(t) = \sum_{j=1}^{k} \varphi_j(\boldsymbol{U})A_{rs}(t,\boldsymbol{U})x_{s,j}(t) + B_r(t,\boldsymbol{U}),\\ \sum_{j=1}^{k} x_{r,j}(0)\varphi_j(\boldsymbol{U}) = X_{0_r}(\boldsymbol{U})\end{gathered} \tag{7.6}$$

and

$$\begin{gathered}\sum_{j=1}^{k} \mathcal{M}_{ij}\frac{d}{dt}x_{r,j}(t) = \sum_{j=1}^{k}\sum_{s=1}^{n} \mathcal{A}_{risj}(t)x_{s,j}(t) + \mathcal{B}_{ri}(t),\\ \sum_{j=1}^{k} \mathcal{M}_{ij}x_{r,j}(0) = \mathcal{N}_{0ri}\end{gathered} \tag{7.7}$$

Here,

$$\mathcal{A}_{risj} = E\left(\varphi_i(\boldsymbol{U})A_{rs}(t,\boldsymbol{U})\varphi_j(\boldsymbol{U})\right), \quad \mathcal{B}_{ri} = E(\varphi_i(\boldsymbol{U})B_r(t,\boldsymbol{U})) \tag{7.8}$$

$$\mathcal{M}_{ij} = E\left(\varphi_i(\boldsymbol{U})\varphi_j(\boldsymbol{U})\right), \quad \mathcal{N}_{0ri} = E(\varphi_i(\boldsymbol{U})X_{0_r}(\boldsymbol{U})) \tag{7.9}$$

By setting

$$\mathcal{M}_{risj} = \mathcal{M}_{ij}\delta_{rs},$$

we have

$$\begin{gathered}\sum_{j=1}^{k}\sum_{s=1}^{n} \mathcal{M}_{risj}\frac{d}{dt}x_{s,j}(t) = \sum_{j=1}^{k}\sum_{s=1}^{n} \mathcal{A}_{risj}(t)x_{s,j}(t) + \mathcal{B}_{ri}(t),\\ \sum_{j=1}^{k}\sum_{s=1}^{n} \mathcal{M}_{risj}(t)x_{s,j}(0) = \mathcal{N}_{0ri}\end{gathered} \tag{7.10}$$

Analogously to algebraical linear systems, we have

$$\mathbb{M}\frac{d\mathbb{X}}{dt} = \mathbb{A}(t)\mathbb{X} + \mathbb{B}(t), \quad \mathbb{M}\mathbb{X}(0) = \mathbb{N}_0, \tag{7.11}$$

$$\mathbb{M}_{pq} = \mathcal{M}_{risj}, \ \mathbb{X}_q = x_{s,j}, \ \mathbb{B}_p = \mathcal{B}_{ri}, \ \mathbb{N}_{0p} = \mathcal{N}_{0ri},$$

$$p = i + (r-1)k, \ q = j + (s-1)k.$$

These equations for a linear ODE for the $k \times n$ components of $\mathbb{X}$. Notice that – in the general situation – the matrices $\mathbb{A}$ and $\mathbb{B}$ depend on time t. If the system is autonomous, id est, if $\boldsymbol{A}$ and $\boldsymbol{B}$ are independent from t, then $\mathbb{A}$ and $\mathbb{B}$ are also independent from t. In such a situation, you can calculate $\mathbb{A}$ and $\mathbb{B}$ once. Otherwise, an evaluation at each time step is requested. An example of VBA code for the solution in the general case is given in Figs. 7.3, 7.4, and 7.5. Function `RKstep` in Fig. 7.5 performs one step of the Runge-Kutta method. It calls function `fsm`, which evaluates $\mathbb{M}^{-1}(\mathbb{A}(t)\mathbb{X} + \mathbb{B}(t))$ – the second member of Eq. (7.11). `aruku, bruku, cruku` contain the coefficients of the method (Sect. 2.4). Function `ABvarode` generates matrices $\mathbb{A}$ and $\mathbb{B}$, analogously to `varlinsys`, page 328.

```
Function RKexec(aruku, bruku, cruku, tspan, x0, mm, basis, k, n, u)
ndim = UBound(x0, 1)
ntp = UBound(tspan, 1)
Dim solx() As Double
ReDim solx(1 To ndim, 0 To ntp)
For i = 1 To ndim
    solx(i, 0) = x0(i)
Next i
Dim x() As Double
ReDim x(1 To ndim)
For j = 1 To ntp
    For i = 1 To ndim
        x(i) = solx(i, j - 1)
    Next i
    tt = tspan(j - 1)
    h = tspan(j) - tt
    aux = RKstep(aruku, bruku, cruku, h, tt, x, mm, u, basis, k, n)
    For i = 1 To ndim
        solx(i, j) = aux(i)
    Next i
Next j
RKexec = solx
End Function
```

Fig. 7.3 Function `RKexec` makes all the steps planned in `tspan` for the solution of (7.11)

```
Function fsm(u, basis, k, n, mm, XX, t)
 resuAB = ABvarode(u, basis, k, n, t)
 aa = resuAB(1)
 bb = resuAB(2)
 nn = k * n
 yy = squeeze(WorksheetFunction.MMult(aa, tocol(XX)))
 Dim zz() As Double
 ReDim zz(1 To nn)
 For i = 1 To nn
   zz(i) = yy(i) + bb(i)
 Next i
 aux = LUSolution(mm, zz)
 fsm = aux
End Function
```

Fig. 7.4 Function fsm evaluates the second member $\mathbb{M}^{-1}(\mathbb{A}(t)\mathbb{X} + \mathbb{B}(t))$ of (7.11). Function `ABvarode` generates matrices $\mathbb{A}$ and $\mathbb{B}$, analogously to `varlinsys`, page 328

```
Function RKstep(aruku, bruku, cruku, h, t, x, mm, u, basis, k, n)
ndim = UBound(x, 1)
nrk = UBound(bruku, 1)
Dim y() As Double
ReDim y(1 To ndim, 0 To nrk)
For i = 1 To ndim
    y(i, 0) = x(i)
    For j = 2 To nrk
        y(i, j) = 0
    Next j
Next i
Dim yy() As Double
ReDim yy(1 To ndim)
For kk = 0 To nrk - 1
    ck = cruku(kk)
    ak = aruku(kk)
    For i = 1 To ndim
        yy(i) = 0
    Next i
    tk = t + ck * h
    For j = 1 To UBound(ak, 1)
        For i = 1 To ndim
            yy(i) = yy(i) + ak(j) * y(i, j - 1)
        Next i
    Next j
    For i = 1 To ndim
        yy(i) = x(i) + h * yy(i)
    Next i
    aux = fsm(u, basis, k, n, mm, yy, t)
    For i = 1 To ndim
        y(i, kk + 1) = aux(i)
    Next i
Next kk
For j = 1 To ndim
    yy(j) = 0
Next j
For i = 1 To nrk
    For j = 1 To ndim
        yy(j) = yy(j) + bruku(i) * y(j, i)
    Next j
Next i
For j = 1 To ndim
    yy(j) = x(j) + h * yy(j)
Next j
RKstep = yy
End Function
```

Fig. 7.5 Function `RKstep` performs one step of Runge-Kutta's method corresponding to coefficients `aruku`, `bruku`, `cruku` of a Runge-Kutta Table (Sect. 2.4)

To use these functions, you must write a `Sub` which collects the data and runs `RKexec`. For instance, let us consider

$$A(t,U) = \begin{pmatrix} 0 & 1 \\ \frac{U^2}{2} & \frac{U}{2} \end{pmatrix}, \quad B(t,U) = \begin{pmatrix} 0 \\ \frac{U^2}{2}(1-tU) \end{pmatrix}, \quad X_0 = \begin{pmatrix} 1 \\ 0 \end{pmatrix}.$$

The exact solution is

$$x_1(t) = Ut + e^{-Ut}, \ \ x_2(t) = U\left(1 - e^{-Ut}\right).$$

We consider U as normally distributed, with mean 2π and standard deviation $\pi/6$. The sample used contains $ns = 20$ variates from U. We generated the matrices $\mathbb{A}$ and $\mathbb{B}$ by using the sample and solved Eq. (7.10) by RK4 solved (7.11) by using the Runge-Kutta method of order 4 presented in Sect. 2.4. The RMS error for 100 time steps on (0,1) ($\Delta t = 0, 01$) was about 3E−2. Examples of solution are given in Figs. 7.6 and 7.7.

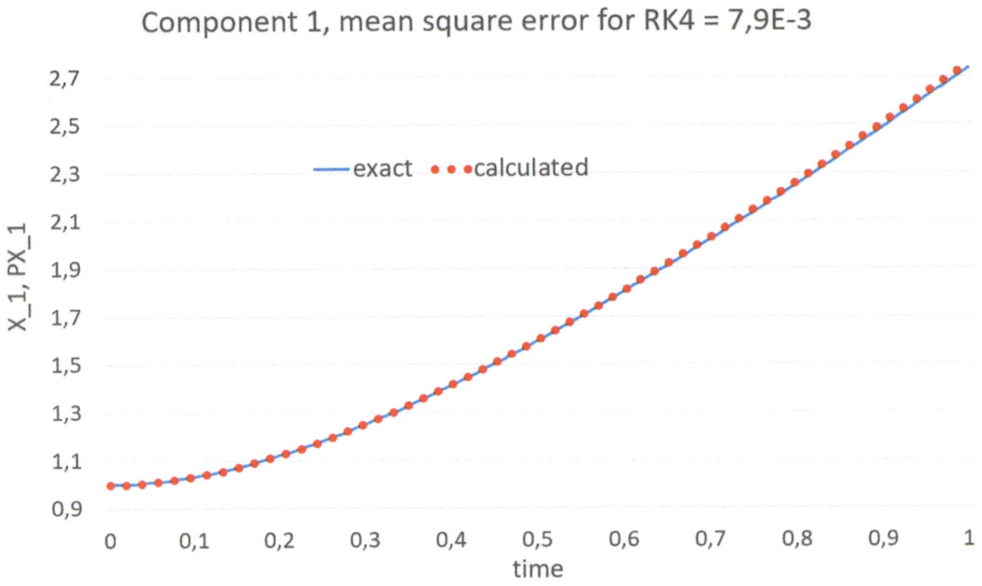

Fig. 7.6 Results obtained by solving Eq. (7.11) by RK4 on (0,1), with a time step 0,01

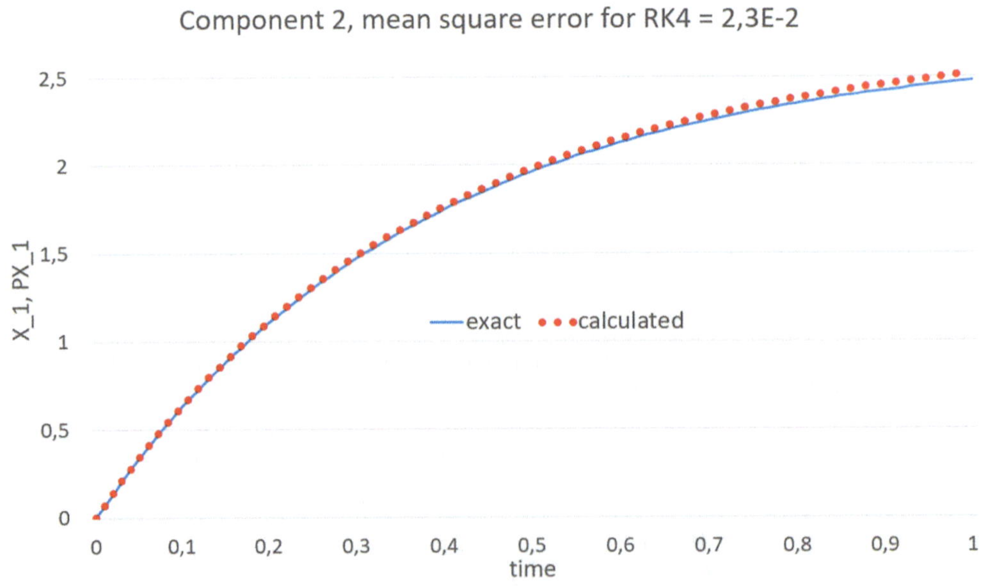

Fig. 7.7 Results obtained by solving Eq. (7.11) by RK4 on (0,1), with a time step 0,01

For autonomous equations, the matrices A and B may be calculated once and passed as parameters. For instance, we may use the code below:

```
Function fsm_auto(k, n, mm, aa, bb, XX)
 nn = k * n
 yy = squeeze(WorksheetFunction.MMult(aa, tocol(XX)))
 Dim zz() As Double
 ReDim zz(1 To nn)
 For i = 1 To nn
   zz(i) = yy(i) + bb(i)
 Next i
 aux = LUSolution(mm, zz)
 fsm_auto = aux
End Function
```

Here, aa corresponds to $\mathbb{A}$ and bb corresponds to $\mathbb{B}$. As an example, let us consider again the system defined in Eq. (6.30), page 346:

$$\begin{pmatrix} x_1' \\ x_2' \end{pmatrix} = \mathbf{A}\begin{pmatrix} x_1 \\ x_2 \end{pmatrix}, \quad \mathbf{A} = \begin{pmatrix} 0 & U \\ -U & 0 \end{pmatrix}, \quad x_1(0) = 0, \quad x_2(0) = 1, t \in (0, \mathrm{T}).$$

The solution is $x_1(t) = \sin(Ut)$, $x_2(t) = \cos(Ut)$. U is normally distributed, with a mean 2π and a standard deviation $\pi/6$. The sample used contains $ns = 20$ variates from U. We generated the matrices $\mathbb{A}$ and $\mathbb{B}$ by using the sample. Then, we solved Eq. (7.10) by RK4, and the result was used to generate the curves corresponding to the elements U in the sample. The RMS error was about 7E−3. Examples of solutions are given in Figs. 7.8 and 7.9. The results are comparable to those obtained by adaptation (see page 347).

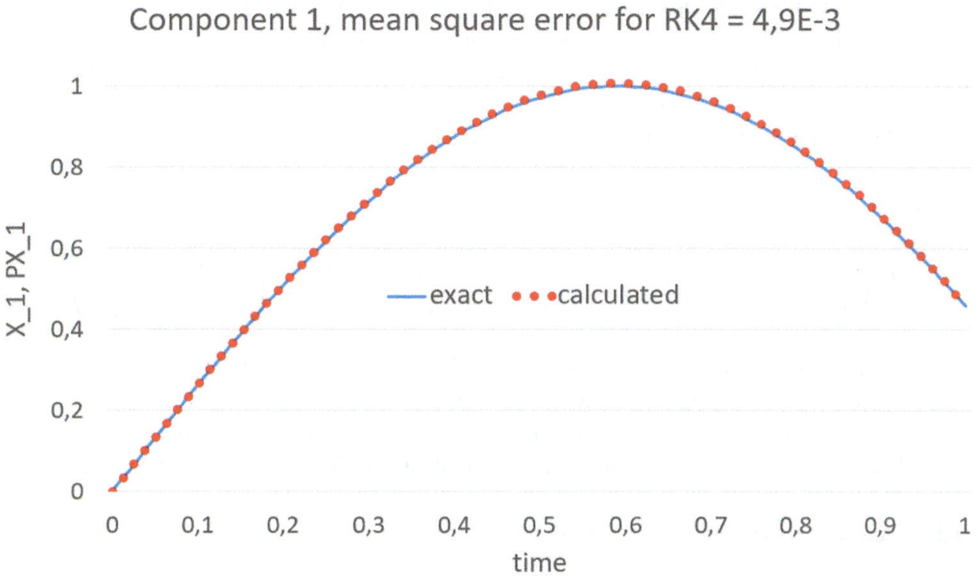

Fig. 7.8 Results obtained by solving Eq. (7.11) by RK4 on (0,1), with a time step 0,01

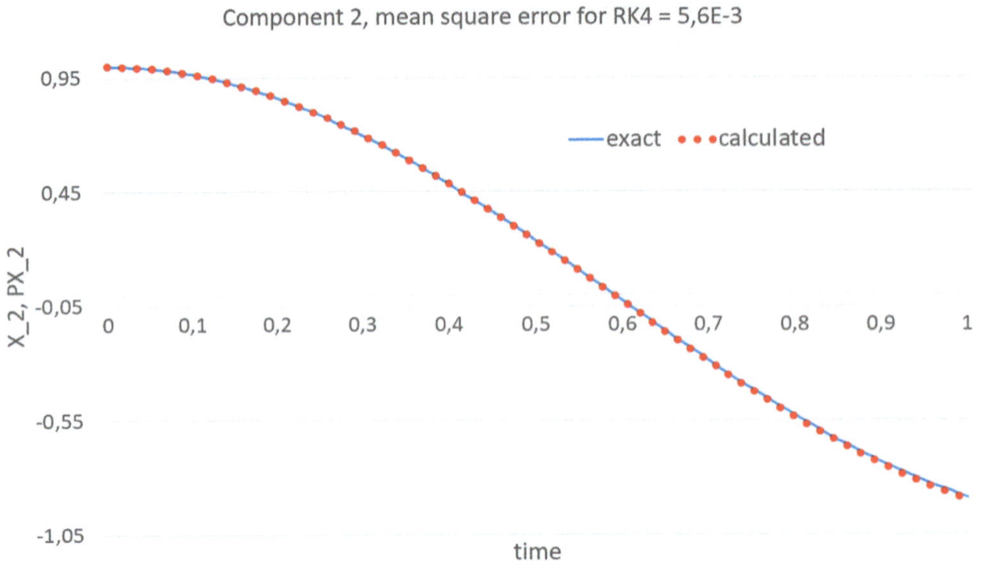

Fig. 7.9 Results obtained by solving Eq. (7.11) by RK4 on (0,1), with a time step 0,01

Exercises

1. Consider the differential equations

$$\frac{\mathrm{d}}{\mathrm{dt}}\begin{pmatrix} x_1 \\ x_2 \end{pmatrix} = \begin{pmatrix} -u_1 & 2 \\ 4 & -u_2 \end{pmatrix}\begin{pmatrix} x_1 \\ x_2 \end{pmatrix}, \quad \begin{pmatrix} x_1(0) \\ x_2(0) \end{pmatrix} = \begin{pmatrix} 2 \\ 5 \end{pmatrix}$$

Assume that the distributions of u_1 and u_2 are triangular: $u_1 \sim T(0, 2, 1)$ and $u_2 \sim T(0, 8, 4)$.

(a) Generate a sample of 6 variates from each random variable u_i.
(b) Use these values to generate a sample of 36 variates from $\boldsymbol{U} = (u_1, u_2)$.
(c) Use the sample to determine a representation of $\boldsymbol{X}(t) = (x_1(t), x_2(t))$ as function of $\boldsymbol{U} = (u_1, u_2)$ for $t \in (0, 1)$.
(d) Determine the CDF and the PDF of $x_1(0.5)$.

2. Consider

$$\frac{\mathrm{d}}{\mathrm{dt}}\begin{pmatrix} x_1 \\ x_2 \end{pmatrix} = \begin{pmatrix} 0.2 - 5u_1 & -2.5 - 5u_1 \\ 1 & 0 \end{pmatrix}\begin{pmatrix} x_1 \\ x_2 \end{pmatrix} + \begin{pmatrix} 8 \\ 3 \end{pmatrix}, \quad \begin{pmatrix} x_1(0) \\ x_2(0) \end{pmatrix} = \begin{pmatrix} 2 \\ 5 \end{pmatrix}$$

Assume that the distributions $u_1 \sim T(-0.1, 0.1, 0)$ and $u_2 \sim N(0, 0.1)$.

(a) Generate a sample of 10 variates from each random variable u_i.
(b) Use these values to generate a sample of 100 variates from $\boldsymbol{U} = (u_1, u_2)$.
(c) Use the sample to determine a representation of $\boldsymbol{X}(t) = (x_1(t), x_2(t))$ as function of $\boldsymbol{U} = (u_1, u_2)$ for $t \in (0, 1)$.
(d) Determine the CDF and the PDF of $x_1(0.5)$.

3. Consider the differential equation

$$\frac{\mathrm{d}}{\mathrm{dt}}\begin{pmatrix} x_1 \\ x_2 \end{pmatrix} = \begin{pmatrix} 2+u_1 & -3+\mathrm{u}_2 \\ 1 & 0 \end{pmatrix}\begin{pmatrix} x_1 \\ x_2 \end{pmatrix} + \begin{pmatrix} 10+5u_1 \\ -3 \end{pmatrix}, \quad \begin{pmatrix} x_1(0) \\ x_2(0) \end{pmatrix} = \begin{pmatrix} 2 \\ 5 \end{pmatrix}$$

Assume that the distributions of u_1 and u_2 are triangular: $u_1 \sim T(1, 3, 2)$ and $u_2 \sim T(0, 2, 1)$.
(a) Generate a sample of 10 variates from each random variable u_i.
(b) Use these values to generate a sample of 100 variates from $\boldsymbol{U} = (u_1, u_2)$.
(c) Use the sample to determine a representation of $\boldsymbol{X}(t) = (x_1(t), x_2(t))$ as function of $\boldsymbol{U} = (u_1, u_2)$ for $t \in (0, 1)$.
(d) Determine the CDF and the PDF of $x_1(0.5)$.

4. Consider the differential equation

$$\frac{\mathrm{d}}{\mathrm{dt}}\begin{pmatrix} x_1 \\ x_2 \end{pmatrix} = \begin{pmatrix} -u_1 & -3-u_2 \\ 0.5+u_2 & -2-u_2 \end{pmatrix}\begin{pmatrix} x_1 \\ x_2 \end{pmatrix} + \begin{pmatrix} 0.2+u_1 \\ 0.1+u_2 \end{pmatrix}, \quad \begin{pmatrix} x_1(0) \\ x_2(0) \end{pmatrix} = \begin{pmatrix} 2 \\ 5 \end{pmatrix}$$

Assume that the distributions of u_1 and u_2 are triangular: $u_1 \sim T(-0.1,0.2,0)$ and $u_2 \sim T(-0.5,0.5,0)$.
(a) Generate a sample of 6 variates from each random variable u_i.
(b) Use these values to generate a sample of 36 variates from $\boldsymbol{U} = (u_1, u_2)$.
(c) Use the sample to determine a representation of $\boldsymbol{X}(t) = (x_1(t), x_2(t))$ as function of $\boldsymbol{U} = (u_1, u_2)$ for $t \in (0, 1)$.
(d) Determine the CDF and the PDF of $x_1(0.5)$.

7.2 Nonlinear Differential Equations

The approach presented in the previous section extends to nonlinear ODE such as

$$\frac{d\boldsymbol{X}}{dt} = \boldsymbol{f}(t, \boldsymbol{X}, \boldsymbol{U}), \qquad \boldsymbol{X}(0) = \boldsymbol{X_0}(\boldsymbol{U}). \tag{7.12}$$

The procedure is analogous and leads to a formulation close to the method of adaptation, presented in 6.2: we use Eq. (7.3) to get

$$\frac{d}{dt}\boldsymbol{PX} = \boldsymbol{f}(t, \boldsymbol{PX}, \boldsymbol{U}), \qquad \boldsymbol{PX}(0) = \boldsymbol{X_0}(\boldsymbol{U}). \tag{7.13}$$

Thus, for $1 \leq r \leq n$

$$\begin{aligned} &\sum_{j=1}^{k} \varphi_j(\boldsymbol{U}) \frac{d}{dt} x_{r,j}(t) = f_r(t, \boldsymbol{PX}, \boldsymbol{U}), \\ &\sum_{j=1}^{k} x_{r,j}(0) \varphi_j(\boldsymbol{U}) = X_{0_r}(\boldsymbol{U}) \end{aligned} \tag{7.14}$$

Then,

$$\sum_{j=1}^{k} \mathcal{M}_{ij}\frac{d}{dt}x_{r,j}(t) = \mathcal{F}_{ri}(t,\boldsymbol{x}),$$
$$\sum_{j=1}^{k} \mathcal{M}_{ij}x_{r,j}(0) = \mathcal{N}_{0ri} \tag{7.15}$$

with $\mathcal{M}_{ij}, \mathcal{N}_{0ri}$ previously defined (Eq. (7.9)) and

$$\mathcal{F}_{ri}(t,\boldsymbol{x}) = E(\varphi_i(\boldsymbol{U})f_r(t,\boldsymbol{PX},\boldsymbol{U})). \tag{7.16}$$

The evaluation of $\mathcal{F}_{ri}$ is made as follows:

1. Determine $\boldsymbol{Z_m} = \boldsymbol{PX}(U_m)$, for $1 \le m \le ns$.
2. Evaluate $\boldsymbol{T_m} = \boldsymbol{f}(t, \boldsymbol{Z_m}, U_m)$, for $1 \le m \le ns$.Let $\boldsymbol{T_m} = (T_{1,\,m}, \ldots, T_{n,\,m})$
3. Determine $V_{m,\,i} = \varphi_i(U_m)$,$1 \le m \le ns$, $1 \le i \le k$.
4. Determine $\mathcal{F}_{ri} \approx \frac{1}{ns}\sum_{m=1}^{ns} V_{m,i}T_{r,m}$

Using the same map $p = i + (r - 1)k$, $q = j + (s - 1)k$ previously introduced, we have

$$\mathbb{M}\frac{d\mathbb{X}}{dt} = \mathbb{F}(t,\mathbb{X},\boldsymbol{U}), \qquad \mathbb{M}\mathbb{X}(0) = \mathbb{N}_0, \tag{7.17}$$

$$\mathbb{M}_{pq} = \mathcal{M}_{risj}, \quad \mathbb{X}_{\mathrm{q}} = x_{s,j}, \quad \mathbb{F}_{\mathrm{p}} = \mathcal{F}_{ri}, \quad \mathbb{N}_{0p} = \mathcal{N}_{0ri}.$$

An example of VBA code for the determination of $\mathbb{F}$ is given below:

```
Function fvarode(u, ap, k, n, mm, XX, t
ns = UBound(u, 1)
nn = k * n
Dim c() As Double
ReDim c(1 To n, 1 To k)
Dim xxxx() As Double
ReDim xxxx(1 To n)
For r = 1 To n
  For i = 1 To k
    pp = i + (r - 1) * k
    c(r, i) = XX(pp)
  Next i
Next r
px = ap.proj(c, u)
tabphi = ap.matval(k, u)
Dim ff() As Double
ReDim ff(1 To nn)
For i = 1 To nn
  ff(i) = 0
Next i
For m = 1 To ns
  uu = u(m)
  For r = 1 To n
    xxxx(r) = px(r, m)
  Next r
  fuu = f(t, xxxx, uu)
  For r = 1 To n
    For i = 1 To k
      pp = i + (r - 1) * k
      aux1 = tabphi(m, i)
      ff(pp) = ff(pp) + fuu(r) * aux1
    Next i
  Next r
Next m
For r = 1 To n
  For i = 1 To k
    pp = i + (r - 1) * k
    ff(pp) = ff(pp) / ns
  Next i
Next r
fvarode = LUSolution(mm, ff)
End Function
```

As an example, let us consider the system defined in Eq. (7.1):

$$\boldsymbol{f}(t, \boldsymbol{x}, u) = \begin{pmatrix} u\sqrt{x_2} \\ 2ux_1^2 \end{pmatrix}, \ X_0 = \begin{pmatrix} 1 \\ 1 \end{pmatrix}.$$

The solution is $x_1(t) = e^{ut}$, $x_2(t) = e^{2ut}$. Using the same data and parameters as those used page 371, we obtain the results in Table 7.2, which are almost identical to those in Table 7.1, page 360. Examples of previsions are given in Figs. 7.10 and 7.11.

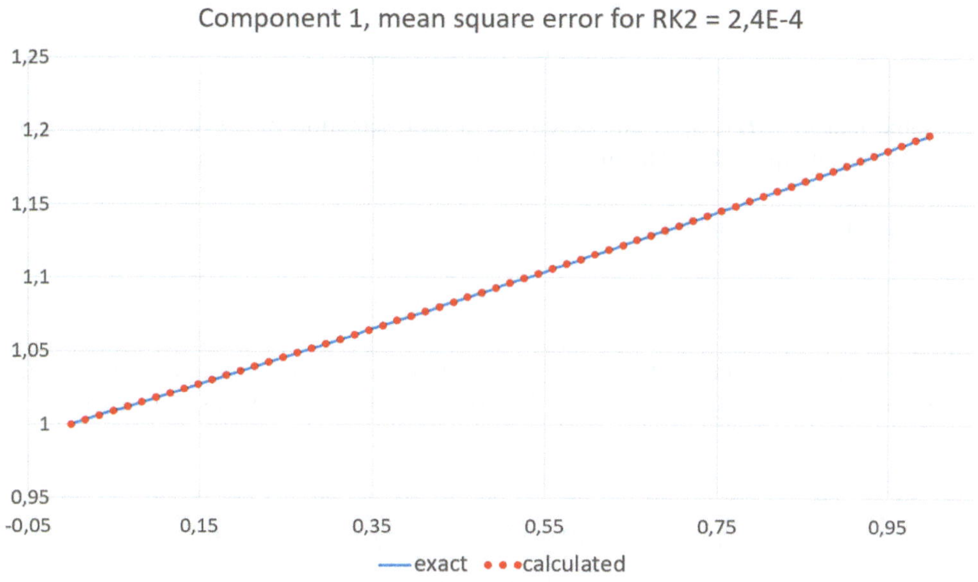

Fig. 7.10 Example of results furnished by Runge-Kutta RK2 for the solution of (7.17). U is uniformly distributed on (0, 1)

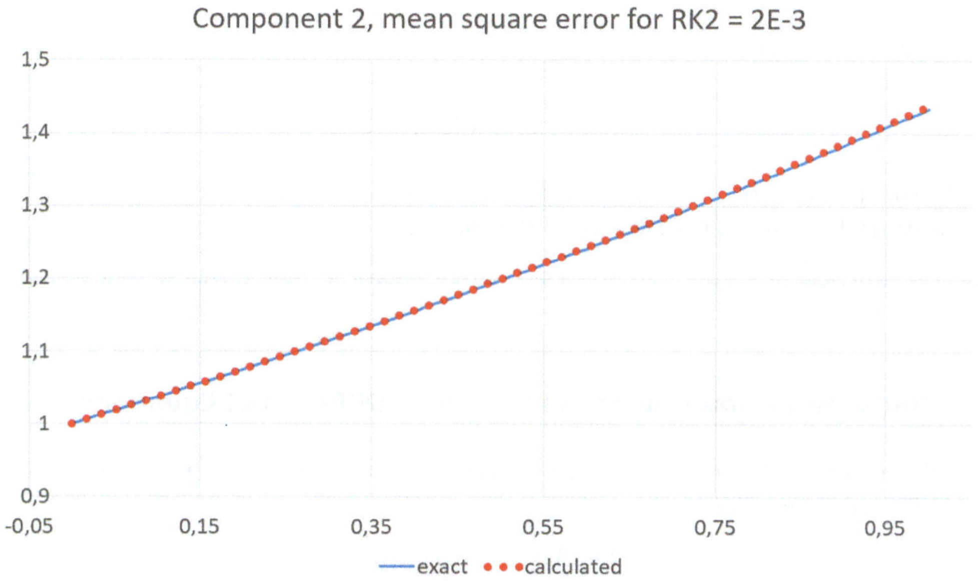

Fig. 7.11 Example of results furnished by Runge-Kutta RK2 for the solution of (7.17). U is uniformly distributed on (0, 1)

Table 7.2 RMS error in the adaptation of RK(nrk) with 100 time steps

nrk	2	3	4	5	6	7	8
uniform	2,5E−3	8,3E−4	8,3E−4	8,2E−4	5,8E−4	5,1E−4	4,4E−4
N(0.5,0.1)	3,1E−3	1,0E−3	1,0E−3	1,0E−3	7,1E−4	6,2E−4	5,4E−4

Exercises

1. Consider the differential equation

$$\frac{dx}{dt} = (1 - t)x^u, \quad x(0) = 1$$

Assume that $u \sim T(1, 3, 2)$ and determine a representation of $x(t)$ as function of u for $t \in (0, 1)$. Determine the CDF and the PDF of $x(0.5)$.

2. Consider the differential equation

$$\frac{dx}{dt} = (1 - t)x^{u_1}, \quad x(0) = u_2$$

Assume that $u_1 \sim T(1, 3, 2)$ and $u_2 \sim N(2, 0.1)$. Determine a representation of $x(t)$ as function of $\boldsymbol{U} = (u_1, u_2)$ for $t \in (0, 1)$. Determine the CDF and the PDF of $x(0.5)$.

3. Consider the differential equation

$$\frac{\mathrm{d}}{\mathrm{d}t}\begin{pmatrix} x_1 \\ x_2 \end{pmatrix} = \begin{pmatrix} (2+u)x_2 \\ -(2+u)x_1 - 4x_1^2 \end{pmatrix}, \quad \begin{pmatrix} x_1(0) \\ x_2(0) \end{pmatrix} = \begin{pmatrix} 0 \\ 2 \end{pmatrix}$$

Assume that $u \sim T(-1, 0, 1)$ and determine a representation of $\boldsymbol{X}(t) = (x_1(t), x_2(t))$ as function of u for $t \in (0, 1)$.

4. Consider the differential equation

$$\frac{dx}{dt} = t^{1-u}(1 + x)^u, \quad x(0) = 1$$

Assume that $u \sim T(0, 1, 0.5)$. and determine a representation of $x(t)$ as function of u for $t \in (0, 1)$. Determine the CDF and the PDF of $x(0.5)$.

7.3 Uncertainties on Curves Connected to Differential Equations

If an ODE involves random parameters, then its solution becomes random – as seen in the preceding sections. Thus, the trajectories

$$t \to \boldsymbol{X}(t) = (x_1(t), x_2(t))$$

and the orbits

$$t \to \left(x_i(t), p_i(t) = \frac{dx_i}{dt}(t) \right)$$

are also random variables. For instance, let us consider the simple ODE:

$$\frac{d}{dt}\begin{pmatrix} x_1 \\ x_2 \end{pmatrix} = \begin{pmatrix} \cos(u) \\ \sin(u) \end{pmatrix}, \quad x_1(0) = x_2(0) = 0.$$

The solution is $x_1(t) = tcos(u)$, $x_2(t) = tsin(u)$. If u is a random variable, x_1 and x_2 are also random. Thus, we may consider the analysis of the variability of these variables and look for statistics associated to them, such as, for instance, their mean and variance.

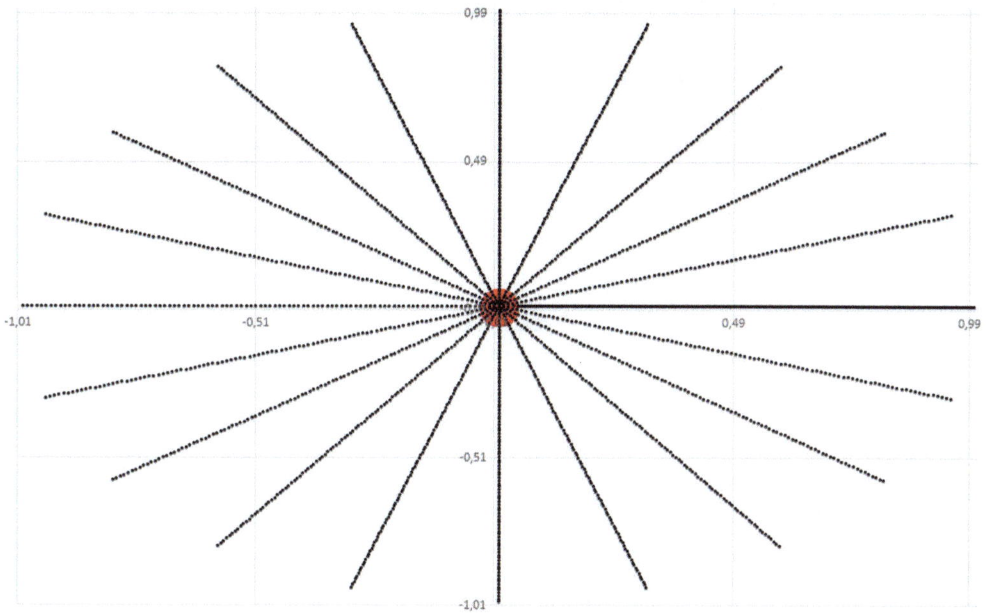

Fig. 7.12 When the mean is taken on the equations, the result is the origin, which is not representative of the family of curves: all are segments of unitary length

However – as shown in Sect. 5.7, page 296 – the analysis of uncertainties in curves may request the use of approaches that are independent from the representation. For instance, let us assume that u is uniformly distributed on $(0, 2\pi)$. Then, taking the mean on the equations leads to $(x_1(t)) = 0$, $E(x_2(t)) = 0$: the resulting curve is not representative of the family (see Fig. 7.12). In such a situation, the approach consisting in searching the most representative member of the family appears as more adequate (see Eq. (5.53), Sect. 5.8, page 308). In this case,

$$dist(C_i, C_j) = 2\left| \sin\left(\frac{1}{2}(u_i - u_j)\right)\right|,$$

so that $J(C_i)$ has a constant value for all the members of the family: any of its members may be considered as a good representative of the family.

Example

Let us consider the ODE

$$\frac{d}{dt}\begin{pmatrix} x_1 \\ x_2 \end{pmatrix} = \begin{pmatrix} 2ut \\ 2u(t-1) \end{pmatrix}, \quad x_1(0) = 0, \quad x_2(0) = u, \quad t \in (0,1).$$

The solution is $x_1(\mathrm{t}) = ut^2$, $x_2(\mathrm{t}) = u(t-1)^2$. Let us assume that u is uniformly distributed on (0,1). The evaluations of the mean trajectory by two different approaches are shown in the Fig. 7.13.

In this case, $p_1(t) = 2ut$ and the orbits of the first variable are the points $(ut^2, 2ut)$. Their mean is shown in Fig. 7.14. The means taken at each time furnish

$$E(x_1(t)) = \frac{1}{2}t^2, \quad E(x_2(t)) = \frac{1}{2}(t-1)^2, \quad E(p_1(t)) = t, \quad t \in (0,1).$$

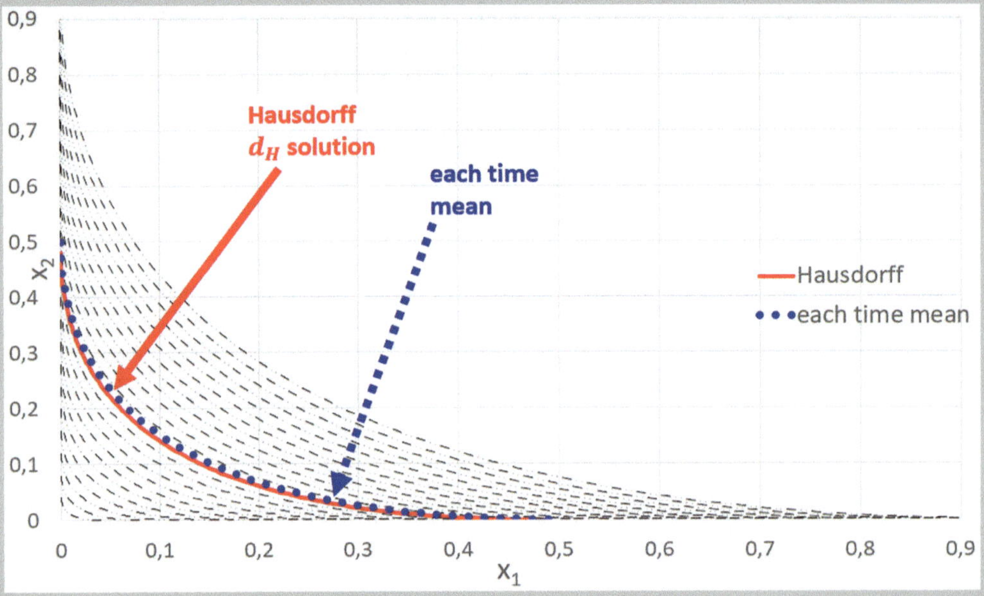

Fig. 7.13 Trajectories of the system. Here, the approach by the Hausdorff distance produces the same result as the mean $E((x_1(t), x_2(t)))$ taken at each time

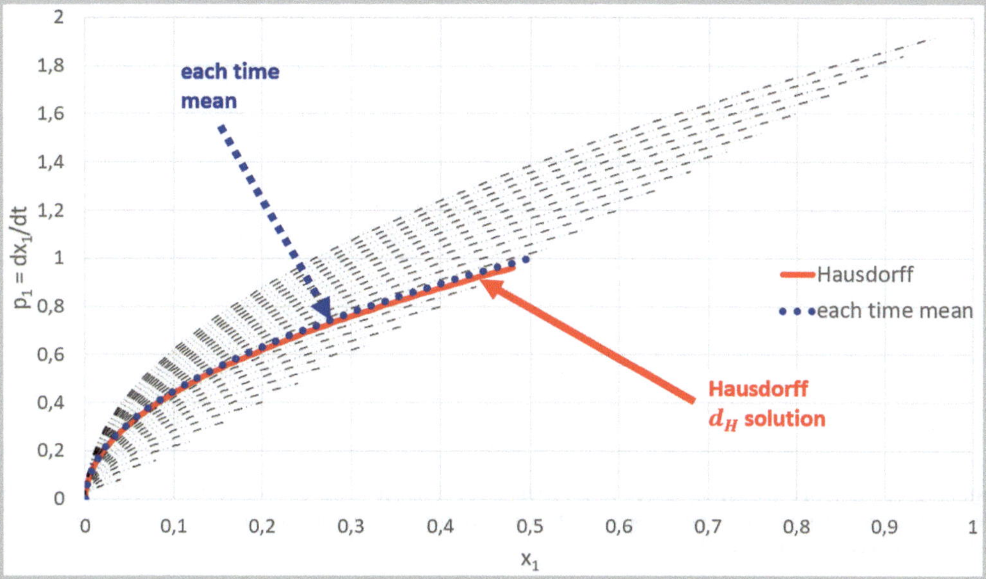

Fig. 7.14 Orbits $(x_1(t), p_1(t))$ of the system. Again, the approach by the Hausdorff distance produces the same result as the mean $E((x_1(t), p_1(t)))$ taken at each time

Example

Let us consider the ODE

$$\frac{d}{dt}\begin{pmatrix} x_1 \\ x_2 \end{pmatrix} = \begin{pmatrix} -(1+|u|)x_2 \\ (1+|u|)x_1 \end{pmatrix}, \quad x_1(0) = 0, \; x_2(0) = u, \quad t \in (0, 2\pi).$$

The solution is $x_1(t) = u \cos((1 + |u|)t)$, $x_2(t) = u \sin((1 + |u|)t)$. Let us assume that u is uniformly distributed on $(-1,1)$. In this case, the means taken at each time furnish

$$E(x_1(t)) = 0, \; E(x_2(t)) = 0, \; E(p_1(t)) = 0, \; E(p_2(t)) = 0.$$

The trajectories are concentric circles: for each u, the corresponding trajectory is the circle of radius $|u|$, centered at the origin (0, 0). Considering that u and -u generate the same circle of radius $|u|$, the family has radius going from 0 to 1. Thus, we expect as mean trajectory the circle corresponding to $|u| = 1/2$.In this case, only the Hausdorff approach furnishes the good result.

The evaluations of the mean trajectory by both the approaches are shown in Fig. 7.15.

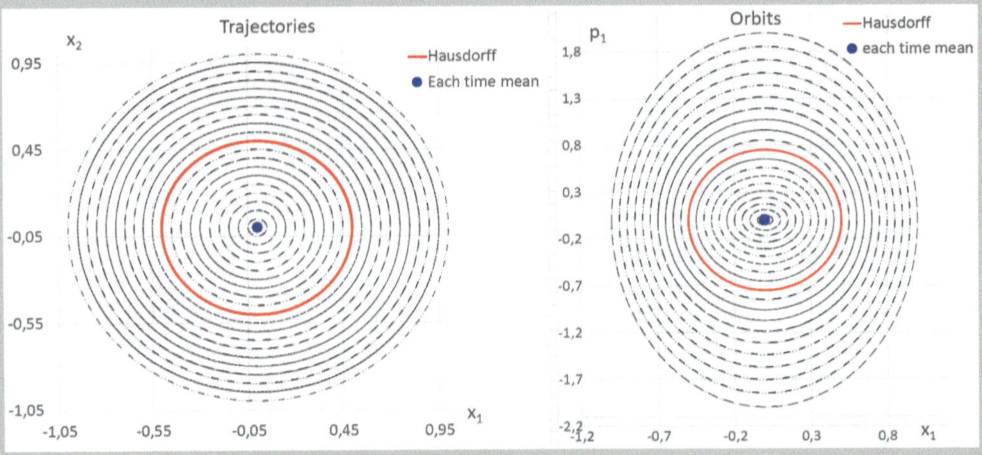

Fig. 7.15 Trajectories and orbits of the system. Here, the approach by the Hausdorff distance does not produce the same result as the mean $\boldsymbol{E}((\boldsymbol{x_1}(\boldsymbol{t}),\ \boldsymbol{x_2}(\boldsymbol{t})))$ taken at each time: means at fixed time are null, while Hausdorff distance produces a correct result

Exercises

1. Consider the differential equation

$$\frac{dx}{dt} = (1-t)x^u, \quad x(0) = 1, \quad t \in (0,1).$$

Assume that $u \sim T(1, 3, 2)$ and determine the most representative history $(t, x(t))$ and the most representative orbit $\left(x(t), \frac{dx}{dt}(t)\right)$.

2. Consider the differential equation

$$\frac{dx}{dt} = (1-t)x^{u_1}, \quad x(0) = u_2, \quad t \in (0,1).$$

Assume that $u_1 \sim T(1, 3, 2)$ and $u_2 \sim N(2, 0.1)$. Determine the most representative history $(t, x(t))$ and the most representative orbit $\left(x(t), \frac{dx}{dt}(t)\right)$.

3. Consider the differential equation

$$\frac{\mathrm{d}}{\mathrm{dt}}\begin{pmatrix} x_1 \\ x_2 \end{pmatrix} = \begin{pmatrix} (2+u)x_2 \\ -(2+u)x_1 - 4x_1^2 \end{pmatrix}, \quad \begin{pmatrix} x_1(0) \\ x_2(0) \end{pmatrix} = \begin{pmatrix} 0 \\ 2 \end{pmatrix}, \quad t \in (0,1).$$

Assume that $u \sim T(-1, 0, 1)$ and determine the most representative trajectory $(x_1(t), x_2(t))$ and the most representative orbit $\left(x_1(t), \frac{dx_1}{dt}(t)\right)$.

4. Consider the differential equation

$$\frac{dx}{dt} = t^{1-u}(1+x)^u, \quad x(0) = 1, \quad t \in (0,1).$$

Assume that $u \sim T(0,\ 1,\ 0.5)$ and determine the most representative history $(t,\ x(t))$ and the most representative orbit $\left(x(t), \frac{dx}{dt}(t)\right)$.

5. Consider the differential equations

$$\frac{\mathrm{d}}{\mathrm{dt}}\begin{pmatrix} x_1 \\ x_2 \end{pmatrix} = \begin{pmatrix} -u_1 & 2 \\ 4 & -\mathrm{u}_2 \end{pmatrix}\begin{pmatrix} x_1 \\ x_2 \end{pmatrix}, \quad \begin{pmatrix} x_1(0) \\ x_2(0) \end{pmatrix} = \begin{pmatrix} 2 \\ 5 \end{pmatrix}, \quad t \in (0,1).$$

Assume that the distributions of u_1 and u_2 are triangular: $u_1 \sim T(0,\ 2,\ 1)$ and $u_2 \sim T(0,\ 8,\ 4)$. Determine the most representative the most representative trajectory $(x_1(t),\ x_2(t))$ and the most representative orbit $\left(x_1(t), \frac{dx_1}{dt}(t)\right)$.

6. Consider $t \in (0,\ 1)$ and

$$\frac{\mathrm{d}}{\mathrm{dt}}\begin{pmatrix} x_1 \\ x_2 \end{pmatrix} = \begin{pmatrix} 0.2 - 5u_1 & -2.5 - 5u_1 \\ 1 & 0 \end{pmatrix}\begin{pmatrix} x_1 \\ x_2 \end{pmatrix} + \begin{pmatrix} 8 \\ 3 \end{pmatrix}, \quad \begin{pmatrix} x_1(0) \\ x_2(0) \end{pmatrix} = \begin{pmatrix} 2 \\ 5 \end{pmatrix}$$

Assume that the distributions $u_1 \sim T(-0.1,\ 0.1{,}0)$ and $u_2 \sim N(0,\ 0.1)$. Determine the most representative trajectory $(x_1(t),\ x_2(t))$ and the most representative orbit $\left(x_1(t), \frac{dx_1}{dt}(t)\right)$.

7. Consider $t \in (0,\ 1)$ and

$$\frac{\mathrm{d}}{\mathrm{dt}}\begin{pmatrix} x_1 \\ x_2 \end{pmatrix} = \begin{pmatrix} 2 + u_1 & -3 + \mathrm{u}_2 \\ 1 & 0 \end{pmatrix}\begin{pmatrix} x_1 \\ x_2 \end{pmatrix} + \begin{pmatrix} 10 + 5u_1 \\ -3 \end{pmatrix}, \quad \begin{pmatrix} x_1(0) \\ x_2(0) \end{pmatrix} = \begin{pmatrix} 2 \\ 5 \end{pmatrix}$$

Assume that the distributions of u_1 and u_2 are triangular: $u_1 \sim T(1,\ 3,\ 2)$ and $u_2 \sim T(0,\ 2,\ 1)$. Determine the most representative trajectory $(x_1(t),\ x_2(t))$ and the most representative orbit $\left(x_1(t), \frac{dx_1}{dt}(t)\right)$.

8. Consider $t \in (0,\ 1)$ and

$$\frac{\mathrm{d}}{\mathrm{dt}}\begin{pmatrix} x_1 \\ x_2 \end{pmatrix} = \begin{pmatrix} -u_1 & -3 - u_2 \\ 0.5 + u_2 & -2 - u_2 \end{pmatrix}\begin{pmatrix} x_1 \\ x_2 \end{pmatrix} + \begin{pmatrix} 0.2 + u_1 \\ 0.1 + u_2 \end{pmatrix}, \quad \begin{pmatrix} x_1(0) \\ x_2(0) \end{pmatrix} = \begin{pmatrix} 2 \\ 5 \end{pmatrix}$$

Assume that the distributions of u_1 and u_2 are triangular: $u_1 \sim T(-0.1{,}0.2{,}0)$ and $u_2 \sim T(-0.5{,}0.5{,}0)$. the most representative trajectory $(x_1(t),\ x_2(t))$ and the most representative orbit $\left(x_1(t), \frac{dx_1}{dt}(t)\right)$

UQ in Game Theory

8

The classic text in Game Theory (GT) is the book by John Von Neumann and Oskar Morgenstern, entitled Theory of Games and Economic Behavior, published in 1944 (Von Neumann & Morgenstern, 1944), what is the culmination of Von Neumann's works on games since the 1920s (Von Neumann, 1928). Previous works may be found in (Cournot, 1838; Bachelier, 1901; Zermelo, 1912; Borel, 1921).

As indicated in the title- and in the introduction of their book – Von Neumann and Morgenstern look for a new approach in Economic Sciences. In their book, they establish the foundations of GT, namely the assumptions about rational behavior of the players, fixed rules, payoffs and outcomes

GT was extensively developed since the works of Von Neumann, including in some of its probabilistic aspects. Nevertheless, Uncertainty Quantification (UQ) in Game Theory (GT) remains rare. In this chapter, we examine some of these applications. Thus, this is not a chapter about GT – although we recall some basic elements - but it concerns the applications of UQ to GT: the reader can find in the literature a large number of presentations of GT provided by the specialists of the domain. In the sequel, we focus on the specific developments concerning the use of UQ in the framework of GT.

8.1 Language from Game Theory

Let us recall some terms often used in the sequel. The first term is "game" itself, which is used as an abbreviation of the complete expression "mathematical game of strategy". Von Neumann and Morgenstern consider a ***game*** as being simply a set of rules (Von Neumann & Morgenstern, 1953, p. 49). They make the difference between this abstract object and one of its concrete achievements - a ***play***. In terms of modern informatics, we may say that a game is an object and a play is an instance of a game. As said, games are defined by ***rules*** – *id est*, the laws governing the game. The rules stipulate what is allowed or forbidden, winnings and losses, the end of the game, the number of participants, etc. In a mathematical game of strategy, the participants have to make to take decisions. A ***move*** is a decision took by or for a player: a ***personal move*** depends only on the decision of the player, while a ***random move*** involves some randomness. Moves may be ***simultaneous*** (when all the players reveal their decision and make their moves at the same time) or ***sequential*** (when only one player is allowed to move at a time). ***Static games*** allow only a unique simultaneous move for all the players, while ***dynamic games*** involve sequential moves or sequences of simultaneous moves. As consequence of

E. Souza de Cursi, *Uncertainty Quantification and Stochastic Modelling with EXCEL*,
Springer Texts in Business and Economics, https://doi.org/10.1007/978-3-030-77757-9_8

the moves, the state of the play (and of the players) change – the ***position*** of the play evolves from its initial state to the actual one by the moves of the players. We say that the players have a ***complete information*** when all the players know all the rules. They have a ***perfect information*** when all the players know the preceding moves and positions. The game is ***cooperative*** when negotiations between players are possible and coalitions of players may be created – otherwise, it is a ***non-cooperative game***. In a game, each player looks for the maximization of a ***utility*** function generated by the payoffs – in general, the payoffs themselves: we say that the players have a ***rational behavior***. A ***zero-sum game*** is a game where the sum of the payoffs is null: a player may have a positive payoff if and only if other players have negative payoffs. In mathematical games of strategy, it is assumed that ***there is no cheating***.

Rational players often use a ***strategy***, *id est*, predefined sequences of moves conditionally determined by the actual position of the game: if the position is P_i then the move is M_i. We say that the strategy is ***pure*** when it consists in choosing a fixed move among the possible ones. We say that the strategy is ***mixed*** when it uses different moves. In general, mixed strategies randomize among the different moves with a given probability. A player may have a ***dominant strategy***, *id est*, a strategy that maximizes his utility for any move of the other players; but it may also have a ***dominated strategy***, *id est*, a strategy that minimizes his utility for any move of the other players.

An ***equilibrium*** – sometimes called ***Nash equilibrium*** -is a set of strategies such that no player may increase his own utility by modifying only his own strategy: if all the other players keep their strategies, he must keep it too – otherwise, he will reduce his own winnings. Analogously to strategies, an equilibrium may be ***pure*** or ***mixed***, according to the use of a single fixed move or several moves with a given probability.

Games are generally represented in a ***normal form***: assume that, at some stage, k players may choose among n possible moves $\boldsymbol{M} = \{M_1, \ldots, M_n\}$: a move is represented by a $k-$dimensional vector $\boldsymbol{m} = (m_1, \ldots, m_k) \in \boldsymbol{M}^k$, where $m_i \in \boldsymbol{M}, \forall\ i$. The result of these move is a payoff for each player, represented by a vector of payoffs $\boldsymbol{u}(\boldsymbol{m}) = (u_1(\boldsymbol{m}), \ldots, u_k(\boldsymbol{m}))$. We may represent such a game by giving the set $\boldsymbol{M}$ and the function $\boldsymbol{u}(\boldsymbol{m})$. In simple situations, the normal form of a game may be represented by a matrix: we say that the game is under **matrix form**. An alternative representation is the ***extended form*** - a tree representing the choices of the players. A second alternative form is the **multimatrix form**, where the payoffs of each player appear in a different matrix.

Uncertainties arrive to Game Theory by several gates – for instance, errors or lacks in the information, randomness involved in the moves (for instance, in mixed strategies), uncertainty in the payoffs and so on. A game involving some uncertainty is referred in the continuation as an ***Uncertain Game (UG)***.

Examples of UG are ***Stochastic Games (SG)*** and ***Bayesian Games (BG)***. A ***SG*** is a dynamical game where the new state is randomly determined at each stage. A ***BG*** is a game where the information of the players is incomplete.

In BG, the assumptions made by one player about the characteristics of the other players are his ***beliefs***. The real set of characteristics of a player – for instance, his real payoffs - is called his ***type***. For instance, if the payoffs are uncertain – then, represented by random variables: the distribution assumed by a player about the payoff of another player is a belief. The real distribution is part of the type of the player.

In UG, it is usual introduce a special supplementary player called ***Nature***, destined to choose randomly the uncertain quantities, such as, for instance, the state of the game, the payoffs or the type of the player.

Concerning UG, GT is mainly interested in the expected payoff of each player, *id est*, in the mean of the payoffs considered as random variables. UQ is mainly interested in the probability distributions of the payoffs – which may be used to evaluate the means but may also be used to implement other decision rules. The equivalent of the Nash Equilibrium is the ***Equal Opportunity*** situation, where all the players have the same probability of obtaining a given non-negative payoff (often zero) - notice that Equal Opportunity does not mean that all the players have the same expected payoff: this last situation referred as ***Egalitarian Solution***. In the sequel, we examine some classical games from both the points of view.

Evolutionary Games(EG) result from the application of GT to biology and look for models describing the evolution of populations having different strategies of survival (see, for instance, (Lewontin, 1961; Maynard-Smith & Price, The logic of animal conflict, 1973; Maynard-Smith, 1982)). In EG, payoffs are interpreted as a measure of the adaptation – a measure of fitness to the environment. The main tool in EG is the ***replicator equation***, which describes the evolution of a population formed by groups having different strategies. Replicator equations are generated from Mathematical Games by considering the mean payoffs: groups having mean payoffs superior to the general mean are more adapted, have a higher fitness and, so, tend to increase, while groups having payoffs inferior to the general mean are less adapted, have a lower fitness and tend to decrease. We shall examine replicator equations connected to the classical games considered. In EG, uncertainty arises by several ways: for instance, initial data is often unknown, and payoffs have to be estimated.

8.2 A Simple Coin Game

Let us consider a game for two players: Player 2 chooses a result and Player 1 toss a coin. If the result agrees to the choice of Player 2, Player 1 wins 1 \$, paid by Player 2. Otherwise, Player 2 wins 1 \$, paid by Player 1. This game is UG, since the result of the toss is uncertain. An analogous game was considered in (Souza de Cursi, 2021)

Denote H a head and T a tail. Let $C_2 \in \{H, T\}$ be the choice of player 2 and $R_1 \in \{H, T\}$ be the result of the coin toss. In this case, the possible moves are the couples (R_1, C_2): $\boldsymbol{m} \in \{(H, H), (H, T), (T, H), (T, T)\}$. We have $\boldsymbol{u}(H, H) = (1, -1)$, $\boldsymbol{u}(H, T) = (-1, 1)$, $\boldsymbol{u}(T, H) = (-1, 1)$, $\boldsymbol{u}(T, T) = (1, -1)$. It is a zero-sum static game with random moves. Let us denote by p the probability of a head: for a fair coin, $p = \frac{1}{2}$, but, in practice, a small deviation may be observed – there is some uncertainty about p. The game may be represented in normal form as shown in Table 8.1 below.

Table 8.1 Matrix representation of the coin game

Player 2 / Player 1	$H\ (p_2)$	$T\ (q_2 = 1 - p_2)$
$H\ (p)$	$(1, -1)$	$(-1, 1)$
$T\ (q = 1 - p)$	$(-1, 1)$	$(1, -1)$

From this Table, we see that:

- If player 2 chooses H, he loses with probability p, so that $P(X_2 = -1 \mid C_2 = H) = p$, $P(X_2 = 1 \mid C_2 = H) = q = 1 - p$.
- If player 2 chooses T, he wins with probability p, so that $P(X_2 = 1 \mid C_2 = T) = p$, $P(X_2 = -1 \mid C_2 = T) = q = 1 - p$.

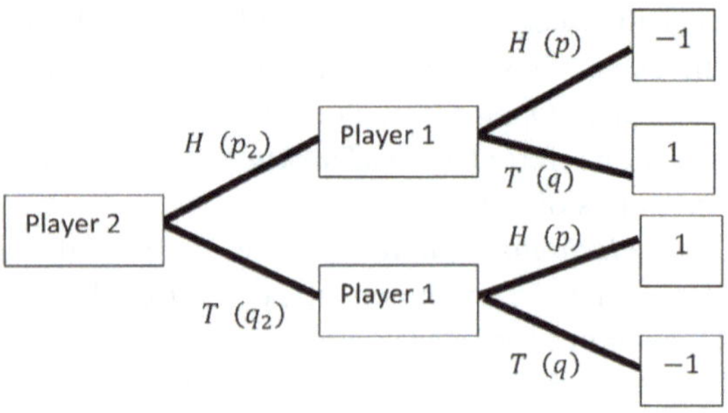

Fig. 8.1 Representation of the coin game by a tree. Point of view of Player 2

8.2.1 GT Strategies When *p* Is Known

In this game, only player 2 may have a strategy since player 1 is dependent on the result of the coin toss. Player 2 may have a strategy defined by the probability of choosing H: we will examine below some possible strategies for its choice. Let p_2 be the probability of choosing H by player 2. The game may be put into a Matrix Form, as indicated in Table 8.1 above. A tree may also represent it. From the point of view of a given player, the tree starts by his own decisions and finishes by the payoffs. For instance, the tree corresponding to the point of view of Player 2 is shown in Fig. 8.1 above.

Tree representation is a practical tool for the analysis of some strategies, by performing a reverse analysis on the tree. For instance, let us consider the classical ***minimax*** and ***maximin*** strategies:

- **Minimax**: the minimax strategy consists in minimizing the maximum loss. From the tree, we quick obtain the result by reverse analysis (see Fig. 8.2). Since the maximum loss is -1 for any decision, it is indifferent to the choice H or T.
- **Maximin**: the maximin strategy consists in maximizing the minimal payoff. Again, this strategy is indifferent to the choice H or T.

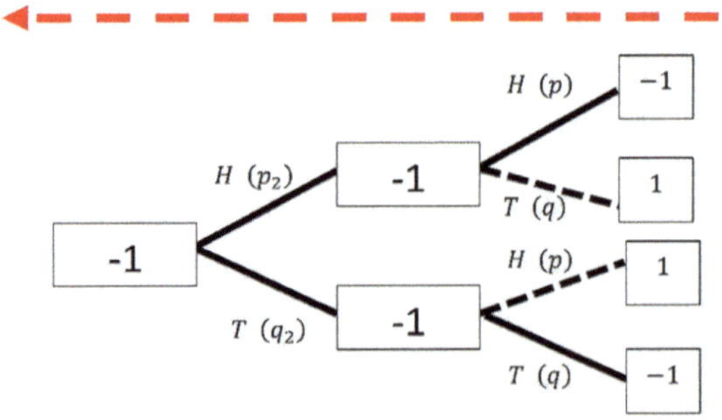

Fig. 8.2 Minimax strategy for the coin game

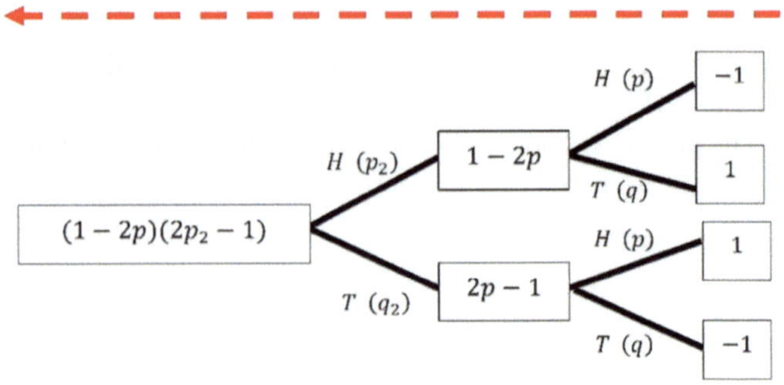

Fig. 8.3 Using the tree to evaluate the expected payoff of player 2

Trees are also useful for the evaluation of means (Fig. 8.3) and probabilities. For instance, from Fig. 8.4, we have

$$P(X_2 = 1) = \pi = p + (1 - 2p)p_2,$$
$$P(X_2 = -1) = 1 - \pi$$

Thus,

$$E(X_2) = \mu = 2\pi - 1 = (2p - 1)(1 - 2p_2)$$
$$E(X_2^2) = 1 \Rightarrow V(X_2) = \beta = 1 - (2p - 1)^2(1 - 2p_2)^2 .$$

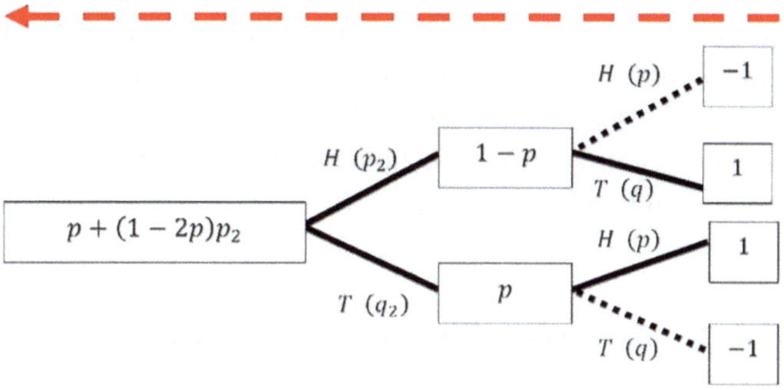

Fig. 8.4 Using the tree to evaluate $P(X_2 = 1)$

Other classical strategies in Game Theory are

- The **fixed *H* strategy**: Player 2 always choose H - then, $p_2 = 1$. The expected value of his payoff is $\mu = 1 - 2p$, with a variance equal to $\beta = 1 - (2p - 1)^2$. For player 2, the probability of winning 1 $ is $\pi = 1 - p$. This strategy is interesting if $p < \frac{1}{2}$.: in this case, it maximizes $P(X_2 = 1)$ and $E(X_2)$.
- The **fixed *T* strategy**: Player 2 always choose T - then, $p_2 = 0$. The expected value of his payoff is $\mu = 2p - 1$, with a variance equal to $\beta = 1 - (2p - 1)^2$. For player 2, the probability of winning 1 $ is $\pi = p$. Such a strategy is interesting if $p > \frac{1}{2}$.: analogously, it maximizes $P(X_2 = 1)$ and $E(X_2)$.
- The **uniformly random strategy**: choose H or T at random, with equal probability. Then, $p_2 = \frac{1}{2}$: the expected value of his payoff is $\mu = 0$, with a variance $\beta = 1$. For player 2, the probability of winning 1 $ is $\pi = \frac{1}{2}$.

Even if this game only allows a strategy for player 2, one can try to build a mixed strategy corresponding to an equilibrium in the sense of Nash. In Game Theory, a mixed Nash equilibrium is determined by making the players indifferent to the choice among the strategies.

For instance, let us denote W_H^1, W_T^1 the expected winnings of Player 1 when the result of the toss is H or T, respectively. We have

$$W_H^1 = p_2 - (1 - p_2) = 2p_2 - 1, W_T^1 = -p_2 + (1 - p_2) = 1 - 2p_2.$$

Player 1 is indifferent to the result when $W_H^1 = W_T^1$, so that $p_2 = \frac{1}{2}$ – this corresponds to the uniformly random strategy. In this case, the expected payoff is zero for all the strategies and all the players. This is not a mixed Nash equilibrium, since Player 1 cannot implement a strategy. If he could choose among different coins corresponding to different values of p, then a mixed Nash equilibrium is found by making the same analysis for Player 2: $W_H^2 = W_T^2$ – what corresponds to $1 - 2p = 2p - 1$, *id est*, $p = \frac{1}{2}$ – a fair coin.

8.2.2 UQ Strategies When *p* Is Known

Now, let us examine this game from the point of view of Uncertainty Quantification. In UQ, the main interest is the determination of probabilities or distributions. In this simple situation, we are interested in analyzing the probability of a positive payoff $P(X_2 = 1) = \pi$. We observe that it may be maximized by choosing $p_2 = 0$ (fixed T strategy) or $p_2 = 1$, if $p < \frac{1}{2}$(fixed H strategy). In both the cases, $\pi > \frac{1}{2}$and $\mu > 0$. The analogous of the Nash equilibrium is the UQ equal opportunity – the situation where the probabilities of winning a non-negative payoff are equal for all the players. In this simple situation, equal opportunity corresponds to $\pi = \frac{1}{2}$, *id est*, $p_2 = \frac{1}{2}$, independently from the value of p – the solution is also egalitarian. Notice that a fair coin – $p = \frac{1}{2}$ – generates equal opportunity egalitarian solution for any value of p_2.

8.2.3 Strategies When *p* Is Unknown

The preceding analysis assume the knowledge of p, what is, in practice, unrealistic. For instance, assume that the coin is chosen at random in a coin deposit D where there are coins having different behaviors: p is a random variable having a distribution – which may be known or unknown. If Player 2 has not any information about this value, how can he define a strategy?

If the game is static, *id est*, if the play consists in a single toss, the definition of a strategy is possible only if the player has some information about p. Otherwise, the only option is a random choice. If the game is dynamic, *id est*, a play consists in several tosses and p is unknown, Player 2 may use the results of the previous moves to define a strategy.

Indeed, this game becomes BG when the value of p is unknown: it is a lack of information and the value of p assumed by Player 2 is his belief. As previously observed, it may be updated from the results of the preceding moves. For instance, Player 2 may use a Bayesian procedure – notice that, as observed in Chap. 1, The Bayesian correction is destined to reduce the uncertainty and is not effective to make any correction if a single value of p is given: to use a Bayesian approach, it is necessary to assume a probability distribution for p – the *prior*. For instance, we may assume a uniform distribution on $\left(\frac{1}{2} - \delta, \frac{1}{2} + \delta\right)$, $\frac{1}{2} > \delta > 0$ and apply the Bayes approach to get a new distribution – the *posterior*. However, in this simple game, it is possible to define other strategies, simpler to implement. Let us examine some of them.

If the game is repeated n times, Player 2 has payoffs $X_2^1, \ldots, X_2^n$ and his total payoff is $S_2^n = X_2^1 + \ldots + X_2^n$. Notice that the possible values of $S_2^n$are $\{n, n-2, \ldots, n \bmod 2, -n \bmod 2, \ldots, -n+2, -n\}$. For instance: if $n = 2$, then the possible values are $\{2, 0, -2\}$. For $n = 3$, the possible values are $\{3, 1, -1, -3\}$. In classical GT, he looks for the maximum of the expected value $E\left(S_2^n\right)$. In the UQ approach, Player 2 appears as interested in maximizing $P\left(S_2^n \geq s\right)$ – the probability of a minimal winning, defined by the Player – in general $s = 0$ (non-negative payoff). The analogous of the Nash equilibrium is the UQ equal opportunity – the situation where the probabilities of winning a non-negative payoff are equal for all the players. In this simple game, it corresponds to $P\left(S_2^n \geq 0\right) = \frac{1}{2}$.

The equal opportunity for a single toss corresponds to $\pi = \frac{1}{2}$, so that $p_2 = \frac{1}{2}$ – the uniformly random strategy.

If Player 2 keeps the same strategy along all the dynamic play, the distribution of S_2^n is analogous to a binomial one. Indeed, let $Y_i = \frac{1+X_2^i}{2}$: Y_i is a Bernoulli variable such that $P(Y_i = 1) = \pi$. Thus, $Z^n = Y_1 + \ldots + Y_n$ is a Binomial variable $\mathcal{B}(n, \pi)$. Since $X_2^i = 2\,Y_i - 1$, we have $S_2^n = 2Z^n - n$, so that

$$P\left(S_2^n = k\right) = P\left(Z = \frac{n+k}{2}\right) = \binom{n}{\frac{n+k}{2}} \pi^{\frac{n+k}{2}} (1-\pi)^{\frac{n-k}{2}}$$

$$E\left(S_2^n\right) = n(2\pi - 1) \;, V\left(S_2^n\right) = 4n\pi(1-\pi)$$

The probability of winning a non-negative payoff (at least 0 $) is

$$P\left(S_2^n \geq 0\right) = P\left(Z^n \geq \frac{n}{2}\right) = 1 - P\left(Z^n < \frac{n}{2}\right)$$

Equal Opportunity means that $P\left(Z^n \geq \frac{n}{2}\right) = P\left(Z^n \leq \frac{n}{2}\right)$, id est, that $\frac{n}{2}$ is a median of Z^n. Thus $n\pi = \frac{n}{2}$, so that $\pi = \frac{1}{2}$ and, again, $p_2 = \frac{1}{2}$ – as in the preceding, a fair coin produces equal opportunity independently of p_2.

Table 8.2 $\left(S_2^n \geq 0\right)$: Anticyclic probability of winning a non-negative payoff for Player 2

p \ n	10	20	30	40	50	100	1000	10000
0,4	0,67	0,66	0,66	0,66	0,66	0,69	0,90	1,00
0,45	0,64	0,61	0,59	0,59	0,58	0,58	0,64	0,84
0,49	0,62	0,59	0,57	0,56	0,56	0,54	0,52	0,52
0,5	0,62	0,59	0,57	0,56	0,56	0,54	0,51	0,50
0,51	0,62	0,59	0,57	0,56	0,56	0,54	0,52	0,52
0,55	0,64	0,61	0,59	0,59	0,58	0,58	0,64	0,84
0,6	0,67	0,66	0,66	0,66	0,66	0,69	0,90	1,00

If Player 2 does not want to offer an equal opportunity to Player 1, he may choose a different strategy. For instance, he may bet on an eventual defect of the coin. A first simple way to define such a strategy consists in using the coin results themselves. For instance, let us assume that Player 2 will take a decision by tossing the coin himself before Player 1. Then, he can choose between the same result observed (cyclic strategy: if the result is *H*, choose *H*; if the result is *T*, choose *T*.) or the inverse result (anticyclic strategy: if the result is *H*, choose *T*; if the result is *T*, choose *H*.). Let us analyze both:

- The cyclic strategy leads to $p_2 = p$, since this is the probability for a result *H* when Player 2 tosses the coin. Then, $\mu = -(2p-1)^2 \leq 0$, with a variance equal to $\beta = 1 - (2p-1)^4$. For Player 2, the probability of winning 1 $ is $\pi = 2p(1-p) \leq \frac{1}{2}$.
- The anticyclic choice corresponds to $p_2 = 1 - p$ (the probability of a result *T* when Player 2 tosses the coin). Then, the expected payoff is $\mu = (2p-1)^2$, with a variance equal to $\beta = 1 - (2p-1)^4$. For Player 2, the probability of winning 1 $ is $\pi = p^2 + (1-p)^2 \geq \frac{1}{2}$.

Table 8.3 (S_2^n): Anticyclic expected payoff for Player 2

p \ n	10	20	30	40	50	100	1000	10000
0,4	0,4	0,8	1,2	1,6	2	4	40	400
0,45	0,1	0,2	0,3	0,4	0,5	1	10	100
0,49	0	0	0	0	0	0	0	4
0,5	0	0	0	0	0	0	0	0
0,51	0	0	0	0	0	0	0	4
0,55	0,1	0,2	0,3	0,4	0,5	1	10	100
0,6	0,4	0,8	1,2	1,6	2	4	40	400

Considering these results, Player 2 must choose the anticyclic strategy: his expected payoff is non-negative, and the probability of a positive payoff is superior to $\frac{1}{2}$. The cyclic strategy leads to a non-positive expected payoff and a probability of positive payoff inferior to $\frac{1}{2}$.

Notice that, if there is some uncertainty on p, there is also some uncertainty on μ, β, π: if p is considered as a random variable, then μ, β, π are random variables. Tables 8.2 and 8.3 show, respectively, the values of $P(S_2^n \geq 0)$ and $E(S_2^n)$ for some values of p and n, when Player 2 uses strategy anticyclic: they increase for unfair coins.

Player 2 may also try to implement the anticyclic strategy by taking his decision from the previous result, without tossing himself the coin. In this case, the game becomes a Markov Chain having four states: $\{s_1 = (H, H), \quad s2 = (H, T), s_3 = (T, H), s_4 = (T, T)\}$. The transition matrix is

$$M = \begin{pmatrix} 0 & 0 & p & p \\ p & p & 0 & 0 \\ 0 & 0 & 1-p & 1-p \\ 1-p & 1-p & 0 & 0 \end{pmatrix}.$$

This chain has a stationary distribution: $P^s = (p(1-p), p^2, (1-p)^2, \quad p(1-p))$. In addition,

$$\forall k \geq 2: \ M^k = \begin{pmatrix} p(1-p) & p^2 & (1-p)^2 & p(1-p) \\ p(1-p) & p^2 & (1-p)^2 & p(1-p) \\ p(1-p) & p^2 & (1-p)^2 & p(1-p) \\ p(1-p) & p^2 & (1-p)^2 & p(1-p) \end{pmatrix}.$$

Table 8.4 $(\widehat{p}_n < \frac{1}{2})$

p \ n	10	20	30	40	50	100	1000	10000
0,4	0,63	0,76	0,82	0,87	0,90	0,97	1,00	1,00
0,45	0,50	0,59	0,64	0,68	0,72	0,82	1,00	1,00
0,49	0,40	0,45	0,47	0,49	0,50	0,54	0,73	0,98
0,5	0,38	0,41	0,43	0,44	0,44	0,46	0,49	0,50
0,51	0,35	0,38	0,38	0,39	0,39	0,38	0,25	0,02
0,55	0,26	0,25	0,23	0,21	0,20	0,13	0,00	0,00
0,6	0,17	0,13	0,10	0,07	0,06	0,02	0,00	0,00

Assume that the initial choice of Player 2 is H with probability p_2. Then, the initial state of the chain is $P^0 = (p_2 p,\ (1-p_2)p,\ \ p_2(1-p),\ \ (1-p_2)(1-p))$. The state at step $k \geq 1$ is

$$P^k = P^0 M^k = P^s.$$

Thus,

$$P\left(X_2^0 = 1\right) = p + (1-2p)p_2;\ \ P\left(X_2^k = 1\right) = p^2 + (1-p)^2 = \pi\ \ (k \geq 1).$$

Hence, this strategy leads to similar results.

A second simple strategy consists in estimating p from the previous results and applying the anticyclic strategy with the estimated value. Let H_n be the number of results H in the first n tosses. Recall that the estimator $\widehat{p}_n$ for p after n tosses is $\widehat{p}_n = \frac{H_n}{n}$. The probability $\alpha_n = P\left(\widehat{p}_n < \frac{1}{2}\right) = P\left(H_n < \frac{n}{2}\right)$ is given in Table 8.4. Notice that H_n is binomially distributed $\mathcal{B}(n,p)$.

Table 8.5 $\left(S_2^n \geq 0\right)$: estimator $\widehat{p}_n$

p \ n	10	20	30	40	50	100	1000	10000
0,4	0,68	0,69	0,75	0,85	0,84	0,95	1,00	1,00
0,45	0,63	0,58	0,63	0,62	0,63	0,71	1,00	1,00
0,49	0,60	0,61	0,58	0,57	0,55	0,54	0,58	0,94
0,5	0,62	0,57	0,58	0,58	0,55	0,55	0,50	0,50
0,51	0,63	0,59	0,58	0,56	0,56	0,56	0,62	0,94
0,55	0,68	0,64	0,64	0,73	0,69	0,73	1,00	1,00
0,6	0,73	0,78	0,83	0,89	0,88	0,96	1,00	1,00

Table 8.6 Expected payoff for Player 2 when using the estimator $\widehat{p}_n$

p \ n	10	20	30	40	50	100	1000	10000
0,4	0,5	1,5	2,8	4,3	5,9	15,0	194,6	1994,6
0,45	0,0	0,3	0,6	1,0	1,4	4,4	89,3	989,3
0,49	0	0	0	0	0	0	6	153
0,5	0,0	0,0	0,0	0,0	0,0	0,0	0,0	0,0
0,51	0	0	0	0	0	0	7	155
0,55	0,4	0,8	1,2	1,7	2,3	5,5	91,1	991,1
0,6	1,1	2,4	3,9	5,5	7,2	16,5	196,2	1996,2

If Player 2 uses the estimator $\widehat{p}_n$, then, at the move $n+1$, he will choose T if $\widehat{p}_n > \frac{1}{2}$, and H if $\widehat{p}_n < \frac{1}{2}$. Thus,

$$P\left(X_2^{n+1} = 1\right) = \Pi_n = p + \alpha_n(1-2p),\ \ P\left(X_2^{n+1} = -1\right) = 1 - \Pi_n,$$

$$E\left(X_2^{n+1}\right) = (1-2p)(2\alpha_n - 1)\ \ , V\left(X_2^{n+1}\right) = 1 - (1-2p)^2(2\alpha_n - 1)^2.$$

For this strategy, the exact evaluation of $P\left(S_2^n \geq 0\right)$ involves an expensive calculation when n increases. In Table 8.5, we show the values furnished by a Monte Carlo simulation with 10^4 runs. The values of $E\left(S_2^n\right)$ are shown in Table 8.6.

An alternative strategy based on the previous results consists in using statistical hypothesis testing. For example, Player 2 can test both the hypothesis $H_< : p \leq \frac{1}{2}$ and $H_> : p \geq \frac{1}{2}$. If both are accepted or both are rejected, Player 2 chooses uniformly random between H and T. If one is accepted and the other is rejected, he chooses the anticyclic strategy: either T if $H_>$ is accepted or H if $H_<$ is accepted. Notice that the decision for this kind of test is based on the number of H previously observed. Indeed, H_n is binomially distributed $\mathcal{B}(n,p)$. Thus, a decision concerning $H_<$ with a significance level $\alpha_<$ may be taken by considering the maximum number $h_<$ such that $P(H_n > h_<) = \alpha_<$. If $H_n \leq h_<$, $H_<$ is accepted; otherwise it is rejected. Analogously, $H_>$ is rejected if $H_n < h_>$, with $P(H_n < h_>) = \alpha_>$ and accepted otherwise. Typical values of $h_<$ and $h_>$ are exhibited in Tables 8.7 and 8.8. These values are furnished by the EXCEL® built-in function `BINOM.INV`.

Table 8.7 Typical values of $h_<$

α \ n	10	20	30	40	50	100	1000	10000
0,01	9	15	21	27	33	62	537	5116
0,05	8	14	19	25	31	58	526	5082
0,1	7	13	19	24	30	56	520	5064
0,25	6	12	17	22	27	53	511	5034

Table 8.8 Typical values of $h_>$

α \ n	10	20	30	40	50	100	1000	10000
0,01	1	5	9	13	17	38	463	4884
0,05	2	6	11	15	19	42	474	4918
0,1	3	7	11	16	20	44	480	4936
0,25	4	8	13	18	23	47	489	4966

This strategy is not as efficient as the previous ones. A Monte Carlo simulation of 10^6 plays has led to a $P\left(S_2^n \geq 0\right) \approx 0.5$ and a expected payoff of zero, for several α from 0.01 to 0.5 and values of n going from 10 to 10^4.

8.2.4 Strategies for the Stochastic Game

The game becomes stochastic (SG) if a different coin is chosen at random in the coin deposit D, with coins having different behaviors: p may be different at each toss. In this case, we say that the special player ***Nature*** chooses a coin and, thus, defines the types of the Players. Analogously to the preceding situation, p is a random variable having a distribution – which may be known or unknown. Since a draw is made at each move, Nature furnishes a different value p^k for each move. In addition, Player 2 may choose a different p_2 for each move. Then,

$$P\left(X_2^k = 1\right) = \pi^k = p^k + \left(1 - 2p^k\right)p_2^k, P(X_2 = -1) = 1 - \pi^k.$$

For a given p^k and a given strategy p_2^k, the expected gain of Player 2 is

$$E\left(X_2^k \,|\, p = p^k, p_2 = p_2^k\right) = \mu^k = 2\pi^k - 1 = \left(2p^k - 1\right)\left(1 - 2p_2^k\right).$$

The expected gain depends on the distribution of p and of the strategy of Player 2. The equivalent of the Nash equilibrium on a single move is determined by setting $\pi^k = \frac{1}{2}$ – a possible solution is $p_2^k = \frac{1}{2}$, i.e., the strategy of a random uniform choice of Player 2. As previously observed, only Player 2 may have a strategy in this game, since Player 1 depends only on the random result of the toss of a randomly chosen coin. If Player 1 may choose the coin with a given value of p^k, a mixed Nash equilibrium is obtained for $p^k = \frac{1}{2}$, *id est*, a fair coin. Analogously, the Equal Opportunity corresponds to $\pi^k = \frac{1}{2}$ and leads to the same results.

In this case, the anticyclic strategy is more difficulty to implement, since p^k varies with k. A solution is to use $\overline{p} = E(p)$ and take $p_2^k = 1 - \overline{p}$. Then,

$$\pi^k = 1 - \overline{p} + (2\overline{p} - 1)p^k, \quad E(X_2^k \mid p = p^k) = (2p^k - 1)(2\overline{p} - 1).$$

Then,

$$E(\pi^k) = 1 - 2\overline{p} + 2\overline{p}^2 \geq \frac{1}{2} \quad , \quad E(X_2^k) = (2\overline{p} - 1)^2 \geq 0$$

If the game is repeated n times, we may consider again the payoff $S_2^n = X_2^1 + \ldots + X_2^n$ of Player 2 and make an analysis similar to the preceding ones. From the preceding, we have $E(S_2^n) = n(2\overline{p} - 1)^2$: analogously to the previous situation, the expected value increases with n when the coin is unfair. The distribution of S_2^n is not binomial-like, since it depends on the distribution of p, what makes difficulty the evaluation of $P(S_2^n \geq 0)$. To solve the difficulty, we may use a Monte Carlo simulation to estimate such a probability. The results obtained this way are shown in Table 8.9 – the calculations assume that p is uniformly distributed on $(\overline{p} - \delta, \overline{p} + \delta)$ $(\delta > 0)$.

In practice, the value of $\overline{p}$ may be unknown: Player 2 assumes a value which is a belief. Analogously to the preceding, Player 2 may use an estimator $\widehat{p}_n = \frac{H_n}{n}$ to take a decision at move $n + 1$. In this case, $\pi^k = p^k + (1 - 2p^k)(1 - \widehat{p}_{k-1})$. If toss k is independent from the past tosses, we have again $E(\pi^k) = 1 - 2\overline{p} + 2\overline{p}^2$ and $E(X_2^k) = (2\overline{p} - 1)^2$.

8.2.5 Replicator Dynamics

As observed in the introduction, Mathematical Games may be used for the analysis of population dynamics: the payoffs give the results of a competition between these populations for the resources and replicator equations describe the evolution of the population (see, for instance, (Cressman & Tao, 2014; Nowak, 2006)). These equations may be interpreted as generalizations of the classical Lotka-Volterra equations (Lotka, 1910, 1920; Volterra, 1926, 1928). A given game may be interpreted in terms of populations dynamics by different ways. For instance, the game represented in Table 8.1 may be interpreted as follows:

- As a first approach, we may consider the population as formed by two subpopulations which adopt the strategies H or D. Such a lecture is possible when the mean payoffs of the strategies are the same for both the players, what is the case of symmetric games. A possible refinement consists in considering that the subpopulations adopt the behavior of each player with a given probability;
- A second approach may consider the population as formed by subpopulations having the behavior of the Players: subpopulation i has the behavior of Player i, including the probabilities of each strategy;

- Finally, we may consider four subpopulations: subpopulations 1 and 2 are formed by internal subpopulations adopting strategies H or D, with the probabilities under consideration.

The replicator equation describes the evolution of a subpopulation by its access to the resources represented by the mean payoffs: the rate of variation of a subpopulation is connected to the difference between its mean payoff and the mean payoff of the whole population. According to this idea, a subpopulation which has a mean payoff larger than the mean payoff of the overall population will increase; in the inverse situation, it will decline. In other words, if the mean payoff of the overall population is $\overline{f}$ and the mean payoff of the subpopulation i is f_i, the rate of variation of the subpopulation i depends on the difference $f_i - \overline{f}$: it declines if $f_i < \overline{f}$ and increases if $f_i > \overline{f}$.

For instance, let us consider the second approach: a population has a proportion x_1 of elements adopting the behavior of Player 1 and a proportion x_2 of elements having the behavior of Player 2. We have $x_1 + x_2 = 1$, with $0 \leq x_1, x_2 \leq 1$ - thus, we may consider x_2 as main variable and determine $x_1 = 1 - x_2$. Using the payoffs presented in Table 8.1, the mean payoff received by an element of the subpopulation x_2 is

$$f_2 = (1 - 2p)(2p_2 - 1)$$

Since the game is a zero-sum game, the mean payoff of an element of the subpopulation x_1 is $f_1 = -f_2$. On the overall population, the mean is

$$\overline{f} = x_1 f_1 + x_2 f_2 = (2x_2 - 1) f_2.$$

The replicator equation reads as

$$\frac{dx_2}{dt} = x_2(f_2 - \overline{f}) = 2x_2(1 - x_2) f_2$$

Such an equation is a logistic equation (Verhulst, 1845; Verhulst, 1847). It has a explicit solution for $0 < x_2(0) < 1$:

$$x_2(t) = \frac{1}{1 + \left(\frac{1}{x_2(0)} - 1\right) e^{-2f_2 t}}$$

We observe that x_2 grows up if $f_2 > 0$ and decreases if $f_2 < 0$. The population is stable if $f_2 = 0$. Indeed, there are 3 steady states: $x_2 = 0$, $x_2 = 1, f_2(x_2) = 0$. $x_2 = 0$ corresponds to the extinction of subpopulation x_2: the whole population is formed by x_1; $x_2 = 1$ corresponds to the extinction of subpopulation x_1: the whole population is formed by x_2; $f_2 = 0$ corresponds to the situation where $p = p_2 = \frac{1}{2}$. When explicit solutions are not available, the stability of these steady states may be analyzed by the standard methods of dynamical systems: for instance, a linearization in a neighborhood of the solution or the determination of the eigenvalues of the matrix of the linearized coefficients. Here, stability depends on the sign of f_2.

Now, let us consider a different lecture of the game: assume that the whole population is formed of individuals applying a fixed strategy H or T. Then let us denote x_H the proportion of individuals applying the strategy H, while $x_T = 1 - x_H$ is the proportion of individuals applying the strategy T. When two individuals compete for a resource, they play as Player 1 with a given probability $\eta > 0$ and with probability $1 - \eta$ as Player 2. Then the expected payoff of an individual adopting the strategy H is

$$f_H = \eta(x_H - x_T) + (1 - \eta)(x_T - x_H) = (2\eta - 1)(2x_H - 1).$$

Since the game is zero-sum, the expected payoff of an individual adopting the strategy T is $f_T = -f_H$. The mean payoff of the whole population is

$$\bar{f} = x_H f_H + x_T f_T = (x_H - x_T) f_H = (2\eta - 1)(2x_H - 1)^2.$$

In this case, the replicator equation reads as

$$\frac{dx_H}{dt} = x_H (f_H - \bar{f}) = 2(2\eta - 1)x_H(1 - x_H)(2x_H - 1)$$

For $\eta = \frac{1}{2}$, the population is stable. For $\eta \neq \frac{1}{2}$, we have three equilibrium states: $x_H = 0$, $x_H = 1$, $x_H = \frac{1}{2}$. Stability depends on the sign of f_H.

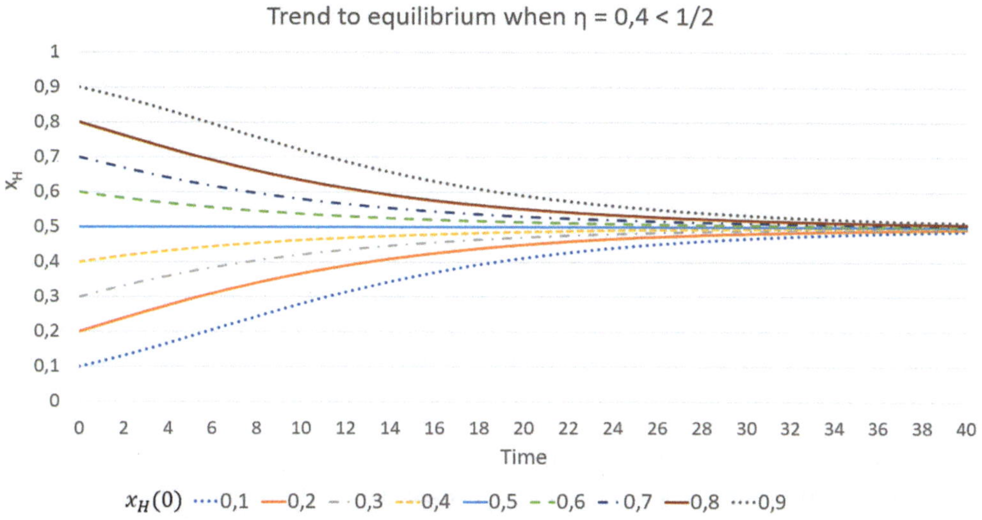

Fig. 8.5 Trend to Equilibrium when $\eta < \frac{1}{2}$

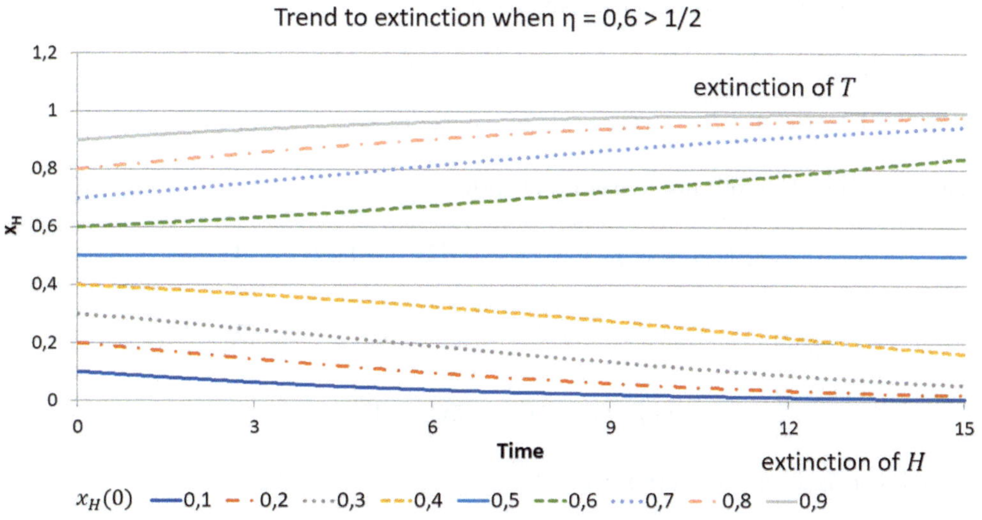

Fig. 8.6 Trend to Extinction when $\eta > \frac{1}{2}$. Equilibrium is possible only for $x_H(0) = \frac{1}{2}$

Indeed, the growth of x_H depends upon the sign of f_H: x_H grows up if $f_H > 0$ $\left(either\ \eta > \frac{1}{2}, x_H > \frac{1}{2} or\ \eta < \frac{1}{2}, x_H < \frac{1}{2}\right)$ and decreases if $f_H < 0$ $\left(either\ \eta < \frac{1}{2}, x_H > \frac{1}{2} or\ \eta > \frac{1}{2}, x_H < \frac{1}{2}\right)$. When $\eta < \frac{1}{2}$, the system tends to the equilibrium with $x_H = \frac{1}{2}$. For $\eta > \frac{1}{2}$, one of the groups goes to extinction, except if there is an exact equilibrium at the start ($x_H = 1/2$). We may simulate this equation under EXCEL®, using the methods previously shown for differential equations. For instance, we use Euler's method to generate the Figs. 8.5, 8.6 and 8.7.

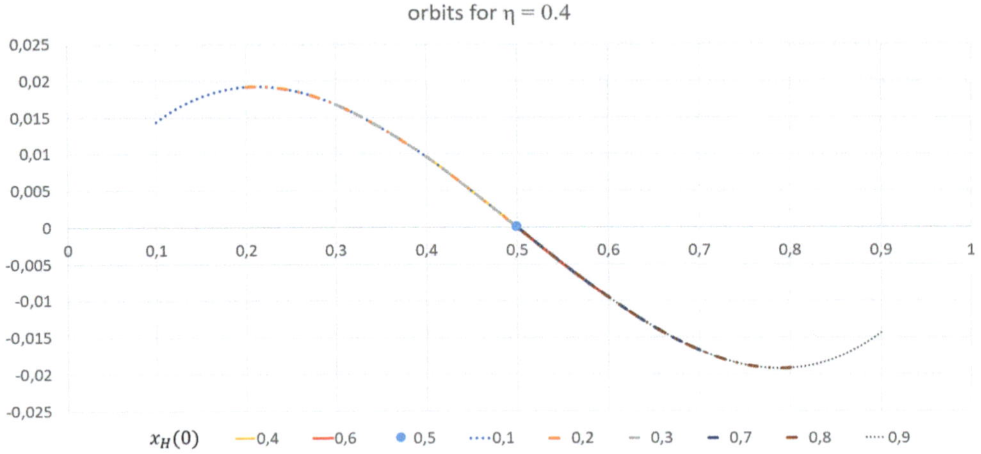

Fig. 8.7 Orbits $(x_H, p_H = \dot{x}_H)$ in the phase space for $\eta = 0.4$. All the curves tend to ($x_H = 0{,}5$, p=0), which corresponds to the equilibrium

Let us introduce other possible lecture of the game: now, we consider two species corresponding to Players 1 and 2 (proportions x_1 and $x_2 = 1 - x_1$, respectively). Each subpopulation is formed of individuals applying a fixed strategy H or T. The internal proportions are denoted x_{iH} (species i, strategy H), $x_{iT} = 1 - x_{i_H}$ (species i, strategy T). Then the global population is divided according proportions $x_{iH}x_i$ (species i, strategy H), $x_{iT}x_i$ (species i, strategy T). We have $x_{1_H}x_1 + x_{1_T}x_1 + x_{2_H}x_2 + x_{2_T}x_2 = 1$. In this case, the mean payoffs are

$$f_{1_H} = x_2(2x_{2_H} - 1), f_{1_T} = x_2(1 - 2x_{2_H}), \overline{f_1} = (2x_{1_H} - 1)(2x_{2_H} - 1)x_2$$

$$f_{2_H} = x_1(1 - 2x_{1_H}), f_{2T} = x_1(2x_{1_H} - 1), \overline{f_2} = (1 - 2x_{1_H})(2x_{2_H} - 1)x_1$$

Then, $\overline{f} = 0$ and the replicator dynamics leads to

$$\frac{d}{dt}(x_1) = x_1\overline{f_1} = x_1(1 - x_1)(2x_{1_H} - 1)(2x_{2_H} - 1),$$

$$\frac{d}{dt}(x_{1H}) = x_{1H}\left(f_{1_H} - \overline{f_1}\right) = 2x_{1_H}(1 - x_1)(1 - x_{1_H})(2x_{2_H} - 1)$$

$$\frac{d}{dt}(x_{2_H}) = x_{2H}\left(f_{2_H} - \overline{f_2}\right) = 2x_{2_H}x_1(1 - x_{2_H})(1 - 2x_{1_H})$$

Analogous analysis may be performed in this case (Figs. 8.8 and 8.9).

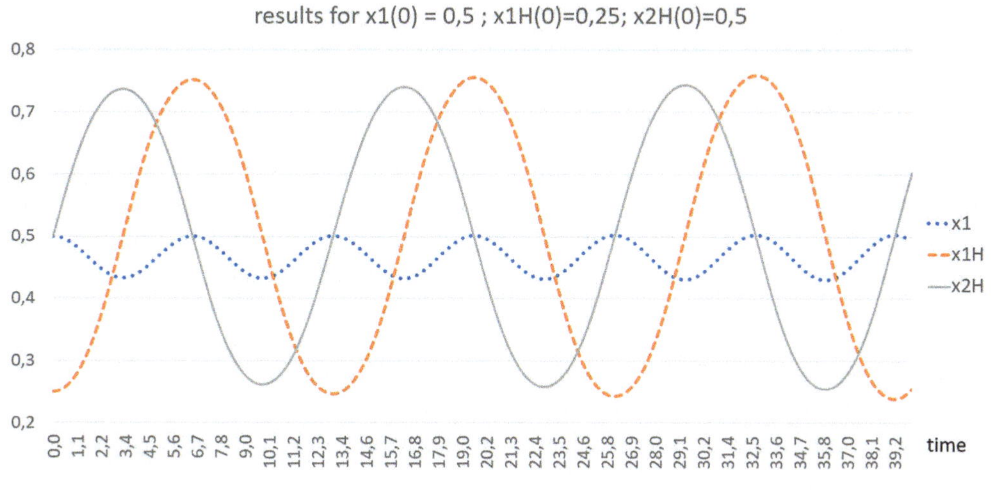

Fig. 8.8 Trajectories of some variables

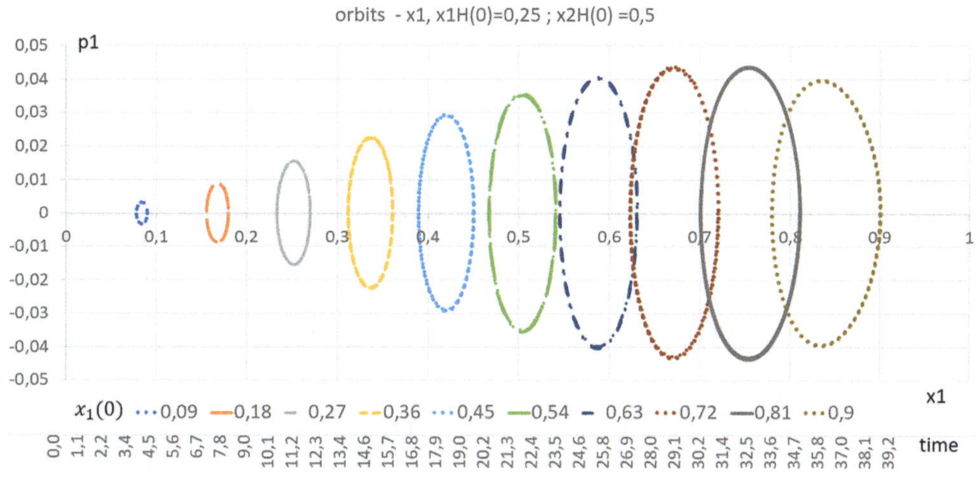

Fig. 8.9 Orbits $(x_1, p_1 = \dot{x}_1)$ for different values of $x_1(0)$

The mean evolution of the system and confidence intervals for orbits or trajectories may be obtained by using the methods presented in Chap. 5.

8.3 A Classical Game: Prisoner's Dilemma

A classical game is prisoner's dilemma: two robbers are arrested and must choose between confess (C) or deny (D). If both confess, they will be sentenced to a light sentence of L years in prison; if one confesses and the other not, the one which has not confessed will be sentenced to a heavy sentence of H years, while the one who has confessed will be sentenced to a very light sentence of V years; if no one confess, both will have a minimal sentence M. In general, $M < V < L < H$. Each prisoner must take a decision. In practice, a prisoner has also to evaluate the probabilities of the decision taken by the other prisoner. This game may be put into a Matrix Form as in Table 8.9 below

Table 8.9 The normal form of the prisoner's dilemma

Player 1 \ Player 2	C_2 (p_2)	D_2 $(q_2 = 1 - p_2)$
C_1 (p_1)	(L, L)	(V, H)
D_1 $(q_1 = 1 - p_1)$	(H, V)	(M, M)

If the prisoners can communicate (cooperative game), they will choose the strategy (D_1, D_2), that ensures to both the minimal sentence. If they do not communicate (non-cooperative game), the situation is more difficulty: they must evaluate the probability p_i that prisoner i confess. If a prisoner is confident in his partner, he evaluates this probability as extremely low; while if he is not so confident in his partner, he will evaluate this probably with higher value.

(D_1, D_2) is a Nash equilibrium for this game: any solitary deviation of a player results in a loss for himself – but its practical implementation cannot be made without communication and strong mutual confidence between the players. A second Nash equilibrium is (C_1, C_2): if one player confesses, the other must confess – otherwise his loss will be larger.

The existence of a mixed Nash equilibrium may be investigated as previously, by making decisions indifferent. For instance, let us consider Player 1. The mean sentence for strategy C_1 is $S_1^C = V + p_2(L - V)$. For strategy D_1, the mean sentence is $S_1^D = M + p_2(H - M)$. Indifference occurs when $S_1^C = S_1^D$, *id est*, $p_2 = p^{mix} = \frac{\Delta_{VM}}{\Delta_{VHML}}$, where

$$\Delta_{VM} = V - M > 0, \Delta_{HL} = H - L > 0, \Delta_{VHML} = \Delta_{VM} + \Delta_{HL} > 0\,.$$

Analogously, indifference between S_2^D and S_2^C occurs for $p_1 = p^{mix}$. If both the players follow that strategy, the mean sentence S^{mix} satisfies $V < S^{mix} < L$. Notice that the mixed strategy is not optimal. Indeed, for Player 1: $P(S_1 = H|\ D_1) = p_2, P(S_1 = M|\ D_1) = q_2, P(S_1 = L|\ C_1) = p_2, P(S_1 = V|\ C_1) = q_2$. So, $P(S_1 = M) = q_1q_2$, $P(S_1 = V) = p_1q_2$, $P(S_1 = L) = p_1p_2$, $P(S_1 = H) = q_1p_2$. Then the mean sentence for Player 1 is

$$S_1 = M + p_1\Delta_{VM} + p_2(H - M) - p_1p_2\Delta_{VHML}\,\,.$$

We observe that the behavior of S_1 depends on the relation between p_2 and the critical value p^{mix}: for $p_2 > p^{mix}$, S_1 decreases when p_1 increases, so that its minimum corresponds to $p_1 = 1$ (strategy C_1); for $p_2 < p^{mix}$, S_1 increases when p_1 increases, so that its minimum corresponds to $p_1 = 0$ (strategy D_1). If $p_2 = p^{mix}$, it is indifferent to choose between C_1 and D_1. Analogously, the mean sentence for Player 2 is

$$S_2 = M + p_2\Delta_{VM} + p_1(H - M) - p_1p_2\Delta_{VHML}\,\,.$$

Thus, the behavior is analogous: for $p_1 > p^{mix}$, the minimum of S_2 is attained at $p_2 = 1$ (strategy C_2); for $p_1 < p^{mix}$, its minimum is attained at $p_2 = 0$ (strategy D_2). If $p_1 = p^{mix}$, it is indifferent to choose between C_2 and D_2.

Then, Player 1 may choose a strategy according to these values:

- Confess: this corresponds to $p_1 = 1$ In this case, $S_1 = S_1^C$. If $p_2 < p^{mix}$, this strategy maximizes S_1. Otherwise, it minimizes S_1.
- Deny: this corresponds to $p_1 = 0$. Then, $S_1 = S_1^D$. If $p_2 < p^{mix}$, this strategy minimizes S_1. Otherwise, it maximizes S_1.
- Apply the minimax strategy: the maximum loss is L for decision C_1 and H for decision D_1. The minimum of the maximal loss is L for decision C_1. Thus, this strategy leads to choose C_1.

- Considering as a payoff the difference between the maximal sentence and the sentence received, the maximin strategy leads to the same result: indeed, the minimal payoff is $H - L$ for C_1 and 0 for D_1. The maximum of the minimal payoff is $H - L$ for decision C_1. Thus, this strategy leads also to choose C_1.

The situation is analogous for Player 2 (the game is symmetric).

Let us adopt the point view of UQ and analyze the probabilities. Let X_1 be the sentence received by Player 1. Then

$$P(X_1 > M) = p_1 + p_2 - p_1 p_2, P(X_1 > V) = p_2, P(X_1 > L) = p_2 - p_1 p_2.$$

Player 1 controls his own decision and may decide the value of p_1. The value of p_2 is chosen independently by Player 2: Player 1 must estimate p_2 and take a decision based on his estimation. The decision will involve a risk of having a sentence superior to a given threshold, considered as acceptable by Player 1. Then the strategies lead to the following results:

- Confess: this corresponds to $p_1 = 1$. If Player 1 chooses this strategy, he is looking for the sentence V or L. We have $(X_1 > M) = p_2$. Thus, Player 1 must have a strong confidence in Player 2: if he does not estimate p_2 as small enough, he will consider the risk as excessive. Then he may choose the betrayal: confess while expecting that Player 2 does not confess – to deceive Player 2 expecting that Player 2 will not betray.
- Deny: this corresponds to $p_1 = 0$. If Player 1 chooses this strategy, he is looking for the minimal sentence M (the other option is the maximal sentence H). Again, Player 1 must have a strong confidence in Player 2 to estimate p_2 as small enough. Otherwise, he will consider the risk as excessive and choose to confess.
- Evaluate the other as himself: the probability of C_2 is the same as C_1, what corresponds to $p_1 = p_2$. Then $P(X_1 > M) = 2p_2 - p_2^2$. Then $P(X_1 > M)$ is an increasing function of p_2 having its maximum at $p_2 = 1$. If player 1 accepts a maximal risk of α, the maximal acceptable value of p_2 is $1 - \sqrt{1 - \alpha}$. For small values of α, this value is close to $\frac{\alpha}{2}$. Thus, if the Player 1 accepts a maximal risk of 0.1%, the maximal acceptable value of p_2 is about 0.05%. For a maximal risk of 0.01%, the value is about 0.005%.

As we may see, either the confidence in the partner is strong either the best strategy is to confess. If both the players make the same analysis, both will confess.

8.3.1 Replicator Dynamics

Let us consider a population formed by two species having the behavior of Players 1 (proportion x_1) and 2 (proportion x_2). Their mean payoffs f_1, f_2 are, respectively, $-S_1$, $-S_2$: notice that S_1, S_2 are positive. To reinterpret them as punitions, we must change their sign. The mean payoff of the whole population is

$$\overline{f} = -x_1 S_1 - (1 - x_1) S_2 = -S_2 + x_1 (S_2 - S_1).$$

Thus, the replicator equation reads as

$$\frac{dx_1}{dt} = x_1 (1 - x_1)(S_1 - S_2).$$

Since

$$S_1 - S_2 = (p_1 - p_2)\underbrace{(V - H)}_{<0},$$

the evolution is driven by $p_1 - p_2$: if $p_1 - p_2 > 0$, then $S_1 - S_2 > 0$ and $x_1(t) \to 1$: thus, $x_2(t) \to 0$ and species 2 goes to extinction. If $p_1 - p_2 < 0$, then $S_1 - S_2 > 0$ and $x_1(t) \to 0$: species 1 goes to extinction. The populations are stable when $p_1 = p_2$. Again, C is the best strategy.

If we consider subspecies applying a fixed strategy C or D, the internal proportions are x_{iC} (species i, strategy C), $x_{iD} = 1 - x_{i_D}$ (species i, strategy D).Then, Analogously to the preceding situation, the mean payoffs are

$$f_{1_C} = -V - (L - V)x_{2C}, f_{1_D} = -M - (H - M)x_{2C}, \overline{f_1} = f_{1_D} + x_{1_C}\left(f_{1_C} - f_{1D}\right)$$

$$f_{2_C} = -V - (L - V)x_{1_C}, f_{2_D} = -M - (H - M)x_{1_C}, \overline{f_2} = f_{2_D} + x_{2_C}\left(f_{2_C} - f_{2_D}\right).$$

The mean payoff is $\overline{f} = \overline{f_2} + x_1\left(\overline{f_1} - \overline{f_2}\right)$ and the replicator dynamics leads to

$$\frac{d}{dt}(x_1) = x_1\left(\overline{f_1} - \overline{f}\right) = x_1(1 - x_1)\left(\overline{f_1} - \overline{f_2}\right),$$

$$\frac{d}{dt}(x_{1C}) = x_{1C}\left(f_{1_C} - \overline{f_1}\right) = x_{1_C}(1 - x_{1_C})\left(f_{1_C} - f_{1D}\right)$$

$$\frac{d}{dt}(x_{2_C}) = x_{2C}\left(f_{2_C} - \overline{f_2}\right) = x_{2_C}(1 - x_{2_C})\left(f_{2_C} - f_{2D}\right)$$

We may simulate this system of differential equations under EXCEL®, using the methods formerly introduced. For instance, using Euler's method, we obtain the results below. Observe that the strategy C is the best: the subpopulations applying strategy D tend to extinction in all the cases. In addition, the system evolves to a stable equilibrium between the species: the orbits $(x_1, p_1 = \dot{x}_1)$ tend to a point where $p_1 = 0$ (Figs. 8.10, 8.11, and 8.12).

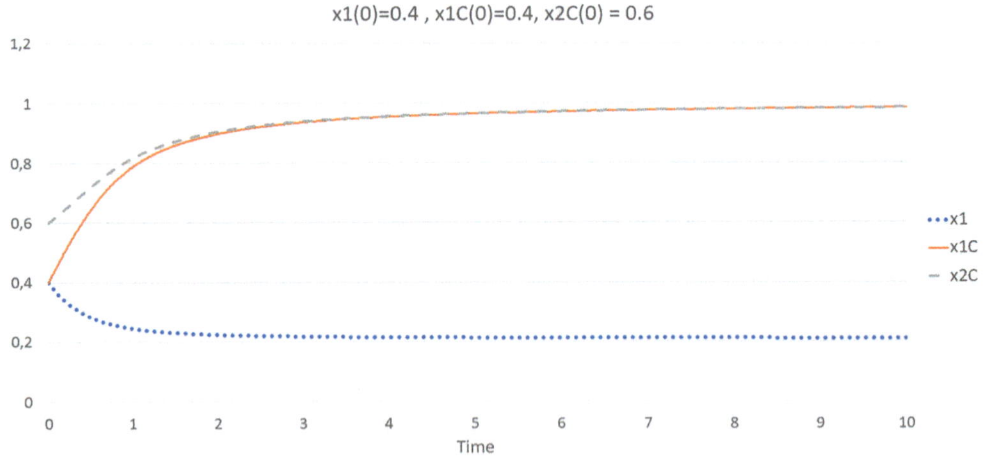

Fig. 8.10 Trajectories of x_1, x_{1_C}, x_{2_C}: x_{1_C}, x_{2_C} both tend to 1, so that x_{1_D}, x_{2_D} both tend to zero. In addition, x_1 tends to a stable value

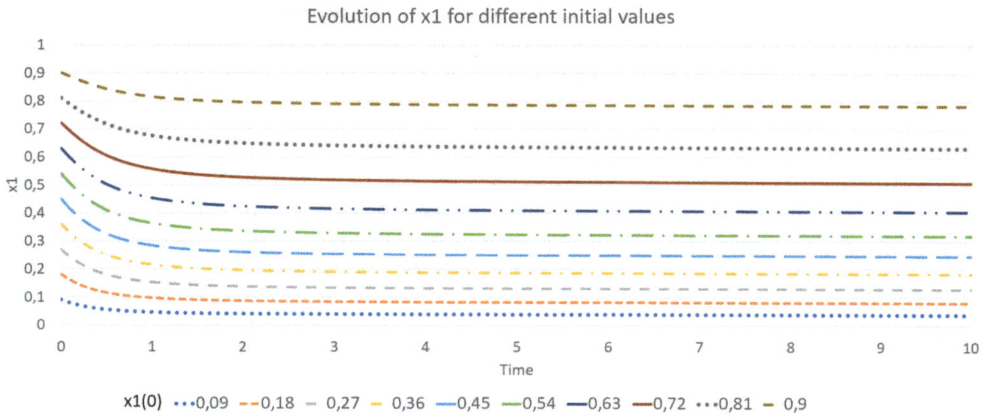

Fig. 8.11 Evolution of x_1 for different initial values. Convergence to stable values is observed in all the cases

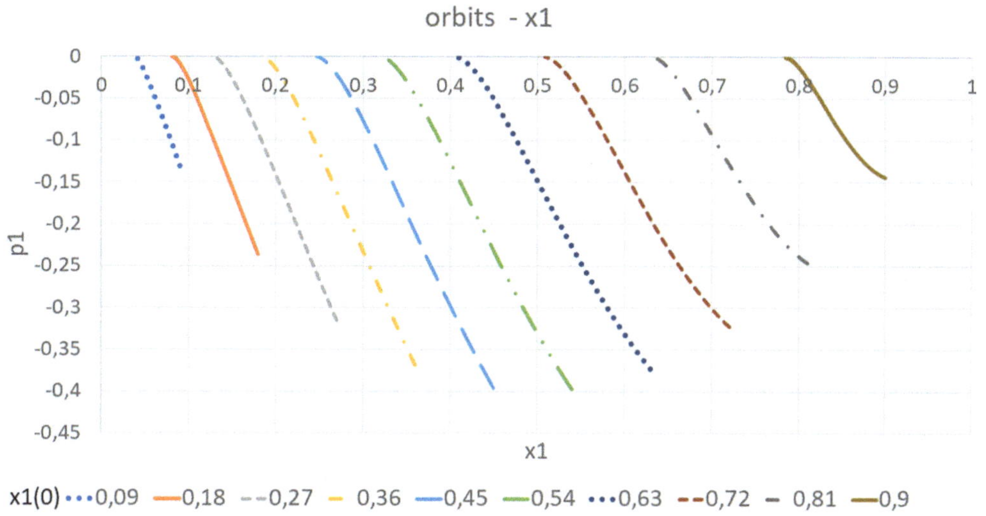

Fig. 8.12 Orbits $(x_1, p_1 = \dot{x}_1)$ for different values of $x_1(0)$. Convergence to a stable value is observed in all the cases

8.4 The Goalie's Anxiety at the Penalty Kick

Penalties in soccer interested researchers in GT since (Levitt et al., 2002). In 2003, an article (Palacios-Huerta, Professionals play minimax, 2003) analyzed penalties in soccer games using GT and a large set of data. The work of Palacios-Huerta kicked off a series of research studies about the same topic: (Garicano et al., 2005; Coloma, 2007; Palacios-Huerta & Volij, 2008; Apesteguia & Palacios-Huerta, 2010; Azar & Bar-Eli, 2011; Coloma & Germán, 2012; Prakash & Garg, 2014; Palacios-Huerta, 2014; Sarkar, 2017; Boczon & Wilson, 2018; Arrondel et al., 2019). In his work, Palacios-Huerta formulates the penalty kick as a game involving the goal and the kicker. Although the paper contains many information, several studies concentrate on the choice of the side to kick for the kicker and the side to plunge for the goalkeeper. In theory, there are many options: left side, center, right side; in addition, the ball can be at ground level, in the middle or high. The paper furnishes information about the influence of the time where the penalty is kicked, about the foot of the kicker and so on. In a simplified model, the study is reduced to the choices L (left-hand side of the goalkeeper) and R (the right-hand side of the goalkeeper). The compilation of data about a large number of matches furnishes estimations of the probability of scoring for the kicker, which is his payoff. The payoff of the goalkeeper is the complement of the payoff of the kicker. Let us consider the kicker K as Player 1 and the goalkeeper G as the Player 2. Let g_L be the probability of the goalkeeper plunging to his left-hand side and k_L be the probability of kicking at the same left-hand side of the goalkeeper for the kicker. Thus, the game may be out into a matrix form as in Table 8.10.

Table 8.10 The penalty game

Goalkeeper G / Kicker K	$L\ (g_L)$	$R\ (1-g_L)$
$L\ (k_L)$	$(\pi_{LL}, 1-\pi_{LL})$	$(\pi_{LR}, 1-\pi_{LR})$
$R\ (1-k_L)$	$(\pi_{RL}, 1-\pi_{RL})$	$(\pi_{RR}, 1-\pi_{RR})$

Typical values for the scoring probabilities are

$$\pi_{LL} = 0,59,\ \ \pi_{RL} = 0,93,\ \ \pi_{LR} = 0,94,\ \ \pi_{RR} = 0,71\ .$$

Since these values result from the analysis of a sample, they are uncertain, and a margin of error must be considered. Typical values for the choices are $g_L = k_L = 0.4$ – but these values may be decision variables.

In the framework of GT, we may determine the Nash mixed strategy by considering the expected gain of each player. Let us denote by EG_{KL}, $EG_{KR}$the expected payoff of the Goalie when the Kicker chooses L or R, respectively. We have

$$EG_{KL} = g_L(\pi_{LR} - \pi_{LL}) + (1 - \pi_{LR}),\ \ EG_{KR} = (1 - \pi_{RR}) - g_L(\pi_{RL} - \pi_{RR}).$$

To make the choice indifferent, we must take $EG_{KL} = EG_{KR}$. Thus,

$$g_L = g_L^* = \frac{\pi_{LR} - \pi_{RR}}{\pi_{LR} + \pi_{RL} - \pi_{RR} - \pi_{LL}}\ .$$

Now, let us denote by EK_{GL}, EK_{GR} the expected payoff of the Kicker when the Goalie chooses L or R, respectively. We have

$$EK_{GL} = \pi_{RL} - k_L(\pi_{RL} - \pi_{LL}),\ \ EK_{GR} = \pi_{RR} + k_L(\pi_{LR} - \pi_{RR}).$$

To make the choice indifferent, we must take $EK_{GL} = EK_{GR}$. Thus,

$$k_L = k_L^* = \frac{\pi_{RL} - \pi_{RR}}{\pi_{LR} + \pi_{RL} - \pi_{RR} - \pi_{LL}}\ .$$

With the values given previously, we have $g_L^* \approx 0.40$, $k_L^* \approx 0.39$ – what is close to the observed values – this was the basic conclusion of (Palacios-Huerta, 2003).

Always in the framework of GT, we may analyze the minimax strategy. If the losses of a given player are evaluated as being the payoffs of the other player, the maximal losses of the Kicker are $1 - \pi_{LL}$ when he chooses L and $1 - \pi_{RR}$ when he chooses R. Thus, his minimax strategy chooses the minimum of these values, which corresponds to the largest value between π_{LL} and π_{RR}. For the given data, the choice is R. The maximin strategy for the Kicker leads to the same result: his minimal payoff are π_{LL} and π_{RR}, so that he will maximize his minimal winning by choosing the maximal value between them. Concerning the Goalie, the maximal losses are π_{LR} and π_{LR}. Let us consider that π_{RL} and π_{LR} are approximately equal: then, the maximum losses are equal for both the sides. Analogously, the minimal winnings of the Goalie are $1-\pi_{LR}$ and $1-\pi_{RL}$, which are approximatively equal. In this case, we obtain the result shown in Table 8.11 below

Table 8.11 Minimax and Maxmin strategies for the penalty kick ($\pi_{RL} \approx \pi_{LR}$)

	Minimax	Maximin
Kicker	R	
Goalie	Uniformly random	

Assuming that the Players choose one of these strategies, the kicker chooses R and the Goalie randomly chooses between R and L. If we assume that the difference of 0.01 between π_{RL} and π_{LR} is significant, then $\pi_{RL} < \pi_{LR}$ and the minimax and maximin strategies for the Goalie are L, instead uniformly random. Notice that these choices do not correspond to a Nash Equilibrium.

Now, let us consider the point of view of UQ: the basis for decision are the probabilities. From the Kicker's point of view, the probability of scoring is S_L or S_R when he chooses to kick L or R, respectively. We have

$$S_L = \pi_{LR} - g_L(\pi_{LR} - \pi_{LL}), \quad S_R = \pi_{RR} + g_L(\pi_{RL} - \pi_{RR}).$$

The global probability of scoring is

$$S = S_R + k_L(S_L - S_R).$$

If the Kicker desires to maximize this probability, so that he will choose either $k_L = 1$, if $S_R < S_L$ or $k_L = 0$, if $S_R > S_L$. When $S_L = S_R$, the Kicker chooses uniformly random between L and R. The result is logical: if the probability of scoring is larger for one of the sides, the Kicker will have the largest probability of scoring by kicking at that side. For instance, if the Goalie is known by always jump to left, $g_L = 1$, so that $S_R = \pi_{RL}$ and $S_L = \pi_{LL}$ and the decision is $k_L = 0$: the kicker strikes at right to maximize his probability of scoring. Analogously, if the Goalie is known by always jump to right, the decision will be to kick at left. Indifference occurs when $S_L = S_R$, id est,

$$g_L = g_L^* \ .$$

From the Goalie's point of view, the probability of not scoring is N_L or N_R when the Kicker chooses L or R, respectively. We have

$$N_L = (1 - \pi_{LR}) + k_L(\pi_{LR} - \pi_{LL}), \quad N_R = (1 - \pi_{RR}) - k_L(\pi_{RL} - \pi_{RR}).$$

The global probability of not scoring is

$$N = N_R + g_L(N_L - N_R).$$

The Goalie desires to maximize N, so that he will choose either $g_L = 0$, if $N_L < N_R$ or $g_L = 1$, if $N_L > N_R$. When $N_L = N_R$, the Goalie chooses uniformly random between L and R. Again, the result is logical: if probability of no scoring is higher at left, the Goalie will choose this side; if the same probability is higher at right, he logically chooses the right side. As an example, if the Kicker is known by always kicking at left, then $k_L = 1$, so that $N_R = 1 - \pi_{RL}$ and $S_L = 1 - \pi_{LL}$ and the decision is $g_L = 1$: the Goalie plunges at left to minimize the probability of scoring. Analogously, if the Kicker is known by always strike at right, the decision will be to jump to right. Indifference occurs when $N_L = N_R$, id est,

$$k_L = k_L^* \ .$$

The indifference values correspond to the Nash equilibrium: in this game, the payoffs are equal to the probabilities. For the data given, $S_L \approx 0.8$, $S_R \approx 0.8$, $S \approx 0.8$, $g_L^* \approx 0.4$, $N_L \approx 0.2$, $N_R \approx 0.2$, $N \approx 0.2$, $k_L^* \approx 0.4$.

In this game, an anticyclic strategy may also be used by the Kicker: he may choose to strike at left with a probability $k_L = 1 - g_L$. In this case,

$$S = S^a = \pi_{LR} + (\pi_{LL} - 2\pi_{LR} + \pi_{RR})g_L + (\pi_{LR} + \pi_{RL} - \pi_{RR} - \pi_{LL})g_l^2$$

This polynomial attains his minimal value at

$$g_L^a = \frac{2\pi_{LR} - (\pi_{LL} + \pi_{RR})}{2(\pi_{LR} + \pi_{RL} - \pi_{RR} - \pi_{LL})}$$

so that

$$S^a \geq S^a_{\min} = \frac{4\pi_{LR}\,\pi_{RL} - (\pi_{LL} + \pi_{RR})^2}{4(\pi_{LR} + \pi_{RL} - \pi_{RR} - \pi_{LL})}$$

The Goalie may use a cyclic strategy and jump to left at probability k_L. Then,

$$N = N^c = 1 - \pi_{RR} + k_L(2\pi_{RR} - \pi_{LR} - \pi_{RL}) + k_L^2(\pi_{LR} + \pi_{RL} - \pi_{RR} - \pi_{LL})$$

This polynomial attains his minimal value at

$$k_L^c = \frac{\pi_{LR} + \pi_{RL} - 2\pi_{RR}}{2(\pi_{LR} + \pi_{RL} - \pi_{RR} - \pi_{LL})}$$

so that

$$N^c \geq N^c_{\min} = \frac{(\pi_{LR} + \pi_{RL})^2 - 4\pi_{LL}\,\pi_{RR} - 4(\pi_{LR} + \pi_{RL} - \pi_{RR} - \pi_{LL})}{4(\pi_{LR} + \pi_{RL} - \pi_{RR} - \pi_{LL})}$$

For the data given, $g_L^a \approx 0.5$, $S^a_{\min} \approx 0.8$, $k_L^c \approx 0.4$, $N^c_{\min} \approx 0.2$ (Figs. 8.13 and 8.14).

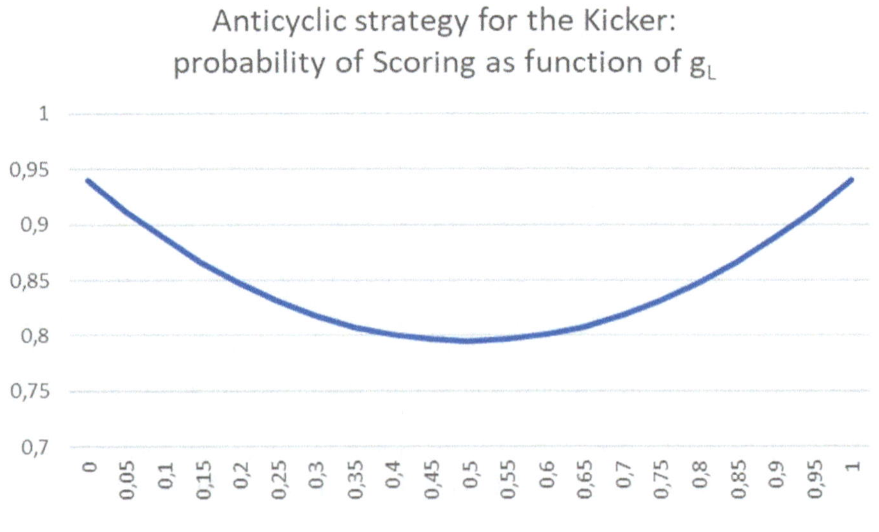

Fig. 8.13 Probability of Scoring as function of g_L in the anticyclic strategy of the Kicker

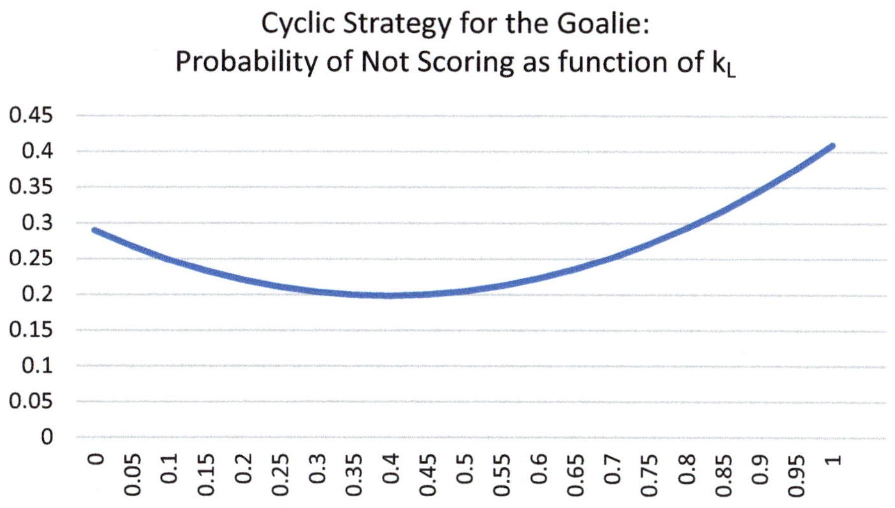

Fig. 8.14 Probability of Not Scoring as function of g_L in the anticyclic strategy of the Goalie

In this game, Equal Opportunity does not exist for the considered values of π_{LL}, π_{LR}, π_{RL}, π_{RR}: indeed, Equal Opportunity gives the same probability to the Goalie and the Kicker, id est, $N = S = 0.5$, but the data corresponds to a minimal S of 0.59 and a maximal N of 0.41.

Exercises

1. A company must decide on the investment on a new method proposed by an inventor. There are two options: risk-based option or contract option. In risk-based option, the company and the inventor will share 50/50 the gains – if any – and the company does not pay anything, except if a gain is obtained. In the contract option, the company pay for the method and will keep 100% of the gains – if any. For the inventor, the risk-based option implies to support the development costs. Assume that these cost represent 1000 money units (but the inventor will ask for a funding $F > 1000$). The expected gain G is a random variable uniformly distributed on (a, b).

   (a) Give the normal form of the game associated to this decision.
   TIP: Two players (inventor and company) and two choices (contract, risk). The payoff is zero for both if there is no agreement. The payoffs of the inventor must take into account the development costs.
   (b) Verify that both choices (contract, contract) and (risk, risk) correspond to Nash equilibria when G and F are large enough.
   (c) Find the relation between F and G for which the strategy is indifferent for the company.
   (d) Find the relation between F and G for which the strategy is indifferent for the inventor.
   (e) Let $F^* = \frac{b+a}{4}$. The inventor asks for a contract with $F = F^* + \delta$ $(\delta > 0)$. What is the company's best choice when considering the mean payoff of the operation?
   (f) The company chooses a contract but considers that the probability of a positive payoff must be 90%. What is the maximum δ acceptable? How this value is modified if the company requests a probability of 99%?
   TIP: Determine the probability of a positive payoff as function of a, b, δ.

2. Two players must choose between two options. If both choose option 1, both win \$1. If both choose option 2, both loose \$2. If a player chooses option 1 and the other chooses option 2, the player choosing option 2 wins \$$\alpha$ ($\alpha > 1$) and the player choosing option 1 looses \$1.

   (a) Give the normal form of the game.
   (b) Verify that there is no pure Nash equilibrium.
   (c) Denoting by p_i the probability of option 1 for Player i, find the mean payoff of each player.
   (d) Determine a mixed Nash Equilibrium. Determine the mean payoffs of the Players in this case. What is the value of α such that the mixed Nash Equilibrium corresponds to a random choice between the two options? What is the mean payoff in this case?
   (e) For each Player, find the probability of not loosing money. Is there an Equal Opportunity Solution for such a minimal payoff?
   (f) Suppose that the Table find in a) represents the payoffs of a population having a proportion x_i of individuals choosing strategy i. Determine the mean payoffs and the replicator equations. Find the steady states.
   TIP: In this case, $p_1 = p_2 = x_1$. Use this equality with the result of c) to find the mean payoff f_i of each subpopulation (use the lines of the Table). Then evaluate the mean payoff of the whole population as $f = x_1 f_1 + (1 - x_1) f_2$ and write the equation for x_1. Steady states corresponds to $f_1 = \bar{f}$ and $f_2 = \bar{f}$.

3. Two people must decide about cleaning the house. If no one makes the task, the house will be dirty and each one will have a negative utility $-d_i$ ($d\_i > 0$). If one cleans the house, the other will pay him a small reward corresponding to the utility of $r > 0$. If both clean the house, the utility is a small positive number for each one c_i ($0 < c\_i < r$). Assume that $d_1 < d_2$ and $c_1 < c_2$. Denote by p_i the probability of cleaning the house for player i,

   (a) Give the normal form of the game.
   (b) Has this game a pure Nash equilibrium?
   (c) Determine the mean payoffs of each player.
   (d) Determine a mixed Nash equilibrium. Is there a Nash equilibrium if one the Players is little bothered by dirtiness ($d_i < c_i$)? Assume that $d_i > c_i$. What are the mean payoffs of each player in this case? What is the minimal value of r such that the mean payoffs are positive?
   (e) Denote by (p_1^*, p_2^*) the probabilities of the mixed Nash Equilibrium found. What is the best decision for Player 1 when $p_2 < p_2^*$?
   TIP: Determine the mean payoff of each strategy and choose the best.
   (f) For each player, find the probability of a positive payoff. Is there a Equal Opportunity Solution for a positive payoff?
   (g) For each Player, find the probability of a minimal payoff of $-r$. Is there a Equal Opportunity Solution for such a minimal payoff?
   (h) Player 2 estimates that p_2 is a random variable uniformly distributed on $(p_2^* - \alpha, p_2 < p_2^* + \beta)$. Determine the probability that cleaning the house is the best strategy.
   TIP: Determine the mean payoff of the strategy as function of p_2.

4. Two Players compete to be the first user of a resource. They must choose between facing the opponent and giving up. When facing the opponent, an injury may be inflicted, and the Player may be defected – these events are synthetized by a negative utility $-b_i$ ($b_i > 0$) for Player i. Assume that the utility of the resource is $u_i > 0$ for the Player i. When giving up, no injury happens, but the utility is reduced to u_i $\left(0 \leq \alpha < \frac{1}{4}\right)$.

   (a) Give the normal form of the game associated.
   TIP: Use the utilities $u_i - b_i$ when facing the opponent and to (0, 0) in the case (give up, give up)
   (b) Find the conditions for (face, give up) to be a Nash equilibrium.
   (c) Find the conditions for (give up, face) to be a Nash equilibrium.
   (d) Find a mixed Nash equilibrium? What are the probabilities and the mean payoffs of each Player?
   (e) Is there an Equal Opportunity Solution with no injury for the Players?
   (f) Assume that b_i is a random variable, uniformly distributed on $(a_i u_i, A_i u_i)$. $0 \leq a_i < A_i \leq 1$. Find the mean payoff of each strategy for each Player. Determine the mean payoff of each Player. Find the probability of facing to be a better option than giving up for each Player.

9 Optimization Under Uncertainty

In Sect. 2.2 (page 50), we introduced the classical mono-objective optimization problem as

$$\boldsymbol{x}^* = \arg\min_C J = \arg\min\{J(\boldsymbol{y}) : \boldsymbol{y} \in C\}. \tag{9.1}$$

According to the terminology introduced in Sect. 2.2, C is the admissible set and J is the objective function. The optimal point is $\boldsymbol{x}^*$ and the optimal value of the objective is $J(\boldsymbol{x}^*)$. In the same section, we observed that, for most of practical applications, $J : \mathbb{R}^n \rightarrow \mathbb{R}$ and $C \subset \mathbb{R}^n$ such that

$$C = \left\{\boldsymbol{y} \in \mathbb{R}^n : \psi_i(\boldsymbol{y}) \leq 0, 1 \leq i \leq p; \psi_{i+p}(\boldsymbol{y}) = 0, 1 \leq i \leq q\right\}, \tag{9.2}$$

with $\psi_j : \mathbb{R}^n \rightarrow \mathbb{R}$, $1 \leq j \leq p + q$. The formulation (9.2) involves Lagrange's multipliers $\boldsymbol{\lambda} = (\lambda_1, \ldots, \lambda_{p+q})^t$ – *shadow prices* in the context of Economics.

Uncertainties may concern the objective function or constraints. If any of these elements becomes a random variable, then both $\boldsymbol{x}^*$ and $J(\boldsymbol{x}^*)$ become random variables. Optimization problems involving random variables are often referred as Stochastic Programming Problems and the reader may find in the literature many works adopting different approaches, such as Robust Optimization, Reliability Analysis, Stochastic Optimization methods, etc. For instance, we may look for the solution of

$$\boldsymbol{x}^* = \arg\min\{E(J(\boldsymbol{y})) : \boldsymbol{y} \in C\}, \tag{9.3}$$

or

$$\boldsymbol{x}^* = \arg\min\{E(J(\boldsymbol{y})) : \boldsymbol{y} \in C, V(J(\boldsymbol{y})) \leq V_{max}\}, \tag{9.4}$$

or

$$\boldsymbol{x}^* = \arg\min\{V(J(\boldsymbol{y})) : \boldsymbol{y} \in C, E(J(\boldsymbol{y})) \leq J_{max}\}. \tag{9.5}$$

We may also look for a solution such that

$$\boldsymbol{x}^* = \arg\min\{E(J(\boldsymbol{y})) : \mathrm{P}(\boldsymbol{y} \in C) \geq 1 - \alpha\}. \tag{9.6}$$

In this Chapter, we do not adopt these points of views. We consider the problem from a different standpoint: we look for the probability distribution of $J(\boldsymbol{x}^*)$ and $\boldsymbol{x}^*$.

E. Souza de Cursi, *Uncertainty Quantification and Stochastic Modelling with EXCEL*,
Springer Texts in Business and Economics, https://doi.org/10.1007/978-3-030-77757-9_9

9.1 Using the Methods of Representation

We can use the approaches presented in Chap. 5 to get a full characterization of the probability distributions of optimal points, Lagrange's multipliers (shadow process) and optimal values.

To fix the ideas, let us consider the following example, adapted from (Kim, 1968):

$$\min \{J(\mathbf{y}) = y_1 - 2y_2 + 3y_3, \mathbf{y} \in C\},$$
$$C = \{\mathbf{y} \in \mathbb{R}^3 : -2y_1 + y_2 + 3y_3 = u_1; 2y_1 + y_2 + 4y_3 = u_2; y_i \geq 0, i = 1, 2, 3\}$$

Assume that u_1 is uniformly distributed on (1, 3), u_2 is uniformly distributed on (3, 5) and these variables are independent. If we look for a solution in the sense of (9.3), we search for the solution of

$$\min \left\{\overline{J(\mathbf{y})} = \overline{y_1} - 2\overline{y_2} + 3\overline{y_3}, \overline{\mathbf{y}} \in \overline{C}\right\},$$
$$\overline{C} = \{\overline{\mathbf{y}} \in \mathbb{R}^3 : -2\overline{y_1} + \overline{y_2} + 3\overline{y_3} = 2; 2\overline{y_1} + \overline{y_2} + 4\overline{y_3} = 4; \overline{y_i} \geq 0, i = 1, 2, 3\}$$

The solution is $\overline{x_1} = 1/2$, $\overline{x_2} = 3$, $\overline{x_3} = 0$, $J(\bar{\mathbf{x}}) = -5.5$. By this method, we do not obtain information about the distribution of $J(\mathbf{x}^*)$.

To get the distribution of these parameters, we can use the approaches presented in Chap. 5. For instance, let us apply the collocation approach: we consider a sample from u_1 formed by the values $u_{1i} = 1 + 0.1i$, $i = 0, \ldots, 20$ and a sample from u_1 formed by the values $u_{2j} = 3 + 0.1j$, $j = 0, \ldots, 20$. We generate a sample from $\mathbf{U} = (u_1, u_2)$ by considering the points (u_{1i}, u_{2j}). The sample is used to generate larger samples from $\mathbf{x}^*$ and $J(\mathbf{x}^*)$, which are used to determine a bidimensional expansion with $k_1 = k_2 = 3$ (see Sect. 5.5.1). Here, we used the representation to generate a sample of $ns = 40401$ variates from $\mathbf{x}^*$, using $u_{1i} = 1 + 0.01i$, $u_{2j} = 3 + 0.01j$, $i, j = 0, \ldots, 200$.

In this simple situation, we may determine the exact values:

$$\mathbf{x}^* = \left(\frac{1}{4}(u_2 - u_1), \frac{1}{2}(u_1 + u_2), 0\right)$$

$$J(\mathbf{x}^*) = -\frac{5}{4}u_1 - \frac{3}{4}u_2$$

A comparison between the approximated and the exact values of $\mathbf{x}^*$ is shown in Fig. 9.1. The global RMS error was 5E-5.

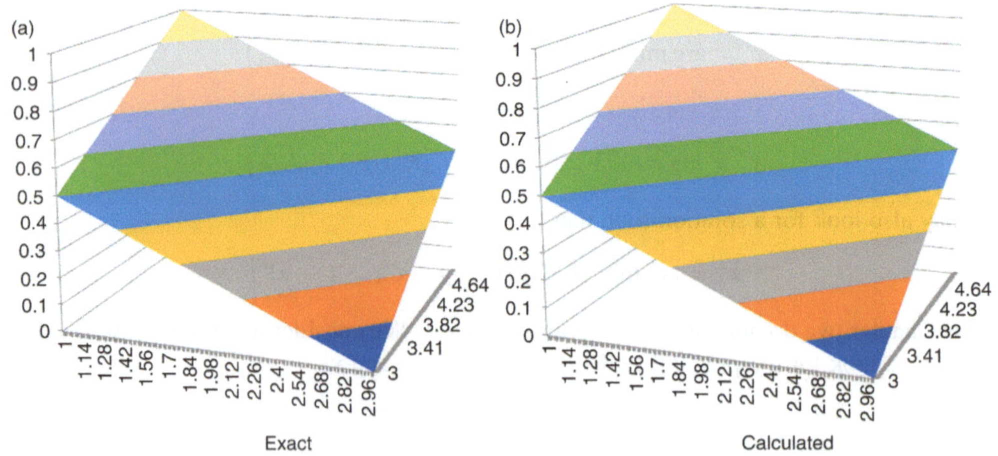

Fig. 9.1 First component x_1^* of $\mathbf{x}^*$ in the introductory example ($\mathbf{U}$ Uniform). (**a**) Exact. (**b**) Calculated

The large sample from x^* can be used to generate a sample of $ns = 40401$ variates from $J(x^*)$, which provides the distribution of $J(x^*)$ shown in Fig. 9.2 – the PDF was evaluated by the SPH approach (see Sect. 2.8). The empirical mean generated by such a large sample furnished the estimations $E(J(x^*)) \approx -5.5$, $\sigma(J(x^*)) \approx 0.85$, identical up to the 5th. decimal for both samples (exact and calculated) – notice that the exact values are $E(J(x^*)) = -11/2 = 5.5$, $\sigma(J(x^*)) = \sqrt{17/24} \approx 0.84$. The RMS error between $J(x^*)$ and its approximation was 2E-4.

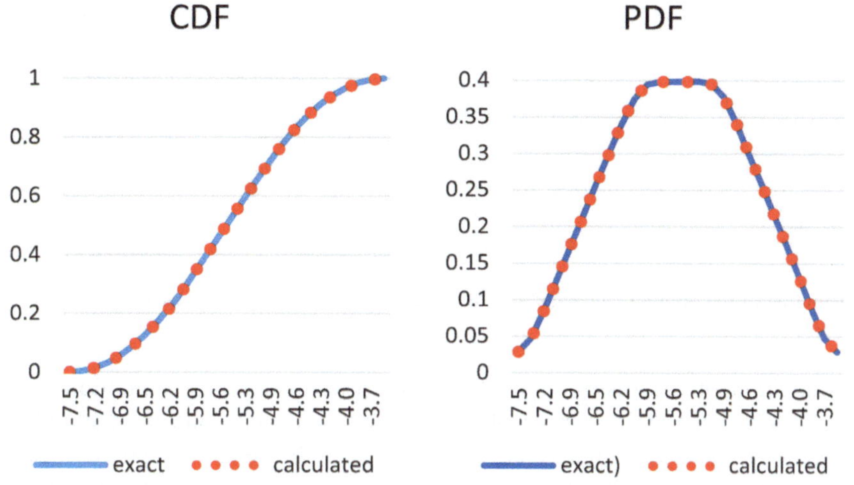

Fig. 9.2 Distribution of the optimal value $J(x^*)$ in the introductory example (U uniform)

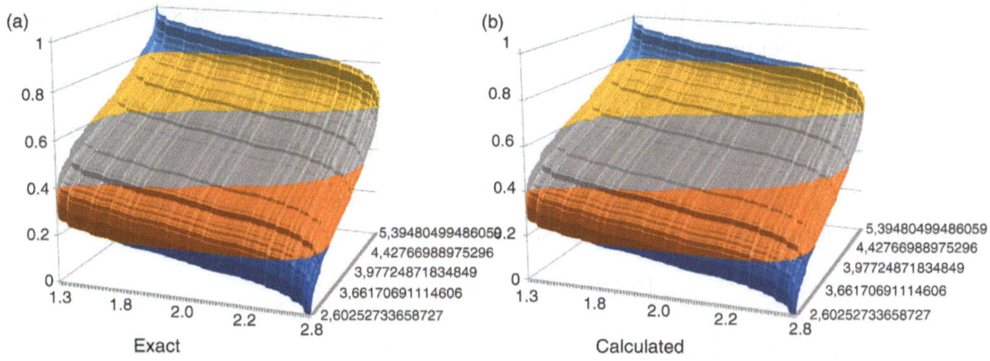

Fig. 9.3 First component x_1^* of x^* in the introductory example (U Gaussian). (**a**) Exact. (**b**) Calculated

Now, assume that $u_1 \sim N(2, 0.25)$, $u_2 \sim N(4, 0.5)$ and these variables are independent. Into an analogous way, we consider a sample of 21 variates from each of these variables and we use again the collocation approach to determine a representation of x^*. The representation was used to generate a large sample analogous to the preceding one, with 201 variates from the gaussian distribution corresponding to each component of U. The CDF and the PDF of $J(x^*)$ were determined analogously. The results are shown in Fig. 9.4. Now, the estimations are $E(J(x^*)) \approx -5.53$, $\sigma(J(x^*)) \approx 0.49$, again identical up to the 5th. decimal for both samples (exact and calculated). Notice that the exact value is $\sigma(J(x^*)) = \sqrt{61}/16 \approx 0.49$. The RMS error was 1E-4 for $J(x^*)$ and 4E-5 for x^*. Although the mean is the same for the two situations, the standard deviations and the distributions differ significantly. A comparison between the approximated and the exact values of x^* is shown in Fig. 9.3, while the distribution of $J(x^*)$ appears in Fig. 9.4.

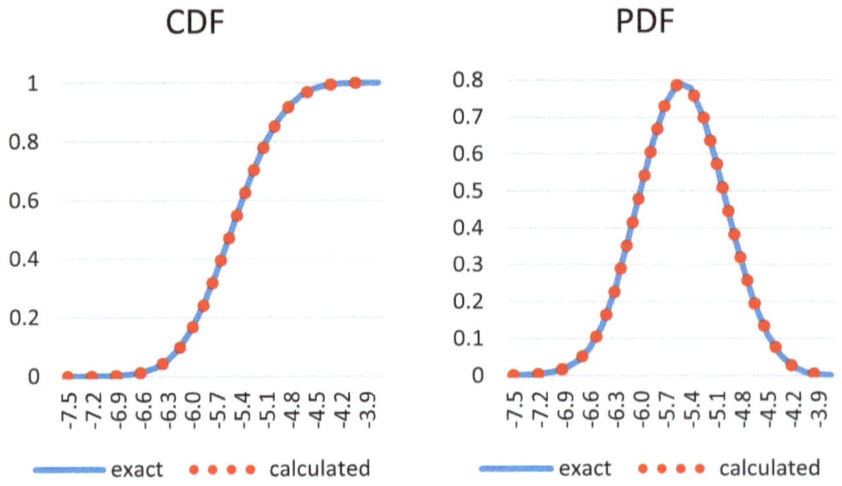

Fig. 9.4 Distribution of the optimal value $J(\boldsymbol{x}^*)$ in the introductory example ($\boldsymbol{U}$ Gaussian)

The determination of the distribution of shadow prices (Lagrange's multipliers) ca be made analogously: we can generate a sample of their values and apply one the methods presented in Chap. 5. The generation of the sample can be made by one of the methods exposed in Sects. 2.2.6 or 2. 2.4 (see pages 61 and 57). For this introductory and simple problem, the objective function and the restrictions are linear, with coefficients independent from $\boldsymbol{U}$, so that their gradients are constant and are not affected by the randomness introduced by $\boldsymbol{U}$. In addition, the active constraints are the same for all the solutions found, so that Lagrange's multipliers are constant. But such is not a general situation: for instance, let us modify the coefficients as follows:

$$\min \{J(\boldsymbol{y}) = (1 - u_1)y_1 + (u_2 - 2)y_2 + (3 - u_1 - u_2)y_3, \boldsymbol{y} \in C\},$$
$$C = \{\boldsymbol{y} \in \mathbb{R}^3 : -2y_1 + y_2 + 3y_3 = u_1; 2y_1 + y_2 + 4y_3 = u_2; y_i \geq 0, i = 1, 2, 3\}$$

In this case, the gradient of the economic function J depends upon $\boldsymbol{U}$, so that the shadow prices also. Here, the solution in the sense of (9.3) satisfies

$$\min \left\{\overline{J(\boldsymbol{y})} = -\overline{y_1} + 2\overline{y_2} - 3\overline{y_3}, \overline{\boldsymbol{y}} \in \overline{C}\right\},$$
$$\overline{C} = \{\overline{\boldsymbol{y}} \in \mathbb{R}^3 : -2\overline{y_1} + \overline{y_2} + 3\overline{y_3} = 2; 2\overline{y_1} + \overline{y_2} + 4\overline{y_3} = 4; \overline{y_i} \geq 0, i = 1, 2, 3\}$$

and the solution is now $\overline{x_1} = 2/7, \overline{x_2} = 0, \overline{x_3} = 6/7, J(\overline{\boldsymbol{x}}) = -20/7 \approx -2.86$. The exact solution is

$$\boldsymbol{x}^* = \left(\frac{1}{14}(-4u_1 + 3u_2), 0, \frac{1}{7}\ (u_1 + u_2)\right)$$

$$J(\boldsymbol{x}^*) = \frac{1}{14}\left(2u_1^2 + 2u_1 - 14u_1u_2 + 9u_2 - 2\ u_2^2\right)$$

We may obtain the distribution of $J(\boldsymbol{x}^*)$ by using the same samples which were considered for the preceding situation. For the uniform distribution, the results are shown in Fig. 9.5. The empirical mean generated by the large sample with 40401 variates furnished the estimations $E(J(\boldsymbol{x}^*)) \approx -2.86$, $\sigma(J(\boldsymbol{x}^*)) \approx 1.16$, identical up to the 5th. decimal for both samples (exact and calculated). The RMS error for the values of $J(\boldsymbol{x}^*)$ was $4E - 5$, while it was of 4E-5 for $\boldsymbol{x}^*$.

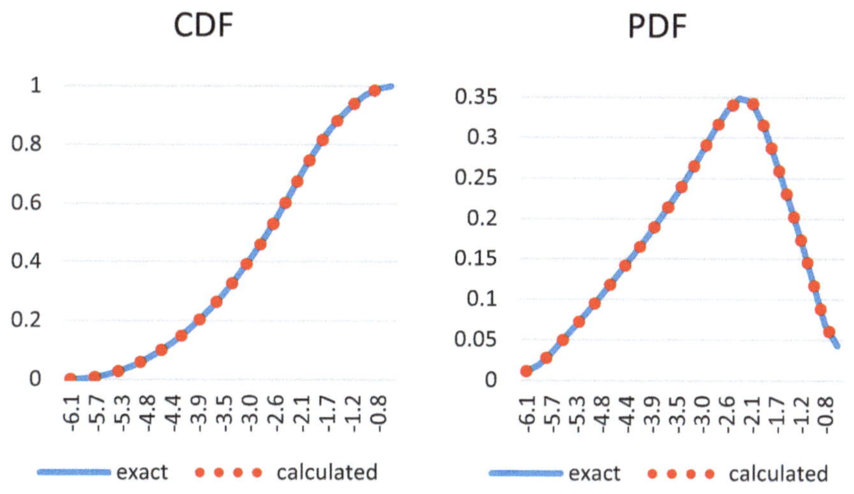

Fig. 9.5 Distribution of the optimal value $J(\boldsymbol{x}^*)$ in the example modified ($\boldsymbol{U}$ uniform)

The shadow prices may be easily derived from the KKT equations (Eq. (2.11), page 51):

$$\lambda_1 = \frac{1}{7}(-1 - u_1 + u_2), \ \ \lambda_2 = \frac{1}{14}(-9 + 5u_1 + 2u_2).$$

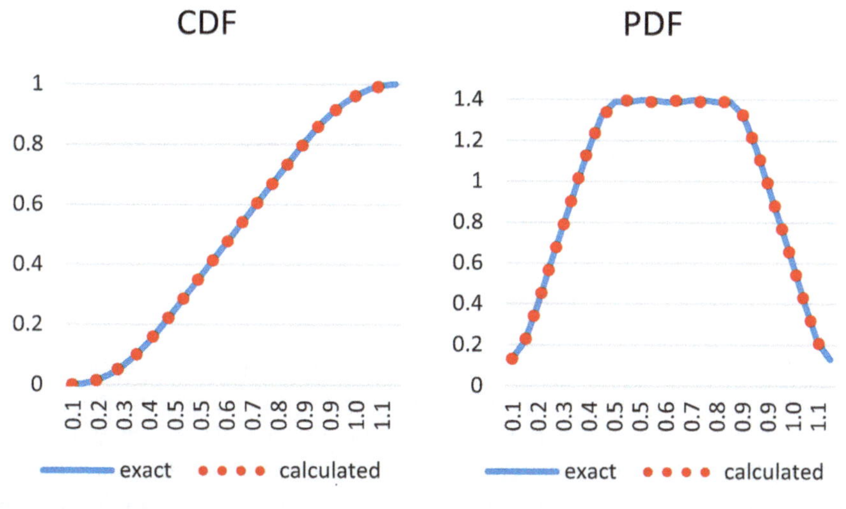

Fig. 9.6 Example of distribution of a shadow price in the second example. Here, we consider λ_2,Lagrange's multiplier associated to the second restriction ($\boldsymbol{U}$ uniform)

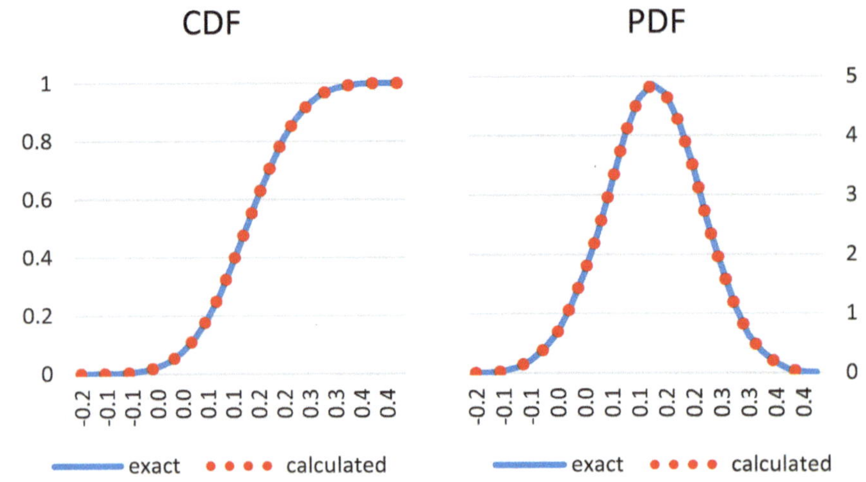

Fig. 9.7 Example of distribution of a shadow price in the second example. Here, we consider λ_1,Lagrange's multiplier associated to the first restriction ($\boldsymbol{U}$ Gaussian)

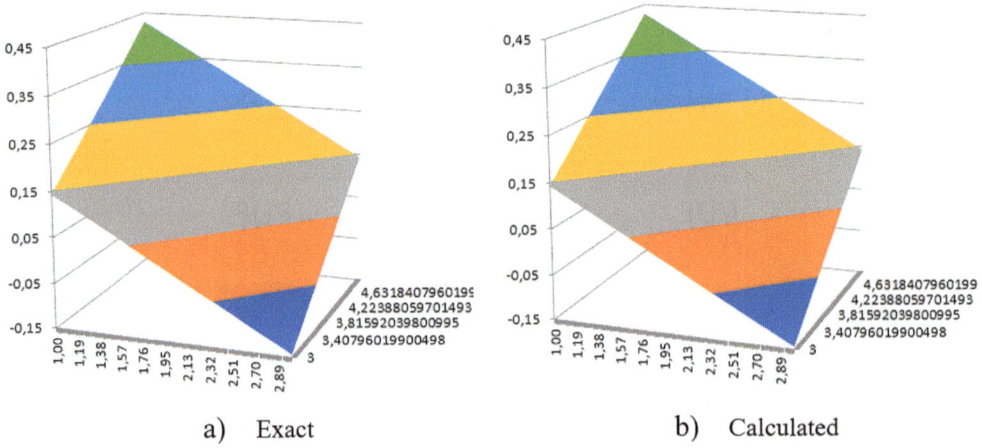

Fig. 9.8 Lagrange's Multiplier λ_1 in the second example ($\boldsymbol{U}$ Uniform). (**a**) Exact. (**b**) Calculated

We may use the samples to estimate the distribution of the Lagrange's multipliers: indeed, we may evaluate the values of λ_1 and λ_2 on the small sample $u_{1i} = 1 + 0.1i$, $u_{2j} = 3 + 0.1j$, $j = 0, \ldots, 20$. The values collected form a sample from the shadow prices, so that we may use them to find representations of these variables. Analogously to $\boldsymbol{x}^*$, we consider collocation with $k_1 = k_2 = 3$. Once the representations are determined, we use the large sample to generate new values of the Lagrange's multipliers, getting a large sample of 40401 values of each multiplier. This sample is used to determine the distribution of the shadow prices. We obtained the estimates $E(\lambda_1) \approx 0.143$, $\sigma(\lambda_1) \approx 0.12$, $E(\lambda_2) \approx 0.643$, $\sigma(\lambda_2) \approx 0.22$, close to the real values $1/7 \approx 0.143$, $\sqrt{2/147} \approx 0.12$, $9/14 \approx 0.643$, $\sqrt{29/588} \approx 0.22$, respectively. The RMS error was 1.6E-5 for λ_1 and 5.3E-5 for λ_2. An example of distribution is shown in Fig. 9.6. A comparison between approximated and exact values of λ_1 is exhibited in Fig. 9.8.

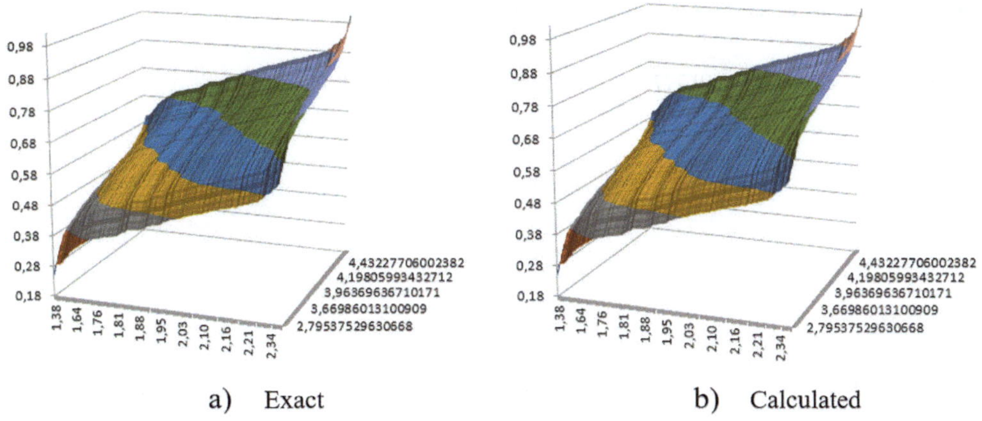

Fig. 9.9 Lagrange's Multiplier λ_2 in the second example ($\boldsymbol{U}$ Gaussian). (**a**) Exact. (**b**) Calculated

The same approach can be applied to the case where $\boldsymbol{U}$ is normally distributed. An example of distribution is shown in Fig. 9.7. We obtained the estimates $E(\lambda_1) \approx 0.15$, $\sigma(\lambda_1) \approx 0.08$, $E(\lambda_2) \approx 0.65$, $\sigma(\lambda_1) \approx 0.11$. Here, the real values are $1/7 \approx 0.143$, $\sqrt{5}/28 \approx 0.08$, $9/14 \approx 0.643$, $\sqrt{41}/56 \approx 0.11$, remarkably close to the real values. The RMS error was 1.6E-5 for λ_1 and 4.5E-5 for λ_2. An example of result appears in Fig. 9.9.

The same approach can be applied for nonlinear optimization. In addition, as shown in Sect. 6.2, iterative descent methods may be adapted (see page 345). For instance, let us consider a regular function η and the modified Rosenbrock's function:

$$f(x) = \sum_{i=1}^{n-1} \left(100 \left(\frac{x_{i+1}}{\eta(u, i+1)} - \left(\frac{x_i}{\eta(u, i)} \right) \right)^2 + \left(\frac{x_i}{\eta(u, i)} - 1 \right)^2 \right). \tag{9.7}$$

The exact solution is $x_i^* = \eta(u, i), 1 \leq i \leq n$, for which $f(x^*) = 0$. Let us consider $n = 5$, $\eta(u, i) = u^2 + i$, $u \sim N(0, 1)$ and a sample of 20 variates from u given in Table 9.1.

Table 9.1 A sample from $N(0, 1)$

−2,274	−1,529	−1,457	−1,324	−1,141	−0,877	−0,746	−0,633	−0,549	−0,461
−0,243	−0,061	0,218	0,234	0,426	0,741	0,929	1,081	1,388	1,915

The SOLVER furnishes numerical results for u taking its values in Table 9.1. As an example, let us consider the results obtained for x_1^*, exhibited in Table 9.2.

Table 9.2 Results furnished by the SOLVER for x_1^* using data in Table 9.1

6,171	3,338	3,122	2,752	2,302	1,770	1,556	1,400	1,301	1,212
1,059	1,004	1,048	1,055	1,181	1,550	1,864	2,168	2,925	4,667

The results in Table 9.2 form a sample from x_1^*: combined with the data in Table 9.1, they furnish a sample from (u, x_1^*), so that this data can be used to determine an expansion Px_1^*. For instance, let us consider a polynomial basis and the degree of approximation 5: collocation furnishes the coefficients in Table 9.3 (see Fig. 9.10).

Table 9.3 Coefficients obtained using Tables 9.1 and 9.2

i	1	2	3	4	5
Coefficient of φ_i	43,44475	19,77749	13,55506	10,71173	9,11702

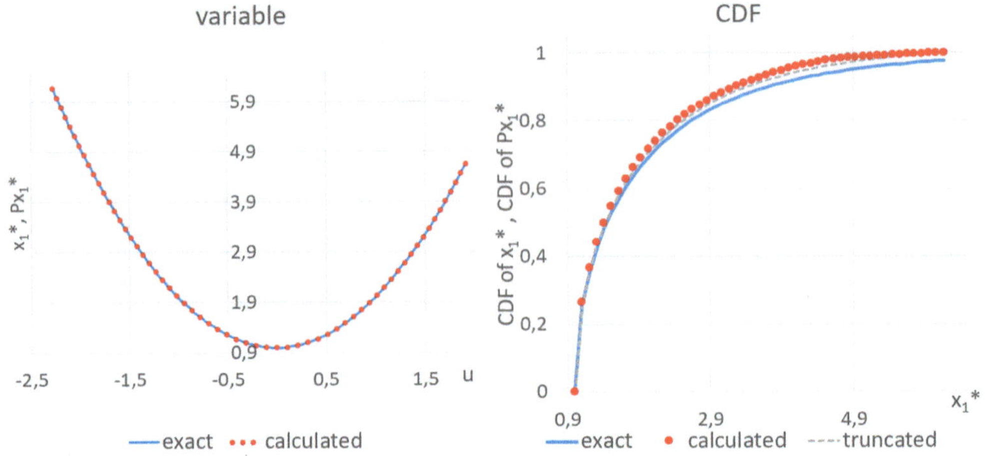

Fig. 9.10 Results corresponding to the coefficients in Table 9.3. At left a comparison between x_1^* and Px_1^*. At right, a comparison between the empirical CDF of Px_1^* and the exact CDF of x_1^*. Notice that the information of the sample is limited to $u \in (-2.274, 1.915)$ and $x_1 \in (1.061, 3.274)$. The CDF of Px_1^* almost coincides with the CDF of the truncation of x_1^* to this interval

Of course, this approach can also be used in constrained optimization: we can generate a sample and determine an expansion by the same way (see examples).

Example
Let us consider again the Cobb-Douglas function with two variables

$$cd(\boldsymbol{x}) = x_1^{\alpha} x_2^{\beta}$$

We are interested in the maximal value of $cd(\boldsymbol{x})$ under the restriction

$$p_1 x_1 + p_1 x_2 = A$$

If the unitary costs are strictly positive: $p_1, p_2 > 0$, then the solution is

$$x_1^* = \frac{\alpha}{\alpha+\beta}\frac{A}{p_1}, x_2^* = \frac{\beta}{\alpha+\beta}\frac{A}{p_2}.$$

The Lagrange's multiplier associated to the restriction is

$$\lambda = -\frac{\alpha+\beta}{A} cd(\boldsymbol{x}^*)$$

We are interested in the situation where uncertainties affect the exponents α, β. Consider a first model where $\beta = 1 - \alpha$ and suppose that $\alpha \in (0.4, 0.6)$. Then

$$E(x_1^*) = \frac{1}{2}\frac{A}{p_1} = 0.5\frac{A}{p_1}, \quad \sigma(x_1^*) = \frac{1}{10\sqrt{3}}\frac{A}{p_1} \approx 0.0577\frac{A}{p_1}$$

$$E(x_2^*) \approx \frac{1}{2}\frac{A}{p_2} = 0.5\frac{A}{p_2}, \quad \sigma(x_2^*) \approx \frac{1}{10\sqrt{3}}\frac{A}{p_2} \approx 0.0577\frac{A}{p_2}$$

For $A = 100, p_1 = 2, p_2 = 4$, we have

$$E(x_1^*) = 25, \sigma(x_1^*) \approx 2.9, E(x_2^*) = 12.5, \sigma(x_2^*) \approx 1.4$$

Table 9.4 Sample from α used in the calculations

0,401836967	0,456233218	0,502981397	0,508512	0,511317
0,511581	0,521121935	0,53794	0,539049	0,596772

Let us apply the method presented: we generate a sample of 10 values of the couple (x_1^*, x_2^*) corresponding to the sample from α shown in Table 9.4. The expansion uses $k = 6$. The RMS errors appear in Table 9.5.

Table 9.5 RMS errors for the first model

x_1^*	x_2^*	λ	$c(\boldsymbol{x}^*)$
2E-2	1E-2	1E-12	5E-3

An example of distribution is shown in Fig. 9.11. The empirical means and standard deviations were

$$E(x_1^*) \approx 25.1, \sigma(x_1^*) \approx 2.7, E(x_2^*) \approx 12.5, \sigma(x_2^*) \approx 1.4$$

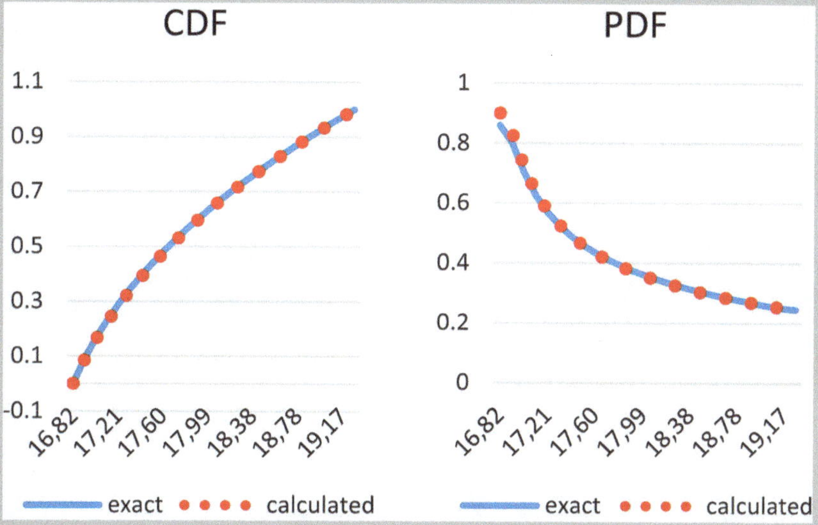

Fig. 9.11 Distribution of $c(\boldsymbol{x}^*)$ with uncertain exponents $\boldsymbol{\beta} = \mathbf{1} - \boldsymbol{\alpha}$

Consider a second model where α and β are independent and suppose that the experts agree that $\alpha \in (0.4, 0.6)$ and $\beta \in (0.3, 0.5)$. In this case

$$E(x_1^*) \approx 0.556\frac{A}{p_1}, \quad \sigma(x_1^*) \approx 0.0459\frac{A}{p_1}$$

$$E(x_2^*) \approx 0.444\frac{A}{p_2}, \quad \sigma(x_2^*) \approx 0.0459\frac{A}{p_2}$$

For $A = 100, p_1 = 2, p_2 = 4$, we have

$$E(x_1^*) \approx 27.8, \quad \sigma(x_1^*) \approx 2.3$$

$$E(x_2^*) \approx 11.1, \quad \sigma(x_2^*) \approx 1.2$$

Let us apply the method presented: we generate a sample of 10 values of the couple $\boldsymbol{x}^* = (x_1^*, x_2^*)$ corresponding to the sample of α previously introduced and the sample from β shown in Table 9.6.

Table 9.6 Sample from β used in the calculations

0,313722041	0,31488304	0,383768292	0,397799173	0,405302848
0,435980503	0,447786821	0,457281824	0,46323338	0,494181001

The expansion uses $k_1 = k_2 = 6$. The expansions determined were used to generate samples of 40421 elements from x^* and λ, by using the values $u_{1,\,i} = 0.2 + 0.002i$, $u_{2,\,j} = 0.3 + 0.002j$, $0 \leq i, j \leq 100$. The RMS errors between the values generated and the exact ones appear in Table 9.7. The empirical means and standard deviations were $E(x_1^*) \approx 27.6$, $\sigma(x_1^*) \approx 2.2$, $E(x_2^*) \approx 11.2$, $\sigma(x_2^*) \approx 1.1$.

Table 9.7 RMS errors for the second model

x_1^*	x_2^*	λ	$c(\boldsymbol{x}^*)$
9E-3	5E-3	1E-10	4E-4

Examples of results are shown in Figs. 9.12 and 9.13.

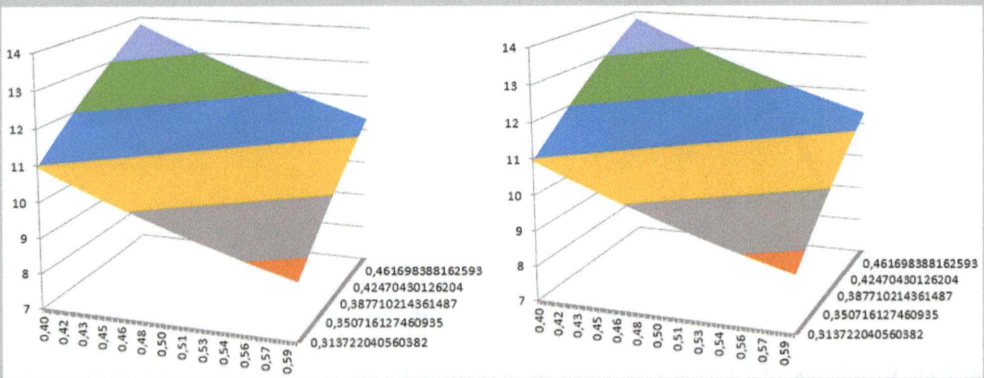

Fig. 9.12 Comparisons between calculated and exact values of x_2^* in the second Cobb-Douglas model with uncertain independent exponents α, β. At left, the exact values, at right, the calculated ones

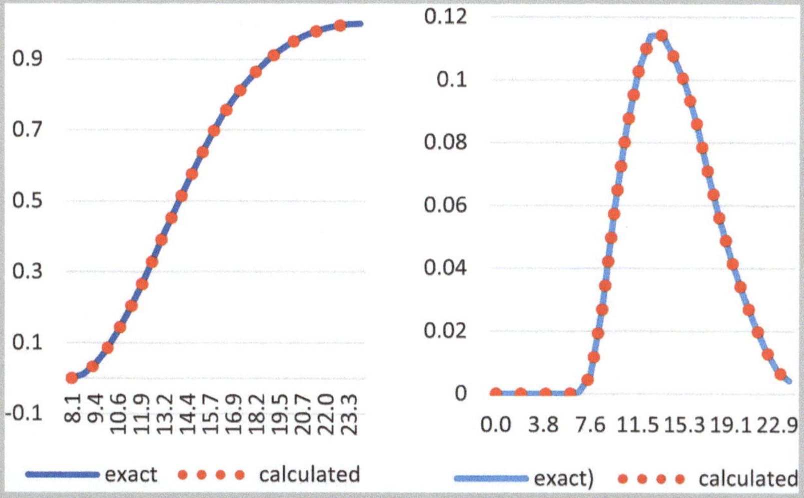

Fig. 9.13 Distribution of $c(\boldsymbol{x}^*)$ in the second Cobb-Douglas model with uncertain independent exponents α, β. At left, the CDF, at right, the PDF

Example

Let us consider the Constant Elasticity Substitution function with two variables

$$ces(\boldsymbol{x}) = \left(\alpha x_1^\beta + (1-\alpha)x_2^\beta\right)^{\frac{1}{\beta}}.$$

We are interested in the maximal value of $ces(\boldsymbol{x})$ under the restriction

$$p_1 x_1 + p_1 x_2 = A$$

Consider again a model where α and β are independent, $\alpha \in (0.4, 0.6)$ and $\beta \in (0.3, 0.5)$, both uniformly distributed. We use the same sample as in the preceding example. Expansions are determined with $k_1 = k_2 = 6$. Once the expansion is found, it is used to generate a sample formed by a family of curves corresponding to na subintervals for α and nb subintervals for β, all equally spaced. The results for $na = nb = 20$ are exhibited in Figs. 9.14 and 9.15. The same analysis is performed with triangular distributions $T(0.45,0.55,0.5)$ and $T(0.38,0.45,0.41)$: the results for $A = 100$, $p_1 = 2$, $p_2 = 4$ are shown in Figs. 9.16 and 9.17.

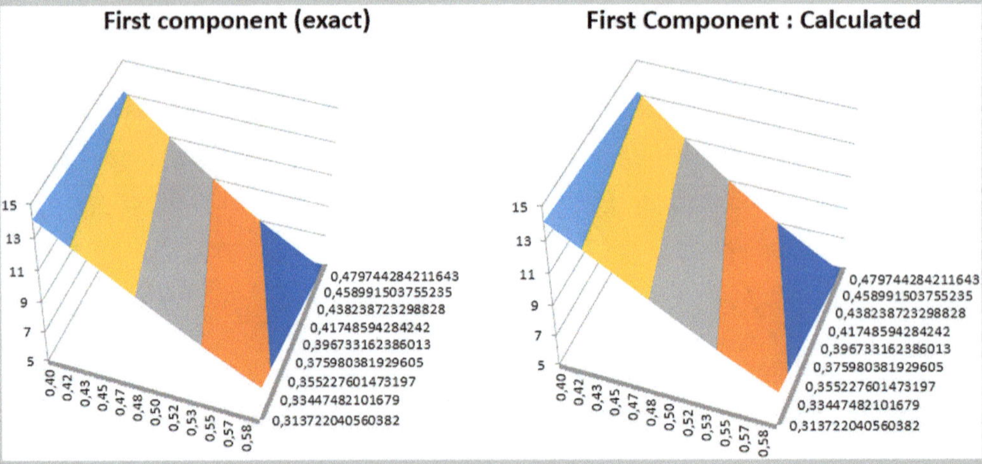

Fig. 9.14 Comparison between calculated and exact values of x_1^* in the Constant Elasticity model with uncertain independent exponents α, β., uniformly distributed. At left, the exact values, at right, the calculated ones

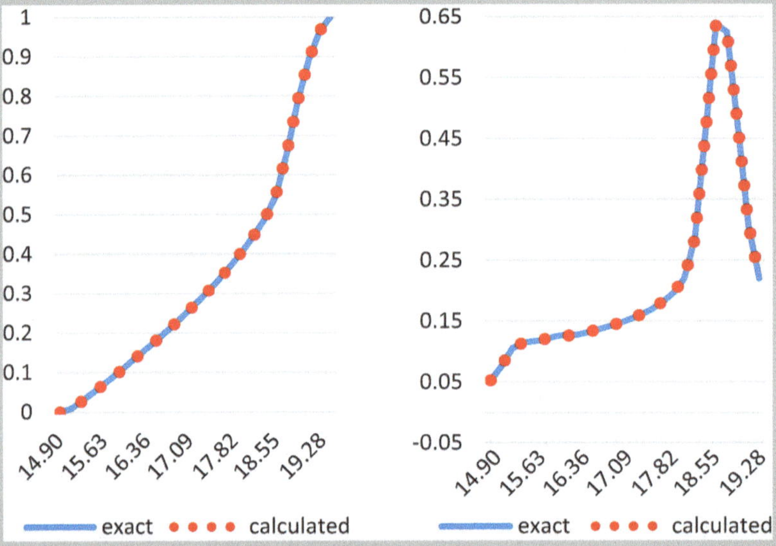

Fig. 9.15 Distribution of $ces(\boldsymbol{x}^*)$ in the Constant Elasticity model with uncertain independent exponents α, β, uniformly distributed. At left, the CDF, at right, the PDF

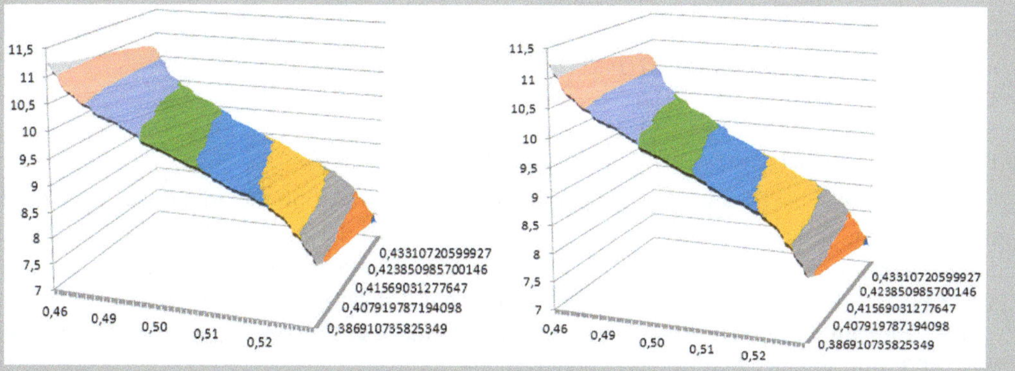

Fig. 9.16 Comparison between calculated and exact values of x_1^* in the Constant Elasticity model with uncertain independent exponents α, β., uniformly distributed. At left, the exact values, at right, the calculated ones

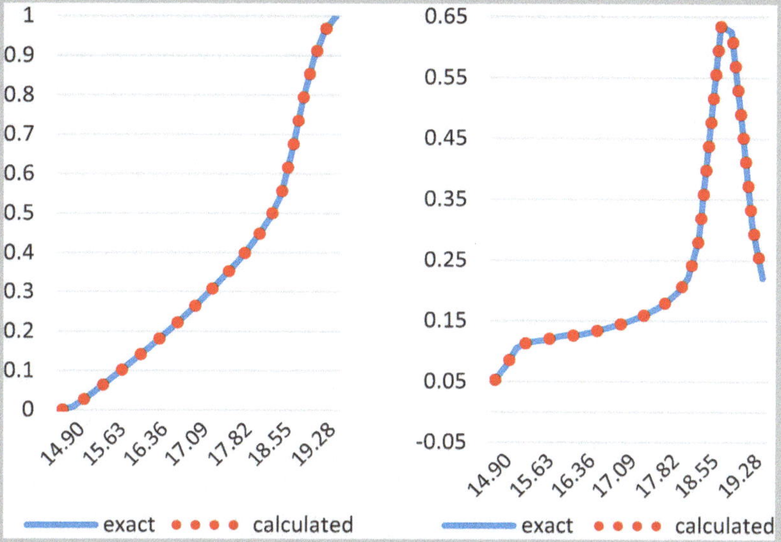

Fig. 9.17 Distribution of $ces(\boldsymbol{x}^*)$ in the Constant Elasticity model with uncertain independent exponents α, β, uniformly distributed. At left, the CDF, at right, the PDF

Exercises

1. Red Brand Canners under uncertainty: a company produces cans of whole tomato, tomato juice and tomato paste, sold by cases. Each case manufactured is sold at a given price, requests a mass of tomatoes, must respect a minimal quality score, and has a variable cost. The forecast demand is evaluated previously. The corresponding values are shown in the Table below:

	Whole tomato	Tomato juice	Tomato paste
Selling price	4	4.5	3.8
mass requested	$T(16,20,18)$	$T(19,21,20)$	$T(22,28,25)$
Quality Score	8	6	5
Variable cost	$T(2.4,2.6,2.52)$	$T(3,3.5,3.18)$	$T(1.8,2.1,1.95)$
Demand	Unlimited	$U(48000, 52000)$	$U(78000, 82000)$

The quality score results from the quantity et the grade of the tomatoes used in the manufacturing process: there tomatoes grade A, which have a quality score of 9 and grade B, which have a quality score of 5. If a product is prepared using fractions x_A of grade A tomatoes and x_B grade B tomatoes, the product has a resulting score $9x_A + 5x_B$.

The company has bought the entire harvest of tomatoes of a panel of producers for a total mass of 3 million mass unities paid 180000 money units.

The harvest is formed of tomatoes grade A and tomatoes grade B - the fraction of tomatoes grade A is a variable $T(0.18,0.22,0.2)$.

The aim is to maximize the profit of the company

(a) Find a deterministic solution where each random variable is taken as its most probable value for triangular distributions and its mean value for the uniform distributions.
(b) Analyze the effects of the variability of the grades in the harvest: consider all the other variables at their most probable (triangular) or mean (uniform) values and determine the distribution of the solution.
(c) Analyze the effects of the fluctuations in the variable costs: consider all the other variables at their most probable (triangular) or mean (uniform) values and determine the distribution of the solution.
(d) Analyze the effects of the manufacturing variabilities: consider all the other variables at their most probable (triangular) or mean (uniform) values and determine the distribution of the solution.
(e) Analyze the effects of the fluctuations in the real demand: consider all the other variables at their most probable values and determine the distribution of the solution.
(f) A producer proposes 80000 mass unities of grade A tomatoes by 6800 money unities. If the company accepts this proposition, how are modified the preceding results?
(g) The Marketing department says that an advertising campaign can modify the distribution of the demand of tomato juice to $U(70000, 740000)$. Determine the new distribution of the solution. How much money units can be invested in the advertising campaign?
TIP: the main variable to be analyzed is the total profit of the company. For question f) and g), you must evaluate the probability of an increase in profit when compared to the situation when you do not make the purchases or the advertising campaign – you consider the result of a) as the basic profit result to simplify the analysis.

2. Consider the economic function

$$f(\boldsymbol{x}) = x_1^{\alpha} + x_2^{1-\alpha}.$$

We are interested in the maximal value of $f(\boldsymbol{x})$ under the restriction

$$p_1 x_1 + p_1 x_2 = A$$

(a) Assume that $A = 100, p_1 = 2, p_2 = 4$ and $\alpha \sim T(0.4,\ 0.6, 0.5)$. Find the distribution of the solution.
(b) Assume that $A = 100, \alpha = 0.5, p_1 \sim T(1.5, 2.5, 2), p_2 \sim T(3,\ 4,\ 5)$. Find the distribution of the solution.

3. Consider the economic function

$$ces(\boldsymbol{x}) = \left(\alpha x_1^{\beta} + (1 - \alpha)x_2^{\beta}\right)^{\frac{1}{\beta}}.$$

We are interested in the maximal value of $p_1x_1 + p_1x_2$, under the restriction $f(\boldsymbol{x}) = 1$.
(a) Assume that $A = 100, p_1 = 2, p_2 = 4, \alpha \sim T(0.4,\ 0.6, 0.5), \beta \sim T(0.3,\ 0.5, 0.4)$. Find the distribution of the solution.
(b) Assume that $A = 100,\ \alpha = 0.5,\ \beta = 0.4,\ p_1 \sim T(1.5, 2.5, 2), p_2 \sim T(3,\ 4,\ 5)$. Find the distribution of the solution.

9.2 Using the Adaptation of a Descent Method

In Sect. 6.2, we introduced the adaptation of iterative methods. Such an approach may be applied to iterative methods of optimization: for instance, in page 345, we introduced the adaptation of the gradient descent with a fixed step. In fact, any iterative method

$$\boldsymbol{X}^{(r+1)} = \boldsymbol{\Psi}\left(\boldsymbol{X}^{(r)}, U\right)$$

can be adapted according to the approach presented in Sect. 6.2 and implemented according to the algorithm presented in page 337.

To illustrate the flexibility of this approach, we shall present below how to adapt the SOLVER: indeed, the SOLVER may be considered as a particular iteration function $\boldsymbol{\Psi}$ which transforms the starting point $\boldsymbol{X}^{(r)}$ into a new point $\boldsymbol{X}^{(r+1)}$. In practice, we can define $\boldsymbol{\Psi}$ as the result of the SOLVER at a predefined small number of iterations - recall that the number of iterations of the SOLVER is defined by an option. For instance, the instruction

```
SolverOptions Iterations:=3
```

fixes the number of iterations of the SOLVER to 3. Thus, we can use the Solver as an iteration function, analogously to the gradient descent iteration function.

To fix the ideas, let us consider again the optimization of the modified Rosenbrock's function given in Eq. (9.7), with $n = 5, k = 5$.

We start by preparing a workbook for the optimization: open a blank workbook, enter the initial data in `B2`, the degree of approximation in `B2`, the stopping conditions: maximum iteration number in `B3`, minimum step in `D3`. Enter the number of iterations of the SOLVER in `F3`.

	A	B	C	D	E	F
1	x(0)	0,2	0,5	0,7	0,4	0
2	k	5				
3	nitmax	100	prec	1,00E-05	SOLVER It	3

Prepare the area for the iterations: the iteration number will appear in B5, the step in D5, the coefficients $\boldsymbol{x}^{(r)}$ will appear in B7:F12, $\boldsymbol{x}^{(r+1)}$ will appear in H7:L12.

	A	B	C	D	E	F	G	H	I	J	K	L
5	r	0	prec	1								
6	coeffs	x1(r)	x2(r)	x3(r)	x4(r)	x5(r)		x1(r+1)	x2(r+1)	x3(r+1)	x4(r+1)	x5(r+1)
7												
8												
9												
10												
11												
12												

Enter the sample from u in line 13 and prepare the area for the evaluation of the values of $\boldsymbol{\Psi}(\boldsymbol{X}^{(r)}, U)$. The values of $\boldsymbol{X}^{(r)}$ will appear in B15:U19. These values will be modified by the SOLVER to generate $\boldsymbol{X}^{(r+1)}$. The corresponding values of the objective function will appear in line 14.

	A	B	C	D	E	F	G	H	I	T	U
13	u	-2,274	-1,529	-1,457	-1,324	-1,141	-0,877	-0,746	-0,633	1,388	1,915
14	fobj										
15	PX										
16											
17											
18											
19											

Give the name u to A13:U13. Then, go to line 21 and enter "= MIN(u)" in B21, "= MAX(u)" in D21.

	A	B	C	D
21	min u	-2,27409	max u	1,915034

The next steps involve VBA: go to *Developer>Visual Basic*, add a module and enter the code below, which evaluates the modified Rosenbrock's function:

```
Function eta(u, i)
eta = u * u + i
End Function
```

```
Function rosemod(x, u)
n = UBound(x, 1)
Dim s1 As Double
Dim s2 As Double
s1 = 0
s2 = 0
For i = 1 To n - 1
  s1 = s1 + ((x(i + 1) / eta(u, i + 1)) - (x(i) / eta(u, i)) ^ 2) ^ 2
  s2 = s2 + ((x(i) / eta(u, i)) - 1) ^ 2
Next i
rosemod = 100 * s1 + s2
End Function
```

Analogously to page 52, a function that evaluates `rosemod` on a Range is required:

```
Function rosemod_on_range(r As Range, u)
x = range2vector(r)
rosemod_on_range = rosemod(x, u)
End Function
```

Function `range2vector` was defined in page 52: it transforms a horizontal or vertical range in a vector. It is implemented in `VectorTools.bas`, which can be imported to the workbook (see page 10) – otherwise, you need to copy the code of `range2vector` page 52 to the module that you are creating.

Function `rosemod_on_range` can be used in a worksheet: go to cell B14 and enter "`=rosemod_on_range(B15:B19;B13)`".

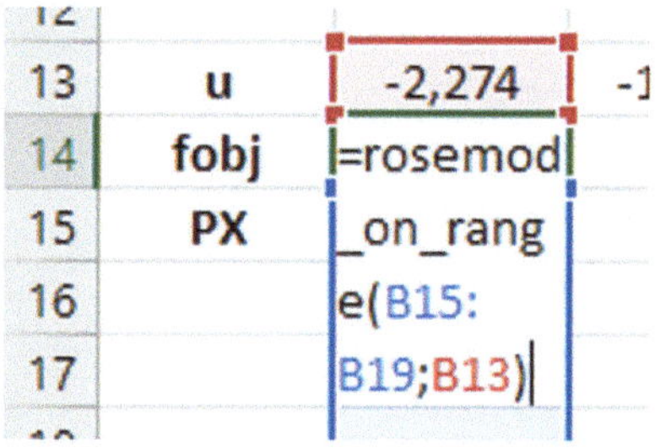

Copy `B14` and paste it on `C14:U14`. All the values appear as 4, since the cells corresponding to $\boldsymbol{X}^{(r)}$ are empty.

	A	B	C	D	E	F	G	H	I	T	U
13	u	-2,274	-1,529	-1,457	-1,324	-1,141	-0,877	-0,746	-0,633	1,388	1,915
14	fobj	4	4	4	4	4	4	4	4	4	4
15	PX										
16											
17											
18											
19											
20											
21	min u	-2,27409	max u	1,915034							

Initialization is made by a Sub that makes a copy of the initial points in the area reserved for $\boldsymbol{x}^{(r)}$ – add the code below to the module:

```
Sub init(k, n)
Range(Cells(7, 2), Cells(11, 12)).ClearContents
For j = 2 To n + 1
  Cells(7, j).value = Cells(1, j).value
  For i = 8 To 6 + k
      Cells(i, j).value = 0
  Next i
Next j
End Sub
```

Since this Sub request parameters, you cannot run it from "*Developer>Macros*": you need to create a new sub that calls it, as the one shown at left.

```
Sub solve()
k = Cells(2, 2).value
n = 5
Call init
End Sub
```

The evaluation of the values of $\boldsymbol{X}^{(r)}$ is made by another Sub – add the code below to the module under construction and modify the Sub solve as below (function readcol is available in VectorTools.bas, which must be imported):

```
Sub calcXr(ap, n, k, colmax)
Dim xr() As Variant
ReDim xr(2 To n + 1)
For i = 2 To n + 1
  xr(i) = readcol(i, 7, 6 + k)
Next i
For j = 2 To colmax
  uu = Cells(13, j).value
  For i = 2 To n + 1
    aux = ap.proj(xr(i), uu)
    Cells(13 + i, j).value = aux
  Next i
Next j
End Sub
```

```
Sub solve()
k = Cells(2, 2).value
n = 5
Call init(k, n)
Dim p As New poly
p.lb = Cells(21, 2).value
p.ub = Cells(21, 4).value
p.zero = 0.001
Dim ap As New expansion
Set ap.phi = p
colmax = Cells(13, Columns.Count).End(xlToLeft).Column
Call calcXr(ap, n, k, colmax)
End Sub
```

Run the macro solve. Now, the workbook appears as in Fig. 9.18. The next step is to use these results to evaluate $X^{(r)}$ and $(X^{(r)}, U)$: here, we desire to use the SOLVER to this task. Thus, our aim is to use the SOLVER to modify lines 15 to 19 – consequently, line 14 will also be modified. This task is achieved by a Sub psi and a callback function cbf – without the callback function cbf, the SOLVER will ask you at each column if you desire to continue or stop the iterations when the maximum number of iterations is reached. psi and cbf are shown in Fig. 9.19.

	A	B	C	D	E			T	U
5	r	0	prec	1					
6	coeffs	x1(r)	x2(r)	x3(r)	x4(r)	x			
7		0,2	0,5	0,7	0,4				
8		0	0	0	0				
9		0	0	0	0				
10		0	0	0	0				
11		0	0	0	0				
12									
13	u	-2,274	-1,529	-1,457	-1,324	-1	)81	1,388	1,915
14	fobj	4,808760243	6,147646	6,34722	6,751655	7,3	7205	6,551577	5,308129336
15	PX	0,2	0,2	0,2	0,2		,2	0,2	0,2
16		0,5	0,5	0,5	0,5		,5	0,5	0,5
17		0,7	0,7	0,7	0,7		,7	0,7	0,7
18		0,4	0,4	0,4	0,4		,4	0,4	0,4
19		0	0	0	0		)	0	0
20									

Fig. 9.18 The workbook after executing `init` and `calcXr`

```
Sub psi(k, n, colmax, nitsolver)
For i = 2 To colmax
  rxad = Range(Cells(15, i), Cells(14 + k, i)).Address
  rfobjad = Range(Cells(14, i), Cells(14, i)).Address
  SolverReset
  SolverOk SetCell:=rfobjad, MaxMinVal:=2, ValueOf:=0, ByChange:=rxad, _
          Engine:=1, EngineDesc:="GRG Nonlinear"
  SolverOptions AssumeNonNeg:=False
  SolverOptions Multistart:=True, PopulationSize:=500
  SolverOptions Iterations:=nitsolver
  SolverOptions StepThru:=True
  resu = SolverSolve(True, "cbf")
  SolverFinish KeepFinal:=1
Next i
End Sub
```

```
Function cbf(reason As Integer)
If reason > 1 Then
   cbf = True
Else
   cbf = False
End If
End Function
```

Fig. 9.19 The macro `psi` and `the callback` function `cbf`. Without the callback function, EXCEL® will open a dialog box at each column calculated

To run psi, you must modify the sub solve again, as shown in Fig. 9.20 – this version of solve will execute psi once. Then, the workbook will appear as shown in Fig. 9.21.

```
Sub solve()
k = Cells(2, 2).value
n = 5
Call init(k, n)
Dim p As New poly
p.lb = Cells(21, 2).value
p.ub = Cells(21, 4).value
p.zero = 0.001
Dim ap As New expansion
Set ap.phi = p
colmax = Cells(13, Columns.Count).End(xlToLeft).Column
Call calcXr(ap, n, k, colmax)
nitsolver = Cells(3, 6).value
Call psi(k, n, colmax, nitsolver)
End Sub
```

Fig. 9.20 The macro `solve` to execute psi exactly one time

	A	B	C	D	E	F	G	H	T	U
13	u	-2,274	-1,529	-1,457	-1,324	-1,141	-0,877	-0,7	1,388	1,915
14	fobj	2,046392606	1,483553	1,962148	1,404448	1,901845	1,551212	1,145	1,749656	2,196232287
15	PX	4,494536818	2,649098	2,2455	2,175505	1,69956	1,337731	1,29	2,270278	3,0727261
16		3,659021288	2,698311	2,065888	2,383833	1,728984	1,603191	1,729	2,254649	2,372220196
17		1,813515087	2,030624	1,290314	1,942205	1,165825	1,191205	1,649	1,549205	1,05339102
18		0,351451663	0,724529	0,197975	0,807476	0,225143	0,526794	0,891	0,456749	0,179192355
19		-0,03548052	-0,02463	-0,08149	-0,02483	0,090107	0,04537	0,142	-0,10933	0,146515682
20										

Fig. 9.21 The workbook after executing `psi` once

The next step is the evaluation of $\boldsymbol{x}^{(r+1)}$: according to the algorithm presented page 337, this step involves the solution of a linear system. The implementation of the steps described page 337 can be made as follows:

```
Sub calcxrplusone(A, V, n, k, colmax)
ns = colmax - 1
Dim aux() As Double
Dim T() As Variant
ReDim T(1 To n, 1 To ns)
For m = 1 To ns
  aux = readcol(m + 1, 15, 14 + n)
  For ele = 1 To n
    T(ele, m) = aux(ele)
  Next ele
Next m
For ele = 1 To n
  ReDim Bele(1 To k)
  For i = 1 To k
    s = 0
    For m = 1 To ns
      s = s + V(m, i) * T(ele, m)
    Next m
    Bele(i) = s / ns
  Next i
  auxl = LUSolution(A, Bele)
  For i = 1 To k
    Cells(6 + i, 7 + ele).value = auxl(i)
  Next i
Next ele
End Sub
```

```
Function calcAV(ap, k, colmax)
u = read_line(13, 2, colmax)
A = ap.meanps(u, k)
V = ap.matrix(k, u)
Dim resu(1 To 2) As Variant
resu(1) = A
resu(2) = V
calcAV = resu
End Function
```

`Sub calcxrplusone` determines the value of $\boldsymbol{x}^{(r+1)}$ by solving the linear systems (6.24), page 336. These linear systems involve matrices $\boldsymbol{A}$ and $\boldsymbol{V}$, described in the algorithm page 337 – both are determined by `Sub calcAV`.

To run this `Sub`, you must modify the `Sub solve` as shown in Fig. 9.22 – it runs `calcxrplusone` exactly one time. Running this new version of solve fills H7:L11 with the values of $\boldsymbol{x}^{(r+1)}$ and the workbook appears as in Fig. 9.23.

```
Sub solve()
k = Cells(2, 2).value
n = 5
Call init(k, n)
Dim p As New poly
p.lb = Cells(21, 2).value
p.ub = Cells(21, 4).value
p.zero = 0.001
Dim ap As New expansion
Set ap.phi = p
colmax = Cells(13, Columns.Count).End(xlToLeft).Column
Call calcXr(ap, n, k, colmax)
nitsolver = Cells(3, 6).value
Call psi(k, n, colmax, nitsolver)
resu = calcAV(ap, k, colmax)
A = resu(1)
V = resu(2)
Call calcxrplusone(A, V, n, k, colmax)
End Sub
```

Fig. 9.22 The macro `solve` to execute `calcxrplusone` exactly one time

	A	B	C	D	E	F	G	H	I	J	K	L
5	r	0	prec	1								
6	coeffs	x1(r)	x2(r)	x3(r)	x4(r)	x5(r)		x1(r+1)	x2(r+1)	x3(r+1)	x4(r+1)	x5(r+1)
7		0,2	0,5	0,7	0,4	0		4,488534	3,656569	1,819481	0,343132	-0,03298
8		0	0	0	0	0		-12,0014	-6,27266	2,025367	2,317627	-0,91873
9		0	0	0	0	0		3,975868	-5,29008	-19,6394	-8,70739	6,824671
10		0	0	0	0	0		14,33111	21,94856	33,91231	11,76306	-12,3063
11		0	0	0	0	0		-7,6819	-11,611	-17,0084	-5,51337	6,555792

Fig. 9.23 The workbook after executing `calcxrplusone` once

The last step consists in executing the iterations: we must update the value of $\boldsymbol{x}^{(r)}$ by copying the values of $\boldsymbol{x}^{(r+1)}$, evaluate the stopping conditions and iterate while the stopping conditions are not satisfied. An example of `Sub updatexr` to update the values of $\boldsymbol{x}^{(r)}$ is given below. `updatexr` also calculates the step RMS and writes the result in `D5`.

```
Sub updatexr(k, n)
s = 0
For i = 1 To k
  For ele = 1 To n
    dx = Cells(6 + i, 7 + ele).value - Cells(6 + i, 1 + ele).value
    s = s + dx * dx
  Next ele
Next i
d = Sqr(s / (k * n))
Cells(5, 4).value = d
For i = 1 To k
  For ele = 1 To n
    Cells(6 + i, 1 + ele).value = Cells(6 + i, 7 + ele).value
  Next ele
Next i
End Sub
```

Finally, we must modify the Sub solve to perform the iterations:

```
Sub solve()
k = Cells(2, 2).value
n = 5
Call init(k, n)
Dim p As New poly
p.lb = Cells(21, 2).value
p.ub = Cells(21, 4).value
p.zero = 0.001
Dim ap As New expansion
Set ap.phi = p
colmax = Cells(13, Columns.Count).End(xlToLeft).Column
Call calcXr(ap, n, k, colmax)
nitsolver = Cells(3, 6).value
resu = calcAV(ap, k, colmax)
a = resu(1)
V = resu(2)
nitmax = Cells(3, 2).value
precdes = Cells(3, 4).value
nit = 0
prec = 1 + precdes
Do Until nit > nitmax Or prec < precdes
  Call psi(k, n, colmax, nitsolver)
  Call calcxrplusone(a, V, n, k, colmax)
  Call updatexr(k, n)
  prec = Cells(5, 4).value
  nit = nit + 1
  Cells(5, 2).value = nit
  Call calcXr(ap, n, k, colmax) 'update PX
Loop
End Sub
```

An example of solution found by this method is shown in Fig. 9.24. In these calculations, the internal number of iterations of the SOLVER was 10. Convergence arises in 7 iterations of the method. When using 3 iterations of the SOLVER, the results are similar, but convergence is slower. At the end of the iterations, the workbook appears as follows:

	A	B	C	D	E	F	K	L
5	r	7	prec	5,18E-06				
6	coeffs	x1(r)	x2(r)	x3(r)	x4(r)	x5(r)	x4(r+1)	x5(r+1)
7		6,171496371	7,171496	8,171495	9,171492	10,17149	9,171492	10,17149
8		-19,0529177	-19,0529	-19,0529	-19,0529	-19,0529	-19,0529	-19,0529
9		17,54876476	17,54875	17,5487	17,5486	17,54836	17,5486	17,54836
10		2,32319E-05	6,18E-05	0,000158	0,000377	0,000888	0,000377	0,000888
11		-1,4043E-05	-3,8E-05	-9,6E-05	-0,00023	-0,00055	-0,00023	-0,00055
12								
13	u	-2,274	-1,529	-1,457	-1,324	-1,141	-0,461	-0,243
14	fobj	3,52474E-13	3,94E-13	4,48E-13	5,56E-13	7,01E-13	1,06E-12	7,56E-13
15	PX	6,171496371	3,337847	3,122079	2,751784	2,302479	1,212296	1,059273
16		7,171495867	4,337846	4,122079	3,751783	3,302478	2,212295	2,059272
17		8,171494684	5,337845	5,122078	4,751783	4,302477	3,212294	3,059272
18		9,17149213	6,337843	6,122076	5,75178	5,302475	4,212292	4,05927
19		10,17148647	7,337839	7,122072	6,751776	6,30247	5,212287	5,059266

Since this approach uses external code for the iterations, it is highly flexible and can be easily used for constrained optimization: if we call the SOLVER with the parameters corresponding to a constrained problem, the method runs with no other modification – thus, only the `Sub psi` must be modified.

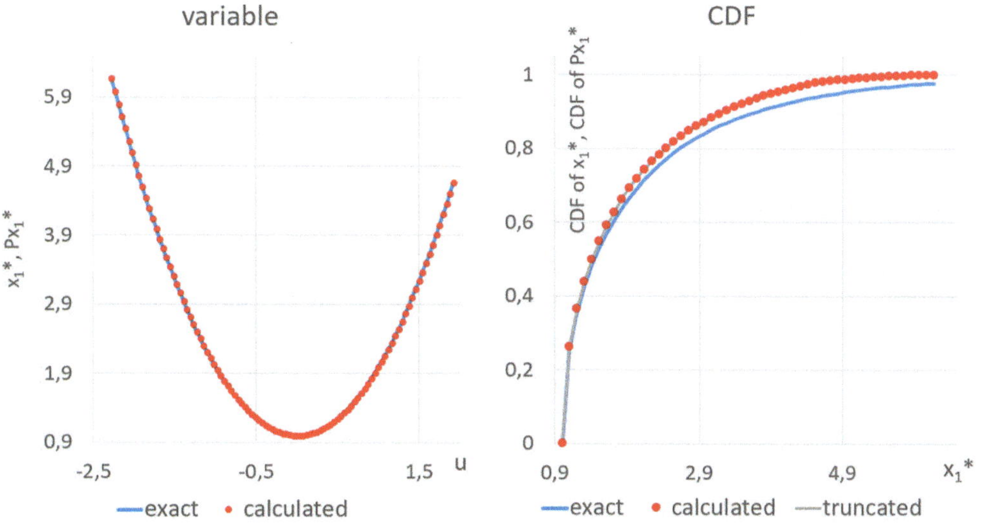

Fig. 9.24 Results furnished by the adaptation of the SOLVER with 10 internal iterations. At left, a comparison between x_1^* and the expansion Px_1^*. At right, a comparison of the empirical CDF of Px_1^* on a large sample and the exact CDF of x_1^*. The sample of 20 variates furnishes information only for $u \in (-2.274, 1.915)$ $x_1 \in (1.061, 3.274)$ – the CDF of Px_1^* coincides with the CDF of x_1^* truncated to this interval

As an example, let us consider the minimization of the modified Rosenbrock's function (9.7) under the constraints

$$\sum_{i=1}^{n} \frac{x_i}{\eta(u,i)} = n, \quad \sqrt{\sum_{i=1}^{n} \left(\frac{x_i}{\eta(u,i)}\right)^2} \leq \sqrt{n}$$

In this case, functions that evaluate the constraints and lines containing their values must be added to the workbook. For instance, we may add the functions below to the module containing `rosemod` and `rosemod_on_range`:

```
Function eql(x, u)
n = UBound(x, 1)
Dim s1 As Double
s1 = 0
For i = 1 To n
  s1 = s1 + (x(i) / eta(u, i))
Next i
eql = s1 - n
End Function
Function eql_on_range(r As Range, u)
x = range2vector(r)
eql_on_range = eql(x, u)
End Function
```

```
Function ineql(x, u)
n = UBound(x, 1)
Dim s1 As Double
s1 = 0
For i = 1 To n
  s1 = s1 + (x(i) / eta(u, i)) ^ 2
Next i
ineql = Sqr(s1) - Sqr(n)
End Function
Function ineql_on_range(r As Range, u)
x = range2vector(r)
ineql_on_range = ineql(x, u)
End Function
```

Now, go to workbook and enter "`=eql_on_range(B15:B19;B13)`" in `B23` and "`=ineql_on_range(B15:B19;B13)`" in `B24` (use it as it was before calculating $X^{(r)}$, page 419). Copy these cells and paste them on `C23 : U24`. The workbook appears as:

	A	B	C	D	E	F		S	T	U
13	u	-2,274	-1,529	-1,457	-1,324	-1,1	!9	1,081	1,388	1,915
14	fobj	4	4	4	4	4		4	4	4
15	PX									
16										
17										
18										
19										
20										
21	min u	-2,274092519	max u	1,915034						
22										
23	inequality	-2,236067977	-2,23607	-2,23607	-2,23607	-2,2:	507	-2,23607	-2,23607	-2,23607
24	equality	-5	-5	-5	-5	-!		-5	-5	-5

Now, we add the constraints to the Sub psi: Relation:=1 defines " $\le$ ", Relation:=2 defines " $=$ ". The new Sub reads as:

```
Sub psi(k, n, colmax, nitsolver)
For i = 2 To colmax
  rxad = Range(Cells(15, i), Cells(14 + k, i)).Address
  rfobjad = Range(Cells(14, i), Cells(14, i)).Address
  SolverReset
  SolverOk SetCell:=rfobjad, MaxMinVal:=2, ValueOf:=0, ByChange:=rxad, _
          Engine:=1, EngineDesc:="GRG Nonlinear"
  SolverAdd Cellref:=Range(Cells(23, i), Cells(23, i)).Address, Relation:=1, FormulaText:="0"
  SolverAdd Cellref:=Range(Cells(24, i), Cells(24, i)).Address, Relation:=2, FormulaText:="0"
  SolverOptions AssumeNonNeg:=False
  SolverOptions Multistart:=True, PopulationSize:=500
  SolverOptions Iterations:=nitsolver
  SolverOptions StepThru:=True
  resu = SolverSolve(True, "cbf")
  SolverFinish KeepFinal:=1
Next i
End Sub
```

Now, run the macro solve. Using 10 internal iterations of the SOLVER, convergence was obtained in 5 iterations of the method. The results obtained are shown in Fig. 9.25. At the end of the iterations, the workbook appears as follows:

	A	B	C	D	E	F	G	H	I	S	T	U
5	r	5	prec	1,75E-11								
6	coeffs	x1(r)	x2(r)	x3(r)	x4(r)	x5(r)		x1(r+1)	x2(r			
7		6,171496777	7,171497	8,171497	9,171497	10,1715		6,171497	7,17			
8		-19,05292116	-19,0529	-19,0529	-19,0529	-19,0529		-19,0529	-19,(			
9		17,54878109	17,54879	17,5488	17,54884	17,54893		17,54878	17,5			
10		-6,57661E-06	-1,8E-05	-4,4E-05	-0,0001	-0,00024		-6,6E-06	-1,8			
11		3,33285E-06	8,82E-06	2,2E-05	5,25E-05	0,000123		3,33E-06	8,82			
12												
13	u	-2,274	-1,529	-1,457	-1,324	-1,141	-0,877	-0,746	-0,(	081	1,388	1,915
14	fobj	4,30087E-16	2,04E-14	1,94E-14	1,76E-14	1,53E-14	8,7E-15	4,15E-15	1,21	E-15	1,39E-15	2,28E-14
15	PX	6,171496777	3,337847	3,12208	2,751784	2,302479	1,769934	1,556378	1,40	8397	2,925184	4,667353
16		7,171496774	4,337847	4,122079	3,751784	3,302479	2,769934	2,556378	2,40	8397	3,925184	5,667354
17		8,171496753	5,337846	5,122079	4,751784	4,302479	3,769934	3,556378	3,40	8397	4,925184	6,667354
18		9,171496704	6,337846	6,122079	5,751783	5,302478	4,769934	4,556378	4,40	8397	5,925184	7,667354
19		10,1714966	7,337845	7,122078	6,751783	6,302478	5,769934	5,556378	5,40	8397	6,925184	8,667356
20												
21	min u	-2,274092519	max u	1,915034								
22												
23	inequality	-1,45926E-08	-2,1E-07	-2,1E-07	-1,9E-07	-1,5E-07	-7,5E-08	-3E-08	7,35	E-08	-1,6E-08	2,25E-07
24	equality	-3,263E-08	-4,8E-07	-4,7E-07	-4,3E-07	-3,5E-07	-1,7E-07	-6,8E-08	1,64	E-08	-3,5E-08	5,02E-07

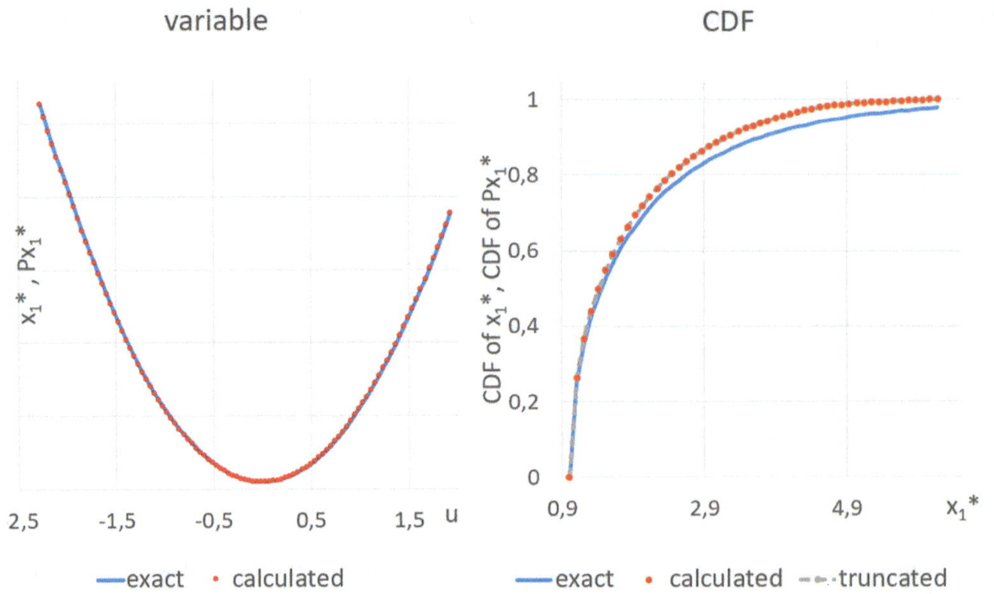

Fig. 9.25 Results furnished by the adaptation of the SOLVER with 10 internal iterations. At left, a comparison between x_1^* and the expansion Px_1^*. At right, a comparison of the empirical CDF of Px_1^* on a large sample and the exact CDF of x_1^*. The sample of 20 variates furnishes information only for $\in(-2.274,\ 1.915)\ x_1^* \in (1.061, 3.274)$: the CDF of Px_1^* coincides with the CDF of the truncation of x_1^* to this interval

Exercises

1. Adapt the SOLVER to the analysis of the Red Brand Canners under uncertainty (page 415).
2. Consider the economic function

$$f(\boldsymbol{x}) = x_1^\alpha + x_2^{1-\alpha}.$$

Adapt the SOLVER to the analysis of the maximal value of $f(\boldsymbol{x})$ under the restriction $p_1x_1 + p_1x_2 = A$ in the following situations:

(a) Assume that $A = 100, p_1 = 2, p_2 = 4$ and $\alpha \sim T(0.4,\ 0.6{,}0.5)$. Find the distribution of the solution.

(b) Assume that $A = 100, \alpha = 0.5, p_1 \sim T(1.5{,}2.5{,}2), p_2 \sim T(3,\ 4,\ 5)$. Find the distribution of the solution.

3. Consider the economic function

$$ces(\boldsymbol{x}) = \left(\alpha x_1^\beta + (1-\alpha)x_2^\beta\right)^{\frac{1}{\beta}}.$$

Adapt the SOLVER to the analysis of the maximal value of $p_1x_1 + p_1x_2$, under the restriction $f(\boldsymbol{x}) = 1$. Solve the following situations:

(a) Assume that $A = 100, p_1 = 2, p_2 = 4, \alpha \sim T(0.4,\ 0.6{,}0.5), \beta \sim T(0.3,\ 0.5{,}0.4)$. Find the distribution of the solution.

(b) Assume that $A = 100,\ \alpha = 0.5,\ \beta = 0.4,\ p_1 \sim T(1.5{,}2.5{,}2), p_2 \sim T(3,\ 4,\ 5)$. Find the distribution of the solution.

9.3 Combining Statistics of the Objective, the Constraints and Expansions

Analogously to Sect. 6.1, we can also use expansions to generate a new optimization problem, involving means of the objective and the constraints. For instance, we can look for the expansion

$$\boldsymbol{Px}^* = \sum_{i=1}^{k} \boldsymbol{x}_i^* \varphi_i(u), \boldsymbol{x}_i^* \in \mathbb{R}^n$$

such that $\boldsymbol{x}^* = (\boldsymbol{x}_1^*, \ldots, \boldsymbol{x}_k^*)$ verifies

$$\boldsymbol{x}^* = \arg\min \left\{ G(\boldsymbol{y}) = E\left(F\left(\sum_{i=1}^{k} \boldsymbol{y}_i \varphi_i(u), u \right) \right), \boldsymbol{y}_i \in \mathbb{R}^n, \boldsymbol{y} = (\boldsymbol{y}_1, \ldots, \boldsymbol{y}_k) \right\}$$

In practice, we may use a sample to estimate the mean:

$$G(\boldsymbol{y}) \approx \frac{1}{ns} \sum_{n=1}^{ns} F\left(\sum_{i=1}^{k} \boldsymbol{y}_i \varphi_i(u_n), u_n \right).$$

For instance, let us consider the minimization of the modified Rosenbrock's function (9.7). The minimization of G may be achieved by adapted methods, such as, for instance, stochastic gradient methods (see, for instance, (Holdorf Lopez et al., 2011)). Under EXCEL®, the SOLVER can handle this minimization, by using workbooks like those introduced in the preceding section. For instance, reopen the workbook created for the solution of the unsconstrained modified Rosenbrock function and clear the contents of cells `H6:L11`. The workbook will appear as follows:

	A	B	C	D	E	F	G	H	I	J	K			T	U
1	x(0)	0,2	0,5	0,7	0,4	0									
2	k	5													
3	nitmax	100	prec	1,00E-05	SOLVER It	10									
4															
5	r	7	prec	0											
6	coeffs	x1(r)	x2(r)	x3(r)	x4(r)	x5(r)									
7															
8															
9															
10															
11															
12															
13	u	-2,274	-1,529	-1,457	-1,324	-1,141	-0,877	-0,746	-0,633	-0,549	-0,461	-0	31	1,388	1,915
14	fobj	4	4	4	4	4	4	4	4	4	4			4	4
15	PX														
16															

Now, select cell `F5` and enter "`=average(B14:U14)`". The workbook appears as

	A	B	C	D	E	F
1	x(0)	0,2	0,5	0,7	0,4	0
2	k	5				
3	nitmax	100	prec	1,00E-05	SOLVER It	10
4						
5	r	7	prec	0	G	4
6	coeffs	x1(r)	x2(r)	x3(r)	x4(r)	x5(r)
7						

Cell `F5` is the objective to be minimized, by changing the values in cells `B7:F11`, which contain the coefficients of the expansion. You can do it manually, as shown in Sect. 2.2.1 or using VBA as shown in Sect. 2.2.2. Let us do it manually: give the name `u` to cells `B13:U13` (see Sect. 1.10). Go to cell `B23` and enter the value `1`. Copy it and paste on `C23:U23`. Go to cell `B24` and enter the formula "`=(u-B21)/(D21-B21)`". The workbook appears as:

	A	B	C	D	E	F	G	H	S	T	U
23	1	1	1	1	1	1	1	1	1	1	1
24	s	0	0,177863	0,195114	0,226907	0,270422	0,333395	0,36479	)0887	0,874073	1

Give the name `s` to `B24:U24`. Go to cell `B25` and enter the formula "`= s^2`". Then enter "`= s^3`" in `B26` and "`= s^4`" in `B27`. The workbook appears as

	A	B	C	D	E	F	G	H	S	T	U
23	1	1	1	1	1	1	1	1	1	1	1
24	s	0	0,177863	0,195114	0,226907	0,270422	0,333395	0,364	0,800887	0,874073	1
25	s^2	0	0,031635	0,038069	0,051487	0,073128	0,111152	0,133	0,64142	0,764004	1
26	s^3	0	0,005627	0,007428	0,011683	0,019775	0,037058	0,048	0,513705	0,667795	1
27	s^4	0	0,001001	0,001449	0,002651	0,005348	0,012355	0,01	0,41142	0,583702	1

Give the name `M` to `B23:U27`. Introduce an initial guess for the coefficients in `B7:F11` and give the name `coeffs` to `B7:F11`. An example of initial guess is given below:

	A	B	C	D	E	F
6	coeffs	x1(r)	x2(r)	x3(r)	x4(r)	x5(r)
7		0,2	0,2	0,2	0,2	0,2
8		0,5	0,5	0,5	0,5	0,5
9		0,7	0,7	0,7	0,7	0,7
10		0,4	0,4	0,4	0,4	0,4
11		0	0	0	0	0

Go to cell B15 and enter "`=MMULT(TRANSPOSE(coeffs);M)`". The workbook appears as:

	A	B	C	D	E	F	G	H	S	T	U
15	PX	0,2	0,313327	0,327177	0,354167	0,394311	0,459327	0,494	.,25492	1,438957	1,8
16		0,2	0,313327	0,327177	0,354167	0,394311	0,459327	0,494	.,25492	1,438957	1,8
17		0,2	0,313327	0,327177	0,354167	0,394311	0,459327	0,494	.,25492	1,438957	1,8
18		0,2	0,313327	0,327177	0,354167	0,394311	0,459327	0,494	.,25492	1,438957	1,8
19		0,2	0,313327	0,327177	0,354167	0,394311	0,459327	0,494	.,25492	1,438957	1,8

Now, open the SOLVER, click on `Reset All`, accept the reinitialization. Then setup it to minimize `F5` by changing `coeffs`:

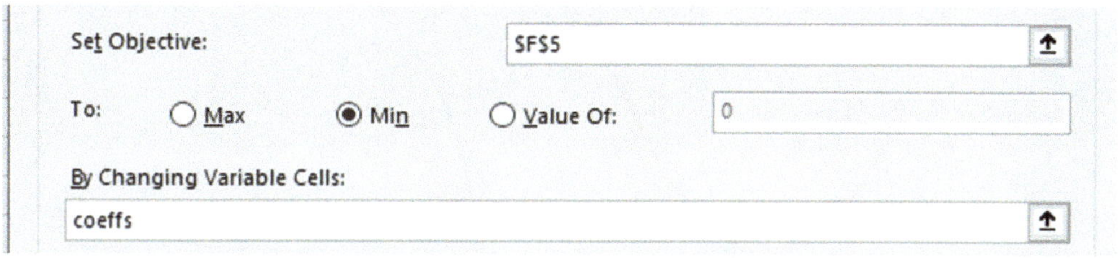

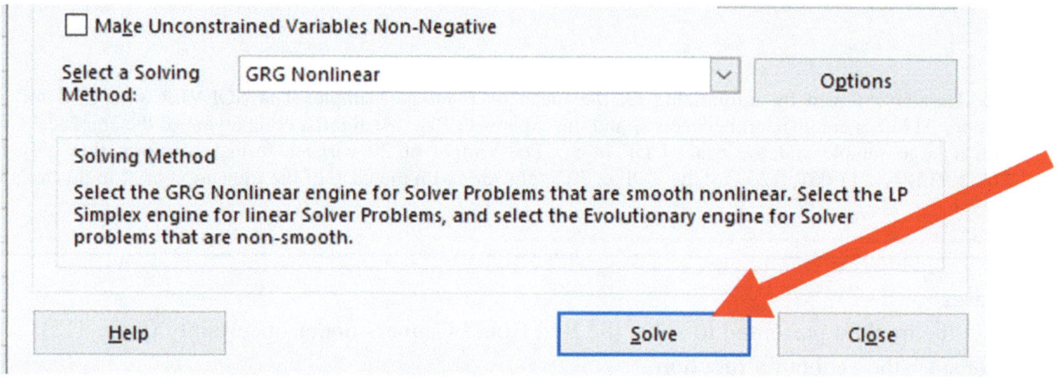

Click on `Solve`, relaunch the SOLVER if necessary. The SOLVER will generate a numerical solution. An example of solution generated by a random starting point is shown in Fig. 9.26. At the end of the calculations, the workbook appears as:

	A	B	C	D	E	F
5	r	1	prec	0	G	2,53E-12
6	coeffs	x1(r)	x2(r)	x3(r)	x4(r)	x5(r)
7		6,171494	7,17149	8,171482	9,171463	10,17142
8		-19,0529	-19,0529	-19,0528	-19,0527	-19,0525
9		17,54874	17,54872	17,54869	17,54863	17,54853
10		2,29E-05	3,95E-06	-6,5E-05	-0,00025	-0,00073
11		-9,7E-07	2,41E-05	9,51E-05	0,000272	0,000695
12						
13	u	-2,274	-1,529	-1,457	-1,324	-1,141
14	fobj	1,86E-11	2,17E-12	1,48E-12	5,84E-13	8,25E-14
15	PX	6.171494	3.337846	3.122079	2.751784	2.302479

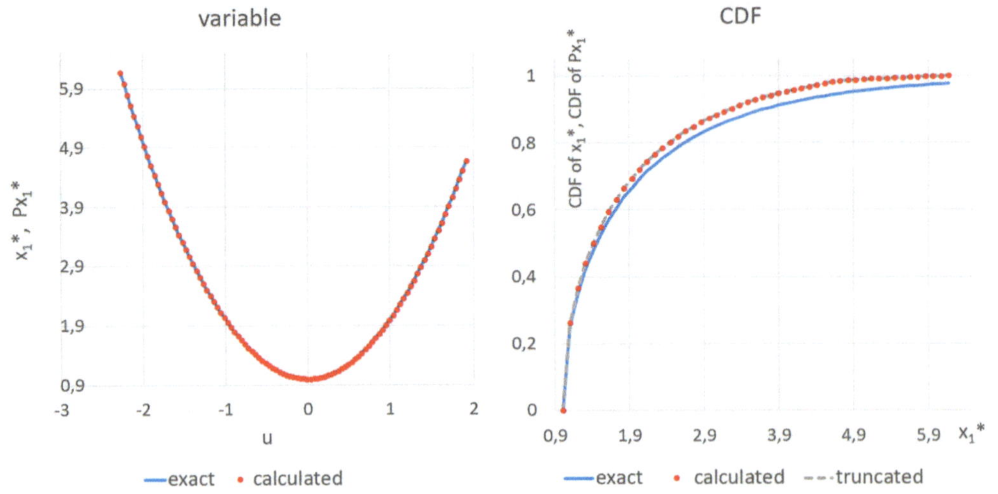

Fig. 9.26 Results obtained by minimizing G, the mean of F on the sample. The SOLVER was used for the minimization. At left, a comparison between x_1 and the expansion Px_1. At right, a comparison of the empirical CDF of Px_1 on a large sample and the exact CDF of x_1. The sample of 20 variates furnishes information only for $\in(-2.274,\ 1.915)\ x_1 \in (1.061,\ 3.274)$): the CDF of Px_1^* coincides with the CDF of the truncation of x_1^* to this interval

Exercises

1. Use the method presented to solve the Red Brand Canners under uncertainty (page 415).
2. Consider the economic function

$$f(\boldsymbol{x}) = x_1^{\alpha} + x_2^{1-\alpha}.$$

Use the method presented to analyze the maximal value of $f(\boldsymbol{x})$ under the restriction $p_1x_1 + p_1x_2 = A$ in the following situations:

(a) Assume that $A = 100$, $p_1 = 2$, $p_2 = 4$ and $\alpha \sim T(0.4,\ 0.6{,}0.5)$. Find the distribution of the solution.

(b) Assume that $A = 100$, $\alpha = 0.5$, $p_1 \sim T(1.5{,}2.5{,}2)$, $p_2 \sim T(3,\ 4,\ 5)$. Find the distribution of the solution.

3. Consider the economic function

$$ces(\boldsymbol{x}) = \left(\alpha x_1^{\beta} + (1-\alpha)x_2^{\beta}\right)^{\frac{1}{\beta}}.$$

Use the method presented to analyze the maximal value of $p_1x_1 + p_1x_2$, under the restriction $f(\boldsymbol{x}) = 1$. Solve the following situations:

(a) Assume that $A = 100$, $p_1 = 2$, $p_2 = 4$, $\alpha \sim T(0.4,\ 0.6{,}0.5)$, $\beta \sim T(0.3,\ 0.5{,}0.4)$. Find the distribution of the solution.

(b) Assume that $A = 100$, $\alpha = 0.5$, $\beta = 0.4$, $p_1 \sim T(1.5{,}2.5{,}2)$, $p_2 \sim T(3,\ 4,\ 5)$. Find the distribution of the solution.

10 Reliability

Reliability is a quite old idea in many areas of human activity facing uncertainties. For example, Civil Engineering faces such a difficulty since its dawn.

In this field, various essential elements may be poorly known, since their variability may be extremely high: external loads, values of physical parameters, soil, not to mention the future climatic events: how to predict future windstorms, rains or floods? In practice, the loads applied to civil engineering structures have a great variability, to the point that some engineers prefer to consider them as unknown in terms of design. Besides this, soils and materials present a great non-homogeneity. These phenomena make mandatory the analysis of uncertainties, especially in certain situations, where the human and economic consequences of a failure can be catastrophic. More recently, industry was incited to reliability analysis, by reasons for such as customer satisfaction, safety, maintenance costs, possible legal liability on failures. Finance also adopted a culture of Risk Analysis connected to Reliability.

Reliability analysis can be performed with the objective of studying and eventually improving an existing system or process. However, it is more efficient to consider reliability since the beginning, taking into account the variability and uncertainties in the design - this is the modern procedure, especially when the correction of errors or the modification of already executed projects can be costly, if not impossible. This is starting point for the modern idea of Reliability Based Design Optimization (RBDO), which consists of seeking optimal solutions that satisfy reliability requirements.

Indeed, design procedures look for the best solution in terms of cost and quality. Thus, design often makes use of optimization tools. As in any optimization procedure, the designer must define the three basic components of any optimization problem: the design variables, the constraints that must be respected and the objective function to be minimized. In general, the design variables are real or integers and denote the parameters for which there is some freedom of choice. The unknowns to be determined define the design. The constraints are the limitations in the freedom of choice, while the objective function gives a numerical evaluation of the design quality, providing a practical method for comparing different solutions, the best ones corresponding to their minima. The constraints and the objective function may involve additional parameters that are not design variables, but supplied data.

Once these basic elements are defined, the designer usually uses a numerical optimization method to obtain an approximate solution. For example, a descent method – deterministic, stochastic or hybrid – can be used. In this procedure, the design variables remain deterministic, even using stochastic descent methods or stochastic modifications of descent methods. This approach is called DDO (Deterministic Design Optimization).

E. Souza de Cursi, *Uncertainty Quantification and Stochastic Modelling with EXCEL*,
Springer Texts in Business and Economics, https://doi.org/10.1007/978-3-030-77757-9_10

However, in many practical situations, the framework includes variability of some parameters or stochastic phenomena, which produces a loss of robustness and reliability. In such a situation, uncertainty must be introduced in the design procedure to obtain robust and reliable solutions, able to face implementation errors and variations in the operational conditions. Different approaches were proposed to include random or uncertain aspects in the design. For instance, robust optimization and RBDO – already mentioned. RBDO looks for a control of the probabilities of some key events, such as failure: the maximal acceptable probability of a key event is introduced as a constraint, which corresponds to requesting a minimum reliability for a key event. In practice, reliability ca be characterized by a reliability index, usually denoted by β and defined for systems that satisfy two basic conditions: on the one hand, the state of the system is fully defined by a vector of parameters $\boldsymbol{x}$; on the other hand, the failure is characterized by the sign of a real variable Z – failure and safe operation correspond to opposite signs of Z – for instance, failure corresponds to $Z > 0$ and safe operation to $Z < 0$.

10.1 Reliability

The reliability of a system is usually defined as the probability that the system ensures its functions without failure. Thus, the reliability is the complement of the probability of failure. Often, the reliability is denoted R and the probability of failure is denoted F. We have $R = 1 - F$. In general, R and F are time-dependent: aging and wear of the components of the system makes that the probability of failure to increase with time. This process is usually summarized in the "bathtub curve" (see Fig. 10.1), which represents a behavior often found in the life of industrial products: at launch, the failure rate is high, due to production errors, gaps in pre-launch testing, unforeseen situations; in a second phase – maturity - the errors have been corrected and the failure rate is approximately constant; in the old age, obsolescence, maintenance failures and wear increase the failure rate.

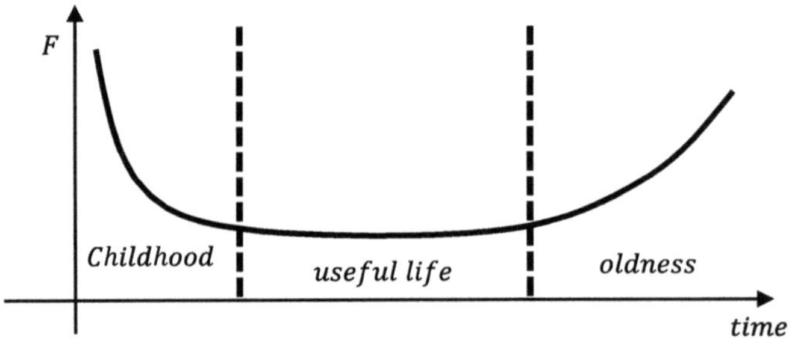

Fig. 10.1 The bathtube curve

The prediction of these periods is one of the concerns of the managers. To make such a prediction, it is necessary to estimate the probability of failure F – that is, the reliability R. Thus, one of the main concerns of reliability is the assessment of the probability of failure and the prediction of its evolution. Another central concern is the determination of probable failure times – whose knowledge allows the design of maintenance policies.

These estimations generally use, in combination, data and models. A simple way to estimate R is to collect data about the number of failures occurring and construct a cumulative distribution of failures up to time t. For instance, let us consider a population of devices, and let $S(t)$ be the number of "survivors" at time t, that is, the devices in operation, without failure, at time t. The number of devices put into operation at the beginning of the process is $S(0)$, and the number of failures at time t is $N(t) = S(0) - S(t)$. Then we have

$$R(t) = \frac{S(t)}{S(0)}, F(t) = 1 - R(t) = \frac{N(t)}{S(0)} \quad .$$

Let T be the instant of time where a failure occurs. Let us consider Δt small enough so that only one failure is possible between t and $t + \Delta t$. Then,

$$P(T \in (t, t + \Delta t)) = R(t) - R(t + \Delta t) = F(t + \Delta t) - F(t).$$

Thus, $F = 1 - R$ is the CDF of the failure time T: we have $P(T < t) = F(t)$ and $P(T \geq t) = R(t)-R(t)$ gives the probability that the system survives at time t and $F(t)$ gives the probability that it does not survive at that same instant of time t.

The failure rate $\lambda(t)$ is the probability of failure per unit of time. It can be estimated in the time interval $(t, t + \Delta t)$ by

$$\lambda(t) \approx \frac{S(t) - S(t + \Delta t)}{S(t)\Delta t} = \frac{N(t + \Delta t) - N(t)}{S(t)\Delta t}$$

Consequently,

$$\lambda(t) \approx \frac{F(t + \Delta t) - F(t)}{R(t)\Delta t} = \frac{R(t) - R(t + \Delta t)}{R(t)\Delta t}$$

Taking the limit for $\Delta t \rightarrow 0+$, we obtain

$$\lambda(t) = -\frac{S'(t)}{S(t)} = \frac{N'(t)}{S(t)} = \frac{F'(t)}{R(t)} = -\frac{R'(t)}{R(t)\Delta t}$$

Notice that

$$P(T \in (t, t + \Delta t) | T \geq t) \approx \lambda(t)\Delta t$$

The expected time between two failures is called Mean Time Between Failures (MTBF). We have

$$MTBF = \frac{1}{\lambda}$$

The expected time from the actual time to the next failure is called Mean Time To Failure (MTTF). We have

$$MTTF = E(T) = \int_0^{+\infty} t dF(t) = -\int_0^{+\infty} t dR(t).$$

To furnish numerical estimations for these values, we need a model for R, which can be a representation generated by observed data. However, if you have some arguments for it, you may also use a theoretical model for R. Popular ones are

- The exponential model: $R(t) = \exp(-\lambda t)$. This model is often used for the middle part of the bathtub curve ("useful life"). Under EXCEL®, `EXPON.DIST(t, λ,True)` returns the CDF $F(t) = 1 - \exp(-\lambda t)$;
- The Weibull model: $R(t) = \exp\left(-\left(\frac{t}{\alpha}\right)^{\beta}\right)$. This model is often used for the last part of the bathtub curve ("oldness"). Under EXCEL®, `WEIBULL.DIST(t,a,b,True)` returns the CDF $F(t) = 1 - \exp\left(-\left(\frac{t}{a}\right)^{b}\right)$;
- The lognormal model: $R(t) = 1 - \Phi\left(\frac{\log(t)-\mu}{\sigma}\right)$, where Φ is the cumulative function of $N(0, 1)$:

$$\Phi(s) = \int_{-\infty}^{s} \phi(t)dt, \ \ \phi(t) = \frac{1}{\sqrt{2\pi}}e^{-\frac{t^2}{2}} \tag{10.1}$$

Under EXCEL®, `LOGNORM.DIST(t, μ, σ,True)` returns $\Phi\left(\frac{\log(t)-\mu}{\sigma}\right)$. In all the situations, replacing `True` by `False` returns the PDF $f(t) = F'(t)$.

In practice, even the theoretical models request data: observations are used to fit the distribution by estimating parameters λ or μ, σ. Estimation can be made by the Maximum of Likelihood (Sect. 3.8.1) or by fitting the empirical CDF. If no model is given, UQ representation can be used (Chap. 5).

Fitting the CDF of the three models above can be made by using affine approximations (see page 106). Indeed, if some data $(t_i, R_i = R(t_i))$, $i = 1, \ldots, n$ is available, we can fit the data to those distributions as follows:

- Exponential model: we have $\log(R_i) = -\lambda t_i$. Thus, we may fit a linear function to the data $y = ax + b$ to the data $x_i = t_i$, $y_i = \log(R_i)$
- Weibull model: we have $\log(-\log R_i) = \beta \log(t_i) - \beta \log(\alpha)$. Thus, we may fit an affine function $y = ax + b$ to the data $x_i = \log(t_i)$, $y_i = \log(-\log R_i)$.
- Lognormal model: we have $\log(t_i) = \sigma\Phi^{-1}(1 - R_i) + \mu$. Thus, we may fit an affine function $y = ax + b$ to the data $x_i = \Phi^{-1}(1 - R_i)$, $y_i = \log(t_i)$

Using the expressions page 106:

$$a = \frac{cov(x, y)}{V(x)} \ , b = E(x) - aE(y).$$

Example

Let us consider the data:

t	1	2	3	4	5	6	7
R	0,7	0,4	0,3	0,2	0,1	0,07	0,05

Open a blank workbook and enter the data:

	A	B	C	D	E	F	G	H
1	t	1	2	3	4	5	6	7
2	R	0,7	0,4	0,3	0,2	0,1	0,07	0,05

Give the name `t` to `B1:H1`, the name `R_` to `B2:H2`. Go to cell `B3` and enter the formula "`=ln(R_)`". Give the name `log_R` to `B3:H3`. The workbook appears as.

3	log R	-0,35667	-0,91629	-1,20397	-1,60944	-2,30259	-2,65926	-2,99573

Go to cell `B4` and enter "`=COVARIANCE.P(t;log_R)/VAR.P(t)`". Then, go to `B5` and enter "`=AVERAGE(log_R)-B4*AVERAGE(t)`". Now, the workbook appears as:

4	a	-0,44649
5	b	0,065396

Thus, $a \approx -0.44649$, id est, $\lambda \approx 0.44649$. We can evaluate the approximated values exp $(-\lambda t_i) =$ exp (at_i): go to cell B6 and enter "`=exp(B4*t)`":

6	R est	0,63987	0,409434	0,261984	0,167636	0,107265	0,068636	0,043918

The Fig. 10.2 below shows the results.

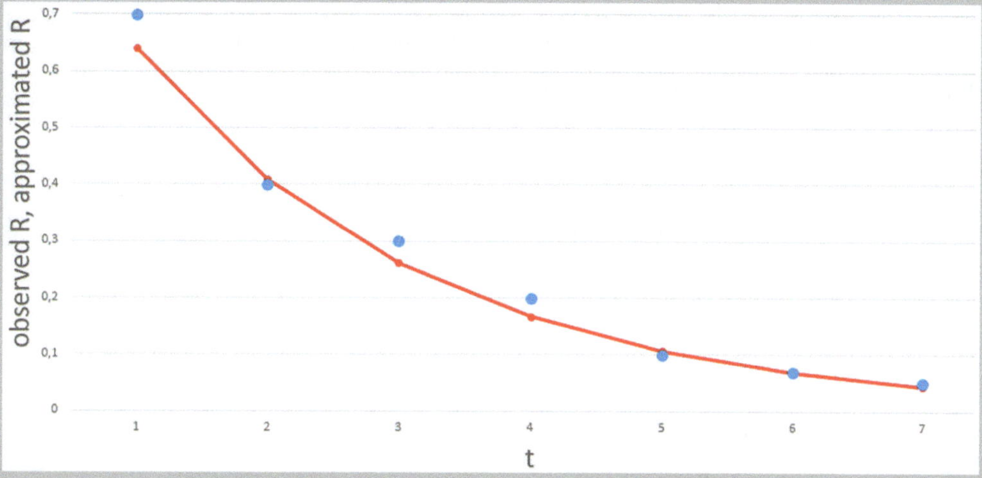

Fig. 10.2 Fitting an exponential distribution

Example

Let us consider the data:

t	1	2	3	4	5	6	7
R	0,8	0,5	0,2	0,06	0,015	0,0024	0,0003

Open a blank workbook and enter the data:

	A	B	C	D	E	F	G	H
1	t	1	2	3	4	5	6	7
2	R	0,8	0,5	0,2	0,06	0,015	0,0024	0,0003

Give the name `t` to `B1:H1`, the name `R_` to `B2:H2`. Go to cell `B3` and enter the formula "`=ln(R_)`". Give the name `log_R` to `B3:H3`. Go to cell `B4` and enter "`=ln(-log_R)`". Give the name `loglog` to `B4:H4`. The workbook appears as.

3	**log R**	-0,22314	-0,69315	-1,60943791	-2,81341	-4,19971	-6,03229	-8,11173
4	**loglog**	-1,49994	-0,36651	0,475884995	1,034398	1,435014	1,797126	2,093311

Enter "`=COVARIANCE.P(LN(t);loglog)/VAR.P(LN(t))`" in `B5`. Then, go to `B6` and enter "`=AVERAGE(loglog)-B4*AVERAGE(LN(t))`". Go to `D6` and enter "`= B5`". Then, go to `D5` and enter "`=EXP(-B6/D6)`". Now, the workbook appears as:

5	a	1,8671	alpha	2,310967787
6	b	-1,56401	beta	1,867100243

Thus, $\alpha \approx 2.311$, $\beta \approx 1{,}867$. We can evaluate the approximated values $\exp\left(-\left(\frac{t_i}{\alpha}\right)^{\beta}\right)$: go to cell `B7` and enter "`=EXP(-POWER(t/D5;D6))`":

7	**R est**	0,811155	0,466029	0,196366612	0,061713	0,014628	0,002637	0,000364

The Fig. 10.3 below shows the results.

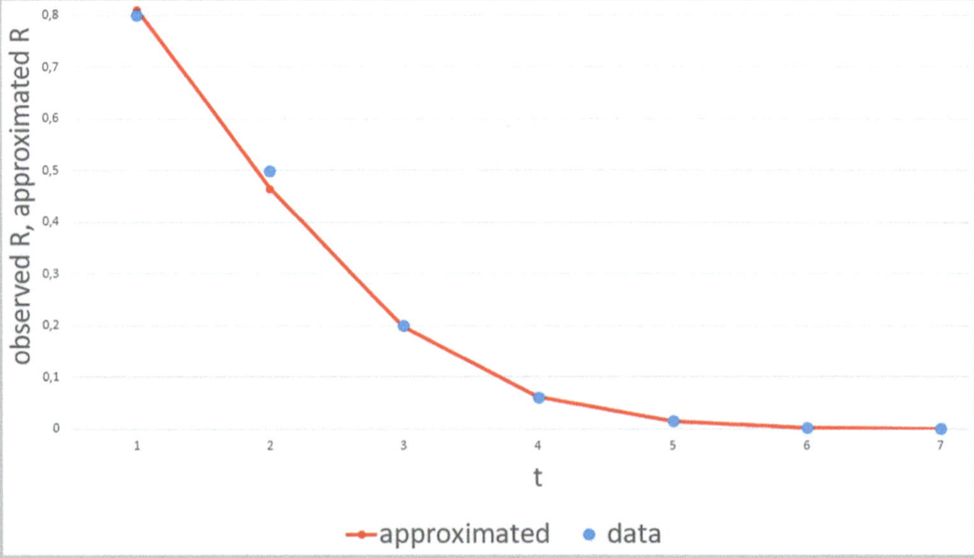

Fig. 10.3 Fitting a Weibull distribution

Example

Let us consider the data:

t	1	2	3	4	5	6	7
R	0,9	0,6	0,4	0,25	0,16	0,1	0,07

Open a blank workbook and enter the data.

	A	B	C	D	E	F	G	H
1	t	1	2	3	4	5	6	7
2	R	0,9	0,6	0,4	0,25	0,16	0,1	0,07

Give the name `t` to `B1:H1`, the name `R_` to `B2:H2`. Go to cell `B3` and enter the formula "`=1-R_`". Give the name `rr1_` to `B3:H3`. Go to cell `B4` and enter "`=NORM.S.INV(rr1_)`". Give the name `phi` to `B4:H4`. The workbook appears as.

3	1-R	0,1	0,4	0,6	0,75	0,84	0,9	0,93
4	phi	-1,28155	-0,25335	0,253347103	0,67449	0,994458	1,281552	1,475791

Enter "`=COVARIANCE.P(rr1;LN(t))/VAR.P(rr1_)`" in `B5`. Then, go to `B6` and enter "`=AVERAGE(LN(t))-B4*AVERAGE(rr1_)`". Now, the workbook appears as:

5	sigma	0,707261
6	mu	0,900144

Thus, $\mu \approx 0.900$, $\sigma \approx 0{,}707$. We can evaluate the approximated values: go to cell `B7` and enter "`=1-LOGNORM.DIST(t;B6;B5;TRUE)`":

7	R est	0,898441	0,615114	0,389502905	0,245925	0,157961	0,103716	0,069622

The Fig. 10.4 below shows the results.

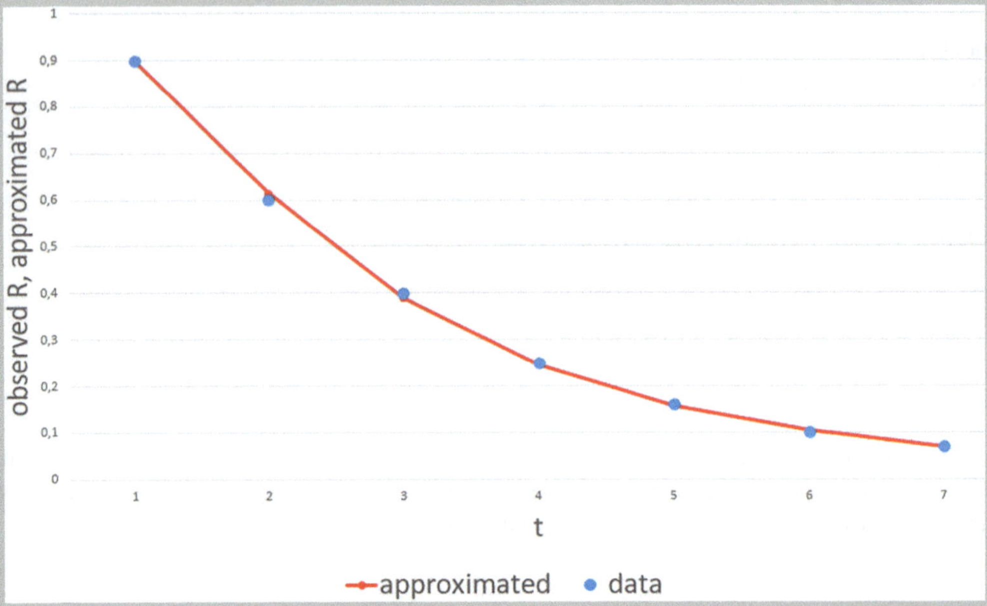

Fig. 10.4 Fitting a Normal distribution

A simple way to analyze the reliability of a system or process is to use fault trees, i.e., diagrams describing the operation of a system or process. These diagrams use elements such as, for instance, those shown in Fig. 10.5.

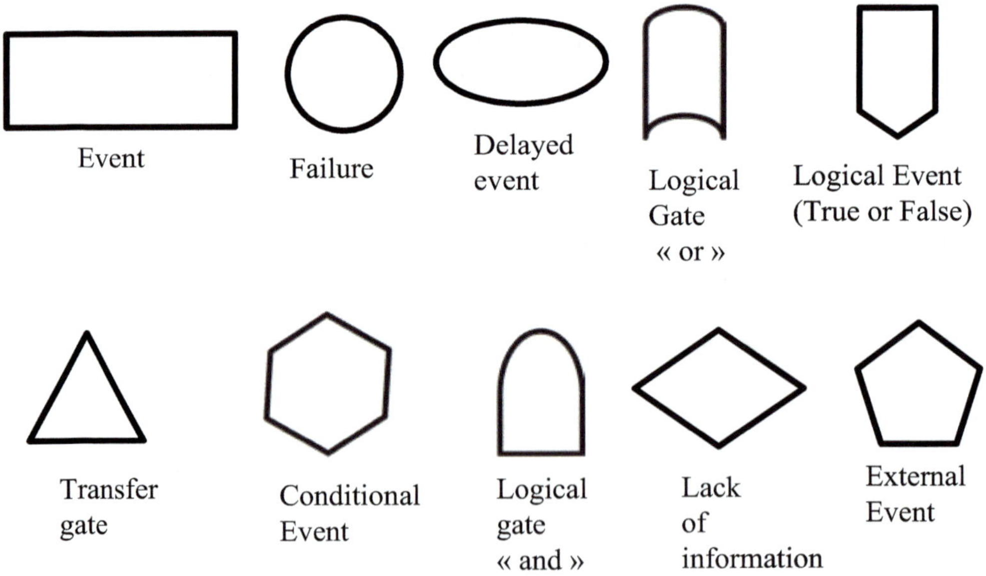

Fig. 10.5 Examples of elements from a Fault Tree

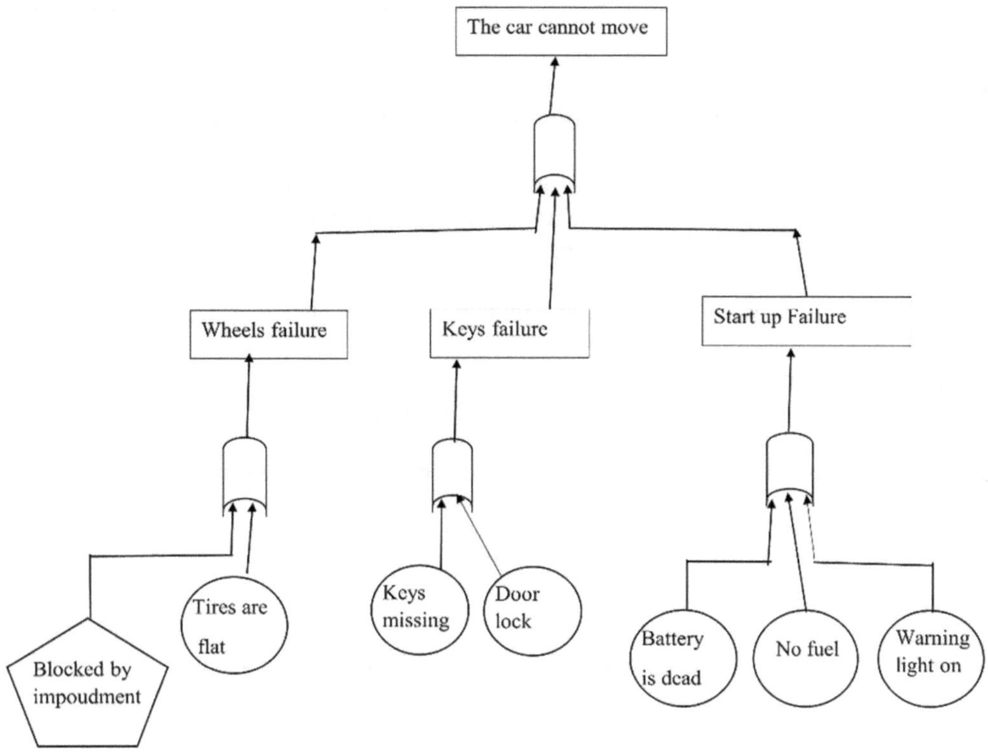

Fig. 10.6 An example of fault tree

By associating probabilities to events, the tree allows the quick calculation of the probabilities of failure. There are two ways to generate a fault tree: bottom-up or top-down. In the bottom-up approach, we start from the basic elements, enumerating the failure possibilities of each one, then, we go to the upper level of the system and we enumerate again the failure possibilities and so on. In the top-down approach, we start from the global system, asking for its possible fault reasons. As an example, let us consider a car and the following elements: keys, tire, fuel, battery, wheels, internal control. In the bottom-up approach, we consider each element. For instance, keys: we look for the possible failures – as an example, consider that failure occurs because keys are missing, or the door does not unlock. The same enumeration is made for each element and we generate the tree by this way. In the top-down approach, we start from the car: if it cannot move, we enumerate the possible reasons: keys, battery, etc. Then, for each reason, we enumerate again the possible reasons: for instance, when considering keys, we list the possibilities "keys missing" and "door does not unlock". that cannot move. For this simple situation, an example of failure tree is shown in Fig. 10.6. If probabilities are assigned to each element in the tree, it is easy to determine the failure probability F. The reader interested in fault trees will find in the literature many books that develop widely this topic – for instance: (Cui & Li, 2020; Ericson, 2011; Limnios, 2007).

Exercises

1. The lifetime of an electronic device is exponentially distributed with a mean of 5 time units.

   (a) Find the reliability at $t = 4$ time units and at $t = 6$ time units.
   (b) Determine the number t of time units such that the reliability is of 5%.
   (c) Determine the MTTF in time units.

2. The lifetime of a device is Weibull distributed with parameters $\alpha = 1, \beta = 2$.

   (a) Find the reliability at $t = 2, 3, 4, 5, 6$.
   (b) Determine the time t such that the reliability is of 5%.

3. The lifetime of a device is Weibull distributed. The observed data is

t	1	2	3	4	5	6	7
R	0,9	0,7	0,5	0,3	0,15	0,08	0,03

   Determine the coefficients α, β.

4. The lifetime of a device is lognormally distributed. The observed data is

t	1	2	3	4	5	6	7
R	0,85	0,75	0,67	0,62	0,57	0,54	0,51

   Determine the coefficients μ, σ.

5. The lifetime of a device is exponentially distributed. The observed data is

t	1	2	3	4	5	6	7
R	0,82	0,67	0,55	0,45	0,37	0,3	0,25

   Determine the coefficient λ.

6. Construct the fault tree for the system

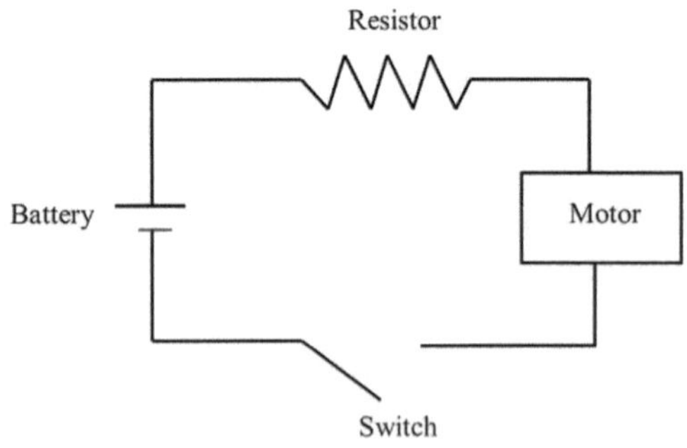

7. Construct a fault tree for the fire on a device considering the following elements: fuel, oxygen, ignition, electricity.

10.2 Limit State Curves

A situation often found in practice is the one where the failure regions are defined by inequalities or, equivalently, the limits of the failure region are defined by equations which separates regions of normal operation and regions of failure. For instance, in dimension 2, the safe and failure regions are separated by curves.

For instance, let us denote by $\boldsymbol{x} \in \mathbb{R}^n$ the design variables of the system: the design space is $\mathbb{R}^n$. Assume that failure occurs when $Z = g(\boldsymbol{x}) > 0$ and the system operates safely when $Z = g(\boldsymbol{x}) < 0$. In such a situation, the safe region and the failure region are separated by the curve$g(\boldsymbol{x}) = 0$, which is called the *limit state curve* or *failure curve*. Then, the space of the design variables splits in two disjoint parts:

- The failure region $\mathcal{F} = \{\boldsymbol{x} \in \mathbb{R}^n : g(\boldsymbol{x}) > 0\}$,
- The safe region $\mathcal{S} = \{\boldsymbol{x} \in \mathbb{R}^n : g(\boldsymbol{x}) \leq 0\}$
  The frontier between these regions corresponds to the limit state curve

$$\mathcal{C} = \{\boldsymbol{x} \in \mathbb{R}^n : g(\boldsymbol{x}) = 0\}.$$

For the sake of convenience, we can bring $\boldsymbol{x}$ to the origin: this is equivalent to describing the points $\boldsymbol{y} \in \mathbb{R}^n$ through the auxiliary variable $\boldsymbol{u}$ such that $\boldsymbol{u} = \boldsymbol{y} - \boldsymbol{x}$. The space of the new variables $\boldsymbol{u}$ is called *centered space*. In principle, such a change of coordinates can be summarized by a simple translation. However, in a more general way, we can consider an enough regular transformation - eventually nonlinear – as, for instance, $\boldsymbol{y} = \phi(\boldsymbol{x}, \boldsymbol{u})$, such that $\phi(\boldsymbol{x}, \boldsymbol{x}) = 0$. Then, the value of z at a point $\boldsymbol{y}$ is $z = h(\boldsymbol{x}, \boldsymbol{u}) = g(\phi(\boldsymbol{x}, \boldsymbol{u})) = g(\boldsymbol{y})$ and

$$\mathcal{F} = \{\boldsymbol{u} \in \mathbb{R}^n : h(\boldsymbol{x}, \boldsymbol{u}) > 0\}, \quad \mathcal{S} = \{\boldsymbol{x} \in \mathbb{R}^n : h(\boldsymbol{x}, \boldsymbol{u}) < 0\}. \tag{10.2}$$

Analogously,

$$\mathcal{C} = \{\boldsymbol{u} \in \mathbb{R}^n : h(\boldsymbol{x}, \boldsymbol{u}) = 0\}. \tag{10.3}$$

Notice that the expression of h can change with $\boldsymbol{x}$. In such a general situation, the design space and the centered space may be different. In the sequel, we assume that the design space is $\mathbb{R}^n$.

Example

Let us consider the situation where $\boldsymbol{x} \in \mathbb{R}^2$ and $g(\boldsymbol{x}) = x_2 - x_1^2 - x_1 - 3$. Consider $\boldsymbol{x} = (x_1, x_2)$, $\phi(\boldsymbol{x}, \boldsymbol{u}) = \boldsymbol{u} + \boldsymbol{x}$. We have $h(\boldsymbol{x}, \boldsymbol{u}) = g(\boldsymbol{u} + \boldsymbol{x}) = u_2 + x_2 - u_1^2 - 2u_1x_1 - x_1^2 - u_1 - x_1 - 3$, id est,

$$h(\boldsymbol{x}, \boldsymbol{u}) = u_2 - u_1^2 - u_1 - 2u_1x_1 + g(\boldsymbol{x}).$$

The curves associated to the points $\boldsymbol{x}^{(1)} = (1, 0)$, $\boldsymbol{x}^{(2)} = (0, 0)$, $\boldsymbol{x}^{(3)} = (1, 1)$, $\boldsymbol{x}^{(4)} = (0, 1)$ are shown at right. Notice that the origin in the centered space corresponds to $\boldsymbol{x}$ in the designs space: the origin is a different **x** for each curve.

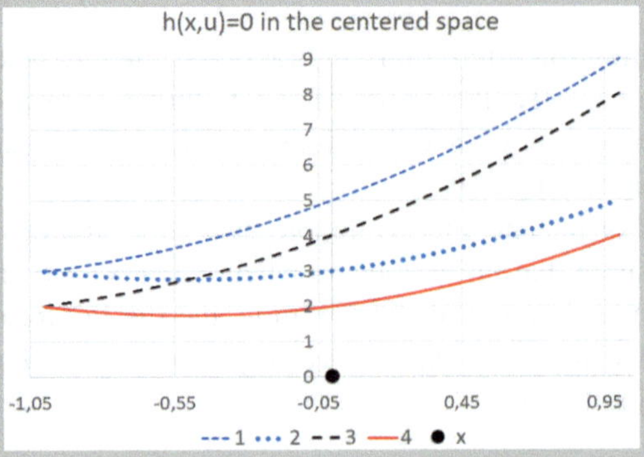

Example

Let us consider the situation where $\boldsymbol{x} \in \mathbb{R}^2$ and $g(\boldsymbol{x}) = x_2 - x_1^3 - x_1^2 - 4$. Consider $\boldsymbol{x} = (x_1, x_2)$, $\phi(\boldsymbol{x}, \boldsymbol{u}) = \boldsymbol{u} + \boldsymbol{x}$. We have.

$$h(\boldsymbol{x},\boldsymbol{u}) = u_2 - u_1^3 - 3u_1^2x_1 - 3u_1x_1^2 - u_1^2 - 2u_1x_1 + g(\boldsymbol{x}).$$

The curves associated to the points $\boldsymbol{x}^{(1)} = (1, 0)$, $\boldsymbol{x}^{(2)} = (0, 0)$, $\boldsymbol{x}^{(3)} = (1, 1)$, $\boldsymbol{x}^{(4)} = (0, 1)$ are shown at right. Notice that the origin in the centered space corresponds to $\boldsymbol{x}$ in the designs space: the origin is a different **x** for each curve.

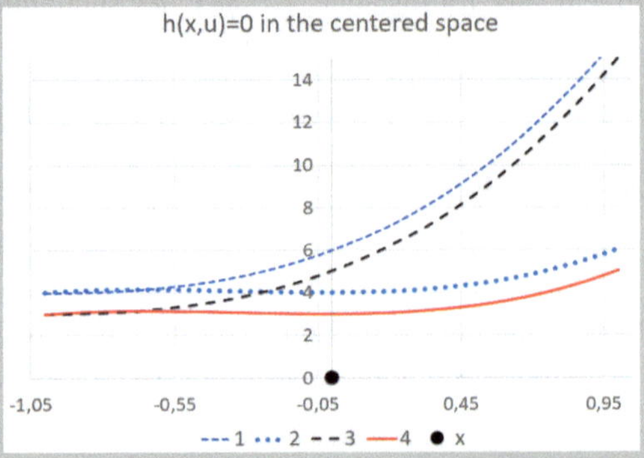

Exercises

1. Let $\boldsymbol{x} \in \mathbb{R}^2$ and $g(\boldsymbol{x}) = x_1x_2 - 2$. Determine $h(\boldsymbol{x}, \boldsymbol{u})$ for $\boldsymbol{x}^{(1)} = (1, 0)$, $\boldsymbol{x}^{(2)} = (0, 0)$, $\boldsymbol{x}^{(3)} = (1, 1)$, $\boldsymbol{x}^{(4)} = (0, 1)$. Draw the limit state curves $h(\boldsymbol{x}, \boldsymbol{u}) = 0$ in the centered space.
2. Let $\boldsymbol{x} \in \mathbb{R}^2$ and $g(\boldsymbol{x}) = x_1^2 + x_2^2 - 4$. Determine $h(\boldsymbol{x}, \boldsymbol{u})$ for $\boldsymbol{x}^{(1)} = (1, 0)$, $\boldsymbol{x}^{(2)} = (0, 0)$, $\boldsymbol{x}^{(3)} = (1, 1)$, $\boldsymbol{x}^{(4)} = (0, 1)$. Draw the limit state curves $h(\boldsymbol{x}, \boldsymbol{u}) = 0$ in the centered space.
3. Let $\boldsymbol{x} \in \mathbb{R}^3$ and $g(\boldsymbol{x}) = x_2 + 1 - 3(2x_3 + 1)^2(3x_1 + 1)$. Determine $h(\boldsymbol{x}, \boldsymbol{u})$ for $\boldsymbol{x}^{(1)} = (1, 0, 0)$, $\boldsymbol{x}^{(2)} = (0, 0, 0)$, $\boldsymbol{x}^{(3)} = (0, 1, 0)$, $\boldsymbol{x}^{(4)} = (0, 0, 1)$. Draw the limit state surfaces $h(\boldsymbol{x}, \boldsymbol{u}) = 0$ in the centered space.

10.3 Design Point

In the same way that we can feel as safer as we are far from the edge of a cliff, we can consider the possible solution $\boldsymbol{x}$ as safer as it is far from the failure region – in fact, the greater the distance between $\boldsymbol{x}$ and $\mathcal{F}$, the larger can be the variations of $\boldsymbol{x}$ that will not generate a failure. Thus, the distance $\beta(\boldsymbol{x}) = d(\boldsymbol{x}, \mathcal{F})$ may be used as a measurement of the reliability of $\boldsymbol{x}$. We have

$$\beta(\boldsymbol{x}) = d(\boldsymbol{x}, \mathcal{F}) = \min\{\|\boldsymbol{y} - \boldsymbol{x}\| : \boldsymbol{y} \in \mathcal{F}\}. \tag{10.4}$$

We can recognize in this formula the definition of the orthogonal projection $\boldsymbol{x}^* = \boldsymbol{P}\boldsymbol{x}$ of $\boldsymbol{x}$ onto $\mathcal{F}$. In regular situations, we may consider the projection of $\boldsymbol{x}$ onto $\mathcal{F} \cup \mathcal{C}$:

$$\beta(\boldsymbol{x}) = \|\boldsymbol{x}^* - \boldsymbol{x}\| = \min\{\|\boldsymbol{y} - \boldsymbol{x}\| : g(\boldsymbol{y}) \geq 0\}. \tag{10.5}$$

In the centered space, we have

$$\beta(\boldsymbol{x}) = \|\boldsymbol{u}^*(\boldsymbol{x})\| = \min\{\|\boldsymbol{u}\| : h(\boldsymbol{x}, \boldsymbol{u}) \geq 0\}. \tag{10.6}$$

In the language of reliability, $\boldsymbol{x}^*$ (or $\boldsymbol{u}^*(\boldsymbol{x})$) is the *design point*, usually denoted DP. $\boldsymbol{x}^*$ (or $\boldsymbol{u}^*(\boldsymbol{x})$) is also called MPP (Most Probable Point) or MPFP (Most Probable Failure Point) . These terminologies are traditional, but their use request some precaution: we should not confuse $\boldsymbol{x}$ - the point in the design space under consideration - and $\boldsymbol{x}^*$- the design point. Moreover, if $\boldsymbol{x}^*$ (or $\boldsymbol{u}^*(\boldsymbol{x})$) defines the failure point which is closest to $\boldsymbol{x}$, this fact does not mean that the failure will occur at that point - the evolution of the values of $\boldsymbol{x}$ may lead to another point. Finally, $\boldsymbol{x}^*$ is the most probable point only for probability distributions decreasing with distance to the failure region: although such a property seems natural, it requests that the distributions satisfy mathematical properties. These considerations make that many authors prefer to use the terminology MCP (Most Central Point) or MCFP (Most Central Failure Point) to avoid misinterpretation. We will use the traditional terminology, and we ask the reader to keep in mind these remarks (Fig. 10.7).

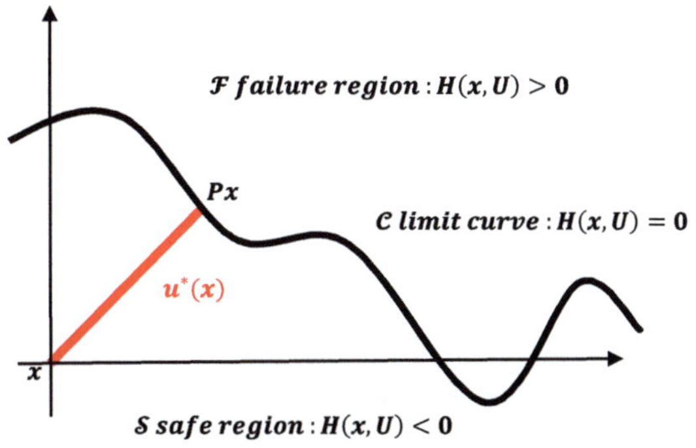

Fig. 10.7 Centered space, Hybrid space and orthogonal projection of a design point $\boldsymbol{x}$

The determination of $\boldsymbol{x}^*$ (or $\boldsymbol{u}^*(\boldsymbol{x})$) can be achieved by the SOLVER: we must solve the optimization problem (10.6).

Example

Let us consider the situation where $\boldsymbol{x} \in \mathbb{R}^2$ and $g(\boldsymbol{x}) = x_2 - x_1^2 - x_1 - 3$. Open a blank workbook and enter the coordinates of the point $\boldsymbol{x}$ and a guess for $\boldsymbol{u}^*$:

	A	B	C
1	x	0	0
2	u*	1	0

Give the name `x` to `B1 : C1` and the name `u` to `B2 : C2`. Go to cell `B3` and enter "=`x + u`". Go to `B4` and enter "=`C3 - B3^2 - B3 - 3`". Go to `B5` and enter "= `B2^2 + C2^2`". The workbook appears as.

	A	B	C
1	x	0	0
2	u*	1	0
3	y	1	0
4	h	-5	
5	||u||^2	1	

Now, run the SOLVER to minimize `B5` by changing `B2 : C2` under the restriction `B4 = 0`.

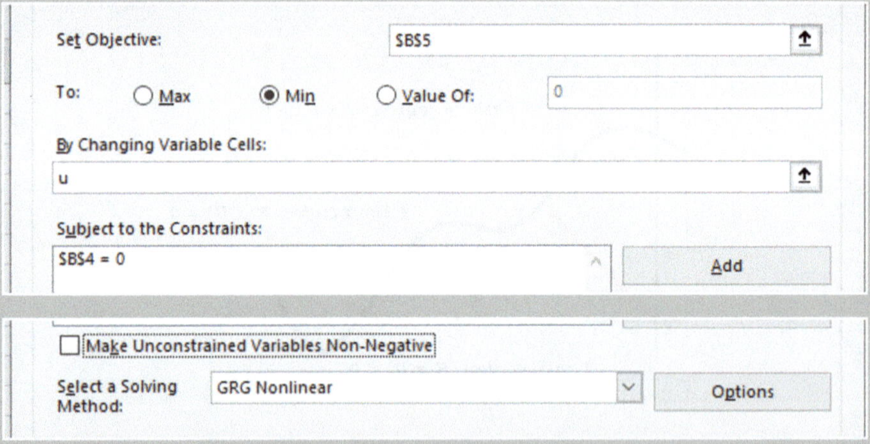

The SOLVER finds a solution:

	A	B	C
1	x	0	0
2	u*	-0,42339	2,755866
3	y	-0,42339	2,755866
4	h	-2,8E-06	
5	||u||^2	7,774058	

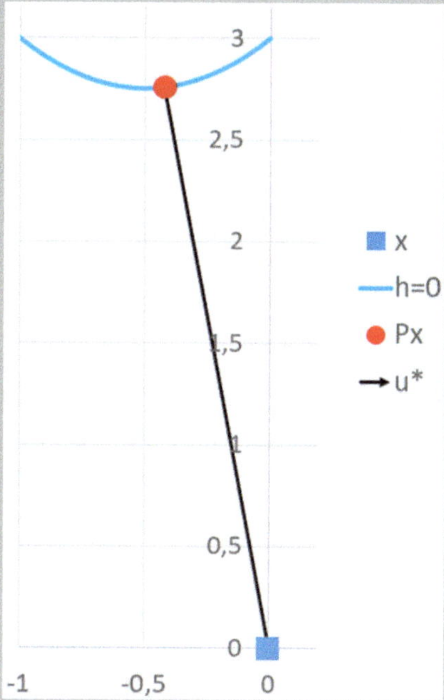

Let us use VBA to gain flexibility: open a new workbook, enter the the coordinates of the point $\boldsymbol{x}$ and the guess for $\boldsymbol{u}^*$. Then, add a module to your workbook (see Sect. 1.5). Enter the code below:

```
Function g(x)
g = x(2) - x(1) ^ 2 - x(1) - 3
End Function
Function h(x, u)
y = plus(x, u)
h = g(y)
End Function
```

```
Function g_on_range(rx As Range)
x = range2vector(rx)
g_on_range = g(x)
End Function
Function h_on_range(rx As Range, ru As Range)
x = range2vector(rx)
u = range2vector(ru)
h_on_range = h(x, u)
End Function
```

This code uses function `range2vector` previously introduced (page 52). It also uses a function `plus`, which evaluates $x + u$. It is implemented in *VectorTools.bas*, which can be imported to the workbook (see Sect. 1.6, page 10). Otherwise, you must add the code below to the module under construction.

```
Function plus(x, y)
nx = UBound(x)
ny = UBound(y)
Dim v() As Variant
ReDim v(1 To nx)
If nx = ny Then
   For i = 1 To nx
         v(i) = x(i) + y(i)
   Next i
Else
   MsgBox "vectors of unequal lengths"
End If
plus = v
End Function
```

Give the name `x` to `B1:C1` and the name `u` to `B2:C2`. Go to cell `B3` and enter "`=h_on_range(x;u)`". Go to cell `B4` and enter "`=SUMSQ(u)`". The workbook appears as.

	A	B	C
1	x	0	0
2	u*	1	0
3	h	-5	
4	||u||^2	1	

Now, you can run the solver to minimize `B4` under the restriction `B3 = 0`. You can do it manually or using a `Sub`:

```
Sub run_solver()
col = Cells(2, Cells.Columns.Count).End(xlToLeft).Column
ruad = Range(Cells(2, 2), Cells(2, col)).Address
rfobjad = Range(Cells(4, 2), Cells(4, 2)).Address
  SolverReset
  SolverOk SetCell:=rfobjad, MaxMinVal:=2, ValueOf:=0, ByChange:=ruad, _
         Engine:=1, EngineDesc:="GRG Nonlinear"
  SolverAdd Cellref:=Range(Cells(3, 2), Cells(3, 2)).Address, Relation:=2, FormulaText:="0"
  SolverOptions AssumeNonNeg:=False
  SolverOptions Multistart:=True, PopulationSize:=500, RequireBounds:=False
  SolverSolve True
End Sub
```

Running this Sub or executing manually the Solver produces the same results at right, which coincide with the preceding ones.

	A	B	C
1	x	0	0
2	u*	-0,42339	2,755866
3	h	-2,8E-06	
4	||u||^2	7,774058	

The use of VBA increases flexibility. For instance, if we desire to change the limit state equation, it is enough to modify function `g`. As an example, let us modify `g` to consider $g(\boldsymbol{x}) = x_2 - x_1^3 - x_1^2 - 4$. Then, we rewrite `g` as follows:

```
Function g(x)
g = x(2) - x(1) ^ 3 - x(1) ^ 2 - 4
End Function
```

With this single modification, we execute the macro `run_solver` and we obtain new results:

	A	B	C
1	x	0	0
2	u*	0,000546	4
3	h	-3,8E-07	
4	||u||^2	16	

10.3.1 Multiple Failure Conditions

In practice, it can be necessary to deal with several failure conditions $g_1(\boldsymbol{x})$, $g_2(\boldsymbol{x})$, . . ., $g_{ng}(\boldsymbol{x})$, so that failure occurs if $g_i(\boldsymbol{x}) > 0$ for at least one of these limit state equations. In the opposite, the safe region verifies $g_i(\boldsymbol{x}) < 0$, $\forall\ i$, $1 \le i \le ng$. Let us denote

$$\mathcal{F}_{\text{i}} = \{\boldsymbol{u} \in \mathbb{R}^n : h_i(\boldsymbol{x}, \boldsymbol{u}) > 0\},\ \ \mathcal{S}_{\text{i}} = \{\boldsymbol{x} \in \mathbb{R}^n : h_i(\boldsymbol{x}, \boldsymbol{u}) < 0\}.$$

Then,

$$\mathcal{F} = \bigcup_{i=1}^{ng} \mathcal{F}_{\text{i}},\ \ \mathcal{S} = \bigcap_{i=1}^{ng} \mathcal{S}_i.$$

Thus,

$$d(\boldsymbol{x}, \mathcal{F}) = \min\{d(\boldsymbol{x}, \mathcal{F}_{\text{i}}), 1 \le i \le ng\}.$$

Consequently, a simple way to extend the preceding ideas to multiple failure conditions consists in using

$$\beta(\boldsymbol{x}) = d(\boldsymbol{x}, \mathcal{F}) = \min\{d(\boldsymbol{x}, \mathcal{F}_{\text{i}}), 1 \le i \le ng\}.$$

Let

$$\beta_i(\boldsymbol{x}) = d(\boldsymbol{x}, \mathcal{F}_{\text{i}}) = \min\{\|\boldsymbol{y} - \boldsymbol{x}\| : \boldsymbol{y} \in \mathcal{F}_{\text{i}}\}.$$

Then

$$\beta(\boldsymbol{x}) = \min\{\beta_i(\boldsymbol{x}), 1 \le i \le ng\}.$$

We have also

$$\beta_i(\boldsymbol{x}) = \|\boldsymbol{u}_i^*\| = \min\{\|\boldsymbol{u}\| : h_i(\boldsymbol{x}, \boldsymbol{u}) \ge 0\}$$

and

$$\beta(\boldsymbol{x}) = \|\boldsymbol{u}^*\| = \min\{\|\boldsymbol{u}_i^*\|, 1 \le i \le ng\}.$$

Thus, a simple way to extend the preceding ideas consists in determining individual design points for each limit state curve and taking the one corresponding to the minimal norm.

Example
Let us consider the situation where $\boldsymbol{x} \in \mathbb{R}^2$ and we have three failure conditions, corresponding to $g_1(\boldsymbol{x}) = x_2 - x_1^2 - x_1 - 3$, $g_2(\boldsymbol{x}) = x_1 - x_2^2 - x_2 - 2$, $g_3(\boldsymbol{x}) = x_1 + x_1x_2 - 1$. Open a blank workbook and enter the coordinates of the point $\boldsymbol{x}$ and the initial guess for $\boldsymbol{u}_1^*$. Give the name `x` to `B1:C1` and the name `u1` to `B2:C2`. Give the name `x` to `B1:C1` and the name `u1_` to `B2:C2`. Go to cell `B3` and enter "`=x + u1_`".

Go to B4 and enter "=C3 - B3^2 - B3-3". Go to D4 and enter "= SQRT(B2^2 + C2^2)". The workbook appears as shown at right.

	A	B	C	D
1	x	0	0	
2	u1*	0	0	
3	x1*	0	0	
4	h1(x,u1*)	-3	||u1*||	0

Copy lines 2 to 4 and paste it in lines 5 to 7 and 8 to 10. Modify the names: u2_ to B5:C7, u3_ to B8:C8. Modify the formulae: "=x + u2_" in B6, "=x + u3_" in B9; "= SQRT(B5^2 + C5^2)" in D7, "= SQRT(B8^2 + C8^2)" in D10. "=B6 - C6^2 - C6-2" in B7, "=B9 + C9*B9-1" in B10. The workbook appears as here.

	A	B	C	D
1	x	0	0	
2	u1*	0	0	
3	x1*	0	0	
4	h1(x,u1*)	-3	||u1*||	0
5	u2*	0	0	
6	x2*	0	0	
7	h2(x,u2*)	-2	||u2*||	0
8	u3*	0	0	
9	x3*	0	0	
10	h3(x,u3*)	-1	||u3*||	0
11				

Now, run the SOLVER three times: to minimize D4 by changing B2:C2 under the restriction B4 = 0; to minimize D7 by changing B5:C5 under the restriction B7 = 0; to minimize D10 by changing B8:C8 under the restriction B10 = 0. The results appear.

	A	B	C	D
1	x	0	0	
2	u1*	-0,50011	2,749997	
3	x1*	-0,50011	2,749997	
4	h1(x,u1*)	-3E-06	||u1*||	2,795101
5	u2*	1,762231	-0,38941	
6	x2*	1,762231	-0,38941	
7	h2(x,u2*)	0	||u2*||	1,804742
8	u3*	0,724535	0,380195	
9	x3*	0,724535	0,380195	
10	h3(x,u3*)	1,06E-08	||u3*||	0,81823

So, $\beta_1(\boldsymbol{x}) = 2.597101$, $\beta_2(\boldsymbol{x}) = 1.804742$, $\beta_3(\boldsymbol{x}) = 0.81823$. The minimal value is $\beta(\boldsymbol{x}) = \beta_3(\boldsymbol{x})$.

Example

Let us consider the situation where $\boldsymbol{x} \in \mathbb{R}^2$ and we have three failure conditions, corresponding to $g_1(\boldsymbol{x}) = x_2 - x_1^2 - x_1 - 3$, $g_2(\boldsymbol{x}) = x_1 - x_2^2 - x_2 - 2$, $g_3(\boldsymbol{x}) = x_1 + x_2 + x_1^2 + x_2^2 - 6$. Let us use VBA to find the design point. Open a blank workbook, add a module and enter the code:

```
Function g_1(x) As Double
Dim aux As Double
g_1 = x(2) - x(1) ^ 2 - x(1) - 3
End Function
Function g_2(x) As Double
g_2 = x(1) - x(2) ^ 2 - x(2) - 2
End Function
Function g_3(x) As Double
g_3 = x(1) + x(2) + x(1) ^ 2 + x(2) ^ 2 - 6
End Function
```

```
Function g_1_on_range(rx As Range) As Double
g_1_on_range = g_1(range2vector(rx))
End Function
Function g_2_on_range(rx As Range) As Double
g_2_on_range = g_2(range2vector(rx))
End Function
Function g_3_on_range(rx As Range) As Double
g_3_on_range = g_3(range2vector(rx))
End Function
```

Prepare the workbook analogously to the preceding one. Give the names `x1_` to `B3:C3`; `x2_` to `B6:C6`; `x3_` to `B9:C9`. Enter the formulae "`=g_1_on_range(x1_)`" in `B4`, "`=g_2_on_range(x2_)`" in `B7`, "`=g_3_on_range(x3_)`" in `B10`. Add to the module the code to run the SOLVER:

```
Sub run_solver()
col = Cells(1, Cells.Columns.Count).End(xlToLeft).Column
ruad = Range(Cells(2, 2), Cells(2, col)).Address
rgad = Range(Cells(4, 2), Cells(4, 2)).Address
rfobjad = Range(Cells(4, 4), Cells(4, 4)).Address
Call rsolv(col, ruad, rgad, rfobjad)

ruad = Range(Cells(5, 2), Cells(5, col)).Address
rgad = Range(Cells(7, 2), Cells(7, 2)).Address
rfobjad = Range(Cells(7, 4), Cells(7, 4)).Address
Call rsolv(col, ruad, rgad, rfobjad)

ruad = Range(Cells(8, 2), Cells(8, col)).Address
rgad = Range(Cells(10, 2), Cells(10, 2)).Address
rfobjad = Range(Cells(10, 4), Cells(10, 4)).Address
Call rsolv(col, ruad, rgad, rfobjad)

n1 = Cells(4, 4).Value
n2 = Cells(7, 4).Value
n3 = Cells(10, 4).Value
If n1 < n2 Then
   If n1 < n3 Then
     ind = 4
   Else
     ind = 10
   End If
Else
  If n2 < n3 Then
    ind = 7
  Else
    ind = 10
  End If
End If
Cells(12, 2).Value = Cells(ind, 4).Value
For i = 2 To col
  Cells(11, i).Value = Cells(ind - 2, i).Value
Next i
End Sub
```

This code uses.

```
Sub rsolv(col, ruad, rgad, rfobjad)
SolverReset
  SolverOk SetCell:=rfobjad, MaxMinVal:=2, ValueOf:=0, ByChange:=ruad, _
         Engine:=1, EngineDesc:="GRG Nonlinear"
  SolverAdd Cellref:=rgad, Relation:=2, FormulaText:="0"
  SolverOptions AssumeNonNeg:=False
  SolverOptions Multistart:=True, PopulationSize:=500, RequireBounds:=False
  SolverSolve True
End Sub
```

Run the Sub run_solver. The results appear:

	A	B	C	D
1	x	0	0	
2	u1*	0	0	
3	x1*	0	0	
4	h1(x,u1*)	-3	||u1*||	0
5	u2*	0	0	
6	x2*	0	0	
7	h2(x,u2*)	-2	||u2*||	0
8	u3*	0	0	
9	x3*	0	0	
10	h3(x,u3*)	-6	||u3*||	0
11	u*			
12	||u*||			

(a) Before running the Sub

	A	B	C	D
1	x	0	0	
2	u1*	-0,50011	2,749997	
3	x1*	-0,50011	2,749997	
4	h1(x,u1*)	-3E-06	||u1*||	2,795101
5	u2*	1,762231	-0,38941	
6	x2*	1,762231	-0,38941	
7	h2(x,u2*)	0	||u2*||	1,804742
8	u3*	1,302775	1,302775	
9	x3*	1,302775	1,302775	
10	h3(x,u3*)	-5,4E-06	||u3*||	1,842402
11	u*	1,762231	-0,38941	
12	||u*||	1,804742		

(b) After running the Sub

The minimal value is $\beta(\boldsymbol{x}) = \beta_2(\boldsymbol{x}) = 1.804742$.

Exercises

1. Let $\boldsymbol{x} \in \mathbb{R}^2$ and $g(\boldsymbol{x}) = x_2 - x_1^2 - x_1 - 3$. Determine the design point for $\boldsymbol{x}^{(1)} = (1, 0)$, $\boldsymbol{x}^{(3)} = (1, 1)$, $\boldsymbol{x}^{(4)} = (0, 1)$. The design point for $\boldsymbol{x}^{(2)} = (0, 0)$ was determined in the example above.
2. Let $\boldsymbol{x} \in \mathbb{R}^2$ and $g(\boldsymbol{x}) = x_2 - x_1^3 - x_1^2 - 4$. Determine the design point for $\boldsymbol{x}^{(1)} = (1, 0)$, $\boldsymbol{x}^{(3)} = (1, 1)$, $\boldsymbol{x}^{(4)} = (0, 1)$. The design point for $\boldsymbol{x}^{(2)} = (0, 0)$ was determined in the example above.
3. Let $\boldsymbol{x} \in \mathbb{R}^2$ and $g(\boldsymbol{x}) = x_1x_2 - 2$. Determine the design point for $\boldsymbol{x}^{(1)} = (1, 0)$, $\boldsymbol{x}^{(2)} = (0, 0)$, $\boldsymbol{x}^{(3)} = (1, 1)$, $\boldsymbol{x}^{(4)} = (0, 1)$.
4. Let $\boldsymbol{x} \in \mathbb{R}^2$ and $g(\boldsymbol{x}) = x_1^2 + x_2^2 - 4$. Determine the design point for $\boldsymbol{x}^{(1)} = (1, 0)$, $\boldsymbol{x}^{(2)} = (0, 0)$, $\boldsymbol{x}^{(3)} = (1, 1)$, $\boldsymbol{x}^{(4)} = (0, 1)$.
5. Let $\boldsymbol{x} \in \mathbb{R}^3$ and $g(\boldsymbol{x}) = x_2 + 1 - 3(2x_3 + 1)^2(3x_1 + 1)$. Determine the design point for $\boldsymbol{x}^{(1)} = (1, 0, 0)$, $\boldsymbol{x}^{(2)} = (0, 0, 0)$, $\boldsymbol{x}^{(3)} = (0, 1, 0)$, $\boldsymbol{x}^{(4)} = (0, 0, 1)$.

6. Let $\boldsymbol{x} \in \mathbb{R}^2$ and $g_1(\boldsymbol{x}) = x_1 - x_2^2 - x_2 - 2$, $g_2(\boldsymbol{x}) = x_2 - x_1^3 - x_1^2 - 4$, $g_2(\boldsymbol{x}) = x_1 + x_2 + x_1^2 + x_2^2 - 6$. Determine the design point for $\boldsymbol{x}^{(1)} = (1, 0)$, $\boldsymbol{x}^{(2)} = (0, 0)$, $\boldsymbol{x}^{(3)} = (1, 1)$, $\boldsymbol{x}^{(4)} = (0, 1)$.

10.4 Reliability Analysis

When uncertainties must be taken into account, $\boldsymbol{x}$ becomes a random variable $\boldsymbol{X}$. For example, we can consider that $\boldsymbol{x}$ is the vector of the nominal values of the design variables and $\boldsymbol{U}$ and the uncertainty – in such a situation, the mean of $\boldsymbol{U}$ is zero. As in the deterministic case, we can consider the more general case where $\boldsymbol{X} = \phi(\boldsymbol{x}, \boldsymbol{U})$ – with a possibly non-linear relationship. In the context of uncertainty, the centered space becomes a space of random variables, while $\boldsymbol{x}$ remains an element of $\mathbb{R}^n$ – the pair $(\boldsymbol{x}, \boldsymbol{U})$ is formed by two objects of distinct natures: in the context of reliability, we say that this pair is an element of the *hybrid space*, that is, a space formed by vectors of $\mathbb{R}^n$ and random variables.

In this case, $Z = H(\boldsymbol{x}, \boldsymbol{U}) = g(\phi(\boldsymbol{x}, \boldsymbol{U}))$ – notice that H is defined on the hybrid space. Z is a random variable and the probability of failure at a point $\boldsymbol{x}$ on the design space is $F = P(Z > 0)$. Similarly, the reliability of the same point is $R = P(Z \leq 0)$: the determination of F and R involve calculations in the centered space, that is, the space of the random variables $\boldsymbol{U}$.

In the context of structural reliability, the terminology *reliability analysis of the point* $\boldsymbol{x}$ of the design space refers to the numerical estimation of the probability of failure associated with $\boldsymbol{x}$. Thus, reliability analysis consists of the numerical determination of $F = P(Z > 0)$ – of course, it is equivalent to determine $R = P(Z \leq 0)$. However, the determination of one of these probabilities involves numerous practical difficulties. For example, we need information about the joint distribution of the components of $\boldsymbol{U}$ (or – equivalently – of $\boldsymbol{X}$). Indeed, if f_X denotes the PDF of X, then we have

$$F = P(Z > 0) = \int_{\mathcal{F}} f_X(\boldsymbol{x}) d\boldsymbol{x}, R = P(Z \leq 0) = \int_{\mathcal{SUC}} f_X(\boldsymbol{x}) d\boldsymbol{x}. \tag{10.7}$$

Thus, the knowledge of f_X is necessary for the determination of these values. Moreover, these values can rarely be determined analytically and must generally be evaluated using numerical approximations, e.g., Monte Carlo simulation – with the additional difficulty that the values of the probabilities of failure probabilities are generally small and that we must use point generation methods adapted to the evaluation of small probabilities, whose order of magnitude is negative: in general, $F \lll 1$. This difficulty makes that, in practice, it is more convenient to use a reliability index to estimate the probability of failure.

Example
Let us consider the situation where $\boldsymbol{x} \in \mathbb{R}^2$ and $g(\boldsymbol{x}) = x_2 - x_1^2 - x_1 - 3$. Consider $\boldsymbol{x} = (x_1, x_2)$, $\phi(\boldsymbol{x}, \boldsymbol{u}) = \boldsymbol{u} + \boldsymbol{x}$. We have $h(\boldsymbol{x}, \boldsymbol{u}) = g(\boldsymbol{u} + \boldsymbol{x}) = u_2 + x_2 - u_1^2 - 2u_1x_1 - x_1^2 - u_1 - x_1 - 3$, id est,

$$h(\boldsymbol{x},\boldsymbol{u}) = u_2 - u_1^2 - u_1 - 2u_1x_1 + g(\boldsymbol{x}).$$

Consider the point $\boldsymbol{x}^{(2)} = (0, 0)$. Then $h(\boldsymbol{x}^{(2)},\boldsymbol{u}) = u_2 - u_1^2 - u_1 - 3$. The probability of failure is

$$F^{(2)} = P\left(h\left(\boldsymbol{x}^{(2)},\boldsymbol{u}\right) > 0\right).$$

Assume that u_1, u_2 are independent random variables having the same distribution $N(0, 1)$. We can evaluate the probability of failure by a Monte Carlo method. For instance, let us generate two samples of ns variates for u_1 and u_2. The, we consider the sample of ns^2 elements formed by $\boldsymbol{u}_{ij} = (u_{1,\,i}, u_{2,\,j})$, $1 \leq i, j \leq ns$. We count the number n_+ of pairs in the sample such that $h(\boldsymbol{x}^{(2)}, \boldsymbol{u}_{ij}) > 0$ and we evaluate the probability of failure as $F \approx n_+/ns^2$. Since the result depends on the sample used, we can consider nr realizations of F, denoted $F_1, \ldots, F_{nr}$, and evaluate $F \approx \overline{F_{nr}}$, the empirical mean of the values obtained. Typical results are shown in the Tables 10.1 and 10.2 below. A run with $ns = 1e5$ furnished $F \approx 0,0011$.

Table 10.1 Monte Carlo Evaluation of the probability of Failure using nr = 10

ns	100	1000	10,000
$\overline{F_{nr}}$	0,002	0,0006	0,0009
s_{nr}	0,004	0,0004	0,0001

Table 10.2 Monte Carlo Evaluation of the probability of Failure using nr = 100

ns	100	1000	10,000
$\overline{F_{nr}}$	0,001	0,0011	0,0009
s_{nr}	0,002	0,0006	0,0002

The main difficulty in this evaluation is the need of the use of exceptionally large samples, due to the smallness of F. Special sampling methods can be used to diminish the computational cost. For instance, importance sampling can be used. (Li & Wu, 2007; Papaioannou et al., 2016).

Example

Let $\boldsymbol{x} \in \mathbb{R}^2$ and $g(\boldsymbol{x}) = x_2 - x_1^3 - x_1^2 - 4$. Consider $\boldsymbol{x} = (x_1, x_2)$, $\phi(\boldsymbol{x}, \boldsymbol{u}) = \boldsymbol{u} + \boldsymbol{x}$. We have

$$h(\boldsymbol{x},\boldsymbol{u}) = u_2 - u_1^3 - 3u_1^2x_1 - 3u_1x_1^2 - u_1^2 - 2u_1x_1 + g(\boldsymbol{x}).$$

Consider the point $\boldsymbol{x}^{(2)} = (0, 0)$. Then $h(\boldsymbol{x}^{(2)},\boldsymbol{u}) = u_2 - u_1^3 - u_1^2 - 4$. Assume that u_1, u_2 are independent random variables having the same distribution $N(0, 1)$.

Analogously to the preceding example, we can evaluate the probability of failure by the Monte Carlo method presented in the preceding example. Typical results are shown in the Tables 10.3 and 10.4 below. A run with $ns = 1e5$ furnished $F \approx 0{,}024$.

Table 10.3 Monte Carlo Evaluation of the probability of Failure using nr = 10

ns	100	1000	10,000
$\overline{F_{nr}}$	0,023	0,026	0,024
s_{nr}	0,014	0,005	0,001

Table 10.4 Monte Carlo Evaluation of the probability of Failure using nr = 100

ns	100	1000	10,000
$\overline{F_{nr}}$	0,023	0,024	0,024
s_{nr}	0,014	0,005	0,001

Exercises

1. Let $\boldsymbol{x} \in \mathbb{R}^2$ and $g(\boldsymbol{x}) = x_2 - x_1^2 - x_1 - 3$. Assuming normality $N(0, 1)$, generate two samples of 1000 variates of u_1, u_2 and use it to evaluate the reliability of each point: $\boldsymbol{x}^{(1)} = (1, 0)$, $\boldsymbol{x}^{(3)} = (1, 1)$, $\boldsymbol{x}^{(4)} = (0, 1)$. $\boldsymbol{x}^{(2)} = (0, 0)$ was analyzed in an example above.
2. Let $\boldsymbol{x} \in \mathbb{R}^2$ and $g(\boldsymbol{x}) = x_2 - x_1^3 - x_1^2 - 4$. Assuming normality $N(0, 1)$, generate two samples of 1000 variates of u_1, u_2 and use it to evaluate the reliability of each point: $\boldsymbol{x}^{(1)} = (1, 0)$, $\boldsymbol{x}^{(3)} = (1, 1)$, $\boldsymbol{x}^{(4)} = (0, 1)$. $\boldsymbol{x}^{(2)} = (0, 0)$ was analyzed in an example above.
3. Let $\boldsymbol{x} \in \mathbb{R}^2$ and $g(\boldsymbol{x}) = x_1 x_2 - 2$. Assuming normality $N(0, 1)$, generate two samples of 1000 variates of u_1, u_2 and use it to evaluate the reliability of each point: $\boldsymbol{x}^{(1)} = (1, 0)$, $\boldsymbol{x}^{(2)} = (0, 0)$, $\boldsymbol{x}^{(3)} = (1, 1)$, $\boldsymbol{x}^{(4)} = (0, 1)$.
4. Let $\boldsymbol{x} \in \mathbb{R}^2$ and $g(\boldsymbol{x}) = x_1^2 + x_2^2 - 4$. Assuming normality $N(0, 1)$, generate two samples of 1000 variates of u_1, u_2 and use it to evaluate the reliability of each point: $\boldsymbol{x}^{(1)} = (1, 0)$, $\boldsymbol{x}^{(2)} = (0, 0)$, $\boldsymbol{x}^{(3)} = (1, 1)$, $\boldsymbol{x}^{(4)} = (0, 1)$.
5. Let $\boldsymbol{x} \in \mathbb{R}^3$ and $g(\boldsymbol{x}) = x_2 + 1 - 3(2x_3 + 1)^2(3x_1 + 1)$. Assuming normality $N(0, 1)$, generate three samples of 100 variates of u_1, u_2, u_3 and use it to evaluate the reliability of each point: $\boldsymbol{x}^{(1)} = (1, 0, 0)$, $\boldsymbol{x}^{(2)} = (0, 0, 0)$, $\boldsymbol{x}^{(3)} = (0, 1, 0)$, $\boldsymbol{x}^{(4)} = (0, 0, 1)$.

10.5 Hasofer-Lind Reliability Index

Among the existing reliability indexes, one of the most popular is the Hasofer-Lind reliability index – the distance between a point in the design space and the failure region, calculated through the design point – the MPP or MPFP.

10.5.1 The General Case

The Hasofer-Lind reliability index of a design point $\boldsymbol{x}$ is $\beta(\boldsymbol{x})$ given by

$$\beta(\boldsymbol{x}) = \|\boldsymbol{u}^*(\boldsymbol{x})\| = \min\{\|\boldsymbol{u}\| : H(\boldsymbol{x},\boldsymbol{u}) \geq 0\}. \tag{10.8}$$

For $x \in C$, an alternative form is

$$\beta(\boldsymbol{x}) = \|\boldsymbol{u}^*(\boldsymbol{x})\| = \min\{\|\boldsymbol{u}\| : H(\boldsymbol{x}, \boldsymbol{u}) = 0\}. \tag{10.9}$$

Three interesting properties of the Hasofer-Lind reliability index are

- The collinearity between the design point $\boldsymbol{u}^* = \boldsymbol{u}^*(\boldsymbol{x})$ and the normal $\boldsymbol{n}^*$ to the limit state curve at the point $\boldsymbol{u}^*$: $\boldsymbol{u}^*(\boldsymbol{x}) = \beta(\boldsymbol{x})\boldsymbol{n}^*$;
- an upper bound for the failure probability F: $F \le P(\|\boldsymbol{u}\| \ge \beta(\boldsymbol{x}))$;
- a nonlinear system of equations defining $\boldsymbol{u}^*$ and β:

$$\boldsymbol{u}^*\|\nabla_u \boldsymbol{H}(\boldsymbol{x}, \boldsymbol{u}^*)\| = \beta \nabla_u \boldsymbol{H}(\boldsymbol{x}, \boldsymbol{u}^*), H(\boldsymbol{x}, \boldsymbol{u}^*) = 0 \tag{10.10}$$

Let us examine the first property: if $\beta(\boldsymbol{x}) > 0$ and H is a differentiable function, then $\boldsymbol{u}^*(\boldsymbol{x})$ is collinear to the unit normal to the limit state curve at $\boldsymbol{u}^*(\boldsymbol{x})$, id est,

$$\boldsymbol{u}^*(\boldsymbol{x}) = \beta(\boldsymbol{x})\boldsymbol{n}^*, \boldsymbol{n}^* = \frac{\boldsymbol{d}^*}{\|\boldsymbol{d}^*\|}, \ \boldsymbol{d}^* = \nabla_u \boldsymbol{H}(\boldsymbol{x}, \boldsymbol{u}^*(\boldsymbol{x})) \ ,$$

$$\nabla_u \boldsymbol{H}(\boldsymbol{x}, \boldsymbol{u}) = \left(\frac{\partial H}{\partial u_1}(\boldsymbol{x}, \boldsymbol{u}), \ldots, \frac{\partial H}{\partial u_n}(\boldsymbol{x}, \boldsymbol{u}) \right) .$$

Indeed, let λ be the Lagrange's multiplier associated to the condition :$H(\boldsymbol{x}, \boldsymbol{u}) \ge 0$. Then, the KKT equations show that (see Sect. 2.2)

$$2\boldsymbol{u}^* - \lambda \boldsymbol{d}^* = 0 \ , \lambda \ge 0, H(\boldsymbol{x}, \boldsymbol{u}^*) \ge 0, \lambda H(\boldsymbol{x}, \boldsymbol{u}^*) = 0.$$

Thus, on the one hand, $\lambda > 0$ (otherwise, $\boldsymbol{u}^* = \boldsymbol{0} \Rightarrow \beta(\boldsymbol{x}) = 0$) and, on the other hand (Fig. 10.8),

$$\boldsymbol{u}^* = \frac{\lambda}{2}\boldsymbol{d}^* \Rightarrow \beta(\boldsymbol{x}) = \|\boldsymbol{u}^*\| = \frac{\lambda}{2}\|\boldsymbol{d}^*\|.$$

Consequently,

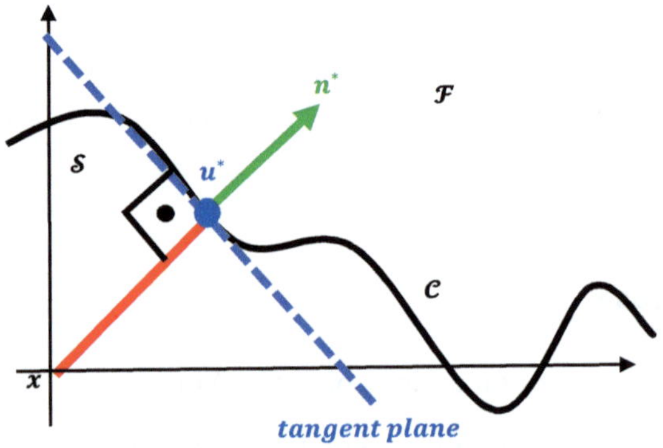

Fig. 10.8 Collinearity between the design point and the normal

$$u^* = \beta(x)\,\frac{d^*}{\|d^*\|} = \beta(x)n^*.$$

The second useful property of the Hasofer-Lind reliability index is an upper bound of the probability of failure: let $A(r)$ be the exterior of the open ball of radius r:

$$A(r) = \{u : \|u\| \geq r\}.$$

We have

$$r_1 > r_2 \Rightarrow A(r_1) \subset A(r_2) \Rightarrow P(A(r_1)) \leq P(A(r_2)).$$

Thus, $r \longrightarrow P(A(r))$ is decreasing. Let us denote $A^*(x) = A(\beta(x))$. We have

$$u \in \mathcal{F} \Rightarrow \|u\| \geq \|u^*\| = \beta(x) \Rightarrow u \in A^*(x).$$

Consequently,

$$\mathcal{F} \subset A^*(x) \Rightarrow F = P(\mathcal{F}) \leq P(A^*(x)).$$

The third property yields from the fact that u^* is a point of the common boundary of $\mathcal{S}$ and $\mathcal{F}$, which is $\mathcal{C}$. Thus,

$$H(x, u^*) = 0 \;\; and\, u^* = \beta n^* = \beta\frac{\nabla_u H(x, u^*)}{\|\nabla_u H(x, u^*)\|}. \quad (10.11)$$

Example

Let us consider the situation where $x \in \mathbb{R}^2$, $g(x) = x_2 - x_1^2 - x_1 - 3$, $\phi(x, u) = u + x$, $h(x, u) = u_2 - u_1^2 - u_1 - 2u_1x_1 + g(x)$.

Then,

$$\nabla_u H(x, u) = (-2u_1 - 2x_1 - 1, 1),$$

Then, u^* is the solution of Eq. (10.11), which read as

$$h(x, u^*) = 0$$

$$u^* = \beta n(x, u^*)$$

We can look for the solution of these equations using the SOLVER. Open a blank workbook, add a module, and enter the code below:

```
Function eqs_on_range(rx As Range, ru As Range, rbeta As Range)
x = range2vector(rx)
u = range2vector(ru)
beta = rbeta.Value
eqs_on_range = eqs(x, u, beta)
End Function
```

```
Function g(x)
g = x(2) - x(1) ^ 2 - x(1) - 3
End Function
Function h(x, u)
y = plus(x, u)
h = g(y)
End Function
Function h_on_range(rx As Range, ru As Range)
x = range2vector(rx)
u = range2vector(ru)
h_on_range = h(x, u)
End Function
Function gradg(x)
Dim gg(1 To 2) As Double
gg(1) = -2 * x(1) - 1
gg(2) = 1
gradg = gg
End Function
```

```
Function gradh(x, u)
y = plus(x, u)
gh = gradg(y)
gradh = gh
End Function
Function vecn(x, u)
gh = gradh(x, u)
s = norm2(gh)
lbda = 1 / s
aux = sm(gh, lbda)
vecn = aux
End Function
Function eqs(x, u, beta)
nn = vecn(x, u)
Dim aux(1 To 3)
aux(1) = u(1) - beta * nn(1)
aux(2) = u(2) - beta * nn(2)
aux(3) = h(x, u)
eqs = aux
End Function
```

This code uses functions `plus(x,u)`, which calculates $x + u$; `sm(x,a)`, which calculates ax; `norm2(x)` which calculates $\|\boldsymbol{x}\| = \sqrt{\sum x_i^2}$. `gradg` evaluates the gradient of g; `gradh` evaluates $\nabla_u H(\boldsymbol{x}, \boldsymbol{u})$; `gradg` evaluates the gradient of g; `vecn` evaluates $\boldsymbol{n}(\boldsymbol{x}, \boldsymbol{u})$; `eqs` evaluates Eq. (10.11).

Go to the worksheet and enter the coordinates of the point $\boldsymbol{x}$ and a guess for $\boldsymbol{u}^*$:

	A	B	C
1	x	0	0
2	u*	1	0

Give the name x to B1:C1 and the name u to B2:C2. Go to cell B3 and enter "=x + u". Go to B4 and enter "=h_on_range(x;u)". Go to B5 and enter a guess for β. Give the name beta to B5. Go to B6 and enter the formula "=eqs_on_range(x;u)". Go to B7 and enter "=SumSq(B6:D6)". The workbook appears as.

Go to B5 and enter a guess for β. Give the name beta to B5. Go to B6 and enter the formula "=eqs_on_range(x;u)". Go to B7 and enter "=SumSq(B6:D6)". The workbook appears as

	A	B	C	D
1	x	0	0	
2	u*	1	0	
3	y	1	0	
4	h	-5		
5	beta	1		
6	eqs	1,948683	-0,31623	-5
7	norm	28,89737		

Now, run the SOLVER to minimize B7 under the constraint B4 = 0. You can do it manually or using a Sub. An example of Sub to achieve the minimization is.

```
Sub run_solver()
col = Cells(2, Cells.Columns.Count).End(xlToLeft).Column
ruad = Range(Cells(2, 2), Cells(2, col)).Address & ";" & Range(Cells(5, 2), Cells(5, 2)).Address
rfobjad = Range(Cells(7, 2), Cells(7, 2)).Address
  SolverReset
  SolverOk SetCell:=rfobjad, MaxMinVal:=2, ValueOf:=0, ByChange:=ruad, _
        Engine:=1, EngineDesc:="GRG Nonlinear"
  SolverAdd Cellref:=Range(Cells(4, 2), Cells(4, 2)).Address, Relation:=2, FormulaText:="0"
  SolverOptions AssumeNonNeg:=False
  SolverOptions Multistart:=True, PopulationSize:=500, RequireBounds:=False
  SolverSolve True
End Sub
```

Running this Sub produces the result at right, which is the same found page 447:

	A	B	C	D
1	x	0	0	
2	u*	-0,42322	2,755895	
3	y	-0,42322	2,755895	
4	h	-3,1E-07		
5	beta	2,7882		
6	eqs	-4,7E-07	2,14E-06	-3,1E-07
7	norm	4,89E-12		

Notice that you can reduce the variables, by eliminating β. Indeed, $\beta = \|\boldsymbol{u}^*\|$, so that we can consider the equations

$$h(\boldsymbol{x}, \boldsymbol{u}^*) = 0, \quad \frac{\boldsymbol{u}^*}{\|\boldsymbol{u}^*\|} = \boldsymbol{n}(\boldsymbol{x}, \boldsymbol{u}^*)$$

In this case, you must modify the code of eqs and eqs_on_range as follows:

```
Function eqs(x, u)
nn = vecn(x, u)
ss = norm2(u)
sss = ss + 0.000000000000001
aux = minus(sm(u, 1 / sss), nn)
eqs = aux
End Function
```

```
Function eqs_on_range(rx As Range, ru As Range)
x = range2vector(rx)
u = range2vector(ru)
eqs_on_range = eqs(x, u)
End Function
```

You must modify also cells B5 and B6. Go to B5 and enter "=SQRT(SUMSQ(B2:C2))". Go to B6 and enter "=eqs(x;u)".
The Workbook appears as at right.

	A	B	C
1	x	0	0
2	u*	1	0
3	y	1	0
4	h	-5	
5	||u*||^2	1	
6	eqs	1,948683	-0,31623
7	norm	3,897367	

Now, the macro to run the SOLVER reads as:

```
Sub run_solver()
col = Cells(2, Cells.Columns.Count).End(xlToLeft).Column
ruad = Range(Cells(2, 2), Cells(2, col)).Address
rfobjad = Range(Cells(7, 2), Cells(7, 2)).Address
  SolverReset
  SolverOk SetCell:=rfobjad, MaxMinVal:=2, ValueOf:=0, ByChange:=ruad, _
          Engine:=1, EngineDesc:="GRG Nonlinear"
  SolverAdd Cellref:=Range(Cells(4, 2), Cells(4, 2)).Address, Relation:=2, FormulaText:="0"
  SolverOptions AssumeNonNeg:=False
  SolverOptions Multistart:=True, PopulationSize:=500, RequireBounds:=False
  SolverSolve True
End Sub
```

Running this Sub or executing manually the Solver produces the same results at right, which coincide with the preceding ones.

	A	B	C
1	x	0	0
2	u*	-0,42322	2,755896
3	y	-0,42322	2,755896
4	h	3,87E-07	
5	||u*||^2	2,788203	
6	eqs	-1E-15	-5,6E-16
7	norm	1,31E-30	

As previously observed, the use of VBA makes the modifications easier. For instance, if we want to consider $g(\boldsymbol{x}) = x_2 - x_1^3 - x_1^2 - 4$, we need just to modify the codes of g and gradg

```
Function g(x)
g = x(2) - x(1) ^ 3 - x(1) ^ 2 - 4
End Function
```

```
Function gradg(x)
Dim gg(1 To 2) As Double
gg(1) = -3 * x(1) ^ 2 - 2 * x(1)
gg(2) = 1
gradg = gg
End Function
```

With this single modification, we execute the macro run_solver and we obtain new results, close to those found page 448.

	A	B	C
1	x	0	0
2	u*	-3,7E-10	4
3	y	-3,7E-10	4
4	h	-1,2E-11	
5	||u*||^2	4	
6	eqs	-8,4E-10	-1,1E-16
7	norm	7,07E-19	

One of the difficulties when using this approach is the convergence to suboptimal solutions. Indeed, you can get convergence to a different point, with a larger value of $\|\boldsymbol{u}^*\|$. Due to nonconvexity, the result may depend on the choice of the starting point. For instance, if the initial point is (1,0) the results are different in this last situation (at right).

	A	B	C
1	x	0	0
2	u*	-0,74716	4,141152
3	y	-0,74716	4,141152
4	h	4,61E-06	
5	||u*||^2	4,208015	
6	eqs	9,99E-16	0
7	norm	9,98E-31	

Notice that the solution of the Eq. (10.11) can be made by other methods if you do not desire or cannot use the SOLVER. For instance, you can simply use a fixed-point procedure which reads as follows:

$$\beta^{(n+1)} = \beta^{(n)} - \omega H\left(\boldsymbol{x}, \boldsymbol{u}^{(\mathrm{n})}\right),$$

$$\boldsymbol{u}^{(\mathrm{n}+1)} = \boldsymbol{u}^{(\mathrm{n})} - \omega\left(\boldsymbol{u}^{(\mathrm{n})} - \beta^{(n)} \frac{\nabla_{\boldsymbol{u}} H(\boldsymbol{x}, \boldsymbol{u}^{(\mathrm{n})})}{\|\nabla_{\boldsymbol{u}} H(\boldsymbol{x}, \boldsymbol{u}^{(\mathrm{n})})\|}\right).$$

Starting from an initial guess $\boldsymbol{u}^{(0)}, \beta^{(0)}$, these iterations are expected to converge to a value close to the exact ones. Iterations are stopped when a maximum iteration number or a desired precision is attained.

Example

Let us consider again the situation where $\boldsymbol{x} \in \mathbb{R}^2$, $g(\boldsymbol{x}) = x_2 - x_1^2 - x_1 - 3$, $\phi(\boldsymbol{x}, \boldsymbol{u}) = \boldsymbol{u} + \boldsymbol{x}$, $h(\boldsymbol{x}, \boldsymbol{u}) = u_2 - u_1^2 - u_1 - 2u_1x_1 + g(\boldsymbol{x})$,

$$\nabla_{\boldsymbol{u}} H(\boldsymbol{x}, \boldsymbol{u}) = (-2u_1 - 2x_1 - 1, 1),$$

Take the workbook created in the preceding example, add the following Sub to the existing module:

```
Sub iter_solver()
om = Cells(9, 2).Value
nitmax = Cells(10, 2).Value
precmin = Cells(11, 2).Value
tolmin = Cells(12, 2).Value
x = read_line(1, 2)
u = read_line(2, 2)
Dim stopping As Boolean
stopping = False
nit = 0
prec = 1 + precmin
tol = 1 + tolmin
While Not stopping
   nit = nit + 1
   eq = eqs(x, u, beta)
   tol = norm2(eq)
   u(1) = u(1) - om * eq(1)
   u(2) = u(2) - om * eq(2)
   beta = beta - om * eq(3)
   prec = norm2(eq)
   Cells(10, 4).Value = nit
   Cells(11, 4).Value = prec
   Cells(12, 4).Value = tol
   stopping = nit > nitmax Or tol < tolmin Or prec < precmin
Wend
Cells(2, 2).Value = u(1)
Cells(2, 3).Value = u(2)
End Sub
```

Select cell B9 and enter the value of ω, go to B10 and enter the maximum iteration number. In B11, enter the minimal step (minimal difference in norm between two iterates):

$$N_1 = \left(\left\| \boldsymbol{u}^{(n+1)} - \boldsymbol{u}^{(n)} \right\|^2 + \left(\beta^{(n+1)} - \beta^{(n)} \right)^2 \right)^{\frac{1}{2}}$$

In B12, enter the desired maximal value for the RMS norm of the equations:

$$N_2 = \left(\left\| \boldsymbol{u}^{(n)} - \beta^{(n)} \frac{\nabla_{\boldsymbol{u}} H\left(\boldsymbol{x}, \boldsymbol{u}^{(n)}\right)}{\left\| \nabla_{\boldsymbol{u}} H\left(\boldsymbol{x}, \boldsymbol{u}^{(n)}\right) \right\|} \right\|^2 + H\left(\boldsymbol{x}, \boldsymbol{u}^{(n)}\right)^2 \right)^{\frac{1}{2}}$$

The iterations are stopped if the iteration number is larger than B10 or N_1 is inferior to B11 or N_2 is inferior to B12. The workbook appears as shown at left. Run the Sub iter_solver. The results appear as at right. The results are close to the preceding ones (Fig. 10.9).

a)

	A	B	C	D
1	x	0	0	
2	u*	0	0	
3	y	0	0	
4	h	-3		
5	beta			
6	eqs	0	0	-3
7	norm	9		
8				
9	om	0,01		
10	nitmax	10000	nit	
11	precmin	1,00E-08	prec	
12	tolmin	1,00E-05	tol	

b)

	A	B	C	D
1	x	0	0	
2	u*	-0,4232163	2,755904	
3	y	-0,4232163	2,755904	
4	h	7,98481E-06		
5	beta	2,788204523		
6	eqs	-2,1E-07	5,94E-06	7,98E-06
7	norm	9,90428E-11		
8				
9	om	0,01		
10	nitmax	10000	nit	2593
11	precmin	1,00E-08	prec	9,99E-06
12	tolmin	1,00E-05	tol	9,99E-06

Fig. 10.9 The workbook before and after running the `Sub`. (**a**) Before running `iter_solver`. (**b**) After running `iter_solver`

Now, modify `g` to consider the case where $g(\boldsymbol{x}) = x_2 - x_1^3 - x_1^2 - 4$ (see preceding examples). The results are shown below (Fig. 10.10):

a)

	A	B	C	D
1	x	0	0	
2	u*	1	0	
3	y	1	0	
4	h	-6		
5	beta			
6	eqs	1	0	-6
7	norm	37		
8				
9	om	0,01		
10	nitmax	10000	nit	
11	precmin	1,00E-08	prec	
12	tolmin	1,00E-05	tol	

b)

	A	B	C	D
1	x	0	0	
2	u*	1,1731E-110	4,000006	
3	y	1,1731E-110	4,000006	
4	h	5,57955E-06		
5	beta	3,999997425		
6	eqs	1,0558E-109	8,15E-06	5,58E-06
7	norm	9,76232E-11		
8				
9	om	0,01		
10	nitmax	10000	nit	2665
11	precmin	1,00E-08	prec	9,95E-06
12	tolmin	1,00E-05	tol	9,95E-06

Fig. 10.10 The workbook before and after running the `Sub`. (**a**) Before running `iter_solver`. (**b**) After running `iter_solver`

Multiple failure conditions can be considered by the same way as shown in Sect. 10.3.1.

Example

Let us consider the situation where $\mathbf{x} \in \mathbb{R}^2$ and we have three failure conditions, corresponding to $g_1(\boldsymbol{x}) = x_2 - x_1^2 - x_1 - 3$, $g_2(\boldsymbol{x}) = x_1 - x_2^2 - x_2 - 2$, $g_3(\boldsymbol{x}) = x_1 + x_2 + x_1^2 + x_2^2 - 6$. Let us prepare a workbook analogous to the one presented in Sect. 10.3.1. Enter x in `B1:C1` and the initial guesses for u_1^* in `B2:C2`, u_2^* in `B6:C6`, u_3^* in `B12:C12`. Give them the names `u1_`, `u2_`, `u3_`. Evaluate their norm in `D3`, `D7`, `D11`. You can use the formula "`=norm2(range2vector(u1_))`" in `D3` and analogous ones in the other cells.

Enter "`=h_on_range(x;u1_;"g_1")`" in `B3` and analogous formula in `B7`, `B11`. Enter "`=eqs_on_range(x;u1_;"gradg_1")`" in `B4` and analogous formulae in `B8`, `B12`.

	A	B	C	D
1	x	0	0	
2	u1*	0	0	
3	h	-3	||u1*||	0
4	eqs	0,707106781	-0,70711	
5	norm	1		
6	u2*	0	0	
7	h	-2	||u2*||	0
8	eqs	-0,707106781	0,707107	
9	norm	1		
10	u3*	0	0	
11	h	-6	||u3*||	0
12	eqs	-0,707106781	-0,70711	
13	norm	1		

Add the code for the evaluation of the gradients:

```
Function gradg_1(x)
Dim gg(1 To 2) As Double
gg(1) = -2 * x(1) - 1
gg(2) = 1
gradg_1 = gg
End Function
```

```
Function gradg_2(x)
Dim gg(1 To 2) As Double
gg(1) = 1
gg(2) = -2 * x(2) - 1
gradg_2 = gg
End Function
```

```
Function gradg_3(x)
Dim gg(1 To 2) As Double
gg(1) = 1 + 2 * x(1)
gg(2) = 1 + 2 * x(2)
gradg_3 = gg
End Function
```

and the for the evaluation of the equations:

```
Function eqs(x, u, dgname As String)
nn = vecn(x, u, dgname)
ss = norm2(u)
aux = minus(sm(u, 1 / (ss + 0.000000000000001)), nn)
eqs = aux
End Function
```

```
Function eqs_on_range(rx As Range, ru As Range, dgname As String)
x = range2vector(rx)
u = range2vector(ru)
eqs_on_range = eqs(x, u, dgname)
End Function
```

Add the code for the evaluation of $h(x, u)$ and its gradient for a variable failure condition:

```
Function h(x, u, gname As String)
y = plus(x, u)
h = Application.Run(gname, y)
End Function
```

```
Function h_on_range(rx As Range, ru As Range, gname As String)
x = range2vector(rx)
u = range2vector(ru)
h_on_range = h(x, u, gname)
End Function
```

```
Function gradh(x, u, dgname As String)
y = plus(x, u)
gh = Application.Run(dgname, y)
gradh = gh
End Function
```

The evaluation of the normal vector is made by the `Sub vecn`.

```
Function vecn(x, u, dgname As String)
gh = gradh(x, u, dgname)
s = norm2(gh)
lbda = 1 / s
aux = sm(gh, lbda)
vecn = aux
End Function
```

Create the `Sub` to run the SOLVER:

```
Sub rsolv(col, lf, lu, lh)
ruad = Range(Cells(lu, 2), Cells(lu, col)).Address
rfobjad = Range(Cells(lf, 2), Cells(lf, 2)).Address
  SolverReset
  SolverOk SetCell:=rfobjad, MaxMinVal:=2, ValueOf:=0, ByChange:=ruad, _
        Engine:=1, EngineDesc:="GRG Nonlinear"
  SolverAdd Cellref:=Range(Cells(lh, 2), Cells(lh, 2)).Address, Relation:=2, FormulaText:="0"
  SolverOptions AssumeNonNeg:=False
  SolverOptions Multistart:=True, PopulationSize:=500, RequireBounds:=False
  SolverSolve True
End Sub
```

```
Sub run_solver()
col = Cells(2, Cells.Columns.Count).End(xlToLeft).Column
Call rsolv(col, 5, 2, 3)
Call rsolv(col, 9, 6, 7)
Call rsolv(col, 13, 10, 11)
End Sub
```

Run the Sub. You will get the results:

	A	B	C	D
1	x	0	0	
2	u1*	-0,423216206	2,755895	
3	h	-3,61396E-07	||u1*||	2,788202
4	eqs	0	-2,2E-16	
5	norm	4,93038E-32		
6	u2*	1,762212561	-0,38949	
7	h	-1,8195E-07	||u2*||	1,804742
8	eqs	-6,66134E-16	1,11E-16	
9	norm	4,5606E-31		
10	u3*	1,302774889	1,302775	
11	h	-5,397E-06	||u3*||	1,842402
12	eqs	-4,44089E-16	-4,4E-16	
13	norm	3,9443E-31		

As we can see, the minimal value is $\|\boldsymbol{u}_2^*\| = 1,804742$. The result is the same as the one obtained in Sect. 10.3.1.

Exercises

1. Let $\boldsymbol{x} \in \mathbb{R}^2$ and $g(\boldsymbol{x}) = x_2 - x_1^2 - x_1 - 3$. Determine $\boldsymbol{u}^*$ for $\boldsymbol{x}^{(1)} = (1, 0)$, $\boldsymbol{x}^{(3)} = (1, 1)$, $\boldsymbol{x}^{(4)} = (0, 1)$. The design point for $\boldsymbol{x}^{(2)} = (0, 0)$ was determined in the example above.
2. Let $\boldsymbol{x} \in \mathbb{R}^2$ and $g(\boldsymbol{x}) = x_2 - x_1^3 - x_1^2 - 4$. Determine $\boldsymbol{u}^*$ for $\boldsymbol{x}^{(1)} = (1, 0)$, $\boldsymbol{x}^{(3)} = (1, 1)$, $\boldsymbol{x}^{(4)} = (0, 1)$. The design point for $\boldsymbol{x}^{(2)} = (0, 0)$ was determined in the example above.
3. Let $\boldsymbol{x} \in \mathbb{R}^2$ and $g(\boldsymbol{x}) = x_1x_2 - 2$. Determine $\boldsymbol{u}^*$ for $\boldsymbol{x}^{(1)} = (1, 0)$, $\boldsymbol{x}^{(2)} = (0, 0)$, $\boldsymbol{x}^{(3)} = (1, 1)$, $\boldsymbol{x}^{(4)} = (0, 1)$.
4. Let $\boldsymbol{x} \in \mathbb{R}^2$ and $g(\boldsymbol{x}) = x_1^2 + x_2^2 - 4$. Determine $\boldsymbol{u}^*$ for $\boldsymbol{x}^{(1)} = (1, 0)$, $\boldsymbol{x}^{(2)} = (0, 0)$, $\boldsymbol{x}^{(3)} = (1, 1)$, $\boldsymbol{x}^{(4)} = (0, 1)$.
5. Let $\boldsymbol{x} \in \mathbb{R}^3$ and $g(\boldsymbol{x}) = x_2 + 1 - 3(2x_3 + 1)^2(3x_1 + 1)$. Determine $\boldsymbol{u}^*$ for $\boldsymbol{x}^{(1)} = (1, 0, 0)$, $\boldsymbol{x}^{(2)} = (0, 0, 0)$, $\boldsymbol{x}^{(3)} = (0, 1, 0)$, $\boldsymbol{x}^{(4)} = (0, 0, 1)$.

10.5.2 The Case of Affine Limit State Equations

Let us consider the situation where $g(\boldsymbol{x}) = \boldsymbol{a}^t\boldsymbol{x} + b$, $\boldsymbol{a} \in \mathbb{R}^n$, $b \in \mathbb{R}$. In this case, $H(\boldsymbol{x}, \boldsymbol{U}) = \boldsymbol{a}^t\boldsymbol{U} + b(\boldsymbol{x})$, $b(\boldsymbol{x}) = b + \boldsymbol{a}^t\boldsymbol{x}$: we have $b(\boldsymbol{x}) < 0$ when $\boldsymbol{x} \in S$. In such a case, C is a hyperplane Π of equation $H(\boldsymbol{x}, \boldsymbol{U}) = 0$ (Fig. 10.11).

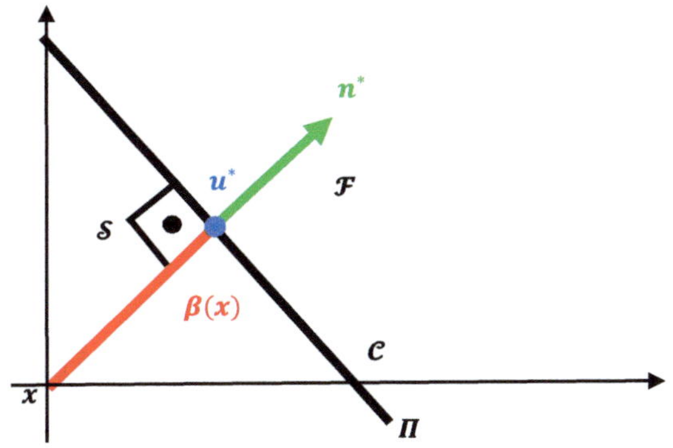

Fig. 10.11 If the limit state is a hyperplane Π, then β is the distance from Π to the origin

Let $\boldsymbol{n}^*$ be the unit normal pointing outwards the safe region: if $\boldsymbol{a} \neq \boldsymbol{0}$, then $\boldsymbol{n}^* = \boldsymbol{a}/\|\boldsymbol{a}\|$. From Eq. (10.1)

$$H(\boldsymbol{x}, \boldsymbol{u}^*) = \boldsymbol{a}^t \boldsymbol{u}^* + b(\boldsymbol{x}) = \beta(\boldsymbol{x}) \frac{\boldsymbol{a}^t \boldsymbol{a}}{\|\boldsymbol{a}\|} + b(\boldsymbol{x}) = 0.$$

Thus,

$$\beta(\boldsymbol{x}) = -\frac{b(\boldsymbol{x})}{\|\boldsymbol{a}\|}, \quad \boldsymbol{u}^* = -\frac{b(\boldsymbol{x})}{\|\boldsymbol{a}\|^2} \boldsymbol{a}. \tag{10.12}$$

Notice that $\beta(\boldsymbol{x}) > 0$ for $\boldsymbol{x} \in \mathcal{S}$.

Example
Let us consider the situation where $\boldsymbol{x} \in \mathbb{R}^2$, $g(\boldsymbol{x}) = x_2 + x_1 - 3$. Let us consider the point $\boldsymbol{x} = (0, 0)$. Here, $\boldsymbol{a} = (1, 1)^t$, $b = -3$. Thus, $b(\boldsymbol{x}) = -3$ and $\|\boldsymbol{a}\| = \sqrt{2}$. Thus,

$$\beta(\boldsymbol{x}) = \frac{3}{\sqrt{2}}, \boldsymbol{u}^*(\boldsymbol{x}) = \left(\frac{3}{2}, \frac{3}{2}\right)^t, \boldsymbol{x}^* = \left(\frac{3}{2}, \frac{3}{2}\right)^t.$$

Let us consider the point $\boldsymbol{x} = (1, 0)$. Now, $b(\boldsymbol{x}) = -2$. Thus,

$$\beta(\boldsymbol{x}) = \sqrt{2}, \boldsymbol{u}^*(\boldsymbol{x}) = (1, 1)^t, \boldsymbol{x}^* = (2, 1)^t.$$

Let us consider the point $\boldsymbol{x} = (0, 1)$. Again, $b(\boldsymbol{x}) = -2$. Thus,

$$\beta(\boldsymbol{x}) = \sqrt{2}, \boldsymbol{u}^*(\boldsymbol{x}) = (1, 1)^t, \boldsymbol{x}^* = (1, 2)^t.$$

These calculations can be made by EXCEL:

Open a blank workbook and enter the coefficients $\boldsymbol{a}$ in `B1:C1`. Give the name `a` to these cells. Enter the value of b in `B2`. Give the name `b` to `B2`. Enter the point $\boldsymbol{x}$ in cells `B3 : C3`. Give the name `x` to `B3 : C3`.

	A	B	C
1	a	1	1
2	b	-3	
3	x	0	0

Enter "`= − (SUMPRODUCT(a;x) + b)/SQRT(SUMSQ(a))`" in `B4`. Give the name `beta` to `B4`. Go to `B5` and enter "`=beta*a/SQRT(SUMSQ(a))`". Give the name u to B5: C5. Go to B6 and enter "`=x + u`".

The workbook appears as at right.

	A	B	C
1	a	1	1
2	b	-3	
3	x	0	0
4	beta	2,12132	
5	u*	1,5	1,5
6	x*	1,5	1,5

Change `B3` in `1`. The results appear:

	A	B	C
1	a	1	1
2	b	-3	
3	x	1	0
4	beta	1,414214	
5	u*	1	1
6	x*	2	1

Example

Multiple linear failure conditions can be considered by taking the minimal β, as shown in Sect. 10.3.1. For instance, let us consider the situation where $\boldsymbol{x} \in \mathbb{R}^2$, $g_1(\boldsymbol{x}) = x_2 + x_1 - 3$, $g_2(\boldsymbol{x}) = x_2 - x_1 - 1$, $g_3(\boldsymbol{x}) = -x_2 - 4$. Let us consider the point $\boldsymbol{x} = (0, 0)$. Here, $\boldsymbol{a_1} = (1, 1)^t$, $b_1 = -3$, $\boldsymbol{a_2} = (-1, 1)^t$, $b_2 = -1$, $\boldsymbol{a_3} = (0, -1)^t$, $b_3 = -4$, Thus, $b_1(\boldsymbol{x}) = -3$, $b_2(\boldsymbol{x}) = -1$, $b_3(\boldsymbol{x}) = -4$, $\|\boldsymbol{a_1}\| = \sqrt{2}$, $\|\boldsymbol{a_2}\| = \sqrt{2}$, $\|\boldsymbol{a_3}\| = 1$. Thus,

$$\beta_1(\boldsymbol{x}) = \frac{3}{\sqrt{2}}, \beta_2(\boldsymbol{x}) = \frac{1}{\sqrt{2}}, \beta_3(\boldsymbol{x}) = 4.$$

Thus, $\beta(\boldsymbol{x}) = \beta_2(\boldsymbol{x})$ and

$$\boldsymbol{u}^*(\boldsymbol{x}) = \left(-\frac{1}{2}, \frac{1}{2}\right)^t, \boldsymbol{x}^* = \left(-\frac{1}{2}, \frac{1}{2}\right)^t.$$

Exercises

1. Let $\boldsymbol{x} \in \mathbb{R}^3$ and $g(\boldsymbol{x}) = x_2 - 2x_1 - x_3 - 3$. Determine $\boldsymbol{u}^*$ for $\boldsymbol{x}^{(1)} = (1, 0, 0), \boldsymbol{x}^{(2)} = (0, 0, 0)$, $\boldsymbol{x}^{(3)} = (1, 1, 1), \boldsymbol{x}^{(4)} = (0, 1, 0)$.
2. Let $\boldsymbol{x} \in \mathbb{R}^3$ and $g(\boldsymbol{x}) = x_3 - x_1 - x_2 - 4$. Determine $\boldsymbol{u}^*$ for $\boldsymbol{x}^{(1)} = (1, 0, 0), \boldsymbol{x}^{(2)} = (0, 0, 0)$, $\boldsymbol{x}^{(3)} = (1, 1, 1), \boldsymbol{x}^{(4)} = (0, 1, 0)$.
3. Let $\boldsymbol{x} \in \mathbb{R}^4$ and $g(\boldsymbol{x}) = x_3 + 2x_4 - 3x_1 - x_2 - 4$. Determine $\boldsymbol{u}^*$ for $\boldsymbol{x}^{(1)} = (1, 0, 0, 1)$, $\boldsymbol{x}^{(2)} = (0, 0, 0, 0)$, $\boldsymbol{x}^{(3)} = (1, 0, 1, 0), \boldsymbol{x}^{(4)} = (0, 1, 0, 1)$.
4. Let $\boldsymbol{x} \in \mathbb{R}^4$ and $g(\boldsymbol{x}) = x_3 + x_4 - x_1 - x_2 - 4$. Determine $\boldsymbol{u}^*$ for $\boldsymbol{x}^{(1)} = (1, 0, 0, 1)$, $\boldsymbol{x}^{(2)} = (0, 0, 0, 0)$, $\boldsymbol{x}^{(3)} = (1, 0, 1, 0), \boldsymbol{x}^{(4)} = (0, 1, 0, 1)$.

10.5.3 The Case of a Convex Failure Region

This property extends to the situation where the failure region is convex: $\beta(\boldsymbol{x})$ corresponds to the distance between the origin and the failure region yet.

Recalling that a regular convex region is the intersection of all the hyperplanes containing it, convexity of the failure region can be exploited to characterize the reliability index as (Fig. 10.12)

$$\beta(\boldsymbol{x}) = \sup\left\{-\frac{\boldsymbol{a}^t\boldsymbol{x} + b}{\|\boldsymbol{a}\|} : \boldsymbol{a}, b; \boldsymbol{a}^t\boldsymbol{y} + b \geq 0, \forall \boldsymbol{y} \in \mathcal{F}\right\}, \tag{10.13}$$

or, equivalently,

$$\beta(\boldsymbol{x}) = \sup\left\{-\frac{\nabla g(\boldsymbol{y})^t(\boldsymbol{x} - \boldsymbol{y})}{\|\nabla g(\boldsymbol{y})\|} : \boldsymbol{y} \in \mathcal{C}\right\}. \tag{10.14}$$

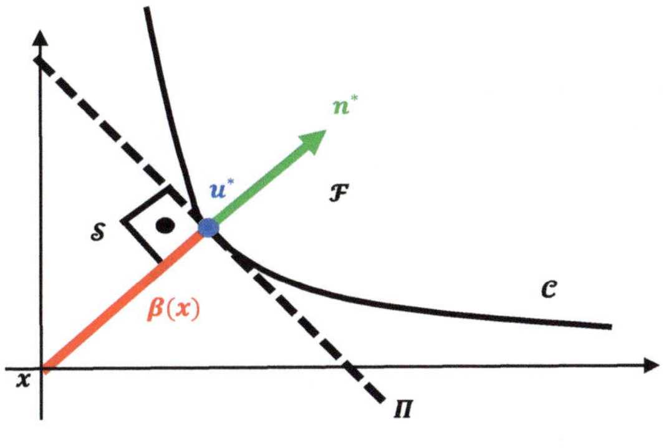

Fig. 10.12 If the failure region is convex, then then β is the distance from $\mathcal{F}$ to the origin, which coincides with the maximal distance between the origin and a hyperplane containing $\mathcal{F}$

Example

Let us consider the situation where $\boldsymbol{x} \in \mathbb{R}^2$, $g(\boldsymbol{x}) = x_2 - x_1^2 - x_1 - 3$. Let us consider the point $\boldsymbol{x} = (0, 0)$ and Eq. (10.14)

Open a blank workbook and enter the point $\boldsymbol{x}$ in B1 : C1. Give the name x to these cells. Enter an initial guess for the point $\boldsymbol{y}$ in cells B2 : C2. Give the name y to B2 : C2.

	A	B	C
1	x	0	0
2	y	0	1

Add a module to the workbook and enter the code for the evaluation of g and its gradient. Add the code for the expression inside the braces in Eq. (10.14).

```
Function g(x)
g = x(2) - x(1) ^ 2 - x(1) - 3
End Function
```

```
Function beta(x, y)
gy = gradg(y)
aux = minus(y, x)
beta = sp(aux, gy) / norm2(gy)
End Function
```

```
Function gradg(x)
Dim gg(1 To 2) As Double
gg(1) = -2 * x(1) - 1
gg(2) = 1
gradg = gg
End Function
```

This code uses sp., minus and norm2, from *VectorTools.bas*, which must be imported. sp.(x,y) evaluates the scalar product between x and y; minus(x,y) evaluates x-y, norm2(x) evaluates $\|x\| = \sqrt{\sum x_i^2}$. To execute these functions in a worksheet, you also need the functions below:

```
Function g_on_range(rx As Range)
x = range2vector(rx)
g_on_range = g(x)
End Function
```

```
Function beta_on_range(rx As Range, ry As Range)
x = range2vector(rx)
y = range2vector(ry)
beta_on_range = beta(x, y)
End Function
```

Go to the worksheet, select B3 and enter "=beta_on_range(x;y)".
Select B4 and enter "=g_on_range(y)"

	A	B	C
1	x	0	0
2	y	0	1
3	beta	0,707107	
4	g	-3	

Now, you are ready to maximize B3 under the restriction B4 inferior to zero, by changing the values of B2 : C2. You can do it manually or running a Sub:

```
Sub run_solver()
col = Cells(1, Cells.Columns.Count).End(xlToLeft).Column
rYad = Range(Cells(2, 2), Cells(2, col)).Address
rfobjad = Range(Cells(3, 2), Cells(3, 2)).Address
  SolverReset
  SolverOk SetCell:=rfobjad, MaxMinVal:=1, ValueOf:=0, ByChange:=rYad, ↵
          Engine:=1, EngineDesc:="GRG Nonlinear"
  SolverAdd Cellref:=Range(Cells(4, 2), Cells(4, 2)).Address, Relation:=2, FormulaText:="0"
  SolverOptions AssumeNonNeg:=False
  SolverOptions Multistart:=True, PopulationSize:=500, RequireBounds:=False
  SolverSolve True
End Sub
```

Running the Sub produces the results at right. The results are the same as in pages 447 and 448.

	A	B	C
1	x	0	0
2	y	-0,42322	2,755896
3	beta	2,788203	
4	g	-4,7E-11	

To determine the solution for $g(\boldsymbol{x}) = x_2 - x_1^3 - x_1^2 - 4$, we modify the evaluations of g and its gradient:

```
Function g(x)
g = x(2) - x(1) ^ 3 - x(1) ^ 2 - 4
End Function
```

```
Function gradg(x)
Dim gg(1 To 2) As Double
gg(1) = -3 * x(1) ^ 2 - 2 * x(1)
gg(2) = 1
gradg = gg
End Function
```

Running the Sub produces the results at right.

	A	B	C
1	x	0	0
2	y	-2,5E-07	4
3	beta	4	
4	g	-4,7E-12	

Example

Multiple limit states can be considered by the simple way introduced in Sect. 10.3.1. Let us consider the situation where $\boldsymbol{x} \in \mathbb{R}^2$, $g_1(\boldsymbol{x}) = x_2 - x_1^2 - x_1 - 3$, $g_2(x) = x_1 - x_2^2 - x_2 - 2$, $g_3(x) = x_1 + x_2 + x_1^2 + x_2^2 - 6$. Let us consider the point $\boldsymbol{x} = (0, 0)$ and Eq. (10.14).

Reopen the workbook of the preceding example and save it under a different name. Copy lines 2 to 4 and paste them in lines 5 to 7 and 8 to 11. Modify the formulae: give the names `y1_`, `y2_`, `y3_` to `B2:C2`, `B5:C5`, `B8:C8`. Enter the formulae "`=beta_on_range(x;y;"gradg_1")`" in `B3`, and analogous expressions in `B6`, `B9`. Enter the initial guesses for `y1_`, `y2_`, `y3_`. Enter `on_range(y_1;"g_1")` in `B4` and analogous expressions in `B7`, `B10`.

The code for `g_1`, `g_2`, `g_3`, `gradg_1`, `gradg_2`, `gradg_3` was previously introduced. The code for `beta`, `beta_on_range`, `on_range` reads as.

```
Function beta(x, y, dgname As String)
gy = Application.Run(dgname, y)
aux = minus(y, x)
beta = sp(aux, gy) / norm2(gy)
End Function
Function on_range(rx As Range, fonc As String)
x = range2vector(rx)
on_range = Application.Run(fonc, x)
End Function
Function beta_on_range(rx As Range, ry As Range, dgname As String)
x = range2vector(rx)
y = range2vector(ry)
beta_on_range = beta(x, y, dgname)
End Function
```

Add a `Sub` to run the SOLVER:

```
Sub rsolv(col, lf, lu, lh)
ryad = Range(Cells(lu, 2), Cells(lu, col)).Address
rfobjad = Range(Cells(lf, 2), Cells(lf, 2)).Address
  SolverReset
  SolverOk SetCell:=rfobjad, MaxMinVal:=1, ValueOf:=0, ByChange:=ryad, _
        Engine:=1, EngineDesc:="GRG Nonlinear"
  SolverAdd Cellref:=Range(Cells(lh, 2), Cells(lh, 2)).Address, Relation:=2, FormulaText:="0"
  SolverOptions AssumeNonNeg:=False
  SolverOptions Multistart:=True, PopulationSize:=500, RequireBounds:=False
  SolverSolve True
End Sub
Sub run_solver()
col = Cells(1, Cells.Columns.Count).End(xlToLeft).Column
Call rsolv(col, 3, 2, 4)
Call rsolv(col, 6, 5, 7)
Call rsolv(col, 9, 8, 10)
End Sub
```

Run the `Sub` `run_solver`. You will get the results.

	A	B	C
1	x	0	0
2	y	0	0
3	beta	0	
4	g	-3	
5	y	0	0
6	beta	0	
7	g	-2	
8	y	0	0
9	beta	0	
10	g	-6	

(a) Before running the Sub

	A	B	C
1	x	0	0
2	y	-0,50019	2,749997
3	beta	2,74981	
4	g	-3E-06	
5	y	1,762225	-0,38943
6	beta	1,804742	
7	g	-6,1E-08	
8	y	1,302775	1,302775
9	beta	1,842402	
10	g	-5,4E-06	
11			

(b) After running the Sub

The minimal value is $\beta(\boldsymbol{x}) = \beta_2(\boldsymbol{x}) = 1.804742$, as in the preceding (Sects. 10.3.1 and 10.5.1).

Exercises

1. Let $\boldsymbol{x} \in \mathbb{R}^2$ and $g(\boldsymbol{x}) = x_2 - x_1^2 - x_1 - 3$. Use the method of this section to determine $\beta(\boldsymbol{x})$ and $\boldsymbol{x}^*$ for $\boldsymbol{x}^{(1)} = (1, 0), \boldsymbol{x}^{(3)} = (1, 1), \boldsymbol{x}^{(4)} = (0, 1)$. The case $\boldsymbol{x}^{(2)} = (0, 0)$ was determined in the example above.
2. Let $\boldsymbol{x} \in \mathbb{R}^2$ and $g(\boldsymbol{x}) = x_2 - x_1^3 - x_1^2 - 4$. Use the method of this section to determine $\beta(\boldsymbol{x})$ and $\boldsymbol{x}^*$ for $\boldsymbol{x}^{(1)} = (1, 0), \boldsymbol{x}^{(3)} = (1, 1), \boldsymbol{x}^{(4)} = (0, 1)$. The case $\boldsymbol{x}^{(2)} = (0, 0)$ was determined in the example above.
3. Let $\boldsymbol{x} \in \mathbb{R}^2$ and $g(\boldsymbol{x}) = x_1x_2 - 2$. Use the method of this section to determine $\beta(\boldsymbol{x})$ and $\boldsymbol{x}^*$ for $\boldsymbol{x}^{(1)} = (1, 0), \boldsymbol{x}^{(2)} = (0, 0), \boldsymbol{x}^{(3)} = (1, 1), \boldsymbol{x}^{(4)} = (0, 1)$.
4. Let $\boldsymbol{x} \in \mathbb{R}^2$ and $g(\boldsymbol{x}) = x_1^2 + x_2^2 - 4$. Use the method of this section to determine $\beta(\boldsymbol{x})$ and $\boldsymbol{x}^*$ for $\boldsymbol{x}^{(1)} = (1, 0), \boldsymbol{x}^{(2)} = (0, 0), \boldsymbol{x}^{(3)} = (1, 1), \boldsymbol{x}^{(4)} = (0, 1)$. For some starting points and some initial points, the results are incorrect. Why?
5. Let $\boldsymbol{x} \in \mathbb{R}^3$ and $g(\boldsymbol{x}) = x_2 + 1 - 3(2x_3 + 1)^2(3x_1 + 1)$. Use the method of this section to determine $\beta(\boldsymbol{x})$ and $\boldsymbol{x}^*$ for $\boldsymbol{x}^{(1)} = (1, 0, 0), \boldsymbol{x}^{(2)} = (0, 0, 0), \boldsymbol{x}^{(3)} = (0, 1, 0), \boldsymbol{x}^{(4)} = (0, 0, 1)$.

10.6 Using the Reliability Index to Estimate the Probability of Failure

When the random variables $\boldsymbol{U}$ form a gaussian vector, the reliability index may be used to estimate the probability of failure. In general, the determination assumes that the distribution of $\boldsymbol{U}$ is $N(\boldsymbol{0}, \boldsymbol{Id})$, id est, the components of $\boldsymbol{U}$ are independent and have the same distribution $N(0, 1)$. For a general Gaussian vector having a mean $\boldsymbol{m}$ and a covariance matrix $\boldsymbol{C}$, id est, a vector $\boldsymbol{U}$ having the distribution $N(\boldsymbol{m}, \boldsymbol{C})$, a transformation $\boldsymbol{U} = \boldsymbol{T}(\boldsymbol{W})$ is used to bring the vector $\boldsymbol{U}$ to a $N(\boldsymbol{0}, \boldsymbol{Id})$ vector $\boldsymbol{W}$.

If $\boldsymbol{U} = (u_1, \ldots, u_n)$ is a gaussian vector, id est, if its components are independent, then $\boldsymbol{C}$ is a diagonal matrix such that $C_{ii} = \sigma_i^2$. Then, each u_i is an independent gaussian variable having as distribution $N(m_i, \sigma_i)$. In such a situation, the transformation reads as

$$w_i = \frac{u_i - m_i}{\sigma_i} \Leftrightarrow u_i = m_i + \sigma_i w_i. \tag{10.15}$$

In the context of Reliability, this transformation is referred as the *transformation of Hasofer-Lind*.

If $\boldsymbol{C}$ is not a diagonal matrix, then start by finding a decomposition $\boldsymbol{C} = \boldsymbol{PDP}^t$, where $\boldsymbol{D}$ is a diagonal matrix and $\boldsymbol{P}^t\boldsymbol{P} = \boldsymbol{PP}^t = \boldsymbol{Id}$. Such a decomposition can be made by determining the eigenvalues and eigenvectors of $\boldsymbol{C}$: $\boldsymbol{D}$ is a diagonal matrix containing the eigenvalues of $\boldsymbol{C}$ as diagonal and $\boldsymbol{P}^t$ is formed by the orthonormalized eigenvectors of $\boldsymbol{C}$, id est,

$$\boldsymbol{P} = \begin{pmatrix} \boldsymbol{P}_1 \\ \dots \\ \boldsymbol{P}_m \end{pmatrix}, \boldsymbol{P}_i\boldsymbol{P}_j^t = \begin{cases} 0, if\ i \neq j \\ 1, if\ i = j \end{cases}, \quad \boldsymbol{CP}_i^t = d_{ii}\boldsymbol{P}_i^t$$

Notice that all the eigenvalues are non-negative, since $\boldsymbol{C}$ is semidefinite: $\forall \boldsymbol{a} \in \mathbb{R}^n$: taking $v = \boldsymbol{a}^t(\boldsymbol{u} - \boldsymbol{m})$, we have $\boldsymbol{a}^t\boldsymbol{C}\boldsymbol{a} = E(vv^t) \geq 0$. Assume that the eigenvalues are ranged in a decreasing order: $d_{11} \geq d_{22} \geq \dots \geq d_{nn} \geq 0$. Let $m = \max\{i : d_{ii} > 0\}$. Let $\boldsymbol{A}$ be the matrix formed by the first m columns of $\boldsymbol{P}$ multiplied by the square root $\sqrt{d_{ii}}$, id est,

$$\boldsymbol{A} = (\boldsymbol{A}_1, \dots, \boldsymbol{A}_m) \quad , \quad \boldsymbol{A}_i = \sqrt{d_{ii}}\boldsymbol{P}_i.$$

Let $\boldsymbol{w} = (w_1, \dots, w_m)$ be a Gaussian vector having as distribution $N(\boldsymbol{0}, \boldsymbol{Id})$. Then $\boldsymbol{U} = \boldsymbol{AW} + \boldsymbol{m}$ has as distribution $(\boldsymbol{m}, \boldsymbol{C})$: indeed, $E(\boldsymbol{U}) = \boldsymbol{A}E(\boldsymbol{W}) + \boldsymbol{m} = \boldsymbol{m}$ and

$$E\left((\boldsymbol{U} - \boldsymbol{m})(\boldsymbol{U} - \boldsymbol{m})^t\right) = \boldsymbol{A}E(\boldsymbol{WW}^t)\boldsymbol{A}^t = \boldsymbol{AA}^t = \boldsymbol{PDP}^t = \boldsymbol{C}$$

```
Public Function cholesky(C)
nnn = UBound(C, 2)
Dim LC() As Double
ReDim LC(1 To nnn, 1 To nnn)
For j = 1 To nnn
    sss = 0
    For kkk = 1 To j - 1
        sss = sss + LC(j, kkk) ^ 2
    Next kkk
    LC(j, j) = C(j, j) - sss
    If LC(j, j) <= 0 Then Exit For
    LC(j, j) = Sqr(LC(j, j))
    For i = j + 1 To nnn
        sss = 0
        For kkk = 1 To j - 1
            sss = sss + LC(i, kkk) * LC(j, kkk)
        Next kkk
        LC(i, j) = (C(i, j) - sss) / LC(j, j)
    Next i
Next j
cholesky = LC
End Function
```

```
Function cholesky_on_range(rC As Range)
C = rC.Value
aux = cholesky(C)
cholesky_on_range = aux
End Function
```

Fig. 10.13 Example of VBA code for Cholesky decomposition

As an alternative, we can determine a decomposition $\boldsymbol{C} = \boldsymbol{A}\ \boldsymbol{A}^t$ – this can be achieved by Cholesky's factorization of $\boldsymbol{C}$. Whichever the case or the method used, the transformation is affine

$$\boldsymbol{U} = \boldsymbol{T}\boldsymbol{W} + \boldsymbol{m},$$

so that the limit state curve can be written as

$$G(\boldsymbol{x}, \boldsymbol{W}) = H(\boldsymbol{x}, \boldsymbol{T}\boldsymbol{W} + \boldsymbol{m}).$$

Consequently, the analysis can be brought to the case where the vector of random variables is formed by independent gaussian variables of same distribution $N(0, 1)$. In the sequel of this subsection, we assume that $\boldsymbol{U}$ satisfies such a condition.

An example of VBA code for Cholesky's decomposition is given in Fig. 10.13.

Example

Let us consider two independent random variables normally distributed: $u_1 \sim N(0, 1)$, $u_2 \sim N(0, 1)$. Let $X_1 = u_1 + u_2$, $X_2 = 3u_1 + 2u_2$, $X_3 = 5u_1 - 4\,u_2$. Let $X = (X_1, X_2, X_3)$. The covariance matrix of X is

$$\boldsymbol{C} = \begin{pmatrix} 2 & 5 & 1 \\ 5 & 13 & 7 \\ 1 & 7 & 41 \end{pmatrix}$$

Open a blank workbook and enter the matrix $\boldsymbol{C}$ in `B1:D3`. Give the name `C_` to these cells. Notice that the matrix is not invertible: Go to B5 and enter "`=MDETERM(C_)`". The determinant of $\boldsymbol{C}$ is near zero.

	A	B	C	D
1	C	2	5	1
2		5	13	7
3		1	7	41
4				
5	Determ	2,93099E-14		

Add a module to the workbook and enter the code for the Cholesky's decomposition. Go to `G1` and enter "`=Cholesky_on_range(C_)`". Give the name `A` to `G1:I3`. To verify the decomposition, go to G5 and enter "`=MMULT(A;TRANSPOSE(A))`". The workbook appears as.

	A	B	C	D	E	F	G	H	I
1	C	2	5	1		A	1,414214	0	0
2		5	13	7			3,535534	0,707107	0
3		1	7	41			0,707107	6,363961	3,86E-07
4									
5	Determ	2,93099E-14				verif	2	5	1
6							5	13	7
7							1	7	41
8									

Look at the last column of A: it is formed by zeros. This indicates that two independent variables are enough to describe the distribution of $\boldsymbol{X}$. For instance, we may consider a gaussian vector $\boldsymbol{U} = (U_1, U_2)$ of two independent variables $N(0, 1)$ and $X_i = A_{i1}U_1 + A_{i2}U_2$. Notice that the variable $A_{i1}U_1 + A_{i2}U_2$ has a distribution $N\left(0, \sqrt{A_{i1}^2 + A_{i2}^2}\right)$. Thus, this approximation leads to $X_1 \sim N(0, \sqrt{2})$, $X_2 \sim N(0, \sqrt{13})$ and $X_3 \sim N(0, \sqrt{41})$ – which corresponds to the exact distributions.

Exercises

1. Consider a vector of 3 gaussian variables (X_1, X_2, X_3) having as covariance matrix

$$C = \begin{pmatrix} 5 & 7 & 1 \\ 7 & 8 & 3 \\ 1 & 3 & 10 \end{pmatrix}$$

What is the minimal number of independent $N(0, 1)$ variables requested to describe this distribution?

2. Consider a vector of 4 gaussian variables (X_1, X_2, X_3, X_4) having as covariance matrix

$$C = \begin{pmatrix} 19 & 19 & 9 & -4 \\ 19 & 29 & -1 & 2 \\ 9 & -1 & 19 & -18 \\ -4 & 2 & -18 & 26 \end{pmatrix}$$

What is the minimal number of independent $N(0, 1)$ variables requested to describe this distribution?

3. Consider a vector of 4 gaussian variables (X_1, X_2, X_3, X_4) having as covariance matrix

$$C = \begin{pmatrix} 25 & -2 & 7 & 9 \\ -2 & 5 & -1 & 5 \\ 7 & -1 & 2 & 2 \\ 9 & 5 & 2 & 10 \end{pmatrix}$$

What is the minimal number of independent $N(0, 1)$ variables requested to describe this distribution?

4. Consider a vector of 5 gaussian variables $(X_1, X_2, X_3, X_4, X_5)$ having as covariance matrix

$$C = \begin{pmatrix} 34 & 11 & 28 & -12 & 7 \\ 11 & 18 & 19 & 0 & -16 \\ 28 & 19 & 33 & -3 & -8 \\ -12 & 0 & -3 & 11 & -9 \\ 7 & -16 & -8 & -9 & 25 \end{pmatrix}$$

What is the minimal number of independent $N(0, 1)$ variables requested to describe this distribution?

5. Consider a vector of 5 gaussian variables $(X_1, X_2, X_3, X_4, X_5)$ having as covariance matrix

$$C = \begin{pmatrix} 5 & 2 & 4 & 6 & -8 \\ 2 & 17 & -2 & -12 & 4 \\ 4 & -2 & 4 & 8 & -8 \\ 6 & -12 & 8 & 20 & -16 \\ -8 & 4 & -8 & -16 & 16 \end{pmatrix}$$

What is the minimal number of independent $N(0, 1)$ variables requested to describe this distribution?

10.6.1 The Case of Affine Limit State Equations

In this section, we examine the situation where $H(\boldsymbol{x}, \boldsymbol{U}) = \boldsymbol{a}^t\boldsymbol{U} + b(\boldsymbol{x})$. As indicated above, we assume that $\boldsymbol{U}$ is a centered Gaussian vector $N(\boldsymbol{0}, \boldsymbol{Id})$, i.e., all its components are independent and have the same distribution $N(0, 1)$ – otherwise the transformation previously introduced must be used to bring the analysis to such a situation.

Under these assumptions, $Z = H(\boldsymbol{x}, \boldsymbol{U})$ is a Gaussian variable such that $E(Z) = b(\boldsymbol{x})$ and $V(Z) = \boldsymbol{a}^t\boldsymbol{a} = \|\boldsymbol{a}\|^2$. Thus,

$$S = \frac{Z - b(\boldsymbol{x})}{\|\boldsymbol{a}\|} \sim N(0, 1).$$

Consequently,

$$F = P(\mathcal{F}) = P(Z > 0) = P\left(S > -\frac{b(\boldsymbol{x})}{\|\boldsymbol{a}\|}\right) = P(S > \beta(\boldsymbol{x}))$$

Let Φ be the CDF associated to $N(0, 1)$ (Cf. Eq. (10.1)). We have

$$F = 1 - \Phi(\beta(x)) = \Phi(-\beta(x)). \tag{10.16}$$

The value of F may be furnished by `NORM.S.DIST(`$-\beta(x)$`;TRUE)`. Typical values are shown in Table 10.5.

Table 10.5 Typical values of the reliability index and the corresponding failure probabilities

β	1	2	3	4	5	6	7	8
F	1.6E-1	2.3E-2	1.3E-3	3.2E-5	2.9E-7	1E-9	1.3E-12	6.2E-16

Example

Let us consider the situation where the failure corresponds to the difference between two variables x_1 and x_2: failure occurs when $x_1 > x_2$. Then, denoting $\boldsymbol{x} = (x_1, x_2)$:

$$g(\boldsymbol{x}) = x_1 - x_2.$$

If $X_1 = x_1 + \sigma_1 u_1$, $X_2 = x_2 + \sigma_2 u_2$, with $u_i \sim N(0, 1)$, $i = 1, 2$. Then, with $\boldsymbol{u} = (u_1, u_2)$:

$$H(\boldsymbol{x}, \boldsymbol{u}) = x_1 - x_2 + \sigma_1 u_1 - \sigma_2 u_2.$$

In this case, $\boldsymbol{a} = (\sigma_1, -\sigma_2)^t$, $b(\boldsymbol{x}) = x_1 - x_2$. Thus,

$$\beta(\boldsymbol{x}) = -\frac{x_1 - x_2}{\sqrt{\sigma_1^2 + \sigma_2^2}} = \frac{x_2 - x_1}{\sqrt{\sigma_1^2 + \sigma_2^2}} \quad .$$

The probability of failure is

$$F = \Phi\left(-\frac{x_2 - x_1}{\sqrt{\sigma_1^2 + \sigma_2^2}}\right) = \Phi\left(\frac{x_1 - x_2}{\sqrt{\sigma_1^2 + \sigma_2^2}}\right).$$ ∎

Exercises

1. Let $\boldsymbol{x} \in \mathbb{R}^2$ and $g(\boldsymbol{x}) = x_2 - 3x_1 - 3$. Let $X_i = x_i + \sigma_i u_i$ with $u_i \sim N(0, 1)$, $i = 1, 2$. Determine the probability of failure for the points (0,0), (1,0), (0,1), when $\sigma_1 = 1$, $\sigma_2 = 2$.
2. Let $\boldsymbol{x} \in \mathbb{R}^3$ and $g(\boldsymbol{x}) = x_3 - 2x_1 - x_2 - 2$. Let $X_i = x_i + \sigma_i u_i$ with $u_i \sim N(0, 1)$, $i = 1, 2, 3$. Determine the probability of failure for the points (0,0,0), (1,0,0), (0,1,0), (0,0,1) when $\sigma_1 = 1$, $\sigma_2 = 2$, $\sigma_3 = 1$.
3. Let $\boldsymbol{x} \in \mathbb{R}^3$ and $g(\boldsymbol{x}) = x_2 - 2x_3 - x_1 - 2$. Let $X_i = x_i + \sigma_i u_i$ with $u_i \sim N(0, 1)$, $i = 1, 2, 3$. Determine the probability of failure for the points (0,0,0), (1,0,0), (0,1,0), (0,0,1) when $\sigma_1 = 1$, $\sigma_2 = 2$, $\sigma_3 = 1$.
4. Let $\boldsymbol{x} \in \mathbb{R}^4$ and $g(\boldsymbol{x}) = x_4 + x_3 - 2x_2 - x_1 - 2$. Let $X_i = x_i + \sigma_i u_i$ with $u_i \sim N(0, 1)$, $i = 1, 2, 3, 4$. Determine the probability of failure for the points (0,0,0,0), (1,0,0,0), (0,1,0,0), (0,0,1,0) and (0,0,0,1) when $\sigma_1 = 1$, $\sigma_2 = 2$, $\sigma_3 = 1$, $\sigma_4 = 2$.
5. Let $\boldsymbol{x} \in \mathbb{R}^4$ and $g(\boldsymbol{x}) = x_4 + 2x_3 - 3x_2 - 2x_1 - 3$. Let $X_i = x_i + \sigma_i u_i$ with $u_i \sim N(0, 1)$, $i = 1, 2, 3, 4$. Determine the probability of failure for the points (0,0,0,0), (1,0,0,0), (0,1,0,0), (0,0,1,0) and (0,0,0,1) when $\sigma_1 = 1$, $\sigma_2 = 2$, $\sigma_3 = 1$, $\sigma_4 = 2$.

10.6.2 The Case of a Convex Failure Region

Assume yet that $\boldsymbol{U}$ is a centered Gaussian vector $N(\boldsymbol{0}, \boldsymbol{Id})$, i.e., all its components are independent and have the same distribution $N(0, 1)$ and that the region $\mathcal{F}$ is convex. In this case (see, for instance, (Souza de Cursi & Sampaio, 2010). See also Fig. 10.12, page 469):

$$\boldsymbol{u} \in \mathcal{F} \Leftrightarrow -(\boldsymbol{u}^*)^t(\boldsymbol{u} - \boldsymbol{u}^*) \le 0 \quad . \tag{10.17}$$

Considering that $\boldsymbol{u}^* = \beta \boldsymbol{n}^*$, we have

$$\boldsymbol{u} \in \mathcal{F} \Leftrightarrow (\boldsymbol{n}^*)^t (\boldsymbol{u} - \boldsymbol{u}^*) \geq 0 \ . \tag{10.18}$$

Let Π^* be the hyperplane having as equation

$$\Pi^* : p^*(\boldsymbol{u}) = (\boldsymbol{n}^*)^t \boldsymbol{u} - (\boldsymbol{n}^*)^t \boldsymbol{u}^* = 0. \tag{10.19}$$

Π^* splits the space in two half spaces, according to the sign of $p^*(\boldsymbol{u})$. We have

$$\boldsymbol{u} \in \mathcal{F} \Rightarrow p^*(\boldsymbol{u}) \geq 0 \Rightarrow F = P(\mathcal{F}) \leq P(p^*(\boldsymbol{u}) \geq 0).$$

$P(p^*(\boldsymbol{u}) \geq 0)$ is the probability of failure associated to the limit state curve defined by Π^*, for which $\boldsymbol{a} = \boldsymbol{n}^*$, $b(\boldsymbol{x}) = -(\boldsymbol{n}^*)^t \boldsymbol{u}^*$. Thus,

$$\beta(x) = \|\boldsymbol{u}^*\| = (\boldsymbol{n}^*)^t \boldsymbol{u}^*; F = P(\mathcal{F}) \leq \Phi(-\beta(\boldsymbol{x})). \tag{10.20}$$

Again, `NORM.S.DIST(`$-\beta(x)$`;TRUE)` can be used to get the values of the upper bound.

Example

Let us consider the situation where $\boldsymbol{x} = (x_1, x_2)$ and the failure occurs when $x_1 x_2 > 1$.

$$g(\boldsymbol{x}) = x_1 x_2 - 1, \mathcal{F} = \{\boldsymbol{x} : x_1, x_2 \geq 0, g(\boldsymbol{x}) > 0\}.$$

In this situation, $\mathcal{F}$ is convex. Let $X_1 = x_1 + \sigma_1 u_1$, $X_2 = x_2 + \sigma_2 u_2$, with $u_i \sim N(0, 1)$, $i = 1, 2$. Then, with $\boldsymbol{u} = (u_1, u_2)$:

$$H(\boldsymbol{x}, \boldsymbol{u}) = x_1 x_2 - 1 + \sigma_1 \sigma_2 u_1 u_2 + \sigma_1 u_1 x_2 + \sigma_2 u_2 x_1.$$

Consider the point $\boldsymbol{x} = (0, 0)^t$. In this case, $H(\boldsymbol{x}, \boldsymbol{u}) = -1 + \sigma_1 \sigma_2 u_1 u_2$. Then

$$u^* = \left(\frac{a}{\sigma_1}, \frac{a}{\sigma_2}\right), a = \frac{1}{\left(\sigma_1^2 + \sigma_2^2\right)^{\frac{1}{4}}}$$

and

$$\beta = \sqrt{\frac{1}{\sigma_1^2} + \frac{1}{\sigma_2^2}} \ .$$

In Table 10.6, we compare the exact value and the upper bound furnished by Eq. (10.20) for the situation where $\sigma_1 = \sigma_2 = \sigma$. The exact values were calculated by numerically integrating the PDF of the couple on the failure region $\mathcal{F}$. ∎.

Table 10.6 Comparison between Eq. (10.20) and the exact value

σ	1	1/2	1/3	1/4	1/5	1/6	1/7
F exact	5.2E-2	1.6E-3	7.7E-6	5.4E-9	5.3E-13	8.5E-18	1.2E-23
Eq. (10.20)	8E-2	2.3E-3	1.1E-5	7.7E-9	7.7E-13	1.1E-17	2.1E-23

Exercises

1. Let $\boldsymbol{x} \in \mathbb{R}^2$ and $g(\boldsymbol{x}) = x_2 - x_1^2 - x_1 - 3$. Let $X_i = x_i + \sigma_i u_i$ with $u_i \sim N(0, 1)$, $i = 1, 2$. Determine an upper bound for the probability of failure for the points (0,0), (1,0), (0,1), when $\sigma_1 = 1, \sigma_2 = 2$.
2. Let $\boldsymbol{x} \in \mathbb{R}^2$ and $g(\boldsymbol{x}) = x_2 - x_1^3 - x_1^2 - 4$. Let $X_i = x_i + \sigma_i u_i$ with $u_i \sim N(0, 1)$, $i = 1, 2$. Determine an upper bound for the probability of failure for the points (0,0), (1,0), (0,1), when $\sigma_1 = 1, \sigma_2 = 2$.
3. Let $\boldsymbol{x} \in \mathbb{R}^2$ and $g(\boldsymbol{x}) = x_1 x_2 - 2$. Let $X_i = x_i + \sigma_i u_i$ with $u_i \sim N(0, 1)$, $i = 1, 2$. Determine an upper bound for the probability of failure for the points (0,0), (1,0), (0,1), when $\sigma_1 = 1, \sigma_2 = 2$.
4. Let $\boldsymbol{x} \in \mathbb{R}^4$ and $g(\boldsymbol{x}) = x_1 x_2 x_3 x_4 - 2$. Let $X_i = x_i + \sigma_i u_i$ with $u_i \sim N(0, 1)$, $i = 1, 2, 3, 4$. Determine an upper bound for the probability of failure for the points (0,0,0,0), (1,0,0,0), (0,1,0,0), (0,0,1,0) and (0,0,0,1) when $\sigma_1 = 1, \sigma_2 = 2, \sigma_3 = 1, \sigma_4 = 2$.
5. Let $\boldsymbol{x} \in \mathbb{R}^4$ and $g(\boldsymbol{x}) = x_4 - x_1 x_2 x_3 - 3$. Let $X_i = x_i + \sigma_i u_i$ with $u_i \sim N(0, 1)$, $i = 1, 2, 3, 4$. Determine an upper bound for the probability of failure for the points (0,0,0,0), (1,0,0,0), (0,1,0,0), (0,0,1,0) and (0,0,0,1) when $\sigma_1 = 1, \sigma_2 = 2, \sigma_3 = 1, \sigma_4 = 2$.

10.6.3 General Failure Regions

Equations (10.16) and (10.20) furnish the most popular estimations of the probability failure using the reliability index. They may be used in the situations where the failure region $\mathcal{F}$ is a subset of the non-negative half space Π_+^* defined by the hyperplane Π^* defined by (10.19), which corresponds to the tangent hyperplane to the failure region at the point $\boldsymbol{u}^*$. Notice that condition (10.18) is sufficient to guarantee that $\mathcal{F} \subset \mathcal{F} \subset \Pi_+^*$.

If $\mathcal{F} \not\subset \Pi_+^*$ (for instance, in the situation of Fig. 10.14) we need to use a different approach. Recall that we continue under the assumption that $\boldsymbol{U}$ is a centered Gaussian vector of covariance $\boldsymbol{Id}$ (identity matrix), i.e., all its components are N (0, 1) independent variables - otherwise, transformations must be used. Thus, the probability density of $\boldsymbol{U}$ is

$$\phi_U(\boldsymbol{u}) = {}_{i=1}^{n} \phi(u_i) = \frac{1}{\sqrt{2^n \pi^n}} \exp\left(-\frac{1}{2}\|\boldsymbol{u}\|^2\right) \tag{10.21}$$

Therefore, the probability density depends only on the distance to the origin $r = \|\boldsymbol{u}\|$ and decreases with r. We will use this property to establish estimates of $F = P(\mathcal{F})$.

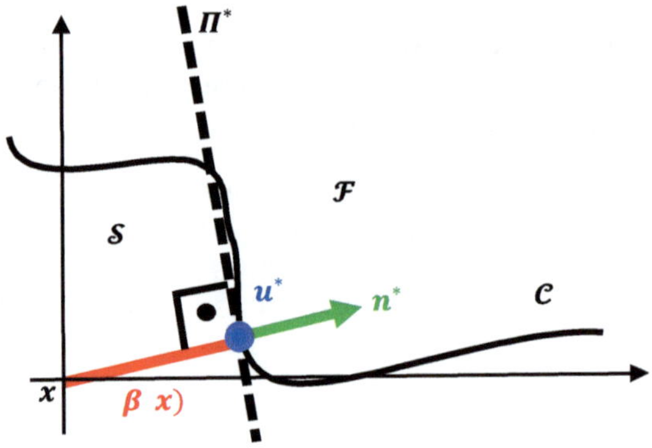

Fig. 10.14 If $\mathcal{F} \not\subset \Pi_+^*$, then a different approach must be used

Let us consider (Fig. 10.15),

$$\mathcal{E} = \{\boldsymbol{e} \in \mathbb{R}^n : \|\boldsymbol{e}\| = 1\}, \quad \mathcal{M}(\boldsymbol{e}) = \{\lambda\boldsymbol{e} : \lambda \geq 0\}. \tag{10.22}$$

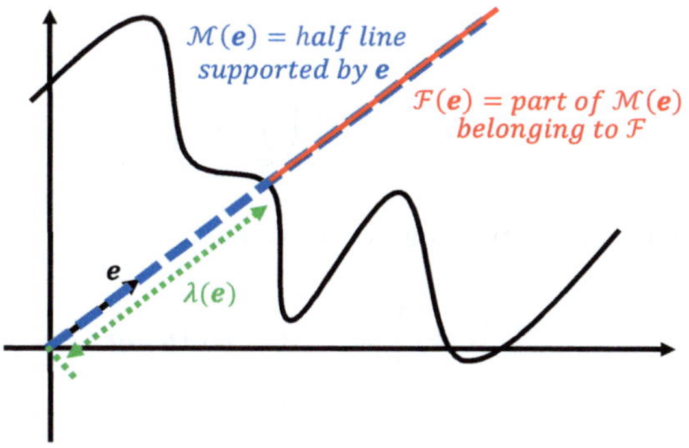

Fig. 10.15 $\mathcal{M}(\boldsymbol{e})$ is the half line oriented by the vector $\boldsymbol{e}$. $\mathcal{F}(\boldsymbol{e})$ is the part of the half line belonging to the failure region $\mathcal{F}$. $\lambda(\boldsymbol{e})$ is the distance between the first point of $\mathcal{F}(\boldsymbol{e})$ and the origin

Then

$$0 \in \mathcal{M}(\boldsymbol{e}), \forall \boldsymbol{e} \in \mathcal{E}, \quad \boldsymbol{x} \neq 0 \Longrightarrow \boldsymbol{x} \in \mathcal{M}\left(\frac{\boldsymbol{x}}{\|\boldsymbol{x}\|}\right),$$

so that

$$\mathbb{R}^n \subset \bigcup_{\boldsymbol{e}\in\mathcal{E}} \mathcal{M}(\boldsymbol{e}) \subset \mathbb{R}^n \Longrightarrow \mathbb{R}^n = \bigcup_{\boldsymbol{e}\in\mathcal{E}} \mathcal{M}(\boldsymbol{e}) \tag{10.23}$$

Let us consider

$$\mathcal{F}(\boldsymbol{e}) = \mathcal{F} \cap \mathcal{M}(\boldsymbol{e}), \quad \Lambda(\boldsymbol{e}) = \{\lambda \in \mathbb{R} : \lambda \geq 0, \lambda\boldsymbol{e} \in \mathcal{F}\}.$$

We have

$$\mathcal{F}(\boldsymbol{e}) = \{\lambda\boldsymbol{e} : \lambda \geq 0, H(x, \lambda\boldsymbol{e}) \geq 0\}, \Lambda(\boldsymbol{e}) = \{\lambda : \lambda \geq 0, H(x, \lambda\boldsymbol{e}) \geq 0\}$$

In addition

$$\mathcal{F} = \bigcup_{\boldsymbol{e}\in\mathcal{E}} \mathcal{F}(\boldsymbol{e}) \ .$$

Let

$$\lambda(\boldsymbol{e}) = \begin{cases} -\infty, \textit{if } \Lambda(\boldsymbol{e}) = \varnothing, \\ \inf \Lambda(\boldsymbol{e}), \textit{if } \ \Lambda(\boldsymbol{e}) \neq \varnothing. \end{cases}$$

Notice that the inf exists for $\Lambda(\boldsymbol{e}) \neq \varnothing$, since $\lambda \geq 0, \forall\, \lambda \in \Lambda(\boldsymbol{e})$. If H is a continuous function, then $\lambda(\boldsymbol{e})$ is the smallest non-negative solution of the equation $H(\boldsymbol{x}, \lambda\boldsymbol{e}) = 0$. Thus,

$$\lambda \geq 0 \ \textit{and}\ H(\boldsymbol{x}, \lambda\boldsymbol{e}) \geq 0 \Longrightarrow \lambda \geq \lambda(\boldsymbol{e}).$$

Let $\mathcal{L}(e) = \{\lambda \boldsymbol{e} : \lambda \geq \lambda(\boldsymbol{e})\}$. We have

$$\forall \boldsymbol{e} \in \mathcal{E} : \mathcal{F}(\boldsymbol{e}) \subset \mathcal{L}(\boldsymbol{e}) \Longrightarrow \mathcal{F} \subset \bigcup_{\boldsymbol{e} \in \mathcal{E}} \mathcal{L}(\boldsymbol{e})$$

and

$$F = P(\mathcal{F}) \leq P\left(\bigcup_{\boldsymbol{e} \in \mathcal{E}} \mathcal{L}(\boldsymbol{e})\right). \tag{10.24}$$

This last value can be determined by integration: we have

$$P\left(\bigcup_{\boldsymbol{e} \in \mathcal{E}} \mathcal{L}(\boldsymbol{e})\right) = \int_{\boldsymbol{u} \in \bigcup_{\boldsymbol{e} \in \mathcal{E}} \mathcal{L}(\boldsymbol{e})} \phi_U(\boldsymbol{u}) d\boldsymbol{u}$$

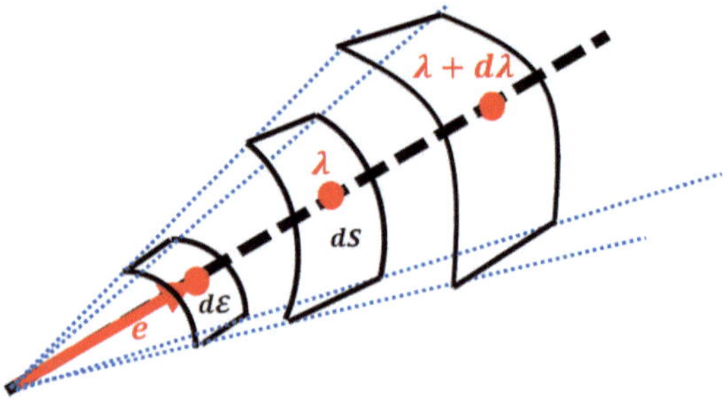

Fig. 10.16 Elements of volume used in the integration

The element of volume is

$$d\boldsymbol{u} = dS d\lambda,$$

where dS is the element of surface for a sphere of radius λ (see Fig. 10.16). The surface of a hyperball of radius r is

$$surf(r) = meas(\|\boldsymbol{u}\| = r) = \frac{2\pi^{\frac{n}{2}} r^{n-1}}{\Gamma\left(\frac{n}{2}\right)}$$

In addition,

$$\frac{dS}{surf(\lambda)} = \frac{d\mathcal{E}}{surf(1)} \Longrightarrow dS = \frac{surf(\lambda)}{surf(1)} d\mathcal{E}$$

and we have

$$P\left(\bigcup_{\boldsymbol{e} \in \mathcal{E}} \mathcal{L}(\boldsymbol{e})\right) = \frac{1}{surf(1)} \int_{\boldsymbol{e} \in \mathcal{E}} d\mathcal{E} \int_{\lambda(\boldsymbol{e})}^{+\infty} surf(\lambda) \phi_U(\boldsymbol{u}) d\boldsymbol{u}.$$

Thus, using that

$$\phi_U(\lambda \boldsymbol{e}) = \frac{1}{\sqrt{2^n \pi^n}} \exp\left(-\frac{\lambda^2}{2}\right),$$

we have

$$P\left(\bigcup_{e \in \mathcal{E}} \mathcal{L}(\boldsymbol{e})\right) = \frac{1}{surf(1)} \frac{1}{2\pi^{\frac{n}{2}}} \int_{e \in \mathcal{E}} d\mathcal{E} \int_{\lambda(e)}^{+\infty} surf(\lambda) \exp\left(-\frac{\lambda^2}{2}\right) d\lambda.$$

Noticing that

$$\int_{\lambda(e)}^{+\infty} surf(\lambda) \exp\left(-\frac{\lambda^2}{2}\right) d\boldsymbol{u} = \frac{2\pi^{\frac{n}{2}}}{\Gamma\left(\frac{n}{2}\right)} \int_{\lambda(e)}^{+\infty} \lambda^{n-1} \exp\left(-\frac{\lambda^2}{2}\right) d\lambda,$$

We have

$$\int_{\lambda(e)}^{+\infty} surf(\lambda) \exp\left(-\frac{\lambda^2}{2}\right) d\boldsymbol{u} = \frac{2^{\frac{n}{2}} \pi^{\frac{n}{2}}}{\Gamma\left(\frac{n}{2}\right)} \int_{\frac{\lambda(e)^2}{2}}^{+\infty} t^{\frac{n}{2}-1} e^{-t} dt$$

and

$$\int_{\lambda(e)}^{+\infty} surf(\lambda) \exp\left(-\frac{\lambda^2}{2}\right) d\boldsymbol{u} = \frac{2^{\frac{n}{2}} \pi^{\frac{n}{2}}}{\Gamma\left(\frac{n}{2}\right)} \Gamma\left(\frac{n}{2}, \frac{\lambda(\boldsymbol{e})^2}{2}\right),$$

where

$$\Gamma(\mathrm{s}, \mathrm{b}) = \int_{\mathrm{b}}^{+\infty} t^{s-1} e^{-t} dt$$

is the incomplete Gamma function. Thus,

$$P\left(\bigcup_{e \in \mathcal{E}} \mathcal{L}(\boldsymbol{e})\right) = \frac{1}{surf(1)} \int_{\mathcal{E}} \frac{\Gamma\left(\frac{n}{2}, \frac{\lambda(e)^2}{2}\right)}{\Gamma\left(\frac{n}{2}\right)} d\mathcal{E} = \mathrm{E}\left(\frac{\Gamma\left(\frac{n}{2}, \frac{\lambda(e)^2}{2}\right)}{\Gamma\left(\frac{n}{2}\right)}\right).$$

and we have

$$F \leq \mathrm{E}\left(\frac{\Gamma\left(\frac{n}{2}, \frac{\lambda(\boldsymbol{e})^2}{2}\right)}{\Gamma\left(\frac{n}{2}\right)}\right). \tag{10.25}$$

Since (see, for instance, (Souza de Cursi, 1992))

$$P\left(\chi^2(n) \geq \lambda(\boldsymbol{e})^2\right) = \frac{\Gamma\left(\frac{n}{2}, \frac{\lambda(\boldsymbol{e})^2}{2}\right)}{\Gamma\left(\frac{n}{2}\right)},$$

we have

$$F \leq E\left(P\left(\chi^2(n) \geq \lambda(\boldsymbol{e})^2\right)\right). \tag{10.26}$$

These values may be estimated by integration or Monte Carlo methods. For instance, let us describe $\mathcal{E}$ using hyperspherical coordinates $\boldsymbol{\theta} = (\theta_1, \ldots, \theta_{n-1})$: let $e(\boldsymbol{\theta}) = (e_1(\boldsymbol{\theta}), \ldots, e_n(\boldsymbol{\theta}))$ be given by

$$e_1(\boldsymbol{\theta}) = {}_{j=1}^{n-1} \sin(\theta_j),$$

$$e_i(\boldsymbol{\theta}) = \cos(\theta_{n-i+1}) {}_{j=1}^{i-1} \sin(\theta_j) \ , 2 \leq i \leq n-1,$$

$$e_n(\boldsymbol{\theta}) = \cos(\theta_1).$$

Then,

$$\mathcal{E} = \left\{e(\boldsymbol{\theta}) : \boldsymbol{\theta} \in (0, \pi)^{n-2} \times (0, 2\pi)\right\}.$$

The Jacobian associated to these coordinates is

$$J(\boldsymbol{\theta}) = (-1)^{[n/2]} {}_{j=1}^{n-2} \sin^{\mathrm{j}}(\theta_{n-j-1}) = (-1)^{[n/2]} {}_{j=1}^{n-2} \sin^{\mathrm{n-j-1}}(\theta_j).$$

Thus,

$$\int_{\mathcal{E}} P\left(\chi^2(n) \geq \lambda(\boldsymbol{e})^2\right) d\mathcal{E} = \int_{\boldsymbol{\theta} \in (0,\pi)^{n-2} \times (0, 2\pi)} \frac{\Gamma\left(\frac{n}{2}, \frac{\lambda(\boldsymbol{e}(\boldsymbol{\theta}))^2}{2}\right)}{\Gamma\left(\frac{n}{2}\right)} J(\boldsymbol{\theta}) d\boldsymbol{\theta}. \tag{10.27}$$

The evaluation of the estimate in Eq. (10.27) can be performed by considering a discretization of $\mathcal{E}$. For instance, in dimension 2, we can consider a given number of steps nsp and a step $\Delta\theta = 2\pi/nsp$. Then, we generate the angles $\theta_i = i\Delta\theta$, $i = 0, \ldots, nsp + 1$ and $\boldsymbol{e}(\theta_i) = (\sin(\theta_i), \cos(\theta_i))$. We obtain the discretized values $F_i = P\left(\chi^2(n) \geq \lambda_i^2\right)$, $\lambda_i = \lambda(\boldsymbol{e}(\theta_i))$ which can be used for a numerical integration. For instance, the trapezoidal rule reads as

$$E\left(P\left(\chi^2(n) \geq \lambda(\boldsymbol{e})^2\right)\right) \approx \frac{\Delta\theta}{2\pi}\left(\frac{F_0 + F_{nsp}}{2} + \sum_{j=1}^{nsp-1} F_j\right). \tag{10.28}$$

Example

Let us consider the situation where $\boldsymbol{x} = (0, 0)$ and the failure occurs when $g(\boldsymbol{x}) = x_2 - x_1^2 - x_1 - 3 > 0$. We have $g(\boldsymbol{x} + \lambda \boldsymbol{e}(\theta_i)) = \lambda \cos(\theta_i) - (\lambda \sin(\theta_i))^2 - (\lambda \sin(\theta_i)) - 3 = -\sin^2(\theta_i)\lambda^2 + (\cos(\theta_i) - \sin(\theta_i))\lambda - 3$. λ_i is a positive root of this equation, if such a root exists. If not any positive solution exists, then $F_i = 0$, what corresponds to $\lambda(\theta_i) = +\infty$.

- Let $\theta_i = 0$: the equation has degree 1 and $\lambda_i = 3$.
- Let $\theta_i = \pi$: the equation has degree 1 and its root is $\lambda = -3$. Thus, there is no positive solution. In this case, we take $F_i = 0$, what corresponds to $\lambda(\theta_i) = +\infty$.
- Let $\theta_i \neq 0$ and $\theta_i \neq \pi$: the equation has degree 2. Its discriminant is

$$\Delta = (\cos(\theta_i) - \sin(\theta_i))^2 - 12\sin^2(\theta_i).$$

Real roots exist only for $\Delta \geq 0$. In this case, the roots are

$$\lambda_\pm = \frac{(\cos(\theta_i) - \sin(\theta_i)) \pm \sqrt{\Delta}}{2\cos^2(\theta_i)}.$$

λ_i corresponds to the smaller positive root. Again, if no positive root exists, we take $F_i = 0$.

We use Eq. (10.28) with different values of *nsp* to evaluate F and we can compare the results in Table 10.7 to those furnished by the Montecarlo approach (page 454).

Table 10.7 Results of Eq. (10.28). The Montecarlo simulation furnished $F \approx 0,0009$ with a standard deviation 0,0002

nsp	1E2	1E3	1E4	1E5	1E6
Eq. (10.28)	0,001088	0,00099	0,00098	0,00098	0,000978

We can obtain a second upper bound by recalling that $\boldsymbol{u}$ is a Gaussian vector formed by independent variables of same distribution $N(0, 1)$. Thus, $u_e = \boldsymbol{u}^t\boldsymbol{e}$ is a Gaussian variable $N(0, 1)$ (Cf., for instance, (Souza de Cursi, 1992)). So, $P(u_e \geq a) = \Phi(-a)$. Since λ is the restriction of u_e to the non-negative values, we have

$$P(\lambda \geq \lambda(\boldsymbol{e})) = 2(1 - \Phi(\lambda(\boldsymbol{e})) = 2\Phi(-\lambda(\boldsymbol{e})).$$

Thus,

$$P\left(\bigcup_{\boldsymbol{e}\in\mathcal{E}} \mathcal{L}(\boldsymbol{e})\right) = \int_{\mathcal{E}} P(\lambda \geq \lambda(\boldsymbol{e}))d\mathcal{E} = 2\int_{\mathcal{E}} \Phi(-\lambda(\boldsymbol{e}))d\mathcal{E}$$

and

$$F \leq 2\int_{\mathcal{E}} \Phi(-\lambda(\boldsymbol{e}))d\mathcal{E}. \tag{10.29}$$

Analogously to the preceding upper bound, this one may be calculated by integration or Monte Carlo methods:

$$\int_{\mathcal{E}} \Phi(-\lambda(\boldsymbol{e}))d\mathcal{E} = \int_{\boldsymbol{\theta}\in(0,\pi)^{n-2}\times(0,2\pi)} \Phi(-\lambda(\boldsymbol{e}(\boldsymbol{\theta})))J(\boldsymbol{\theta})d\boldsymbol{\theta}. \tag{10.30}$$

Notice that the PDF of λ is

$$\phi_\lambda(\lambda) = 0, \textit{if } \lambda < 0; \quad \phi_\lambda(\lambda) = 2\phi(\lambda), \textit{if } \lambda > 0 \ .$$

Example
Let us consider the situation where $\boldsymbol{x} = (x_1, x_2)$ and the failure occurs when $(x_1 - u_1)^2 + (x_2 - u_2)^2 > \alpha^2 (\alpha > 0)$. Let $\boldsymbol{x} = (0, 0)^t$: then, $\boldsymbol{u}^*$ is not unique since any point $\boldsymbol{u}^*$ such that $\|\boldsymbol{u}^*\| = \alpha$ is a design point. Then, $\beta(\boldsymbol{x}) = \alpha$ and the probability of failure is

$$F = \frac{1}{2\pi} \int_{\|\boldsymbol{u}\|\geq\alpha} \exp\left(-\frac{1}{2}\|\boldsymbol{u}\|^2\right) d\boldsymbol{u},$$

id est,

$$F = \frac{1}{2\pi} \int_{\alpha}^{+\infty} 2\pi r \exp\left(-\frac{1}{2}r^2\right) dr = \exp\left(-\frac{\alpha^2}{2}\right).$$

Table 10.8 presents a comparison between the exact value above and the upper bound (10.27). ∎

Table 10.8 Comparison between Eq. (10.27) and the exact value

α	1	2	3	4	5	6	7
F exact	0.61	0.14	1.1E-2	3.4E-4	3.7E-6	1.5E-8	2.3E-11
Eq. (10.27)	0.61	0.14	1.1E-2	3.4E-4	3.7E-6	1.5E-8	2.3E-11

Example
Let us consider the situation where $\boldsymbol{x} = (x_1, x_2)$ and the failure occurs when $x_1x_2 > 1$ (see Example page 479). Let $\boldsymbol{x} = (0, 0)^t$: in Table 10.9, we compare the exact value and the upper bound furnished by Eq. (10.27) for the situation where $\sigma_1 = \sigma_2 = \sigma$. ∎

Table 10.9 Comparison between Eq. (10.27) and the exact value

σ	1	1/2	1/3	1/4	1/5	1/6	1/7
F exact	5.2E-2	1.6E-3	7.7E-6	5.4E-9	5.3E-13	8.5E-18	1.2E-23
Eq. (10.27)	9.2E-2	2.3E-3	1.1E-5	7.7E-9	7.7E-13	1.1E-17	2.1E-23

Exercises

1. Let $\boldsymbol{x} \in \mathbb{R}^2$ and $g(\boldsymbol{x}) = x_2 - x_1^2 - x_1 - 3$. Let $X_i = x_i + u_i$ with $u_i \sim N(0, 1)$, $i = 1, 2$. Determine an upper bound for the probability of failure at the points (1,0), (0,1), (1,1).

2. Let $\boldsymbol{x} \in \mathbb{R}^2$ and $g(\boldsymbol{x}) = x_2 - x_1^3 - x_1^2 - 4$. Let $X_i = x_i + u_i$ with $u_i \sim N(0, 1)$, $i = 1, 2$. Determine an upper bound for the probability of failure at the points (0,0), (1,0), (0,1).
3. Let $\boldsymbol{x} \in \mathbb{R}^2$ and $g(\boldsymbol{x}) = x_1 x_2 - 2$. Let $X_i = x_i + u_i$ with $u_i \sim N(0, 1)$, $i = 1, 2$. Determine an upper bound for the probability of failure at the points (0,0), (1,0), (0,1).

10.7 The Transformations of Rosenblatt and Nataf

The preceding analysis were limited to Gaussian variables $\boldsymbol{U}$. When $\boldsymbol{U}$ is not gaussian, it is necessary to use more complex transformations, such as the *Rosenblatt Transformation* (Rosenblatt, 1952), which is defined as follows: let F_1 be the marginal CDF of U_1 and, for $i > 1$, F_i be the CDF of U_i conditional to $(U_1, \ldots, U_{i-1})$:

$$F_i(u_i|U_1 = u_1, \ldots, U_{i-1} = u_{i-1}) = P(U_i < u_i|U_1 = u_1, \ldots, U_{i-1} = u_{i-1})$$

Let us consider the CDF Φ of a Gaussian variable $N(0, 1)$ and introduce new variables $\boldsymbol{W} = (W_1, \ldots, W_n)$ such that $P(W_i < w_i) = \Phi(w_i)$ with

$$\begin{aligned} w_1 &= \Phi^{-1}(F_1(u_1)), \\ w_2 &= \Phi^{-1}(F_2(u_2|U_1 = u_1)), \\ w_3 &= \Phi^{-1}(F_3(u_3|U_1 = u_1, U_2 = u_2)), \\ &\vdots \\ w_n &= \Phi^{-1}(F_n(u_n|U_1 = u_1, U_2 = u_2, \ldots, U_{n-1} = u_{n-1})) \end{aligned}$$

Let Φ_i be the CDF of W_i conditional to $(W_1, \ldots, W_{i-1})$: we have

$$\Phi_i(w_i|W_1 = w_1, \ldots, W_{i-1} = w_{i-1}) = F_i(u_i|U_1 = u_1, \ldots, U_{i-1} = u_{i-1}) = \Phi(w_i),$$

so that $\boldsymbol{W}$ is a gaussian vector of independent variables. Notice that Rosenblatt transformation may generate non gaussian variables if $\boldsymbol{U}$ contains dependent variables (see example below)

Example

Let us consider the situation where $\boldsymbol{U} = (U_1, f(U_1))$, where f is a regular bijective function and U_1 is uniformly distributed on $(-1, 1)$. Here,

$$F_1(u_1) = \frac{1 + u_1}{2} \Rightarrow w_1 = \Phi^{-1}\left(\frac{1 + u_1}{2}\right).$$

In addition,

$$F_2(u_2|U_1 = u_1) = 0, \textit{if } u_2 \leq f(U_1);\ \ F_2(u_2|U_1 = u_1) = 1, \textit{otherwise}.$$

Thus,

$$w_2 = \Phi^{-1}(0) = -\infty, \textit{if } u_2 \leq f(U_1);\ \ w_2 = \Phi^{-1}(1) = +\infty, \textit{otherwise}.$$

In this situation, the variable U_2 can be eliminated and the description of the uncertainties can be made by using only U_1 and w_1: we have

$$\boldsymbol{U} = (2\Phi(w_1) - 1, f(2\Phi(w_1) - 1)) \qquad \blacksquare$$

The use of the Rosenblatt transformation requests the complete knowledge of the distribution of the vector $\boldsymbol{U}$, so that its use may result difficulty in practice - in general, only the marginal laws are known. An alternative is to use an approximation: the *Nataf transformation* (Nataf, 1962), which uses only the marginal distributions of the variables. The Nataf transformation generates a vector $\boldsymbol{W}$ of independent Gaussian variables in two steps: first, we generate a vector $\boldsymbol{V} = (v_1, \ldots, v_n)$ by using the marginal laws:

$$v_i = \Phi^{-1}(F_i(u_i)), \quad F_i(u_i) = P(U_i < u_i).$$

In the second step, we look for a vector $\boldsymbol{W} = (W_1, \ldots, W_m), m \leq n$ formed by independent Gaussian variables of same distribution $N(0, 1)$, such that $\boldsymbol{V} = \boldsymbol{MW}$, with $\boldsymbol{M}$ a $n \times m$ matrix, which may be determined by the methods presented in pages 474 and 475. For instance, by solving the equation

$$cov\,(\boldsymbol{V}, \boldsymbol{V}) = \boldsymbol{MM}^t.$$

As previously indicated page 474, $\boldsymbol{M}$ can be determined by Cholesky's decomposition. In practice, it may be useful to avoid the evaluation of $cov\,(\boldsymbol{V}, \boldsymbol{V})$, using an approximation: for instance, $cov\,(\boldsymbol{V}, \boldsymbol{V}) \approx cov\,(\boldsymbol{U}, \boldsymbol{U})$. As an alternative, the procedure of factorization can be applied to the correlation matrix $\boldsymbol{R_V} = \rho(\boldsymbol{V}, \boldsymbol{V})$, which may be approximated into an analogous way: $\boldsymbol{R_V} \approx \boldsymbol{R_U} = \rho(\boldsymbol{U}, \boldsymbol{U})$.

Example

Let us consider again the situation where $\boldsymbol{U} = (U_1, f(U_1))$, where f is a regular bijective function and U_1 is uniformly distributed on $(-1, 1)$. Then,

$$F_1(u_1) = \frac{1 + u_1}{2}, F_2(u_2) = \frac{1 + f^{-1}(u_2)}{2}.$$

Thus,

$$v_1 = \Phi^{-1}\left(\frac{1 + u_1}{2}\right), \quad v_2 = \Phi^{-1}\left(\frac{1 + f^{-1}(u_2)}{2}\right).$$

Let $r = \rho(U_1, U_2)$: the correlation matrix is

$$\boldsymbol{R_U} = \begin{pmatrix} 1 & r \\ r & 1 \end{pmatrix}$$

and

$$\boldsymbol{M} = \begin{pmatrix} 1 & 0 \\ r & \sqrt{1 - r^2} \end{pmatrix} \Rightarrow \boldsymbol{MM}^t = \boldsymbol{R_U}.$$

Thus, using the approximation $R_V \approx R_U$, we have

$$u_1 = 2\Phi(w_1) - 1, u_2 = f\left(2\Phi\left(rw_1 + w_2\sqrt{1-r^2}\right) - 1\right) \qquad \blacksquare$$

One of the effects of the transformations of Rosenblatt and Nataf is the modification of the geometry of the failure region and of the limit state curve. For instance, let us consider the situation where $n = 2$ and the boundary curve is linear: $g(\boldsymbol{x}) = x_1 - x_2$. Consider the point $\boldsymbol{x} = (3, 4)$ and $X_i = x_i + u_i$, where u_1, u_2 are independent variables uniformly distributed on $(-1, 1)$. The associated Rosenblatt transformation modifies the geometry as shown in Fig. 10.17. Notice that the original failure region is convex.

In this same situation, let us consider the case where X_1 is log-normal distributed with mean x_1: we can use $u_1 = (\log(X_1) - \mu_1)/\sigma_1$, id est, $X_1 = \exp(\mu_1 + \sigma_1 u_1)$. Since $u_1 = 0$ for $X_1 = x_1$(centered space), we have $\mu_1 = \log(x_1)$. The Rosenblatt transformation is illustrated in Fig. 10.18, where we consider $\sigma_1 = 1$ – the other data remain identical. In this case, the failure region becomes non-convex.

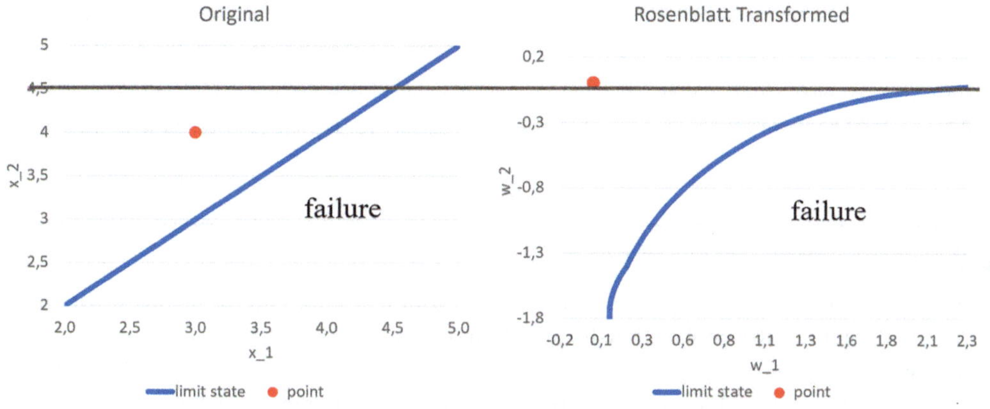

Fig. 10.17 Example of modification of the geometry by the Rosenblatt Transformation. The failure region is convex

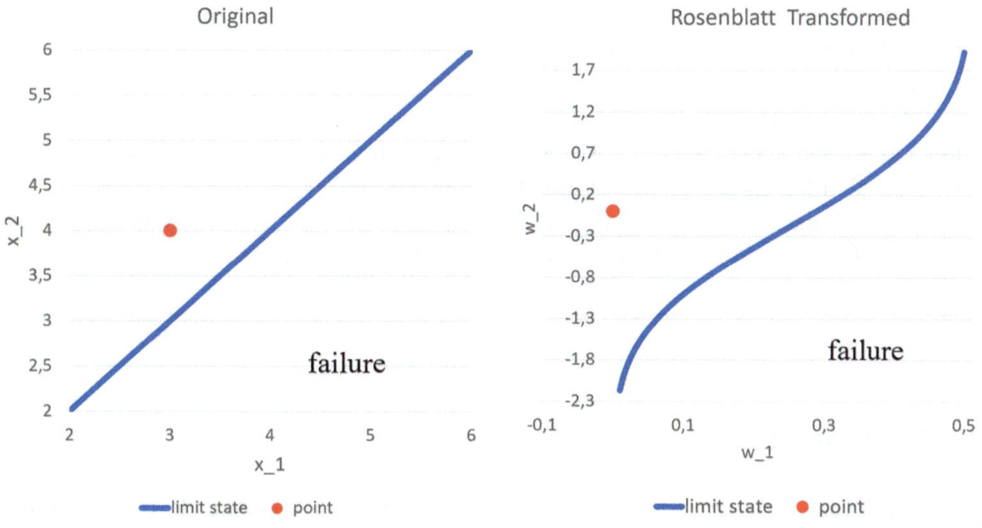

Fig. 10.18 A second example of modification of the geometry by the Rosenblatt Transformation. The failure region becomes non-convex

As a last example, let us consider the situation where X_1, X_2, X_3 are uniformly distributed and $g(\boldsymbol{x}) = x_3 - x_1^2 x_2$, with $X_1 \sim U(3,\ 5)$, $X_2 \sim U(250{,}350)$, $X_3 \sim U(2E3,\ 9E3)$. The Rosenblatt transformation is illustrated in Fig. 10.19. In this case, the failure region becomes non-convex.

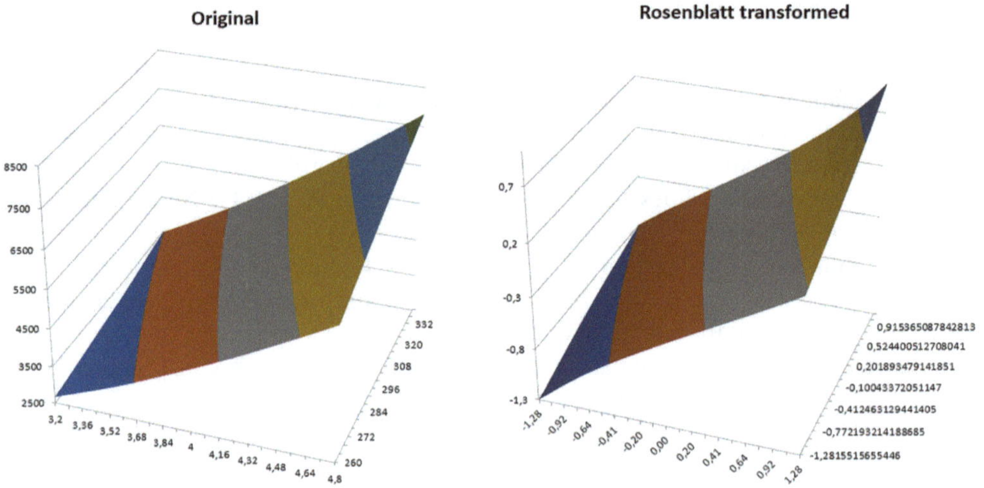

Fig. 10.19 A third example of modification of the geometry by the Rosenblatt Transformation. The failure region becomes non-convex

Exercises

1. Let $\boldsymbol{x} \in \mathbb{R}^2$ and $g(\boldsymbol{x}) = x_2 - x_1^2$. Let $X_i = x_i + u_i$ with u_i lognormally distributed with $\sigma_i = 1$, $i = 1, 2$. Plot the Rosenblatt transformed limit curve.
2. Let $\boldsymbol{x} \in \mathbb{R}^2$ and $g(\boldsymbol{x}) = x_2 - x_1^3 - x_1^2 - 4$. Let $X_i = x_i + u_i$ with u_i lognormally distributed with $\sigma_i = 1$, $i = 1, 2$. Plot the Rosenblatt transformed limit curve.
3. Let $\boldsymbol{x} \in \mathbb{R}^2$ and $g(\boldsymbol{x}) = x_1 x_2 - 2$. Let $X_i = x_i + u_i$ with u_i lognormally distributed with $\sigma_i = 1$, $i = 1, 2$. Plot the Rosenblatt transformed limit curve.
4. Let $\boldsymbol{x} \in \mathbb{R}^2$ and $g(\boldsymbol{x}) = x_2 - x_1^2$. Let $X_i = x_i + u_i$ with u_i uniformly distributed on $(-1{,}1)$, $i = 1, 2$. Plot the Rosenblatt transformed limit curve.
5. Let $\boldsymbol{x} \in \mathbb{R}^2$ and $g(\boldsymbol{x}) = x_2 - x_1^3 - x_1^2 - 4$. Let $X_i = x_i + u_i$ with u_i uniformly distributed on $(-1{,}1)$,$i = 1, 2$. Plot the Rosenblatt transformed limit curve.
6. Let $\boldsymbol{x} \in \mathbb{R}^2$ and $g(\boldsymbol{x}) = x_1 x_2 - 2$. Let $X_i = x_i + u_i$ with u_i uniformly distributed on $(-1{,}1)$, $i = 1, 2$. Plot the Rosenblatt transformed limit curve.
7. Let $\boldsymbol{x} \in \mathbb{R}^3$ and $g(\boldsymbol{x}) = x_3 - x_1^2 x_2$. Let $X_i = x_i + u_i$ with u_i lognormally distributed with $\sigma_i = 1$, $i = 1, 2, 3$. Plot the Rosenblatt transformed limit curve.

10.8 FORM and SORM

As shown in Sect. 10.5, the reliability index of Hasofer-Lind corresponds to the value of an optimization problem:

$$\beta(\boldsymbol{x}) = \|\boldsymbol{u}^*(\boldsymbol{x})\|, \boldsymbol{u}^*(\boldsymbol{x}) = \arg\min\left\{\|\boldsymbol{u}\|^2 : h(\boldsymbol{x}, \boldsymbol{u}) \geq 0\right\}.$$

The solution of the optimization problem above provides $\boldsymbol{u}^*$, which furnishes the value of the reliability index. We presented methods of solution in Sect. 10.5, but the reader will find in the literature about optimization many methods, algorithms, and software for its numerical solution - eventually, global optimization methods must be considered, if the safe or the failure region is non-convex. As shown in the preceding, we may also determine its value by solving a system of algebraic equations:

$$\boldsymbol{u}^*\|\nabla_u h(\boldsymbol{x}, \boldsymbol{u}^*)\| = \beta \nabla_u h(\boldsymbol{x}, \boldsymbol{u}^*), h(\boldsymbol{x}, \boldsymbol{u}^*) = 0.$$

Other these standard methods, there are methods specific to Reliability, which are examine below.

10.8.1 First Order Reliability Method (FORM)

The first - and extremely popular method - is the First Order Reliability Method (FORM), which consists in solving a sequence of problems with affine limit state curves: at each step, the limit state curve is approximated by a hyperplane and we use the expression given in Eq. (10.12) to get an approximated value of β. FORM can be interpreted as an adaptation of the sequential quadratic programming procedure (SQP) to reliability analysis. Assuming that all the variables are independent and $N(0, 1)$ – otherwise, transformations must be considered - the algorithm of the method reads as (see Fig. 10.20)

Algorithm of the method FORM

Input: the point $\boldsymbol{x}$, starting point $\boldsymbol{u}^{(0)} \in \mathcal{S}$, stopping conditions (for instance, maximal iteration number, minimal step at each iteration). Programs evaluating $h(\boldsymbol{x}, \boldsymbol{u})$ and $\nabla_u h(\boldsymbol{x}, \boldsymbol{u})$, a program for the correction of a tentative design point $\boldsymbol{u}$ by determining a point $\overline{\boldsymbol{u}}(\boldsymbol{u})$ such that $|h(\boldsymbol{x}, \overline{\boldsymbol{u}}(\boldsymbol{u}))| < tol$, where tol is a predefined tolerance.

Output: estimations of $\boldsymbol{u}^*(\boldsymbol{x})$ and $\beta(\boldsymbol{x})$.

Initialization: If the initial point does not verify $|h(\boldsymbol{x}, \boldsymbol{u}^{(0)})| < tol$, then correct it: $\boldsymbol{u}^{(0)} \longleftarrow \overline{\boldsymbol{u}}(\boldsymbol{u}^{(0)})$. Initialize $\beta^{(0)} = \|\boldsymbol{u}^{(0)}\|$. Set the iteration number to zero : $i \longleftarrow 0$.

1. Approximate the limit state curve by a hyperplane $\Pi^{(i)}: p^{(i)}(\boldsymbol{u}) = (\boldsymbol{a}^{(i)})^t \boldsymbol{u} + b^{(i)}(\boldsymbol{x})$. For instance, use a Taylor-MacLaurin series: $\boldsymbol{a}^{(i)} = \nabla_u h(\boldsymbol{x}^{(i)}, \boldsymbol{u}^{(i)}), b^{(i)}(\boldsymbol{x}) = h(\boldsymbol{x}, \boldsymbol{u}^{(i)}) - (\boldsymbol{a}^{(i)})^t \boldsymbol{u}^{(i)}$.
2. Set $\boldsymbol{n}^{(i)} = \boldsymbol{a}^{(i)}/\|\boldsymbol{a}^{(i)}\|, \beta^{(i+1)} = -b^{(i)}(\boldsymbol{x})/\|\boldsymbol{a}^{(i)}\|$, $\boldsymbol{u}^{(i+1)} = \beta^{(i+1)}\boldsymbol{n}^{(i)}$.
3. If $|h(\boldsymbol{x}, \boldsymbol{u}^{(i+1)})| \geq tol$ then $\boldsymbol{u}^{(i+1)} \longleftarrow \overline{\boldsymbol{u}}(\boldsymbol{u}^{(i+1)})$.
4. Set $\beta^{(i+1)} = \|\boldsymbol{u}^{(i+1)}\|$.
5. Increment the iteration number $i \longleftarrow i + 1$.
6. Test for the stopping conditions. If the iterations continue, then go to 1, else estimate $\beta(\boldsymbol{x}) \approx \beta^{(i)}, \boldsymbol{u}(\boldsymbol{x}) \approx \boldsymbol{u}^{(i)}$.

To make this algorithm to work correctly, it is necessary to choose the starting point in the safe region. We must also ensure that the generated sequence of points maintains $b^{(i)}(\boldsymbol{x}) < 0$ during the iterations - notice that $\mathbf{0} \in \mathcal{S}$, so that $\boldsymbol{u}^{(i)}$ generally satisfies the condition $(\boldsymbol{a}^{(i)})^t\boldsymbol{u}^{(i)} < 0$. When the candidate point $\boldsymbol{u}^{(i+1)}$ does not belong to the safe region $\mathcal{S}$, it is necessary to perform the correction $\bar{\boldsymbol{u}}\left(\boldsymbol{u}^{(i+1)}\right)$, in order to bring it to the safe region (with a small enough tolerance *tol*). For this purpose, optimization methods can be used, such as, for example, those presented in (Souza de Cursi et al., 2004). The SOLVER can also be used for such a correction.

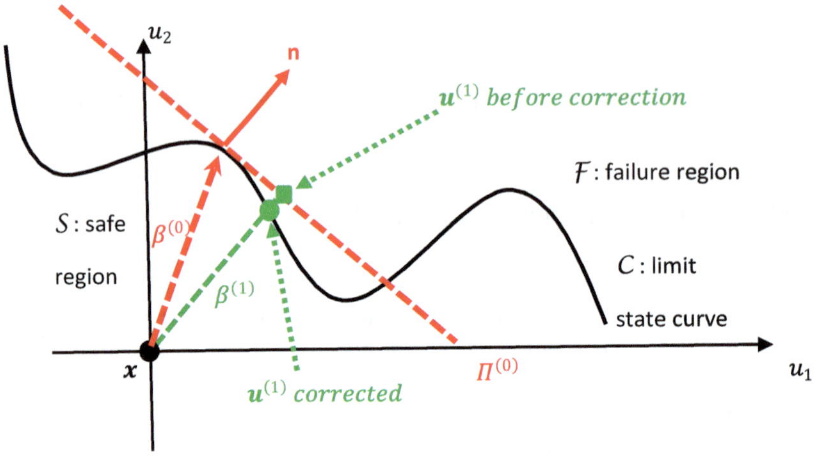

Fig. 10.20 Iterative determination of the reliability index by FORM

In general, FORM converges quickly and can be used manually in a worksheet. The method can also be used in VBA mode. The reader will find in the literature many works on FORM, with implementations available on the net. Below, we illustrate this method with a simple implementation. Notice that this is not an optimal implementation and that you can find on internet sophisticated software implementing FORM, but we focus on other aspects, more pedagogical and coherent with the spirit of this book, which concerns the use of EXCEL.

Example

Let us consider the case where $\boldsymbol{x} = (0, 0)$ and $g(\boldsymbol{x}) = x_2 - x_1^2 - x_1 - 3$. This problem was previously solved, and we found $\beta(\boldsymbol{x}) \approx 2.8$ (see page 447).

Open a blank workbook and add the modules evaluating `g, h`, etc. page 458. You will need also function `g_on_range` introduced page 460. Enter the coordinates of the point $\boldsymbol{x}$ and a guess for $\boldsymbol{u}^*$. Give the name `x` to `B1:C1` and the name `u` to `B2:C2`. Go to `B3` and enter "`=x + u`". Give the name `Px` to `B3:C3`. Go to `B4` and enter "`=g_on_range(Px)`".

Go to D4 and enter "= ABS(B4)". Go to B5 and enter "=Sqrt(SumSq(u))".

	A	B	C	D
1	x	0	0	
2	u	1	0	
3	Px	1	0	
4	g(x)	-5	abs(g)	5
5	beta	1		

Observe that $g(\boldsymbol{Px}) < 0$: the point $\boldsymbol{Px}$ is in the interior of the safe region S, not on the limit state curve. To bring it on the limit state curve, we can use the SOLVER. Go to the tab *Data* and launch the SOLVER to get a value of $\boldsymbol{Px}$ such that $g(\boldsymbol{Px}) = 0$- you can choose between two options:

either to minimize cell D4 by changing B3:C3:

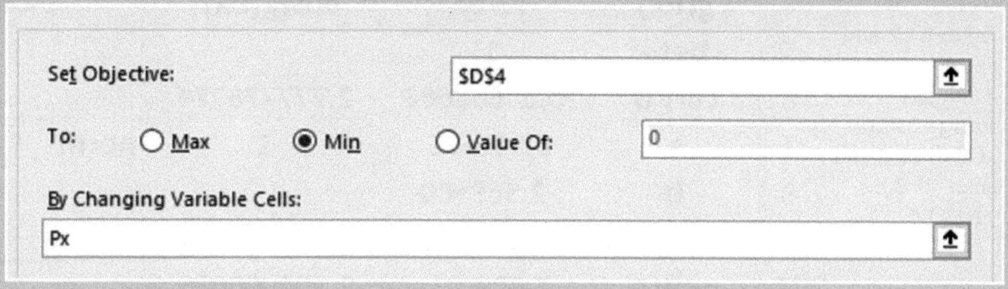

or to find a root of $g(\boldsymbol{x})$ starting from $\boldsymbol{Px}$:

Let us choose this last option: we look for a vector such that $(\boldsymbol{Px}) = 0$. Running the SOLVER furnishes

	A	B	C	D
1	x	0	0	
2	u	1	0	
3	Px	-0,66666666	2,777778774	
4	g(Px)	1E-06	abs(g(Px))	1E-06
5	beta	1		

Now, ***Px*** is on the limit state curve. Go to `B6` and enter "=`Px-x`". The corrected value of ***u*** appears. Give the name `uc` to `B6:C6`. Now, let us make one step of the algorithm. Go to `B7` and enter the formula "= `gradh_on_range(x;uc)`". This furnishes the vector ***a***. In E7, enter "=`SQRT(B7*B7 + C7*C7)`" (the value of $\|\boldsymbol{a}\|$). Go to `B8` and enter "= `B4 - B7*B6-C7*C6`" – this is the value of $b^{(i)}(\boldsymbol{x})$. Go to `B9` and enter the initial estimation for the new value of $\beta(\boldsymbol{x})$: "= `B8/E7`". Go to `B10` and enter "= `B9*B7/E7`". Go to `C10` and enter "=`B9*C7/E7`".

Give the name `un` to `B10:C10`. Go to `B11` and enter "=`x + un`". Give the name `Pxn` to `B11:C11`. Go to `B12` and enter "=`g_on_range(Pxn)`". Go to `D12` and enter "=`Abs(B12)`".

	A	B	C	D	E
1	x	0	0		
2	u	1	0		
3	Px	-0,66666666	2,777778774		
4	g(Px)	1E-06	abs(g(Px))	1E-06	
5	beta	1			
6	corr u	-0,66666666	2,777778774		
7	a	3,33E-01	1	norm	1,054093
8	b	-2,56E+00			
9	new beta	2,4244129			
10	new u	7,67E-01	2,300000039		
11	new Px	1	0		
12	g(new Px)	-5	abs(g(new Px))	5	

The value in B12 is negative. Again, we need to run the SOLVER to bring it to zero by modifying B11:C11.

After running the SOLVER, go to `B13` and enter "=`Pxn-x`". Go to `B14` and enter "=`SQRT((B13-B3)^2 + (C13-C3)^2)`".

	A	B	C	D	E
1	x	0	0		
2	u	1	0		
3	Px	-0,66666666	2,777778774		
4	g(Px)	1E-06	abs(g(Px))	1E-06	
5	beta	1			
6	corr u	-0,66666666	2,777778774		
7	a	3,33E-01	1	norm	1,054093
8	b	-2,56E+00			
9	new beta	2,4244129			
10	new u	7,67E-01	2,300000039		
11	new Px	-0,15267914	2,870632536		
12	g(new Px)	7,52955E-07	abs(g(new Px))	7,53E-07	
13	corr u	-0,15267914	2,870632536		
14	prec	0,522307374			

Cells B13:C13 give the new estimate for u. B14 gives the norm of the difference between two iterates. To iterate, we must restart the operations using this new value of u: copy B13:C13 and paste **its values** in B2:C2. Then you can make the same operations again, starting from entering "=x + u" in B3. You can use a Macro to do it:

```
Sub autoiterate()
    Range("B3:C3").Select
    Selection.ClearContents
    Range("B3").Select
    ActiveCell.Formula2R1C1 = "=x+u"
    Range("B4").Select
    SolverOk SetCell:="$B$4", MaxMinVal:=3, ValueOf:=0, ByChange:="$B$3:$C$3", _
        Engine:=1, EngineDesc:="GRG Nonlinear"
    SolverSolve True

    Range("B11:C11").Select
    Selection.ClearContents
    Range("B11").Select
    ActiveCell.Formula2R1C1 = "=x+un"
    SolverOk SetCell:="$B$12", MaxMinVal:=3, ValueOf:=0, ByChange:="$B$11:$C$11", _
        Engine:=1, EngineDesc:="GRG Nonlinear"
    SolverSolve True
End Sub
```

With the macro autoiterate, you have only to copy B13:C13 and paste its values in B2:C2.

Then run autoiterate: the result is

13	corr u	-0,22058765	2,828071259
14	prec	0,080143804	

Run autoiterate again. The results are

13	corr u	-0,68853904	2,785547515
14	prec	0,469879528	

Run autoiterate yet. The results become

13	corr u	-0,6874208	2,785126557
14	prec	0,001194852	

We stop the iterations here – if you continue, the iterations will oscillate around these values. The estimation of the reliability index is $\|\boldsymbol{u}\| \approx 2.8$.

In this situation too, the use of VBA is more efficient. For instance, open a blank workbook, add a module and enter the VBA code below:

```
Sub form()
col = Cells(1, Columns.Count).End(xlToLeft).Column
n = col - 1
x = read_line(1, 2, col)
ut = read_line(2, 2, col)
aux = h(x, ut)
If Abs(aux) > tol Then
   Call correction
End If
uac = read_line(2, 2, col)
xac = plus(x, uac)
beta = norm2(uac)
nitmax = Cells(4, 2).Value
tol = Cells(5, 2).Value
prec = 1 + tol
Cells(5, 4).Value = prec
While nit <= nitmax And prec > tol
  a = gradh(x, uac)
  b = h(x, uac) - sp(a, uac)
  beta = -b / norm2(a)
  ut = sm(a, beta / norm2(a))
  aux = h(x, ut)
  Cells(6, 2).Value = aux
  Cells(7, 2).Value = Abs(aux)
  If Abs(aux) > tol Then
     Call correction
  End If
  un = read_line(2, 2, col)
  xn = plus(x, un)
  beta = norm2(un)
  nit = nit + 1
  prec = norm2(minus(un, uac))
  Cells(5, 4).Value = prec
  coef = 1
  uac = plus(sm(un, coef), sm(uac, 1 - coef))
  For i = 2 To col
    Cells(2, i).Value = uac(i - 1)
  Next i
  Cells(3, 2).Value = norm2(uac)
  Cells(4, 4).Value = nit
Wend
End Sub
```

Sub `Correction` reads as

```
Sub correction()
col = Cells(1, Cells.Columns.Count).End(xlToLeft).Column
ruad = Range(Cells(2, 2), Cells(2, col)).Address
Cells(6, 2).Formula = "=h_on_range(B1:C1,B2:C2)"
Cells(7, 2).Formula = "=abs(B6)"
rfobjad = Range(Cells(7, 2), Cells(7, 2)).Address
  SolverReset
  SolverOk SetCell:=rfobjad, MaxMinVal:=2, ValueOf:=0, ByChange:=ruad, _
         Engine:=1, EngineDesc:="GRG Nonlinear"
  SolverOptions AssumeNonNeg:=False
  SolverOptions Multistart:=True, PopulationSize:=500, RequireBounds:=False
  SolverSolve True
End Sub
```

The code of the `Sub form` uses `read_line, plus, norm2, minus, sp., sm`. These functions are available in *VectorTools.bas*. `read_line` reads the values on a line into a vector, `plus` evaluates the sum of two vectors, `norm2` calculates the norm of a vector, `minus` evaluates the difference of two vectors, `sp.` calculates the scalar product of two vectors and `sm` multiplies a vector by a real number. The method contains a coefficient of relaxation coef to update $\boldsymbol{u}^{(i+1)} \leftarrow coef * \bar{u}\left(\boldsymbol{u}^{(i+1)}\right) + (1 - coef) * \boldsymbol{u}^{(i)}$.

The code uses different cells of the Active Worksheet, as shown here:

	A	B	C	D
1	x	0	0	
2	u	1	0	
3	beta			
4	nitmax	10	i	
5	tol	1,00E-10	prec	
6	h			
7	abs			

`B1:C1` contains the point $\boldsymbol{x}$; `B2:C2` is the initial guess for $\boldsymbol{u}$. `B3` will exhibit the value of $\beta(\boldsymbol{x})$. The maximal iteration number is given in `B4` and the tolerance *tol* is given in `B5`. `D4` will contain the iteration number, `D5` will exhibit the precision between two iterates of $\boldsymbol{u}$. `B6` will contain the value of $h(\boldsymbol{x}, \boldsymbol{u})$, `B7` its absolute value.

Run the macro `form`. You will get the results: the method converges in 2 iterations and estimates $\beta(\boldsymbol{x}) \approx 2.9$.

	A	B	C	D
1	x	0	0	
2	u	-0,66666572	2,777777	
3	beta	2,856657278		
4	nitmax	10	i	2
5	tol	1,00E-10	prec	0
6	h	-1,7399E-12		
7	abs	1,73994E-12		

Example

Multiple failure conditions can be treated by the simplified previously introduced in sect. 10.3.1 and exploited in the sequel.

To do this, we need to run form with different parameters. Let us illustrate the procedure by using the same example considered in the preceding: $\mathbf{x} \in \mathbb{R}^2$ and three failure conditions, corresponding to $g_1(\boldsymbol{x}) = x_2 - x_1^2 - x_1 - 3$, $g_2(\boldsymbol{x}) = x_1 - x_2^2 - x_2 - 2$, $g_3(\boldsymbol{x}) = x_1 + x_2 + x_1^2 + x_2^2 - 6$. We consider a workbook as shown at right. Each group of 4 lines surrounded by black corresponds to a function g_i. We used h_on_range as in page 465.

	A	B	C	D
1	x	0	0	
2	nitmax	10	tol	1,00E-10
3	u	1	0	
4	beta			
5	h	-5	abs	5
6	i		prec	
7	u	1	0	
8	beta			
9	h	-1	abs	1
10	i		prec	
11	u	1	0	
12	beta			
13	h	-4	abs	4
14	i		prec	

For instance, we may use the code

```
Sub form(col, lx, lu, lh, nitmax, tol, gname As String, dgname As String)
n = col - 1
x = read_line(lx, 2, col)
ut = read_line(lu, 2, col)
aux = h(x, ut, gname)
If Abs(aux) > tol Then
   Call correction(col, lx, lu, lh, gname)
End If
uac = read_line(lu, 2, col)
xac = plus(x, uac)
beta = norm2(uac)
nit = 0
prec = 1 + tol
Cells(lh + 1, 2).Value = nit
Cells(lh + 1, 4).Value = prec
While nit <= nitmax And prec > tol
  a = gradh(x, uac, dgname)
  b = h(x, uac, gname) - sp(a, uac)
  beta = -b / norm2(a)
  ut = sm(a, beta / norm2(a))
  aux = h(x, ut, gname)
  Cells(lh, 2).Value = aux
  Cells(lh, 4).Value = Abs(aux)
  If Abs(aux) > tol Then
     Call correction(col, lx, lu, lh, gname)
  End If
  un = read_line(lu, 2, col)
  xn = plus(x, un)
  beta = norm2(un)
  nit = nit + 1
  Cells(lh + 1, 2).Value = nit
  prec = norm2(minus(un, uac))
  Cells(lh + 1, 4).Value = prec
  coef = 1
  uac = plus(sm(un, coef), sm(uac, 1 - coef))
  For i = 2 To col
    Cells(lu, i).Value = uac(i - 1)
  Next i
  Cells(lu + 1, 2).Value = norm2(uac)
Wend
End Sub
```

```
Sub form_multiple()
col = Cells(1, Columns.Count).End(xlToLeft).Column
nitmax = Cells(2, 2).Value
tol = Cells(2, 4).Value
Call form(col, 1, 3, 5, nitmax, tol, "g_1", "gradg_1")
Call form(col, 1, 7, 9, nitmax, tol, "g_2", "gradg_2")
Call form(col, 1, 11, 13, nitmax, tol, "g_3", "gradg_3")
End Sub
```

Run the `Sub form_multiple`. You will get the results

	A	B	C	D
1	x	0	0	
2	nitmax	10	tol	1,00E-10
3	u	-0,66666572	2,777777	
4	beta	2,856657278		
5	h	-1,7399E-12	abs	1,74E-12
6	i	2	prec	0
7	u	2	-1	
8	beta	2,236067977		
9	h	0	abs	0
10	i	1	prec	0
11	u	1,918677465	0,306225	
12	beta	1,942960933		
13	h	1,24631E-08	abs	1,25E-08
14	i	1	prec	0
15				

So, $\beta(\boldsymbol{x}) \approx 1.9$. As previously observed, the results may vary with the starting point. For instance:

	A	B	C	D
1	x	0	0	
2	nitmax	10	tol	1,00E-10
3	u	0	0	
4	beta			
5	h	-3	abs	3
6	i	1	prec	0
7	u	0	0	
8	beta			
9	h	-2	abs	2
10	i	1	prec	0
11	u	0	0	
12	beta			
13	h	-6	abs	6
14	i	1	prec	0

a) Initial data

	A	B	C	D
1	x	0	0	
2	nitmax	10	tol	1,00E-10
3	u	-1	3	
4	beta	3,16227766		
5	h	0	abs	0
6	i	1	prec	0
7	u	2	-1	
8	beta	2,236067977		
9	h	0	abs	0
10	i	1	prec	0
11	u	1,302777443	1,302774	
12	beta	1,842402982		
13	h	4,9738E-13	abs	4,97E-13
14	i	1	prec	0

b) Final result: $\beta(\boldsymbol{x}) \approx 1.8$.

	A	B	C	D
1	x	0	0	
2	nitmax	10	tol	1,00E-10
3	u	0	1	
4	beta			
5	h	-2	abs	2
6	i	1	prec	0
7	u	0	1	
8	beta			
9	h	-4	abs	4
10	i	1	prec	0
11	u	0	1	
12	beta			
13	h	-4	abs	4
14	i	1	prec	0

a) Initial data

	A	B	C	D
1	x	0	0	
2	nitmax	10	tol	1,00E-10
3	u	-1	3	
4	beta	3,16227766		
5	h	0	abs	0
6	i	1	prec	0
7	u	1,777779686	-0,66667	
8	beta	1,898671295		
9	h	1,58936E-10	abs	1,59E-10
10	i	1	prec	0
11	u	0,306225364	1,918677	
12	beta	1,942960932		
13	h	1,13709E-08	abs	1,14E-08
14	i	1	prec	0

d) Final result: $\beta(\boldsymbol{x}) \approx 1.9$.

Exercises

1. Let $\boldsymbol{x} \in \mathbb{R}^2$ and $g(\boldsymbol{x}) = x_2 - x_1^2 - x_1 - 3$. Let $X_i = x_i + u_i$ with $u_i \sim N(0, 1)$, $i = 1, 2$. Use FORM to evaluate the reliability index of the points (1,0), (0,1).
2. Let $\boldsymbol{x} \in \mathbb{R}^2$ and $g(\boldsymbol{x}) = x_2 - x_1^3 - x_1^2 - 4$. Let $X_i = x_i + u_i$ with $u_i \sim N(0, 1)$, $i = 1, 2$. Use FORM to evaluate the reliability index of the points (0,0), (1,0), (0,1).

3. Let $\boldsymbol{x} \in \mathbb{R}^2$ and $g(\boldsymbol{x}) = x_1 x_2 - 2$. Let $X_i = x_i + u_i$ with $u_i \sim N(0, 1)$, $i = 1, 2$. Use FORM to evaluate the reliability index of the points (0,0), (1,0), (0,1).
4. Let $\boldsymbol{x} \in \mathbb{R}^4$ and $g(\boldsymbol{x}) = x_1 x_2 x_3 x_4 - 2$. Let $X_i = x_i + u_i$ with $u_i \sim N(0, 1)$, $i = 1, 2, 3, 4$. Use FORM to evaluate the reliability index of the points (0,0,0,0), (1,0,0,0), (0,1,0,0), (0,0,1,0) and (0,0,0,1).
5. Let $\boldsymbol{x} \in \mathbb{R}^4$ and $g(\boldsymbol{x}) = x_4 - x_1 x_2 x_3 - 3$. Let $X_i = x_i + u_i$ with $u_i \sim N(0, 1)$, $i = 1, 2, 3, 4$. Use FORM to evaluate the reliability index of the points (0,0,0,0), (1,0,0,0), (0,1,0,0), (0,0,1,0) and (0,0,0,1).

10.8.2 Second Order Reliability Method (SORM)

The second – and less popular than FORM - is the Second Order Reliability Method (SORM), which is a variant from FORM, where a quadratic approximation is used instead the affine one. In SORM, the limit state curve is approximated by a quadratic equation and we must solve at each step an optimization problem involving a quadratic objective function and quadratic constraints – such a problem is said a Quadratically Constrained Quadratic Problem (QCQP). Analogously to FORM, the reader will find many works and implementations of SORM. WE give below a simple implementation intended to a pedagogical aim.

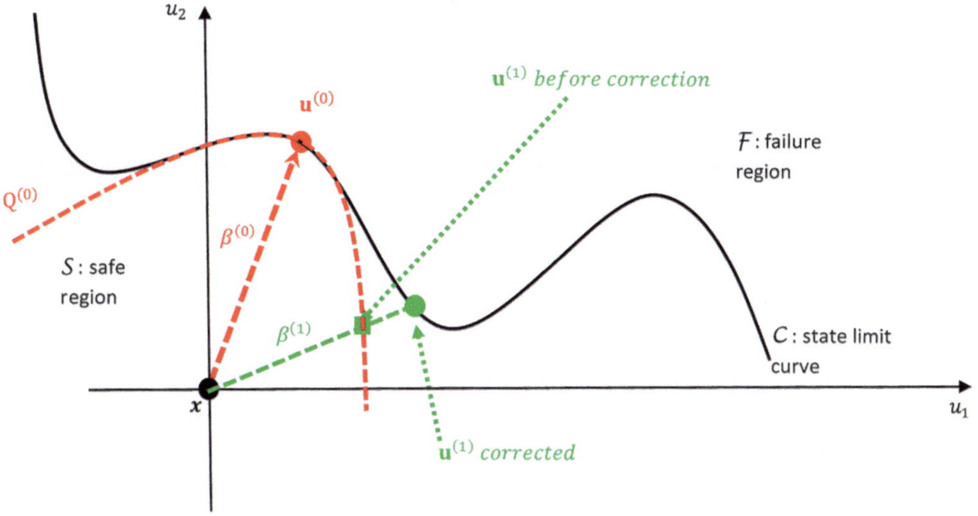

Fig. 10.21 Iterative determination of the reliability index by SORM

The algorithm of the method reads as (Fig. 10.21)

Algorithm of the method SORM

Input: the point $\boldsymbol{x}$, starting point $\boldsymbol{u}^{(0)} \in \mathcal{S}$, stopping conditions (for instance, maximal iteration number, minimal step at each iteration). Programs evaluating $h(\boldsymbol{x},\boldsymbol{u})$, $\nabla_u h(\boldsymbol{x},\boldsymbol{u})$ and the Hessian $\nabla_u^2 h(\boldsymbol{x},\boldsymbol{u})$, a program for the correction of a tentative design point $\boldsymbol{u}$ by determining a point $\overline{\boldsymbol{u}}(\boldsymbol{u})$ such that $|h(\boldsymbol{x},\overline{\boldsymbol{u}}(\boldsymbol{u}))| < tol$, where tol is a predefined tolerance.

Output: estimations of $\boldsymbol{u}^*(\boldsymbol{x})$ and $\beta(\boldsymbol{x})$.

Initialization: If the initial point does not verify $|h(\boldsymbol{x},\boldsymbol{u}^{(0)})| < tol$, then correct it: $\boldsymbol{u}^{(0)} \longleftarrow \overline{\boldsymbol{u}}(\boldsymbol{u}^{(0)})$. Initialize $\beta^{(0)} = \|\boldsymbol{u}^{(0)}\|$. Set the iteration number to zero : $i \longleftarrow 0$.

1. Approximate the limit state curve by a quadratic form $Q^{(i)}: q^{(i)}(\boldsymbol{u}) = \boldsymbol{u}^t\boldsymbol{A}^{(i)}\boldsymbol{u}/2 + \boldsymbol{u}^t\boldsymbol{B}^{(i)} + C^{(i)}(\boldsymbol{x})$. For instance, use a Taylor-MacLaurin series: $\boldsymbol{A}^{(i)} = \nabla_u^2 h(\boldsymbol{x}^{(i)},\boldsymbol{u}^{(i)})$, $\boldsymbol{B}^{(i)} = \nabla_u h(\boldsymbol{x}^{(i)},\boldsymbol{u}^{(i)}) - \boldsymbol{A}^{(i)}\boldsymbol{u}^{(i)}$, $C^{(i)}(\boldsymbol{x}) = h(\boldsymbol{x},\boldsymbol{u}^{(i)}) - (\boldsymbol{B}^{(i)})^t\boldsymbol{u}^{(i)} + (\boldsymbol{u}^{(i)})^t\boldsymbol{A}^{(i)}(\boldsymbol{u}^{(i)})^t/2.$
2. Find the solution of the QCQP: $\boldsymbol{u}^{(i+1)} = \arg\min\{\boldsymbol{u}^t\boldsymbol{u} : q^{(i)}(\boldsymbol{u}) \geq 0\}$
3. If $|h(\boldsymbol{x},\boldsymbol{u}^{(i+1)})| \geq tol$ then $\boldsymbol{u}^{(i+1)} \longleftarrow \overline{\boldsymbol{u}}(\boldsymbol{u}^{(i+1)})$.
4. Set $\beta^{(i+1)} = \|\boldsymbol{u}^{(i+1)}\|$.
5. Increment the iteration number $i \longleftarrow i + 1$.
6. Test for the stopping conditions. If the iterations continue, then go to 1, else estimate $\beta(\boldsymbol{x}) \approx \beta^{(i)}, \boldsymbol{u}(\boldsymbol{x}) \approx \boldsymbol{u}^{(i)}$.

Example

Let us exemplify how to implement SORM in dimension 2. We use as example the case where $\boldsymbol{x} = (0, 0)$ and $g(\boldsymbol{x}) = x_2 - x_1^2 - x_1 - 3$. This problem was previously solved, and we found $\beta(\boldsymbol{x}) \approx 2.8$ (see page 447).

Open a blank workbook and add the modules evaluating g, h, etc. page 458. You will need also function g_on_range introduced page 460. Add a module evaluating $\nabla_u^2 h$:

```
Function d2g(x)
Dim gg(1 To 2, 1 To 2) As Double
gg(1, 1) = -2
gg(1, 2) = 0
gg(2, 1) = 0
gg(2, 2) = 0
d2g = gg
End Function
```

```
Function d2h(x, u)
y = plus(x, u)
gh = d2g(y)
d2h = gh
End Function
```

Now, prepare the worksheet: `B1:C1` contains the point $\boldsymbol{x}$; `B2:C2` is the initial guess for $\boldsymbol{u}$. `B3` will exhibit the value of $\beta(\boldsymbol{x})$. The maximal iteration number is given in `B4` and the tolerance *tol* is given in `B5`. `D4` will contain the iteration number, `D5` will exhibit the precision between two iterates of $\boldsymbol{u}$. `B6` will contain the value of $h(\boldsymbol{x}, \boldsymbol{u})$, `B7` its absolute value. The matrix $\boldsymbol{A}^{(i)}$ will appear in `B8:C9`, the vector $\boldsymbol{B}^{(i)}$ will be exhibited in `B10:C10`, $C^{(i)}(\boldsymbol{x})$ will appear in `B11`. The value of $q^{(i)}(\boldsymbol{u})$ will be in `B12`. Give the names `a` to `B8:C9`, `b` to `B10:C10`, `c_` to `B11`. The formulae in `B6` and `B7` will be written by the macro. You must enter the formula in `B12`, which reads as "`=MMULT(u;MMULT(a; TRANSPOSE (u)))/2-MMULT(u; TRANSPOSE (b)) + c_`".

The workbook appears as shown at right. EXCEL® produces an error `#VALUE`, due to the fact that the cells defining `a,b,c` are empty – the error will disappear when executing the method.

	A	B	C
1	x	0	0
2	u	1	0
3	beta		
4	nitmax	100	i
5	tol	1,00E-10	prec
6	h	-5	
7	abs	5	
8	a		
9			
10	b		
11	c		
12	q(u)	#VALUE!	

Otherwise, you can enter random values in these cells and the error will disappear:

	A	B	C
1	x	0	0
2	u	1	0
3	beta		
4	nitmax	100	i
5	tol	1,00E-10	prec
6	h	-5	
7	abs	5	
8	a	0	0
9		0	0
10	b	0	0
11	c	0	
12	q(u)	0	

An example of `Sub` that implements SORM in dimension 2 is given in the sequel (Fig. 10.22). It uses functions `read_line`, `plus`, `norm2`, `sp.`, `sm`, `minus`, which are implemented in `VectorTools.bas` – this module must be imported to the workbook (see page 10, Sect. 1.6).

```
Sub sorm()
col = Cells(1, Columns.Count).End(xlToLeft).Column
ruad = Range(Cells(2, 2), Cells(2, col)).Address
n = col - 1
x = read_line(1, 2, col)
ut = read_line(2, 2, col)
aux = h(x, ut)
If Abs(aux) > tol Then
   Call correction
End If
uac = read_line(2, 2, col)
xac = plus(x, uac)
beta = norm2(uac)
nitmax = Cells(4, 2).Value
tol = Cells(5, 2).Value
prec = 1 + tol
Cells(5, 4).Value = prec
While nit <= nitmax And prec > tol
  a = d2h(x, uac)
  Range("$B$8:$C$9").Value = a
  b = minus(gradh(x, uac), prodmat(a, uac))
  Range("$B$10").Value = b(1)
  Range("$C$10").Value = b(2)
  c = h(x, uac) - sp(b, uac) + sp(uac, prodmat(a, uac)) / 2
  Range("$B$11").Value = c
  aux = WorksheetFunction.Transpose(u)
  SolverReset
  SolverOk SetCell:="$B$12", MaxMinVal:=2, ValueOf:=0, ByChange:=ruad, _
          Engine:=1, EngineDesc:="GRG Nonlinear"
  SolverOptions AssumeNonNeg:=False
  SolverAdd cellRef:="$B$12", relation:=3, formulaText:=0
  SolverOptions Multistart:=True, PopulationSize:=500, RequireBounds:=False
  SolverSolve True
  un = read_line(2, 2, col)
  aux = h(x, un)
  If Abs(aux) > tol Then
     Call correction
     un = read_line(2, 2, col)
  End If
  xn = plus(x, un)
  beta = norm2(un)
  nit = nit + 1
  prec = norm2(minus(un, uac))
  Cells(5, 4).Value = prec
  coef = 1
  uac = plus(sm(un, coef), sm(uac, 1 - coef))
  Cells(2, 2).Value = uac(1)
  Cells(2, 3).Value = uac(2)
  Cells(3, 2).Value = norm2(uac)
  Cells(4, 4).Value = nit
Wend
End Sub
```

Fig. 10.22 An example of implementation of SORM

Running the Sub SORM produces the results at right

	A	B	C	D
1	x	0	0	
2	u	-0,65811384	2,775	
3	beta	2,851970973		
4	nitmax	100	i	7
5	tol	1,00E-10	prec	1,27E-11
6	h	-5,6391E-08		
7	abs	5,63905E-08		
8	a	-2	0	
9		0	0	
10	b	-1	1	
11	c	-3,86622765		
12	q(u)	-7,73245524		

Thus, we obtain the estimation $\beta(\boldsymbol{x}) \approx 2.85$.

Example

Multiple failure conditions can be treated by the simplified approach previously presented. Analogously to FORM, we need to run form with different parameters. Let us illustrate the procedure by using the same example considered in the preceding: $\mathbf{x} \in \mathbb{R}^2$ and three failure conditions, corresponding to $g_1(\boldsymbol{x}) = x_2 - x_1^2 - x_1 - 3$, $g_2(\boldsymbol{x}) = x_1 - x_2^2 - x_2 - 2$, $g_3(\boldsymbol{x}) = x_1 + x_2 + x_1^2 + x_2^2 - 6$.

We consider a workbook as shown at right. Each group of 4 lines surrounded by black corresponds to a function g_i. We used h_on_range as in page 465.

	A	B	C	D
1	x	0	0	
2	nitmax	10	tol	1,00E-10
3	u	1	0	
4	beta			
5	h	-5	abs	5
6	i		prec	
7	u	1	0	
8	beta			
9	h	-1	abs	1
10	i		prec	
11	u	1	0	
12	beta			
13	h	-4	abs	4
14	i		prec	

For instance, we may use the code

```
Sub sorm_multiple()
col = Cells(1, Columns.Count).End(xlToLeft).Column
nitmax = Cells(2, 2).Value
tol = Cells(2, 4).Value
Call sorm(col, 1, 3, 5, nitmax, tol, "g_1", "gradg_1", "hgone", 1)
Call sorm(col, 1, 7, 9, nitmax, tol, "g_2", "gradg_2", "hgtwo", 2)
Call sorm(col, 1, 11, 13, nitmax, tol, "g_3", "gradg_3", "hgthree", 3)
End Sub
```

Functions g_x, gradg_x are identical to those introduced in the preceding. Hgone, hgtwo, hgthree determine the matrix of second derivatives $\nabla_u^2 h_i$. The new function sorm reads as

```
Sub sorm(col, lx, lu, lh, nitmax, tol, gname As String, _
        dgname As String, d2gname As String, ind)
inda = 20 + 10 * ind - 10
ra = Range(Cells(inda, 2), Cells(inda + 1, 3)).Address
rh = Range(Cells(inda + 4, 2), Cells(inda + 4, 2)).Address
x = read_line(lx, 2, col)
ut = read_line(lu, 2, col)
aux = h(x, ut, gname)
If Abs(aux) > tol Then
   Call correction(col, lx, lu, lh, gname)
End If
uac = read_line(lu, 2, col)
xac = plus(x, uac)
beta = norm2(uac)
nit = 0
prec = 1 + tol
Cells(lh + 1, 2).Value = nit
Cells(lh + 1, 4).Value = prec
While nit <= nitmax And prec > tol
  a = d2h(x, uac, d2gname)
  Range(ra).Value = a
  b = minus(gradh(x, uac, dgname), prodmat(a, uac))
  Cells(inda + 2, 2).Value = b(1)
  Cells(inda + 2, 3).Value = b(2)
  c = h(x, uac, gname) - sp(b, uac) + sp(uac, prodmat(a, uac)) / 2
  Cells(inda + 3, 2).Value = c
  aux = WorksheetFunction.Transpose(u)
  SolverReset
  SolverOk SetCell:="$B$24", MaxMinVal:=2, ValueOf:=0, ByChange:=ruad, _
         Engine:=1, EngineDesc:="GRG Nonlinear"
  SolverOptions AssumeNonNeg:=False
  SolverAdd cellRef:="$B$24", relation:=3, formulaText:=0
  SolverOptions Multistart:=True, PopulationSize:=500, RequireBounds:=False
  SolverSolve True
  un = read_line(lu, 2, col)
  aux = h(x, un, gname)
  If Abs(aux) > tol Then
     Call correction(col, lx, lu, lh, gname)
  End If
  un = read_line(lu, 2, col)
  xn = plus(x, un)
  beta = norm2(un)
  nit = nit + 1
  prec = norm2(minus(un, uac))
  Cells(lu + 3, 4).Value = prec
  coef = 1
  uac = plus(sm(un, coef), sm(uac, 1 - coef))
  For i = 2 To col
    Cells(lu, i).Value = uac(i - 1)
  Next i
  Cells(lu + 1, 2).Value = norm2(uac)
  Cells(lu + 3, 4).Value = nit
Wend
End Sub
```

Run the `Sub sorm_multiple`. The results appear:

	A	B	C	D
1	x	0	0	
2	nitmax	10	tol	1,00E-10
3	u	-0,66666572	2,777777	
4	beta	2,856657278		
5	h	-1,7399E-12	abs	1,74E-12
6	i	0	prec	2
7	u	2	-1	
8	beta	2,236067977		
9	h	0	abs	0
10	i	0	prec	1
11	u	1,918677465	0,306225	
12	beta	1,942960933		
13	h	1,24631E-08	abs	1,25E-08
14	i	0	prec	1

As shown in the preceding, the results depend on the starting point and the parameters used. For instance, we can get also:

	A	B	C	D
1	x	0	0	
2	nitmax	10	tol	1,00E-10
3	u	0	0	
4	beta			
5	h	-3	abs	3
6	i		prec	
7	u	0	0	
8	beta			
9	h	-2	abs	2
10	i		prec	
11	u	0	0	
12	beta			
13	h	-6	abs	6
14	i		prec	

a) Initial data

	A	B	C	D
1	x	0	0	
2	nitmax	10	tol	1,00E-10
3	u	-1	3	
4	beta	3,16227766		
5	h	0	abs	0
6	i	0	prec	1
7	u	2	-1	
8	beta	2,236067977		
9	h	0	abs	0
10	i	0	prec	1
11	u	1,302777443	1,302774	
12	beta	1,842402982		
13	h	3,15745E-08	abs	3,16E-08
14	i	0	prec	1

b) Final result : $\beta(\boldsymbol{x}) \approx 1.8$

	A	B	C	D
1	x	0	0	
2	nitmax	10	tol	1,00E-10
3	u	0	1	
4	beta			
5	h	-2	abs	2
6	i		prec	
7	u	0	1	
8	beta			
9	h	-4	abs	4
10	i		prec	
11	u	0	1	
12	beta			
13	h	-4	abs	4
14	i		prec	

a) Initial data

	A	B	C	D
1	x	0	0	
2	nitmax	10	tol	1,00E-10
3	u	-1	3	
4	beta	3,16227766		
5	h	0	abs	0
6	i	0	prec	1
7	u	1,777779686	-0,66667	
8	beta	1,898671295		
9	h	1,58936E-10	abs	1,59E-10
10	i	0	prec	1
11	u	0,306225364	1,918677	
12	beta	1,942960932		
13	h	1,13709E-08	abs	1,14E-08
14	i	0	prec	1

b) Final result : $\beta(\boldsymbol{x}) \approx 1.9$

Exercises

1. Let $\boldsymbol{x} \in \mathbb{R}^2$ and $g(\boldsymbol{x}) = x_2 - x_1^2 - x_1 - 3$. Let $X_i = x_i + u_i$ with $u_i \sim N(0, 1)$, $i = 1, 2$. Use SORM to evaluate the reliability index of the points (1,0), (0,1).
2. Let $\boldsymbol{x} \in \mathbb{R}^2$ and $g(\boldsymbol{x}) = x_2 - x_1^3 - x_1^2 - 4$. Let $X_i = x_i + u_i$ with $u_i \sim N(0, 1)$, $i = 1, 2$. Use SORM to evaluate the reliability index of the points (0,0), (1,0), (0,1).
3. Let $\boldsymbol{x} \in \mathbb{R}^2$ and $g(\boldsymbol{x}) = x_1x_2 - 2$. Let $X_i = x_i + u_i$ with $u_i \sim N(0, 1)$, $i = 1, 2$. Use SORM to evaluate the reliability index of the points (0,0), (1,0), (0,1).
4. Let $\boldsymbol{x} \in \mathbb{R}^4$ and $g(\boldsymbol{x}) = x_1x_2x_3x_4 - 2$. Let $X_i = x_i + u_i$ with $u_i \sim N(0, 1)$, $i = 1, 2, 3, 4$. Use SORM to evaluate the reliability index of the points (0,0,0,0), (1,0,0,0), (0,1,0,0), (0,0,1,0) and (0,0,0,1).
5. Let $\boldsymbol{x} \in \mathbb{R}^4$ and $g(\boldsymbol{x}) = x_4 - x_1x_2x_3 - 3$. Let $X_i = x_i + u_i$ with $u_i \sim N(0, 1)$, $i = 1, 2, 3, 4$. Use SORM to evaluate the reliability index of the points (0,0,0,0), (1,0,0,0), (0,1,0,0), (0,0,1,0) and (0,0,0,1).

10.9 Reliability Based Design Optimization

As we have already observed, optimization problems may involve uncertainty. For instance, when considering

$$\boldsymbol{x}^* = \arg\min\{\, f(\boldsymbol{x}) : \boldsymbol{x} \in C \subset \mathbb{R}^n\} \tag{10.31}$$

both the objective f and the admissible set C may contain uncertainty, arising from different variability sources.

Thus, variability and uncertainties should be considered in the optimization procedure to generate a robust result. Indeed,

- The effects of variations of the parameters involved in the objective and constraints must be analyzed, to prevent the situation where small variations of some parameters can cause significant variations of the result (loss of robustness)
- The impacts of implementation errors of the solution found must be considered, to avoid situations where small errors in the implementation produce substantial variations in the real result (variability and uncertainty)
- the effects of fluctuations of the external environment must be evaluated to ensure minimal safety conditions (unreliability).

To avoid these difficulties, it is necessary to introduce uncertainties in the design process. A possible aim is to get a complete characterization of the variability of the solutions: as shown in the preceding, UQ furnishes tools for such a characterization.

However, in the case where we are interested in the determination of robust solutions without the full determination of the variability of the solution, alternatives exist in the literature.

For instance, we can look for solutions having a small sensitivity to the variations of some parameters (*Robust Optimization*): in such a situation, we must introduce an index of robustness

$\beta(\boldsymbol{x})$, analogous to the reliability index, which gives to each possible solution $\boldsymbol{x}$ a value measuring the robustness. Assuming that robustness increases with β, we may either look for the solution that minimizes the objective on the admissible set and ensures a minimal safety:

$$\boldsymbol{x}^* = \arg\min\{ f(\boldsymbol{x}) : \boldsymbol{x} \in C \subset \mathbb{R}^n, \beta(\boldsymbol{x}) \geq \beta_{min}\} \tag{10.32}$$

or determine the solution that ensures the maximal safety for a maximal value of the objective:

$$\boldsymbol{x}^* = \arg\min\{\beta(\boldsymbol{x}) : \boldsymbol{x} \in C \subset \mathbb{R}^n, f(\boldsymbol{x}) \leq f_{max}\}. \tag{10.33}$$

In these formulations, the robustness index may be replaced by the reliability index: in this case, we refer to these optimization problems as being *Reliability Based Design Optimization* (RBDO) problems. More generally, we can look for the solution that controls the probability of failure:

$$\boldsymbol{x}^* = \arg\min\{ f(\boldsymbol{x}) : \boldsymbol{x} \in C \subset \mathbb{R}^n, F = P(\mathcal{F}) \leq p_{max}\} \tag{10.34}$$

or

$$\boldsymbol{x}^* = \arg\min\{F = P(\mathcal{F}) : \boldsymbol{x} \in C \subset \mathbb{R}^n, f(\boldsymbol{x}) \leq f_{max}\}. \tag{10.35}$$

10.9.1 The Bilevel or Double Loop Approach for a Desired β

The formulation (10.32) leads to a bi-level optimization problem, since it deals with nested optimization problems: on the one hand, the determination of $\beta(\boldsymbol{x})$ and, on the other hand, the minimization of $f(\boldsymbol{x})$. Indeed, typical optimization iterations start from an initial guess $\boldsymbol{x}^{(0)}$ to generate a sequence of points $\boldsymbol{x}^{(1)}, \boldsymbol{x}^{(2)}, \ldots$. At each step, the generation of $\boldsymbol{x}^{(i+1)}$ requests the evaluation of $\beta(\boldsymbol{x})$ for the candidate points $\beta(\boldsymbol{x})$ – so, it request the solution of an internal optimization problem. A typical algorithm is given below:

Bi-level (or double loop) algorithm for RBDO (desired $\boldsymbol{\beta}$)

Input: the starting point $\boldsymbol{x}^{(0)}$, stopping conditions (for instance, maximal iteration number, minimal step at each iteration), a program determining $\boldsymbol{u}^*(\boldsymbol{x})$, a program for the correction of $\boldsymbol{x}$ for a given $\boldsymbol{u}^*$.

Output: estimations of $\boldsymbol{x}^*$, $\boldsymbol{u}^*(\boldsymbol{x}^*)$ and $\beta(\boldsymbol{x}^*)$.

Initialization: Set the iteration number to zero: $k \leftarrow 0$.

1. **Inferior level (interior loop)**: find $(\boldsymbol{u})^k = \boldsymbol{u}^*(\boldsymbol{x}^{(k)})$. Set $\beta^{(k)} = \|(\boldsymbol{u}^*)^k\|$.
2. **Superior level (external loop)**: Determine $\boldsymbol{x}^{(k+1)}$ as a correction of $\boldsymbol{x}^{(k)}$ using $(\boldsymbol{u})^k$.
3. Increment the iteration number $k \leftarrow k + 1$.
4. Test for the stopping conditions. If the iterations continue, then go to 1, else estimate $\boldsymbol{x}^* \approx \boldsymbol{x}^{(k)}, \boldsymbol{u}^*(\boldsymbol{x}^*) \approx \boldsymbol{u}^*(\boldsymbol{x}^{(k)}), \beta(\boldsymbol{x}^*) \approx \|\boldsymbol{u}^*(\boldsymbol{x}^{(k)})\|$.

Examples of methods for the correction of $\boldsymbol{x}^{(k)}$ can be found in the literature. We presented some methods in Sect. 2.2 The determination of $(\boldsymbol{u})^k$ is usually called the *reliability level*, while the determination of $\boldsymbol{x}^{(k+1)}$ is called the *optimization level*. The numerical solution by iterative methods may request a large number of calls to each of these levels. Thus, simplification methods have been proposed in the literature - such as, for example, the use of FORM at the reliability level.

A popular approach consists in generating $\boldsymbol{x}^{(k+1)}$ by determining successively: at the internal loop,

$$\boldsymbol{u}^{(k)} = \arg\min\left\{\|\boldsymbol{u}\| : h\left(\boldsymbol{x}^{(k)}, \boldsymbol{u}\right) \geq 0 \; and \; \|\boldsymbol{u}\| \geq \beta_{min}\right\}$$

and, at the external loop (correction of $\boldsymbol{x}^{(k)}$):

$$\boldsymbol{x}^{(k+1)} = \arg\min\left\{ f(x) : h\left(\boldsymbol{x}, \boldsymbol{u}^{(k)}\right) \leq 0 \right\}.$$

The main difficulty in the implementation of the double loop is the fact that there are two problems of optimization intertwined: the reliability analysis and the optimization of the objective function. Nested calls to the SOLVER are a difficulty task. The best solution consists in the use of the SOLVER only for one of these optimization problems and the use of another method to the other problem. For instance, reliability analysis can be performed by FORM with an iterative procedure other than the SOLVER.

Example

Let us consider the simple objective function $f(x) = x_1^2 + (x_2 - 1)^2$ and the failure region defined by $g(x) = x_1 + x_2 - 3$. We look for a RBDO solution with $\beta(x) \geq 5$. The global minimum of f is (0, 1) and does not satisfy this condition.

Since the limit state curve is affine, we can evaluate the reliability index by Eq. (10.12).

Open a blank workbook, add a module and enter the code for the evaluation of functions f, g, including their versions f_on_range, g_on_range

```
Function fobj(x)
fobj = (x(1)) ^ 2 + (x(2) - 1) ^ 2
End Function
Function fobj_on_range(rx As Range)
x = range2vector(rx)
fobj_on_range = fobj(x)
End Function
Function g(x)
a = read_line(10, 2, 3)
b = Cells(11, 2).Value
g = sp(a, x) + b
End Function
Function g_on_range(rx As Range)
x = range2vector(rx)
g_on_range = g(x)
End Function
```

Add the code for the evaluation of β:

```
Function hlind(x)
a = read_line(10, 2, 3)
b = Cells(11, 2).Value
hlind = -(sp(a, x) + b) / norm2(a)
End Function
```

```
Function hlind_on_range(rx As Range)
x = range2vector(rx)
hlind_on_range = hlind(x)
End Function
```

Go to the workbook. Enter the initial guess for x in `B1:C1`. Give the name `x` to `B1:C1`. Enter the formulae "`=fobj_on_range(x)`" in `B2`, "`=g_on_range(x)`" in `B3`, "`=hlind_on_range(x)`" in `B4`. Enter the desired β in `B5` and the formula "`=B5 - B4`" in `B6`. Enter the values of a and b such that $g(x) = a^t x + b$ in `B10:C10` for a, `B11` for b. The workbook appears as shown at right.

	A	B	C
1	x	0	0
2	f(x)	1	
3	g(x)	-3	
4	beta(x)	2,12132	
5	beta min	5	
6	inequality	2,87868	
7			
8			
9			
10	a	1	1
11	b	-3	

Now, run the solver to minimize `B2` by changing `x` under the restrictions `B3`, `B6` negative:

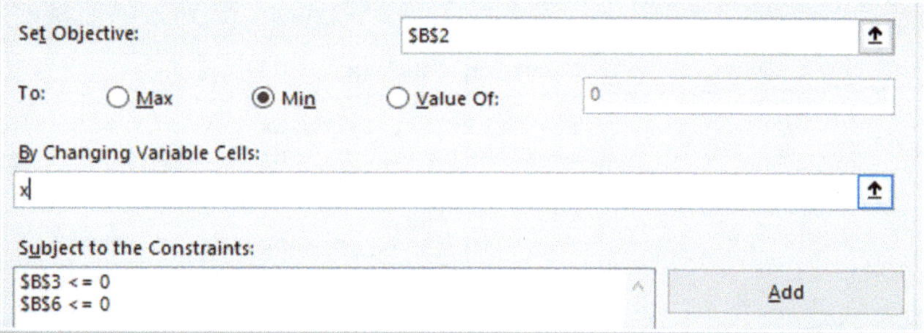

Click on "`Solve`". The results appear:

	A	B	C
1	x	-2,53553	-1,53553
2	f(x)	12,85786	
3	g(x)	-7,07107	
4	beta(x)	5	
5	beta min	5	
6	inequality	0	
7			
8			
9			
10	a	1	1
11	b	-3	

If you desire, you can use a Sub to solve the problem:

```
Sub runsolver()
col = Cells(1, Cells.Columns.Count).End(xlToLeft).Column
rxad = Range(Cells(1, 2), Cells(1, col)).Address
rfobjad = Range(Cells(2, 2), Cells(2, 2)).Address
rgad = Range(Cells(3, 2), Cells(3, 2)).Address
rineqad = Range(Cells(6, 2), Cells(6, 2)).Address
  SolverReset
  SolverOk SetCell:=rfobjad, MaxMinVal:=2, ValueOf:=0, ByChange:=rxad, _
          Engine:=1, EngineDesc:="GRG Nonlinear"
  SolverAdd Cellref:=rineqad, Relation:=1, FormulaText:="0"
  SolverAdd Cellref:=rgad, Relation:=1, FormulaText:="0"
  SolverOptions AssumeNonNeg:=False
  SolverOptions Multistart:=True, PopulationSize:=500, RequireBounds:=False
  SolverSolve True
End Sub
```

Running this Sub produces the same results. If you are also interested in the value of $\boldsymbol{u}(\boldsymbol{x})$, add the code for its evaluation in the module

```
Function findu(x)
a = read_line(10, 2, 3)
b = Cells(11, 2).Value
beta = -(sp(a, x) + b) / norm2(a)
findu = sm(a, beta / norm2(a))
End Function
Function findu_on_range(rx As Range)
x = range2vector(rx)
findu_on_range = findu(x)
End Function
```

Go to cell B7 and enter the formula "=findu_on_range(x)". You will get the result:

7	u	3,535534	3,535534

Example

Let us consider again the simple objective function $f(x) = x_1^2 + (x_2 - 1)^2$ and the failure region defined by $g(x) = x_2 - x_1^2 - x_1 - 3$. We look for a RBDO solution with $\beta(x) \geq 5$. To prevent from running nested instances of the SOLVER, we evaluate $\beta(x)$ by the fixed-point iterations previously introduced (page 461). To do this, we modify the function hlind as shown in Fig. 10.23.

```
Function hlind(x)
om = Cells(9, 2).Value
nitmax = Cells(10, 2).Value
precmin = Cells(11, 2).Value
tolmin = Cells(12, 2).Value
Dim u(1 To 2) As Double
u(1) = 0
u(2) = 0
beta = norm2(u)
Dim stopping As Boolean
stopping = False
nit = 0
prec = 1 + precmin
tol = 1 + tolmin
While Not stopping
   nit = nit + 1
   eq = eqs(x, u, beta)
   tol = norm2(eq)
   u(1) = u(1) - om * eq(1)
   u(2) = u(2) - om * eq(2)
   beta = beta - om * eq(3)
   prec = norm2(eq)
   stopping = nit > nitmax Or tol < tolmin Or prec < precmin
Wend
hlind = beta
End Function
```

Fig. 10.23 Fixed-point evaluation of $\beta(x)$

Recall that the evaluation of eqs requests the gradient of g, function h(x,u) and its gradient. It also requests the function vecn (see page 458). The workbook must furnish the parameters for the iterations, as shown at right.

	A	B	C
1	**x**	0	0
2	f(x)	1	
3	g(x)	-3	
4	beta(x)	2,788205	
5	beta min	5	
6	inequality	2,211795	
7			
8			
9	**om**	0,01	
10	**nitmax**	10000	
11	**precmin**	1,00E-08	
12	**tolmin**	1,00E-05	
13			

Run the SOLVER to minimize `B2` by changing `x` under the restriction `B3`, `B6` negative – for instance, you can run the `Sub runsolver` previously introduced. The results appear.

	A	B	C
1	x	0,68917	-2,12006
2	f(x)	10,20971	
3	g(x)	-6,28418	
4	beta(x)	4,999999	
5	beta min	5	
6	inequality	6,76E-07	

Example

Let us consider again the simple objective function $f(x) = x_1^2 + (x_2 - 1)^2$ and the failure region defined by $g(x) = x_2 - x_1^3 - x_1^2 - 4$. We look for a RBDO solution with $\beta(x) \geq 5$.

Run the `Sub runsolver` previously introduced. The results appear.

	A	B	C
1	x	0	-1
2	f(x)	4,000015	
3	g(x)	-5	
4	beta(x)	5	
5	beta min	5	
6	inequality	3,33E-10	

Exercises

1. Let $\boldsymbol{x} \in \mathbb{R}^2, f(x) = x_1^2 + x_2^2$ and $g(\boldsymbol{x}) = x_1 x_2 - 1$. Let $X_i = x_i + u_i$ with $u_i \sim N(0, 1)$, $i = 1, 2$. Find a RBDO solution with $\beta = 3, 4, 5, 6$.
2. Let $\boldsymbol{x} \in \mathbb{R}^2, f(x) = x_1^2 + x_2^2$ and $g(\boldsymbol{x}) = x_1 x_2^2 - 1$. Let $X_i = x_i + u_i$ with $u_i \sim N(0, 1)$, $i = 1, 2$. Find a RBDO solution with $\beta = 3, 4, 5, 6$.
3. Let $\boldsymbol{x} \in \mathbb{R}^2, f(x) = (x_1 + x_2 - 6)^2 + (x_1 - x_2 - 10)^2$ and $g(\boldsymbol{x}) = x_1 - 3x_2^2 + 2$. Let $X_i = x_i + u_i$ with $u_i \sim N(0, 1)$, $i = 1, 2$. Find a RBDO solution with $\beta = 3, 4, 5, 6$.
4. Let $\boldsymbol{x} \in \mathbb{R}^2, f(x) = x_1^2 + x_2^2$ and $g(\boldsymbol{x}) = x_1^2 + 8x_2 - 2$. Let $X_i = x_i + u_i$ with $u_i \sim N(0, 1)$, $i = 1, 2$. Find a RBDO solution with $\beta = 3, 4, 5, 6$.
5. Let $\boldsymbol{x} \in \mathbb{R}^2, f(x) = (x_1 + x_2 - 6)^2 + (x_1 - x_2 - 10)^2$ and $g(\boldsymbol{x}) = x_1 + x_2^3 - 2$. Let $X_i = x_i + u_i$ with $u_i \sim N(0, 1)$, $i = 1, 2$. Find a RBDO solution with $\beta = 3, 4, 5, 6$.

10.9.2 The Bilevel or Double Loop Approach for a Desired Objective

As observed, an alternative formulation is furnished by (10.33): we can exchange the rules of β and f, to look for the maximal β for a prescribed maximal value of f. In this case, we run the SOLVER to maximize β under some restrictions.

Example

Let us consider the simple objective function $f(x) = x_1^2 + (x_2 - 1)^2$ and the failure region defined by $g(x) = x_1 + x_2 - 3$. We look for a RBDO solution with $f(x) \leq 3$. As previously indicated, the limit state curve is affine, and we can evaluate the reliability index by Eq. (10.12).

Go to the workbook. Enter the initial guess for x in `B1:C1`. Give the name `x` to `B1:C1`. Enter the formulae "`=fobj_on_range(x)`" in `B2`, "`=g_on_range(x)`" in `B3`, "`=hlind_on_range(x)`" in `B4`. Enter the maximal f in `B5` and the formula "`=B2 - B5`" in `B6`. Enter the values of a and b such that $g(x) = a^t x + b$ in `B10:C10` for a, `B11` for b. The workbook appears as shown at right.

	A	B	C
1	x	0	0
2	f(x)	1	
3	g(x)	-3	
4	beta(x)	2,12132	
5	f max	3	
6	inequality	-2	
7			
8			
9			
10	a	1	1
11	b	-3	
12			

Now, run the solver to maximize `B4` by changing `x` under the restrictions `B3, B6` negative:

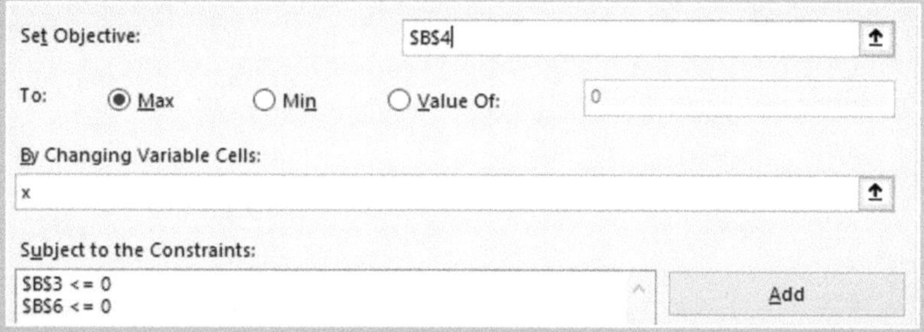

Click on "`Solve`". The results appear:

	A	B	C
1	x	-1,22432	-0,22517
2	f(x)	3	
3	g(x)	-4,44949	
4	beta(x)	3,146264	
5	f max	3	
6	inequality	3,86E-07	
7			

An example of Sub to solve the problem is given below:

```
Sub runsolver()
col = Cells(1, Cells.Columns.Count).End(xlToLeft).Column
rxad = Range(Cells(1, 2), Cells(1, col)).Address
rfobjad = Range(Cells(4, 2), Cells(4, 2)).Address
rgad = Range(Cells(3, 2), Cells(3, 2)).Address
rineqad = Range(Cells(6, 2), Cells(6, 2)).Address
  SolverReset
  SolverOk SetCell:=rfobjad, MaxMinVal:=1, ValueOf:=0, ByChange:=rxad, _
          Engine:=1, EngineDesc:="GRG Nonlinear"
  SolverAdd Cellref:=rineqad, Relation:=1, FormulaText:="0"
  SolverAdd Cellref:=rgad, Relation:=1, FormulaText:="0"
  SolverOptions AssumeNonNeg:=False
  SolverOptions Multistart:=True, PopulationSize:=500, RequireBounds:=False
  SolverSolve True
End Sub
```

Analogously to the preceding examples, running this Sub produces the same results. If you are also interested in the value of $\boldsymbol{u}(\boldsymbol{x})$, proceed as indicated page 512.

Example

Let us consider again the simple objective function $f(x) = x_1^2 + (x_2 - 1)^2$ and the failure region defined by $g(x) = x_2 - x_1^2 - x_1 - 3$. We look for a RBDO solution with $f(x) \leq 3$. As in the previous examples, we evaluate $\beta(x)$ by the fixed-point iterations (see pages 461 and 513). Running the Sub runsolver produces the results below (Fig. 10.24):

a)

	A	B	C
1	x	0	0
2	f(x)	1	
3	g(x)	-3	
4	beta(x)	2,788205	
5	f max	3	
6	inequality	-2	
7			
8			
9	om	0,01	
10	nitmax	10000	
11	precmin	1,00E-08	
12	tolmin	1,00E-05	

b)

	A	B	C
1	x	0,373683	-0,69126
2	f(x)	3	
3	g(x)	-4,20458	
4	beta(x)	3,53679	
5	f max	3	
6	inequality	-4,8E-10	
7			
8			
9	om	0,01	
10	nitmax	10000	
11	precmin	1,00E-08	
12	tolmin	1,00E-05	

Fig. 10.24 Results furnished by runsolver. (**a**) Initial data. (**b**) results

Example

Let us consider again the simple objective function $f(x) = x_1^2 + (x_2 - 1)^2$ and the failure region defined by $g(x) = x_2 - x_1^3 - x_1^2 - 4$. We look for a RBDO solution with $f(x) \leq 3$. As in the previous examples, we evaluate $\beta(x)$ by the fixed-point iterations (see pages 461 and 513). Running the `Sub runsolver` produces the results below (Fig. 10.25):

a)

	A	B	C
1	x	0	0
2	f(x)	1	
3	g(x)	-4	
4	beta(x)	3,999999	
5	f max	3	
6	inequality	-2	
7			
8			
9	om	0,01	
10	nitmax	10000	
11	precmin	1,00E-08	
12	tolmin	1,00E-05	

b)

	A	B	C
1	x	0	-0,73205
2	f(x)	3	
3	g(x)	-4,73205	
4	beta(x)	4,732048	
5	f max	3	
6	inequality	3,85E-07	
7			
8			
9	om	0,01	
10	nitmax	10000	
11	precmin	1,00E-08	
12	tolmin	1,00E-05	

Fig. 10.25 Results furnished by `runsolver`. (**a**) Initial data. (**b**) Results

Exercises

1. Let $\boldsymbol{x} \in \mathbb{R}^2, f(x) = x_1^2 + x_2^2$ and $g(\boldsymbol{x}) = x_1 x_2 - 1$. Let $X_i = x_i + u_i$ with $u_i \sim N(0, 1), i = 1, 2$. Find a solution that maximizes β for $f(x) \leq 2, f(x) \leq 3, f(x) \leq 4, f(x) \leq 5$.
2. Let $\boldsymbol{x} \in \mathbb{R}^2, f(x) = x_1^2 + x_2^2$ and $g(\boldsymbol{x}) = x_1 x_2^2 - 1$. Let $X_i = x_i + u_i$ with $u_i \sim N(0, 1), i = 1, 2$. Find a solution that maximizes β for $f(x) \leq 2, f(x) \leq 3, f(x) \leq 4, f(x) \leq 5$.
3. Let $\boldsymbol{x} \in \mathbb{R}^2, f(x) = (x_1 + x_2 - 6)^2 + (x_1 - x_2 - 10)^2$ and $g(\boldsymbol{x}) = x_1 - 3x_2^2 + 2$. Find a solution that maximizes β for $f(x) \leq 2, f(x) \leq 3, f(x) \leq 4, f(x) \leq 5$.
4. Let $\boldsymbol{x} \in \mathbb{R}^2, f(x) = x_1^2 + x_2^2$ and $g(\boldsymbol{x}) = x_1^2 + 8x_2 - 2$. Let $X_i = x_i + u_i$ with $u_i \sim N(0, 1)$, $i = 1, 2$. Find a solution that maximizes β for $f(x) \leq 2, f(x) \leq 3, f(x) \leq 4, f(x) \leq 5$.
5. Let $\boldsymbol{x} \in \mathbb{R}^2, f(x) = (x_1 + x_2 - 6)^2 + (x_1 - x_2 - 10)^2$ and $g(\boldsymbol{x}) = x_1 + x_2^3 - 2$. Find a solution that maximizes β for $f(x) \leq 2, f(x) \leq 3, f(x) \leq 4, f(x) \leq 5$.

Bibliography

Akaike, H. (1974, December). A new look at the statistical model identification. *IEEE Transactions on Automatic Control, 19*(6), 716–723. https://doi.org/10.1109/TAC.1974.1100705

Apesteguia, J., & Palacios-Huerta, I. (2010, December). Psychological pressure in competitive environments: Evidence from a randomized natural experiment. *American Economic Review, 100*(5), 2548–2564.

Arrondel, L., Duhautois, R., & Laslier, J.-F. (2019). Decision under psychological pressure: The shooter's anxiety at the penalty kick. *Journal of Economic Psychology, 70*(C), 22–35.

Arrous, J. (2000). *Algèbre linéaire et Science économique, un cas exemplaire.* Retrieved 1 13, 2021, from Research Gate page of Jean Arrous: https://www.researchgate.net/publication/242301525_ALGEBRE_LINEAIRE_ET_SCIENCE_ECONOMIQUE_UN_CAS_EXEMPLAIRE

Autrique, L., & Souza de Cursi, E. (1997). On stochastic modification for global optimization problems: An efficient implementation for the control of the vulcanization process. *International Journal of Control, 67*(1), 1–22.

Azar, O. H., & Bar-Eli, M. (2011). Do soccer players play the mixed-strategy Nash equilibrium? *Applied Economics, 43* (25), 3591–3601. https://doi.org/10.1080/00036841003670747

Bachelier, L. (1900). Théorie de la spéculation. *Annales scientifiques de l'École Normale Supérieure, série 3, 17,* 21–86. https://doi.org/10.24033/asens.476

Bachelier, L. (1901). Théorie mathématique du jeu. *Annales scientifiques de l'E.N.S., 18*(3eme série), 143–209.

Bassi, M., Souza de Cursi, E., Pagnacco, E., & Ellaia, R. (2018). Statistics of the Pareto front in multi-objective optimization under uncertainties. *Latin American Journal of Solids and Structures, 15*(11), e130. https://doi.org/10.1590/1679-78255018

Black, F., & Scholes, M. (1973). The pricing of options and corporate liabilities. *Journal of Political Economy, 81*(3), 637–654. https://doi.org/10.1086/260062

Blankenship, G., & Baras, J. (1981). Accurate evaluation of stochastic Wiener integrals with applications to scattering in random media and to nonlinear filtering. *SIAM Journal of Applied Mathematics, 41*(3), 518–552. https://doi.org/10.1137/0141043

Boczon, M., & Wilson, A. J. (2018). *Goals, constraints, and public assignment: A field study of the UEFA champions league.* University of Pittsburgh. Retrieved June 2, 2020, from https://www.econ.pitt.edu/sites/default/files/working_papers/Working%20Paper.18.16.pdf

Borel, E. (1921, July 11). La théorie du jeu et les équations intégrales à noyau symétrique. *Comptes rendus hebdomadaires des séances de l'Académie des Sciences de Paris, 173,* 1304–1308.

Bouhadi, M., Ellaia, R., & Souza de Cursi, E. (2004a). Global optimization under nonlinear restrictions by using stochastic perturbations of the projected gradient. In C. A. Floudas & P. N. Pardalos (Eds.), *Frontiers in global optimization* (pp. 541–561). Kluwer Academic Press. https://doi.org/10.1007/978-1-4613-0251-3_29

Bouhadi, M., Ellaia, R., & Souza de Cursi, E. (2004b). Stochastic perturbation methods for affine restrictions. In C. A. Floudas & P. Pardalos (Eds.), *Advances in convex analysis and global optimization* (pp. 487–499). Kluwer Academic Press.

Boukhetala, K., & Guidoum, A. (2011). *Sim.DiffProc: A Package for Simulation of Diffusion Processes in R.* HAL. Retrieved 1 04, 2021, from https://hal.archives-ouvertes.fr/hal-00629841/en/

Box, G. E. (1958). A note on the generation of random normal deviates. *Annals of Mathematical Statistics, 29*(2), 610–611.

Brown, R. (1828). A brief account of Microscopical observations made in the months of June, July, and August, 1827, on the particles contained in the pollen of plants; and on the general existence of active molecules in organic and inorganic bodies. *The Philosophical Magazine, 4*(21), 161–173. https://doi.org/10.1080/14786442808674769

Chen, F. (2008, December 18). *Random Music Generator.* Retrieved from OpenStax CNX: http://cnx.org/contents/62219fba-96df-418a-b9aa-14fddd7c30fa@2

Choi, B. (1992). *ARMA model identification.* Springer. https://doi.org/10.1007/978-1-4613-9745-8

E. Souza de Cursi, *Uncertainty Quantification and Stochastic Modelling with EXCEL*,
Springer Texts in Business and Economics, https://doi.org/10.1007/978-3-030-77757-9

Choleski, A.-L. (1910). *Sur la résolution numérique des systèmes d'équations linéaires.* manuscript.

Chorin, A. J. (1973). Accurate evaluation of Wiener integrals. *Mathematics of Computation, 27*(121), 1–15. https://doi.org/10.2307/2005242

Coloma, G. (2007). Penalty kicks in soccer: An alternative methodology for testing mixed-strategy equilibria. *Journal of Sports Economics, 8*(5), 530–545.

Coloma, & Germán. (2012). *The penalty-kick game under incomplete information.* Universidad del CEMA.

Coletti G., & Scozzafava, R. (2000). Zero probabilities in stochastic independence. In B. Bouchon-Meunier R. R. Yager & L. A. Zadeh (Eds.) *Information, Uncertainty and Fusion.* The Springer International Series in Engineering and Computer Science, vol 516. Springer, Boston, MA. https://doi.org/10.1007/978-1-4615-5209-3_14

Cournot, A. (1838). *Recherches sur les principes mathématuiques de la théorie des richesses.* Hachette.

Cowles, A. (1933). Can stock market forecasters forecast? *Econometrica, 1*(3), 309–324. https://doi.org/10.2307/1907042

Cox, J., Ingersoll, J., & Ross, S. (1985). A theory of the term structure of interest rates. *Econometrica, 53*(2), 385–407. https://doi.org/10.2307/1911242

Cramer, G. (1750). *Introduction à l'Analyse des Lignes Courbes Algébriques.* Genève. https://doi.org/10.3931/e-rara-4048

Cressman, R., & Tao, Y. (2014). The replicator equation and other game dynamics. *Proceedings of the National Academy of Sciences of the United States of America, 111*(Supplement 3), 10810–10817. https://doi.org/10.1073/pnas.1400823111

Cui, T., & Li, S. (2020). *Space fault tree theory and system reliability analysis.* EDP Sciences.

Dautray, R., & Lions, J.-L. (2012). *Mathematical analysis and numerical methods for science and technology* (Vol. 3). Springer.

Dickey, D. A., & Fuller, W. A. (1979). Distribution of the estimators for autoregressive time series with a unit root. *Journal of the American Statistical Association, 366a*, 427–431. https://doi.org/10.1080/01621459.1979.10482531

Einstein, A. (1905). Über die von der molekularkinetischen Theorie der Wärme geforderte Bewegung von in ruhenden Flüssigkeiten suspendierten Teilchen. *Annalen der Physik, 17*, 549–560.

El Mouatasim, A., Ellaia, R., & Souza de Cursi, E. (2006). Random perturbation of the variable metric method for unconstrained nonsmooth nonconvex optimization. *Applied Mathematics and Computer Science, 16*(4), 463–474.

enseignements. (2020, December 13). Retrieved from page personnelle de V. Lemaire: https://www.lpsm.paris/pageperso/lemaire/docs/stats-ENPC/series-chrono.pdf

Ericson, C. A. (2011). *Fault tree analysis primer.* CreateSpace Incorporated.

Fisk, D. L. (1963). *Quasi-martingales and stochastic integrals.* Michigan State University, Department of Statistics. Office of Naval Research. Retrieved 12 27, 2020, from https://apps.dtic.mil/dtic/tr/fulltext/u2/414838.pdf

Föllmer, H. (1981). Calcul d'Ito sans probabilités. *Séminaire de probabilités (Strasbourg), 15*, 143–150. Retrieved from http://www.numdam.org/item?id=SPS_1981__15__143_0

Garicano, L., Palacios-Huerta, I., & Prendergast, C. (2005). Favoritism under social pressure. *The Review of Economics and Statistics, 87*(2), 208–216.

Gauss, C. F. (1811). Disquisitio de elementis ellipticis Palladis ex oppositionibus annorum 1803, 1804, 1805, 1807, 1808, 1809. *Commentationes societatis regiæ scientiarum Gottingensis recentiores – Commentationes mathematicae*, pp. 1–26.

Graham, C., & Talay, D. (2013). *Stochastic simulation and Monte Carlo methods.* Springer. https://doi.org/10.1007/978-3-642-39363-1

Hannan, E. J., & Quinn, B. G. (1979). The determination of the order of an autoregression. *Journal of the Royal Statistical Society. Series B (Methodological), 41*(2), 190–195. Retrieved from http://www.jstor.org/stable/2985032

Holdorf Lopez, R., Souza de Cursi, J. E., & Lemosse, D. (2011). Approximating the probability density function of the optimal point of an optimization problem. *Engineering Optimization, 43*(3), 281–303. https://doi.org/10.1080/0305215X.2010.489607

Itô, K. (1944). Stochastic integral. *Proceedings of the Imperial Academy, 20*(8), 519–524. https://doi.org/10.3792/pia/1195572786

Itô, K. (1946). On a stochastic integral equation. *Proceedings of the Japan Academy, 22*(2), 32–35. https://doi.org/10.3792/pja/1195572371

Itô, K. (1950). Stochastic differential equations in a differentiable manifold. *Nagoya Mathematical Journal, 1*, 35–47. https://projecteuclid.org/euclid.nmj/1118764702

Itô, K. (1951). On a formula concerning stochastic differentials. *Nagoya Mathematical Journal, 3*, 55–65. Retrieved from https://projecteuclid.org/euclid.nmj/1118799221

Jafari, M. A., & Abbasian, S. (2017). The moments for solution of the Cox-Ingersoll-Ross interest rate model. *Journal of Finance and Economics, 5*(1), 34–37. https://doi.org/10.12691/jfe-5-1-4

Jordan, W. (1962). *Handbook of Geodesy* (Vol. 1). (M. W. Carta, Trans.)

Kharroubi, I., & Pham, H. (2015). Feynman–Kac representation for Hamilton–Jacobi–Bellman IPDE. *The Annals of Probability, 43*(4), 1823–1865. https://doi.org/10.1214/14-AOP920

Kim, N. J. (1968). *Linear Programming with Random Requirements.* Logan, UT 84321: Utah Water Research Laboratory. Retrieved February 7, 2021, from https://digitalcommons.usu.edu/water_rep/272?utm_source=digitalcommons.usu.edu%2Fwater_rep%2F272&utm_medium=PDF&utm_campaign=PDFCoverPages

Kloeden, P., & Platen, E. (1989). A survey of numerical methods for stochastic differential equations. *Stochastic Hydrology and Hydraulics, 3*, 155–178. https://doi.org/10.1007/BF01543857

Kolmogorov, A. N. (1950). *Foundations of the theory of probability.* Chelsea Publishing Company.

Kress, R. (1998). *Numerical analysis.* Springer. https://doi.org/10.1007/978-1-4612-0599-9

Kuhn, H. W. (1957). Extensive games and the problem and information. In H. W. Kuhn & A. W. Tucker (Eds.), *Contributions to the theory of games, annals of mathematical studies* (pp. 193–216). Princeton University Press.

Kwiatkowski, D., Phillips, P. C., Schmidt, P., & Shin, Y. (1992). Testing the null hypothesis of stationarity against the alternative of a unit root: How sure are we that economic time series have a unit root? *Journal of Econometrics, 54* (1–3), 159–178. https://doi.org/10.1016/0304-4076(92)90104-Y

Langevin, P. (1908). Sur la théorie du mouvement brownien. *Comptes-Rendus de l'Académie des Sciences, 146*, 530–532.

Leontieff, W. W. (1936). Quantitative input and output relations in the economic systems of the United States. *The Review of Economics and Statistics, 18*(3), 105–125. https://doi.org/10.2307/1927837

Leontieff, W. W. (1937). Interrelation of prices, output, savings, and investment. *The Review of Economics and Statistics, 3*, 109–132. https://doi.org/10.2307/1927343

Leontieff, W. W. (1986). *Input-output economics.* Oxford University Press.

Levitt, S., Chiappori, P., & Groseclose, T. (2002). Testing mixed-strategy equilibria when players are heterogeneous: The case of penalty kicks in soccer. *American Economic Review, 92*, 1138–1151. https://doi.org/10.1257/00028280260344678

Lévy, P. (1922, November 13). Sur la détermination des lois de probabilité par leurs fonctions caractéristiques. *Comptes Rendus de l'Académie des Sciences de Paris*, pp. 854–856.

Lewontin, R. C. (1961, July). Evolution and the theory of games. *Journal of Theoretical Biology, 1*(3), 382–403.

Leybourne, S. J., & McCabe, B. P. (1994). A consistent test for a unit root. *Journal of Business and Economic Statistics, 12*(2), 157–166. https://doi.org/10.1080/07350015.1994.10510004

Li, F., & Wu, T. (2007). An Importance Sampling Based Approach for Reliability Analysis. *2007 IEEE International Conference on Automation Science and Engineering* (pp. 956–961). Scottsdale, AZ: IEEE. https://doi.org/10.1109/COASE.2007.4341815.

Limnios, N. (2007). *Fault Trees.* ISTE/Wiley.

Lotka, A. J. (1910, March 1). Contribution to the theory of periodic reactions. *The Journal of Physical Chemistry, 14*(3), 271–274. https://doi.org/10.1021/j150111a004

Lotka, A. J. (1920, July 1). Analytical note on certain rhythmic relations in organic systems. *Proceedings of the National Academy of Sciences of the USA, 6*(7), 410–415. https://doi.org/10.1073/pnas.6.7.410

Maynard-Smith, J. (1982). *Evolution and the theory of games.* Cambridge University Press.

Maynard-Smith, J., & Price, G. R. (1973, November 2). The logic of animal conflict. *Nature, 246*, 15–18.

Merton, R. C. (1973). Theory of rational option pricing. *The Bell Journal of Economics and Management Science, 4*(1), 141–183. https://doi.org/10.2307/3003143

Milstein, G., & Tretyakov, M. (2004). *Stochastic numerics for mathematical physics.* Springer. https://doi.org/10.1007/978-3-662-10063-9

Milstein, G., & Tretyakov, M. (2012). Solving the Dirichlet problem for Navier–Stokes equations by probabilistic approach. *BIT Numerical Mathematics, 52*, 141–153. https://doi.org/10.1007/s10543-011-0347-z

Milstein, G., & Tretyakov, M. (2013). Probabilistic methods for the incompressible Navier–Stokes equations with space periodic conditions. *Adv. in Appl. Probab., 45*(3), 742–772. https://doi.org/10.1239/aap/1377868537

Milstein, G., & Tretyakov, M. (2020). Mean-Square approximation of Navier-Stokes equations with additive noise in vorticity-velocity formulation. *Numerical Mathematics: Theory, Methods and Applications, 14*(1), 1–30. https://doi.org/10.4208/nmtma.OA-2020-0034

Milstein, G. N. (1973). Approximate integration of stochastic differential equations. *Theory of Probability and its Applications, 19*(3), 557–562. https://doi.org/10.1137/1119062

Morillon, J.-P. (1997). Numerical solutions of linear mixed boundary value problems using stochastic representations. *International Journal for Numerical Methods in Engineering, 40*, 387–405. https://doi.org/10.1002/(SICI)1097-0207(19970215)40:3%3C387::AID-NME69%3E3.0.CO;2-D

Muller, M. E. (1958). An inverse method for the generation of random normal deviates on large-scale computers. *Mathematical Tables and Other Aids to Computation, 12*, 167–174.

Muller, M. E. (1959). A comparison of methods for generating normal deviates on digital computers. *Journal of the Association for Computing Machinery, 6*, 376–383.

Muller, M. E. (n.d.). Generation of normal deviates. In *Princeton University, statistical techniques research group*. Princeton University.

Nash, J. F., Jr. (1950a). Equilibrium points in n-person games. *Proceedings of the National Academy of Sciences of the United States of America, 36*(1), 48–49.

Nash, J. F., Jr. (1950b, April). The bargaining problem. *Econometrica, 18*(2), 155–162.

Nataf, A. (1962). Détermination des distributions de probabilités dont les marges sont données. *Comptes Rendus de l'Académie des Sciences Paris, 225*, 42–43.

Nowak, M. (2006). *Evolutionary dynamics: Exploring the equations of life*. Harvard University Press.

Organisation for Economic Co-operation and Development. (2021, January 13). *Input-Output Tables (IOTs)*. Retrieved from OECD Homepage: https://www.oecd.org/sti/ind/input-outputtables.htm

Palacios-Huerta, I. (2003). Professionals play minimax. *Review of Economic Studies, 70*, 395–415.

Palacios-Huerta, I. (2014). *Beautiful game theory: How soccer can help economics*. Princeton University Press.

Palacios-Huerta, I., & Volij, O. (2008, January). Experientia Docet: Professionals play minimax in laboratory experiments. *Econometrica, 76*(1), 71–115.

Paley, R., Wiener, N., & Zygmund, A. (1933). Notes on random functions. *Mathematische Zeitschrift, 37*, 647–668. https://doi.org/10.1007/BF01474606

Papaioannou, I., Papadimitriou, C., & Straub, D. (2016). Sequential importance sampling for structural reliability analysis. *Structural Safety, 62*, 66–75. https://doi.org/10.1016/j.strusafe.2016.06.002

Pardoux, E., & Talay, D. (1985). Discretization and simulation of stochastic differential equations. *Acta Applicandae Mathematicae, 3*, 23–47. https://doi.org/10.1007/BF01438265

Pearson, K. (1905). The problem of random walk. *Nature, 294*.

Phillips, P. C., & Perron, P. (1988, June). Testing for a unit root in time series regression. *Biometrika, 75*(2), 335–346. https://doi.org/10.1093/biomet/75.2.335

Pogu, M., & Souza de Cursi, E. (1994). Global optimization by random perturbation of the gradient method with a fixed parameter. *Journal of Global Optimization, 5*(2), 159–180.

Pourahmadi, M. (2001). *Foundations of time series analysis and prediction theory*. Wiley.

Prakash, P., & Garg, R. (2014, September). Preferred side of the penalty kick. *International Journal of Science, Technology and Management, 3*(9), 7–12.

Regnault, J. (1863). *Calcul des chances et philosophie de la bourse*. Castel, Mallet-Bachelier.

Robert, C. P. (2006). *Le choix bayésien*. Springer.

Romeo, G. (2020). *Elements of mathematical economics with EXCEL*. Academic Press. https://doi.org/10.1016/C2018-0-02476-7

Rosenblatt, M. (1952). Remarks on a multivariate transformation. *Annals of Mathematical Statistics, 23*(3), 470–472. https://doi.org/10.1214/aoms/1177729394

Sarkar, A. (2017). *The Gambler's fallacy and hot outcome: Cognitive biases or adaptive thinking for goalkeepers' decisions on dive direction during penalty shootouts. Master Thesis*. Ohio, USA: Bowling Green State University.

Schwarz, G. (1978). Estimating the dimension of a model. *The Annals of Statistics, 6*(2), 461–464. https://doi.org/10.1214/aos/1176344136

Shapley, L. S. (1953). Stochastic games. *Proceedings of the National Academy of Sciences of the United States of America, 39*(10), 1095–1100.

Sinclair, B. (2005, June 9). *Machine Repair Model*. Retrieved from OpenStax CNX: http://cnx.org/contents/56f1bed0-bd34-4c28-a2ec-4a3f9ded8e18@3

Souza de Cursi, E. (1992). *Introduction aux Probabilités et Statistiques*. Ecole Centrale Nantes.

Souza de Cursi, E. (1994). Numerical methods for linear boundary value problems based on Feynman-Kac representations. *Mathematics and Computers in Simulation, 36*(1), 1–16.

Souza de Cursi, E. (2015). *Variational methods for engineers with Matlab®*. Hoboken, NJ: John Wiley & Sons, Inc.

Souza de Cursi, E. (2021). Uncertainty quantification in game theory. *Chaos, Solitons & Fractals, Elsevier, 143*. https://doi.org/10.1016/j.chaos.2020.110558

Souza de Cursi, E., & Sampaio, R. (2010). *Modelling and convexity*. ISTE-Wiley.

Souza de Cursi, J. E., Ellaia, R., & Bouhadi, M. (2004). Global optimization under nonlinear restrictions by using stochastic perturbations of the projected gradient. In C. A. Floudas & P. Pardalos (Eds.), *Ftontiers in global optimization* (pp. 541–561). Springer.

Stratonovich, R. L. (1966). A new representation for stochastic integrals and equations. *SIAM Journal on Control, 4*(2), 362–371. https://doi.org/10.1137/0304028

Talay, D. (2015). Simulation of stochastic differential equations. In B. Engquist (Ed.), *Encyclopedia of applied and computational mathematics*. Springer. https://doi.org/10.1007/978-3-540-70529-1_346

Vasicek, O. (1977). An equilibrium characterization of the term structure. *Journal of Financial Economics, 5*(2), 177–188. https://doi.org/10.1016/0304-405X(77)90016-2

Verhulst, P. (1845). Recherches mathématiques sur la loi d'accroissement de la population. *Nouveaux mémoires de l'Académie Royale des Sciences et Belles-Lettres de Bruxelles, 18*, 14–54. Retrieved August 15, 2020, from http://eudml.org/doc/182533

Verhulst, P. (1847). Deuxième Mémoire sur la Loi d'Accroissement de la Population. *Mémoires de l'Académie Royale des Sciences, des Lettres et des Beaux-Arts de Belgique, 20*, 1–32. Retrieved August 20, 2020, from http://eudml.org/doc/178976

Volterra, V. (1926). *Variazioni e fluttuazioni del numero d'individui in specie animali conviventi* (Vol. 2). (http://www.liberliber.it, Ed.) Memoria della Reale Accademia Nazionale dei Lincei. Ser. VI. Retrieved August 12, 2020, from https://www.europeana.eu/pt/item/2022117/urn_axmedis_00000_obj_bd05ae74_d168_4c92_9a65_4f461377f7bd

Volterra, V. (1928, April). Variations and fluctuations of the number of individuals in animal species living together. *ICES Journal of Marine Science, 3*(1), 3–51. https://doi.org/10.1093/icesjms/3.1.3

Von Neumann, J. (1928). Zur Theorie der Gesellschaftsspiele. *Mathematische Annalen, 100*(1), 295–320.

Von Neumann, J., & Morgenstern, O. (1944). *Theory of games and economic behavior*. Princeton University Publishing.

Von Neumann, J., & Morgenstern, O. (1953). *Theory of games and economic behavior*. Princeton University Press.

Zermelo, E. (1912). Über eine Anwendung der Mengenlehre auf die Theorie des Schachspiels. In *Proceedings of the Fifth International Congress of Mathematicians* (pp. 501–504). Cambridge University Press.

Zhou, Y., & Cai, W. (2016). Numerical solution of the Robin problem of Laplace equations with a Feynman-Kac formula and reflecting Brownian motions. *Journal of Scientific Computing, 69*(1). https://doi.org/10.1007/s10915-016-0184-y

Zidani, H. (2013). *Représentation de solution en optimisation continue, multi-objectif et applications*. INSA Rouen, France and EMI, Rabat, Maroc.

Index

E. Souza de Cursi, *Uncertainty Quantification and Stochastic Modelling with EXCEL*,
Springer Texts in Business and Economics, https://doi.org/10.1007/978-3-030-77757-9

S

FSC
www.fsc.org

MIX
Papier aus verantwortungsvollen Quellen
Paper from responsible sources
FSC® C105338

If you have any concerns about our products,
you can contact us on
ProductSafety@springernature.com

In case Publisher is established outside the EU,
the EU authorized representative is:
Springer Nature Customer Service Center GmbH
Europaplatz 3, 69115 Heidelberg, Germany

Printed by Libri Plureos GmbH
in Hamburg, Germany